Techniques and
New Developments in
Photosynthesis Research

NATO ASI Series

Advanced Science Institutes Series

A series presenting the results of activities sponsored by the NATO Science Committee, which aims at the dissemination of advanced scientific and technological knowledge, with a view to strengthening links between scientific communities.

The series is published by an international board of publishers in conjunction with the NATO Scientific Affairs Division

A	**Life Sciences**	Plenum Publishing Corporation
B	**Physics**	New York and London
C	**Mathematical**	Kluwer Academic Publishers
	and Physical Sciences	Dordrecht, Boston, and London
D	**Behavioral and Social Sciences**	
E	**Applied Sciences**	
F	**Computer and Systems Sciences**	Springer-Verlag
G	**Ecological Sciences**	Berlin, Heidelberg, New York, London,
H	**Cell Biology**	Paris, and Tokyo

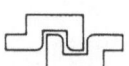

Series A: Life Sciences

Techniques and New Developments in Photosynthesis Research

Edited by

J. Barber

Imperial College of Science, Technology and Medicine
London, United Kingdom

and

R. Malkin

University of California, Berkeley
Berkeley, California

Plenum Press
New York and London
Published in cooperation with NATO Scientific Affairs Division

Proceedings of a NATO Advanced Study Institute on
Techniques and New Developments in Photosynthesis Research,
held July 31–August 13, 1988,
on the Island of Spetsai, Greece

Library of Congress Cataloging in Publication Data

NATO Advanced Study Institute on Techniques and New Developments in
Photosynthesis Research (1988: Nísos Spétsai, Greece)
  Techniques and new developments in photosynthesis research / edited by J.
Barber and R. Malkin
     p.   cm. — (NATO ASI series. Series A, Life sciences: vol. 168.)
  "Proceedings of a NATO Advanced Study Institute on Techniques and New
Developments in Photosynthesis Research, held July 31–August 13, 1988, on the
Island of Spetsai, Greece"—T.p. verso.
  "Published in cooperation with NATO Scientific Affairs Division."
  Bibliography: p.
  Includes index.
  ISBN 978-1-4684-8573-8   ISBN 978-1-4684-8571-4 (eBook)
  DOI 10.1007/978-1-4684-8571-4

  1. Photosynthesis—Research—Methodology—Congresses. I. Barber, J.
(James). II. Malkin, R. III. North Atlantic Treaty Organization. Scientific Affairs
Division. IV. Title. V. Series: NATO ASI series. Series A, Life sciences; v. 168.
QK882.N37   1988   89-8460
581.1'3342'072—dc20   CIP

© 1989 Plenum Press, New York
Softcover reprint of the hardcover 1st edition 1989

A Division of Plenum Publishing Corporation
233 Spring Street, New York, N.Y. 10013

Dedicated to
George Akoyunoglou
(1927-1986)

We wish to dedicate this book, as we did the ASI, to our friend and colleague, George Akoyunoglou (1927-1986). It was George who organised the very first photosynthesis meeting to be held in the Anargyrios and Korgialenios School in 1978 on the Island of Spetsai. Precisely ten years later we returned to the same beautiful venue to hold this NATO sponsored advanced school on photosynthesis research. Although George can not be with us any longer we were honoured that his wife Joan attended and actively participated in this meeting and was kind enough to contribute to this book.

George started his scientific career as a Ph.D. student with Melvin Calvin in Berkeley working on Rubisco. In 1963 he returned to Greece as head of the photobiology laboratory at the Nuclear Research Centre, Demokritos in Athens. Gradually his interest shifted towards the biogenesis of photosystem one and two. Together with Joan he pioneered the use of intermittent light treatments to manipulate the synthesis of thylakoid membrane components. In this way he and Joan were able to separate temporally the assembly of reaction centre complexes from that of their associated light-harvesting complexes, permitting the step-wise synthesis and interconnection of these chlorophyll-proteins to be studied. The approach which they developed and used is now forming the basis for tackling many intriguing questions of gene expression and regulation.

Both of us met George many times at various international gatherings and of course it was he who organised the 5th International Congress on Photosynthesis in 1980 at Halkidiki, Greece. Just a few months before George was taken ill he visited Berkeley in California. By coincidence JB was also visiting and the three of us spent a delightful but challenging question and answer session with Berkeley students.
During that session George demonstrated his gift for communication and for passing on the enthusiasm which he had for his subject and for which we will always remember him.

Jim Barber
Dick Malkin

v

PREFACE

From July 31 to August 13 a NATO Advanced Study Institute on Photosynthesis was held at the Anargyrios and Korgialenios School on the Island of Spetsai, Greece. The Institute focused on techniques and recent advances in photosynthesis research and brought together teachers and students with a wide range of interest and experience. It was a very stimulating occasion which allowed cross-fertilization to occur between biophysicists, biochemists, molecular biologists and physiologists. Lectures and discussions ranged from the description of the molecular structure of the photosynthetic bacterial reaction centre and of tobacco Rubisco through to the regulation of carbon metabolism and the application of genetic engeering. This book is comprised of the contents of the major lectures and a selection of relevant posters displayed at the Institute. Taken together the book is an excellent representation of the most up to date thoughts and activities in photosynthesis research across a wide, but interlocking, spectrum of topics. The papers presented here are a written record of the high quality of both the lecturers and students alike and emphasises the value of the NATO ASI series as a reference source.

The successful organisation of the Institute and the production of this book would not have been possible without the support of our colleagues. We therefore wish to thank Pam Cook, Lyn Barber, Niki Gounaris, Alison Telfer, Sotiria Nikolaidon, David Chapman, Steven Mayes and Wei Qiu Wang for all their help during the course of the Institute.

October 1988   J. Barber and R. Malkin
                                                       London and Berkeley

CONTENTS

INTRODUCTION

SECTION I: STRUCTURE AND FUNCTION OF PROTEIN COMPLEXES

SECTION II: MEMBRANES AND ELECTRON TRANSPORT

SECTION III: CARBON FIXATION: MEASUREMENT AND REGULATION

SECTION IV: MOLECULAR BIOLOGY, GENETIC ENGINEERING AND BIOTECHNOLOGY

SECTION V: STRESS PHENOMENA AND ADAPTATION

INTRODUCTION

SOLAR ENERGY FROM PHOTOCHEMISTRY

Sir George Porter PRS

Department of Pure & Applied Biology
Imperial College of Science, Technology & Medicine
Prince Consort Road
London SW7 2BB UK

INTRODUCTION

This book is about Nature's storage and utilisation of solar energy, achieved by photosynthesising chemical fuels, with the sun as energy source, and storing those fuels as chemical potential.

The two essential requirements of man for survival on Earth are food and fuel; both are products of photosynthesis. Food is not now a production problem, although its distribution presents difficulties that are partly geographical but mainly economic. For example the U.S.S.R. and E. Europe on present projections, will have a 20% deficit of food production in 1990 whilst the developed countries of Europe will have a 25% surplus. World wide a 10% surplus over requirements is expected.

Fuel, on the other hand, presents big problems. Even today a large proportion of people rely mainly on biomass, such as wood and dung. The amount of these materials available is diminishing and energy is the bottleneck to development in many underdeveloped countries. Not only is fuel essential to life - rice and wheat have to be cooked before they are edible - but it is essential to all aspects of modern life from transport, manufacturing processes and communications to lighting, refrigeration, and irrigation. A modern city community would die quickly without its energy lifeline.

All this energy, except for nuclear, comes from the sun and apart from a relatively small amount of hydro-electric power, the sun's energy is stored through photochemistry and photobiology. We have a vast store of fossil fuels which have been accumulated over the last three hundred million years or so and the process continues today on a renewable basis and a grand scale.

The total solar energy falling on earth in two weeks is equivalent to all stored fossil fuels. The annual fixation as carbohydrates is 2×10^{11} tons; this is 10 x man's annual energy use and 200 x his food consumption. Turnover of the carbon dioxide in the atmosphere occurs every 300 years, of oxygen every 2000 years and even the water is turned over every 2 million years. But the overall efficiency of photosynthesis is low, only about 0.2%.

The Time Scale of Photosynthesis

The Earth was formed 4.6 billion years ago and life has been present on Earth for most of this time with the sun as the energy source for the endergonic evolution of life. The early photochemical processes in the upper atmosphere involved the photosynthesis of larger organic molecules from smaller ones such as water, ammonia and perhaps methane, for which ultra-violet light would be necessary. Then more than three billion years ago, pigment molecules able to absorb in the visible region, like chlorophyll, were developed so that photosynthesis, more or less as we know it today, became possible.

The storage of the products of photosynthesis as coal, oil and gas has occurred only over the last three hundred million years and man's use of these fossil fuels on a large scale occupies only a few generations. In the United States, for example, the renewable fuel, wood, predominated up to 1850 then, by 1900, coal was supplying 85% of the energy and, today, coal has itself been overtaken by oil and gas as the principal fuels.

There is no immediate danger of our fuels running out. Although even the most optimistic forecasts do not expect oil resources to last more than 60 years, coal will be available (at a price, because it will become more and more expensive to recover) for a few hundred years. Nevertheless, looked at over a time scale of one or two thousand years our consumption of, and dependence on, the store of fossil fuels is a mere hiccup in history.

The shortages, particularly of oil, will clearly be felt within our lifetime. Furthermore, these projections are based on the present pattern of consumption and, if the standard of living of the developing countries were raised to that of Western Europe, the rate of consumption of energy of the World as a whole would have to be increased by a factor of about eight. If the population doubles, as seems inevitable, the energy requirement would then be sixteen times the present.

It is doubtful whether this rate of energy consumption could be borne by our planet, especially from renewable resources in a steady state, and it is important that all the possible energy sources should be investigated now. Apart from the non-renewable fossil fuels, only two energy sources have the potential to supply these large energy needs - nuclear reactors and photosynthesis. We would be wise not to rely on either of these alone and, fortunately they are in some ways complimentary. Nuclear reactors are more suitable for providing electricity to the grids of the world whilst chemical products of photosynthesis provide the forms of energy that are essential for transport and storage.

METHODS FOR SOLAR ENERGY CONVERSION AND STORAGE

There are four main methods of conversion of solar energy:

Thermal - This is extensively used in a passive way - as we sit in the sun in Spetses for example. Most of our energy for heating buildings is obtained in this way and always will be. On the other hand because of thermodynamic restrictions, conversion of solar heat to work is not promising economically.

Mechanical - Wind, hydro power and waves. The first two of these are as ancient as technology itself and they are excellent for conversion to work and electricity. The amount available is limited and like nuclear, they are most suitable for non-storable energy and direct supply to the grid.

Photoelectric - Photovoltaic cells are the most efficient (up to 20%) and convenient means of converting solar radiation into electricity and are being rapidly developed for remote places and for small, portable generators. They are too expensive at present for large scale applications but amorphous materials look promising.

Photosynthesis - This is already the source of all our chemical fuels which unlike the above, have the great advantage of storage and transportability. Biomass is the main source of fuel in many developing countries where it is merely burned. It is also used in the form of sugar cane for alcohol fuel production, in Brazil for example, though this is not yet economic without subsidy. Problems are the low

4

efficiency of solar conversion, the need for large amounts of water and fertiliser, and the expensive harvesting and processing needed to make a convenient liquid fuel from plant material.

Photochemical production of fuels - In principle a purely chemical, in vitro, method of photosynthesis might provide a higher efficiency and a more convenient fuel. In spite of very active research over the last decade or so, and successful mimicking of most of the photoconversion processes of natural photosynthesis, no viable overall process has yet emerged.

The physical, chemical and biological methods are very closely related: all use an absorber of photons which utilises electronic excitation to move electrons to higher potential. Furthermore the possibilities and theoretical efficiencies of all are based on the same considerations of thermodynamics. These have been discussed in detail (1) and will not be repeated here.

We need only note that thermodynamic restrictions, with the Sun as the hot source and the Earth as the reservoir, coupled with the losses due to irreversibility at finite power extraction, the polychromatic nature of the radiation and the need for storage, set an upper limit of 27% for the efficency of conversion of solar radiation energy to stored chemical potential. This could be increased by raising the operating temperature nearer to the 6000K of the sun (concentrators allow the use of more than one "sun"). A further improvement is possible by using a sandwich type of absorber, with more than one threshold wavelength, so as to overcome some of the inefficiencies of the use of polychromatic radiation. This principle is already used in some production photovoltaic cells, it can almost double efficiencies and does not appear to have been used by the plant in natural photosynthesis.

There is clearly no theoretical reason why efficiencies much greater than those of natural photosynthesis should not be attained, either by chemical models of photosynthesis or by genetic engineering of photosynthesising organisms.

If our studies of photosynthesis are to provide the means to more effective storage of solar energy, what changes should we be seeking? In principle we could use any elements and any chemical reaction which stores solar energy, but in practice we are much more restricted. If we are to collect radiation over a very large area, and transport the products for combustion or some similar reaction elsewhere, it is clear that taking the raw materials from the environment and returning them after use is far and away the most cost efficient method - that is we run a cyclic process which leaves the environment unchanged but richer in energy.

The energy producing reaction that is used both in the metabolism of food and the combustion of a fuel is combination of the fuel/food with the oxygen of the atmosphere and the task of photosynthesis is to complete the cycle by making oxygen from water. Other cycles, such as those of the photosynthetic bacteria, cycle sulphur rather than oxygen but store little energy.

The photodissociation of water into oxygen and some reduced products is the essence of photosynthesis and most models are based on this reaction:

$$2H_2$$
$$\uparrow$$

$$2H_2O = O_2 + (4H^+ + 4e) \qquad \Delta G = 470 \text{ kJ/M}$$

$$\downarrow + CO_2$$

$$(CH_2O) + H_2O$$

The reducing equivalent may produce hydrogen directly or it may be used to reduce other components of the environment of which carbon dioxide and nitrogen are the most generally available. The reduction of these to carbohydrate and ammonia provides convenient and useful products without changing significantly the overall energetics of the reaction.

The solar photolysis of water can be carried out in many ways. We may use photovoltaic cells, followed by electrolysis; or we may employ photoelectrochemical devices based on semiconductors, pioneered by the titanium dioxide cell of Fujishima and Honda. We might design new heterogeneous photochemical systems, such as have already been used successfully to generate hydrogen and oxygen separately, but not at the same time, with the help of sacrificial donors and acceptors.

Unfortunately a complete system would probably involve at least ten components and these would have to be spacially arranged with respect to each other to prevent back reactions. It soon becomes tempting to return to the living plant, which does all this complex arranging for us, and try to improve on what is, after all, already a working system.

SUMMARY

In summary photovoltaics and photosynthesis, modified perhaps by genetic engineering, seem more promising at present than photochemical or photoelectrochemical methods for solar energy utilisation and the hope of achieving overall efficiencies better than 10% are good in both cases. But what are the prospects of solar energy providing power on a significant scale, by any method which gives 10% efficiency or better? How much land would be required? The average, day and night, winter and summer, insolation over the Earth's surface is typically 200 w/m^2 in temperate latitudes. To supply man's current power demands would require an area of the Earth's surface only one thirtieth of that already under cultivation.

Even to supply projected future energy needs of the increased population with a "developed" standard of living would need less land than we already use for agriculture. It is possible that desert or arid land could be used; the amount of water needed stoichiometrically for photosynthesis is far less than a plant normally demands.

One of the first scientists to have this vision was Ciamician, a pioneer photochemist in Bologna. In 1912 he expressed it eloquently, as follows: "Where vegetation is rich, photochemistry may be left to the plants and, by rational cultivation ... solar energy may be used for industrial purposes. In the desert regions, unadapted to any kind of cultivation, photochemistry will artificially put their solar energy to practical uses. On the arid lands there will spring up industrial colonies without smoke and without smokestacks; forests of glass tubes will extend over the plains, and glass buildings will rise everywhere; inside of these will take place the photochemical processes that hitherto have been the guarded secret of the plants, but that will have been mastered by human industry which will know how to make them bear even more abundant fruit than nature, for nature is not in a hurry and mankind is. And if in a distant future the supply of coal becomes completely exhausted, civilisation will not be checked by that, for life and civilisation will continue as long as the sun shines! If our black and nervous civilisation, based on coal, shall be followed by a quieter civilisation based on the utilisation of solar energy, that will not be harmful to progresss and to human happiness. The photochemistry of the future should not however be postponed to such distant times; I believe that history will do well in using from this very day all the energies that nature puts at its disposal."

Many of us share Ciamician's enthusiasm, others are more sceptical. One of the early sceptics was Dean Swift, in Gulliver's Travels. You may remember that he describes how, when Gulliver visited the Lagado Academy, (which I fear was probably modelled on the Royal Society), he met "a man who had been eight years upon a project for extracting sun-beams out of cucumbers, which were to be put into vials hermetically sealed, and let out to warm the air in inclement summers. "He told me," says Gulliver, "he did not doubt in eight years more, he should be able to supply the Governors gardens with sunshine at a reasonable rate; but he complained that his stock was low and entreated me to give him something as an encouragement to ingenuity, especially since this had been a very dear season for cucumbers." He clearly had funding problems, like many of us poor scientists today.

I think Swift was just having a bit of fun here and that he was really a solar energy enthusiast because, through the mouth of the King of Brobdingnag he "gives it as his opinion that he who made two ears of corn or two blades of grass grow where only one grew before would deserve better of mankind and do a more essential service for the country than a whole race of politicians put together."

Whilst the study of photosynthesis for its own sake, as the most important chemical process on earth, needs little justification, the deliberations which took place at the NATO-ASI held in Spetses and their record in written form presented in this book also have the real possibility of leading towards a practical goal of immense importance.

REFERENCES

1. Porter, G. (1983) J. Chem. Soc. Faraday Trans. 2. 79, 473-482

SECTION I

STRUCTURE AND FUNCTION OF PROTEIN COMPLEXES

CRYSTALLIZATION OF MEMBRANE PROTEINS

Hartmut Michel

Max-Planck-Institut für Biophysik
Abteilung Molekulare Membranbiologie
Heinrich-Hoffmann-Str. 7
6000 Frankfurt/M 71, West Germany

INTRODUCTION

Prerequisite to understand the function and mechanism of action of any biological macromolecule is the knowledge of its structure. The only way to obtain detailed structural knowlege of large molecules is X-ray crystallography. Of course, X-ray crystallography needs crystals and the main area, where no crystals had been obtained until recently was the area of the membrane proteins.

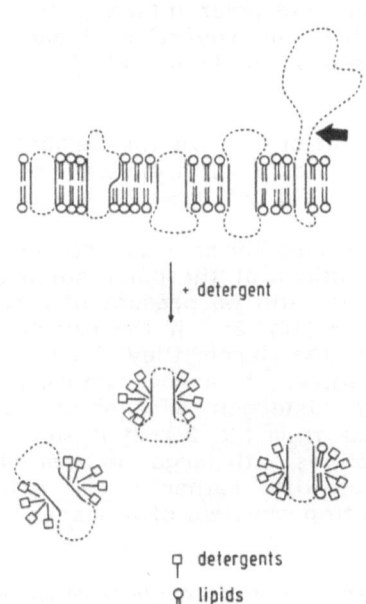

+ detergent

detergents
lipids

Fig. 1. Model of a biological membrane. The polar surface of the membrane proteins is indicated by the dashed lines (taken from reference 1)

The difficulties to handle and to crystallize membrane proteins reside in the amphiphilic surface of the membrane proteins. Membrane proteins are highly hydrophilic where they are in contact with the aqueous phases on both sides of the membranes whereas they are hydrophobic where they are in contact with the alkane chains of the membrane lipids. This is schematically depicted in fig. 1.

In order to solubilize, to isolate and to characterize membrane proteins one has to use detergents. Detergents are amphiphilic molecules, which form micelles above a certain concentration, which is called the critical micellar concentration (CMC). If an excess of detergents is added to biological membranes, the detergent micelles take up the lipids and the membrane proteins, thereby disrupting the membrane. One must add a large excess of detergent in order to make sure that most of the detergent micelles contain only one protein molecule. The membrane proteins within the detergent micelles can then be isolated by various chromatographic methods.

RESULTS AND DISCUSSION

If one thinks on how to bring these membrane proteins into real crystalline order, one ends up with two principal possibilities (1). Examples for both are shown in fig. 2. One should be able to form two-dimensional crystals of the membrane protein in the plane of the membrane first and then order these two-dimensional crystals in the third dimension with respect to translation, rotation and up and down orientation ("type I"). One would then have a perfect three-dimensional crystal and nothing artificial like a detergent would be present. In addition one would be able to make statements about lipid protein interaction. However, we have to consider that there are hydrophobic and polar interactions in the planes of the membranes, whereas there are only polar interactions perpendicular to the membranes. As a result, in a reasonable crystallization procedure one would have to increase both hydrophobic and polar interactions at the same time. This is difficult to achieve. There are several crystals of this type. They are all small, disordered and only useful for electron-microscopic structural investigations.

The alternative is to crystallize the membrane proteins within their detergent micelles. How such a crystal could be built up, is shown in the lower part of fig. 2 ("type II"-crystal).

The crystal lattice will be formed preferentially by the membrane protein via polar interactions at the polar surface of the membrane proteins. The detergent will still be present in a micelle-like manner. The following is clear from fig. 2: if the micelles surrounding the membrane proteins are too large, they would prevent the desired attractive interactions between the polar surface parts of the membrane proteins. Therefore the detergent micelles should be as small as possible. It is also clear from fig. 2 that it seems much easier to crystallize membrane proteins with large extramembrane surface domains compared to those ones with rather small extramembranous surface domains. All well diffracting crystals of membrane proteins belong to type II.

In general one has to use detergents with a short alkyl chain. However, an experience by people working with membrane proteins is that membrane proteins tend to be more instable in detergents of short alkyl chain lengths. Normally, an increase of the alkyl chain by one methylene group leads to an increase in stability of a factor of about three. One therefore has to find a compromise.

12

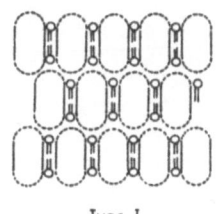

Type I

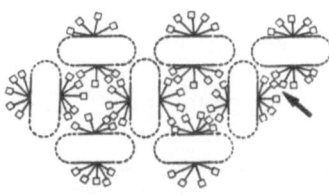

Type II

Fig. 2. The two basic types of membrane protein crystals. Type I: stacks of two-dimensionally crystalline membranes ordered in the third dimension. Type II: A possible arrangement of a membrane protein when crystallized with the detergent micelle bound to its hydrophobic surface. The arrow indicates an area of close, critical contact between detergent micelles.

Of great importance has been the addition of small amphiphilic molecules like heptane-1,2,3-triol (1). Such compounds are not detergents themselves, since their alkyl chains are too short to form micelles. They form mixed micelles with the detergents and these mixed micelles are smaller than the pure detergent micelles. They also might act by displacing a few detergent molecules which are too large to fit perfectly into the proteins crystal lattice. Especially a large head-group of a detergent, like the glucose moiety of octylglucoside may lead to steric hindrance . One also has to consider specific interactions between the protein and the small amphiphilic molecules, since for the crystallization of the photosynthetic reaction center from the purple bacterium Rhodopseudomonas viridis crystals can only be obtained in the presence of one of the various isomers of heptane-1,2,3-triol.

An additional benefit of the small amphiphilic molecules is that they are able to suppress phase separation into a detergent-enriched phase and into a detergent-depleted phase to a large extend. The detergent-rich phase also contains the membrane proteins. The phase separation is induced by the precipitant. It is largly due to an attractive interaction between the detergent micelles and seems to be a "salting out" phenomenon, very much like the "salting out" of proteins. Crystals of membrane proteins are frequently obtained at precipitant concentrations just below the start of the phase separation. Therefore, attractive interactions of detergent micelles may help to form nuclei for membrane protein crystallization (2).

Table 1 lists those detergents which so far could be used to crystallize one or the other membrane protein. With highly stable membrane proteins like the porins from E. coli all the detergents listed in the table have been succesfully used.

TABLE I. Detergents which have been used in membrane protein crystallization

detergent (abbreviation)	molecular weight	CMC* (mM)	examples**
N-decyl-N,N-dimethylamine-N-oxide (DDAO)	201	10.4	bLH
N-dodecyl-N,N-dimethylamine-N-oxide (LDAO)	229	1.2	bLH, RC
octyl(hydroxyethyl)sulfoxide (C8-HESO)	206	29.9	Po, bLH
octyl(dihydroxypropyl)sulfoxide (C8-DOPSO)	246	30	bLH
octyl-ß-D-glucopyranoside (OG)	292	30	various
nonyl-ß-D-glucopyranoside (NG)	306	6.5	various
decanoyl-N-methylglucamide (MEGA-10)	350	6-7	bR
octyl-ß-D-thioglucopyranoside (OTG)	308	9	various
N-dodecyl-N,N-dimethyl-ammonio-propane sulfonate	336	2-4	Po
dodecyl-ß-D-maltoside (DM)	511	1.6	PS I-RC

For very large membrane protein complexes also Triton X 100 (PS I-RC, see ref.3) and Brij 35 (Cytochrome oxidase, see ref. 4) could be used.

* The values for the critical micellar concentration (CMC) are as given by the manufacturers. They are for low ionic strength. The CMC is lower at higher ionic strength.

** abbreviations: bLH = bacterial light-harvesting complex, bR =

bacteriorhodopsin, RC = photosynthetic reaction centres from purple bacteria, Po = Porin, PSI-RC = photosynthetic reaction centre of photosystem I.

For less stable membrane proteins preferentially those with large polar head groups like glucose or maltose residues and longer alkyl chains are needed. Detergents with a zwitterionic head group denature many membrane proteins. In the cases of bacteriorhodopsin and the bacterial light-harvesting complexes crystals were obtained with all detergents which did not denature the protein. A special case seems to be the reaction center from Rhodopseudomonas viridis where membrane protein crystals were only obtained with LDAO as detergent. The need for LDAO seems to be due to the fact that one LDAO molecule is firmly bound by three neighbouring alpha helices and therefore visible in the electron density map.

In order to achieve supersaturation and to precipitate the membrane protein the same precipitants as with soluble proteins can be used: salts like ammonium sulphate, sodium or potassium phosphate and citrate, as well as polymeric precipitants like polyethyleneglycol. As technique to achieve supersaturation both vapor diffusion using the sitting drop method as well as dialysis can be used. It is technically easier to perform vapor diffusion experiments, whereas in dialysis experiments the relevant parameters, like detergent concentration, concentrations of additives and precipitating agent can be varied individually.

The most important parameter in the crystallization of membrane proteins however is the membrane protein itself. The protein preparation should be stable and homogeneous with respect to purity and conformation. It should be well characterized. Membrane proteins from the photosynthetic apparatus seem to be especially suited for crystallization since they are available in large amounts. In addition, they frequently contain bound pigments and conformational changes can be monitored by simply recording absorption spectra.

REFERENCES

1. H. Michel (1983)
   Crystallization of Membrane Proteins
   Trends Biochem. Sci. 8, 56-59

2. R. M. Garavito, Z. Markovic-Housley and J. A. Jenkins (1986)
   The Growth and Characterization of Membrane Protein Crystals
   Journal of Crystal Growth 76, 701-709

3. R. C. Ford, D. Picot and R. M. Garavito (1987)
   Crystallization of the Photosystem I Reaction Centre
   EMBO J. 6, 1581-1586

4. S. Yoshikawa, T. Tera, Y. Takahashi, T. Tsukihara and W. S. Caughey (1988)
   Crystalline Cytochrome c Oxidase of Bovine Heart Mitochondrial Membrane: Composition and X-ray Diffraction Studies
   Proc. Natl. Acad. Sci. USA 85, 1354-1358

FTIR INVESTIGATIONS ON ORIENTATION OF PROTEIN SECONDARY STRUCTURES AND PRIMARY REACTIONS IN PHOTOSYNTHESIS

E. Nabedryk, W. Mäntele*, J. Breton

Service de Biophysique, Département de Biologie
CEN Saclay 91191 Gif-sur-Yvette cedex, France
*Institut für Biophysik und Stralhenbiologie der Universität
Freiburg, D7800 Freiburg, FRG

1. INTRODUCTION

In the field of photosynthesis, a variety of spectroscopic techniques have been used to investigate the physical properties of the components (pigments, charge carriers, proteins) in the intact photosynthetic membrane and in the isolated complexes as well as in model systems which were thought to mimic the in vivo environment. These techniques have revealed a very high level of organization of the membrane components in vivo (1-4). Although a wealth of spectroscopic data has been collected over the years, the problem of inferring from these measurements alone the most probable configuration of the pigments and proteins in vivo is extremely complex.

In the case of the bacterial photosynthetic reaction center (RC), absorption of light leads to the transfer of an electron from the primary donor P, a dimer of bacteriochlorophyll (BChl) a or of BChl b, to the intermediary acceptor I, a bacteriopheophytin (BPheo) a or BPheo b monomer and subsequently to a quinone (Q) acceptor. Only recently, the X-ray analysis of the RC crystal (5,6) has allowed to visualize the electron transfer pathway and has yielded a detailed description of the molecular architecture of the complex. X-ray analysis of the RC, by demonstrating the fixed arrangement of the chlorophyll (Chl) molecules and by giving informations on the pigment binding sites provides a unique opportunity to test the results of the previous spectroscopic investigations done on the same systems. It thus constitutes an invaluable reference for analyzing the structural properties of the other systems such as antenna complexes, photosystems from higher plants or other membrane proteins (for example purple membrane and porin) for which no high resolution X-ray structure is available. It is with this perspective in mind that we will discuss some results recently obtained on the structural and functional aspects of the organization of the pigments and proteins within the RC of Rps. viridis and Rb. sphaeroides. More specifically, a comparison will be made between X-ray and infrared (IR) data both for the global structure of the RC and for the localized cofactors-protein interactions which are affected by primary electron transfer reactions.

2. CHARACTERIZATION OF PROTEIN SECONDARY STRUCTURES IN THE REACTION CENTER: COMPARISON WITH OTHER MEMBRANE PROTEINS

There is now a large body of evidence to support the concept of α-helices as the membrane-spanning components of intrinsic proteins in photosynthetic membranes. This includes studies by IR spectroscopy (2,4), X-ray diffraction (7,8) as well as calculation of the distribution of hydrophobic residues in the amino acid sequences (9). In the absence of high resolution structural data, the degree of orientation of the secondary structures in a membrane protein can be investigated by polarized IR spectroscopy of oriented multilayer films of the native membrane or of liposomes in which the protein has been incorporated. It thus offers a starting point for model building of membrane proteins.

The most useful spectral range for an analysis by IR spectroscopy of the secondary structure of proteins is the so-called amide I (mainly peptide C=O stretching vibration) and amide II (in part NH bending vibration) absorption regions between 1700 and 1500 cm^{-1}. The frequency of the amide bands depends on the peptide backbone conformation : the amide I band, particularly, absorbs at higher wavenumbers for the main component in an α-helix than it does for a β-sheet (10). In addition, IR amide vibrations from an ordered structure are anisotropic and polarized predominantly parallel or perpendicular to the polypeptide chain axis. Accordingly, by measuring the IR dichroism of the amide absorption bands, it is possible to estimate the degree of orientation of the α-helices or β-sheets in intrinsic proteins relative to the membrane plane (11). Furthermore, using Fourier transform IR (FTIR) spectroscopy, second-derivative spectra can be obtained leading to a resolution-enhancement by band narrowing of IR spectra (12).

For a quantative determination of the orientation of protein secondary structures, the extent of each type of secondary structure present in the protein must be known. This can be evaluated from analysis of UV circular dichroism (CD) spectra of isolated or reconstituted complexes. UVCD spectra can be analyzed as linear combinations of conformationally related subspectra (13).

Using these spectroscopic approaches, a transmembrane organization of α-helical segments has been demonstrated in various reaction centers (RCs) and antenna complexes from both bacteria and green plants. Here, a comparison is made with i) the bacteriorhodopsin in purple membrane (PM) from Halobacterium halobium, a high α-helical-containing membrane protein for which transmembrane arrangement of α-helices has been demonstrated (11,14-16) and ii) porin (OmpF), a high β-sheet-containing integral protein (trimer) forming transmembrane channels across E. coli outer membranes.

The FTIR absorbance spectra of air-dried multilayers of PM, Rb. sphaeroides RC and porin are compared in Fig. 1a. For PM and RC, the IR spectrum shows major components at 1660 cm^{-1} (PM) and 1656 cm^{-1} (RC) for the amide I band and at 1545 cm^{-1} (PM) 1547 cm^{-1} (RC) for the amide II band, indicating a high proportion of α-helical structure in both membrane proteins. In contrast, for porin, the major components at 1631 cm^{-1} and 1530 cm^{-1} as well as several shoulders on the amide I band at 1650-1675 cm^{-1} and at 1696 cm^{-1} are indicative of antiparallel β-sheet structure as the predominant type of protein secondary structure in this protein (17,18). In both RC and porin IR spectra, the band at ~ 1738 cm^{-1} is due to the absorption of the ester carbonyl groups from the lipids in which the proteins have been incorporated.

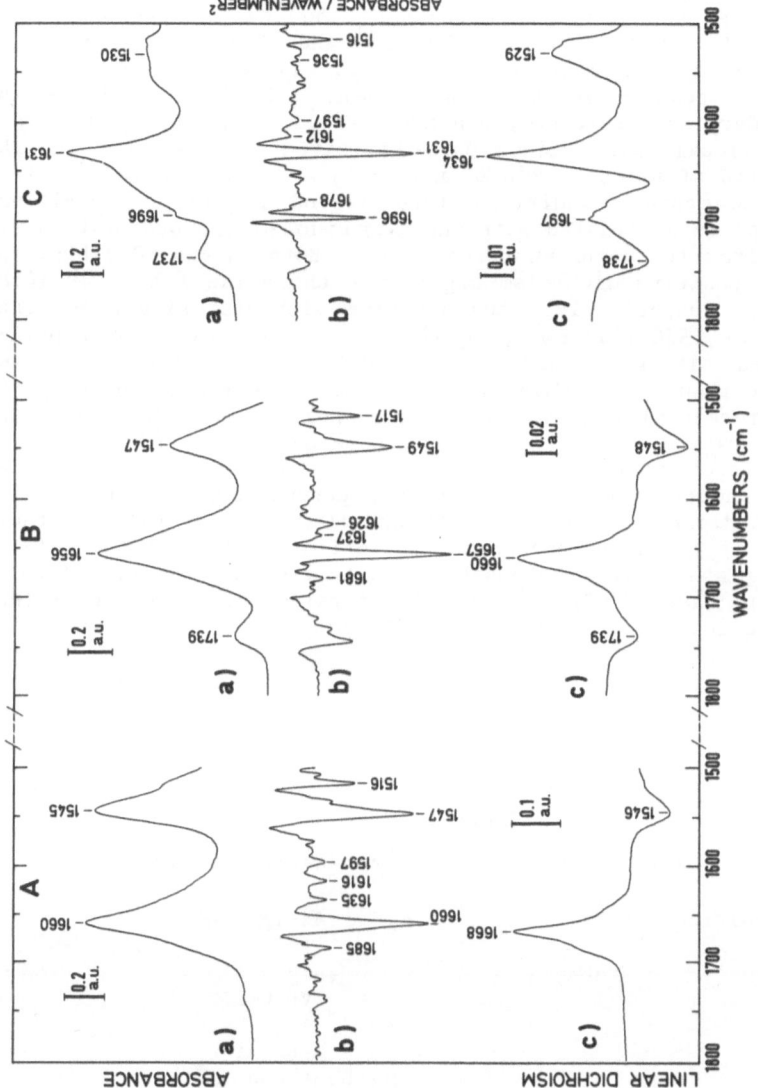

FIGURE 1. FTIR spectra of oriented air-dried membrane multilayers of A) purple membrane from Halobacterium halobium B) Rb. sphaeroides RC C) porin. In order to remove the major source of reflection occuring at the membrane/air interface, the films were covered with Nujol (18,28). The films were tilted with respect to the IR beam which was linearly polarized parallel (to record $A_{//}$) or perpendicular (to record $A_{\perp}$) to the plane of incidence (11). a) absorbance spectra calculated from $(2A_{\perp} + A_{//})/3$. b) Second derivative spectra. c) Linear dichroism spectra $(A_{//} - A_{\perp})$. Measuring conditions : 256 scans. Resolution : 4 cm^{-1}, T = 20°C.

In agreement with IR spectroscopy, UVCD spectra of sonicated PM and isolated RC show a typical α-helix pattern (data not shown) leading to a 69% and 51% α-helical content in PM (16,19) and RC (4,20) respectively. For porin, a 68% β-sheet content was estimated by UVCD (18).

A more detailed analysis of the IR peak frequencies of the characteristic secondary structure components can be made by second-derivative analysis of FTIR spectra (12,21). The second-derivative spectra generated from data in Fig. 1a are displayed on Fig. 1b. They show three main peaks at 1660 cm^{-1}, 1547 cm^{-1} and 1516 cm^{-1} for PM, at 1657 cm^{-1}, 1549 cm^{-1} and 1517 cm^{-1} for RC, at 1696 cm^{-1}, 1631 cm^{-1} and 1516 cm^{-1} for porin, as well as several weak but reproducible peaks. Tentative assignments of the main bands to specific secondary structure components (α-helix, β-sheet, β-turn, etc.) and individual amino acid groups are given in Table I. The IR dichroism spectra of films of PM, RC and porin are compared in Fig. 1c. Under our experimental conditions (see legend Fig.1), a positive dichroism signal is associated with the alignment of the oscillators at less than 55° from the membrane normal (11). Both PM and RC exhibit a characteristic positive dichroism signal for the amide I band at 1668 and 1660 cm^{-1}, respectively, and negative dichroism signal for the amide II band at 1546-1548 cm^{-1}, qualitatively indicating of peptide groups oriented rather perpendicular to the membrane plane. In an α-helix, the carbonyl stretching vibration of the polypeptide chain is polarized preferentially parallel to the helix axis while the NH deformation vibration is mostly polarized perpendicular to this axis (11 and references therein). An opposite situation is found in a β-sheet where the peptide bonds are rather perpendicular to the chain axis (18). IR dichroism spectra of PM and RC therefore indicate that the α-helical segments are preferentially oriented along the normal to the membrane plane, as previously demonstrated for PM (11,14-16). However, in the case of RC, this was the first direct evidence of the transmembrane orientation of α-helices (11,20).

TABLE I

Tentative bands assignments from the second-derivative IR spectra.

IR Band position (cm^{-1})	Assignment
1516	Tyr C=C
Amide II	
1530	β-sheet
1536	β-sheet and β-turns
1547-1549	α-helix
Amide I	
1631-1637	β-sheet
1657-1660	α-helix
1678-1681	β-turns and antiparallel β-sheets
1685-1696	β-turns

In contrast, the IR dichroism spectrum of porin shows positive signals for the amide I and amide II bands indicating that the transition moments of these vibrations are oriented at less than 55° with respect to the membrane normal. Furthermore, the small magnitude of the dichroism signals suggests an average orientation of the corresponding transitions at a value not too far from the magic angle (φ =55°). Qualitatively, from our IR dichroism data (18), it thus appears that in porin a β-sheet arrangement with all strands strictly perpendicular or parallel to the membrane can be excluded.

It must be noticed that among these three membrane proteins for which structural and functional properties are extensively studied, the RC is the only one the 3-D structure of which has been determined by X-ray analysis of the crystal at high resolution. Structural models of bacteriorhodopsin and porin have been proposed from electron or neutron diffraction and image reconstruction on 2-D sheets or 3-D crystals at a resolution of about 6 Å. In the case of PM, our IR and CD data (11,16) imply a large amount of transmembrane α-helices in good agreement with the generally accepted model derived from diffraction experiments (14, 22,23).

In the case of porin, a new model for the packing of β-strands has been proposed from both polarized IR and diffuse X-ray data (18). On the one hand, X-ray diffraction studies of Kleffel et al. (17) on 3-D crystals of porin indicate that a significant fraction of β-strands are oriented at a very small angle to the membrane normal. On the other hand, from our IR dichroism data, it can be calculated that the average angle of inclination away from the membrane normal for the β-strands is about 45°. This applies severe constraints on the β-sheet orientations that could exist in porin. Only one arrangement of strands does satisfy both IR and X-ray requirements. In this model, each porin monomer consists of at least two β-sheet domains, both with their plane perpendicular to the membrane. One sheet has its strands direction lying nearly parallel to the membrane normal while the other sheet has its strands inclined at a small angle away from the membrane plane. Although the kind of β-sheet arrangement we are proposing makes it difficult to envisage a pore built within single monomers, created only by β-sheet interactions, the question of how the pores are constructed is still open. A 3-D structure of this pore protein would thus add a great deal of information concerning the variety of polypeptide configurations that can exist in biological membranes.

3. ORIENTATION OF α-HELICES IN THE BACTERIAL REACTION CENTER. COMPARISON WITH X-RAY STRUCTURAL MODEL

In 1982, from our visible and IR dichroism data (20), we concluded that the average orientation for all the α-helical segments in Rb. sphaeroides RC is 30° with respect to the membrane normal (Table II). Such a transmembrane organization of α-helices was later corroborated by the sequence analysis of RC polypeptides which predicted five hydrophobic domains for each L and M polypeptide (24,25) that could fit to transmembrane α-helices. Later, X-ray analysis of the RC crystal has beautifully shown the orientation of eleven α-helices (including the one from the H subunit) preferentially parallel to the C2 symmetry axis of the RC core (7,8). Comparable α-helical contents were estimated from X-ray and UVCD (see Table II), 42% (J. Deisenhofer and H. Michel, personal communication) and 47% (26) in Rps. viridis RC, 51% (8) and 51% (4,20) in Rb. sphaeroides RC.

In Rps. viridis RC crystals, the electron density map clearly indicates that the ten α-helices observed in the LM subunit are tilted on the average at a small angle with respect to the 2-fold axis with eight of them tilted at less than 25° and the remaining two at about 38° from this axis (7). In Rb. sphaeroides RC, recent X-ray data indicate an average tilt angle of 22° (27). These data are quite in agreement with IR dichroism results of RC (4,20) and LM (28) films of Rb. sphaeroides leading to an average orientation for all the α-helices at less than 30° from the membrane normal and at 20-25° for the purely transmembrane α-helical segments in the LM core (4,26). Taken together, IR dichroism and X-ray results imply that the C2 symmetry axis evident from the X-ray map coincides with the normal to the membrane. This conclusion is in agreement with the original proposal of Deisenhofer et al. (5) and has been recently confirmed from the energy minimization calculations of Yeates et al. (27) on Rb. sphaeroides RC. In the RC of Rps. viridis, an experimental average tilt angle of 38° has been estimated from IR dichroism for all the α-helices including those of the bound cytochrome c (26). The cytochrome part of the RC protein accounting for about one third of the molecular weight of the total complex and the X-ray structure showing that the α-helical segments of this unit have almost no preferred orientation with respect to the C2 symmetry axis, the exact tilt for the only α-helices of the LM complex of Rps. viridis RC should be lower than 38° and closer to the angle observed by X-ray or by IR dichroism of RC and LM films of Rb. sphaeroides. It must be emphasized that the tilt angle calculated from our CD and IR data represents an average value for all the α-helices contained in the system under investigation. The possibility that some α-helical segments are either rather parallel to the membrane (i.e. joining transmembrane -helices) or random has to be taken into account. Indeed, the RC X-ray model clearly shows that in L, M and H subunits, several α-helical segments do not span the membrane (7,8).

TABLE II

Estimation of the α-helix content from UVCD data and of the average orientation of α-helices (φ) from IR dichroism. Comparison with X-ray data. *The φ value is given for the purely transmembrane α-helices.

	% α-helix		φ	
	UVCD	X-ray	IR	X-ray
Rb. sphaeroides				
LMH	51	51	30°	22°*
LM	55	–	20-25°*	
H	31[x]			
Rps. viridis				
LMH-Cyt	47	42[+]	38°	
LM				25°* for 8 α-helices
				38°* for 2 α-helices

[x]The sample was obtained from R. Debus, M. Okamura and G. Feher.
[+]J. Deisenhofer and H. Michel, personal communication.

They are generally shorter (6 to 18 residues length) than the transmembrane ones (22-29 residues length). In Rb. sphaeroides RC, five of them are found in L, six in M and one in H (8). Thus, our conclusion that the transmembrane α-helices in LM of Rb. sphaeroides are tilted on the average at 20-25° from the normal to the membrane is fairly consistent with the X-ray data. This gives confidence for interpreting similar spectroscopic data on other Chl-protein complexes as in the case, for example, of the antenna complexes (2,4,29,30), photosystem I from higher plants (31) and other membrane systems for which no X-ray structure is available. For the organization of proteins in antenna complexes, the reader is referred to (4).

4. LIGHT-INDUCED FTIR DIFFERENCE SPECTROSCOPY OF THE PRIMARY REACTIONS IN PHOTOSYNTHESIS

This experimental approach is aimed at gaining informations on i) the molecular interactions between the pigments involved in the charge separation and their anchoring sites in the protein cage and ii) the changes which affect these molecular interactions during the charge separation and the subsequent steps of stabilization.

In contrast to resonance Raman (RR) spectroscopy where selectivity allows the pigment vibrations in a pigment-protein complex to be studied, IR spectroscopy is sensitive to all bonds of the complex i.e., bonds of Chl (conjugated or side groups), protein (peptide and side chain), lipids and even bound water molecules. IR spectroscopy has proved to be an extremely useful tool for the study of Chl-Chl interactions in vitro since it is not limited to the bonds involved in the π electron system. However, in the case of a large pigment-protein complex, the IR vibrational bands of the chromophores are superimposed on the much stronger absorption bands of the protein and water. This non-selectivity makes the in vivo investigation of single pigments molecules in a large Chl-protein complex a rather difficult enterprise with conventional IR. Using FTIR difference spectroscopy, the sensitivity is high enough to detect perturbations in the vibrational modes of single groups of Chls and protein in a RC. Only modes that change in intensity or position during the external perturbation will appear in the difference spectrum. This is an important advantage since the rest of the signals originating from functional groups not participating in the change are subtracted out. Thus, a FTIR difference spectrum provides a mean of determining molecular interactions at the level of single chemical groups.

This approach has been successfully applied to the study of the primary electron donor photooxidation (PIQ $\longrightarrow$ P$^+$IQ$^-$) and of the intermediary electron acceptor photoreduction (PIQ$^-$ $\longrightarrow$ PI$^-$Q$^-$) in a variety of BChl a- and BChl b- containing purple photosynthetic bacteria (32-35) as well as in green plants (36,37). The light-induced difference spectra consist of a number of characteristic bands highly reproducible in frequency and in amplitude. For a detailed interpretation of the IR difference bands, however, a comparison to model compounds spectra as well as investigation of isotope-labelled complexes is necessary. Previous IR spectroscopic studies from the last years have covered only the neutral species (38). In order to obtain appropriate models for the photooxidation of P and the photoreduction of I, the cation of (B)Chl a (b) and the anion of (B)Pheo a (b) have been generated in an electrochemical cell specially designed for IR measurements (39,40). The selectivity of FTIR spectroelectrochemistry of isolated pigments can be as high as the selectivity for the light-induced FTIR spectroscopy of large Chl-protein complexes.

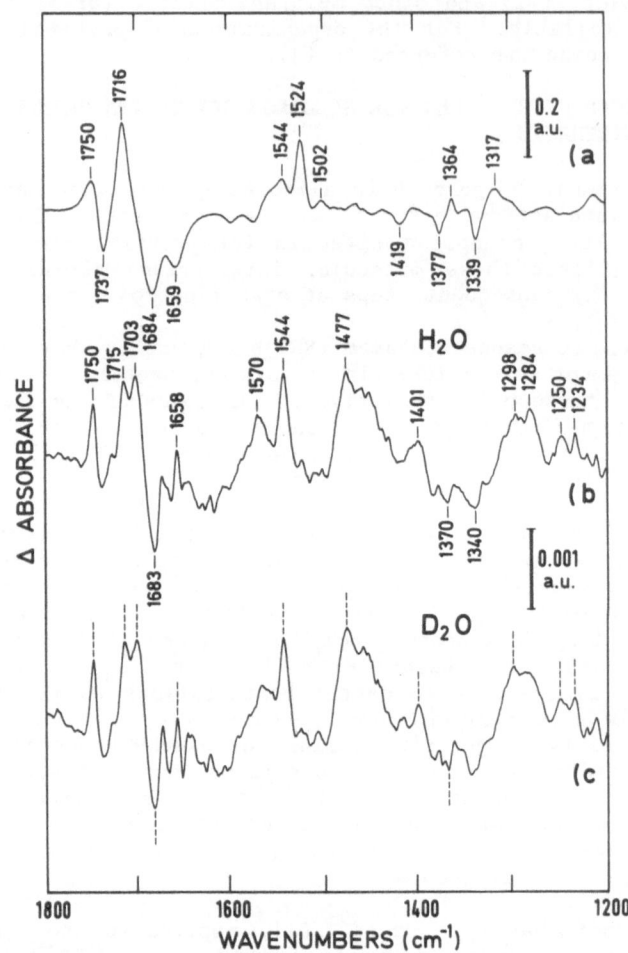

FIGURE 2. Comparison of a) the FTIR difference spectrum of the BChl a cation formation (BChl a$^+$ - BChl a) in deuterated THF (U=+0.5V, T=295K, 64 scans) with the light-induced FTIR difference spectra of the primary donor photooxidation in films of <u>Rb. sphaeroides</u> RC prepared from b) H_2O suspension c)D_2O suspension after three days incubation in D_2O at room temperature. 715 nm $\langle$ λ excitation $\langle$ 1100 nm, T=255K, 512 scans, Resolution : 4 cm^{-1}.

This new application of FTIR spectroscopy combined with electro-chemistry for the characterization of pigment radicals is described in (40) and in this volume (M. Leonhard et al). Here, we will compare the molecular changes associated in vivo with the light-induced oxidation of the primary donor and reduction of the intermediary acceptor to those obtained in vitro with pigments radicals generated electro-chemically.

Fig. 2 shows the light-minus-dark FTIR spectra of the primary donor photooxidation (termed P^+ spectra) in Rb. sphaeroides RC and the difference FTIR spectrum of the BChl a cation formation, BChl a^+-minus-BChl a (termed BChl a^+ spectrum) generated in the electro-chemical cell in deuterated tetrahydrofuran (THF). The BChl a cation can also be generated in protic solvents such as methanol (35,39,41). The FTIR difference spectrum of the photoreduction of the intermediary acceptor (also termed I^- spectrum) in Rps. viridis RC is displayed on Fig. 3 together with the BPheo b anion -minus- BPheo b spectrum (termed BPheo b^- spectrum) for comparison.

Such FTIR difference spectra contain contributions arising from the appearance of radicals (positive bands) and disappearance of neutral species (negative bands). In general, spectral band features reflect absorption changes of specific chemical groups. They may be either due to an intrinsic change in oscillator strength upon radical formation or to a change of environment, e.g., H-bonding. At this level of sensitivity, the changes observed in vivo are of the order of 10^{-3}-10^{-4} absorbance units. Any large conformational change of the protein backbone concomitant with these photoreactions would give rise to differential bands in the amide I and amide II regions of the IR spectrum. Our results (Fig. 2b and Fig. 3b) clearly demonstrate that such large changes of the protein backbone do not occur upon photooxidation of P or photoreduction of I. However, specific band features are observed, especially in the carbonyl frequency region (1620-1760 cm^{-1}). Using model compound spectra, absorption changes detected in vivo can be related to the bond strength of the four carbonyls, namely the 7c and 10a ester, 9 keto and 2 acetyl groups for the pigments in P and I. In particular, interactions assumed by these groups may be tentatively described from our FTIR data for both the neutral and radical states of P and I.

4.1. The environment of the carbonyl groups of the BChl molecules constituting the primary donor in Rb. sphaeroides and Rps. viridis RC

X-ray analysis of RC crystals demonstrated that several C=O groups of the two BChls (BCLP, BCMP) constituting P show specific interactions with amino acid side chains of the L and M protein subunits. Moreover, several asymmetries in the distribution of the protein polar groups surrounding P is clearly exhibited in Rps. viridis RC (42).

4.1.1. The ester carbonyls
In 1986, Michel et al. (42) suggested that the 10a ester C=O group at ring V of BCMP in Rps. viridis could interact with the side chain of SER M203. However, a refined 2.3 Å map (J. Deisenhofer and H. Michel, personal communication) favours a free 10a C=O. No information is, at the present time, available for this group in Rb. sphaeroides. In P^+ and BChl a^+ spectra (Fig. 2), the highest frequency band in the C=O region could be in principle assigned to the 7c and/or 10a C=O ester groups. For bonds not involved in conjugation such as these ester carbonyls, only inductive coulombic effects from the π electron system would be expected (35). Comparing the distance of these groups to the

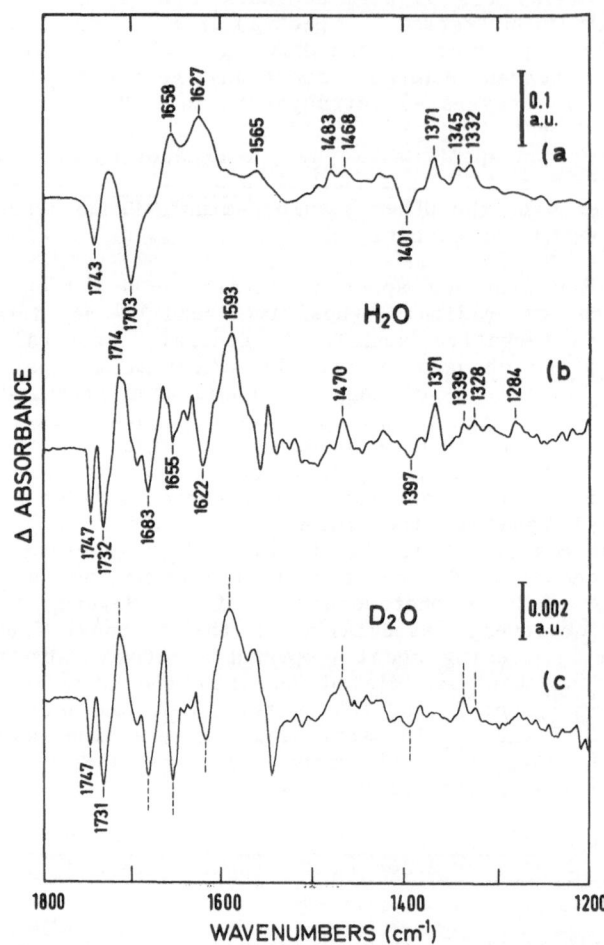

FIGURE 3. Comparison of a) the FTIR difference spectrum of the BPheo b anion formation (BPheo b⁻ – BPheo b) in deuterated THF at U=-0.8V (64 scans, T=295K) with the light-induced FTIR difference spectra of the intermediary acceptor photoreduction in films of <u>Rps. viridis</u> RC prepared from b) H_2O suspension and covered with dithionite redox H_2O buffer c) D_2O suspension and covered with dithionite redox D_2O buffer. 715 nm < λ excitation < 1100 nm, T=240K, 1024 scans, Resolution : 4 cm⁻¹.

conjugated system, the 10a ester C=O should be predominantly affected, as demonstrated using pyroBChl a (M. Leonhard et al., this volume). In the BChl a$^+$ spectrum, it is shown that when free of molecular interaction, the 10a C=O is shifted from 1737 cm^{-1} to 1750 cm^{-1} upon cation formation in THF. When the BChl a cation is generated in methanol in which the ester C=O are expected to be H-bonded, only a positive band at 1756 cm^{-1} is observed without a negative counterpart (35,41). This was interpreted as due to the breakage of the H-bond on the 10a C=O during cation formation (35). In the P$^+$ spectrum of Rb. sphaeroides RC, a band feature closely corresponding to the BChl a+ spectrum in methanol is observed at 1750 cm^{-1}. This suggests that in Rb. sphaeroides, the 10a ester C=O is H-bonded in the neutral state of the BChl dimer. This H-bond would be essentially broken in the P$^+$ state. In contrast, in Rps. viridis RC, our FTIR data suggest non-interacting ester C=O group(s) in the neutral dimer (35), in good agreement with X-ray data. The possibility that amino acid side chain carboxyl groups might contribute in the 1700-1760 cm^{-1} region was investigated by H-D exchange with isolated RC (Fig. 2c). No band shift or amplitude effect was observed upon exchange, indicating the absence of a contribution from accessible protonated carboxylic groups.

4.1.2. The keto carbonyls

The X-ray structure of Rps. viridis RC indicates differences in H-bonding to the 9 keto C=O of P for the L and M side monomers. On the L side, the 9 keto C=O of BCLP is close to THR L248 while on the M side, the THR is replaced by a non H-bonding ILE. In contrast, recent X-ray data at 3.2 Å resolution on Rb. sphaeroides RC suggest that the amino acid environment surrounding P is more symmetric than in Rps. viridis. In particular, both 9 keto C=O are not H-bonded (43). In the P+ spectra of Rb. sphaeroides (Fig. 2), a clear negative band at 1683 cm^{-1} as well as two split positive signals at 1715 cm^{-1} and 1703 cm^{-1} are observed. When BChl a$^+$ is formed in THF, a shift from 1684 cm^{-1} to 1716 cm^{-1} close to that observed in vivo is found. We thus assign the 1683 cm^{-1} band in P$^+$ spectra to a free keto C=O in the neutral special pair. An additional negative band at 1669 cm^{-1} might account for a weakly interacting group assigned to the keto of the second BChl monomer. However, a possible contribution of C=O from the protein or quinones in the P$^+$ spectrum (the P$^+$ spectrum is actually a P$^+$Q$^-$-minus-PQ spectrum) cannot be excluded. The 1683 cm^{-1} and 1669 cm^{-1} C=O groups in P would give rise to the 1715 cm^{-1} and 1703 cm^{-1} positive signals in P$^+$ and thus point to a non symmetric environment of the two keto C=O in both the neutral and the cation form of the dimer. The IR bands positions are in good agreement with RR data (44). However, the X-ray structural model of Rb. sphaeroides points to an equivalent non H-bonded form for 9 keto C=O of both monomers (43). In this respect, a detailed structural analysis of the Rb. sphaeroides RC would help to corroborate the description derived from the FTIR data. Indeed, the P$^+$ spectra of Rps. viridis RC only suggests a weakly H-bonded keto C=O group absorbing at 1674 cm^{-1} (35) in agreement with RR and X-ray data. A large positive band at 1712 cm^{-1} would account for free keto C=O in the cation dimer, as in Rb. sphaeroides.

4.1.3. The acetyl carbonyls

The 2a acetyl C=O group of BChl a is expected to absorb in the 1620-1665 cm^{-1} region. In the BChl a$^+$ spectrum in THF (Fig. 2a), a negative band at 1659 cm^{-1} indicates a free acetyl C=O in the neutral BChl a state. However, in P$^+$ spectra, a clear assignment as in the case of the ester and keto C=O vibrations does not seem justified due to the possible contribution from peptide, quinones C=O groups or of OH bending vibration of bound water. In Rps. viridis RC, both acetyls of the dimer are close to amino acid groups. The 1636 cm^{-1} and 1624 cm^{-1} IR bands (32) observed in Rps. viridis P$^+$ spectra, in agreement with RR bands at 1634 cm^{-1} and 1628 cm^{-1} (45), could thus correspond to the two H-bonded acetyl C=O observed in the RC crystal.

4.2. The environment of the carbonyl groups of the intermediary acceptor in Rps. viridis and C. vinosum

4.2.1. The ester carbonyls

In Rps. viridis RC, the 10a ester C=O group in both BPheo L and BPheo M appears to be close to the side chains of TRP L100 and TRP M127, respectively (42). The Rps. viridis I$^-$ spectrum shows two negative bands at 1747 cm^{-1} and 1732 cm^{-1}. Instead a single band at 1743 cm^{-1} is observed in the BPheo b$^-$ spectrum (Fig. 3). As discussed in (33,35,46), the two bands observed in vivo might be assigned to the 10a and 7c ester C=O groups of the neutral intermediary acceptor, provided that no contribution of amino acid side chains appears in this spectral region. Comparing with the negative 1743 cm^{-1} band in the BPheo b$^-$ spectrum, the 1747 cm^{-1} and 1732 cm^{-1} negative bands in Rps. viridis I$^-$ spectrum would then correspond to a free and a weakly bonded ester group, respectively. Taking together the IR and X-ray data, we assign the band at 1732 cm^{-1} to the 10a C=O group whose frequency is downshifted by H-bonding to TRP L100. As equivalent signals are also observed in the I$^-$ spectra of C. vinosum (at 1729 cm^{-1}) and of various photosystem II preparations (at 1720 cm^{-1}), we thus suggest that the 10a ester C=O of the BPheo a (C. vinosum) and Pheo a (photosystem II) is also H-bonded in its neutral state (46).

In contrast to P$^+$ spectra, an hydrogen isotope effect is detected in Rps. viridis I$^-$ spectra when the photoreduction of the intermediary acceptor is performed in D$_2$O (Fig. 3). Under these conditions, a ~ 30% decrease of the band at 1747 cm^{-1} relative to the 1732 cm^{-1} band is observed (with almost no shift of the 1732 cm^{-1} band) which favours the assignment of at least part of this band to a protein side chain carboxylic group protonated in the neutral state of the intermediary acceptor. The corresponding COOD band displaced to lower frequency would be masked by the 1731 cm^{-1} band. The assignment of this protein group will be discussed below in connection with the description of the keto C=O binding site.

4.2.2. The keto carbonyl

From the Rps. viridis X-ray structural model, the most important difference in the binding sites of BPheo L and BPheo M is the presence of a GLU residue (L104) close to the keto C=O of BPheo L, with the correct distance to form an H-bond (42). It is replaced in the M branch by a non-polar VAL M131 residue. Michel et al. (42) have proposed that GLU L104 is protonated. In Rb. sphaeroides RC, this GLU residue is conserved in the sequence of the L subunit (as well as in the sequence of the D1 polypeptide of photosystem II) and is also positioned near the BPheo L ring V keto C=O (43,47). Indeed, our FTIR data on Rps. viridis RC indicate an interacting keto C=O group at 1683 cm^{-1} (downshifted with respect to the 1703 cm^{-1} keto C=O band of the BPheo b model compound in THF) for the intermediary acceptor. In C. vinosum and photosystem II, the 9 keto C=O of (B)Pheo a is assigned to a negative band at 1675-1676 cm^{-1} (46) suggesting an H-bonded C=O in the neutral state of these intermediary acceptors.

Furthermore, in view of the isotope effect on the 1747 cm^{-1} band in Rps. viridis I$^-$ spectra which suggests that this mode is in part due to a protonated carboxylic group of an amino acid, we recently proposed

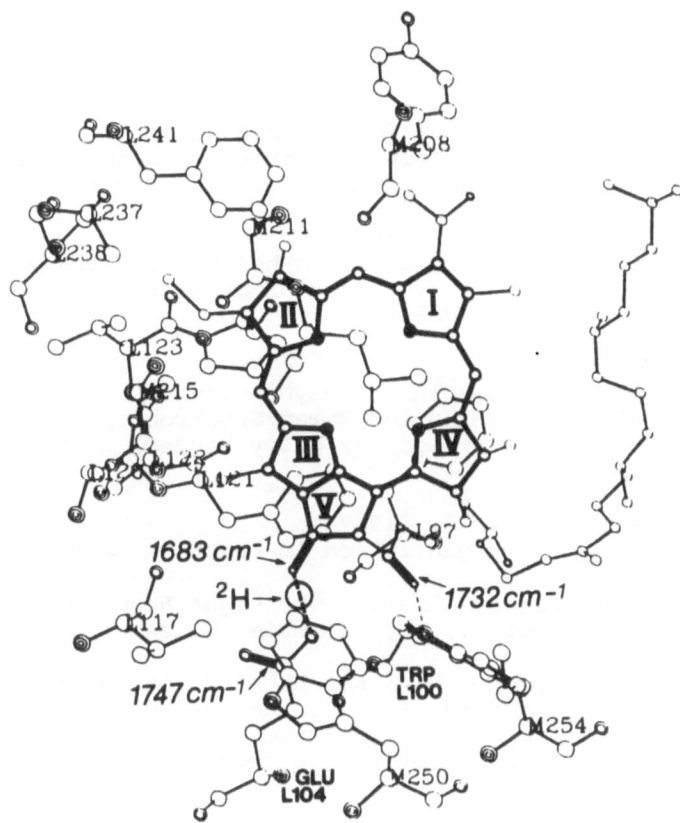

FIGURE 4. The protein residues surrounding BPheo L. Possible hydrogen bonds are indicated with dashed lines. From Michel et al., ref. 42. Tentative IR bands assignments (Fig. 3) to specific C=O groups are also indicated.

(46) as a working hypothesis that the exchangeable proton is the one from GLU L104 side chain (Fig. 4). Such H-D exchange of a proton has also been recently identified by Feher et al. (48) from ENDOR studies of the reduced intermediary acceptor in <u>Rb. sphaeroides</u> RC and an H-bond was postulated between GLU L104 and the 9 keto C=O of BPheo L. As the difference spectra obtained from the intermediary acceptor reduction in <u>C. vinosum</u> (BPheo a) and in photosystem II (Pheo a) show similarities to that observed from the <u>Rps. viridis</u> RC, we suggest that the bonding interactions of the intermediary acceptor might thus be comparable.

ACKNOWLEDGEMENTS

We would like to thank S. Andrianambinintsoa, Dr. G. Berger and J. Kléo for the preparation of the purified reaction centers and chlorophylls, A. Wollenweber for her help with the electrochemistry, A.-M. Bardin for the UVCD measurements. Part of this work was funded by the European Community (Contract ST2J - 0118 - 2-D).

REFERENCES

1 Breton J. and Verméglio A. (1982)
Orientation of photosynthetic pigments in vivo.
in: 'Photosynthesis : Energy Conversion by Plants and Bacteria'
(Govindjee, Ed.), Academic Press, New York, pp.153-194.

2 Breton J. and Nabedryk E. (1984)
Transmembrane orientation of alpha-helices and the organization
of chlorophylls in photosynthetic pigment-protein complexes.
FEBS Lett. 176(2), 355-359.

3 Lutz M. (1984)
Resonance Raman studies in photosynthesis.
in: 'Advances in Infrared and Raman Spectroscopy,'
(Clark R.J.H, Hester R.E eds.), Wiley-Heyden, Londres, Vol. 11,
pp. 211-300.

4 Breton J. and Nabedryk E. (1987)
Pigment and protein organization in reaction center and antenna
complexes.
in: 'Topics in Photosynthesis - The Light Reactions',
pp. 159-195, (Barber J., ed), Elsevier Science Publishers B.V.

5 Deisenhofer J., Epp O., Miki K., Huber R. and Michel H. (1984)
X-ray structure analysis of a membrane protein complex.
Electron density map at 3A resolution and a model of the
chromophores of the photosynthetic reaction center of
Rhodopseudomonas viridis.
J. Mol. Biol. 180, 385-398.

6 Allen J.P., Feher G., Yeates T.O., Komiya H. and Rees D.C.
(1987)
Structure of the reaction center from Rhodobacter sphaeroides
R-26: The cofactors.
Proc. Natl. Acad. Sci. USA 84, 5730-5734.

7 Deisenhofer J., Epp O., Miki K., Huber R. and Michel H. (1985)
Structure of the protein subunits in the photosynthetic
reaction center of Rhodopseudomonas viridis at 3A
resolution.
Nature (Lond.) 318, 618-624.

8 Allen J.P., Feher G., Yeates T.O., Komiya H. and Rees D.C.
(1987)
Structure of the reaction center from Rhodobacter sphaeroides
R-26: The protein subunits.
Proc. Natl. Acad. Sci. USA 84, 6162-6166.

9 Zuber H. (1987)
The structure of light-harvesting pigment-protein complexes.
in: 'Topics in Photosynthesis. The Light Reactions',
pp. 197-259. (J. Barber, ed.). Elsevier Science Publishers B.V.

10 Susi H. (1969)
Infrared spectra of biological macromolecules and related
systems.
in: 'Structure and Stability of Biological Macromolecules',
(Timasheff S.N. and Fasman G.D., eds.) pp. 575-663, Dekker, New
York.

11 Nabedryk E. and Breton J. (1981)
    Orientation of intrinsic proteins in photosynthetic membranes.
    Polarized infrared spectroscopy of chloroplasts and
    chromatophores.
    Biochim. Biophys. Acta 635, 515-524.

12 Susi H. and Byler D.M. (1983)
    Protein structure by Fourier transform infrared spectroscopy :
    second derivative spectra.
    Biochem. Biophys. Res. Commun. 115, 391-397.

13 Chen Y.H., Yang T. and Chau K.H. (1974)
    Determination of the alpha-helix and beta form of proteins in
    acqueous solution by circular dichroism.
    Biochemistry 13, 3350-3359.

14 Henderson R. and Unwin P.N.T. (1975)
    Three-dimensional model of purple membrane obtained by electron
    microscopy.
    Nature (Lond.) 257, 28-32.

15 Rothschild K.J. and Clark N.A. (1979)
    Polarized infrared spectroscopy of oriented purple membrane.
    Biophys. J. 25, 473-488.

16 Nabedryk E., Bardin A.-M. and Breton J. (1985)
    Further characterization of protein secondary structures in
    purple membrane by circular dichroism and polarized infrared
    spectroscopies.
    Biophys. J. 48, 873-876,

17 Kleffel B.R., Garavito M., Baumeister W. and Rosenbusch J.P.
    (1985)
    Secondary structure of a channel-forming protein : porin from
    E. coli outer membranes.
    EMBO J. 4, 1589-1592.

18 Nabedryk E., Garavito R.M. and Breton J. (1988)
    The orientation of beta-sheets in porin. A polarized Fourier
    transform infrared spectroscopic investigation.
    Biophys. J. 53, 671-676.

19 Mao D. and Wallace B.A. (1984)
    Differential light scattering and absorption flattening optical
    effects are minimal in the circular dichroism spectra of small
    unilamellar vesicles.
    Biochemistry 23, 2667-2673.

20 Nabedryk E., Tiede D.M., Dutton P.L. and Breton J. (1982)
    Conformation and orientation of the protein in the bacterial
    photosynthetic reaction center.
    Biochim. Biophys. Acta 682, 273-280.

21 Byler D.M. and Susi H. (1986)
    Examination of the secondary structure of proteins by
    deconvolved FTIR spectra.
    Biopolymers. 25, 469-487.

22 Trewhella J., Popot J.-L., Zaccai G. and Engelman D.M. (1986)
    Localization of two chymotryptic fragments in the structure of
    renatured bacteriorhodopsin by neutron diffraction.
    EMBO J. 5, 3045-3049.

23 Tsygannik I.N. and Baldwin J.M. (1987)
Three-dimensional structure of deoxycholate-treated purple
membrane at 6A resolution and molecular averaging of three
crystals forms of bacteriorhodopsin.
Eur. Biophys. J. 14, 263-272.

24 Williams J.C., Steiner L.A., Ogden R.C., Simon M.I. and
Feher G. (1983)
Primary structure of the M subunit of the reaction center from
Rhodopseudomonas sphaeroides.
Proc. Natl. Acad. Sci., USA 80, 6505-6509.

25 Williams J.C., Steiner L.A., Feher G. and Simon M.I. (1984)
Primary structure of the L subunit of the reaction center from
Rhodopseudomonas sphaeroides.
Proc. Natl. Acad. Sci USA 81, 7303-7307.

26 Nabedryk E., Berger G., Andrianambinintsoa S. and
Breton J. (1985)
Comparison of alpha-helix orientation in the chromatophore,
quantasome and reaction centre of Rhodopseudomonas viridis by
circular dichroism and polarized infrared spectroscopy.
Biochim. Biophys. Acta 809, 271-276.

27 Yeates T.O, Komiya H., Rees D.C., Allen J.P. and
Feher G. (1987)
Structure of the reaction center from Rhodobacter sphaeroides
R-26 : membrane-protein interactions.
Proc. Natl. Acad. Sci. USA 84, 6438-6442.

28 Nabedryk E., Tiede D.M., Dutton P.L. and Breton J. (1984)
Polarized infrared spectroscopy of bacterial reaction centers;
the LMH and LM complexes in reconstituted membranes
in: 'Advances in Photosynthesis Research', Vol. II, 177-180.
(Sybesma C, ed.), Nijhoff M, Junk D.W. Publ.

29 Nabedryk E., Andrianambinintsoa S. and Breton J. (1984)
Transmembrane orientation of alpha-helices in the thylakoid
membrane and in the light-harvesting complex. A polarized
infrared spectroscopy study.
Biochim. Biophys. Acta 765, 380-387.

30 Hiller R.G., Bardin A.-M. and Nabedryk E. (1987)
The secondary structure content of pigment-protein complexes
from the thylakoids of two chromophyte algae.
Biochim. Biophys. Acta 894, 365-369.

31 Nabedryk E., Biaudet P., Darr S., Arntzen C.J. and Breton J.
Breton J. (1984)
Conformation and orientation of chlorophyll-proteins in
Photosystem-I by circular dichroism and polarized infrared
spectroscopies.
Biochim. Biophys. Acta 767, 640-647.

32 Mäntele W., Nabedryk E., Tavitian B.A., Kreutz W. and
Breton J. (1985)
Light-Induced Fourier transform infrared (FTIR) spectroscopic
investigations of the primary donor oxidation in bacterial
photosynthesis.
FEBS Lett. 187, 227-232.

33 Nabedryk E., Mäntele W., Tavitian B.A. and Breton J. (1986)
   Light-induced Fourier transform infrared (FTIR) spectroscopic
   investigations of the intermediary electron acceptor reduction
   in bacterial photosynthesis.
   Photochem. Photobiol. 43, 461-465.

34 Nabedryk E., Tavitian B.A., Mäntele W., Kreutz W. and
   Breton J. (1987)
   Fourier transform infrared (FTIR) spectroscopic investigations
   of the primary reaction in purple photosynthetic bacteria.
   in 'Progress in Photosynthesis Research', Vol. I, pp. 177-180.
   (Biggins J., ed.). Martinus Nijhoff Publ., Dordrecht.

35 Mäntele W.G., Wollenweber A., Nabedryk E. and Breton J. (198
   Infrared spectroelectrochemistry of bacteriochlorophylls and
   bacteriopheophytins : implications for the binding of the
   pigments in the reaction center from photosynthetic bacteria.
   Proc. Natl. Acad. Sci. USA. 8468-8472.

36 Tavitian B.A., Nabedryk E., Mäntele W. and Breton J. (1986)
   Light-induced Fourier transform infrared (FTIR) spectroscopic
   investigations of primary reactions in Photosystem I and
   Photosystem II.
   FEBS Lett. 201, 151-157.

37 Tavitian B.A., Nabedryk E., Wollenweber A., Mäntele W. and
   Breton J. (1988)
   FTIR spectroscopy of primary photosynthetic reactions in
   the two photosystems of green plants. Comparison with
   spectroelectrochemistry of chlorophyll models.
   in: 'Spectroscopy of Biological Molecules New Advances',
   (Schmidt E.D., Schneider F.W. & Siebert F., eds.), Wiley &
   Sons Chichester, UK, pp. 297-300.

38 Ballschmiter K. and Katz J.J. (1969)
   An infrared study of chlorophyll-chlorophyll and chlorophyll-
   water interactions.
   J. Am. Chem. Soc. 91, 2661-2677.

39 Wollenweber A. (1986)
   Spektroelektrochemische Untersuchungen an den
   Bakteriochlorophyllen a und b, sowie den
   Bakteriophäophytinen a und b.
   Diplomarbeit Universität Freiburg, FRG.

40 Mäntele W., Wollenweber A., Rashwan F., Heinze J., Nabedryk
   Berger G. and Breton J. (1988)
   Fourier transform infrared spectroelectrochemistry of the
   bacteriochlorophyll a anion radical.
   Photochem. Photobiol. 47, 451-455.

41 Mäntele W., Wollenweber A., Nabedryk E., Breton J., Rashwan
   Heinze J. and Kreutz W. (1987)
   Fourier-transform infrared (FTIR) spectroelectrochemistry of
   bacteriochlorophylls.
   in 'Progress in Photosynthesis Research'. Vol. I, pp. 329-332.
   (Biggins J., ed.). Martinus Nijhoff Publ., Dordrecht.

42 Michel H., Epp O. and Deisenhofer J. (1986)
   Pigment-protein interactions in the photosynthetic reaction
   centre from Rhodopseudomonas viridis.
   EMBO J. 5, 2445-2451.

43 Tiede D.M., Budil D.E., Tang J., El-Kabbani O., Norris J.R.
Chang C.H. and Schiffer M. (1988)
Symmetry breaking structures involved in the docking of
cytochrome c and primary electron transfer in reaction centers
of Rhodobacter sphaeroides.
in: 'The Photosynthetic Bacterial Reaction Center:
Structure and Dynamics', (Breton J. and Vermeglio A., eds.)
Vol. 149, pp. 13-20. NATO ASI Series, Plenum, N.Y.

44 Robert B. and Lutz M. (1986)
Structure of the primary donor of Rhodopseudomonas sphaeroides:
difference resonance Raman spectroscopy of reaction centers.
Biochemistry 25(9), 2303-2309.

45 Zhou Q., Robert B. and Lutz M. (1987)
Interspecific structural variations of the primary donor in
bacterial reaction centers.
in: 'Progress in Photosynthesis Research', Vol. 1, pp. 395-398.
(Biggins J., ed.), Martinus Nijhoff.Publ., Dordrecht.

46 Nabedryk E., Andrianambinintsoa S., Mäntele W.
and Breton J. (1988)
FTIR spectroscopic investigations of the intermediary electron
acceptor photoreduction in purple photosynthetic bacteria and
green plants.
in: 'The Photosynthetic Bacterial Reaction Centers : Structure
and Dynamics',
(Breton J. and Verméglio A., eds.), Vol. 149, pp. 237-250. NATO
ASI Series, Series A: Life Sciences, Plenum, New York.

47 Allen J.P., Feher G., Yeates T.O., Komiya H. and
Rees D.C. (1988)
Structure of the reaction center from Rhodobacter sphaeroides
R-26 and 2.4.1.
in: 'The Photosynthetic Bacterial Reaction Center: Structure
and Dynamics', (Breton J. and Verméglio A., eds.) Vol. 149, pp.
5-11. NATO ASI Series, Plenum, N.Y.

48 Feher G., Isaacson R.A., Okamura M.Y. and Lubitz W. (1988)
Endor of exchangeable protons of the reduced intermediate
acceptor in reaction cent rs from Rhodobacter sphaeroides R-26.
in: 'The Photosynthetic Bacterial Reaction Center: Structure
and Dynamics', (Breton J. and Verméglio A., eds.) Vol. 149, pp.
229-235. NATO ASI Series, Plenum, N.Y.

THE STRUCTURE AND FUNCTION OF PHOTOSYSTEM II

Wim Vermaas

Arizona State University
Department of Botany
Tempe, AZ 85287-1601   USA

INTRODUCTION

Photosystem II (PS II) is a pigment-protein complex in thylakoid membranes from all oxygenic photosynthetic organisms (cyanobacteria and photosynthetic eukaryotes). It catalyzes the light-induced reduction of plastoquinone by water through a number of redox reactions. The electron transport chain in PS II is composed of various protein-bound components, which are held in close proximity and suitable orientation with respect to each other by the protein environment, so that a rapid and efficient electron transport is feasible. The PS II complex consists of at least five integral membrane proteins in the thylakoid (together forming the PS II "core complex"), in addition to several peripheral proteins. Many of these polypeptides interact with one or more components of the electron transport chain or with light-harvesting pigments, such that these protein ligands can fulfill a specific function in PS II.

Over the last few years, the understanding of the interactions between the proteins and the cofactors involved in various aspects of PS II activity has rapidly advanced. One of the major driving forces towards a better insight into these interactions has been the crystallization and X-ray diffraction analysis of reaction centers from photosynthetic purple bacteria[1-5], along with the realization that there is an extensive functional and structural homology between parts of PS II and of the reaction center complex from purple bacteria[6-10]. It has become clear how much the PS II proteins contribute to the particular properties and orientation the PS II cofactors need to optimally function in energy transfer and electron transport through the Photosystem. For this reason, protein structure and cofactor function need to be considered in conjunction, even though generally biochemical and biophysical aspects of PS II have been treated separately. A simultaneous and integrated understanding of structural and functional aspects is expected to be critical to the design of further experimentation geared towards an elucidation of the mechanism by which PS II can successfully utilize light energy to initiate a well-orchestrated series of redox reactions that is one of the corner stones of the photosynthesis process.

This review is designed to provide a synthesis of structural and functional information available on PS II, while we have attempted to minimize overlap with other recent reviews[7,10-14] in the area. Space limitations preclude the review to be anywhere close to comprehensive. For more information, the reader is referred to any of the recent reviews, or to one of the other chapters in this volume.

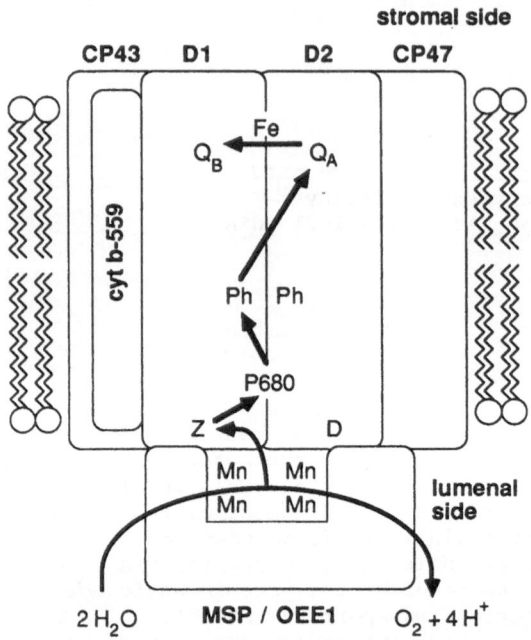

Fig. 1. Model of the Photosystem II core complex organization in the thylakoid membrane as it occurs in all oxygenic photosynthetic organisms. The site(s) of interaction of the 33,000 M_r peripheral protein with the core complex are still the subject of speculation; one of the more probable scenarios for this interaction has been depicted in this model. P680: the "primary donor" chlorophyll(s) of the reaction center; Ph: the primary acceptor in the reaction center, pheophytin a; Q_A: the first electron-accepting plastoquinone; Q_B: the second electron accepting plastoquinone; Fe: a non-heme iron of as yet unknown function; Z: the physiological donor to P680; D: an electron donor with slow redox kinetics; Mn: the manganese involved in water splitting.

To serve as a frame of reference for the following sections of the review, a model of most of the proteins and their cofactors in PS II is shown in Fig. 1. The genes coding for these polypeptides are summarized in Table I. The PS II core complex is composed of at least five proteins: (a) the chlorophyll-binding proteins CP47 and CP43, which function as light-harvesting antenna proteins, (b) the 32-34 kDa proteins D1 and D2, which provide the environment necessary for many of the PS II redox reactions to function properly, and (c) cytochrome b-559. In addition to these integral PS II proteins, there are several peripheral proteins, the most important of which perhaps is the

Table I. Genes and proteins of the Photosystem II core complex

| Gene | Product | Approximate mol. weight (kDa) | Apparent number of gene copies per genome | |
			spinach	Synechocystis 6803*
psbA	D1	32	1	3
psbB	CP47	47	1	1
psbC	CP43	43	1	1
psbD	D2	32	1	2
psbE	cyt b-559	10	1	1
psbF	cyt b-559	4	1	1

*: Synechocystis 6803 is a cyanobacterium that is frequently used for specific mutagenesis (site-directed mutagenesis, etc.) because of its spontaneous transformation and homologous recombination with foreign DNA.

"33 kDa" manganese-stabilizing protein (MSP[15], or OEE-1[16]), since this polypeptide is indispensable for stable oxygen evolution in vivo[16,17]. As will be discussed in more detail in the next sections, the functional center of PS II is P680, the primary donor in the reaction center, which, upon excitation by light, transfers an electron to the primary acceptor pheophytin $\underline{a}$ (Pheo). Pheo$^-$ reduces the plastoquinone Q_A, which in turn reduces the secondary plastoquinone Q_B, which, upon double-reduction to the quinol, can exchange with free plastoquinone in the thylakoid. Oxidized P680 is rapidly rereduced by Z (probably a Tyr residue in D1; discussed in a subsequent section), which then obtains an electron from a (presumably tetranuclear) Mn cluster. A molecule of O_2 is produced when four "electron holes" have accumulated in the Mn cluster.

The functional homology with the reaction centers from purple bacteria extends from either the primary acceptor or the primary donor in the reaction center to Q_B[7-10]. The PS II donor side shares no homology with that of the anoxygenic bacterial system. Structurally, the homology between PS II and the bacteria is limited to regions of D1 and D2, which are significantly homologous to parts of the bacterial L and M subunits, respectively. Also the topology of the D1 protein in the thylakoid has been shown to be homologous to that of the L subunit[18]. However, CP47, CP43, cyt b-559 and the MSP have no significant homology with the reaction center or antenna proteins in purple bacteria.

ELECTRON TRANSPORT THROUGH PHOTOSYSTEM II

In this section, first primary charge separation will be discussed, followed by the events that occur at the electron acceptor side. In the last part of this section, reactions at the donor side of PS II will be briefly reviewed.

P680

PS II electron transport is initiated by light excitation of a specific chlorophyll $\underline{a}$ (P680), the primary donor in the PS II reaction center. The absorption maximum of this pigment in the red region of the spectrum is close to 680 nm[19]. P680 excitation results either from excitation transfer from PS II antenna pigments (most notably chlorophylls, carotenoids and -in cyanobacteria- phycobilins) to P680, or from direct excitation of this reaction center pigment. Since in most preparations the antenna pigments greatly outnumber the P680's (by a factor of several hundred in intact systems or thylakoids[20]), light energy usually reaches the reaction center through the antenna pigments. The precise supermolecular structure of P680 is as yet unknown: some data are suggestive of a chlorophyll a monomer and some can be interpreted to indicate a dimeric structure[21] (the latter would resemble, to some extent, the "special pair" in photosynthetic purple bacteria[1-3]). However, it should also be kept in mind that there are many "intermediate" forms between pure monomeric and dimeric structures: two chlorophylls may be moved with respect to each other in a lateral and/or vertical direction, thus having the pigments behave neither as pure monomers nor as a dimer. The situation may be complicated by the observation that the strength of the interaction between the chlorophylls appears dependent on the state (singlet, triplet, oxidation state) of the special pair[21].

P680 is assumed to be localized close to the lumenal side of the thylakoid[21]. Upon excitation, P680 rapidly (in the ps time range) transfers an electron to the "primary acceptor", Pheo[22,23]. In purple bacteria, two bacteriopheophytin a's are found in relatively symmetrical orientations with respect to the primary donor (one bound to L, and one to M)[1-3], and only the one that interacts with the L subunit is an intermediate for further electron transfer[24]. In PS II, the Pheo/reaction center stoichiometry appears also to be two[25] (in fact, the composition of reaction center preparations has been normalized towards the Pheo content, assuming that to be two per reaction center[26]), but it should be kept in mind that it has not yet been experimentally proven that both Pheos are located at sites homologous to the bacteriopheophytins in purple bacteria.

Even though the functional and structural similarities between PS II and bacterial reaction centers have led to the development of an attractive homology concept, one should realize that part of this concept has not yet been confirmed experimentally, and there are even experimental results that hint to potential differences between the acceptor sides of PS II and reaction centers from purple bacteria. The most "serious" potential difference between the bacterial and PS II systems may be at the level of primary charge separation. In the first place, it is possible that the orientation of P680 is different from that of the special pair in bacteria: the spin-polarized chlorophyll triplet state, originating from a back reaction between Pheo and P680$^+$, is oriented radically differently in the PS II complex as compared to in the bacterial reaction center[27]. Moreover, the bacterial reaction center proteins provide ligands (His residues) to the central Mg^{2+} in bacteriochlorophylls (voyeur pigments) that are localized in between the primary donor and the two bacteriopheophytins[1-3]. The function of these bacteriochlorophylls may be to facilitate rapid and efficient electron transfer between the primary donor and acceptor rather than to act as redox intermediates. However, in PS II Mg^{2+} ligands for voyeur chlorophylls cannot be found in homologous positions compared to the situation in purple bacteria[2,7], thus casting some doubt on a striking similarity between the bacterial and PS II systems in this respect. In spite of these differences, the functional homology between the subsequent reactions at the acceptor sides of the two systems is very strong.

It should be pointed out that in the cyanobacterium <u>Synechocystis</u> 6803 a specific mutation of the His-197 of D2 into Tyr results in a loss of assembly (and, thus, of function) of the PS II complex[28,29]. The

His-197 residue in D2 is in a homologous position to that of the M-subunit His that binds one of the bacteriochlorophylls of the special pair. This indicates that the His-197 residue in D2 has an important function in the PS II complex, even though it has not been proven that it contributes to the binding of P680.

Pheophytin a

Pheo seems to be close to "half way" between P680 and Q_A[21] in terms of redox midpoint potential (and from comparison with bacterial systems also perhaps in terms of localization), and presumably functions to facilitate efficient and rapid electron transfer resulting in stable charge separation between P680 and Q_A. Reduction of Q_A by Pheo$^-$ occurs well within a ns[23].

From comparison with the reaction center from purple bacteria, electron transfer between Pheo (bound to D1) and Q_A may well involve Trp-255 in D2; this residue is homologous to Trp-250 of the bacterial M subunit, which is acting as a "bridge" between bacteriopheophytin in the L-branch and Q_A[2,3,7]. However, the aromatic groups that may interact with the bacteriopheophytins[2,3] do not seem to be very much conserved in D1 and D2, indicating that the interaction between pheophytin and the protein complex may be somewhat different in PS II compared to in the bacterial system. However, at the acceptor side the functional homology between bacterial reaction centers and PS II is so striking that the D1/D2 complex very likely contains two Pheos in positions similar to those in the bacterial system, with the Pheo bound to D1 being involved in linear electron transport to Q_A. Obviously, for more precise insight either X-ray diffraction studies after crystallization of PS II, or (as a next-best alternative) site-directed mutagenesis in psbA and psbD (followed by mutant analysis) will be required. (This review will not cover the various procedures for specific mutagenesis, which have been developed for use in certain cyanobacterial species and which have been recently reviewed elsewhere[30,31].)

Q_A

This plastoquinone is in proximity of the stromal side of the thylakoid membrane[21]. Thus, the redox reactions between P680 and Pheo and between Pheo and Q_A together have resulted in transporting an electron virtually across the thylakoid. Q_A in PS II has two relatively anomalous properties. First, in contrast to the situation in photosynthetic bacteria[32], Q_A in PS II cannot be easily extracted; however, recently replacement of Q_A by ubiquinone has been reported[33]. Second, in contrast to plastoquinones that are free in the thylakoid, Q_A can only be reduced to Q_A^- under physiological conditions[34]. Limitations imposed by its environment presumably prevent the formation of Q_A^-. A possible explanation for the lack of double-reduction of Q_A is that the Q_A^-/Q_A^- midpoint redox potential is lower than that of Pheo/Pheo$^-$. The amino acid residues or cofactors that induce this anomalous behavior have not yet been identified, although in bacterial systems removal of the non-heme iron between Q_A and Q_B (to be discussed later) facilitates double-reduction of Q_A[35], possibly by allowing protons access to the (semi)reduced Q_A.

The binding site of Q_A may include the His-214 residue of D2 (the one that is in a homologous position to the His residue binding Q_A in the bacterial M subunit): after mutagenesis of this D2 residue to Asn, the PS II complex no longer assembles, indicating an important function of the His-214 residue in D2[28,29].

The plastosemiquinone Q_A^- reduces a second plastoquinone, Q_B. The kinetics of this redox reaction (100-200 us[36]) are almost 6 orders of magnitude slower than that of the electron transport reactions between P680 and Q_A. The reason why PS II can afford the redox reaction between the two quinones to be slow is that re-reduction of oxidized P680 by the physiological donor Z (to be discussed later) occurs on the ns time scale, whereas a back reaction between P680$^+$ and Q_A^- is slower (150 us[37]); in other words, the electron on Q_A is far enough removed from P680$^+$ to effectively prevent a back reaction (which would result in a loss of the light energy to heat or fluorescence).

Q_B

Q_B is reversibly bound to D1, with its binding affinity to the protein being much higher when it is in the semiquinone state than when it is fully oxidized or reduced[38,39]. Q_B is a plastoquinone molecule, coming from the plastoquinone "pool" in the thylakoid membrane, only distinguishing itself from its kin in the pool in that it occupies the Q_B "site" in the D1 protein, where it can be reduced by Q_A^-. The apparent equilibrium constant between $Q_A^-.Q_B$ and $Q_A.Q_B^-$ is approximately 15-25 (equilibrium strongly towards $Q_A.Q_B^-$) at neutral pH under normal conditions[36,40]. Once Q_B^- is formed, the semiquinone remains at the site until it has been further reduced by Q_A^-, thus forming the quinol. This can leave the site, freeing it up for an oxidized plastoquinone[38,39].

PS II-directed herbicides (such as atrazine and diuron) have long been known to inhibit electron transport between Q_A and Q_B, and to induce a reversed electron transport from Q_B^- to Q_A[41]. This can be explained by the concept of a common binding environment for the quinone and the herbicides[38,39,42,43]. The phenomena associated with herbicide binding and inhibition of electron transport could be explained by either of two mechanisms. (a) Since Q_B has a low binding affinity, and the herbicides and Q_B^- have a much higher affinity for binding to D1, herbicides cannot bind while Q_B^- occupies the site, but will "jump on" once the electron resides at Q_A^-. (b) Alternatively, it may be suggested that both the herbicide and Q_B^- can occupy the binding environment simultaneously (even though with a lower affinity as when only one of the two occupies the binding environment), but that the semiquinone is "forced out" either by being actually induced to transfer the electron back to Q_A (due to a herbicide-induced decrease of the redox midpoint potential of the Q_B/Q_B^- couple), or by losing the competition with the herbicide at the moment that it becomes oxidized (when the system is in the $Q_A^-.Q_B$ state). Either of the two alternatives mentioned in (b) are more probable than the situation of either a quinone or a herbicide (but not both) fitting into the binding pocket. It was found that after covalent linkage of a (relatively small) quinone to the binding environment, atrazine bound with a lower affinity but had the same number of binding sites compared to a sample that did not contain covalently linked quinone[44]. Another argument for different (but potentially related or overlapping) binding sites for Q_B as compared to herbicides is that the kinetics of binding and release of plastoquinone are greatly different from that of atrazine or diuron: whereas Q_B binding and plastoquinol release must be in the ms time range, binding and release of atrazine or diuron takes many seconds[45].

Both Q_B properties and herbicide binding appear very sensitive to the protein environment. Mutants (or biotypes) of several plants (for example, _Amaranthus hybridus_, _Senecio vulgaris_, _Solanum nigrum_ and others) were observed to be resistant to atrazine, but not to a variety of other PS II-directed herbicides; the atrazine resistance was found to be also present in thylakoids from these mutants, indicating

Table II. Mutations of D1 and L causing herbicide resistance

	Organism	Residue	Resistant against	Reference
D1:	Synechococcus sp. PCC 7002	Phe-211 to Ser	A	54
	Synechococcus sp. PCC 7002	Val-219 to Ile	D	54
	Chlamydomonas reinhardtii	Val-219 to Ile	D	56
	Chlamydomonas reinhardtii	Ala-251 to Val	A,M	57
	Chlamydomonas reinhardtii	Phe-255 to Tyr	A	56
	Chlamydomonas reinhardtii	Gly-256 to Asp	A	11
	Chlamydomonas reinhardtii	Ser-264 to Ala	A	58
	Amaranthus hybridus and others	Ser-264 to Gly	A	47
	Anacystis nidulans R2	Ser-264 to Ala	A,D	59
	Chlamydomonas reinhardtii	Leu-275 to Phe	D	11
L:	Rhodopseudomonas viridis	Phe-216 to Ser	T	55
	Rhodopseudomonas viridis	Arg-217 to His and Ser-223 to Ala	T	55
	Rhodobacter sphaeroides	Tyr-222 to Gly	T	60
	Rhodobacter sphaeroides	Ser-223 to Pro	T	60
	Rhodobacter capsulatus	Leu-229 to various residues	A	61
	Rhodobacter sphaeroides	Ile-229 to Met	T	60,62

A: atrazine; D: diuron; M: metribuzin; T: terbutryn. Note that the
numbering of D1 residues is not identical to that of L in this area
of the proteins: Phe-216 in L may be homologous to Phe-255 in D1, and
Ser-223 in L is assumed to be homologous to Ser-264 in D1. All
triazine-resistant biotypes from higher plants of which psbA has
been sequenced have been shown to contain a Ser-264 to Gly mutation.
In green algae (Chlamydomonas reinhardtii and Euglena gracilis
(U. Johanningmeier and R.B. Hallick, Curr. Genet. 12 (1987) 465-470))
and in cyanobacteria, the Ser residue appears preferentially mutated
to Ala.

that PS II electron transport was no longer sensitive to the herbicide[46]. Sequencing of the psbA gene of these mutants revealed that the
mutant D1 protein was different in a single amino acid residue (Gly
instead of Ser at position 264) compared to wild type[11,47]. In addition
to an altered spectrum of herbicide sensitivity, the mutants also showed
modified Q_B properties. It was observed that the semiquinone equilibrium (between $Q_A^-.Q_B$ and $Q_A.Q_B^-$) was about in the middle, causing the
equilibrium Q_A^- concentration to be anomalously high[40]. This has a
marked effect on fluorescence induction as well as on flash-induced
oxygen evolution patterns[40,48,49]. In contrast to the interpretation
given in some reports[49,50], the rate of forward electron transfer from
Q_A to Q_B (until equilibrium has been reached) has not dramatically
changed in the mutant compared to in wild type[50]: it is the semiquinone
equilibrium that has shifted[40,48,51].

From labeling studies using radioactive azidoatrazine, the azido
group of the herbicide analog was concluded to interact mainly with
Met-214 in D1 (which is the residue next to the His residue that
presumably interacts with Q_B), as well as with residue(s) between
His-215 and Arg-225[52]. These sites are localized between the fourth and
fifth membrane-spanning helices in D1[18,43,53] (as is Ser-264), but are
some 40-50 residues removed from the residue in which a mutation led to

herbicide resistance. This indicates that several parts of the region between these two helices are involved in creating the herbicide-binding environment. Further evidence for the amino acid residues involved in quinone and herbicide binding to be localized in the region between these two membrane-spanning regions in D1 comes from sequencing of herbicide-resistant mutants of higher plants, green algae and cyanobacteria[11,54]. These data have been summarized in Table II. The sites at which a mutation can lead to herbicide resistance (and in some cases also to changes in Q_B properties) are all located between the

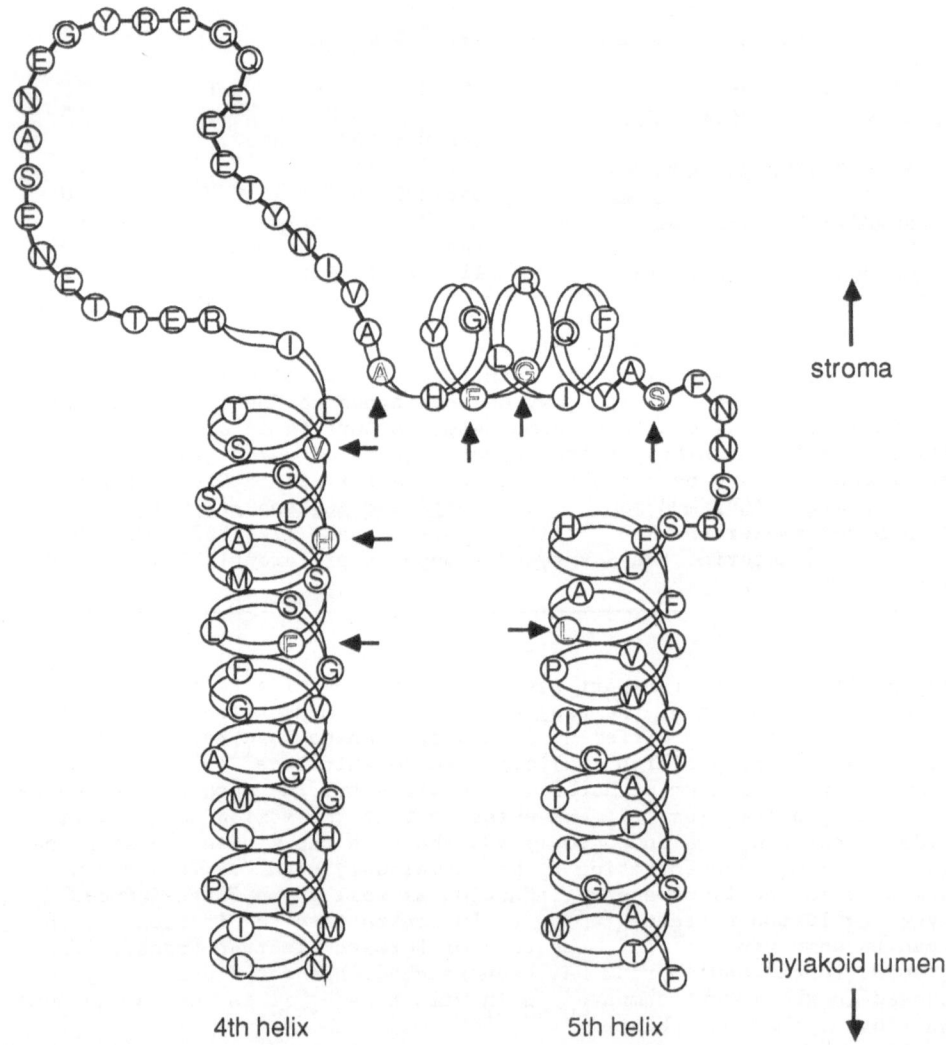

stroma

4th helix   5th helix

thylakoid lumen

Fig. 2. Hypothetical model of the herbicide-binding environment in the D1 protein in photosystem II. For clarity, only a region of D1 has been shown: from the beginning of the fourth membrane-spanning helix until the end of the fifth one. The region between the two helices is at the stromal side of the thylakoid. Arrows point to residues that can be mutated to give a herbicide-resistant phenotype (Table II). One arrow points to the His residue that is assumed to be involved in Q_B binding. Modified after [43].

last part of the fourth membrane-spanning helix in D1 and the first part of the fifth one (Fig. 2). It should be noted that not all of the mutations that lead to a herbicide-resistant phenotype have a marked effect on Q_B (and Q_B^-) properties[11].

In reaction centers from purple bacteria, herbicide binding has significant homologies with that in PS II. Even though herbicide affinity in general is several orders of magnitude lower in bacteria than in PS II, some of the sites at which mutations induce herbicide resistance are homologous in both systems (Table II). This implies that several details regarding herbicide binding to PS II can be learned by comparing it to the herbicide-binding environment in purple bacteria for which a high-resolution structure is available. A detailed crystal structure of <u>Rhodopseudomonas</u> <u>viridis</u> reaction centers with the herbicide terbutryn or the PS II inhibitor o-phenanthrolin has been published[2]. The two inhibitors appear to interact with different parts of the Q_B/herbicide binding environment[2], thus confirming the concept of close, but non-identical binding sites for different types of inhibitors or quinones. Crystallographic analysis in conjunction with the creation and characterization of herbicide-resistant[55] or site-directed mutants appears a powerful combination to lead to a further understanding of quinone and herbicide binding in bacterial reaction centers, and a significant part of the insight obtained presumably can be extrapolated to PS II.

The non-heme iron

Between Q_A and Q_B, a non-heme Fe(II) is located. The function of the iron in PS II electron transfer is not yet understood. In photosynthetic bacteria, reaction center activity with another divalent cation replacing the iron appears to be virtually unchanged[63,64]. However, in the absence of any divalent cation occupying the site the rate of electron transfer from Q_A to Q_B is decreased by a factor of two[63], and, perhaps more importantly, electron transfer from bacteriopheophytin to Q_A is significantly impaired[64]: the rate of forward electron transfer is decreased by a factor of 20, which cuts the quantum yield of stable charge separation about in half: in this case reduced pheophytin can back-react with the oxidized donor at a rate comparable to that of Q_A reduction.

Even though the non-heme iron does not appear to undergo redox changes during normal electron transport, there are conditions under which oxidation and/or subsequent rereduction can be detected in PS II. It was observed that a compound in the thylakoid could be oxidized by ferricyanide and could rapidly oxidize Q_A[65,66]. The midpoint redox potential of this component was found to be about 350-400 mV[65,66], and this (at the time mysterious) component was dubbed Q400. Later, it was elegantly shown by EPR and Mossbauer spectroscopy that Q400 was the non-heme iron near Q_A[67]. In photosynthetic bacteria, no conditions have been found to date under which redox reactions involving the iron can be observed to occur.

In spite of its lack of obvious involvement in electron transport between Q_A and Q_B, the non-heme iron has significant effects on various parameters involving the Q_B/herbicide binding environment. It was observed that oxidation of the non-heme iron by ferricyanide markedly altered the binding affinity of diuron[68,69] and o-phenanthrolin[69], but not of atrazine[68,69] in PS II. Perhaps even more surprisingly, it was found that selected high-potential quinones (such as phenyl-p-benzoquinone), when bound to the Q_B site, could first accept an electron in the normal way from Q_A, to form a semiquinone, and then could oxidize the non-heme iron to form a quinol and Fe^{3+}[70,71].

The previous discussion indicates that the non-heme iron appears to be in redox equilibrium with both Q_A and Q_B. The observation of reductant-induced oxidation of the non-heme iron provides a unique opportunity to "measure" the in situ midpoint redox potential of the semiquinone/quinol redox couple at the Q_B site, and can thus provide an independent measure of the stabilization of the semiquinone at the Q_B site.

The bicarbonate effect

It has been well-established that after CO_2/bicarbonate depletion (generally carried out in the presence of formate which presumably competes with bicarbonate for binding) electron transport between Q_A and Q_B has been inhibited[72,73]. This inhibition can be reversed by addition of bicarbonate. There are various indications that this "bicarbonate effect" is localized at or near the non-heme iron and/or Q_B. In the first place, the affinity of PS II-directed herbicides is generally affected by bicarbonate depletion[42]. Secondly, the ESR signal due to the Q_A^-/Fe^{2+} complex is greatly increased upon incubation with formate[74]; this increase is reversible upon incubation with bicarbonate. In the third place, after incubation with ferricyanide the oxidation of Q_A^- by the oxidized non-heme iron was reported to be not observed after bicarbonate depletion[75].

The bicarbonate effect is specific for PS II and has not been observed in photosynthetic bacteria. It has been suggested that in PS II bicarbonate provides two ligands for the non-heme iron, which in photosynthetic bacteria have been provided by a glutamate residue (Glu-232 in the M subunit)[7]. However, it is difficult to visualize why formate cannot provide ligands for the iron when glutamate can. Even though it is obvious that the formate/bicarbonate antagonism has rather dramatic effects on PS II function, its mechanism, its possible in vivo role, as well as its relationship to the mechanism of herbicide/quinone binding require further experimentation.

Photoinhibition

Upon exposure of thylakoids or more intact photosynthetic systems to high light intensity, PS II is specifically inactivated[76]. This phenomenon, referred to as photoinhibition, has often been linked to the rapid turnover of D1 in the light, at a rate where breakdown may be larger than synthesis, thus causing a depletion of D1 in PS II[77,78]. Other PS II proteins (including D2) turn over much less rapidly. Since D1 is a crucial component of a functional PS II, centers without D1 (photoinhibited centers) will be inactive. Originally the mechanism of photoinhibition was thought to be related to radical reactions involving the semiquinone and quinol anion forms of Q_B[78]. However, this hypothesis has not been substantiated by a great deal of experimental evidence, and is also conceptually unattractive because of (1) the long lifetime of Q_B^- in darkness, and (2) the short residence time of fully reduced Q_B at the Q_B site. What appears to be clear, though, is that the effect of photoinhibition also involves the reaction center itself[79,80] (which is not surprising in view of the fact that D1 is involved in primary reactions and perhaps also in the water-splitting process). It should be pointed out, however, that "photoinhibition" can be observed under many different experimental conditions and after various pretreatments, and it is not yet clear that the various types of photoinhibition all pertain to the same phenomenon (such as depletion of the D1 protein in PS II complexes)[81,82].

The rapid turnover of D1 in PS II poses a number of fascinating and challenging questions. (1) Can D1, the "heart" of the PS II complex, be removed and another protein be put in without the whole complex falling apart, or does PS II disassemble, and can it subsequently reassemble around another D1 protein? (2) What is the primary reason of the rapid D1 turnover? (3) What is the mechanism by which damaged D1 gets removed from the complex? There do not seem to be satisfactory answers to these questions yet. However, over the last few years, hypotheses regarding the origin of photoinhibition and the primary effects (such as D1 turnover) have been developed that may contribute to an answer. With regard to the origin of photoinhibition it has been suggested that a chlorophyll close to P680 can be oxidized by P680$^+$ [83], and the oxidized chlorophyll generates cofactor and/or protein damage. This appears to be a relatively reasonable hypothesis: the oxidized pigment could be around for a relatively long time, since it does not have a specific electron donor connected to it. The redox potential of an oxidized chlorophyll certainly is high enough to react with many molecules close by. It could be envisioned that a residue in D1 has a reasonable probability reacting with the oxidized chlorophyll, potentially resulting in a radical within the protein, and subsequently in protein damage. In support of this hypothesis, in systems where the oxidized primary donor has a lower potential (PS I, photosynthetic bacteria) no photoinhibition is observed. However, this scenario would imply that the oxidized chlorophyll has to be bound to D1 and not to, for example, CP47 or CP43. This presumably means that a D1 homolog of a "voyeur" bacteriochlorophyll of the bacterial reaction center complex would be the most likely candidate as the pigment that could be oxidized by P680. It should be pointed out that at this moment the hypothesis is still highly speculative.

The question now is how the damage near the reaction center is "translated" to attracting the attention of a specific protease (presumably located in the soluble phase around the thylakoid) to remove the damaged D1. One would have to postulate the existence of a conformational change in D1, either by damage to one or more amino acid residues or to the loss of a cofactor, resulting in susceptibility to protease attack. A somewhat more specific hypothesis has been formulated regarding the region of D1 recognized by the protease. It was observed[84] that the D1 sequence contained a region in between the 4th and 5th membrane-spanning helices that was rich in specific amino acid residues that can make up a so-called PEST sequence[85]. Such a sequence has been found to be a potential region where proteolysis occurs[84]. Neither any other PS II protein, nor the L and M subunit of bacteria have such a sequence (and neither of these turn over rapidly).

Thus, a relatively viable hypothesis for the mechanism of photoinhibition and the rapid turnover of D1 is that oxidized P680 has a certain probability to oxidize a nearby chlorophyll (which may be reduced in turn by cytochrome b-559[83]; see a subsequent section). The chlorophyll radical formed can oxidize amino acid residues and/or PS II cofactors, which triggers a "conformational change" in D1 so that the PEST sequence becomes exposed and can be recognized by a protease, thus initiating the breakdown of the damaged protein. However, whether and how a new D1 can be put in into the PS II complex that lacks D1 without disassembly of the entire complex remains an unanswered question.

Z and D

The physiological donor to P680 is called Z. This entity donates an electron to P680$^+$ in 20-250 ns when the oxygen-evolving system is intact[86]. Until recently, Z was supposed to be a plastoquinol, since upon oxidation of Z an EPR signal (Signal II$_f$ or Signal II$_{vf}$) was seen that was interpreted as that of a plastoquinol cation radical[87,88] (f and vf stand for "fast" and "very fast", respectively, depending on how rapidly the signal decayed, which is dependent on the intactness of the oxygen-evolving system).

An EPR signal of virtually identical shape, but with much slower kinetics, originates from the oxidized form of a component called D. It is as yet unknown what the physiological role of this component is, but it is known that it slowly communicates with the water-splitting system[40,89,90], and is stable for minutes or hours in its oxidized form in spite of its high redox potential.

A major problem in attributing Z and D to be plastoquinols was that the amount of plastoquinone per PS II reaction center did not appear to be sufficient to account for both Z and D, in addition to the quinones at the acceptor side[91,92]. Moreover, the results of iodination experiments[93,94] could be interpreted to indicate that Tyr residues in D1 and D2 might be directly or indirectly involved in electron donation (via Z and D, respectively) to the PS II reaction center. While experiments were set up to probe the function of conserved Tyr residues near the presumable P680 binding site in D1 and D2 using site-directed mutagenesis, it was shown that selective deuteration of Tyr residues in cells had a large effect on EPR Signal II_s whereas plastoquinone deuteration had not[95]. This was the first experimental evidence that D (and, by analogy, Z) is not a plastoquinol, but rather a Tyr residue in the D2 or D1 protein that can undergo redox reactions.

That D indeed was a Tyr residue in the D2 protein was convincingly shown by site-directed mutagenesis experiments in psbD from the cyanobacterium Synechocystis 6803[96,97]. From this organism, both copies of psbD were deleted by transformation with the appropriate plasmids (this facultative photoheterotrophic cyanobacterium shows homologous recombination between its genome and foreign DNA that can be taken up spontaneously into the cell. This recombination process leads to an insertion of foreign DNA into the cyanobacterial genome, while deleting native DNA that was between the recombination sites[13,30,31,98]). Subsequently, using oligonucleotide-directed mutagenesis in cloned psbD, a site-directed mutation was introduced in one of the bases that codes for Tyr-160 in D2, a conserved residue that appears close to P680 in topological models of D2[53]. The mutation induced was chosen such that the Tyr-160 would be replaced by Phe, which is identical to Tyr except that it lacks the hydroxy group on the ring (this hydroxy group is expected to be crucial for D (or Z) to participate in the redox reaction). After introduction of the plasmid with the site-directed mutation into the cyanobacterium, it was observed that in the cyanobacterial mutant having a Phe residue (instead of Tyr) at position 160 of D2, no EPR Signal II_s could be detected[96,97]. This indicates indeed that Tyr-160 in D2 is the species D. However, it is necessary to do the appropriate control experiments to make sure that the loss of EPR Signal II_s is specifically linked to a mutation of Tyr-160, and is not caused by a disturbance of the protein environment leading to, for example, a decreased stability of D^+, thus inducing the disappearance of the EPR Signal. As a control, Met-159 (next to the Tyr-160) was mutated into Arg (this mutation is expected to have a far larger effect on the structure of the D2 protein than the Tyr-to-Phe mutation)[97]. It was observed that in the Arg-159 mutant EPR Signal II_s was still present[97], thus implying that the disappearance of the EPR Signal in the Phe-160 mutant was a specific effect, and that indeed Tyr-160 in D2 is the component D.

One of the perhaps surprising features of the mutant that contains a Phe residue at position 160 of D2 is that it grows photoautotrophically, indicating that function of D is not crucially required for PS II electron transport[96,97]. However, under photoautotrophic conditions the growth rate of the mutant is 2-3 fold slower than that of the wild type. This was found not to be caused by a decrease in the total number of PS II centers (as measured by herbicide binding stoichiometry[97]). It is possible that in the mutant a number of PS II centers is inactive, but it is currently unclear why in the absence of D in all centers some PS II centers would function normally while others would be inactive.

Since the EPR Signals originating from Z^+ and D^+ are so similar, it can be hypothesized that Z is also a Tyr residue, in a very similar environment to that of Tyr-160 in D2. In view of (a) the symmetry between D1 and D2, and (b) the results of iodination experiments that D1 gets iodinated under conditions where Z turns over[93,94], Z is almost certainly the Tyr-160 residue in D1. This residue is in an almost perfectly symmetrical location in comparison with D. However, experimental evidence supporting this hypothesis is still lacking.

The water-splitting system

The process of oxygen evolution and water splitting is currently the least understood part of PS II, in spite of intensive efforts in many laboratories to elucidate its mechanism and the protein components involved in the process. The complexity of this aspect of PS II does not allow oxygen evolution to be covered in any detail in this review, and therefore only some of the basic generalities will be discussed.

In dark-adapted systems, oxygen evolution in PS II was found to oscillate with flash number when exposed to a series of brief (≤ 10 us) saturating flashes[99,100]. The oscillation periodicity of four was explained by the existence of five "S-states", S_0 through S_4, that differed from one another by the number of electrons (or "electron holes") in the oxygen-evolving system. Once the state S_4 is reached by four subsequent donations of an electron to the reaction center (via Z), the oxygen-evolving system spontaneously relaxes to the state S_0 while producing an oxygen molecule[100]. The fact that after dark adaptation the first burst of oxygen evolution is observed in the third flash was explained by assuming that the states S_0 and S_1 are stable in darkness, while S_2 and S_3 decay to S_1[100]. Even though there have been some refinements of this model over the years (including proton production as a function of flash number, and the realization that S_0 can react slowly with D^+), the principles of the model put forward almost two decades ago are still considered to be valid.

The redox reactions associated with the changes in S states involve manganese necessary for stable oxygen evolution[88,101]. There are four manganese per PS II center[88]. Even though a wealth of measurements is available that pertain to the questions how the four manganese are localized with regard to each other or with regard to specific residues in the protein environment, and which valency changes in one or several of the manganese may occur during S-state transitions[88,102-104], the problem of how manganese is exactly involved in oxygen evolution has not yet been solved.

In addition to the manganese, also calcium and chloride ions have an important function in oxygen evolution[105]. However, in contrast to the manganese, the calcium and chloride do not appear to act as redox reagents during the water splitting process; the latter ions may fulfill structural roles. Various other monovalent anions have been shown to be able to functionally replace chloride[106], even though these anions induce changes in some of the properties of the water-splitting system[107]. The specificity for calcium is high: only strontium has been found to be somewhat effective in functionally replacing calcium at the oxygen-evolving site[108], whereas other divalent cations are not.

Some phenomenological mechanisms of action of Ca and Cl ions as well as of the manganese-stabilizing protein and two other peripheral proteins, OEE-2 and OEE-3 (of about 23 and 16 kDa, respectively), have been described. A full coverage of all the evidence is far beyond the scope of this review. However, other papers[105,106,108] provide a relatively comprehensive treatment of most of these aspects of oxygen evolution. Regarding the function of the various PS II proteins in water splitting, it has been established that OEE-2 and OEE-3 are not required for oxygen evolution activity provided that sufficient calcium and chloride are present[105]. Upon extraction of MSP, the Mn and - in some cases - part of the oxygen evolution activity can be retained[109,110], and the structure of the manganese site appears to be unaffected[111]. Therefore, it is obvious that Mn binding does not crucially depend on either of the three peripheral proteins involved in oxygen evolution. The question then arises which protein(s) then bind Mn. There is evidence for the involvement of D1 (and perhaps D2) in manganese binding and oxygen evolution. The first clue in this direction came from the realization that a mutant inhibited at the donor side of PS II and containing a low Mn content was altered in D1[112]. Moreover, it was observed that iodination of D1 could be observed after chloride depletion without removal of Mn, and a binding of iodide to a site normally occupied by chloride was suggested[113]. This would put at least one of the chloride binding sites on the D1 protein. The localization of D (and probably Z) in the D1/D2 complex, and the close interaction that needs to exist between Mn and Z make an involvement of the D1/D2 complex in Mn binding very attractive. Detailed models incorporating this concept have already appeared[114]. However, it should be kept in mind that at this point a direct interaction between the MSP and D1 and D2 proteins have not yet been shown. In contrast, a crosslinking between MSP and CP47 has been reported[115,116], even though it remains to be seen whether the close contact between these two proteins is of any functional significance with respect to water oxidation: for example, Fig. 1 depicts a scenario where CP47 is in direct contact with MSP, but the Mn is not associated closely with any of these two chlorophyll proteins.

The attempts to solve the mechanism of oxygen evolution have become perfect examples of the multidisciplinary nature of the current research in photosynthesis. Experiments using EPR, ENDOR, absorption spectroscopy, and biochemistry have all contributed to elucidating the mechanism of the water-splitting process. It is likely that these disciplines will be joined relatively soon by molecular genetics: the gene for the OEE-1 protein (MSP) has recently been cloned and sequenced from different organisms[15,17,117], and since it is probable that D1 and/or D2 are involved in aspects of water splitting, site-directed mutagenesis of residues in these proteins that may be involved in oxygen evolution are planned in various groups.

Chlorophyll binding

The obvious function of CP43 and CP47 is light harvesting by the chlorophyll bound to these proteins, and transferring the energy to eventually reach the reaction center. There are presumably between 20 and 50 chlorophylls in total bound to CP43 and CP47[118-120]. The sites where the pigments are bound to the proteins are still unknown. In

48

various other (bacterio)chlorophyll-binding systems, His has been shown or implied to function as a ligand to the Mg in the pigment[2,3,121,122]. By analogy, His residues also are good candidates to bind chlorophyll in the PS II antenna proteins. However, the total number of conserved His residues in CP43 and CP47 is about 25 (of which some are localized in relatively hydrophilic regions)[123,124], and is therefore perhaps lower than the number of chlorophylls that need to be accomodated. It should be kept in mind, however, that His residues are not the only candidates for chlorophyll binding. For example, in the bacteriochlorophyll c antenna of <u>Chloroflexus aurantiacus</u> the number of pigments to be accomodated is equal to or larger than that of the weight of antenna apoprotein.[125] The high pigment/protein ratio was explained by assuming that also other residues (such as Asn and Gln) could bind pigments[126], and/or that groups in the pigments themselves provide ligands for the Mg in neighboring pigment molecules[127,128]. Evidence that pigments indeed might be bound to Asn residues came from the sequence determination of a bacteriochlorophyll b-binding antenna protein from <u>Ectothiorhodospira halochloris</u>, which has an Asn residue at a site homologous to the His residue that binds the pigment in other bacterial antennas[129]. Furthermore, X-ray diffraction analysis of crystals of a bacteriochlorophyll a-binding protein from a green bacterium has suggested that the Mg of two of the seven bacteriochlorophylls bound to the protein are not liganded by His residues[130].

The function of the chlorophyll-binding proteins in PS II is not limited to light harvesting. It appears that these proteins also have a function in assembly of the PS II complex. It was observed that upon interruption of the <u>psbB</u> gene (encoding CP47) not only CP47, but also D1 and D2 disappeared from the thylakoid membrane[29]. This indicates that, at least in cyanobacterial systems, CP47 is needed for a stable assembly of D1 and D2. However, once the complex has assembled in the presence of CP47, this protein can be extracted while retaining a (partially) functional D1/D2/cyt b-559 complex[26,119,131]. It can be excluded that the loss of D1 and D2 from thylakoids in mutants with interrupted <u>psbB</u> is due to the loss of proteins encoded in the same operon as <u>psbB</u>, since an interruption in the DNA just downstream of the gene does not lead to loss of photoautotrophic growth (and therefore not to a loss of D1 and D2, of course)[123].

CP43 appears to be less closely involved in the assembly of the reaction center than CP47 is. Upon alteration of <u>psbC</u>, mutants are obligatory photoheterotrophic (indicating that they lack PS II activity)[29,132], but D1, D2 and CP47 can still be detected in the thylakoid, even though in greatly decreased amounts[29]. This notion of CP43 being somewhat less directly involved in assembly and stability of the PS II reaction center is strengthened by the observation that CP43 can be extracted from the PS II complex without apparent loss of PS II electron transport intermediates between (and including) Z and Q_A[118,133].

Cytochrome b-559

The role of cyt b-559 in PS II is still obscure. This protein obviously is closely related with the reaction center, since it is copurified with D1 and D2[26,131]. However, there is no clear evidence for a direct involvement of cyt b-559 in the regular electron transport chain[134], although interruption or deletion of the genes for this cytochrome leads to a loss of PS II activity[135] It has been suggested that cyt b-559 may function in photoactivation[134], or in protecting the reaction center against photoinhibition: it may serve as a donor to an oxidized chlorophyll[83]. However, until further experimental evidence has been obtained this hypothesis still is quite speculative.

Other PS II proteins

In addition to the five PS II core proteins shown in Fig. 1, there are at least three other proteins that are presumably associated with the PS II complex. These proteins, of 73, 39 and 41 amino acids, are encoded by psbH, psbI and psbJ, respectively[136,137]. These proteins all contain a hydrophobic region that is long enough to span the thylakoid membrane, and are quite conserved[136,137]. The psbH gene product is the "10 kDa phosphoprotein" in PS II, which is cotranscribed in chloroplasts with psbB[138]. psbI and psbJ are cotranscribed with the cytochrome b-559 genes[137]. The role of the three small proteins in PS II still remains to be determined.

CONCLUDING REMARKS

The last several years have marked a significant progress in the understanding of the function and structure of PS II. Among the most important factors contributing to this progress was the crystallization of the reaction center from purple non-sulfur bacteria, along with the realization that some of the information elucidated for the bacterial system could be extrapolated to PS II. Also the development of preparation procedures for protein complexes (perhaps most notably the D1/D2/cyt b-559 complex) of various compositions and various functions has helped the characterization of PS II.

It may be anticipated that over the next few years specific mutagenesis will begin to play a more prominent role in PS II research, since it provides a direct route for the design of mutants with changes in one or several specific amino acids of a single protein. The characterization of these mutants will presumably contribute significantly to gaining insight into important features of PS II that are not yet understood. One of the best examples to date of how specific mutagenesis can aid research on PS II may be the use of this technique for the identification of Tyr-160 in D2 as the component D. Even though generation and characterization of specific PS II mutants requires equipment and expertise to cover both molecular genetics and biochemistry or biophysics, there is a clear trend towards integration of these disciplines. Such a synthesis may prove to be very fruitful to elucidate important aspects of PS II function, structure and assembly.

ACKNOWLEDGEMENTS

Support by a grant from the National Science Foundation (DMB 87-16055) is gratefully acknowledged.

REFERENCES

1. Deisenhofer, J., Epp, O., Miki, K., Huber, R. and Michel, H. (1985) Structure of the protein subunits in the photosynthetic reaction centre of Rhodopseudomonas viridis at 3Å resolution. Nature 318, 618-624
2. Michel, H., Epp, O. and Deisenhofer, J. (1986) Pigment-protein interactions in the photosynthetic reaction centre from Rhodopseudomonas viridis. EMBO J. 5, 2445-2451

3. Allen, J.P., Feher, G., Yeates, T.O., Komiya, H. and Rees, D.C.
    (1987) Structure of the reaction center from _Rhodobacter
    sphaeroides_ R-26: The cofactors. Proc. Natl. Acad. Sci. USA
    84, 5730-5734

4. Allen, J.P., Feher, G., Yeates, T.O., Komiya, H. and Rees, D.C.
    (1987) Structure of the reaction center from _Rhodobacter
    sphaeroides_ R-26: The protein subunits. Proc. Natl. Acad.
    Sci. USA 84, 6162-6166

5. Chang, C.-H., Tiede, D., Tang, J., Smith, U., Norris, J. and
    Schiffer, M. (1986) Structure of _Rhodopseudomonas sphaeroides_
    R-26 reaction center. FEBS Lett. 205, 82-86

6. Rochaix, J.-D., Dron, M., Rahire, M. and Malnoe, P. (1984)
    Sequence homology between the 32K dalton and the D2
    chloroplast membrane polypeptides of _Chlamydomonas
    reinhardii_, Plant Mol. Biol. 3, 363-370

7. Michel, H. and Deisenhofer, J. (1988) Relevance of the
    photosynthetic reaction center from purple bacteria to
    the structure of photosystem II, Biochem. 27, 1-7

8. Rutherford, A.W. (1986) How close is the analogy between the
    reaction centre of Photosystem II and that of purple
    bacteria? Biochem. Soc. Trans. 14, 15-17

9. Rutherford, A.W. (1987) How close is the analogy between the
    reaction centre of PS II and that of purple bacteria? 2.
    The electron acceptor side. _In_: Progress in Photosynthesis
    Research, Vol. I (J. Biggins, ed.), pp. 277-283, Martinus
    Nijhoff, Dordrecht

10. Barber, J. (1987) Photosynthetic reaction centres: a common link.
    Trends in Biochem. Sci. 12, 321-326

11. Rochaix, J.-D. and Erickson, J. (1988) Function and assembly of
    photosystem II: genetic and molecular analysis. Trends in
    Biochem. Sci. 13, 56-59

12. Satoh, K. (1988) Reality of P-680 chlorophyll protein.
    Identification of the site of primary photochemistry in
    oxygenic photosynthesis. Physiol. Plant. 72, 209-212

13. Vermaas, W.F.J. (1988) Photosystem II function as probed by
    mutagenesis. _In_: Light energy transduction in
    photosynthesis: higher plants and bacterial models (D.A.
    Bryant and S.E. Stevens, Jr., eds.), Waverley Press, in press

14. Mathis, P. and Rutherford, A.W. (1987) The primary reactions of
    photosystems I and II of algae and higher plants. _In_:
    Photosynthesis (J. Amesz, ed.), pp. 63-96, Elsevier,
    Amsterdam

15. Kuwabara, T., Reddy, K.J. and Sherman, L.A. (1987) Nucleotide
    sequence of the gene from the cyanobacterium _Anacystis
    nidulans_ R2 encoding the Mn-stabilizing protein involved in
    photosystem II water oxidation. Proc. Natl. Acad. Sci. USA
    84, 8230-8234

16. Mayfield, S.P., Bennoun, P. and Rochaix, J.-D. (1987) Expression
    of the nuclear encoded OEE1 protein is required for oxygen
    evolution and stability of photosystem II particles in
    Chlamydomonas reinhardtii. EMBO J. 6, 313-318

17. Philbrick, J.B. and Zilinskas, B.A. (1988) Cloning, nucleotide
    sequence and mutational analysis of the gene encoding the
    photosystem II manganese-stabilizing polypeptide of
    Synechocystis 6803. Mol. Gen. Genet., in press

18. Sayre, R.T., Andersson, B. and Bogorad, L. (1986) The topology of
    a membrane protein: the orientation of the 32 kd Qb-binding
    chloroplast thylakoid membrane protein. Cell 47, 601-608

19. Doring, G., Stiehl, H.H. and Witt, H.T. (1967) A second chlorophyll reaction in the electron chain of photosynthesis - registration by the repetitive excitation technique. Z. Naturforsch. 22b, 639-644

20. Govindjee and Govindjee, R. (1975) Introduction to photosynthesis. In: Bioenergetics of Photosynthesis (Govindjee, ed.), pp. 1-50, Academic Press, New York

21. Diner, B.A. (1986) Photosystems I and II: structure, proteins and cofactors. In: Photosynthesis III, Encyclopedia of Plant Physiology, New Series, Vol. 19 (L.A. Staehelin and C.J. Arntzen, eds.), pp. 422-436, Springer Verlag, Berlin

22. Klevanik, A.V., Klimov, V.V., Shuvalov, V.A. and Krasnovskii, A.A. (1977) Reduction of pheophytin in the photoreaction of photosystem II of higher plants. Dokl. Akad. Nauk SSSR 236, 241-244

23. Shuvalov, V.A., Klimov, V.V., Dolan, E., Parson, W.W. and Ke, B. (1980) Nanosecond fluorescence and absorbance changes in photosystem II at low redox potential. FEBS Lett. 118, 279-282

24. Michel-Beyerle, M.E., Plato, M., Deisenhofer, J., Michel, H., Bixon, M. and Jortner, J. (1988) Unidirectionality of charge separation in reaction centers of photosynthetic bacteria. Biochim. Biophys. Acta 932, 52-70

25. Omata, T., Murata, N. and Satoh, K. (1984) Quinone and pheophytin in the photosynthetic reaction center II from spinach chloroplasts. Biochim. Biophys. Acta 765, 403-405

26. Nanba, O. and Satoh, K. (1987) Isolation of a photosystem II reaction center consisting of D-1 and D-2 polypeptides and cytochrome b-559. Proc. Natl. Acad. Sci. USA 84, 109-112

27. Rutherford, A.W. (1985) Orientation of EPR signals arising from components in Photosystem II membranes. Biochim. Biophys. Acta 807, 189-201

28. Vermaas, W.F.J., Williams, J.G.K. and Arntzen, C.J. (1987) Site-directed mutations of two histidine residues in the D2 protein inactivate and destabilize Photosystem II in the cyanobacterium Synechocystis 6803. Z. Naturforsch. 42c, 762-768

29. Vermaas, W.F.J., Ikeuchi, M. and Inoue, Y. (1988) Protein composition of the photosystem II core complex in genetically engineered mutants of the cyanobacterium Synechocystis sp. PCC 6803. Photosynth. Res., in press

30. Williams, J.G.K. (1988) Construction of specific mutations in the photosystem II photosynthetic reaction center by genetic engineering methods in the cyanobacterium Synechocystis 6803. Meth. Enzymol., in press

31. Vermaas, W.F.J., Carpenter, S. and Bunch, C. (1988) Specific mutagenesis as a tool for the analysis of structure/function relationships in Photosystem II. In: Applications of Molecular Biology in Bioenergetics of Photosynthesis (G.S. Singhal et al., eds.), Narosa Publishing House, New Delhi, in press

32. Woodbury, N.W., Parson, W.W., Gunner, M.R., Prince, R.C. and Dutton, P.L. (1986) Radical-pair energetics and decay mechanisms in reaction centers containing anthraquinones, naphthoquinones or benzoquinones in place of ubiquinones. Biochim. Biophys. Acta 851, 6-22

33. Diner, B.A., de Vitry, C. and Popot, J.-L. (1988) Quinone exchange in the Q_A binding site of photosystem II reaction

center core preparations isolated from <u>Chlamydomonas reinhardtii</u>. Biochim. Biophys. Acta 934, 47-54

34. Van Gorkom, H.J. (1974) Identification of the reduced primary electron acceptor of photosystem II as a bound semiquinone anion. Biochim. Biophys. Acta 347, 439-442

35. Dutton, P.L., Prince, R.C. and Tiede, D.M. (1978) The reaction center of photosynthetic bacteria. Photochem. Photobiol. 28, 939-949

36. Robinson, H.H. and Crofts, A.R. (1983) Kinetics of the oxidation-reduction reactions of the photosystem II quinone acceptor complex, and the pathway for deactivation. FEBS Lett. 153, 221-226

37. Glaser, M., Wolff, Ch. and Renger, G. (1976) Indirect evidence for a very fast recovery kinetics of chlorophyll-a$_{II}$ in spinach chloroplasts. Z. Naturforsch. 31c, 712-721

38. Velthuys, B.R. (1981) Electron-dependent competition between plastoquinone and inhibitors for binding to photosystem II. FEBS Lett. 126, 277-281

39. Wraight, C.A. (1981) Oxidation-reduction physical chemistry of the acceptor quinone complex in bacterial photosynthetic reaction centers: evidence for a new model of herbicide activity. Isr. J. Chem. 21, 348-354

40. Vermaas, W.F.J., Renger, G. and Dohnt, G. (1984) The reduction of the oxygen-evolving system in chloroplasts by thylakoid components. Biochim. Biophys. Acta 764, 194-202

41. Velthuys, B.R. and Amesz, J. (1974) Charge accumulation at the reducing side of system 2 of photosynthesis. Biochim. Biophys. Acta 333, 85-94

42. Vermaas, W.F.J., Renger, G. and Arntzen, C.J. (1984) Herbicide/quinone binding interactions in photosystem II. Z. Naturforsch. 39c, 368-373

43. Trebst, A. (1987) The three-dimensional structure of the herbicide-binding niche on the reaction center polypeptides of Photosystem II. Z. Naturforsch. 42c, 742-750

44. Vermaas, W.F.J., Arntzen, C.J., Gu, L.-Q. and Yu, C.A. (1983) Interactions of herbicides and azidoquinones at a photosystem II binding site in the thylakoid membrane. Biochim. Biophys. Acta 723, 266-275

45. Vermaas, W.F.J., Dohnt, G. and Renger, G. (1984) Binding and release kinetics of inhibitors of Q_A^- oxidation in thylakoid membranes. Biochim. Biophys. Acta 765, 74-83

46. Pfister, K. and Arntzen, C.J. (1979) The mode of action of photosystem II-specific inhibitors in herbicide-resistant weed biotypes. Z. Naturforsch. 34c, 996-1009

47. Hirschberg, J. and McIntosh, L. (1983) Molecular basis of herbicide resistance in <u>Amaranthus hybridus</u>. Science 222, 1346-1349

48. Vermaas, W.F.J. and Arntzen, C.J. (1983) Synthetic quinones influencing herbicide binding and photosystem II electron transport; the effect of triazine resistance on quinone binding properties in thylakoid membranes. Biochim. Biophys. Acta 725, 483-491

49. Holt, J.S., Stemler, A.J. and Radosevich, S.R. (1981) Differential light responses of photosynthesis by triazine-resistant and triazine-susceptible <u>Senecio vulgaris</u> biotypes. Plant Physiol. 67, 744-748

50. Bowes, J., Crofts, A.R. and Arntzen, C.J. (1980) Redox reactions on the reducing side of Photosystem II in chloroplasts with

altered herbicide binding properties. Arch. Biochem.
Biophys. 200, 303-308

51. Ort, D.R., Ahrens, W.H., Martin, B. and Stoller, E.W. (1983)
Comparison of photosynthetic performance in triazine-
resistant and susceptible biotypes of _Amaranthus hybridus_.
Plant Physiol. 72, 925-930

52. Wolber, P.K., Eilmann, M. and Steinback, K.E. (1986) Mapping of
the triazine binding site to a highly conserved region of the
Q_B protein. Arch. Biochem. Biophys. 248, 224-233

53. Trebst, A. (1986) The topology of the plastoquinone and
herbicide-binding polypeptides of photosystem II in the
thylakoid membrane. Z. Naturforsch. 41c, 240-245

54. Gingrich, J.C., Buzby, J.S., Stirewalt, V.L. and Bryant, D.A.
(1988) Genetic analysis of two new mutations resulting in
herbicide resistance in the cyanobacterium _Synechocystis_ sp.
PCC 7002. Photosynth. Res., in press

55. Sinning, I. and Michel, H. (1987) Sequence analysis of mutants
from _Rhodopseudomonas viridis_ resistant to the herbicide
terbutryn. Z. Naturforsch. 42c, 751-754

56. Erickson, J.M., Rahire, M., Rochaix, J.-D. and Mets, L. (1985)
Herbicide resistance and cross-resistance: changes at three
distinct sites in the herbicide-binding protein. Science
228, 204-207

57. Johanningmeier, U., Bodner, U. and Wildner, G.F. (1987) A new
mutation in the gene coding for the herbicide-binding protein
in _Chlamydomonas_. FEBS Lett. 211, 221-224

58. Erickson, J.M., Rahire, M., Bennoun, P., Delepelaire, P., Diner,
B. and Rochaix, J.-D. (1984) Herbicide resistance in
Chlamydomonas reinhardtii results from a mutation in the
chloroplast gene for the 32-kilodalton protein of photosystem
II. Proc. Natl. Acad. Sci. USA 81, 3617-3621

59. Golden, S.S. and Haselkorn, R. (1985) Mutation of herbicide
resistance maps within the _psbA_ gene of _Anacystis nidulans_
R2. Science 229, 1104-1107

60. Paddock, M.L., Williams, J.C., Rongey, S.H., Abresch, E.C.,
Feher, G. and Okamura, M.Y. (1987) Characterization of three
herbicide-resistant mutants of _Rhodopseudomonas sphaeroides_
2.4.1: structure-function relationship. _In:_ Progress in
Photosynthesis Research, Vol. III (J. Biggins, ed.), pp.
775-778, Martinus Nijhoff, Dordrecht

61. Bylina, E.J. and Youvan, D.C. (1987) Genetic engineering of
herbicide resistance: saturation mutagenesis of isoleucine
229 of the reaction center L subunit. Z. Naturforsch. 42c,
769-774

62. Gilbert, C.W., Williams, J.G.K., Williams, K.A.L. and Arntzen,
C.J. (1985) Herbicide action in photosynthetic bacteria. _In:_
Molecular Biology of the Photosynthetic Apparatus (K.E.
Steinback, S. Bonitz, C.J. Arntzen and L. Bogorad, eds.), pp.
67-71, Cold Spring Harbor Laboratory, Cold Spring Harbor

63. Debus, R.J., Feher, G. and Okamura, M.Y. (1986) Iron-depleted
reaction centers from _Rhodopseudomonas sphaeroides_ R-26.1:
characterization and reconstitution with Fe^{2+}, Mn^{2+}, Co^{2+},
Ni^{2+}, Cu^{2+} and Zn^{2+}. Biochem. 25, 2276-2287

64. Kirmaier, C., Holten, D., Debus, R.J., Feher, G. and Okamura,
M.Y. (1986) Primary photochemistry of iron-depleted and
zinc-reconstituted reaction centers from _Rhodopseudomonas
sphaeroides_. Proc. Natl. Acad. Sci. USA 83, 6407-6411

65. Ikegami, I. and Katoh, S. (1973) Studies on chlorophyll fluorescence in chloroplasts. II. Effect of ferricyanide on the induction of fluorescence in the presence of 3-(3,4-dichlorophenyl)-1,1-dimethylurea. Plant Cell Physiol 14, 829-836

66. Bowes, J.M. and Crofts, A.R. (1980) Binary oscillations in the rate of reoxidation of the primary acceptor of photosystem II. Biochim. Biophys. Acta 590, 373-384

67. Petrouleas, V. and Diner, B.A. (1986) Identification of Q_{400}, a high-potential electron acceptor of photosystem II, with the iron of the quinone-iron acceptor complex. Biochim. Biophys. Acta 849, 264-275

68. Wraight, C.A. (1985) Modulation of herbicide binding by the redox state of Q_{400}, an endogenous component of Photosystem II. Biochim. Biophys. Acta 809, 320-330

69. Diner, B.A. and Petrouleas, V. (1987) Light-induced oxidation of the acceptor-side Fe(II) of photosystem II by exogenous quinones acting through the Q_B binding site. II. Blockage by inhibitors and their effects on the Fe(III) EPR spectra. Biochim. Biophys. Acta 893, 138-148

70. Zimmermann, J.-L. and Rutherford, A.W. (1986) Photoreductant-induced oxidation of Fe^{2+} in the electron-acceptor complex of photosystem II. Biochim. Biophys. Acta 851, 416-423

71. Petrouleas, V. and Diner, B.A. (1987) Light-induced oxidation of the acceptor-side Fe(II) of photosystem II by exogenous quinones acting through the Q_B binding site. I. Quinones, kinetics, and pH-dependence. Biochim. Biophys. Acta 893, 126-137

72. Govindjee and van Rensen, J.J.S. (1978) Bicarbonate effects on the electron flow in isolated broken chloroplasts. Biochim. Biophys. Acta 505, 183-213

73. Vermaas, W.F.J. and Govindjee (1982) Bicarbonate or carbon dioxide as a requirement for efficient electron transport on the acceptor side of photosystem II. In: Photosynthesis, Vol. II (Govindjee, ed.), pp. 541-558, Academic Press, New York

74. Vermaas, W.F.J. and Rutherford, A.W. (1984) EPR measurements on the effects of bicarbonate and triazine resistance on the acceptor side of photosystem II. FEBS Lett. 175, 243-248

75. Radmer, R. and Ollinger, O. (1980) Isotopic composition of photosynthetic O_2 flash yields in the presence of $H_2^{18}O$ and $HC^{18}O_3^-$. FEBS Lett. 110, 57-61

76. Powles, S.B. (1984) Photoinhibition of photosynthesis induced by visible light. Ann. Rev. Plant Physiol. 35, 15-44

77. Nedbal, L., Setlikova, E., Masojidek, J. and Setlik, I. (1986) The nature of photoinhibition in isolated thylakoids. Biochim. Biophys. Acta 848, 108-119

78. Kyle, D.J. and Ohad, I. (1986) The mechanism of photoinhibition in higher plants and green algae. In: Photosynthesis III. Encyclopedia of Plant Physiology, New Series, Vol. 19 (L.A. Staehelin and C.J. Arntzen, eds.), pp. 468-475, Springer Verlag, Berlin

79. Tytler, E.M., Whitelam, G.C., Hipkins, M.F. and Codd, G.A. (1984) Photoinactivation of photosystem II during photoinhibition in the cyanobacterium _Microcystis_ _aeruginosa_. Planta 160, 229-234

80. Demeter, S., Neale, P.J. and Melis, A. (1987) Photoinhibition:
     impairment of the primary charge separation between P-680 and
     pheophytin in photosystem II of chloroplasts. FEBS Lett.
     214, 370-374

81. Arntz, B. and Trebst, A. (1986) On the role of the Q_B protein of
     PS II in photoinhibition. FEBS Lett. 194, 43-49

82. Callahan, F.E., Becker, D.W. and Cheniae, G.M. (1986) Studies on
     the photoactivation of the water-oxidizing enzyme. II.
     Characterization of weak light photoinhibition of PS II and
     its light-induced recovery. Plant Physiol. 82, 261-269

83. Thompson, L.K. and Brudvig, G.W. (1988) Cytochome b_{559} may
     function to protect photosystem II from photoinhibiiton.
     Biopys. J. 53, 269a.

84. Greenberg, B.M., Gaba, V., Mattoo, A.K. and Edelman, M. (1987)
     Identification of a primary in vivo degradation product of
     the rapidly-turning-over 32 kd protein of photosystem II.
     EMBO J. 6, 2865-2869

85. Rogers, S., Wells, R. and Rechsteiner, M. (1986) Amino acid
     sequences common to rapidly degraded proteins: the PEST
     hypothesis. Science 234, 364-368

86. Brettel, K., Schlodder, E. and Witt, H.T. (1984) Nanosecond
     reduction kinetics of photooxidized chlorophyll-a_{II} ($P680$) in
     single flashes as a probe for the electron pathway, H^+
     release and charge accumulation in the O_2-evolving complex.
     Biochim. Biophys. Acta 766, 403-415

87. O'Malley, P.J. and Babcock, G.T. (1984) EPR properties of
     immobilized quinone cation radicals and the molecular origin
     of Signal II in spinach chloroplasts. Biochim. Biophys. Acta
     765, 370-379

88. Babcock, G.T. (1987) The photosynthetic oxygen-evolving process.
     In: Photosynthesis (J. Amesz, ed.), pp. 125-158, Elsevier,
     Amsterdam

89. Velthuys, B.R. and Visser, J.W.M. (1975) The reactivation of EPR
     signal II in chloroplasts treated with reduced
     dichlorophenol-indophenol: evidence against a dark
     equilibrium between two oxidation states of the oxygen
     evolving system. FEBS Lett. 55, 109-112

90. Styring, S. and Rutherford, A.W. (1987) In the oxygen-evolving
     complex of Photosystem II the S_0 state is oxidized to the S_1
     state by D^+ (Signal II_{slow}). Biochem 26, 2401-2405

91. De Vitry, C., Carles, C. and Diner, B.A. (1986) Quantitation of
     plastoquinone-9 in photosystem II reaction center particles.
     Chemical identification of the primary quinone, electron
     acceptor Q_A. FEBS Lett. 196, 203-206

92. Takahashi, Y. and Katoh, S. (1986) Numbers and functions of
     plastoquinone molecules associated with photosystem II
     preparations from Synechococcus sp.. Biochim. Biophys. Acta
     848, 183-192

93. Takahashi, Y., Takahashi, M. and Satoh, K. (1986) Identification
     of the site of iodide photooxidation in the photosystem II
     reaction center complex. FEBS Lett. 208, 347-351

94. Ikeuchi, M. and Inoue, Y. (1987) Specific ^{125}I labeling of D1
     (herbicide-binding protein). An indication that D1 functions
     on both the donor and acceptor sides of photosystem II. FEBS
     Lett. 210, 71-76

95. Barry, B.A. and Babcock, G.T. (1987) Tyrosine radicals are involved in the photosynthetic oxygen-evolving system. Proc. Natl. Acad. Sci. USA 84, 7099-7103

96. Debus, R.J., Barry, B.A., Babcock, G.T. and McIntosh, L. (1988) Site-directed mutagenesis identifies a tyrosine radical involved in the photosynthetic oxygen-evolving system. Proc. Natl. Acad. Sci. USA 85, 427-430

97. Vermaas, W.F.J., Rutherford, A.W. and Hansson, O. (1988) Site-directed mutagenesis in Photosystem II of the cyanobacterium Synechocystis sp. PCC 6083: the donor D is a tyrosine residue in the D2 protein. Proc. Natl. Acad. Sci. USA, in press

98. Porter, R.D. (1986) Transformation in cyanobacteria. Crit. Rev. Micorbiol. 13, 111-132

99. Joliot, P., Barbieri, G. and Chabaud, R. (1969) Un nouveau modele des centres photochimiques du systeme II. Photochem. Photobiol. 10, 309-329

100. Kok, B., Forbush, B. and mcGloin, M. (1970) Cooperation of charges in photosynthetic oxygen evolution. I. A linear four-step model. Photochem. Photobiol. 11, 457-475

101. Cheniae, G.M. and Martin, I.F. (1970) Sites of function of manganese within photosystem II. Roles in O_2 evolution and system II. Biochim. Biophys. Acta 197, 219-239

102. Cole, J.L., Yachandra, V.K., McDermott, A.E., Guiles, R.D., Britt, R.D., Dexheimer, S.L., Sauer, K. and Klein, M.P. (1987) Structure of the manganese complex of photosystem II upon removal of the 33-kilodalton extrinsic protein: an X-ray absorption spectroscopy study. Biochem. 26, 5967-5973

103. Brudvig, G.W. and Crabtree, R.H. (1986) Mechanism for photosynthetic O_2 evolution. Proc. Natl. Acad. Sci. USA 83, 4586-4588

104. Dismukes, G.C. (1986) The metal centers of the photosynthetic oxygen-evolving complex. Photochem. Photobiol. 43, 99-115

105. Homann, P.H. (1987) The relations between the chloride, calcium and polypeptide requirements of photosynthetic water oxidation. J. Bioenerg. Biomembr. 19, 105-123

106. Critchley, C. (1985) The role of chloride in photosystem II. Biochim. Biophys. Acta 811, 33-46

107. Ono, T.-A., Nakayama, H., Gleiter, H., Inoue, Y. and Kawamori, A. (1987) Modification of the properties of S2 state in photosynthetic O_2-evolving center by replacement of chloride by other anions. Arch. Biochem. Biophys. 256, 618-624

108. Boussac, A. and Rutherford, A.W. (1988) Nature of the inhibition of the oxygen evolving enzyme of photosystem II induced by NaCl washing and reversed by the addition of Ca^{2+} or Sr^{2+}. Biochem. 27, 3476-3483

109. Ono, T.-A. and Inoue, Y. (1983) Mn-preserving extraction of 33-,24- and 16-kDa proteins from O_2-evolving PS II particles by divalent salt-washing. FEBS Lett. 164, 255-260

110. Miyao, M. and Murata, N. (1985) The Cl^- effect on photosynthetic oxygen evolution: interaction of Cl^- with 18-kDa, 24-kDa and 33-kDa proteins. FEBS Lett. 180, 303-308

111. Miller, A.-F., De Paula, J.C. and Brudvig, G.W. (1987) Formation of the S_2 state and structure of the Mn complex in photosystem II lacking the extrinsic 33 kilodalton polypeptide. Photosynth. Res. 12, 205-218

112. Metz, J.G., Pakrasi, H.B., Seibert, M. and Arntzen, C.J. (1986) Evidence for a dual function of the herbicide-binding D1 protein in photosystem II. FEBS Lett. 205, 269-274

113. Ikeuchi, M., Koike, H. and Inoue, Y. (1988) Iodination of D1 (herbicide-binding protein) is coupled with photooxidation of $^{125}I^-$ associated with Cl^--binding site in photosystem II water oxidation system. Biochim. Biophys. Acta 932, 160-169

114. Dismukes, G.C. (1988) The spectroscopically derived structure of the manganese site for photosynthetic water oxidation and a proposal for the protein-binding sites for calcium and manganese. Chem. Scripta, in press

115. Enami, I., Satoh, K. and Katoh, S. (1987) Crosslinking between the 33 kDa extrinsic protein and the 47 kDa chlorophyll-carrying protein of the PS II reaction center core complex. FEBS Lett. 226, 161-165

116. Bricker, T.M., Odom, W.R. and Queirolo, C.B. (1988) Close association of the 33 kDa extrinsic protein with the apoprotein of CPal in photosystem II. FEBS Lett. 231, 111-117

117. Tyagi, A., Hermans, J., Steppuhn, J., Jansson, Ch., Vater, F. and Herrmann, R.G. (1987) Nucleotide sequence of cDNA clones encoding the complete "33 kDa" precursor protein associated with the photosynthetic oxygen-evolving complex from spinach. Molec. Gen. Genet. 207, 288-293

118. Yamagishi, A. and Katoh, S. (1985) Further characterization of the two photosystem II reaction center complex preparations from the thermophilic cyanobacterium Synechococcus sp.. Biochim. Biophys. Acta 807, 74-80

119. Akabori, K., Tsukamoto, H., Tsukihara, J., Nagatsuka, T., Motokawa, O. and Toyoshima, Y. (1988) Disintegration and reconstitution of photosystem II reaction center core complex. I. Preparation and characterization of three different types of subcomplex. Biochim. Biophys. Acta 932, 345-357

120. Yamaguchi, N., Takahashi, Y. and Satoh, K. (1988) Isolation and characterization of a photosystem II core complex depleted in the 43 kDa-chlorophyll-binding subunit. Plant Cell Physiol. 29, 123-129

121. Youvan, D.C. and Ismail, S. (1985) Light-harvesting II (B800-B850 complex) structural genes from Rhodopseudomonas capsulata. Proc. Natl. Acad. Sci. USA 82, 58-62

122. Zuber, H. (1985) Structure and function of light-harvesting complexes and their polypeptides. Photochem. Photobiol. 42, 821-844

123. Vermaas, W.F.J., Williams, J.G.K. and Arntzen, C.J. (1987) Sequencing and modification of psbB, the gene encoding the CP-47 protein of photosystem II, in the cyanobacterium Synechocystis 6803. Plant Mol. Biol. 8, 317-326

124. Chisholm, D. and Williams, J.G.K. (1988) Nucleotide sequence of psbC, the gene encoding the CP-43 chlorophyll a-binding protein of photosystem II, in the cyanobacterium Synechocystis 6803. Plant Mol. Biol. 10, 293-301

125. Feick, R.G. and Fuller, R.C. (1984) Topography of the photosynthetic apparatus of Chloroflexus aurantiacus. Biochem. 23, 3693-3700

126. Zuber, H. (1987) Structural principles of the antenna system of photosynthetic organisms. In: Progress in Photosynthesis Research, Vol. II (J. Biggins, ed.), pp. 1-8, Martinus Nijhoff, Dordrecht

127. Blankenship, R.E., Brune, D.C. and Wittmershaus, B.P. (1988)

Chlorosome antennas in green photosynthetic bacteria. In: Light energy transduction in photosynthesis: higher plants and bacterial models (D.A. Bryant and S.E. Stevens, Jr., eds.), Waverley Press, in press

128. Brune, D.C., King, G.H. and Blankenship, R.E. (1988) Interactions between bacteriochlorophyll c molecules in oligomers and in chlorosomes of green photosynthetic bacteria. In: Photosynthetic Light-harvesting Systems (H. Scheer and S. Schneider, eds.), pp. 141-151, Walter de Gruyter, Berlin

129. Wagner-Huber, R., Brunisholz, R.A., Bissig, I., Frank, G. and Zuber, H. (1988) A new possible binding site for bacteriochlorophyll b in a light-harvesting polypeptide of the bacterium Ectorhodospira halochloris. FEBS Lett. 233, 7-11

130. Tronrud, D.E., Schmid, M.F. and Matthews, B.W. (1986) Structure and X-ray amino acid sequence of a bacteriochlorophyll a protein from Prosthecochloris aestuarii refined at 1.9Å resolution. J. Mol. Biol. 188, 443-454

131. Barber, J., Chapman, D.J. and Telfer, A. (1987) Characterization of a PS II reaction centre isolated from the chloroplasts of Pisum sativum. FEBS Lett. 220, 67-73

132. Dzelzkalns, V.A. and Bogorad, L. (1988) Molecular analysis of a mutant defective in photosynthetic oxygen evolution and isolation of a complementing clone by a novel screening procedure. EMBO J. 7, 333-338

133. Boska, M., Yamagishi, A. and Sauer, K. (1986) EPR signal II in cyanobacterial Photosystem II reaction-center complexes with and without the 40 kDa chlorophyll-binding subunit. Biochim. Biophys. Acta 850, 226-233

134. Cramer, W.A., Theg, S.M. and Widger, W.R. (1986) On the structure and function of cytochrome b-559. Photosynth. Res. 10, 393-403

135. Pakrasi, H.B., Williams, J.G.K. and Arntzen, C.J. (1988) Targeted mutagenesis of the psbE and psbF genes blocks photosynthetic electron transport: evidence for a functional role of cytochrome b_{559} in photosystem II. EMBO J. 7, 325-332

136. Hird, S.M., Dyer, T.A. and Gray, J.C. (1986) The gene for the 10 kDa phosphoprotein of photosystem II is located in chloroplast DNA. FEBS Lett. 209, 181-186

137. Cantrell, A. and Bryant, D.A. (1988) Nucleotide sequence of the genes encoding cytochrome b-559 from the cyanelle genome of Cyanophora paradoxa. Photosynth. Res., in press

138. Westhoff, P., Farchaus, J.W. and Herrmann, R.G. (1986) The gene for the M_r 10,000 phosphoprotein associated with photosystem II is part of the psbB operon of the spinach plastid chromosome. Curr. Genet. 11, 165-169

STRUCTURAL ORGANIZATION AND FUNCTION OF

POLYPEPTIDE SUBUNITS IN PHOTOSYSTEM I

Barry D. Bruce, Richard Malkin, R. Max Wynn and April Zilber

Division of Molecular Plant Biology
University of California
Berkeley, CA   94720

INTRODUCTION

Photosystem I (PSI) is one of three multi-subunit complexes found in thylakoids of higher plants, algae and cyanobacteria.[1] In conjunction with PSII and the cytochrome $\underline{b}_6$-$\underline{f}$ complex, electrons are transferred from H_2O to NADP and accompanying this process, protons are translocated across the thylakoid membrane. This non-cyclic electron transfer system yields NADPH, O_2 and a proton gradient, of which the latter is used for the synthesis of ATP. In the overall reaction from water to NADP, PSI catalyzes a light-dependent transfer of electrons between two mobile proteins, plastocyanin and ferredoxin.[2,3] The PSI complex differs from PSII and most other photochemical reaction center complexes found in non-O_2 evolving organisms in that PSI utilizes a bound electron complex which includes low-potential iron-sulfur centers.

The characterization of the electron transfer components associated with PSI has been an intensive area of investigation for over 20 years,[3,4] starting with the identification of the reaction center chlorophyll, P700, by Kok.[5] In the light, P700 undergoes photooxidation, yielding a chlorophyll cation radical. The electron lost from P700 is transferred to a series of bound electron carriers within the PSI complex.[3,4] The first of these, denoted A_o, is the least well characterized but has been proposed to be a monomeric chlorophyll species. The second electron acceptor, denoted A_1, is also poorly characterized. Recent studies have focused on the possible identity of A_1 as vitamin K_1 as the PSI complex is known to contain two molecules of vitamin K_1 per P700.[4] Following electron transfer to A_1, a series of bound Fe-S centers function as electron carriers. Using EPR spectroscopy, three different centers, $Fe-S_X$, $Fe-S_A$ and $Fe-S_B$, have been identified[2] and while the latter two are proposed to be [4Fe-4S] centers, the molecular composition of $Fe-S_X$ remains more controversial although we will present data which indicates this center must contain [4Fe-4S]. The properties and approximate midpoint redox potentials of these electron carriers are shown in Table I and this table documents that in the light, PSI produces a strong reductant which is capable of reducing ferredoxin (E_m = -420 mV) and NADP (E_m = -320mV).

Table I. Properties of Bound Electron Carriers in the PSI Complex

Electron carrier	Chemical Identity	Midpoint Redox Potential
P700	Chlorophyll a dimer	+ 450 mV
A_o	Chlorophyll a monomer	?
A_1	vitamin K_1 (?)	?
$Fe-S_X$	[4Fe-4S] cluster	- 705 mV
$Fe-S_B$	[4Fe-4S] cluster	- 580 mV
$Fe-S_A$	[4Fe-4S] cluster	- 530 mV

While the analysis of the primary reactants of PSI is fairly extensive, the study of the protein components associated with the PSI complex is less complete and is only a more recent area of investigation.[2] This was due in part to the absence of well-defined PSI complexes which were free from other membrane complexes. With the recent isolation of resolved PSI complexes from a variety of different thylakoid membrane systems,[6-8] it has become possible to associate protein subunits with PSI and to ask questions about the role of these subunits in specific functions. This article will review recent work in this area and will attempt to synthesize a number of results that will ultimately describe the nature of the function and organization of protein subunits in PSI.

PSI-200, PSI-120 and LHCPI, CPI

A native PSI complex, denoted PSI-200, which contains approximately

200 Chl/P700 has been isolated from higher plant thylakoids after solubil-ization with low concentrations of Triton X-100.[6] A native complex, which contains approximately 120 Chl/P700, has been described from cyano-bacterial membranes.[7] The larger number of chlorophyll molecules in the higher plant PSI complex is due to presence of Chl a/b binding proteins which are absent from the cyanobacterial preparation. The presence of Chl b in the higher plant PSI complex suggested the presence of an antenna complex which might function in a similar manner to the LHCPII complex of PSII. The identification of LHCPI, a Chl a/b-containing antenna complex, in PSI was facilitated by the fractionation of the PSI-200 complex.[9,10] This was accomplished by treatment of PSI-200 with dodecylmaltoside plus Zwittergent 16, followed by sucrose density gradient centrifugation. Two chlorophyll-containing fractions could be separated by this procedure. The lower band, denoted PSI-120, was photochemically active and contained P700 in a ratio of 120 Chl/P700. Little or no Chl b was found in this fraction. The upper green band, which was photochemically inactive, had a Chl a/b ratio of 3-4. Studies on the reconstitution of the two fractions indicated that the Chl a/b fraction functioned as a light-harvesting and antenna complex capable of transferring energy to the PSI reaction center.[11] This antenna complex has been designated LHCPI to indicate its function in PSI and to distinguish it from LHCPII.

Examination of the fractionated PSI reaction center and antenna complex has shown that the two functionally different fractions contain different polypeptide subunits. This allowed a first tentative association of specific subunits with either reaction center or antenna activities. The results of an SDS-PAGE analysis of these complexes is shown in Figure 1. PSI-200 contains approximately 10 subunits (lane 1). The broad band at 62 kDa is actually composed of two subunits of similar molecular weight[12-14]; 3-4 subunits between 24 and 27 kDa are present and 5-7 sub-units below 22 kDa can be detected. Also shown in the figure (lane 2) is the subunit composition of the PSI-120 core complex. This complex is almost totally deficient in the 24-27 kDa subunits as well as lacking several low molecular weight subunits. Previous work from our laboratory has shown the LHCPI complex contains three or more subunits, yet only one of these, denoted LHCPIb was shown to be chlorophyll-containing.[9,15] Other groups have confirmed and extended these findings[16-18] and it is now generally accepted that the LHCPI complex consists of several chloro-phyll-binding subunits in the molecular weight range from 24-27 kDa. These subunits appear to be distinct from those associated with LHCPII although antibody cross-reactivity has indicated some structural similarities between these two major groups of antenna complexes.[19,20]

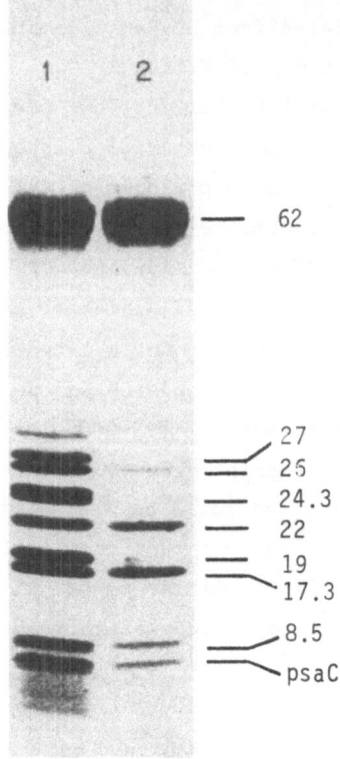

Fig. 1. SDS-PAGE analysis of spinach PSI complexes. Lane 1- PSI-200; Lane 2- PSI-120.

Both PSI-200 and PSI-120 are active in transferring electrons from
plastocyanin to ferredoxin/NADP. Thus, there does not appear to be a loss
of any essential electron transfer components in the separation of the
LHCPI complex from the PSI-120 complex. The PSI-120 complex is one of the
simplest complexes in terms of polypeptide composition which can still
catalyze the overall PSI electron transfer activity, and we have focused
our efforts on the properties of these subunits in terms of PSI electron
transfer functions.

One additional fractionation procedure that is pertinent to the
discussion of PSI function is the isolation of a chlorophyll-containing
complex known as CPI. This complex is isolated after treatment of complexes such as PSI-200 or PSI-120 with SDS[12] or chaotropic agents, such
as urea.[21] Under denaturing SDS-PAGE conditions, CPI has been found to
contain only the two high molecular weight subunits of the complex and
none of the low molecular weight subunits.[12-14] CPI contains approxi-

mately 100 Chl $\underline{a}$ molecules and retains P700 and two early electron acceptors $(A_o, A_1)^{21}$ while Golbeck and co-workers have recently shown it is possible to isolate CPI with Fe-S_x intact.[22] The preparation and characterization of the CPI complex therefore allows an identification of the subunits which bind the PSI reaction center chlorophyll and some of the early electron acceptors. The properties of CPI also suggest it is likely to be the only chlorophyll-binding component in the PSI-120 reaction center core complex, indicating that none of the low molecular weight subunits in PSI-120 have a role in binding chlorophyll.

PSI Subunit Stoichiometry

One important point of controversy relating to the PSI complex is its subunit stoichiometry. In particular, there are several reports which have given different values for the number of copies of the high molecular weight subunits per PSI complex.[7,8] Some of these discrepancies may arise from the different techniques used to determine these stoichiometries and also to the uncertainty of exact molecular weights for individual subunits. We have reinvestigated this problem using the aquatic higher plant, Lemna, and the green alga, Dunaliella. Both of these organisms can be uniformly labeled by growth on ^{14}C-substrates to obtain uniformly ^{14}C-labeled cell material for determination of subunit stoichiometries. PSI complexes were then isolated from ^{14}C-labeled membranes and stoichiometries calculated on the basis of radio-labeling instead of relying on alternative methods, such as protein staining on gels, which depends on the amount of stain bound to a particular subunit.

An analysis of the results of several determinations of subunit stoichiometries for PSI-120 complexes from Lemna and Dunaliella are summarized in Table II. Three different methods were used to analyze the stoichiometry, and as can be seen from the table, there is substantial agreement among all the methods. The results indicate that the PSI core complex in both these organisms contains two copies of the high molecular weight subunits per approximately one copy of several lower molecular weight subunits. Similar results have been obtained for the native PSI-200 complex from each organism. It is to be noted that the actual molecular weight of the high molecular weight subunit (84 kDa as deduced from cDNA analysis) has been used in the calculations and that the migration of these subunits under SDS-PAGE conditions is highly anomalous. Further work with the Lemna PSI complex has shown that a single copy of each high molecular weight subunit is present in PSI.

In the case of uniformly ^{14}C-labeled <u>Dunaliella</u> membranes, the cytochrome $\underline{b}_6$-$\underline{f}$ complex was isolated from the same membranes used to isolate PSI. The cytochrome complex was then used an as internal standard for specific activity based on protein concentration since the molecular weight of the individual subunits is known with great accuracy from cDNA gene sequencing for the three chloroplast-encoded subunits of this complex and the concentration of cytochrome $\underline{f}$ can be accurately determined by absorption difference spectroscopy. This allowed a calculation of the minimal molecular weight of PSI based on the ratio of protein/P700. A value of approximately 320 kDa per P700 was calculated using this method. This compares well with a "theoretical" molecular weight based on the experimentally determined subunit stoichiometry of the PSI complex and its known pigment composition. Based on two copies of the high molecular weight subunits and single copies of several low molecular weight sub-units, along with 75 chlorophylls, this calculation yields an expected molecular weight of approximately 290 kDa/P700. The expected value and the experimentally determined value based on uniform labeling differ by only 10°/o, and this general agreement supports the conclusion that the PSI complex contains single copies of the high molecular weight subunits per P700.

Table II. Subunit Stoichiometry of Core PSI Complexes from <u>Lemna</u> and <u>Dunaliella</u>

Subunit MW	Lemna PSI-120				Dunaliella PSI-120		
(kDa)				Method			
	1	2	3		1	2	3
84	2.6	2.2	2.2		2.1	2.4	2.1
22	1.0	1.0	1.0		1.0	1.0	1.0
19	0.5	0.4	0.5		0.8	0.3	0.5
16.5	2.3	1.5	1.9		ND	ND	ND
10.3	0.8	0.5	0.7		ND	ND	ND
9.0	ND	ND	0.3		ND	ND	ND
8.5	1.4	0.8	1.3		1.8	1.1	0.8

Method 1. Counting 150 individual gel slices

   2. Autoradiography

   3. Counting of individual stained bands

 ND. Not Determined

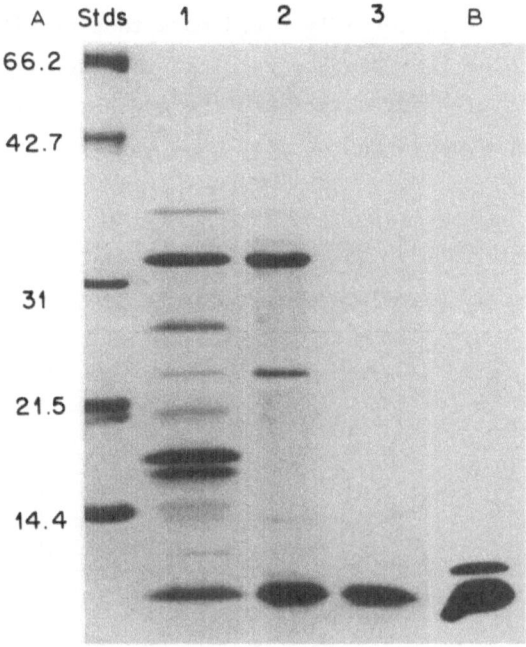

Fig. 2. SDS-PAGE and immunoblot analysis of the
purified membrane Fe-S protein: (A)
Coomassie stained SDS-PAGE analysis of (1)
extract, (2) DEAE-Biogel fraction (3)
Sephadex G-50 fraction. (B) Immunoblot
using an antibody against the PSI psaC gene
product. From ref. 28.

Identification of the Fe-S$_A$/Fe-S$_B$ Binding Protein in PSI. The
recent determination of the complete sequence of chloroplast DNA led to
the identification of two open reading frames which were proposed to code
for iron-sulfur apoproteins.[23,24] The frxA or psaC gene product was
suggested to code for a 9 kDa protein containing two 4Fe-4S clusters.
However, direct evidence that such a protein existed in chloroplasts was
lacking until Moller and co-workers[25] and Matsubara and colleagues[26]
identified a subunit in PSI with an amino acid sequence that corresponded
to the psaC gene product. This biochemical linkage provided the first
evidence that a low molecular weight subunit in PSI could be the apoprotein of an iron-sulfur protein. However, since both groups characterized
the protein in its fully denatured state, it was not possible to draw
conclusions concerning the native state of the protein, and in particular,

to define if the apoprotein actually bound more than one [4Fe-4S] centers. The resolution of this problem required the isolation of the Fe-S protein with its Fe-S clusters intact and the demonstration that the isolated protein corresponded to the psaC gene product.

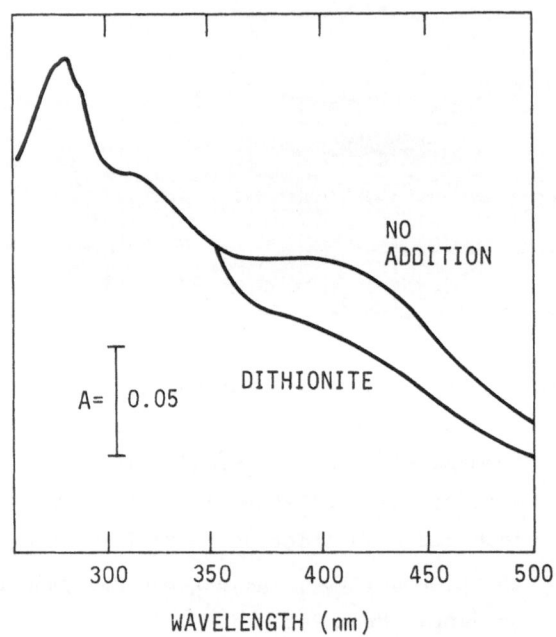

Fig. 3. Absorbance spectra of the isolated membrane
Fe-S protein from ref. 28.

A membrane-bound Fe-S protein had previously been isolated in our laboratory in 1974,[27] but the relationship of this protein to the components of PSI was not defined. We recently reconsidered this problem by isolating the Fe-S protein using an improved procedure and have carried out a more detailed examination of the properties of the isolated Fe-S protein.[28] An SDS-PAGE analysis of the purified protein, shown in Figure 2, indicates the denatured protein has a molecular weight of approximately 9 kDa and also shown in this figure is an immunoblot with an antibody raised against the 9 kDa PSI subunit in Dr. Matsubara's laboratory. The strong cross reaction of the antibody with our isolated 9 kDa protein confirms the identity of the two proteins. Shown in Figure 3 is

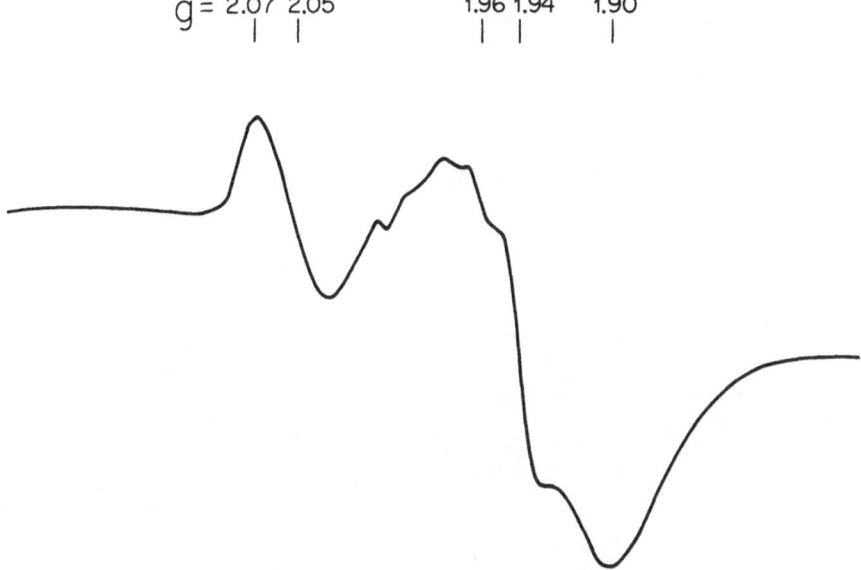

Fig. 4. EPR spectrum of the reduced isolated membrane
Fe-S protein. From ref. 28.

the optical absorbance spectrum of the purified protein, and this spectrum
is indicative of an Fe-S protein containing at least one [4Fe-4S] center.
The low temperature EPR spectrum. shown in Figure 4, is reminiscent of the
EPR spectrum of clostridial-type ferredoxins which contain two [4Fe-4S]
centers.[24] That the isolated protein contained 8Fe and 8S$^=$ was confirmed by analyses of non-heme iron and sulfide--approximately 7-8 atoms
of Fe and S$^=$ per mole of protein were found in the isolated protein.
The identity of the isolated Fe-S protein as the psaC gene product was
further confirmed by N-terminal amino acid sequencing of the isolated
protein: the N-terminal sequence of the psaC gene product, the isolated
Fe-S protein and the apoprotein subunit of PSI were identical.

While the EPR spectrum of the isolated Fe-S protein is not identical
to that of Fe-S$_A$ and Fe-S$_B$ in PSI, the spectrum does suggest that the
protein contains two interacting Fe-S clusters. This EPR spectrum and the
observed stoichiometry of iron and sulfide in the isolated protein would
argue that the isolated protein contains two different clusters which
presumably would correspond to Fe-S$_A$ and Fe-S$_B$. It is probable that
the properties of these centers have been altered as a result of the

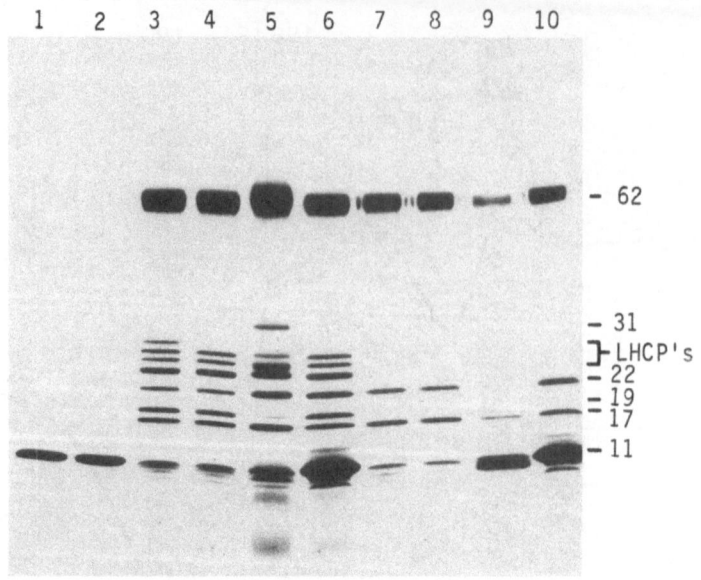

Fig. 5. SDS-PAGE of EDC-crosslinked PSI preparations. Lane 1--plastocyanin (PC); lane 2--
PC + EDC; lane 3--PSI-200; lane 4--PSI-200 +
EDC; lane 5--PSI-200 + EDC + PC; lane 6--
PSI-200 + EDC + carboxyl-modified PC;
lane 7--PSI-120; lane 8--PSI-120 + EDC;
lane 9--PSI-120 + EDC + PC;
lane 10--PSI-120 + EDC + carboxyl-modified PC

purification procedure or that the membrane environment of the protein is
important in conferring these properties. Further characterization of the
isolated protein and its interaction with PSI complexes may resolve some
of these questions.

Plastocyanin and Ferredoxin Binding Proteins in PSI. As previously
described, one of the unusual properties of the PSI reaction center complex is that it interacts with two soluble, mobile electron carriers,
plastocyanin and ferredoxin. In intact chloroplasts, these two proteins
are localized in the lumen and stroma, respectively and thus their sites
of interaction with PSI must be on opposite sides of the membrane. The
nature of the interaction of these two soluble proteins with the membrane-bound PSI complex remains poorly defined, particularly in relation
to the identity and nature of their binding sites within the complex.

The interaction of plastocyanin and ferredoxin with PSI has been recently investigated in our laboratory using chemical crosslinking approaches. This has involved the water soluble zero-length chemical crosslinking reagent, N-ethyl-3-(3-dimethylaminopropryl) carbodiimide (EDC). EDC is known to form a covalent bond between carboxyl and amino groups on proteins and because both plastocyanin and ferredoxin are acidic proteins, one would anticipate crosslinking of carboxyl groups on these proteins to amino groups on PSI subunits. Crosslinked products could then be detected by immunoblotting procedures utilizing specific antibodies for the two respective soluble proteins as well as for specific PSI subunits.

Shown in Figure 5 are the results of an SDS-PAGE analysis of the crosslinking of plastocyanin to PSI-200 and PSI-120. The incubation of plastocyanin with PSI-200 and EDC produced two major effects: a new band with an apparent molecular weight of 31 kDa appeared (lane 5) and a band of approximately 19 kDa disappeared relative to the other subunits in the complex (lane 5). PSI-120 treated in a similar manner showed comparable results (lane 9) except that the 19 kDa subunit was depleted in this pre-paration and therefore less of the 31 kDa protein was detected. No new bands were detected in control experiments with EDC and PSI complexes as compared with the PSI preparations alone (lanes 3 and 4 as well as lanes 7 and 8). These results suggest a specific interaction of plastocyanin with

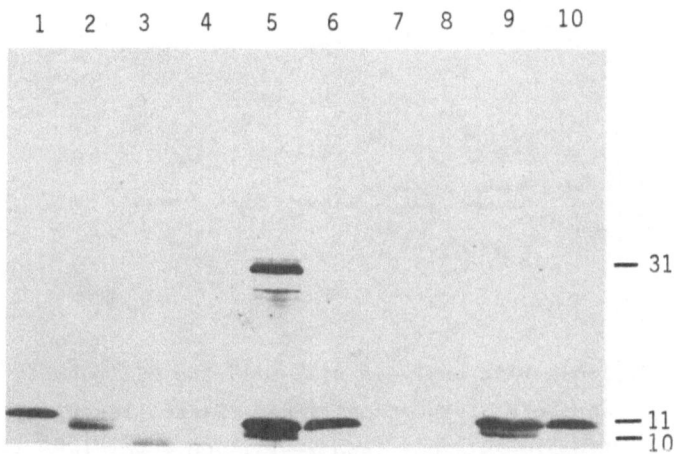

Fig. 6. Immunoblot analysis with a plastocyanin antibody of EDC-crosslinked PSI complexes Lanes
are as in Figure 5.

the 19 kDa subunit of the PSI complex. No crosslinked product of 31 kDa was observed in experiments in which PSI was reacted with a plastocyanin sample which had been modified to block carboxyl groups (lanes 6 and 10). These results support the view that ionic interactions are critical for the stabilization of the plastocyanin-PSI complex prior to covalent crosslinking with EDC.

In order to confirm directly that the product at 31 kDa is a crosslinked product between plastocyanin and the PSI 19 kDa subunit, immunoblotting with specific antibodies was done. The results shown in Figure 6 were obtained when the blot was probed with an antibody to plastocyanin. As can be seen from the EDC-treated samples of PSI-200 (lane 5) and PSI-120 (lane 9), the 31 kDa band showed a positive crossreaction with the plastocyanin antibody. The remaining lanes are appropriate controls which indicate that only when PSI, EDC and plasto-cyanin were present was the crosslinked product detected. An experiment using an antibody to the PSI 19 kDa subunit is shown in Figure 7. In this case, several positive crossreacting bands were detected. The band at approximately 17 kDa, probably arises from a contamination in the 19 kDa

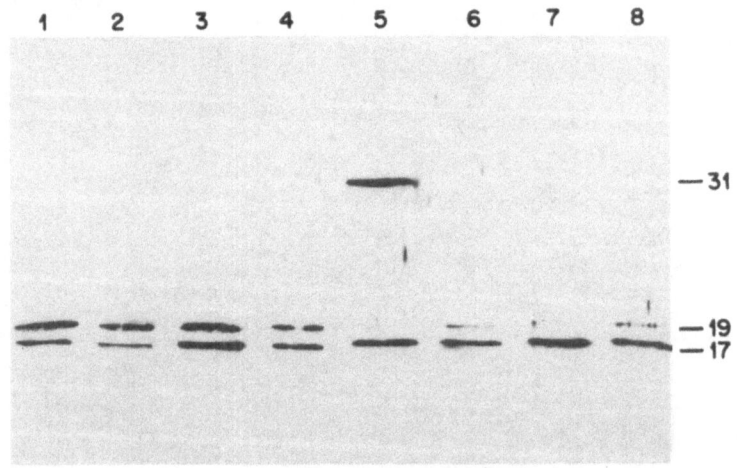

Fig. 7. Immunoblot analysis with a 19 kDa PSI subunit
antibody of EDC-crosslinked PSI complexes.
Lane 1--PSI-200; lane 2--PSI-200 + PC; lane 3--
PSI-200 + EDC; lane 4--PSI-200 + EDC + carboxyl-modified PC; lane 5--PSI-200 + EDC + PC;
lane 6--PSI-120 + EDC +PC; lane 7--PSI-120 +
EDC; lane 8--PSI-120.

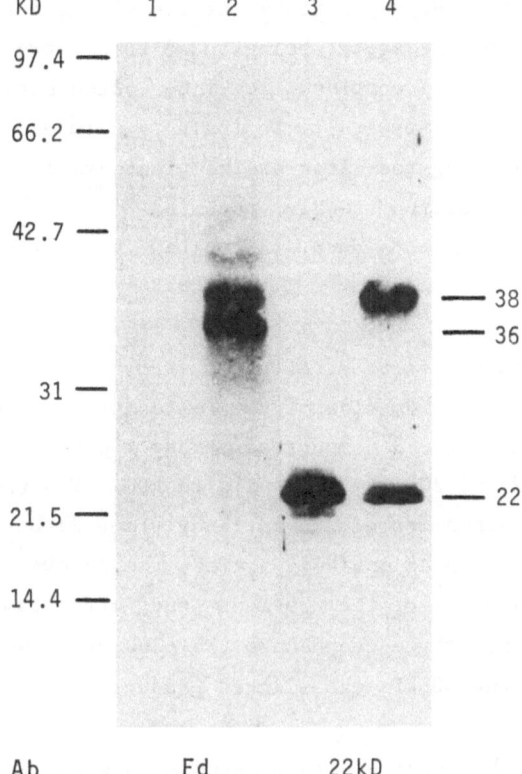

Fig. 8. Immunoblot analysis of EDC-crosslinked PSI
complexes. Lanes 1 and 3: PSI-200 + EDC;
lanes 2 and 4: PSI-200 + EDC + ferredoxin
(Fd). Lanes 1 and 2 were probed with a Fd
antibody and 3 and 4 with an antibody to
the 22 kDa PSI subunit.

subunit antibody. Lanes 5 and 6 show that the 31 kDa product also
crossreacts with the 19 kDa subunit antibody when tested with PSI-200 and
the PSI-120. Again, the remaining lanes show appropriate controls that
document the specificity of this reaction.

These results identify a specific subunit in the PSI complex to which
plastocyanin binds. We believe the 19 kDa subunit forms part of the
"docking" site for plastocyanin at the donor side of PSI. There is no
evidence that the 19 kDa subunit contains any prosthetic groups so we

visualize this interaction as facilitating the electron transfer between plastocyanin and P700, the latter being bound to the two high-molecular weight subunits of the PSI complex. The interaction between the 19 kDa subunit and plastocyanin appears to be ionic in nature. Several groups are currently considering the sites on the plastocyanin molecule which interact with other electron transfer proteins[30,31] and it will be of interest to extend these considerations to the interaction of plastocyanin with PSI.

We have utilized a similar strategy to study the interaction of ferredoxin with the reducing side of the PSI complex. The immunoblot experiment shown in Figure 8 demonstrates that two new products are formed when ferredoxin and PSI-200 are treated with EDC. The two products, at 36 and 38 kDa, react with a ferredoxin antibody (lane 2) but only the 38 kDa subunit crossreacts with an antibody against the 22 kDa subunit of PSI (lane 4). As in the case of the previous study with plastocyanin, controls show that all three components (PSI-200, EDC and ferredoxin) must be present to form the 38 kDa crosslinked product.

In contrast to the results with plastocyanin where only a single crosslinked product was observed, two products were detected in the crosslinking experiment with ferredoxin. When a similar experiment was done with the PSI-120 core complex, bands at 36 and 38 kDa also appeared under crosslinking conditions. It is also possible to study the interaction of ferredoxin with PSI in untreated membranes since the site of binding is on the stromal side of the membrane. With membranes, two major crosslinked products were also detected (Figure 9, lane 4) with molecular weights of 38 and 48 kDa, respectively. Immunoblot analysis identified the former as a crosslinked product between the 22 kDa PSI subunit and ferredoxin while the latter was a product between ferredoxin-NADP reductase and ferredoxin (lanes 2 and 4). A complex between the reductase and ferredoxin has been known to form, and crosslinking with EDC has been used to characterize this complex.[32,33] It was, therefore, not surprising to detect this product in membranes under our crosslinking conditions. However, the finding that only the 38 kDa crosslinked product is detected in membranes would argue that the second product of 36 kDa seen with isolated PSI complexes probably arises from a binding of ferredoxin to a second subunit in PSI which has become accessible due to detergent treatment.

The results on the interaction of ferredoxin with PSI strongly argue for a specific PSI subunit which facilitates binding to the complex. Hoffman et al[34] have recently reported the amino acid sequence of a 21 kDa PSI subunit from tomato which they believe is identical to the spinach 22 kDa subunit. These workers point out several positively charged regions on the protein which could serve to bind the negatively charged ferredoxin molecule. Further studies of the crosslinked product may provide details on the mode of interaction of these two proteins.

Model for the Structural Organization of PSI. A model which incorporates our recent results as well as those of others on the function and

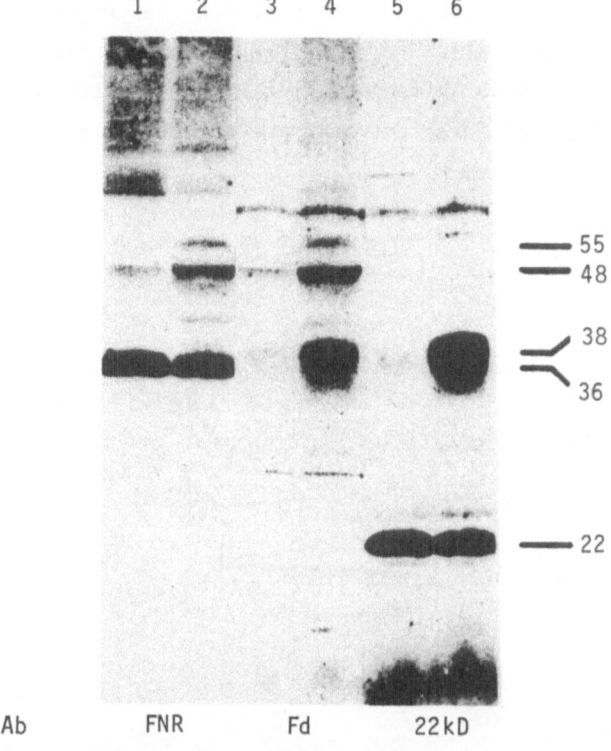

Fig. 9. Immunoblot analysis of EDC-crosslinked
membranes. Lanes 1,3 and 5-membranes +
EDC; lanes 2,4 and 6- membranes + EDC +
Fd. Lanes 1-2 were probed with an antibody to Fd:NADP reductase (FNR), 3-4
with a Fd antibody and 5-6 with the
22 kDa PSI subunit antibody.

organization of polypeptide subunits in the PSI complex is shown in
Figure 10. Single copies of the two high molecular weight subunits (psaA
and psaB gene products) are proposed, based on the described subunit stoichiometry measurements. Because each of these subunits has 2 conserved
cysteine residues, this stoichiometry dictates that $Fe-S_X$ be a single
[4Fe-4S] cluster that is coordinated by both subunits. This view is in
contrast to one presented by Golbeck and co-workers[35,36] which proposed
that $Fe-S_X$ contains two [2Fe-2S] clusters. The latter would require at
least two copies of each high molecular weight subunit per complex in
order to provide the eight cysteine ligands for the two Fe-S clusters.
The organization of the remaining electron carriers (A_o, A_1, P700),
associated with the high molecular weight subunits, is shown in a manner
analogous to that found for the bacterial reaction center complex[37] and
recently proposed for PSII.[38] In this regard, it is striking that PSI
contains two molecules of vitamin K_1 per P700 [39-41] but only one
appears to be required for electron transport to the stable electron

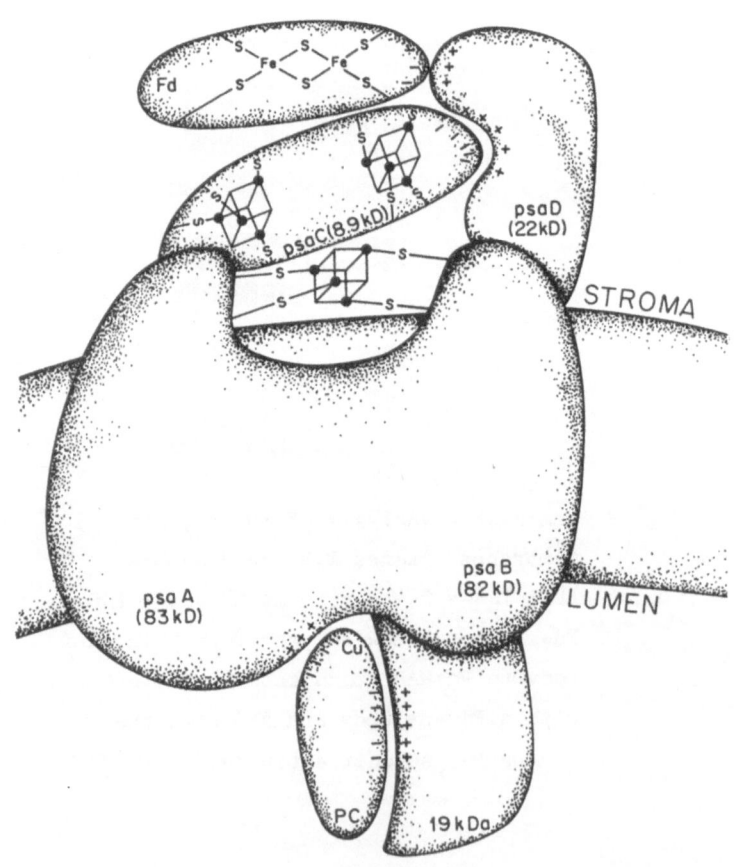

Fig. 10 Model for the organization of Photosystem I.

acceptors.[39,41] Although this interpretation is not yet fully agreed upon (see refs. 40 and 42), it suggests a symmetrical prosthetic group organization, as shown in the figure, but also indicates that only one "arm" of the two branches is functional in transferring electrons to Fe-S$_X$. Again, the analogy with other reaction center complexes would seem appropriate although there is much that still requires substantiation.

Plastocyanin and ferredoxin are known to interact with the lumenal and stromal surfaces of the PSI complex, respectively. This would dictate that the two specific subunits found to crosslink to these proteins would have respective lumenal (19 kDa) and stromal (22 kDa) orientation. The 22 kDa subunit has been shown by Hoffman et al[34] to be a peripheral PSI protein and to have several "patches" of positive charge which could interact with ferredoxin. We would expect that the lumenal 19 kDa subunit would also contain areas of positive charge which could interact with the negatively charged plastocyanin molecule.

It is important to stress, however, that neither the 19 kDa nor 22 kDa subunits contain prosthetic groups and that plastocyanin must also interact with the high molecular subunits which bind P700 while ferredoxin must interact with the 8.9 kDa psaC subunit which binds Fe-S$_A$ and Fe-S$_B$. Presumably these interactions are also ionic in nature although they have not been investigated in detail. The interaction of both plastocyanin and ferredoxin in the PSI complex can, however, occur in the absence of the 19 kDa and 22 kDa subunits. This conclusion is based on two lines of evidence: recent studies have shown that the maize PSI complex, which apparently lacks the 19 kDa subunit, can still rapidly transfer electrons from reduced plastocyanin to P700,[43-45] and have shown that studies of a Lemna mutant which has a PSI complex lacking the 22 kDa subunit[46] still shows substantial rates of ferredoxin-dependent NADP reduction.[47,48] These results indicate that the two binding proteins are not essential for electron transport activity but they have a role in facilitating these reactions.

While the advances in our understanding of the function of specific carriers and components in PSI have greatly advanced in recent years, this article has also raised questions which remain to be answered. The recent description of preparations from cyanobacteria which have been crystallized[49,50] is an advance that offers the potential of a three dimensional structure for this complex. The tools of molecular genetics are also being applied to studies of the function of the PSI complex and recent work with genes for PSI from cyanobacteria[51,52] allow the possibility of using site-directed mutagenesis to examine molecular details of function. It would therefore appear we are entering a new and exciting era of investigation of structure-function relationships in PSI.

REFERENCES

1. N. Nelson, Structure and function of protein complexes in the photosynthetic membrane, in: "Photosynthesis," J. Amesz, ed., Elsevier Science Publishers, Amsterdam (1987).
2. R. Malkin, Photosystem I, in: "The Light Reactions," J. Barber, ed., Elsevier Science Publishers, Amsterdam (1987).
3. A. W. Rutherford, and P. Heathcote, Primary photochemistry in photosystem I., Photosyn. Res. 6:295 (1985).
4. P. Mathis, and A. W. Rutherford, The primary reactions of photosystems I and II of algae and higher plants, in: "Photosynthesis," J. Amesz, ed., Elsevier Science Publishers, Amsterdam (1987).
5. B. Kok, Absorption changes induced by the photochemical reaction of photosynthesis, Nature 179:583 (1957).
6. J. E. Mullet, J. J. Burke, and C. J. Arntzen, Chlorophyll proteins of photosystem I, Plant Physiol. 65:814 (1980).
7. D. J. Lundell, A. N. Glazer, A. Melis and R. Malkin, Characterization of a cyanobacterial PSI complex, J. Biol. Chem. 260:646 (1985).
8. C. Bengis, and N. Nelson, Purification and properties of the photosystem I reaction center from chloroplasts, J. Biol. Chem. 250:2783 (1975).
9. E. Lam, W. Ortiz, S. Mayfield, and R. Malkin, Isolation and characterization of a light-harvesting chlorophyll a/b protein complex associated with photosystem I, Plant Physiol. 74:650 (1984).
10. P. Haworth, J. L. Watson, and C. J. Arntzen, The detection, isolation and characterization of a light-harvesting complex which is specifically associated with photosystem I, Biochim. Biophys. Acta 724:151 (1983).
11. W. Ortiz, E. Lam, M. Ghirardi, and R. Malkin, Antenna function of a chlorophyll a/b protein complex of photosystem I, Biochim. Biophys. Acta 766, 505-509.
12. E. Vierling, and R. S. Alberte, P700 chlorophyll a protein. Purification, characterization and antibody production, Plant Physiol. 72:625 (1983).
13. L. E. Fish, and L. Bogorad, Identification and analysis of the maize P700 chlorophyll a apoproteins PSI-A1 and PSI-A2 by high pressure liquid chromatography analysis and partial sequence determination, J. Biol. Chem. 261:8134 (1986).
14. B. D. Bruce, and R. Malkin, Subunit stoichiometry of the chloroplast photosystem I complex, J. Biol. Chem. 263:7302 (1988).
15. E. Lam, W. Ortiz, and R. Malkin, Chlorophyll a/b proteins of photosystem I, FEBS Lett. 168:10 (1984).
16. R. Bassi, and D. Simpson, Chlorophyll protein complexes of barley photosystem I, Eur. J. Biochem. 163:221 (1987).
17. J. M. Anderson, A chlorophyll a/b-protein complex of photosystem I, Photosyn. Res. 8:221 (1984).
18. R. Nechustai, C. C. Peterson, G. F. Peter, and J.-P. Thornber, Purification and characterization of a light-harvesting chlorophyll a/b-protein of photosystem I of Lemna gibba, Eur. J. Biochem. 164:345 (1987).

19. P. K. Evans, and J. M. Anderson, The chlorophyll a/b-proteins of PSI and PSII are immunologically related, FEBS Lett. 199:227 (1986).

20. M. J. White, B. R. Green, Antibodies to the photosystem I and chlorophyll a + b antenna cross-react with polypeptides of CP29 and LHCII, Eur. J. Biochem. 163:545 (1987).

21. P. Mathis, K. Sauer, and R. Remy, Rapidly reversible flash-induced electron transfer in a P700 chlorophyll-protein complex isolated with SDS, FEBS Lett. 88:275 (1978).

22. J. H. Golbeck, K. G. Parrett, T. Mehari, K. L. Jones, and J. J. Brand, Isolation of the intact photosystem I reaction center core containing P700 and iron-sulfur center F_X, FEBS Lett. 228:268 (1988).

23. K. Ohyama, H. Fukuzawa, T. Kohchi, H. Shirai, T. Sano, S. Sano, K. Umesono, Y. Shiki, M. Takeuchi, Z. Chang, S. Aota, H. Inokuchi, and H. Ozeki, Chloroplast gene organization deduced from complete sequence of liverwort Marchantia polymorpha chloroplast DNA, Nature 322:572 (1986).

24. K. Shinozaki, M. Ohme, M. Tanaka, T. Wakasugi, N. Hayashida, T. Matsubayashi, N. Zaita, J. Chun-Wonyse, J. Obokata, K. Yamaguchi-Shinozaki, C. Ohto, K. Torazawa, B. Y . Meng, M. Sugita, H. Deno, T. Kamogashira, K. Yamada, J. Kusuda, F. Takaiwa, A. Kato, N. Tohdoh, H. Shimada, and M. Suguira, The complete nucleotide sequence of the tobacco chloroplast genome: its gene organization and expression, EMBO J. 5:2043 (1986).

25. P. B. Hoj, I. Svendsen, H. V. Scheller, and B. L. Moller, Identification of a chloroplast-encoded 9-kDa polypeptide as a 2[4Fe-4S] protein carrying centers A and B of photosystem I, J. Biol. Chem. 262:12676 (1987).

26. H. Oh-oka, Y. Takahashi, K. Wada, H. Matsubara, K. Ohyama, and H. Ozeki, The 8 kDa polypeptide in photosystem I is a probable candidate of an iron-sulfur protein coded by the chloroplast gene frxA, FEBS Lett. 218:52 (1987).

27. R. Malkin, P. J. Aparacio, and D. I. Arnon, The isolation and characterization of a new iron-sulfur protein from photosynthetic membranes, Proc. Natl. Acad. Sci. USA 71, 2362 (1974).

28. R. M. Wynn, and R. Malkin, Characterization of an isolated chloroplast membrane Fe-S protein and its identification as the photosystem I Fe-S_A/Fe-S_B binding protein, FEBS Lett. 229:293 (1988).

29. W. H. Orme-Johnson, and H. Beinert, Heterogeneity of paramagnetic species in two iron-sulfur proteins: Clostridium pasteuranium ferredoxin and milk xanthine oxidase, Biochem. Biophys. Res. Commun. 36:337 (1969).

30. D. J. Davis, and K. Hough, Preparation of a covalently linked adduct between plastocyanin and cytochrome f, Biochem. Biophys. Res. Commun. 116:1000 (1983).

31. L. M. Geren, J. Stonehuerner, D. J. Davis, and F. Millett, The use of a water-soluble carbodiimide to cross-link cytochrome c to plastocyanin, Biochim. Biophys. Acta 724:62 (1983).

32. B. J. Vieira, K. K. Colvert, and D. J. Davis, Chemical cross-linking as probes of regions on ferredoxin involved in its interaction with ferredoxin:NADP reductase, Biochim. Biophys. Acta 851:109 (1986).

33. G. Zanetti, A. Oliverti, and B. Curti, A cross-linked complex between ferredoxin and ferredoxin-NADP$^+$ reductase, J. Biol. Chem. 259:6153 (1984).

34. N. E. Hoffman, E. Pichersky, V. S. Malik, K. Ko, and A. R. Cashmore, Isolation and sequence of a tomato cDNA clone encoding subunit II of the photosystem I reaction center, Plant Mol. Biol. 10:435 (1988).

35. J. H. Golbeck, K. G. Parrett, and A. E. McDermott, Photosystem I charge separation in the absence of center A and B. III. Biochemical characterization of a reaction center particle containing P700 and F_X, Biochim. Biophys. Acta 893:149 (1987).

36. J. H. Golbeck, A. E. McDermott, and W. K. Jones, and D. M. Kurtz, Evidence for the existence of [2Fe-2S] as well as [4Fe-4S] clusters among F_A, F_B and F_X. Implications for the structure of the photosystem I reaction center, Biochim. Biophys. Acta 891:94 (1987).

37. J. Deisenhofer, O. Epp, K. Miki, R. Huber, and H. Michel, Structure of the protein subunits in the photosynthetic reaction centre of Rhodopseudomonas viridis, Nature 318, 618 (1985).

38. H. Michel, and J. Deisenhofer, Relevance of the photosynthetic reaction center from purple bacteria to the structure of photosystem II, Biochemistry 27:1 (1988).

39. R. Malkin, On the function of two vitamin K_1 molecules in the PSI electron acceptor complex. FEBS Lett. 208:343 (1986).

40. K. Ziegler, W. Lockau, and W. Nitschke, Bound electron acceptors of photosystem I. Evidence against the identity of redox center A_1 with phylloquinone. FEBS Lett. 217:16 (1987).

41. P. Setif, I. Ikegami, and J. Biggins, Light-induced charge separation in photosystem I at low temperature is not influenced by vitamin K_1, Biochim. Biophys. Acta 894:146 (1987).

42. G. P. Palace, J. E. Franke, and J. T. Warden, Is phylloquinone an obligate electron carrier in photosystem I?, FEBS Lett. 215:58 (1987).

43. R. Nechustai, G. Schuster, N. Nelson, and I. Ohad, Photosystem I reaction centers from maize bundle sheath and mesophyll chloroplasts lack subunit III, Eur. J. Biochem. 159:157 (1986).

44. W. Haehnel, V. Hesse, and A. Propper, Electron transfer from plastocyanin to P700. Function of a subunit of photosystem I reaction center, FEBS Lett. 111:79 (1980).

45. R. Ratajczak, R. Mitchell, and W. Haehnel, Properties of the oxidizing site of photosystem I. Biochim. Biophys. Acta 933:306 (1988).

46. B. D. Bruce, and R. Malkin. Structure-function studies of the higher plant photosystem I complex. in: "Plant Membranes. Structure, Function, Biogenesis," C. Leaver and H. Sze, eds., Alan R. Liss, Inc., New York (1987).

47. Y. Shahak, H. B. Posner, and M. Avron, Evidence for a block between plastoquinone and cytochrome f in a photosynthetic mutant of Lemna with abnormal flowering behavior, Plant Physiol. 57:577 (1976).

48. E. Lam, and R. Malkin, Characterization of a photosynthetic mutant of Lemna lacking the cytochrome b_6-f complex. Biochim. Biophys. Acta 810:106 (1985).

49. R. C. Ford, D. Picot, and G. M. Garavito, Crystallization of the photosystem I reaction centre, EMBO J. 6:1581 (1987).

50. I. Witt, H. T. Witt, S. Gerken, W. Saenger, J. P. Dekker, and M. Rogner, Crystallization of reaction center I of photosynthesis, FEBS Lett. 221:260 (1987).

51. A. Cantrell, and D. A. Bryant, Molecular cloning and nucleotide sequence of the psaA and psaB genes of the cyanobacterium Synechococcus sp. PCC 7002. Plant Mol. Biol. 9:453 (1987).

52. E. Rhiel, and D. A. Bryant, Preliminary results concerning the psaC, psaD, psaE and psaF genes and their products in the cyanobacteria Synechococcus sp. PCC 7002 and Nostoc sp. PCC 8009, in: "Light-Energy Transduction in Photosynthesis: Higher Plants and Bacterial Models," D. A. Bryant, and S. E. Stevens, Jr., eds., Waverly Press, (Baltimore), in press.

STRUCTURAL AND FUNCTIONAL PROPERTIES OF THE ISOLATED PHOTOSYSTEM TWO REACTION CENTRE

J. Barber, D.J. Chapman, K. Gounaris, J.B. Marder and A. Telfer

AFRC Photosynthesis Research Group, Department of Pure & Applied Biology, Imperial College of Science, Technology & Medicine, London SW7 2BB UK

INTRODUCTION

The sequencing of the genes for the L and M subunits of the purple bacterial reaction centre and the D1 and D2 polypeptides of photosystem two (PS2) revealed striking homologies in localized areas (1,2). The likely meaning of these homologies emerged from the X-ray crystallography work of Deisenhofer et al (3) using crystals of the isolated reaction centre of Rhodopseudomonas viridis. Such a comparison indicated that like the L and M polypeptides, the D1 and D2 polypeptides probably form the heart of the PS2 reaction centre (4-6). This prediction was supported by work of Nanba and Satoh (7) who isolated a chlorophyll binding complex from spinach chloroplasts consisting only of the D1, D2 and cytochrome b559 polypeptides which were able to catalyse photoreactions indicative of the PS2 reaction centre. Using pea rather than spinach, we also isolated a D1/D2/cyt b559 complex which we have subjected to detailed structural and functional analysis. Here we summarise our findings.

ISOLATION OF REACTION CENTRE COMPLEX

Isolation of the reaction centre complex has been documented in detail in ref.(8). Briefly, membrane fragments enriched in PS2 were isolated from greenhouse grown pea plants (Pisum sativum var. Feltham First), essentially according to the method of Berthold et al. (9). These membrane fragments were suspended to 4 mg $Chl.ml^{-1}$ in 5 mM $MgCl_2$, 15 mM NaCl, 50 mM Mes (pH 6.5) and 10% (w/v) glycerol, frozen at 77 K and stored at 190 K. PS2 reaction centres were prepared from a 50 mg sample of the stored membrane fragments using a method developed from the approach suggested by Nanba and Satoh (7) in which solubilization in Triton X-100 is followed by ion-exchange chromatography. The procedure was firstly to remove extrinsic membrane polypeptides by incubating the thawed sample on ice in the dark in 50 mM Tris, pH 9.0 (0.8 mg $Chl.ml^{-1}$) for 10 min, followed by centrifugation at 40,000 xg at 4°C for 20 min. The pellets were resuspended in 50 mM Tris, pH 7.2 (to about 50 ml) and 8 ml of 30% Triton X-100 added to give a final Chl concentration of 0.8 $mg.ml^{-1}$. A 60 min incubation with stirring, in the dark and on ice, was followed by centrifugation at 100,000 xg for 60 min. The supernatant was then applied to a chromatography column (20 x 120 mm) containing Fractogel TSK DEAE-650(S) (Merck-BDH). The column was then extensively washed with 30 mM NaCl in a running buffer containing 0.2% Triton X-100 and 50 mM Tris-Cl, pH 7.2, until no more chlorophyll could be eluted. This procedure removed more than 98% of the chlorophyll applied, taking up to 3 h at 1.0 $ml.min^{-1}$. The column was then subjected to a linear NaCl gradient from 30 to 200 mM in the same running buffer and the fraction which eluted at about 110 mM NaCl collected as 2.0 ml samples. These samples were pooled, diluted 4-fold in running buffer and loaded onto a smaller column (10 x 60 mm) of the same DEAE-Fractogel. As before, the column was washed extensively with 30 mM NaCl (about 2 h at 0.5 ml.min^{-1}) and a linear NaCl gradient applied to obtain the PS2 reaction centre as a sharp peak at about 110 mM NaCl. The absorption spectra and ratio of Chl to cytochrome b559 were measured immediately to judge the success of the isolation. Samples were stored at 190 K after addition of glycerol to 10% (w/v) if not required for immediate use.

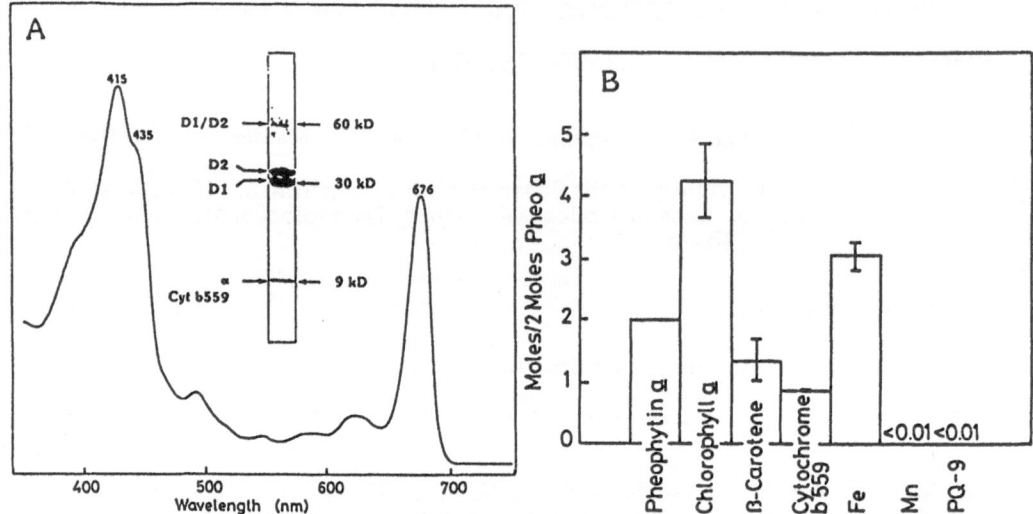

Fig. 1A   Absorption spectrum and SDS-PAGE profiles of purified pea PS2 reaction centre. The absorption spectrum was measured after suspending in 60 mM Tris-Cl (pH 8.5) buffer at room temperature. The SDS-PAGE track was obtained using 10-20% gel containing 6 M urea.

B.   Composition of isolated PS2 reaction centres based on two pheophytin molecules as the reference level.

PROPERTIES

i) Composition and Structure

This isolated complex was shown to be composed of the D1 and D2 polypeptide by immunoblotting of SDS-PAGE profile (see Fig. 1A) using antibodies raised to the products of psbA and psbD genes (10). These genes were expressed in E.coli as fused products with β-galactosidase using a pUR-type expression vector (11). This blotting procedure also showed that a higher molecular weight band at about 60 kD was an aggregate (presumably a dimer) of D1 and D2 (10). Nanba and Satoh (7) identified a band at about 9 kD as one of the apoproteins of cytochrome b559 (the product of the psbE gene) while we have recently used N-terminus analysis to show that a smaller band having a molecular weight on our SDS-PAGE system less than 4 kD is the product of psbF gene (12). In the same work it was possible to investigate the origin of a fifth polypeptide in the isolated reaction centre. This polypeptide has low molecular mass which is located by SDS-PAGE between the cytochrome b559 α- and β-bands. N-terminal analysis of this polypeptide indicates that it is the product of a gene located upstream of psbD gene in the wheat chloroplast genome and which has been called the psbI gene (12). This gene encodes a polypeptide of 36 amino acid residues of which 20 (between 6 and 26) are hydrophobic and may constitute a membrane-spanning region.

The chemical analysis of the isolated complex indicates that it binds one haem, four chlorophyll a, one β-carotene and at least one non-haem iron per two pheophytin molecules (see Fig.1B). Since the bacterial reaction centre contains two bacteriopheophytins then it is at present assumed that the basic PS2 unit would also contain two pheophytins.

Again by analogy with the bacterial system it seems highly likely that two of the chlorophylls form a special pair and our preliminary C.D. spectroscopy supports this contention (13). The other two chlorophylls may be bound as monomeric species but unlike the bacterial system, the appropriate histidine residues are not obvious (4). The isolated PS2 reaction centre complex also differs from isolated bacterial reaction centres in that it does not bind the quinones which would constitute the secondary electron acceptors Q_A and Q_B. Nevertheless inspection of those regions of the D1 and D2 sequences which are likely to be involved in the binding of plasto-quinone reveal similarities with the Q_A and Q_B binding sites on the L and M subunits (4).

L.D. spectroscopy supports the finding that there is only one carotenoid in the complex and that this lies approximately parallel to the long axis of the reaction centre complex and at right angles to the plane of the membrane (13).

The isolated D1/D2/cyt b559 complex normally does not bind other proteins and is free of CP43 and CP47. However, it will bind specifically to an affinity column which has incorporated into its matrix the 33 kD protein involved in stabilizing the Mn cluster necessary for water splitting (14). In contrast it was found that CP43 and CP47 did not bind to this affinity column. The elution of the D1/D2/cyt b559 complex could be accomplished by flushing the column with 1 M $CaCl_2$, a treatment which, interestingly, can be used to release the 33 kD from the thylakoid membrane surface.

ii) Functional properties

Without any additions the isolated complex shows primary photochemistry expected for a PS2 reaction centre devoid of secondary electron donors and acceptors. Flash spectroscopy studies by Takahashi et al (15) and by Danielius et al (16) have monitored the formation and decay of the radical pair $P680^+Pheo^-$. The back reaction between $P680^+$ and $Pheo^-$ has also been detected as a spin polarized triplet signal at low temperatures (17,18). We have shown that this light induced EPR signal is only detected in the redox potential range between 400 and -530 mV. The loss of the EPR signal at negative potential is almost certainly due to chemical reduction of the pheophytin acceptor while at potentials more oxidising than 400 mV the reason for the loss of the spin-polarized triplet is not clear. Under these conditions our EPR work indicated that P680 could be photooxidised which led to the suggestion that at 400 mV and above, the non-haem iron in the complex is oxidised to Fe^{3+} and therefore can act as an electron acceptor with reduced pheophytin acting as the donor (18). Kinetic analysis of the forward and backward reactions indicate that the radical pair is formed within a few picoseconds (19) while recombination at room temperature shows a half time of approximately 36 ns. The charge separation can be stabilised if various acceptors and donors are introduced to the isolated complex. In the presence of dithionite, electrons can be donated rapidly to the photooxidised P680 so as to yield the state $P680Pheo^-$. The photoaccumulation of $Pheo^-$ is shown by its characteristic spectrum (7,20). If instead of adding an electron donor, the artificial electron acceptor, silicomolybdate (SiMo) is added to the isolated reaction centre, the $P680^+Pheo$ state is created (20). Under these conditions the reaction centre seems to be stable and addition of artificial electron donors, such as diphenylcarbazide or manganese(II) chloride (8,20), facilitate a net electron flow to SiMo. The quantum yield, however, for this light induced electron flow is rather low with the highest rates being only about 1000 μequiv.mg $Chl^{-1}.h^{-1}$. Nevertheless this reaction has served as a very useful monitor of the activity of the reaction centre especially in relation to its rate of thermal and photo-degradation.

Initial experiments to replace SiMo by other electron acceptors were not successful. More recently, however, conditions have been found were quinones can be used as electron acceptors including plastoquinone-9 (8,21). The presence of these quinones give rise to a light induced signal at 430 nm. The light-dark difference spectrum has shown that this signal is due to the photoreduction of cytochrome b559

(8). A series of experiments have been conducted using different detergent concentrations which clearly indicate that the primary cause of the redox change of the cytochrome is the photoreduction of the quinone. In addition to the above findings it was found that with quinone present, the isolated reaction centre could support the photoreduction of 2,6-dichlorophenol indophenol (DCPIP) (21).

Of particular note is that quinone mediated processes are partially inhibited by the herbicide 3-(3,4-dichlorophenyl)-1,1-dimethyl urea (DCMU) with a $C_{1/2}$ value in the region of 50 μM (8). With the aid of radiolabelled herbicide we have been able to confirm that the isolated D1/D2/cyt b559 complex can bind these compounds (22). It was found that there is approximately one binding site for DCMU and for ioxynil but two for i-dinoseb. In all cases, however, the binding constants were much higher than for intact thylakoids indicating that there has been some modification of the organisation of the complex after its isolation. It seems highly likely that it is this structural modification which is responsible for the poor binding of plasto-quinone to the vacant Q_A and Q_B sites.

We have found that if the reaction centre complex is isolated from pea thylakoids treated so as to activate a membrane bound kinase, (e.g. in the presence of a reducing agent such as NADPH or dithionite) then the D1 and D2 polypeptides become phosphorylated (23,24). The phosphorylation occurs at threonine residues and by using proteolytic mapping techniques we have located the precise sites to be the conserved threonines at the N-terminus of both polypeptides. We have no explanation for the role of this protein phosphorylation, but it will undoubtedly create a significant change in the electrostatic properties of the exposed surface of the PS2 complex.

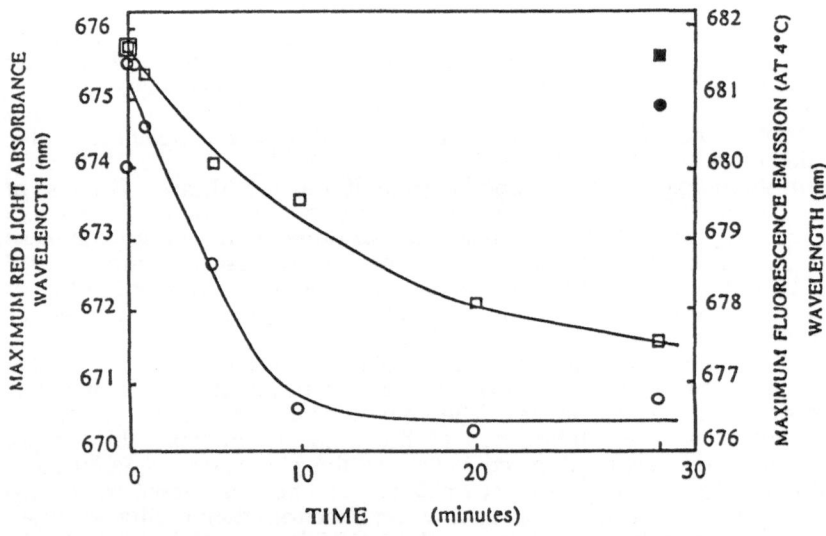

Fig.2   Effect of light treatment at 4°C on wavelength (nm) maximum red light absorbance (□), maximum fluorescence emission (at 4°C) (O) of PS2 reaction centre preparations.

It could be that the phosphorylation of the D1 and D2 polypeptides plays a role in destabilization of the PS2 reaction centre complex prior to its degradation. It is well established that in vivo the D1 polypeptide is rapidly turning over (25) and that this phenomenon is closely related to the vulnerability of the PS2 reaction centre to photodamage (26). When isolated the D1/D2/cyt b559 complex is also very easily damaged by light. Exposure to 500 μE m^{-2} s^{-1} of white light at 4°C results in changes in the absorption spectrum. After a 30 minute illumination period the red absorption peak shifts from 676 nm to 670 nm with a reduction in its magnitude.

A similar blue shift also occurs for the peak of chlorophyll fluorescence from the isolated complex but, in this case, there is a concomitant increase in the yield of emission. The time courses for these changes are shown in Fig.2. Also during the 30 minute illumination there was a decline in the electron transport capacity of the isolated complex. An example of this decline is shown in Fig.3 where the data has come from monitoring the quinone mediated photoreduction of DCPIP. If the Triton X-100 associated with the isolated complex was exchanged with β-lauryl dodecyl maltoside then it becomes significantly more stable especially to dark degradation at temperatures above 4°C (see Fig.4).

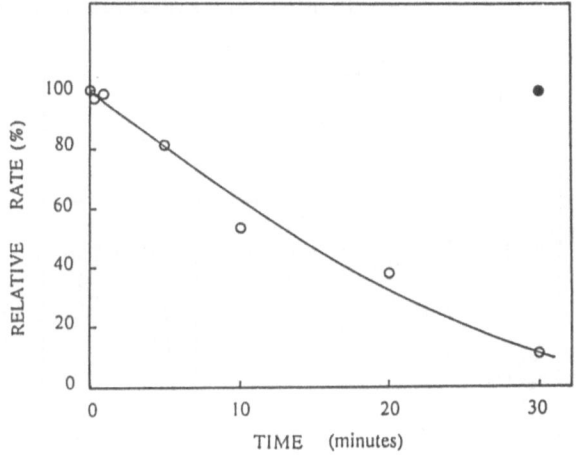

Fig.3   Effect of light treatment at 4°C on the rate of reduction of DCPIP by the isolated PS2 reaction centre preparation plus decylplastoquinone.

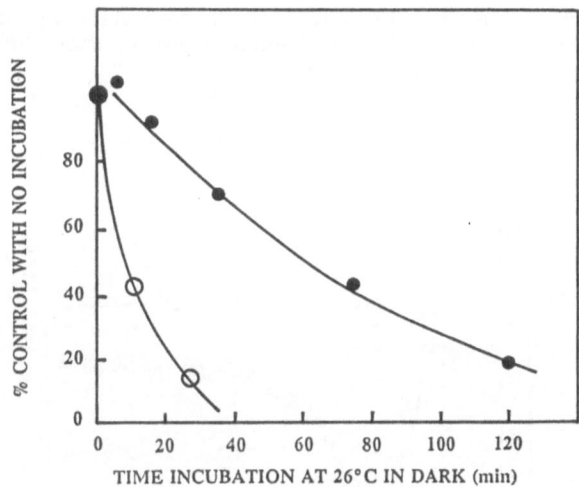

Fig.4   Effect of incubation at 26°C in the dark on relative rates of light dependent reduction of silicomolybdate in PS2 reaction centre preparations in Triton X-100 (○) or dodecylmaltoside (●).

Freshly isolated PS2 reaction centre complexes fluoresce at room temperature with its maximum about 683 nm. As can be seen in Fig.5 when a modulated fluorescence measuring technique is employed the yield of chlorophyll fluorescence is dramatically reduced if the isolated reaction centre is exposed to actinic light in the presence of sodium dithionite (20,27). Under such conditions the P680Pheo$^-$ state photo-accumulates and indeed the simultaneous measurement of absorption and emission shown in Fig.6 indicates the link between the two phenomena. Thus it is concluded that the P680Pheo$^-$ state is a very effective quencher of chlorophyll fluorescence. Similar experiments carried out in the presence of SiMo, also showed that the P680$^+$Pheo state is an effective quencher of emission from the isolated complex. These results could be interpreted as support for the Klimov model (28) in which it is argued that chlorophyll fluorescence results from a recombination of P680$^+$ and Pheo$^-$. Because of quantum yield considerations, however, it seems unlikely that this model is correct (see 27,29). It is more likely that a decline in the yield of fluorescence due to the photoaccumulation of P680Pheo$^-$ or P680$^+$Pheo is due to the fact that both of these species are effective quenchers since their singlet states are low lying compared to those of chlorophyll (30).

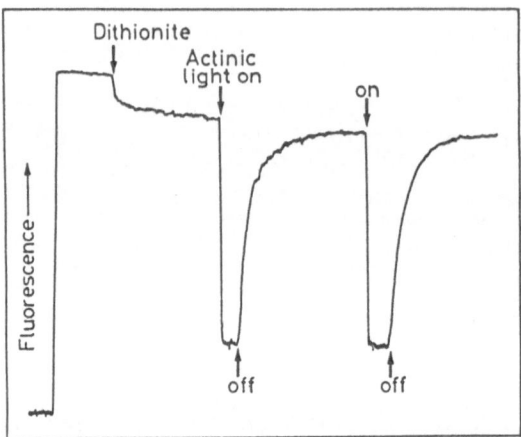

Fig.5   Light induced quenching of chlorophyll fluorescence from PS2 reaction centres in the presence of excess sodium dithionite (see ref.27). The fluorescence was excited by 1 μs pulses of light at 1.6 kHz from a light emitting diode passing through a short pass filter (λ < 670 nm). The fluorescence was detected at wavelengths greater than 700 nm. The actinic illumination was saturating white light.

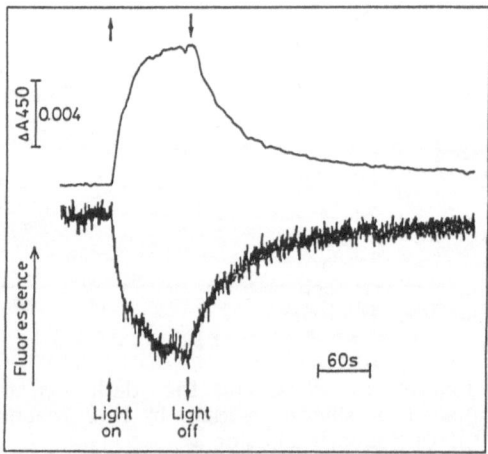

Fig.6   Simultaneous recording of the formation and decay of reduced pheophytin due to light on (upward arrows) and light off (downward arrows) measured as an absorption change at 430 nm and as chlorophyll fluorescence in a PS2 reaction centre preparation treated with sodium dithionite.

CONCLUSION

Our understanding of the structure and function of the PS2 reaction centre is growing rapidly, a developemnt which would not have been possible without the outstanding results which have been attained with isolated reaction centres of purple photosynthetic bacteria (see 31). We now have in our hands a comparable complex isolated from higher plants which is amenable to a wide range of experiments. To date we have collected sufficient data to draw the cartoon shown in Fig.7. This diagrammatic representation relies heavily on analogies with the bacterial system. Without doubt some of these analogies are legitimate but there must, and will be, significant differences. Obvious differences will be on the oxidising side, simply because the PS2 reaction centre has the capacity to extract electrons from water. Already data is accumulating which suggests that the D1 and D2 polypeptides are important for the oxidising reactions as well as harbouring the component involved in stabilizing the reducing potential (see article in this book by Vermaas). Our work implies that the 33 kD and the associated Mn cluster, are intimately associated with the D1 and D2 polypeptides. There are however many unanswered questions. What precisely is the role of cytochrome b559? How does this reaction centre disassemble and then reassemble in order to accommodate turnover of the D1 polypeptide? What are the molecular mechanisms behind the photoinactivation? By what means does the reaction centre create the strong oxidising potential necessary to split water? During the course of the next two or three years these questions will be tackled and we can look forward to some very exciting developments. Perhaps the most important goal will be obtaining a complete physical and chemical description of the unique molecular processes which bring about the oxidiation of water to its elemental components.

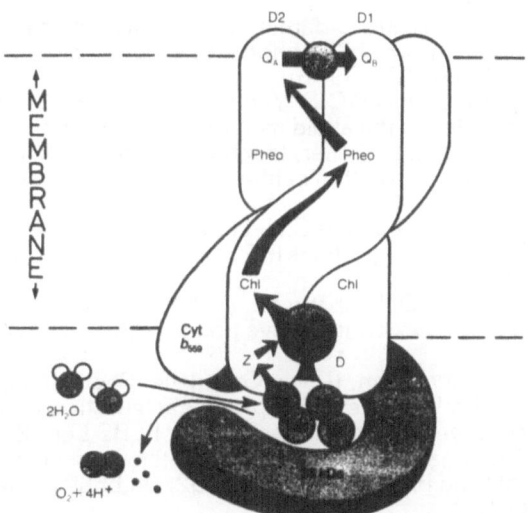

Fig.7   A working model for the organisation of the PS2 reaction centre based on its analogy with the reaction centre structure of purple bacteria. Note that cytochrome b559 consists of two polypeptides encoded for by psbE and psbF genes. There may also be an additional intrinsic polypeptide encoded by psbI gene. An extrinsic 33 kDa protein probably binds to the D1 and D2 polypeptide and in so doing stabilizes a cluster of four manganese atoms necessary for oxidising water. Z and D are oxidised by the primary donor P680 (which is likely to be a special pair of chlorophyll a molecules) and are almost certainly tyrosine residues at position 160 in the D1 and D2 sequences (see ref.32 and article by Vermaas in this book).

ACKNOWLEDGEMENTS

We wish to thank the Agricultural and Food Research Council and Science and Engineering Research Council for financial support. We also acknowledge the help and support of our colleagues, John DeFelice, Ken Davis, Jenny Nicolson, William Newell and Peter Nixon.

REFERENCES

1. J.E. Hearst and K. Sauer, Protein sequence homologies between positions of the L and M subunits of reaction centres of Rhodopseudomonas capsulata and the Q_B protein of chloroplast thylakoid membranes: proposed relation to quinone-binding sizes, Z. fur Naturforsch 39:421-424 (1984)

2. J.C. Williams, L.A. Steiner, G. Feher and M.I. Simon, Primary structure of the L subunit of the reaction cnetre from Rhodopseudomonas sphaeroides. Proc. Natl. Acad. Sci USA 81:7303-7307 (1984)

3. J. Deisenhofer, O. Epp, K. Miki, R. Huber and H. Michel, Structure of the protein subunits in the photosynthetic reaction centre of Rhodopseudomonas viridis at 3 A° resolution, Nature 318:618-624 (1985)

4. H. Michel and J. Deisenhofer, The relevance of the photosynthetic reaction centre from purple bacteria to the structure of photosystem II, Biochemistry 27:1-7 (1988)

5. A. Trebst, The topology of the plastoquinone and herbicide binding peptides of photosystem two in the thylakoid membrane, Z. fur Naturforsch 41c:240-245 (1986)

6. J. Barber and J.B. Marder, Application of molecular genetics for determining photosynthetic membrane structure, in: Biotech. Genet. Eng. Revs. Vol.4, pp355-404, Intercept Ltd., Newcastle, UK (1986)

7. O. Nanba and K. Satoh, Isolation of a photosystem two reaction centre containing D1 and D2 polypeptides and cytochrome b559, Proc. Natl. Acad. Sci. USA, 84:109-112 (1987)

8. D.J. Chapman, K. Gounaris and J. Barber, Electron transport properties of the isolated D1/D2 cytochrome b559 photosystem two reaction centre, Biochim. Biophys. Acta 933:423-431 (1988)

9. D.A. Berthold, G.T. Babcock and C.F. Yocum, A highly resolved O_2 evolving PSII preparation from spinach thylakoid membranes, FEBS Lett. 134:231-234 (1981)

10. J.B. Marder, D.J. Chapman, A. Telfer, Identification of psb A and psb B gene products, D1 and D2, as reaction centre proteins of photosystem two, Plant Mol. Biol. 9:325-333 (1987)

11. P.J. Nixon, T.A. Dyer, J. Barber and C.N. Hunter, Immunological evidence for the presence of the D1 and D2 proteins in PS2 cores of higher plants, FEBS Lett. 209:83-86 (1986)

12. A.N. Webber, L. Packman, D.J. Chapman, J. Barber and J.C. Gray, The photosystem II reaction centre complex contains five polypeptides, FEBS Lett. (1989) in press

13. W.R. Newell, H.van Amerongen, R. van Grondelle, R. Aalnerts, J.W. Drake, P. Udvarhelyi and J. Barber, Spectroscopic characterisation of the reaction centre of photosystem two from higher plants, FEBS Lett. 228:162-166 (1988)

14. K. Gounaris, D.J. Chapman and J. Barber, The interaction between the 33 kD manganese stabilising protein and the D1/D2/cyt b559 complex, FEBS Lett. 234:374-378 (1988)

15. Y. Takahashi, P. Hansson and P. Mathis, Primary radical pair in the photosystem II reaction centre, Biochim. Biophys. Acta 893:49-59 (1988)

16. R.V. Danielius, K. Satoh, P.J.M. Van Kan, J.J. Plijter, A.M. Nuijs and H.J. Van Gorkom, The primary reaction of photosystem two in the D1-D2-cytochrome b559 complex, FEBS Lett. 213:241-244 (1987)

17. M.Y. Okamura, K. Satoh, R.A. Isaacson, and G. Feher, Evidence of the primary charge separation in the D1/D2/ complex of photosystem II in spinach: EPR of the triplet state, in: Progress in Photosynthesis Research, J. Biggins, ed., Vol.1 pp379-381, Martinus Nijhoff Publ., The Hague (1987)

18. A. Telfer, M.C.W. Evans and J. Barber, Oxidation-reduction potential dependence of reaction centre triplet formation in the isolated D1/D2/cytochrome b559 photosystem two complex, FEBS Lett. 232:209-213 (1988)

19. D. Klug, Excitons and charge separation in higher plant chromophore protein complexes. Ph.D. Thesis, University of London (1988)

20. J. Barber, D.J. Chapman and A. Telfer, Characterisation of a photosystem two reaction centre isolated from the chloroplast, FEBS Lett. 220:67-73 (1987)

21. K. Gounaris, D.J. Chapman and J. Barber, Reconstitution of plastoquinone in the D1/D2/cyt b559 photosystem two reaction centre, FEBS Lett. 240:143-147 (1988)

22. M.T. Giardi, J.B. Marder and J. Barber, Herbicide binding to the isolated photosystem two reaction centre, Biochim. Biophys. Acta 934:64-71 (1988)

23. A. Telfer, J.B. Marder and J. Barber, Photosystem two reaction centres isolated from phosphorylated pea thylakoids carry phosphate on the D1 and D2 polypeptide subunits, Biochim. Biophys. Acta 893:557-563 (1987)

24. J.B. Marder, A. Telfer and J. Barber, The D1 polypeptide subunits of the photosytem II reaction centre has a phosphorylation site at its amino terminus, Biochim. Biophys. Acta 932:362-365 (1988)

25. M. Edelman, P. Goloubinoff, J.B. Marder, H. Fromm, R. Fluhr and A.K. Matoo, Structure-function relationships and regulation of the 32 kD protein in the photosynthetic membranes, in: Molecular form and function of the plant genome, L. Vloten-Doting, G.S.P. Groot and T.C. Hall, eds.) pp291-300, Plenuum Press (1985)

26. D.J. Kyle, The 32000 Dalton Q_B protein of photosystem II, Photochem. Photobiol. 41:107-116 (1985)

27. J. Barber, S. Malkin and A. Telfer, The origin of chlorophyll fluorescence in vivo and its quenching by the photosystem two reaction centre, Proc. Roy. Soc. (London) in press (1988)

28. V.V. Klimov, A.V. Klevonik, V.A. Shuvalov and A.A. Krasnovsky, Reduction of pheophytin in the primary light reactions of photosystem II, FEBS Lett. 82:183-186 (1977)

29. G.H. Schatz, H. Brock and A.R. Holzwarth, Picosecond kinetics of fluorescence and absorbance changes in photosystem II particles excited at low photon density, Proc. Natl. Acad. Sci. USA, 84:8414-8418 (1987)

30. I. Fujita, M.S. Davis and J. Fajer, Anion radicals of pheophytin and chlorophyll a. Their role in the primary charge separation of plant photosynthesis, J. Am. Chem. Soc. 100:6280-6282 (1978)

31. J. Barber, Photosynthetic reaction centres: A common link, Trends Biochem. Sci. 12:321-326 (1987)

32. R.J. Debus, B.A. Barry, G.T. Babcock and L. McIntosh, Site directed mutagenesis identifies a tyrosine radical involved in photosynthetic oxygen evolution system. Proc. Natl. Acad. Sci (USA) 85:427-430 (1988)

PIGMENT-PROTEIN COMPLEX ORGANIZATION AND MODULATION OF THE F730/F685 RATIO AT 77°K IN CHLOROPLAST THYLAKOIDS

J. H. Argyroudi-Akoyunoglou and C. Vakirtzi-Lemonias

NRCPS "Demokritos", Athens, Greece

INTRODUCTION

It is generally accepted that in chloroplasts the fluorescence emission at room temperature comes from the antennae Chla of PSII (685 nm), while at 77°K it originates from PSII (685, 695 nm) as well as from PSI (735, 720 nm). The Chla fluorescence yield changes, as well as the 77°K fluorescence F730/F685 ratio changes, observed in chloroplasts upon manipulation of their ionic environment or quality of light exposure (PSI or PSII light), have been considered to monitor changes in the excitation energy transfer (spillover) from the antennae of PSII to the reaction center of PSI[1-3,5] or changes in the initial distribution of the absorbed light energy between PSI and PSII[4-9]. These fluorescence changes are thought to reflect randomization of PS units[10], upon "low-salt" grana unstacking[11]-facilitating spillover and thus enhancing the F730/F685 ratio- and segregation of PS units upon cation-induced grana restacking (PSII in grana, PSI in stroma lamellae)-blocking spillover and enhancing the F685/F730 ratio-; or changes in the organization of Chl among the pigment-protein complexes within each photosynthetic unit, causing changes in the effective absorption cross section of the units (especially of PSI)[4-9].

To establish whether the modulation of the fluorescence parameters reflect indeed changes in the organization of the pigment-protein complexes in the PS units, we studied the kinetics of the F730/F685 ratio increase upon cation depletion, or decrease upon addition of the ether phospholipid PAF[12], and tried to correlate it with any changes in the distribution of Chl among the pigment-protein complexes. Our results show that there is a very good correlation between the two parameters, suggesting that the changes in the F730/F685 ratio may indeed reflect changes in the organization of Chl among the pigment-protein complexes comprising the photosynthetic units.

MATERIALS AND METHODS

Chloroplasts isolated from spinach, pea or bean leaves (pellet at 1,000 x g x 10 min) after homogenization in 0.3 M sucrose-0.01 M KCl-0.05 M Phosphate buffer, pH 7.2, were resuspended in homogenization buffer; aliquots of known Chl content were repelleted, and Tricine-NaOH (0.05 M, pH 7.2) or phosphate-KCl (0.05 M-0.01 M, pH 7.2) were added to make 30 ug Chl/ml (for fluorescence measurements at 77°K)[7,11] or 180 ug Chl/0.1 ml (for SDS-PAGE of pigment-proteins)[13]. At various time intervals after Tricine addition, aliquots were withdrawn and immediately frozen in liquid Nitrogen and their fluorescence recorded at 77° K , or solubilized in SDS (SDS/Chl=10) and analyzed by SDS-PAGE as before[13].

Chloroplasts isolated as above and then washed and suspended in 0.05 M Tricine-NaOH, pH 7.2, at 30 ug Chl/ml were incubated for 30 min with a freshly prepared film of AGEPC (acetyl glyceryl ether phosphocholine, PAF) or analogues, in an ice bath; the thylakoid suspension was removed into a capillary tube, frozen in liquid Nitrogen and its fluorescence spectrum recorded at 77°K[7]. For Chla fluorescence yield measurements, 50 ul aliquots containing 13 ug Chl were incubated with the phospholipid film as above, in the dark, then transferred with the aid of a syringe into a 2.5 ml Tricine containing cuvette, and the Fmax at 685 nm was recorded[11].

PSII activity was determined as the DPC-DCIP photoreduction[14] after incubation of 50 ml thylakoid samples (1.5 ug Chl) with PAF as above.

Grana and stroma lamellae were isolated by differential centrifugation of French press disrupted chloroplasts suspended in 0.05 M phosphate-0.15 M KCl, pH 7.2, at 400 ug Chl/ml[15]. The 10K and 240K pellet fractions (grana and stroma lamellae, respectively) were washed and suspended in Tricine at 30 ug Chl/ml, and 50 ul aliquots were incubated with the film of PAF as above. All phospholipids were products of NOVA, Switzerland.

SDS-PAGE was done at SDS/Chl=10, at 4°C[13]. The resolved pigment-protein complexes were scanned in a Joyce-Loeble Chromoscan with a 620 nm cutoff filter, and the area under each peak was estimated by weight.

RESULTS AND DISCUSSION

Suspension of chloroplasts in "low-salt" Tricine buffer

Suspension of spinach or pea chloroplasts in Tricine induces a drastic increase in the F730/F685 ratio (Fig. 1, left). In contrast, suspension of bean plastids into Tricine does not affect the ratio. Parallel to the F730/ /F685 ratio increase in spinach and pea, drastic changes are observed in the distribution of Chl among the pigment-protein complexes (Table 1, Fig. 1, right): the CPIa complex of PSI, which is completely missing in the resolution patterns of phosphate-KCl samples prior to Tricine addition, is gradually organized from its components (CPI and LHC-I) upon suspension in Tricine, and the CPIa/CPI ratio is drastically increased; the percent Chl bound on the LHCP[2] dimer form of LHC-II, known to be enriched in LHC-I[16], is greatly increased; and the Chl distribution between PSII and PSI complexes increases in favor of PSI. All changes are completed within 5 min from Tricine addition Thylakoids isolated in phosphate-KCl (180 ug Chl) were pelleted and then suspended in 0.1 ml of either phosphate-KCl or Tricine buffer. At various time intervals after Tricine addition SDS was added (0.3 M Tris-HCl, pH 8.8, 1% SDS, 10% glycerol) to make SDS/Chl=10, and SDS-PAGE followed imediately. PSI complexes: CPIa+CPI+LHCP2 (LHC-I). PSII complexes: LHCP1+CPa+LHCP3.

and the time course of the F730/F685 ratio increase is closely correlated with that of Chl organization into the PSI complexes.

The inability of "low-salt" to induce changes in the F730/F685 ratio is similarly quite well correlated with the lack of drastic changes in the organization of Chl in the various pigment-protein complexes (Table 1, Fig. 1). The difference between spinach/pea and bean plastids may be attributed to the greater stability of the complexes in bean thylakoids, not allowing their dissociation upon manipulation of the ionic environment. This is supported by the finding that the cation or SDS concentration required for dissociation of the oligomeric supramolecular structures of the pigment-protein complexes are by far greater in bean than in spinach (Fig. 2).

The results, therefore suggest that depletion of salt induces redistribution of Chl among pigment-protein complexes in favor of the PSI unit, and thus higher F730/F685, resulting from its greater absorption cross section.

92

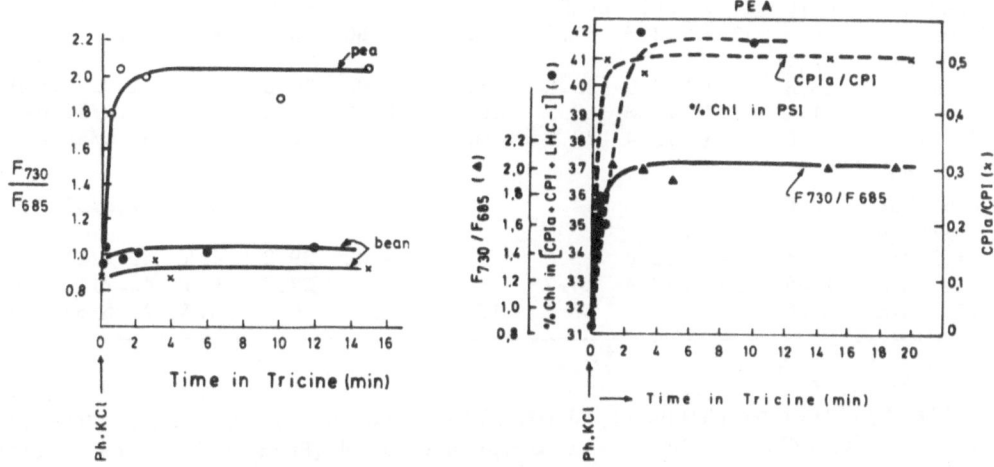

Fig. 1 Left: The kinetics of the F730/F685 ratio increase at 77°K in bean and pea plastids isolated in phosphate-KCl upon their suspension in Tricine at 30 ug Chl/ml. Right: Correlation of the F730/F685 increase with the organization of Chl into the PSI complexes in pea plastids. (o-o), (x-x): etiolated bean exposed to continuous light for 92h or 192h, respectively.

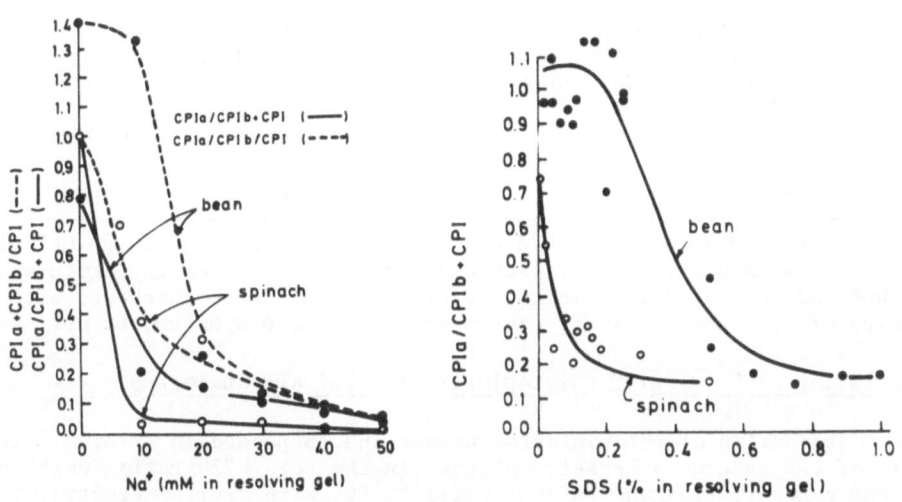

Fig. 2. The dissociation of the CPIa complex into CPI, as affected by SDS or cation concentration in the resolving gel, in spinach and bean thylakoids suspended in Tricine at 180 ug Chl/0.1 ml. (SDS/Chl=10).

Table 1. The distribution of Chl among the pigment-protein complexes as affected by suspension of plastids into 0.05 M Tricine-NaOH, pH 7.2.

BEAN-192h	CPIa	CPIb	CPI	LHCP1	LHCP2	CPa	LHCP3	FP	PSI / PSII
				Chlorophyll Distribution (%)					
Phos-KCl	14.3	3.4	12.3	31.3	11.3	4.86	18.3	4.2	41.3/54.5=0.75
Tricine									
1 min	13.8	3.7	12.5	25.8	11.2	4.98	20.8	7.2	41.2/51.6=0.79
5 min	11.1	3.7	12.3	27.5	10.8	5.30	19.2	10.0	38.0/52.0=0.73
20 min	13.5	2.3	12.4	29.6	9.4	7.0	20.2	5.5	37.6/56.6=0.67
SPINACH									
Phos-KCl	0.5	-	18.3	14.2	12.7	15.20	27.6	11.4	31.5/57.0=0.55
Tricine									
1.5 min	9.34	-	12.0	13.6	19.0	9.02	29.2	7.5	40.3/51.8=0.78
3 min	11.55	-	12.4	14.7	18.8	8.49	27.7	6.1	42.7/50.9=0.84
15 min	10.00	-	14.1	17.7	17.5	7.84	26.8	5.3	41.6/52.3=0.80

Table 2. Effect of phospholipid molecular structure on the F685/F730 ratio at 77°K and the Chla fluorescence yield (Fmax/Chl) in pea plastids

	F685/F730* ratio	Fmax/Chl** (arbitrary units)
(plastids in Tricine-NaOH)	0.36	48
(plastids in phosphate-KCl)	1.10	100
+Phospholipid		
1-0-C16-2-acetyl-glycero-3-PC,(PAF 16)	3.00	130
1-0-C18-2-acetyl-glycero-3-PC,(PAF 18)	0.45	43
1-0-C16-glycero-3-PC, (L-PAF 16)	1.37	160
1-0-C18-glycero-3-PC, (L-PAF 18)	0.60	80
1,2-di-C16-glycero-3-PC, (PC 16)	0.43	42
1,2-di-C18-glycero-3-PC, (PC 18)	0.37	-
1-0-C16-2-acetyl-glycerol	0.36	52
1-0-C16-glycerol	0.40	-
1-0-C16-2-acetyl-glycero-3-phospho N(CH3) - -hexanolamine	2.05	59***
1-C16-glycero-3-PC, (L-PC 16)	0.53	60
1,2-0-0-di-C16-glycero-3-PC	0.41	46
1,2-0-0-di-C16-glycerol	0.37	-

*50 ul thylakoid aliquots washed and suspended in Tricine (30 ug Chl/ml) were incubated with 0.5 umoles of each phospholipid . **50 µl thylakoid aliquots washed and suspended in Tricine (13 ug Chl/50 ul) were incubated with 0.25 umoles of each phospholipid, except in (***) where 0.05 umoles were used.

Incubation of "low-salt" chloroplasts with PAF and analogues

Incubation of chloroplasts, washed and suspended in Tricine, with the film of PAF induces a drastic increase in the F685/F730 ratio, which depends on PAF concentration and PAF/Chl ratio[12]. PC, with similar electrical properties, is completely ineffective. As shown in Table 2, the molecular structure requirement for the effect to be observed is the ether bond and C-16 chain length at C-1 position of the glycerol moiety, a short C-tail at C-2 position and a large polar head at C-3. Since all compounds are zwitterionic with similar electrical properties, the effect can not be attributed to neutralization of thylakoid surface charges, in a way similar to cation action, nor to lateral movement -leading to segregation of PSII in grana and PSI in stroma lamellae-. This is also suggested from the finding that isolated grana or stroma lamellae incubated with PAF behave in a similar way (Fig. 3). Instead it seems that the phospholipids act by incorporation into the thylakoid bilayer, as suggested by the molecular shapes of the active molecules: only the cone-like PAFs are active , but not the rod-like PCs. Furthermore, the difference in activity between PAF C-16 and PAF C-18 may be attributed to the easier incorporation of the former into the thylakoid as compared to the latter, because of its larger critical micellar concentration (an increase by one methylene group of the C-1 alkyl chain in PAF results in a 3-fold decrease in CMC)[17], and probably its smaller micellar volume (data for 1-O-C16- and 1-O-C18-2-deoxy-PC gives 140 and 4,900 molecules/micelle, respectively)[18].

Upon incorporation of PAF into the thylakoid the organization of the photosynthetic units is perturbed: in addition to the increase in the F685/ /F730 ratio at 77°K, the Fmax/Chl is greatly enhanced, suggesting that PSII is affected (Table 2). This is also suggested from the finding that the F685/F730 ratio increase is larger in the PSII-rich grana fractions than in the PSI-rich stroma lamellar ones (Fig. 3). Indeed when thylakoids incubated with PAF are analyzed for pigment-protein complexes, one observes that Chl binding on PSII complexes (LHCP1 and CPa) is greatly increased. In contrast, incubation of thylakoids under similar conditions with PC has no effect on the distribution of Chl among the pigment-protein complexes (Table 3).

Table 3. Chlorophyll distribution among the pigment-protein complexes in bean or pea thylakoids, as affected by incubation with PAF or PC.

Sample	Chlorophyll distribution (%)									PSII / PSI
	CPIa	CPIb	CPI	X2	LHCP1	LHCP2	CPa	LHCP3	FP	
BEAN-72h CL										
-PAF	3.8	3.5	9.9	0.4	26.0	12.7	3.4	19.5	20	1.65
+PAF C-16	9.9	6.7	7.4	3.5	38.6	9.2	6.1	13.1	5.5	1.85
+PC	4.3	5.0	10.5	1.1	30.1	15.7	4.6	19.7	8.8	1.55
PEA										
-PAF	7.9	2.1	14.2	2.0	20.5	17.1	4.3	22.3	9.6	1.19
+PAF C-16	9.7	6.0	9.9	4.5	26.8	12.3	6.4	18.0	6.3	1.47
+PC	6.8	-	16.1	1.9	19.4	16.6	4.2	25.9	9.0	1.29

50 ul aliquots of thylakoids washed and suspended in 0.05 M Tricine-NaOH, (25 ug Chl/50 µl) were incubated with a film of PAF or PC (0.25 µmoles) in ice for 30 min. SDS was added (SDS/Chl=10) and PAGE followed immediately.

Finally, PAF incorporation into the thylakoid seems also to affect the PSII activity of thylakoids (measured as DPC-DCIP reduction) (Fig. 4). Preliminary experiments have shown that PAF mediates electron transport from DPC to DCIP even in the absence of thylakoid Chl, or in the presence of DCMU, suggesting that PAF acts as a mediator of electron transport in thylakoids as well (see Fig. 4).

In conclusion, the F730/F685 (77°K) ratio increase upon suspension of pea and spinach chloroplasts in "low-salt"-buffer (Tricine) correlates with the Chl organization into the PSI complexes CPIa and LHC-I (LHCP[2]), while the F685/F730 ratio increase in bean and pea plastids upon addition of PAF C-16 correlates with the Chl organization mainly in the PSII unit complexes LHCP1 and CPa. Similarly, the lack of observable changes in the F730/F685

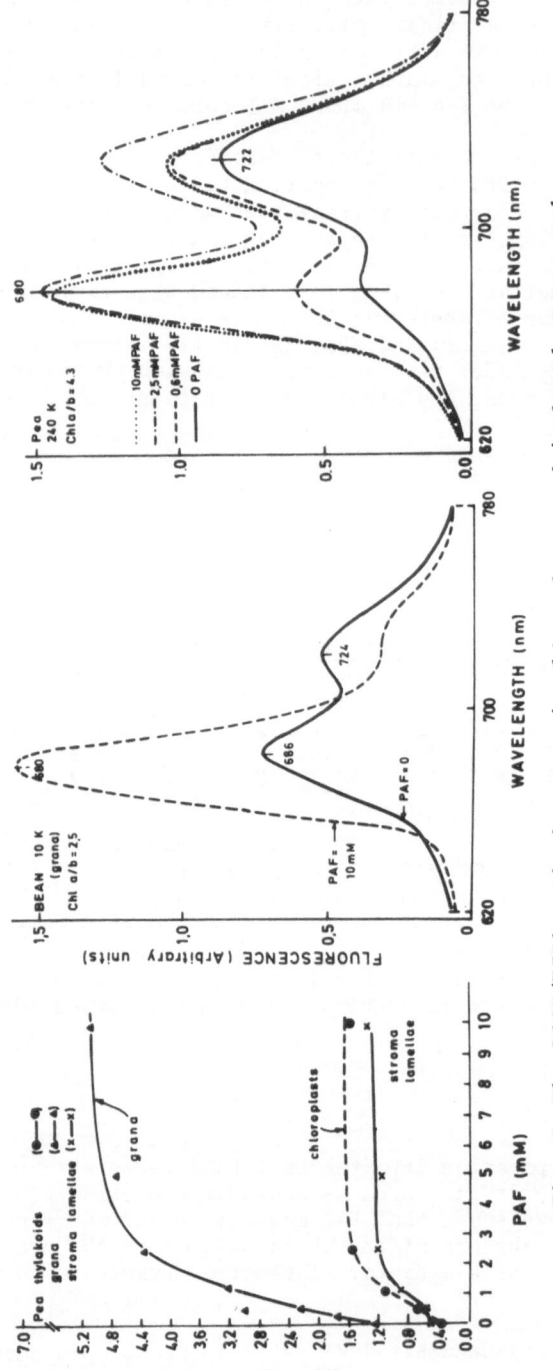

Fig. 3. The F685/F730 ratio increase in chloroplasts and isolated stroma and grana lamellae after their incubation with PAF C-16. Chloroplasts and thylakoid fractions were washed twice with 0.05 M Tricine-NaOH, pH 7.2, and were suspended in Tricine at 30 ug Chl/ml. 50 ul aliquots were incubated with the freshly prepared phospholipid film for 30 min at 0°C; then they were transferred to a capillary tube, immersed in liquid Nitrogen and their fluorescence recorded at -196°C.

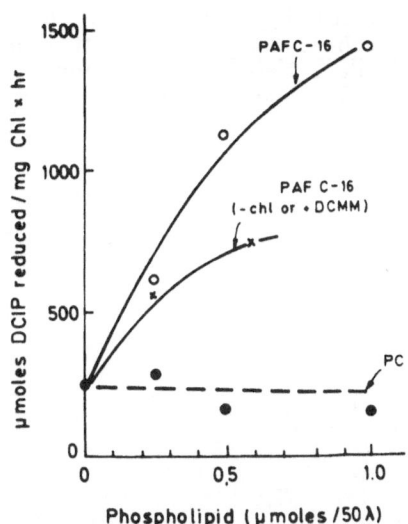

Fig. 4. PAF C-16-induced mediation of electron transport (DPC-DCIP reduction) Chloroplasts washed and suspended in 0.05 M Tricine-NaOH, pH 7.2, to have 1.5 μg Chl/50 ul were used. PSII activity determined as the DPC-DCIP reduction. Incubation with the film of PAF was for 30 min at 0°C prior to PSII activity determination.

ratio in bean correlates with the lack of changes in the distribution of Chl between PSI and PSII complexes. The effect of PAF C-16 on the F685/F730 ratio of plastids, as well as of isolated stroma lamellar and grana fractions, is believed to be elicited through its incorporation into the thylakoid bilayer because of its wedge-shaped structure, rather than to charge neutralization processes leading to lateral movement of complexes.

Acknowledgment:We thank Spyro Daoussis for his expert technical assistance.

REFERENCES
1. P. Homann, Plant Physiol. 44:932 (1969)
2. N. Murata, Biochim. Biophys. Acta 189:171 (1969)
3. N. Murata, Biochim. Biophys. Acta 172:242 (1969)
4. C. Bonaventura and J. Myers, Biochim. Biophys. Acta 189:366 (1969)
5. Govindjee and G. Papageorgiou, in: "Photophysiology", A. C. Giese, ed., Academic Press, N.Y., pp 1-46.
6. J. H. Argyroudi-Akoyunoglou and G. Akoyunoglou, Arch. Biochem. Biophys. 227:469 (1983)
7. J. H. Argyroudi-Akoyunoglou, A. Castorinis and G. Akoyunoglou, Photo-biochem Photobiophys. 4:201 (1982)
8. A. Castorinis, G. Akoyunoglou and J. H. Argyroudi-Akoyunoglou, Photo-biochem. Photobiophys. 4:283 (1982)
9. G. Akoyunoglou and J.H. Argyroudi-Akoyunoglou, in:"Ion Interactions in energy transfer biomembranes", G. C. Papageorgiou and J. Barber, eds, Plenum, N.Y. pp. 211-222.
10. J. Barber, FEBS Lett. 118:1 (1980)
11. J. H. Argyroudi-Akoyunoglou and G. Akoyunoglou, Arch. Biochem. Biophys. 179:370 (1977)
12. J. H. Argyroudi-Akoyunoglou and C. Vakirtzi-Lemonias, Arch. Biochem. Biophys. 253:38 (1987)
13. J. M. Anderson, J. C. Waldron and S. W. Thorne, FEBS Lett. 99:227 (1978)
14. L. P. Vernon and E. R Shaw, Plant Physiol. 44:1645 (1969)
15. P. V. Sane, D. J. Goodchild and R. B. Park, Biochim. Biophys. Acta 216:162 (1970)

16. T. Y. Kuang, J.H. Argyroudi-Akoyunoglou, H. Nakatani, J. Watson and C.J. Arntzen, <u>Arch. Biochem. Biophys.</u> 235:618 (1984)

17. W. J. Kramp et. al., <u>Fed. Proc.</u> 42:1909 (1983)

18. H. U. Weltzien, B. Arnold and R. Reuther, <u>Biochim. Biophys. Acta</u> 466: 411 (1977)

PHOTOREGULATION OF LHC-II ACCUMULATION IN THYLAKOIDS DURING CHLOROPLAST DEVELOPMENT

J.H. Argyroudi-Akoyunoglou

NRCPS "Demokritos", Athens, Greece

The biogenesis of the photosynthetic units depends on close cooperation of the chloroplast and cytoplasm protein synthesizing machinery: the reaction center (RC) core proteins are coded by the plastid DNA, synthesized on thylakoid-bound ribosomes, and cotranslationaly inserted into the developing membrane; on the other hand, the light-harvesting antennae apoprotein LHC-II, is coded by nuclear DNA, synthesized on cytoplasmic ribosomes, as higher molecular weight hydrophilic precursor, it is post-translationally imported through the plastid envelope, and after its processing to its mature size, by cleavage of its transient sequence, it is incorporated into the thylakoid. No precursor or mature form have been detected in the envelope or stroma of the chloroplast, suggesting a rapid transport through the various compartments. A small amount of precursor LHC-II apoprotein has been found in the thylakoid recently, suggesting that processing may also take place in the thylakoid[1-3]. Stabilization of the mature hydrophobic protein in the thylakoid seems to require concomitant Chlorophyll synthesis, which allows its stabilization via Chl-protein complex formation.

It is clear, therefore, that LHC-II incorporation in thylakoids involves a number of steps: transcription, translation, transport, processing and stabilization; of these, the transcription and post-translational controls are the most important.

Transcriptional control: Circadian rhythm in LHC-II gene transcription and its regulation by phytochrome

The transcription of the LHC-II gene is controlled in a positive way by light, via the phytochrome photoreceptor protein[4]. Phytochrome affects LHC-II mRNA accumulation at all stages of etioplast development[5]. Five to 17 day old etiolated plants show an increase in the amount of the translatable LHC-II mRNA in the dark following a 2 min Red-Light(R) pulse, which is reversed by Far-Red(FR) light immediately following the R pulse: the initial increase in the dark after the R pulse, is lower in 5 and 17 d old plants than in 9 to 13 (Fig.1). The kinetics of the LHC-II mRNA accumulation in the dark, following the R pulse, show rhythmical oscillations with a period of 24 hours, after an initial 12h rise and decay (Fig.2). In these experiments etiolated bean leaves were exposed to the R pulse and then kept in the dark for 70h. At various time intervals the poly A^+-mRNA was isolated, translated on a wheat germ system in the presence of S^{35}-met, and aliquots of translation products, containing equal TCA precipitable radioactivity, were either analyzed as such by SDS-PAGE and the electropherograms visualized by fluorography, or they

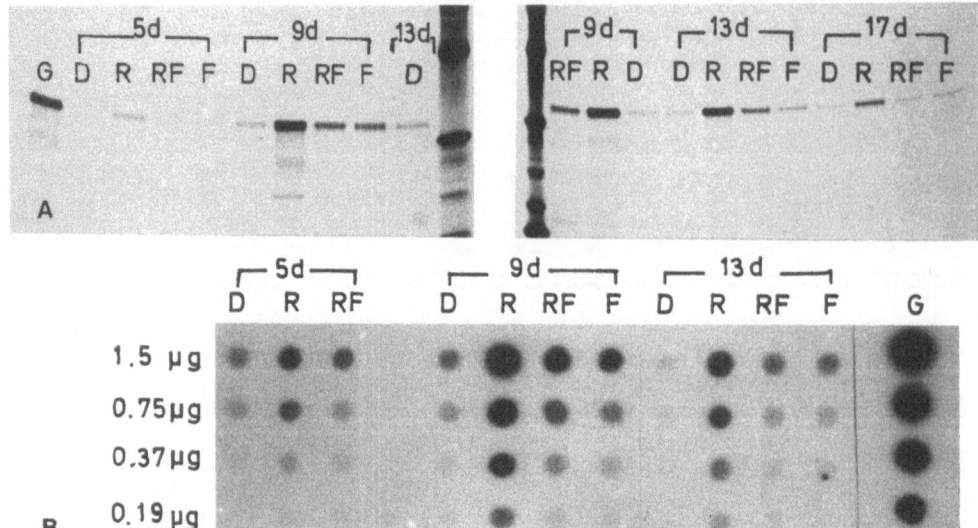

Fig. 1. Phytochrome effect on the level of LHC-II mRNA in 5 to 17 day etio-
lated bean leaves exposed to a 2 min R pulse (R), or to 2 min R+5 min
FR (RF) or to 5 min FR (F), and then kept in the dark for 4 hours.
A: Fluorogram of immunoprecipitates of total in vitro translation
products (samples contained equal counts). B: Dot-Hybridization with
LHC-II cDNA. (Tavladoraki, P., Kloppstech, K., Argyroudi-Akoyunoglou,
J.H., 1988, in press).

were immunoprecipitated with antiserum against LHC-II, and the immunoprecipitates electrophoresed and detected as above. The results show the relative
amount of LHC-II mRNA present in the total translatable mRNA of each sample.
A similar rhythm in LHC-II mRNA accumulation was observed in plants exposed
to continuous light, or transferred to the dark after various periods in the
light (Fig.2). A second R pulse, applied many hours after initiation of the

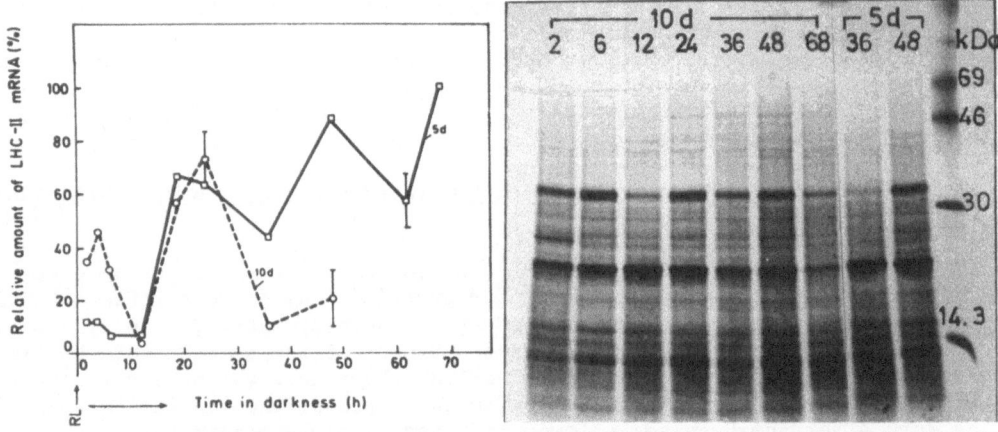

Fig. 2. Circadian rhythm in LHC-II mRNA accumulation in etiolated bean leaves
exposed to 2min R and then kept in the dark (left),or to continuous
light (CL) (right). Relative LHC-II mRNA calculated after scanning
of the 32kDa LHC-II precursor band from fluorograms of total in
vitro translation products containing equal counts.Data expressed as
% maximal value.(Right):In vitro translation products; plants exposed
to CL for 2 to 48h (Tavladoraki,P., Kloppstech, K., Argyroudi-Akoyunoglou, J.H., 1988, in press).

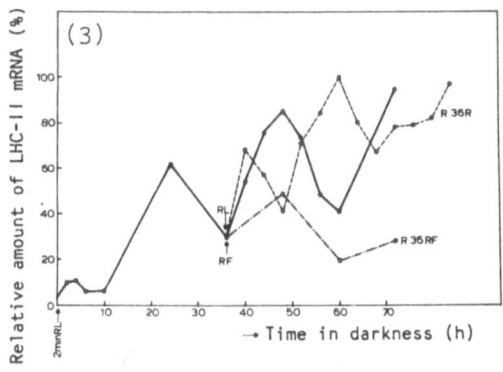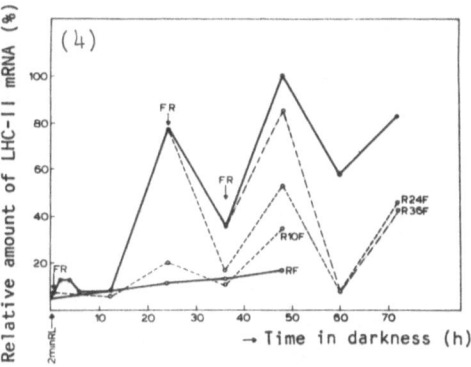

Fig. 3. Phase-shift in the R-induced circadian rhythm in LHC-II mRNA accumulation, as induced by a second R-pulse, applied 36h after initiation of the rhythm (R36R) and its prevention by FR immediately following the second R pulse (R36RF). Solid line shows the initial rhythm.

Fig. 4. Persistent FR reversibility of the R-induced LHC-II mRNA increase. FR applied 0 time (RF), 10 min (R10F), 24h (R24F) or 36h (R36F) after the R pulse (Tavladoraki, Kloppstech, Argyroudi-Akoyunoglou, 1988, in press).

rhythm, induced a phase-shift,while FR, applied immediately after the second R pulse, prevented the phase-shift and reduced the LHC-II mRNA level accumulating thereafter (Figs 3 and 4). FR reversibility on the LHC-II mRNA level increase induced by a R pulse, persisted for at least 36 hours in the dark after initiation of the rhythm, but it was gradually reduced as time from the R pulse was prolonged.

The finding that a 2 min R pulse is sufficient to induce the rhythm, which persists even under constant light or dark conditions, while a second R pulse applied hours after initiation of the rhythm induces a phase-shift, which can be prevented by FR, suggests that the LHC-II gene transcription follows a circadian endogenous rhythm, which is under phytochrome control. Phytochrome seems to act as a synchronizer of the oscillator. Finally, the persistent FR-reversibility of the R-induced LHC-II mRNA increase,suggests that a stable Pfr form responsible for sustaining transcription exists for a long period of time (at least 36h).

Post translational control/Competition between LHC and RC apoproteins for Chlorophyll under conditions of limited Chl accumulation

The LHC-II mRNA accumulation, however, can not be correlated with the stabilization of the LHC-II apoprotein in thylakoids: eventhough a 2 min R pulse is adequate for LHC-II mRNA transcription, LHC-II apoprotein cannot be detected in thylakoids -but only in minute amout-, no matter how long the dark period is prolonged thereafter. In addition, up to date, no rhythmical oscillation in the accumulation of LHC-II apoprotein has ever been observed in thylakoids. Furthermore, inspite of the presence of translatable LHC-II mRNA in ImL plants,no LHC-Chl-protein complexes can be detected in their thylakoids (Fig. 5). The thylakoids under these conditions contain only Chl$\underline{a}$, a tenth of the Chl content of plants exposed to continuous light, and small-sized PS units with core pigment-protein complexes and no LHC components[7-10]. LHC complexes are formed only after transfer of the ImL plants to Continuous light. Main reason for the discrepancy seems to be on one hand the requirement for concomittant Chl synthesis, needed for stabilization of LHC apoprotein through Chl-protein complex formation, and on the other, the competition which seems to exist between RC and LHC apoproteins for Chl, under conditions of limited Chl availability. This became evident from experiments with plants exposed briefly to continuous light and then transferred to the dark -where Chl synthesis ceases-[11-13]. In

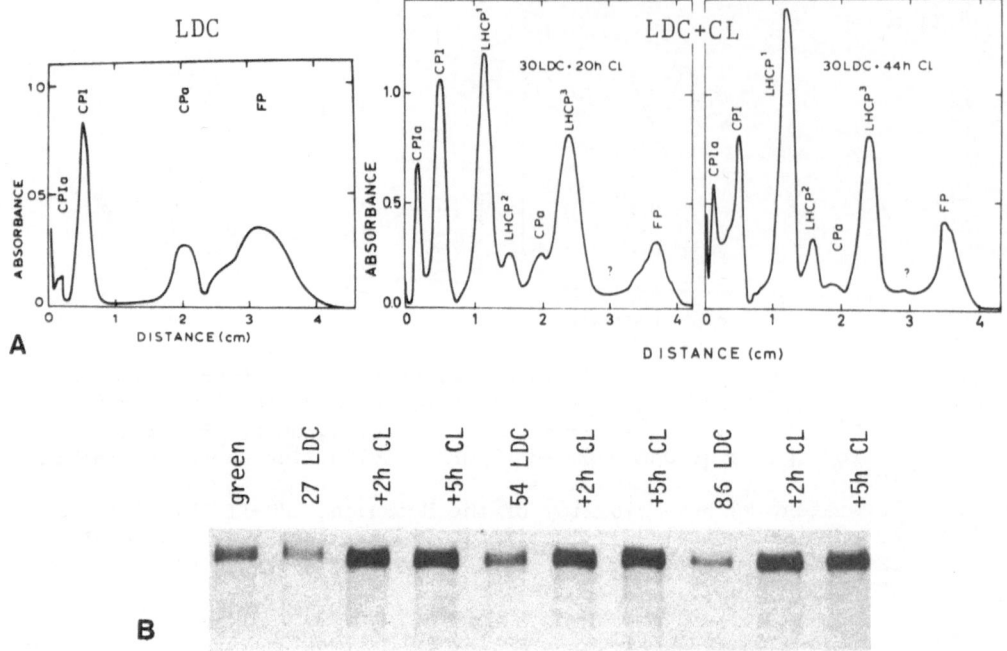

Fig. 5. Chlorophyll-protein complex and LHC-II mRNA accumulation in etiolated bean leaves exposed to ImL (2minL+98minD in cycles, LDC) and
then transferred to continuous light (CL). (Ref. 8 and Tavladoraki
and Argyroudi-Akoyunoglou, unpublished results). Data in B show
immunoprecipitates of translation products.

this case, the Chla/leaf remains constant in the dark (see Table 1), the
Chla/Chlb ratio increases gradually and parallel to LHC-protein degradation,
and new small-sized units are formed, lacking H_2O-splitting activity[12,13].
Degradation of LHC-II in the dark, however, occurs only after brief preexposure to continuous light and parallel to new RC-protein synthesis (Fig.6).

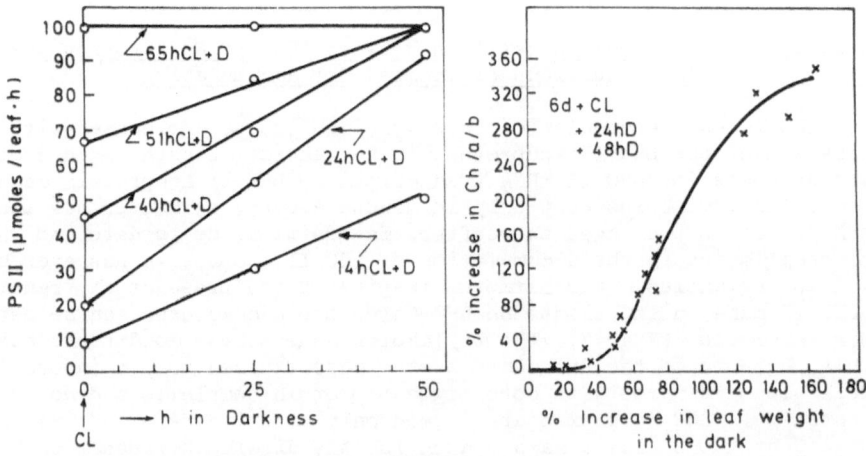

Fig. 6. Left: Photosystem II content (% of green) in etiolated leaves exposed to continuous light (CL) for 8 to 65h and then transferred to
dark for 48h. Right: The increase in Chla/b ratio in plants transferred to the dark, after preexposure to CL for 24 or 48h, as affected by leaf fresh weight increase.

As shown in Fig. 6, upon exposure of etiolated bean leaves to continuous
light, the PSII content increases to reach that of the green control after
65h. After transfer to the dark, new units are formed only in cases where
the preexisting unit content prior to the dark transfer was lower than that
of the green control. New PSII unit formation is thus more extensive in
plants transferred to the dark after brief preexposure to light, and so is
the degradation of the preexisting LHC-II polypeptide[12,13]. Since new PS
unit formation parallels leaf growth, the increase in Chl_a/_b ratio, monitoring destruction of LHC-II apoproteins, is also more extensive when
growth of the leaf is more extensive (Fig. 6, right).

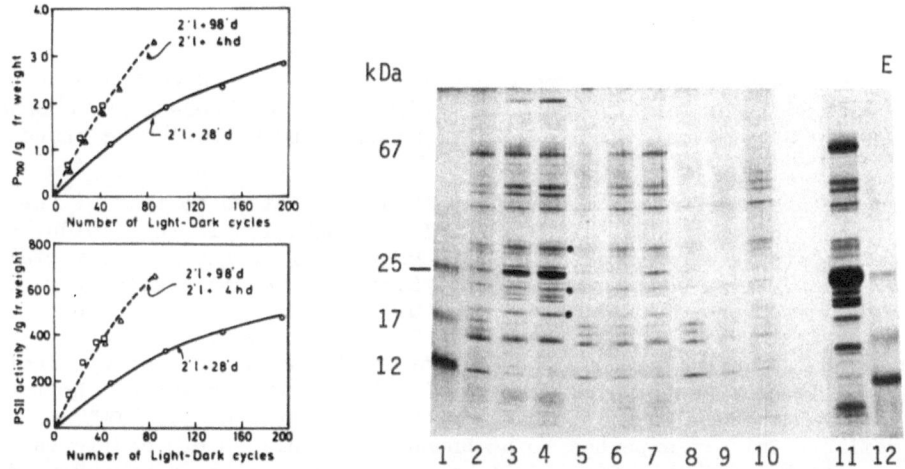

Fig. 7. Left. PS unit content in thylakoids of leaves exposed to ImL with
        dark intervals of varying duration (2minL+28minD, 2minL+98minD,
        2minL+4hD), vs the number of light-dark cycles (LDC). Right:SDS-
        -PAGE of thylakoids in plants exposed to LDC of 2minL+28minD (slots
        2 to 4), 2minL+98minD (slots 5 to 7), and 2minL+4hD (slots 8 to
        10), for 2, 4, 6 days respectively. 11:green control; 1,12:Stds.

        Since Chla is required for the stabilization of a great number of
thylakoid polypeptides -through binding and Complex formation-these results
have suggested to us that LHC-II degradation in the dark may involve the
removal of its stabilizing Chl by the new RC proteins formed in the dark;
we have thus proposed that whenever Chl_a is the limiting factor, it is the
Chl_a synthesis, relative to that of the other thylakoid components, which
regulates the assembly of the complexes and the fate of the polypeptides
which require Chl for their stabilization. It seems that whenever the rate
of Chl_a synthesis is low, relative to that of other components, the RC apoproteins formed compete with those of the LHC for the small amount of Chl
available[14,15]. The affinity of the RC proteins for Chl_a seems to be
higher than that of the LHC, and thus, (a) in ImL plants only RC proteins
are stabilized, and (b) in dark-transferred plants the new RC proteins
formed in the dark, remove Chl from preexisting LHC, rendering it unstable
and subject to proteolytic digestion. This proposal has been tested by a
variety of experimental conditions designed to modulate either the relative rate of RC protein formation, or the rate of Chl accumulation.

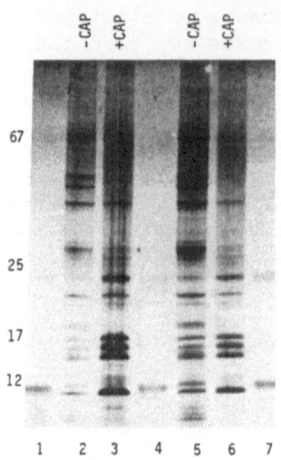

Fig. 8. SDS-PAGE of thylakoid polypeptides in etiolated bean leaves exposed
to 35 light-Dark Cycles (2minL+98minD) in the presence or absence
of Chloramphenicol. Slots 2,3:4 μg Chl; slots 5,6:140 μg protein.

To test whether the rate of Chl synthesis, relative to that of the
other thylakoid components, affects LHC-II accumulation in thylakoids, e-
tiolated plants were exposed to ImL with shorter or longer dark intervals
(2minL+28minD, 2minL+98minD, 2minL+4hD). Since Chl accumulates only during
the 2min light phase of the cycle, more Chl is formed per day in the shorter
dark interval ImL than in the longer[16]. Thus, for the same amount of Chl
accumulated, and assuming that the RC protein formation remains unaffected by ImL, less RC proteins are expected to be present in ImL of shorter
dark interval than in longer. Thus, in the long dark interval ImL most of
the available Chl was expected to be bound on RC Chl-protein complexes, and
much less to be available for LHC-II stabilization, while in the short dark
interval ImL some Chl was expected to be available for LHC-stabilization, as
well. Fig. 8 shows that this was indeed the case. By increasing the rate of
Chl synthesis i.e. by decreasing the dark interval in ImL, Chl becomes available to bind on LHC apoproteins, resulting in their stabilization in the
thylakoid. Thus, an equal amount of Chl, formed under the same total irradiation received, can be used either for the stabilization of few large-in-
-size PS units, containing LHC components (short dark interval ImL) or many
small-sized units with no LHC components (long dark interval ImL) (see also
Table 2).

To test whether the rate of RC protein synthesis affects stabilization
of LHC-II in thylakoids, we administered chloramphenicol (CAP) to plants exposed to ImL (2minL+98minD), or plants transferred to the dark (after brief
preexposure to CL), to inhibit plastid-coded RC protein synthesis. It was
found that indeed LHC-II accumulates in CAP treated ImL plants (Fig. 8) and
it is not degraded in plants transferred to the dark in the presence of CAP
after preexposure to CL[17]. This suggests that under conditions where RC-
core-protein synthesis is inhibited, (i.e., in absence of RC-protein-Chl
binding) the LHC-II apoprotein can be stabilized through binding with the
Chl available.

The results suggest that the rate of Chl synthesis relative to that of
the other thylakoid components (polypeptides) regulates the assembly of LHC
and that LHC accumulation depends on competition between RC and LHC apoproteins for Chl. This can explain why in young ImL plants no LHC is stabilized,
while it can be accumulated in the older (see Fig. 9 right, compare sample
at 5d+86 Light-Dark Cycles-slot 4- with that at 11d+86 LDC-slot 8-). As
shown in Fig. 9 (left), this is reflected by the low rate of PSII unit formation in the 11d plant (the PSII activity per g never reaches that of a

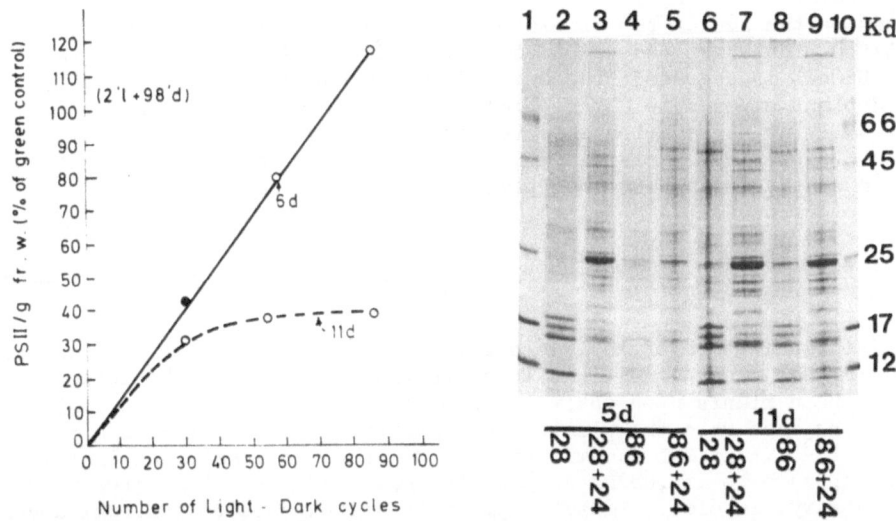

Fig. 9. Left: Photosystem II unit content (expressed as % activity/g fr w of a green control) in 5 and 11d etiolated bean leaves exposed to ImL (2minL+98minD). Calculation based on Chl/g fr w. Right:SDS-PAGE of their thylakoid polypeptides prior to or after transfer to continuous light for 24h. 140 μg protein loaded per slot.

green control in ImL) and the high rate of PSII unit formation in the 5d plant (the PSII activity per g reaches and overpasses that of a green control in ImL). LHC-II stabilization in the older ImL plant, but not in the younger,in spite their similar Chl content (see Table 3), suggests that LHC--II stabilization is due to the lower rate of RC-protein formation in the older plant, permitting some of the Chl present to bind on LHC apoprotein and stabilize it. Similarly, this hypothesis can also explain the finding that degradation of LHC-II in the dark following a brief exposure to CL,

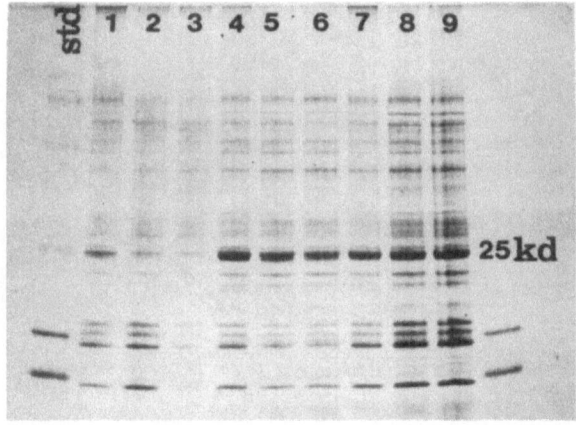

Fig. 10. SDS-PAGE of thylakoid polypeptides in 4-11d etiolated bean leaves exposed to CL for 14h and then transferred to the dark for 24 or 48h Slots 1-3:4d (14CL, +24hD, +48hD,respectively). Slots 4-6:7d (14hCL, +24hD, +48hD). Slots 7-9:11d (14hCL, +24hD, +48hD).

depends on the age of the etiolated tissue used (Fig. 10). Thus, in 4-5 d etiolated plants exposed to CL (slot 1) and transferred to the dark (slots 2, 3) LHC-II is degraded; the degradation in the dark is less extensive in the 7d plant exposed to CL (slot 4) and transferred to the dark (slots 5, 6); finally no degradation is observed in the 11d plant under similar conditions (slots 7-9). This finding is probably due to the higher rate of RC protein synthesis in the young plants transferred to the dark as compared to the older ones.

LHC-II accumulation in continuous light/Control by preexisting PSII unit content

The final state of thylakoid development depends greatly on initial developmental stage/conditions.Thus,eventhough massive accumulation of Chl occurs in 5d plants upon their transfer to CL following 28 LDC of ImL, this is not the case when the plants are transferred to CL after long preexposure to ImL[18]. Thus, Chl accumulation in CL is greatly affected in this case by preexposure to ImL,being completely inhibited after long preexposure. On the contrary, in older etiolated plants (11d) exposed to ImL, Chl accumulates in CL irrespective of duration of preexposure to ImL[19]. The accumulation of LHC-II in thylakoids closely parallels the accumulation of Chl: in young etiolated plants briefly exposed to ImL and then transferred to CL, LHC-II accumulates with no problem (see Fig. 9, right), but in those transferred to CL after long preexposure to ImL, no LHC-II is accumulated, in spite the fact that the LHC-II mRNA is present and can be translated under all conditions (see Fig. 5). On the contrary, in the 11d plant, LHC-II accumulates in CL irrespective of preexposure time in ImL (Fig. 9, right). These results can be explained on the basis of the photosynthetic activity (Fig.11).

As shown (Fig. 11), when ImL plants are transferred to CL, new PSII units are formed: this occurs, however, only in the cases where the preexisting PSII unit content reached in ImL is lower than that of the fully green plant. This suggests that new PSII unit formation depends on the need of the plant to reach the content of the fully green plant and perform full photosynthetic electron transport. In the young plant new units are formed in CL only after short preexposure to ImL, where the PSII unit content is still low, but not after long preexposure, where the PSII unit content has reached already that of the fully green plant (Fig.11, left). In contrast,

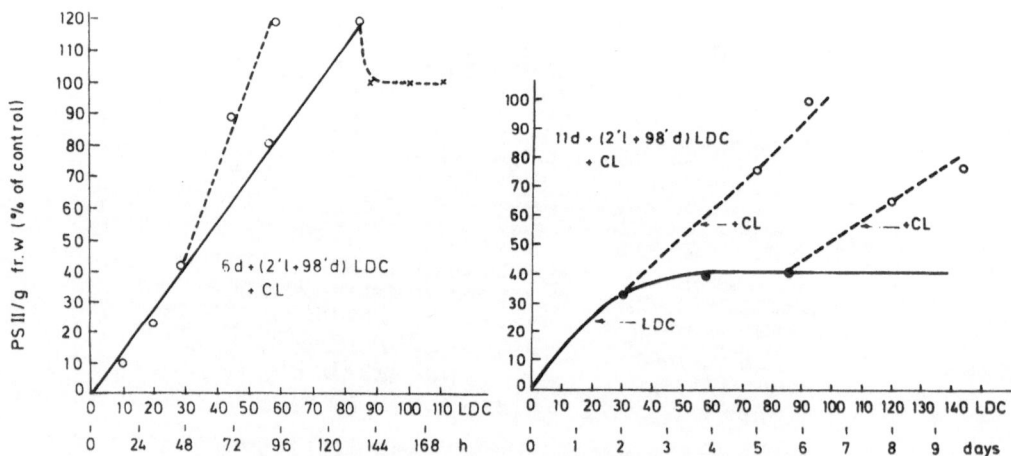

Fig. 11. The PSII unit content/g fr w expressed as % of that found in a green control plant. 6d and 11d etiolated bean plants were exposed to ImL (solid line) and after 28 or 86 Light-Dark Cycles (LDC)were transferred to CL (---).

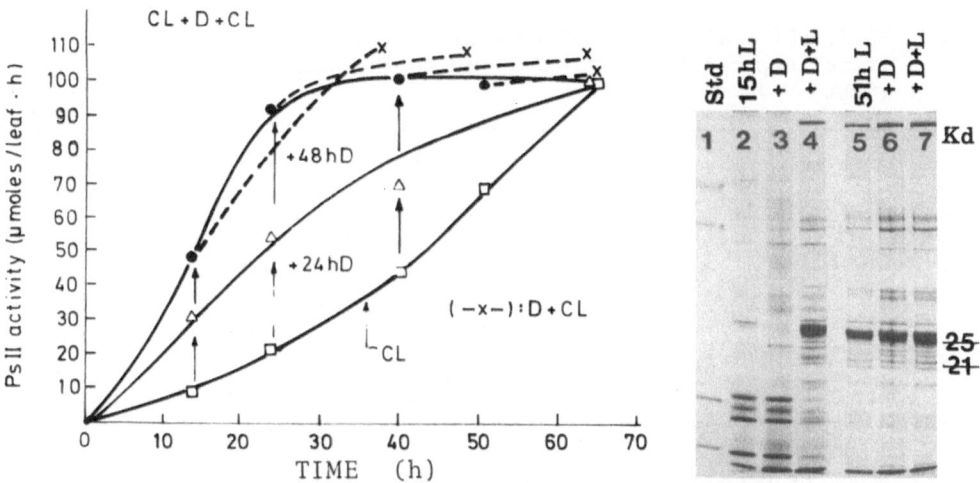

Fig. 12. Left: Photosystem II content/leaf in 6d etiolated bean leaves exposed to continuous light for 70h (CL, □) and at various time intervals transferred to the Dark for 24h (+24hD, △), or 48h (+48hD, ●). After 48h in the dark, the plants were transferred again to CL (--x--x). Right:SDS-PAGE of thylakoids of leaves exposed to 15 or 51h CL, then transferred to the dark for 48h (+D) and again to Light for 24h (+D+L).

in the older plant (Fig. 11,right) new PSII units are formed irrespective of preexposure to ImL, since the PSII unit content in ImL is lower than that of the green control throughout ImL exposure. Thus,it seems that LHC-II, as well as Chl, accumulation in CL follows closely the formation of new PSII units, and therefore depend on preexisting PSII unit content (relative to that of the green control).

Similarly,in plants exposed briefly to CL, then transferred to the dark, and then again transferred to CL, LHC-II accumulates in thylakoids only in cases where the PSII unit content reached in the dark prior to reexposure to light is lower than that of the green control (Fig. 12). Plants exposed to CL for 15h, then transferred to dark for 48h have only 50% of the activity of a green control plant (50% of the PSII units). Therefore, when exposed to CL have the capacity to form new PSII units to reach the content of the green control; and new units are indeed formed, which in CL are large in size and contain LHC components. Under these conditions, therefore, the thylakoids accumulate LHC-II (Fig. 12 right, 15h-D+L). On the contrary, plants exposed to CL for 50h and then transferred to the dark for 48h contain 100% already of the PSII unit content of a green plant, and in the following exposure to CL no changes in activity or LHC-II incorporation are observed (Fig. 12, right, 51h-D+L).

LHC-II accumulation in CL, therefore seems to depend on PSII unit formation and thus on preexisting PSII unit content.

Control of LHC accumulation on the PSII/PSI ratio (stoichiometry)

LHC-II accumulation in thylakoids during chloroplast development increases the PSII unit size, as suggested by the parallel decrease in the half-rise time of the fluorescence induction kinetics ($t\frac{1}{2}$) of isolated plastids and the decrease in the light-intensity requirement for saturation of PSII activity[9,10] (Tables 1-3). In all cases studied (young and old ImL plants, plants exposed to CL, then transferred to the dark and again to light, plants exposed to ImL with dark interval of varying duration), an inverse relationship was found between PSII unit size and PSII/PSI ratio (the larger the size, the lower the ratio) (see Tables 1, 2, 3)[9,10,12,13,16,19].

Table 1. PSII, PSI/leaf, PSI/P700 and PSII/PSI ratio in 5d etiolated bean leaves exposed to CL, transferred to the dark and reexposed to light[12,13].

	Chla/l (μg)	Chla/b	Activity/1 PSII	PSI	PSI/ P700	PSII (%)	PSI	PSII/PSI
14hCL	14.4	3.0	9.8	4.2	1.1	10	9	1.1
14hCL+48hD	16.1	14.6	49.1	9.0	1.1	50	19	2.7
14hCL+48hD+24hCL	131.5	3.0	102.0	40.0	1.1	104	83	1.2
51hCL	119.6	2.4	67.8	34.5	1.0	69	72	1.0
51hCL+48hD	125.4	3.3	95.9	41.0	1.1	98	85	1.2
51hCL+48hD+24hCL	175.9	2.4	99.7	50.8	1.0	102	106	1.0
65h CL control	172.9	2.4	98.0	48.3	1.0	100	100	1.0

The rate of PSII and PSI unit formation in plants exposed to CL remains constant, and thus, the PSII/PSI ratio found is constant (assumed to be 1,0); this ratio increases after transfer to the dark to about 2.5(under conditions of LHC-II degradation) and it is reduced again upon reexposure to CL (in cases where new PSII units with LHC complexes are formed) (Table 1). In plants exposed to ImL, the PSII/PSI ratio increases to a value of 2.5 (under conditions of no LHC-II stabilization); this value drops to about 1.0 upon transfer of plants to CL (in cases where new PSII units are formed) or to intermediate values (when no extensive accumulation of LHC-II occurs) (Tables 2, 3). Thus, in 5d plants exposed to 85 (2minL+98minD) cycles and transferred to CL, where no new PSII units are formed-nor any LHC-II is stabilized, no change in the PSII/PSI ratio is observed in CL. Similarly, in plants exposed to 50h CL and transferred to darkness, the PSII unit content reaches that of the control in the dark; thus, upon reexposure to CL there is no new PSII formation, nor LHC-II stabilization, and no change in the PSII/PSI ratio.

The change in the PSII/PSI ratio occurs via modulation of the rate of the PSI unit synthesis (as suggested by Tables 1-3). This may indicate that the PSI unit formation is drastically reduced in the dark, so that the ratio is increased in ImL and in dark-transferred plants. It this were the case, however, the ratio would be expected to decrease upon transfer of plants to CL, irrespective of preexisting PSII unit size and content. This does not seem to be the case, however (see Table 2; 6d plant +86 LDC+24h CL). Thus,

Table 2. Growth of PS units in etiolated bean leaves exposed to ImL with dark intervals of varying duration (Tzinas et al, 1988)

	Chl/g	Chla/b	Activity/g PSII	PSI	t½ (ms)	PSII (%)	PSI	PSII/PSI
(2minL+28minD)								
48 LDC	385	11.5	193	133	45	36	22	1.68
48 LDC+70h CL	2740	2.2	525	605	10	100	100	1.00
95 LDC	670	6.0	335	231	40	64	38	1.68
95 LDC+70h CL	2730	2.2	524	601	10	100	100	1.00
(2minL+98minD)								
28 LDC	218	high	251	120	100	47	20	2.40
28 LDC+70h CL	2750	2.2	528	605	12	100	100	1.00
85 LDC	630	12.1	662	334	70	125	52	2.30
85 LDC+70h CL	1230	3.3	547	283	32	103	47	2.20
(2minL+4hD)								
35 LDC	320	high	375	205	85	71	34	2.10
35 LDC+70h CL	1195	3.2	849	569	47	160	94	1.70

Table 3. Growth of the PS units in young and old etiolated bean leaves exposed to ImL (2minL+98minD in cycles) and transferred to cont. light[19].

	Chl/g (µg)	Chla/b	Activity/g PSII	PSI	t½ ·(ms)	PSII (%)	PSI	PSII/PSI
6d+28 LDC	210	high	200	102	98	43	18	2.4
+24h CL	1960	3.2	334	384	11	72	68	1.0
6d+85 LDC	525	16.0	543	289	76	117	51	2.3
+24h CL	735	6.5	460	275	30	100	49	2.0
6-day control	2900	3.0	464	570	11	100	100	1.0
11d+28 LDC	460	12.1	230	127	28	33	22	1.5
+90h CL	2895	3.2	695	579	15	101	101	1.0
11d+85 LDC	670	6.0	275	188	25	40	33	1.2
+90h CL	2050	3.3	515	423	15	75	74	1.0
11-day control	2920	3.2	685	570	17	100	100	1.0

PSI unit formation is probably controlled by another mechanism. Tables 1-3 suggest that PSI formation stops in ImL or CL as soon as PSII unit formation is stopped. Thus, the content (as well as the size) of the PSII unit seems to control the PSI unit content, and the PSII/PSI ratio.

The modulation of the PSII/PSI ratio may be the means by which the plant regulates balanced absorption of light energy between PSI and PSII. When the PSII unit absorption cross section is increased (via LHC-II stabilization), the plant responds by making more PSI units to equalize the distribution of absorbed light energy between the photosystems. When the PSII unit size, its absorption cross section and LHC-II stabilization are reduced (ImL or transfer to the dark), the rate of PSI unit formation is reduced, since less PSI are required for equal light absorption.

REFERENCES

1. R. J.Ellis, Chlorophyll-proteins: Synthesis, transport and assembly, Ann Rev Plant Physiol., 32:111 (1981).
2. N. H.Chua and G.W. Schmidt, Transport of proteins into mitochondria and chloroplasts, J. Cell Biol. 81:461 (1979).
3. P. R.Chitnis et al., Assembly of the precursor and processed Light-harvesting Chla/b protein of Lemna into the Light-harvesting Complex II of barley etiochloroplasts, J. Cell Biol. 102:982 (1986).
4. K. Apel, Phytochrome-induced appearance of mRNA activity for the apoprotein of the light-harvesting chlorophyll a/b protein of barley (Hordeum vulgare), Eur. J. Biochem. 97:183 (1979).
5. P. Tavladoraki, G. Akoyunoglou, A. Bitsch, G. Meyer and K. Kloppstech, Age and Phytochrome-induced changes at the level of the translatable mRNA coding for the LHC-II apoprotein of Phaseolus vulgaris leaves, in:"Regulation of Chloroplast Differentiation", G. Akoyunoglou and H. Senger eds., Liss, N.Y. pp. 559-564 (1986).
6. P. Tavladoraki, K. Kloppstech and J.H. Argyroudi-Akoyunoglou, Circadian rhythm in the expression of the mRNA coding for the apoprotein of the light-harvesting complex of Photosystem II: Phytochrome control and persistent Far Red reversibility, Plant Physiol. in press (1988).
7. J. H.Argyroudi-Akoyunoglou and G. Akoyunoglou, Photoinduced changes in the Chlorophyll a to Chlorophyll b ratio in young bean plants, Plant Physiol., 46:247 (1970).
8. J. H.Argyroudi-Akoyunoglou and G. Akoyunoglou, The Chlorophyll-protein complexes of thylakoids in greening plastids of Phaseolus vulgaris, FEBS Lett., 104:78 (1979)

9. G. Akoyunoglou, Development of the Photosystem II unit in plastids of bean leaves greened in periodic light, Arch. Biochem. Biophys., 183:571 (1977).

10. G. Akoyunoglou, Assembly of functional components in chloroplast photosynthetic membranes, in: Photosynthesis, G. Akoyunoglou ed., Vol. V, Balaban, Pa., pp. 353-366 (1981).

11. J. Bennett, Biosynthesis of the light-harvesting Chla/b protein. Polypeptide turnover in darkness. Eur. J. Biochem. 118:61 (1981).

12. J. H.Argyroudi-Akoyunoglou, A. Akoyunoglou, K. Kalosakas and G. Akoyunoglou, Reorganization of the PSII unit in developing thylakoids of higher plants after transfer to darkness: Changes in Chl b, Light-harvesting Chl-protein content, and grana stacking. Plant Physiol., 70:1242 (1982).

13. A. Akoyunoglou, and G. Akoyunoglou, Reorganization of thylakoid components during chloroplast development in higher plants after transfer to darkness. Changes in PSI unit components and in cytochromes, Plant Physiol. 79:425 (1985).

14. G. Akoyunoglou, Biosynthesis, Assembly and properties of the pigment-protein Complexes, in: Proc. 16th FEBS Congress, VNU Sci:Press, Utrecht, pp 35-46 (1985).

15. G. Akoyunoglou and J.H. Argyroudi-Akoyunoglou, Post-translational regulation of chloroplast differentiation, in:Regulation of Chloroplast Differentiation, G. Akoyunoglou and H. Senger eds., Liss, N.Y., pp. 571-582 (1986).

16. G. Tzinas, G. Akoyunoglou and J.H. Argyroudi-Akoyunoglou, The effect of the dark interval in intermittent light on thylakoid development: photosynthetic unit formation and light-harvesting protein accumulation, Photosynthesis Res., 14:241 (1987).

17. G. Tzinas and J.H. Argyroudi-Akoyunoglou, Chloramphenicol-induced stabilization of light-harvesting complexes in thylakoids during development, FEBS Lett., 229:135 (1988).

18. G. Akoyunoglou and J.H. Argyroudi-Akoyunoglou, Control of thylakoid growth in Phaseolus vulgaris, Plant Physiol. 61:834 (1978).

19. S. Tsakiris and G. Akoyunoglou, Formation and growth of Photosystem I and II units in old bean leaves, Plant Sci. Lett., 35:97 (1984).

A NOVEL METHOD FOR THE VISUALIZATION OF OUTER SURFACES FROM STACKED REGIONS OF THYLAKOID MEMBRANES

Jenny E. Hinshaw and Kenneth R. Miller

Division of Biology and Medicine, Brown University, Providence, Rhode Island 02912, U.S.A.

INTRODUCTION

Freeze-etch studies of thylakoid membranes reveal outer and inner surfaces that are rich in structural detail. The inner surface is covered with tetrameric particles that have been associated with the extrinsic oxygen-evolving proteins of PS II[1,2] and are concentrated in the stacked (grana) regions of the membrane[3]. The outer surface of non-stacked regions contains large particles identified as the CF_1 subunit of the ATP synthase[4] and smaller particles which may represent PS I. The only thylakoid surface that has not been studied in detail is the outer surface from the stacked region. In the past this surface has been inaccessible by the stacking process itself. In this paper we describe a method to expose the previously hidden outer surface. The structure of this surface is remarkably smooth, with no apparent surface particles. This striking lateral heterogeneity of the outer surface and the extreme smoothness of the stacked region suggest new restrictions on the lateral and transverse distribution of thylakoid membrane components.

MATERIALS AND METHODS

A suspension of spinach thylakoid membranes, isolated as described by Miller & Staehelin[4], were placed on alcian blue (1mg/ml) treated glass coverslips in a low ionic strength buffer, containing 5mM MES (2[N-Morpholino]ethanasulfonic Acid]) and 2mM $MgCl_2$. After the membranes were attached to the glass coverslips by a 5 min. incubation in this buffer they were mechanically disrupted by a squirt of buffer (10ml) from a syringe (fig. 1). The coverslips were quickly frozen in liquid freon 22 and then placed on a stage in a Balzers BAF 400 freeze-etching device, and freeze-dried, without prior fracturing, for 15-60 minutes at -80° C. After freeze-drying, the membranes were rotary shadowed at an angle of 23° with platinum followed by a coat of carbon. The resulting replica was washed in bleach and examined in a Philips 410 electron microscope (fig. 2.).

The distance between the stacked membranes was calculated from measurements of replicas prepared by single-sided shadowing at an angle of 45° (fig. 3). This was calculated using the height of the tetramer, reported by Seibert et al.[1] as 8.0 nm, as an internal standard and the assumption that the minimum possible membrane width is 5.0 nm. The shadow length of the edge of the inner surface fragment, located on the outer stacked surface, was found to be identical to the tetramer shadow length. This suggests the distance from the PSs surface to the top of the membrane fragment is also 8.0 nm and therefore the distance between the membranes is only 3.0 nm.

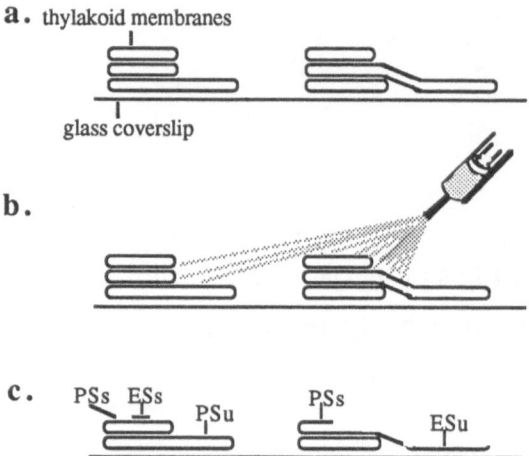

a. thylakoid membranes

glass coverslip

b.

c. PSs ESs PSu PSs ESu

Figure 1. Schematic diagram of the technique used to expose the outer stacked surface. **a.** Membranes are adhered to an alcian blue treated glass coverslip. **b.** A stream of buffer from a syringe mechanically disrupts these membranes. **c.** The result is the exposure of all four surfaces, including the newly exposed outer stacked surface. The inner surfaces are labeled according to the convention of Branton *et al* [24]: ESu denotes the inner surface in a non-stacked membrane region, and ESs in a stacked region. Similarly, the outer surfaces are labeled PSs and PSu.

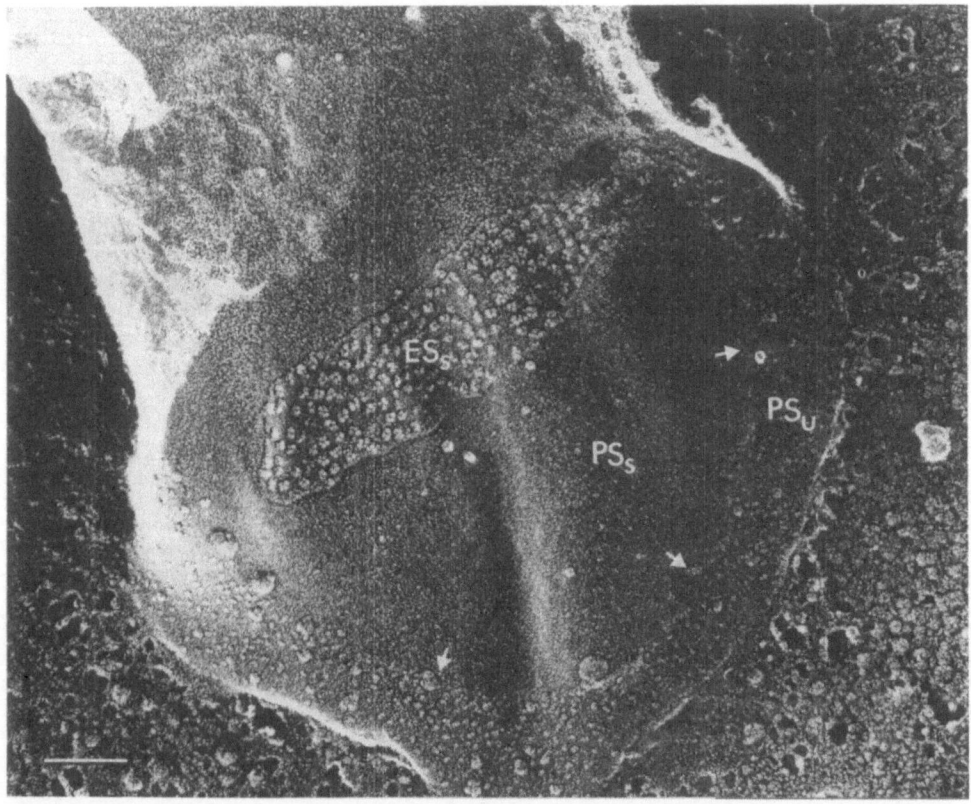

Figure 2. This Electron micrograph enables us to identify the stacked outer surface, PSs, by its positioning among the the other membrane surfaces. A fragment of the inner surface (ESs) remains attached to this smooth PSs surface. Figure 1c illustrates how such an inner membrane fragment may remain attached. In addition, this PSs surface is surrounded by and continuous with the non-stacked regions (PSu). The numerous particles of the non-stacked regions abruptly end at the edge of the stacked region (see arrows). Bar = 100nm.

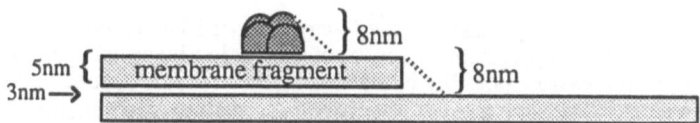

Figure 3. Diagram of the single-sided shadowing method used to determine the distance between the membranes. The height of the tetramer and the distance form the membrane fragment to the PSs surface are both 8nm. Assuming a minimum membrane width of 5nm, the distance separating the outer surfaces of adhering membranes cannot exceed 3nm.

RESULTS AND DISCUSSION

The techniques which we have reported have enabled us to examine the previously hidden outer stacked surface of thylakoid membranes. This new surface was identified by its positioning among other membrane surfaces. For example, occasionally a piece of an inner stacked surface (ESs), identifiable by its numerous tetramers, remains attached to the smooth outer surface (fig. 2). Also, this smooth surface is frequently continuous with and surrounded by the distinct non-stacked outer surface (PSu in Fig. 2). There is a clear delineation between the extremely smooth outer stacked surface (PSs) and the particle-rich outer non-stacked surface (PSu).

The most striking feature of the PSs surface is its lack of any surface particle. This is especially interesting since the rest of the thylakoid surfaces are highly structured. The surface topology of the PSs surface is comparable to freeze-etched liposomes. This suggests the PSs surface lacks large extrinsic proteins and bulky protrusions from intrinsic protein complexes. Although no extrinsic proteins have been localized to this surface, both the intrinsic PS II[5] and LHC II[6,7] complexes are exposed at the outer (PSs) surface, and the cytochrome b6/f complex may also be exposed at this surface[8]. In addition, there is good evidence LHC II is phosphorylated on the PSs surface[9] by a 64kDa protein that has been localized to the stacked regions of thylakoid membranes [10]. Our micrographs indicate that the extent of surface exposure of these three complexes and the LHC II kinase is limited to peptide regions which are smaller than the resolution limit of platinum shadowing. In addition, the extreme smoothness of the PSs surface suggests the individual polypeptide loops of these proteins that are exposed on this surface do not cluster to the extent that would result in a surface particle. For example, if all the surface exposed loops of the four PS II core proteins were clustered at a single point they would form an aggregate structure of approximately 500 amino acids[11,12] which would be visible on this surface. The lack of apparent surface structure on the PSs surface is also consistent with the recent crystalline data of LHC II. The thickness of frozen hydrated two dimensional crystals of LHC II is only 4.8 nm[13]. The structure of LHC II in such crystals would easily fit into the membrane (5.0 nm width) without significant surface exposure.

The striking exclusion of both large and small particles from the PSs surface clearly illustrates lateral heterogeneity on the outer surface of thylakoid membranes. The absence of the large particle on the PSs surface supports earlier conclusions that ATP synthase is excluded from the stacked region[14,15]. If the smaller particle on the PSs surface is associated with PS I, then PS I is also restricted to the non-stacked regions. This correlates with thin-section immunolabeling[14] and fractionation studies[16] that have shown PS I to be exclusively located to non-stacked regions.

Single-sided shadowing indicates the distance between the stacked membranes is at most 3nm. Considering the minimal surface exposure of LHC II on the PSs surface, it would be necessary for the appressed membranes to be fairly close for LHC II molecules on one membrane to interact with adjacent membranes. The fact that pieces of inner surface remain attached to the outer surface after mechanical disruption (fig. 2), suggests such a physical attraction between adjacent stacked membranes does exist. Barber[17] has suggested the physical forces involved in thylakoid stacking include Van der Waals interactions and electrostatic repulsion. There is good evidence that LHC II molecules are responsible for this membrane stacking[18]. More specifically, the N-terminus of LHC II has been shown to play a major role in membrane appression. Removal of this region by amino-peptidases severely disrupts membrane appression, possibly by increasing electrostatic repulsion between the adjacent membranes[17]. For such a small N-terminal segment to affect membrane adhesion, the distance between the membranes must be minimal.

This small distance would also limit the accessibility of soluble proteins to the space between stacked membranes. In addition, protein complexes with large outer surface protrusions would be prevented from migrating into the stacked region. In fact, it has been suggested that the exclusion of ATP synthase and PS I from the stacked regions might be due to steric hindrance created by these closely appressed membranes[19]. According to this theory, ATP synthase and PSI would be excluded from the stacked region for steric reasons. The CF1 subunit of ATP synthase extends into the stromal space about 12nm[4] while the ferredoxin NADP-reductase of PS I extends at least 5nm. This exclusion of ATP synthase and PS I from stacked regions is confirmed by thin section immunolabeling which localizes these complexes only in non-stacked regions[14]. On the other hand, the cytochrome b6/f complex does not have any large protrusions on the outer surface[19] and therefore would be free to migrate into the stacked region. This is supported by fractionation and immunological studies that localize this complex to both stacked and non-stacked regions[20,21].

The two remaining complexes, LHC II and PS II, do not have large exposed regions on the outer surface but have been predominantly localized within the stacked regions. It has been suggested LHC II freely migrates into the stacked regions and then is held there by attractive interactions with other LHC II molecules. LHC II can migrate back to non-stacked regions only when these attractive forces are disrupted by phosphorylation[9]. After migrating into the stacked regions, PS II may also become concentrated in this region by lateral attractive forces between PS II and LHC II. The small amount of PS II that is found in non-stacked regions is believed to have fewer of these closely associated LHC II molecules[22,23]. In conclusion, the smoothness of the PSs surface and the small distance between adjacent membranes suggests new restrictions for the organization of the thylakoid membrane.
[This work was supported by NIH grant GM 28799.]

REFERENCES

1. Seibert, M., DeWitt, M., and Staehelin, L.A. (1987). J. Cell Biol. *105*, 2257-2265
2. Simpson, D.J., and Andersson, B. (1986). Carlsberg Res. Commun. *51*, 467-474.
3. Staehelin, L. A. (1976). J. Cell Biol. *71*, 136-158.
4. Miller, K.R. and Staehelin, L.A. (1976). J. Cell Biol. *68*, 30-47.
5. Sayre, R.T., Andersson, B., and Bogorad, L. (1986). Cell *47*, 601-608.
6. Mullet, J.E. (1983). J. Biol. Chem. *258*, 9941-9948.
7. Andersson, B., Anderson, J.M., and Ryrie, J. (1982). Eur. J. Biochem. *123*, 465-472.
8. Mansfield, R.W., and Anderson, J.M. (1985). Biochim. Biophys. Acta. *809*, 435-444.
9. Bennett, J. (1983). Biochem. J. *212*, 1-13.
10. Coughlan, S.J. and Hind, G. (1987). Biochemistry *26*, 6515-6521.
11. Alt, J., Morris, J., Westoff, P. and Herrmann, R.G. (1984). Curr. Gen. *8*, 597-606.
12. Deisenhofer, J., Epp, O., Miki, K., Huber, R., and Michel, H. (1985). Nature *318*, 618-624.
13. Lyon, M.K., and Unwin, N.T. (1988). J. Cell Biol. *106*, 1515-1523.
14. Vallon, O., Wollman, F.A., and Olive, J. (1986). Photobiochem. Photobiophys. *12*, 203-220.
15. Allred, D.R., and Staehelin, L.A. (1985). Plant. Physiol. *78*, 199-202.
16. Andersson, B., and Anderson, J.M. (1980). Biochim. Biophys. Acta. *593*, 427-440.
17. Barber, J. (1982). Ann. Rev. Plant Physiol. *33*, 261-295.
18. McDonnel, A., and Staehelin, L.A. (1980). J. Cell Biol. *84*, 40-56.
19. Murphy, D.J. (1986). Biochim. Biophys. Acta, *864*, 33-94.
20. Anderson, J. (1982). FEBS. LETTS. *138*, 62-66.
21. Olive, J., Vallon, O., Wollman, F., Recouvreur, M., and Bennoun, P. (1986). Biochim. Biophys. Acta *851*, 239-248.
22. Melis, A., and Duysens, L.M.N. (1979). Photochem. Photobiol. *29*, 373-382.
23. Thielen, A.P.G.M., and Van Gorkum, H.J. (1981). Biochim. Biophys. Acta *645*, 111-120.
24. Branton, D., Bullivant, S., Gilula, N.B., Karnovsky, M.J., Moor, H., Muhlethaler, K., Northcote, D.H., Packer, L., Satir, B., Satir, P., Speth, V., Staehelin, L.A., Steere, R.L., and Weinstein, R.S. (1975). Science *190*. 54-56.

INFRARED SPECTROSCOPY AND ELECTROCHEMISTRY OF CHLOROPHYLLS: MODEL COMPOUND STUDIES ON THE INTERACTION IN THEIR NATIVE ENVIRONMENT

M. Leonhard*, A. Wollenweber*, G. Berger[s], J. Kléo[s], E. Nabedryk[s], J. Breton[s] and W. Mäntele*

*Institut für Biophysik und Strahlenbiologie der Universität Freiburg, Albertstraße 23, D-7800 Freiburg
[s]Service de Biophysique, Département de Biologie CEN Saclay, F-91191 Gif-sur-Yvette Cédex

INTRODUCTION

In the primary reactions of photosynthesis, the light-induced charge separation and stabilization involves the generation of a radical cation and of radical anions at the various steps of electron transfer. The basic units that can perform these processes – called photosynthetic reaction centers (RC) – are well-characterized for bacterial photosynthesis in their structure[1,2] and function (for a review, see Parson[3]). In bacterial RC, the primary electron donor (P) is a bacteriochlorophyll (BChl) a or b dimer and the first electron acceptor a bacteriopheophytin (BPheo) a or b molecule. They are embedded in the protein matrix and provided with very specific interactions that seem to be responsible for the spectral and redox properties of the pigments in their native environment as well as for the efficiency and specificity of electron transfer. X-ray structure analysis[1] has demonstrated the close proximity of protein residues to carbonyl groups of the pigments, thus suggesting H-bonding.

Resonance Raman and infrared (IR) spectroscopy have proven to be useful tools to investigate the interaction of pigments in organic solvents or in their native environment[4,5]. In previous work[6-8], we have used light-induced IR difference spectroscopy with photosynthetic membranes and reaction centers to study the molecular changes upon primary donor photooxidation and acceptor reduction. With the sensitivity of Fourier transform infrared (FTIR) spectroscopy, highly structured difference spectra were obtained that are governed to a large extent by absorption changes of individual pigment bonds. By comparison with the spectra of model compounds, information on the bonding of the pigments in the reaction center can be obtained.

In order to get appropriate model spectra, we have combined the sensitivity of FTIR spectroscopy with electrochemical techniques for the reversible and well-defined generation of cations and anions of isolated pigments in H-bonding and non-H-bonding solvents. Cells and procedures for this "IR-Spectroelectrochemistry" have been described[9]. The cation and anion model spectra reveal the partial change of bond order of individual bonds of the π-electron system due to addition or abstraction of an electron, but also strong inductive effects of a net charge on bonds not involved in conjugation and on H-bonds formed between the pigments and

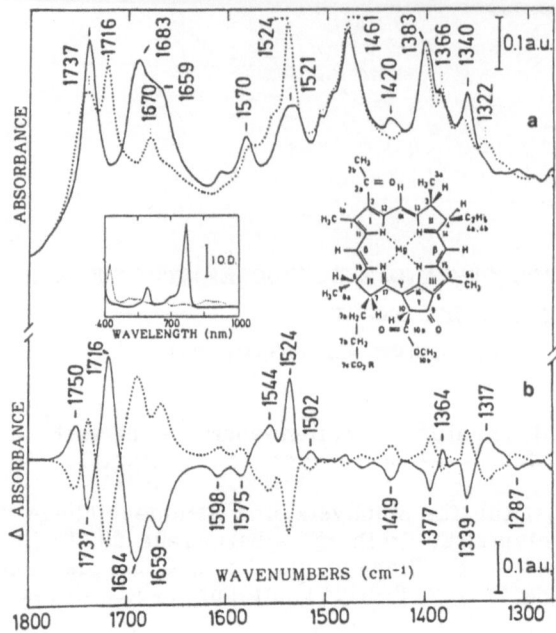

Fig. 1. Insets: Structure of BChl a, spectra of neutral BChl
a (full line) and of the cation (dashed line) in THF.
a) IR absorbance spectra of neutral BChl a (full line)
and of the cation (dashed line). b) IR difference
spectra of BChl a cation formation at U=+0.5V (BChl
a⁺−BChl a, full line) and of the reverse reaction
(BChl a−BChl a⁺, dashed line) at U=−0.5V.

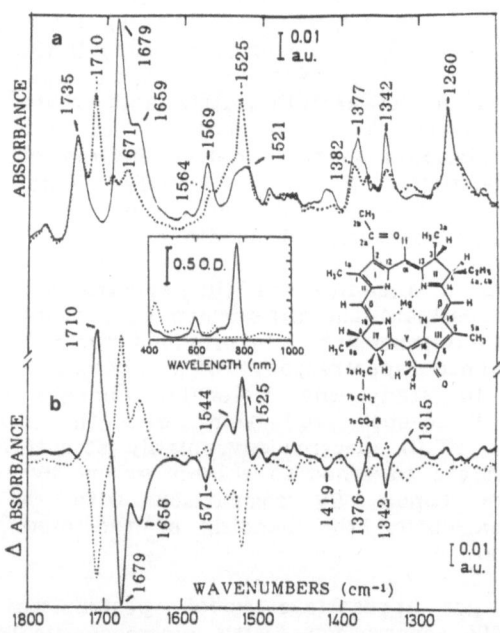

Fig. 2. Insets: Structure of PBChl a, spectra of neutral
PBChl a (full line) and of the cation (dashed line) in
THF. a) IR absorbance spectra of neutral PBChl a (full
line) and of the cation (dashed line). b) IR difference
spectra of PBChl a cation formation at U=+0.8V
(PBChl a⁺−PBChl a, full line) and of the reverse
reaction (PBChl a−PBChl a⁺,dashed line) at U=0.27V.

external ligands. Among these effects upon cation formation in RC, a change from H-bonded to free of the ester C=O group(s) of one of the primary donor BChl a molecules in *Rb. sphaeroides* RC was observed[6]. In *Rp. viridis* RC, however, a non-interacting BChl b ester C=O was found[6].

The ester groups at the 7c and 10a position of the BChl molecule can, in general, not be distinguished. For the ester C=O group contributing to the difference spectra, we have favoured the one at 10a position due to its proximity to the π-electron system. Using modified pigments, we can now present conclusive evidence that it is this group which changes its bond strength and, in BChl a containing RC, its H-bonding state during primary charge separation.

MATERIALS AND METHODS

BChl a from *Rs. rubrum* was prepared as described[10]. After the extraction of pelleted cells with methanol-water the BChl a was separated from polar and less polar compounds using octadecyl silica cartridges. The following high performance liquid chromatography on reversed phase columns (rp-HPLC) in ethanol-water yielded highly purified BChl a. Pyrobacteriochlorophyll a (PBChl a) was obtained by adapting a relatively mild pyrolysis method first described by Pennington[11] for chlorophyll. BChl a in pyridine was kept at 100°C in evacuated and sealed tubes for 12 hours, leading to the pyro derivative in good yield. The PBChl a was separated from BChl a and other impurities by rp-HPLC. The solvent was evaporated under vacuum and the PBChl a stored under N_2 at -20°C. All operations were carried out in minimum light.

Cyclic voltammetry, coulometry and constant potential electrolysis were performed using a home-built potentiostat. Solvent and electrolyte were prepared as described[9]. IR spectra were recorded as described[9], with a measuring cell adjusted to 90 μm pathlength. Spectra in the 400-1000 nm region were recorded with the same cell using a home-built spectrophotometer.

RESULTS AND DISCUSSION

The IR absorbance spectra of unmodified BChl a in tetrahydrofuran (THF) in the neutral state (full line) and in the fully evolved cation state (dashed line) is shown in fig. 1a. The corresponding absorbance spectra (inset) clearly indicate the disappearance of the main absorbance peak at 772 nm and the formation of a near-infrared band typical for the cation. The cation-minus-neutral IR difference spectrum is shown in fig. 1b, with the bands of the cation chosen to appear positive. In previous work[12], such difference spectra of BChl a and b generated in a H-bonding and a non-H-bonding solvent were compared to light-induced difference spectra of the primary donor photooxidation. In the 1750 cm^{-1} to 1620 cm^{-1} spectral region, changes of band positions and intensity can, to a large extent, be attributed to changes of bond order and H-bonding of the C=O groups at the periphery of the BChl molecule[12]. The band shift from 1684 cm^{-1} to 1716 cm^{-1} can be interpreted as due to the 9 keto C=O group, which shows a small change of bond order upon the BChl radical formation. Since this differential band was found in an almost identical position for the pigment in the RC, we concluded that this group, in the neutral and cation state, is non-interacting. The difference bands in the 1730 cm^{-1} - 1750 cm^{-1} region can be assigned to either one of the ester C=O groups. In RC containing BChl a, only a positive band appears at 1750 cm^{-1}. It was best modelled by he cation formation in H-bonding solvents and was thus attributed to an ester C=O group strongly H-bonded in the neutral state (resulting in a

117

shift of its absorption to lower wavenumbers, possibly connected with broadening and weakening), but free in the cation state[6]. In *Rp. viridis* RC, a shift of the ester C=O absorption from 1745 cm^{-1} to 1755 cm^{-1} was observed. It was best modelled by BChl b in THF (1731 cm^{-1} -> 1747 cm^{-1}) and was interpreted in terms of a non-interacting ester group.

Pyrolysis of BChl a allows to distinguish between the 7c and the 10a ester C=O group, since the 10a ester group is replaced by a hydrogen (see fig 2, inset). Fig. 2a shows an IR absorbance spectrum of neutral PBChl a in THF (full line). The band at 1735 cm^{-1} appears reduced by 50% as compared to the corresponding band for BChl a. The 9 keto and the 2a acetyl C=O absorption remain essentially unchanged. In the IR absorbance spectrum of the fully evolved PBChl a cation (fig. 2a ,dashed line) and in the redox-induced IR difference spectrum, the shift of the 9 keto C=O absorption to 1710 cm^{-1} and the absorbance decrease of the 2a acetyl group closely resemble the spectral changes observed upon BChl a cation formation. In the ester C=O frequency region, however, the band shift, which was observed for BChl a (fig 1b), has disappeared. The absorbance changes can thus be attributed to the 10a ester C=O group.

CONCLUSIONS

The comparison of the redox-induced IR difference spectra of normal BChl a and PBChl a clearly indicates that the bond order of the 10a ester roup is altered upon cation formation. Since it is assumed that this group is not involved in conjugation, orbital interactions with the π-electron system have to be accounted for[13]. As suggested by X-ray structure analysis[1] and light-induced FTIR difference spectra[6-8,12], this group, in RC, can form a H-bond with appropriate protein side chain residues. An altered bond order, in connection with a change in H-bonding, may thus be part of the mechanisms that are responsible for charge stabilization.

REFERENCES

1. H. Michel, O. Epp, and J. Deisenhofer, The EMBO Journal 5:2445 (1986)
2. J. P. Allen, and G. Feher, Proc. Natl. Acad. Sci. USA 81:4795 (1987)
3. W. W. Parson, Annu. Rev. Biophys. Bioeng. 11:57 (1982)
4. K. M. Ballschmiter, and J. J. Katz, J. Am. Chem. Soc. 91:2661 (1969)
5. M. Lutz, in: "Advances in Infrared and Raman Spectroscopy", R. J. H. Clark, and R. E. Hester, eds., Wiley, Heyden 11:211 (1984)
6. W. Mäntele, E. Nabedryk, B. A. Tavitian, W. Kreutz, and J. Breton, FEBS Lett. 187:227 (1985)
7. B. A. Tavitian, E. Nabedryk, W. Mäntele, and J. Breton, FEBS Lett. 201:151 (1986)
8. E. Nabedryk, W. Mäntele, B.A. Tavitian, and J. Breton, Photochem. Photobiol. 43:461 (1986)
9. W. Mäntele, A. Wollenweber, F. Rashwan, J. Heinze, E. Nabedryk, and J. Breton, Photochem. Photobiol. 47:451 (1988)
10. G. Berger, A. Wollenweber, J. Kléo, S. Andrianambinintsoa, and W. Mäntele, J. Liquid Chrom. 10:1519 (1987)
11. F. C. Pennington, H. H. Strain, W. A. Svec, and J. J. Katz, J. Am. Chem. Soc. 86:1418 (1964)
12. W. Mäntele, A. Wollenweber, E. Nabedryk, and J. Breton, Proc. Natl. Acad. Sci. USA, (1988) manuscript in press
13. H. Wolf, H.jr. Brockmann, H. Biere, and H. H. Inhoffen, Liebigs Ann. Chem. 704:208 (1967)

ENZYMATIC CHLORINATION OF CHLOROPHYLL A IN VITRO

Mathias Senge* and Horst Senger

FB Biologie/Botanik, Philipps-Universität, Lahnberge
D-3550 Marburg/Lahn, Federal Republic of Germany

INTRODUCTION

Chlorinated chlorophylls (chl) came into focus in recent years with reports suggesting the participation of 13^2-HO-20-Cl-chl a in photosystem I [1,2]. However, studies revealed the artificial formation of this compound [3,4]. Nevertheless chlorinated chlorophylls deserve further attention, since their unique physical and chemical properties make them promising model compounds for comparative chl studies.

(1)   (2)

Although the great susceptibility of the methine bridge positions of chl a (1) to nucleophilic attack is known, so far attempts to synthetize chlorinated chl's were sucessful only on the level of Mg-free porphyrin systems [5,6]. H_2O_2/HCl was employed in those studies as halogenating agent, which inevitably led to pheophytinization of the chl's, thus required always remetallation as a second synthetic step with onyl low yields. In this contribution we report the synthesis of 20-Cl-chl a (2), which circumvents the remetallation step and uses a commercially available chloroperoxidase (CPO) (EC 1.11.1.10). This enzyme is capable of the H_2O_2/KCl-dependent chlorination of a variety of substrates [7]. The chlorination of chl a proceeds regioselective at the C-20-postion with a yield of 75 %.

MATERIALS AND METHODS

Chl a (1) was prepared from <u>Scenedesmus obliquus</u> mutant C-6E and purified by rp-HPLC <4>. Pheophytins were prepared according to the method of
Lötjönen and Hynninen <8>. Absorption spectroscopy was performed using an
Uvikon 820 spectrophotometer (Kontron, Munich, F.R.G.). Fluorescence emission spectra were recorded with a Shimadzu RF-540 spectrofluorophotometer
with excitation and emission bandwidth of 5 nm, each. ^{1}H-NMR was performed
in d_6-acetone using a Nicolet NT 360 MHz instrument. Fast atom bombardment
mass spectra were measured with a Kronos MS 50 instrument. TLC was performed
according to Dörnemann and Senger <1> and the localization of radioactive
bands was tested with a Rita-3200 radio-thin-layer analysator (Raytest,
Straubenhardt, F.R.G.).

The optimum assay for chlorination of chl a consisted of the following
components: 100 ml KH_2PO_4-buffer (pH 5.0, 0.3 M), 20 ml KCl (0.3 M), 50 ml
H_2O_2 (12 mM), 90 ml H_2O, 10 mg chl a in 10 ml acetone (11 µmole) and 640
units CPO. CPO was purchased from Sigma (Munich, F.R.G.; C-0278).

RESULTS AND DISCUSSION

Following the method of chlorination of monochlorodimedone to dichlo-
rodimedone by Hager et al. <7>, we optimized the system for chlorination
of chl a. The optimum pH for the chlorination of chl a had to be balanced
between pheophytinization of chl (below pH 4.5) and the pH optimum of CPO
reaction at low pH (2.75). A kinetic analysis of chl a chlorination at
different pH was performed and showed sufficient reaction and no pheophytinization at pH 5.0.

For preparation of chlorinated chl's and analysis of the reaction products chl a was mixed with the assay components given in the methods section. After different incubation times the chl's were extracted with petrol
ether (b.p. 40-60 $^{\circ}$C)/diethylether = 2/1 (v/v), dried over $MgSO_4$, the solvent was evaporated under reduced pressure and the products analyzed by
HPLC. The maximum amount of products was obtained with 60 min incubation
time. Variation of the incubation temperature showed an optimum at 27 $^{\circ}$C.
Under these conditions chl a was converted with 75 % yield.

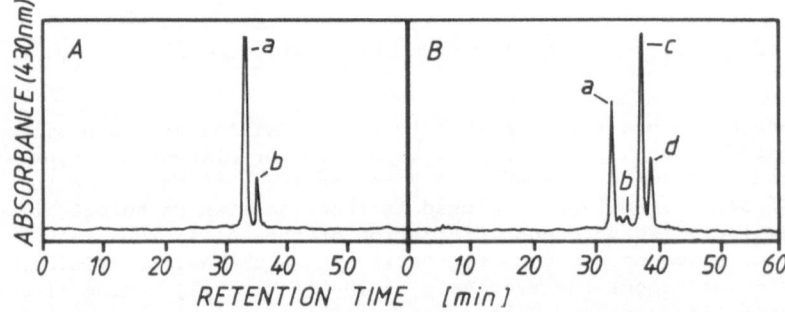

Fig. 1. HPLC elution profiles of a chl a sample (A) and the same
sample after incubation with CPO and cosubstrates for 60 min.
a: chl a; b: chl a'; c: 20-Cl-chl a; d: 20-Cl-chl a'.

Table 1. Spectroscopic properties of chl a, 20-Cl-chl a and their corresponding pheophytins (pheo). All spectra were measured in acetone.

Compounds	Absorption Maxima (nm)					Fluorescence (nm)	
						Excitation	Emission
chl a	661	617	428			430	670
pheo a	665	610	531	503	408	410	674
20-Cl-chl a	668	624	432			433	675
20-Cl-pheo a	675	617	542	512	412	412	682

HPLC analysis (Fig. 1.) showed that chl a (peak a) was converted by CPO into two compounds eluting at higher retention times (peaks c and d) compared to chl a. Fractionation of these peaks and subsequent extraction showed chl a, chl a' and two additional compounds with identical UV/vis and fluorescence emission spectra. Since the reaction conditions were favourable for epimerization, these two compounds were assumed to be the 13^2-epimers of one new compound, whereas the $(13^2 S)$-epimer was not a product of the enzymatic reaction but was formed during the incubation and extraction process. This was proven by their identical absorption spectra and CD-spectroscopy (data not shown). The maxima of the UV/vis-spectra of both compounds shifted to longer wavelengths compared to chl a (Table 1.). The fluorescence emission maximum was also shifted to a longer wavelength. Preparation of the corresponding pheophytins and spectroscopic analysis showed a typical pheophytin spectrum with maxima shifted to longer wavelengths, when compared to the spectrum of pheophytin a.

The first hint that chlorination had taken place was given by the fact that the spectroscopic properties of the product compound were nearly iden-

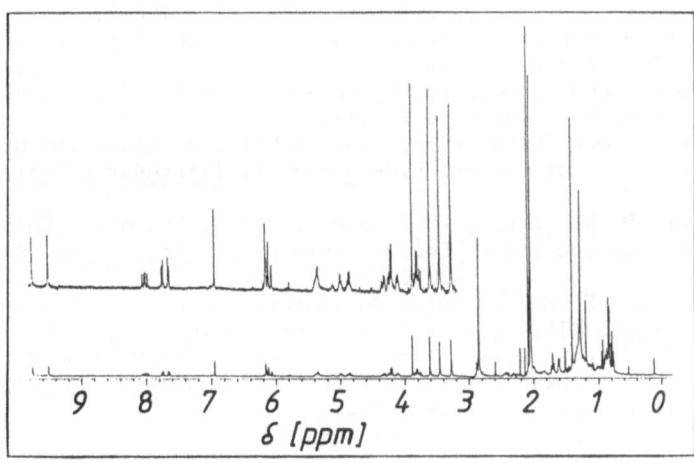

Fig. 2. ^{1}H-NMR-spectrum of 20-Cl-chl a in d_6-acetone.

tical to those of 13^2-HO-20-Cl-chl a (Chl RC I) reported by Dörnemann and Senger <1>. A comparison of the retention time of the reaction product of the enzymatic chlorination with those of chemically synthetized standards revealed its identity with 20-Cl-chl a (2) <9>.

To prove that chlorination really had taken place, chl a was incubated with CPO in the presence of ^{36}chloride. After incubation and extraction the pigments were separated by TLC and scanned for radioactivity. With the enzyme present a labeled band comigrated with a sample of authentic 20-Cl-chl a. The control experiment without enzyme gave no labeled bands.

For structural analysis of the reaction product mass spectrometry and ^{1}H-NMR spectroscopy were employed. Fast atom bombardment mass spectrometry gave a M^+-peak at 927 mass units. This corresponds to the theoretically calculated peak with the most abundant isotopes of 20-Cl-chl a.

The ^{1}H-NMR spectrum of the reaction product is shown in Fig. 2. It is the typical spectrum of a chl a derivative with some exceptions. Most peaks are shifted slightly low field, especially the signals associated with the central ring system. This is in accordance with the introduction of an electronegative group. The signal of the C-20 proton, which in chl a occurs at 8.5 ppm, is missing. This shows that substitution has taken place at this position. The fact that signal integration accounts for all other protons in the correct stoichiometry proves the regioselectivity of the chlorination reaction. The compilation of all spectroscopic data identifies the structure of the reaction product to be 20-Cl-Chl a.

Further investigations of the exact reaction mechanism and detailed characterisations of the spectroscopy and chemistry of chlorinated chl's are currently in progress.

REFERENCES

1. D. Dörnemann and H. Senger, The structure of Chlorophyll RC I, a chromophore of the reaction center of photosystem I, Photochem. Photobiol. 43: 573 (1986).
2. H. Scheer, E. Groß, B. Nitsche, E. Cmiel, S. Schneider, W. Schäfer, H.-M. Schiebel, and H.-R. Schulten, The Structure of Methylpheophorbide RC I. Photochem. Photobiol. 43: 559 (1986).
3. M. Senge, D. Dörnemann, and H. Senger, The chlorinated chlorophyll RC I, a preparation artefact, FEBS Lett. 234: 215 (1988).
4. M. Senge and H. Senger, Chlorination of Chlorophyll in vitro, Photochem. Photobiol., in press (1988).
5. R. D. Woodward and V. Scarric, A New Aspect of the Chemistry of Chlorins, J. Am. Chem. Soc. 83: 4676 (1961).
6. P. H. Hynninen and S. Lötjönen, Electrophilic Substitution at the δ-Methine Bridge of Pheophorbide a and a', Tetrahedron Lett. 22: 1845 (1981).
7. L. P. Hager, D. R. Morris, F. S. Brown, and H. Eberwein, Chloroperoxidase II. Utilization of Halogen Anions, J. Biol. Chem. 241: 1769 (1966).
8. S. Lötjönen and P. H. Hynninen, An improved method for the preparation of (10R)- and (10S)-pheophytins a and b, Synthesis : 708 (1983).
9. M. Senge, A. Struck, D. Dörnemann, H. Scheer, and H. Senger, Hydroxylation of Chlorinated and Unchlorinated Chlorophylls in vitro, Z. Naturforsch. 43c: in press (1988).

PROPERTIES OF THE 47KDa PROTEIN AND

THE D1-D2-CYT B559 COMPLEX

D.F. Ghanotakis(1), G.T. Babcock(2) and C.F. Yocum(3)

(1) Department of Chemistry, University of Crete
    Iraklion, Crete, Greece
(2) Department of Chemistry, Michigan State University
    East Lansing, MI 48824, USA
(3) Departments of Biology and Chemistry, The University
    of Michigan, Ann Arbor, MI 48109, USA

ABSTRACT

In the present paper we describe a new method for the isolation of the 47 kDa protein and the D1-D2-Cyt b559 complex that uses the non-ionic detergent dodecylmaltoside in combination with high concentrations of lithium perchlorate to dissociate polypeptides of the PSII complex, followed by FPLC ion-exchange chromatography to separate these polypeptides. By carrying out a low temperature EPR study of the two systems we support the proposal that the D1-D2 complex, and not the 47 kDa polypeptide, is the species which binds the reaction center of PSII. The EPR signal from the spin-polarized triplet was also used as a probe of the stability of the D1-D2-Cyt b559 preparation. It was found that more than 80% of the EPR signal intensity from the spin-polarized triplet could still be observed after incubation of our D1-D2-Cyt b559 complex in the dark at room temperature for five hours.

MATERIALS AND METHODS

Subchloroplast PSII membranes and the PSII oxygen-evolving reaction center complex were prepared as described in ref. 1. EPR spectra were recorded at helium temperature on a Bruker ER200D spectrometer operating at X-band. A Hewlett Packard 5245L electronic counter with a 5255A frequency converter plug-in and a Bruker ER035M gaussmeter were used to measure microwave frequency and magnetic field respectively. Optical spectroscopy was carried out on a Perkin-Elmer Lambda 5 UV/Vis spectrophotometer.

RESULTS

A Tris-treated preparation of the PSII reaction center complex was treated with 15% dodecylmaltoside, 1.5% Taurine and 4M $LiClO_4$. The solubilized complex was desalted, and subsequently loaded onto a Mono-S Pharmacia column (attached to a Pharmacia FPLC system); the fraction that did not bind on the Mono-S column was loaded onto a Mono-Q column and was subjected to a gradient of $LiClO_4$. The elution profile from the Mono-Q column consists of five main fractions (see ref. 2 for details). Table I

Table I. Polypeptide Content of various fractions from the Mono-Q column
----

p. 1	43 kDa
p. 2	28 kDa
p. 3	47 kDa
p. 4	D1 - D2 - Cyt b_{559}
p. 5	D1 - D2 - 47 kDa -Cyt b_{559}

----

lists the polypeptides found in each fraction, as determined from SDS polyacrylamide gel electrophoresis of individual fractions. As shown in the Table, fraction p. 1 contains the Chl-binding polypeptide of apparent molecular mass 43 kDa while fraction p. 2 contains the 28 kDa species, another Chl-binding protein that was isolated and characterized in ref. 1. The fractions designated as p. 3 and p. 4 are of primary interest because they contain the 47 kDa and the D1, D2, Cyt b559 species respectively. The fraction denoted as p. 4 was further purified by repeating the solubilization and isolation procedure described above

From the absorption spectra of the isolated fractions p. 3 and p. 4 it is apparent that none of these fractions contains optically detectable amounts of Chl b (Figure 1). An enrichment in the D1-D2 complex is accompanied by the corresponding appeearence of a discrete peak at 417 nm in the absorption spectrum of this fraction. The 437 nm Soret band is due to the absorption by Chl-a, whereas the 417 nm peak arises from absorption contributions from Pheo a as well as from the Soret band of oxidized Cyt b559. Reduction of the D1-D2Cyt b559 complex with dithonite produced a distinctive red shift of the Soret band (data not shown); as reported in ref. 6, reduction of Cyt b559 by dithionite is accompanied by a 14 nm red shift in the Soret region and a weak absorption increase at 559 nm. The presence of absorption bands due to Chl a makes it difficult to discriminate between contributions to the 417 nm band which arise from Cyt b559 as well as from Pheo a.

To investigate further the question of which protein(s) bind(s) the primary donor and acceptor of PSII, we carried out a low temperature EPR study of the purified 47 kDa species and of the D1-D2-Cyt b559 complex. As shown in Figure 2, a spin polarized triplet was observed in the D1-D2-Cyt b559 complex, but no such triplet was observed in the 47 kDa protein. Reduction of both preparations with dithionite had no significant effect on the observed signals (data not shown)

The P_{680} triplet EPR signal was used as an assay of the stability of our D1-D2-Cyt b559 preparation. We determined the intensity of the EPR signal as the sum of the intensities of the two highest-field and the two lowest-field peaks, as they are the farthest removed from the spectrally dense g=2.0 region. We found that even after a dark incubation period of five hours at room temperature the intensity of the P_{680} triplet EPR spectrum decreased by only 15.5% (see ref. 2). Akabori et al. (3) have measured optically the formation of the P680 triplet in D1-D2-Cyt b559 complexes; they found that complexes prepared according to Nanba and Satoh (4, 5) with Triton X-100 as detergent were relatively unstable, as they lost about 85% of the triplet signal after five hours of incubation at 25 °C. Akabori et al. (3) also found that use of milder detergents, such as Octyl-glucopyranoside and Octyl-thioglucopyranoside afforded more stable complexes.

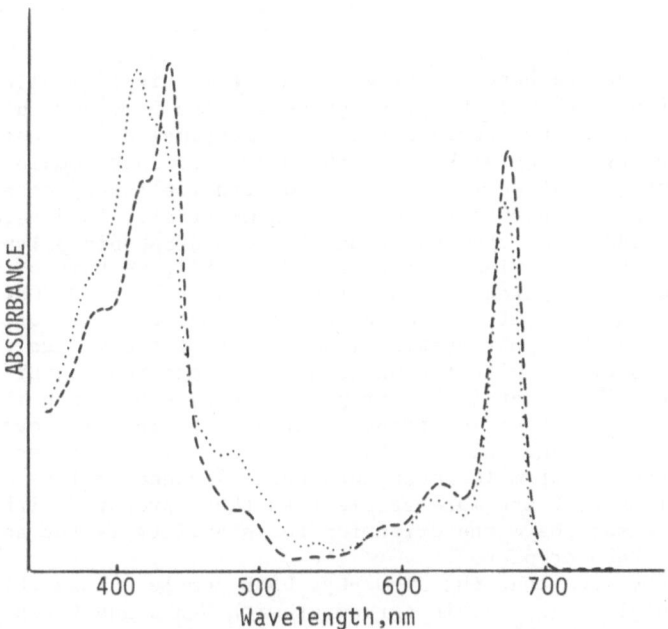

Figure 1. Room temperature absorption spectra of the 47 kDa (- - -) species and the D1-D2-Cyt b559 complex (.......).

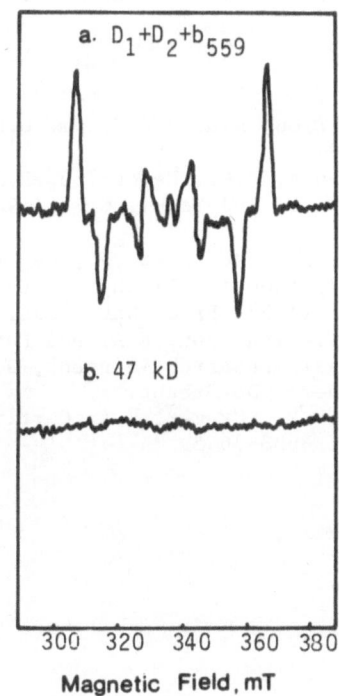

Figure 2. (a) EPR spectrum of the light induced triplet in the D1-D2-Cyt b559 complex. The experiment was carried out at 4.2 K
(b) The same as in a. but with the 47 kDa species.

DISCUSSION

The procedure we have reported in this communication allows the isolation of both the 47 kDa polypeptide and the D1–D2–Cyt b559 complex. Our method uses the non-ionic detergent dodecylmaltoside instead of Triton and the recovery of the 47 kDa and the D1–D2–Cyt b559 complex under these conditions of separation is accomplished with yields approaching 80% based on the chlorophyl content of the starting material. In developing a stepwise procedure for the isolation of the hydrophobic polypeptide components of the reaction center "core" of PSII, we have noted that under our conditions of separation and purification, the 43 kDa Chl-binding protein behaves as if it were more loosely bound to the "core" than the 47 kDa species. This observation indicates that the 47 kDa protein is more intimately associated with the photochemical reaction center than the 43 kDa; it is possible that in higher plants this association is essential for some aspects of electron transfer within the reaction center.

From this work and that of Akabori et al. (3), it is becoming more clear that the use of mild detergents such as dodecylmaltoside results in preparations that are more stable than those prepared with Triton X-100 at least in cases where the criterion for stability is the ability to generate the P680 triplet.

The observation that the D1–D2–Cyt b559 complex shows the spin polarized triplet strongly suggests, in agreement with Nanba and Satoh (4) that this complex contains the binding sites for P680 and Pheo a.

ACKNOWLEDGEMENTS

This work was supported by grants from NSF (DMB-8515932) to CFY and from USDA and NIH (GM25480) to GTB. We thank Dr. Julio de Paula for helpful discussions.

REFERENCES

1. Ghanotakis, D.F., Demetriou, D.M. and Yocum, C.F. (1987) Biochim. Biophys. Acta 891, 15–21
2. Ghanotakis, D.F., de Paula, J.C., Demetriou, D.M., Bowlby, N.R., Petersen, J., Babcock, G.T. and Yocum, C.F. (1988) Biochim. Biophys. Acta, submitted
3. Akabori, K., Tsukamoto, H., Tsukihara, J., Nagatsuka, T., Motokawa, O. and Toyoshima, Y. (1988) Biochim. Biophys. Acta 932, 345–357.
4. Nanba, O. and Satoh, K. (1987) Proc. Nat. Acad. Sci. 84, 109–112.
5. Okamura, M.Y., Satoh, K., Isaacson, R.A. and Feher, G. (1987) in Progress in Photosynthesis Research (Biggens, J. ed.) Vol. I, 379–381 Martnus Nijhoff publishers, Dordrecht.
6. Babcock, G.T., Widger, W.R., Cramer, W.A, Oertling, W.A. and Metz, J.G. (1985) Biochemistry 24, 3638–3645.

A HYPOTHETICAL MODEL FOR THE STRUCTURE AND

FUNCTION OF THE PHOTOSYSTEM II REACTION CENTER

Zoran G. Cerovic

Institute of Botany and Botanical Gardens
Faculty of Sciences, University of Belgrade
Takovska 43, 11000 Belgrade, Yugoslavia

Chlorophyll fluorescence has extensively been used to investigate the function of Photosystem II reaction centers (PSII RC) of higher plants (see Ref. 1.). The assignment of variable chlorophyll fluorescence to the redox state of PSII RC[2] has been of prime importance for these investigations. However, the interpretation of fluorescence signals is still encumbered by two major problems: the discrepancy between the measured and predicted Fm/Fo ratios[3] and the presence under physiological conditions[4] of the "energy-dependent" quenching[5] (q_E) in addition to photochemical quenching[2] (q_Q). To account for these phenomena, the existence of a radiationless deactivation pathway in "closed" reaction centers was postulated[3]. It is considered[4] to be of non-photochemical nature and to depend on ultrastructural changes of the thylakoid membrane.

The investigation of the relationship between the complex changes of variable chlorophyll fluorescence and carbon metabolism during photosynthetic induction under well defined in vitro conditions, led to the conclusion that complex changes in fluorescence yield reflect the changes in the proportion of PSII RC present in the four major possible states[6]: DPA, $^+$DPA$^-$, DPA$^-$ and $^+$DPA, where D = secondary donor(s), A = secondary acceptor(s), P = "primary" donor of the reaction center. The DPA and DPA$^-$ states correspond to the open and closed reaction centers of Duysens[2], respectively. The $^+$DPA state, very rare under physiological conditions, has been shown[7] to be associated with an elevated fluorescence yield. The state $^+$DPA$^-$ accumulates when a high proton gradient is present across the thylakoid membrane because it controls both the oxidation of the acceptor side and the reduction of the donor side of PSII (the oxidation of water). This state is associated with a lower fluorescence yield[7], even though the acceptor is reduced and "normal" photochemistry is blocked. In this paper it is proposed that the population of PSII RC in this state is responsible for the q(E) type of fluorescence quenching and the Fm/Fo discrepancy.

The fluorescence quenching by "closed" reaction centers in the state $^+DPA^-$ would be of photochemical nature, but instead of leading to a new stable charge separation, it would represent the photochemical neutralization of the pre-existing charges (on the donor and acceptor). It can be visualized as radical pair generation in a reaction center system of chlorin rings and subsequent oxido-reduction of this radical pair with the existing charged secondary acceptor and donor, or the excited reaction center system of chlorin rings serving as bridge for charge recombination, the vacant HOMO accepting one electron from the reduced acceptor and the electron from the LUMO reducing the oxidized donor (figure 1.).

Figure 1 is an energetic scheme depicting the photochemistry and oxido-reduction in the states DPA and $^+DPA^-$. The given values for standard redox potentials of the appropriate orbitals, although being for chlorophyll $\underline{a}$ and pheophytin $\underline{a}$ in solution[8], accommodates well to the expected macro-energetic picture of the reaction $\underline{in\ vivo}$. It was proposed[9] recently that the rate constant for radical pair formation in PSII RC is decreased, and thus the fluorescence yield increased, as a consequence of the presence of a negative charge on the reduced acceptor A^-. A simple mechanism for radical pair formation with one chlorophyll oxidized and one pheophytin reduced[9,10] would implicate an even smaller rate constant, higher fluorescence yield, in the state $^+DPA^-$ because of the presence of both a positive charge on one side of the thylakoid membrane and a negative one on the opposite side. To account for a low fluorescence yield associated with this state[5,7], the occurrence of primary charge separation in a symmetrical structure of PSII RC has to be postulated (figure 1. and 2.)

The molecular structure (figure 2.) of the proposed model is based on the homology between the D1-D2 proteins of PSII and M-L of photosynthetic bacteria[11], and on the existing data for the position and function of the chromophores in bacteria and PSII[10,11]. The model accounts for the role of quinones as both acceptors and donors[12] of PSII, and their orientation in the membrane[13]. A dual function is proposed for Q_B acting both as primary donor and secondary acceptor. The recent finding[14] that D1 is the site of both the oxidation and reduction in PSII RC is in line with this proposal. A new function for cytochrome $\underline{b}$-559 is postulated. It would perform the communication between the oxygen evolving complex and the primary donor to the system of chlorin rings (figure 2.). The presence of two hemes per PSII, the transmembrane structure of the apoproteins[15] and the detected perpendicular orientation of the cytochrome hemes to the plane of the membrane[13], all argue in favor of the proposed function. This assignment for an obligatory physiological role of cytochrome $\underline{b}$-559 in oxygen evolution of PSII is in accordance with the close correlation found between the appearance of its high potential form and the capacity of greening barley leaves to photo-oxidize water[16], and the recent evidence for a significant structural role of this cytochrome in oxygen evolution[17].

The presented model of PSII RC was constructed in order to explain the phenomenon of variable chlorophyll fluorescence inherent to this Photosystem, and is based on the new comparative data on both structure and function of its constituents. It conceives the decreased efficiency of photochemistry (higher fluorescence yield) as a consequence of the presence of surplus charge (positive or negative) that remains at the reaction center when an imbalance in the consumption of the products of the primary reactions (ATP, NADPH) is present[6]. Moreover the same structure enables an efficient consumption of excitons from the antenna chlorophylls (lower fluorescence yield) independent of light saturation. This is accomplished either through the generation of reducing equivalents and oxygen by the reaction centers in the state DPA, or the generation of heat by the reaction centers in the state $^+DPA^-$, when photosynthesis is over-saturated with light, as long as the "consuming reactions" of the carbon metabolism are well balanced.

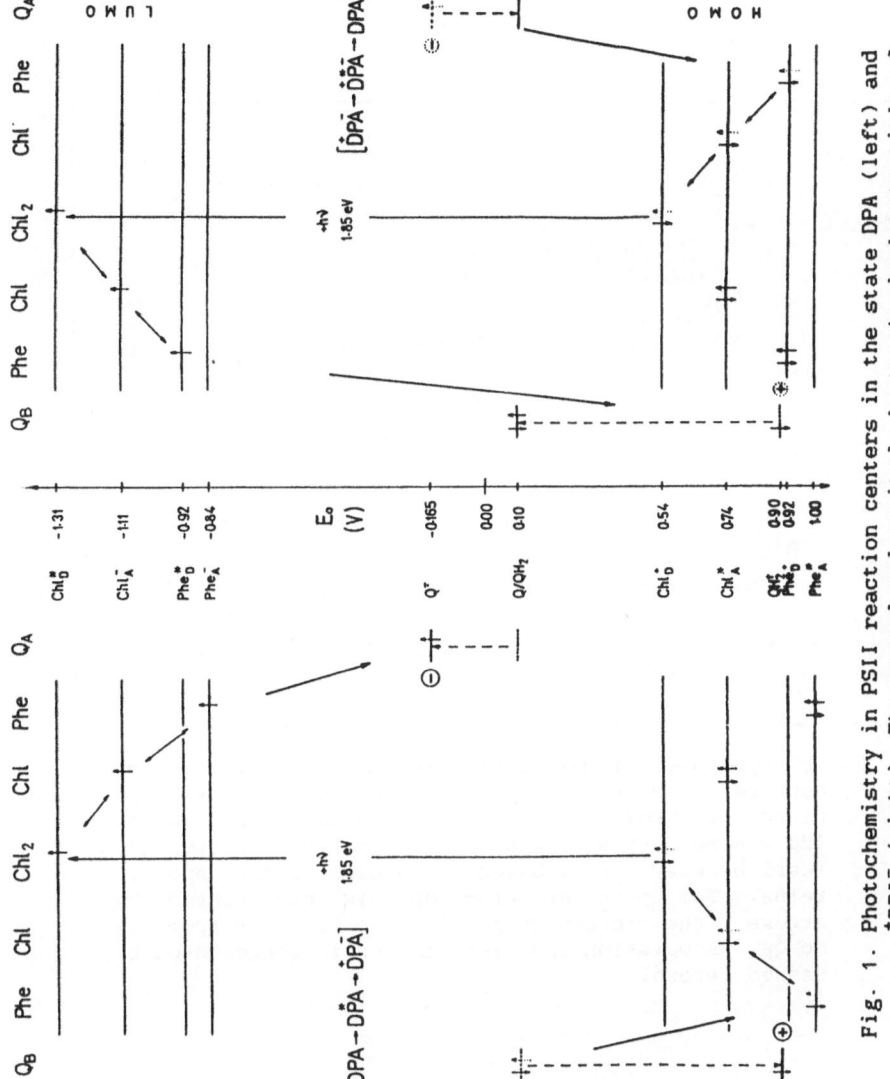

Fig. 1. Photochemistry in PSII reaction centers in the state DPA (left) and $^+$DPA$^-$ (right). The ground and excited electrochemical potentials of chlorophyll $\underline{a}$ (Chl) and pheophytin $\underline{a}$ (Phe) are taken from Ref. 8, and the potentials of plastoquinone (Q) pairs are estimates from Ref. 18. Only one of several possible distributions of electrons, represented by little arrows, is presented to depict the possible route for charge separation and oxido-reduction.

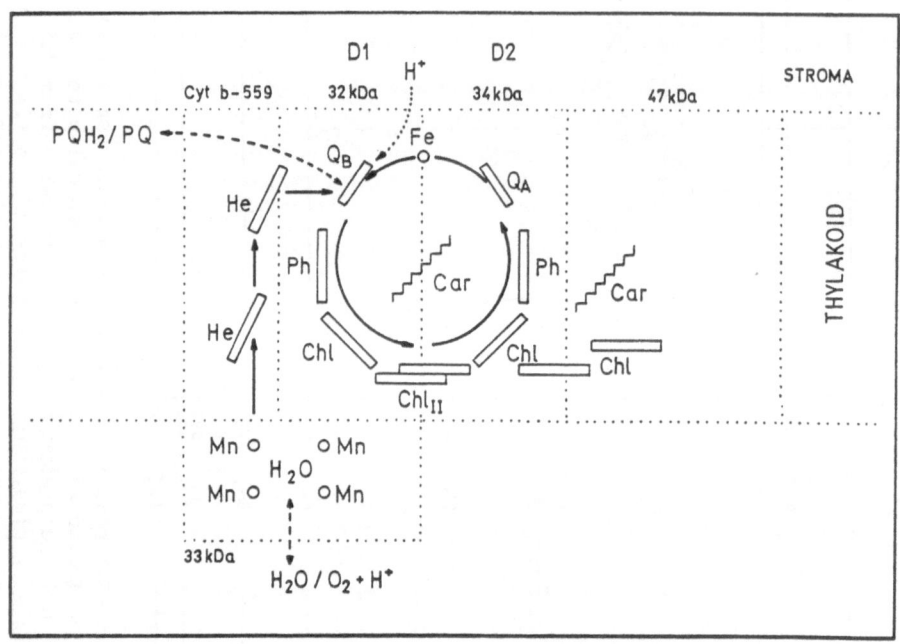

Fig. 2. The hypothetical molecular structure of PSII reaction
centers enabling the proposed photochemistry. The
rings of chromophores are represented by rectangles
(He = heme), free metals by circles, and carotenoids
(Car) by wavy lines. Doted lines delimit the apoproteins. The pass of electrons is represented by
arrows, the slowest step being the transfer from Q_A
to Q_B. Association and dissociation is represented by
dashed arrows.

References

1. Govinjee, J. Amesz and D. C. Fork, eds., "Light Emission by Plants and Bacteria", Academic Press, New York (1986)
2. L. N. M. Duysens and H. E. Sweers, Mechanism of two photochemical reactions in algae as studied by means of fluorescence, in: "Microalgae and Photosynthetic Bacteria", Jap. Soc. Plant Physiol., eds., University of Tokyo Press, Tokyo, pp. 353 (1963)
3. W. L. Butler and M. Kitajima, Fluorescence quenching in Photosystem II of chloroplast, Biochim. Biophys. Acta 376:116 (1975)
4. G. H. Krause, C. Vernotte and J. -M. Briantais, Photoinduced quenching of chlorophyll fluorescence in intact chloroplasts and algae. Resolution into two components, Biochim. Biophys. Acta 679:116 (1982)
5. N. Murata and K. Sugahara, Control of excitation transfer in photosynthesis - III. Light-induced decrease of chlorophyll $\underline{a}$ fluorescence related to photophosphorylation system in spinach chloroplasts, Biochim. Biophys. Acta 189:182 (1969)
6. Z. G. Cerović, Photosynthesis and chlorophyll fluorescence in isolated chloroplast, Doctoral Thesis, Belgrade University, Yugoslavia (1988)
7. R. Delosme, Variations du rendement de fluorescence de la chlorophylle in vivo sous l'action d'éclairs de forte intensité, C. R. Acad. Sci. (Paris) 272:2828 (1971)
8. G. R. Seely, The energetics of electron-transfer reactions of chlorophyll and other compounds, Photochem. Photobiol. 27:639 (1978)
9. G. H. Schatz and A. R. Holzwarth, Mechanism of chlorophyll fluorescence revised: prompt or delayed emission from Photosystem II with closed reaction centers?, Photosynth. Res. 10:309 (1986)
10. B. A. Diner, The reaction center of Photosystem II, in: "Encyclopedia of Plant Physiology, New Series Vol. 19, Photosynthesis III", L. A. Staehelin and C. J. Arntzen, eds., Springer Verlag, Berlin (1986)
11. J. Deisenhofer, O. Epp, K. Miki, R. Huber and H. Michel, Structure of the protein subunits in the photosynthetic reaction center of Rhodopseudomonas viridis at 3A resolution, Nature 318:618 (1985)
12. P. J. O'Malley, G. T. Babcock and R. C. Prince, The cationic plastoquinone radical of the chloroplasts water splitting complex. Hyperfine splitting from a single methyl group determines the EPR spectral shape of signal II, Biochim. Biophys. Acta 766:283 (1984)
13. A. W. Rutherford, Orientation of EPR signals arising from components in Photosystem II membranes, Biochim. Biophys. Acta 807:189 (1985)
14. M. Ikeuchi and Y. Inoue, Specific ^{125}I labeling of D1 (herbicide-binding protein). An indication that D1 functions on both the donor and acceptor sides of photosystem II, FEBS Lett. 210:71 (1987)
15. W. R. Widger, W. A. Cramer, M. Hermodson and R. G. Herrmann, Evidence for a hetero-oligomeric structure of the chloroplast cytochrome $\underline{b}$-559, FEBS Lett. 191:186 (1985)
16. M. Plesničar and D. S. Bendall, The photochemical activities and electron carriers of developing barley leaves, Biochem. J. 136:803 (1973)
17. L. K. Thompson, J. M. Sturtevant and G. W. Brudvig, Differential scanning calorimetric studies of Photosystem II: Evidence for a structural role for cytochrome b_{599} in the oxygen-evolving complex, Biochemistry 25:6161 (1986)
18. P. Rich, Electron and proton transfers through quinones and cytochrome bc complexes, Biochim. Biophys. Acta 768:53 (1984)

BINDING OF [45]Ca BY PHOTOSYSTEM II POLYPEPTIDES

Andrew N. Webber, Richard Wales and John C. Gray

Botany school, University of Cambridge, Downing Street
Cambridge, CB2 3EA

Introduction

It is now well established that photosystem II reactions have an absolute requirement for calcium ion(s). The calcium ions are required for both the normal functioning of the water oxidising complex[1,2] and for electron flow from the secondary electron donor, Z, to the reaction center[3]. The number of calcium binding sites is controversial and is reported to range between 1 and 3 per reaction center[4,5]. The role of calcium in photosystem II and the location of the calcium binding sites is unknown. Exogenously added calcium can partially reverse the inhibition of oxygen evolution due to the removal of the hydrophilic polypeptides (33, 23, 16 and 10kDa) by various salt treatments[2,6,7,8]. Such observations have led to the suggestion that these hydrophilic polypeptides provide a binding site(s) for calcium ion required for water oxidation. However, conflicting results suggest that light is required for the release of calcium ion from photosystem II membranes depleted of the hydrophilic polypeptides[9] and indicate that the calcium binding site is associated with an intrinsic photosystem II component. In order to understand the role of calcium in photosystem II it is clearly important to identify the calcium binding sites. Here we report the identification of individual polypeptides in photosystem II which are able to bind calcium ions when immobilised to nitrocellulose membranes. The putative calcium binding site for one of these polypeptides has been further identified by comparison of the derived amino acid sequence to known calcium binding proteins.

Materials and methods

Photosystem II membranes were prepared from wheat thylakoid membranes by the method of Berthold et al.[10]. Following electrophoresis on SDS-polyacrylamide gels, polypeptides were electroblotted to either nitrocellulose membrane for calcium binding studies or to polyvinylidene difluoride (PVDF) membrane for protein sequencing. Nitrocellulose membrane bearing the photosystem II polypeptides was incubated for 10 min with 10mM imidizole-HCl (pH 6.8), 60mM KCl, 5mM $MgCl_2$ and 30µCi [45]$CaCl_2$ (20µCi/mg Ca^{2+}) and then washed for 5 mins with 50% ethanol[11]. Polypeptides able to retain [45]Ca after washing were visualised by autoradiography. N-terminal amino acid sequence analysis was performed by gas phase sequencing on an Applied Biosystems 470A Gas-phase Sequencer coupled to a 120A PTH-analyser under the control of a 900A data module.

Results and discussion

Calcium binding polypeptides in photosystem II were identified by their ability to bind ^{45}Ca. Following SDS-polyacrylamide gel electrophoresis and transfer to nitrocellulose, photosystem II polypeptides were incubated with ^{45}CaCl$_2$ and washed with 50% ethanol. Polypeptides that were able to bind ^{45}Ca, and retain it after subsequent washings were visualised by autoradiography. Figure 1 shows that calmodulin and three photosystem II polypeptides were able to bind ^{45}Ca. The ability of calmodulin to bind ^{45}Ca indicates the usefulness of this technique[11]. Calcium binding was not affected by varying the concentration of Mg^{2+} in the incubation medium between 0-5mM (data not shown).

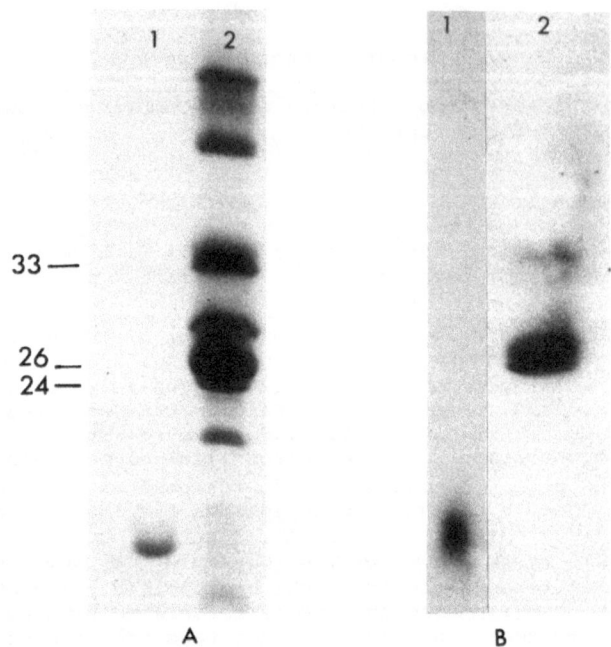

Figure 1. A. Polypeptide profile of calmodulin (lane 1) and wheat photosystem II membranes fractionated on a 15% acrylamide gel. B. Autoradiography of the nitrocellulose membrane bearing photosystem II polypeptides following incubation with ^{45}Ca.

```
L - - L L - - L * - * - * G - I * - - * L - - L L - - L    Test sequence

L D E I E G P F E V S A D G S V K F E E K D G I D Y A A    Pea 33kDa sequence
167                                             195
L D E L F E E L D K N G D G E V S F E E F Q V L V K K I    Bovine ICaBP
46                                              73
L D D L F Q E L D K N G N G E V S F E E F Q V L V K K I    Pig ICaBP
49                                              76
L D N L F E E L D K N D D G E V S Y E E F E V F F K K L    Rat ICaBP
40              X   Y   Z   -Y  -X   -Z            67
```

Figure 2.Comparison of the predicted calcium binding domain of the pea 33kDa polypeptide with the calcium binding domains of bovine[13], porcine[14] and rat[15] intestinal calcium-binding proteins, and the EF hand test sequence of Tufty and Kretsinger[16]. In the test sequence, * indicates an oxygen containing residue, L indicates a hydrophobic residue, I indicates a conserved aliphatic hydrophobic residue, G is glycine and - is any amino acid. X, Y, Z, -Y, -X, -Z refer to vertices of the calcium co-ordination octahedron.

134

The calcium-binding polypeptide of 33kDa co-migrates with the 33kDa polypeptide of the oxygen-evolving complex. The amino acid sequence of the 33kDa polypeptide, deduced from a full-length cDNA clone for the pea polypeptide[12] has been screened for calcium binding sites by computer aided comparison to known calcium binding proteins. As shown in figure 2, the pea 33kDa polypeptide contains a region highly homologous to mammalian intestinal calcium binding proteins, and contains all the essential amino acids required to form a calcium binding 'EF-hand' structure. A prediction of the 3-D tertiary 'EF-hand' structure of the 33kDa polypeptide calcium binding region is shown in figure 3.

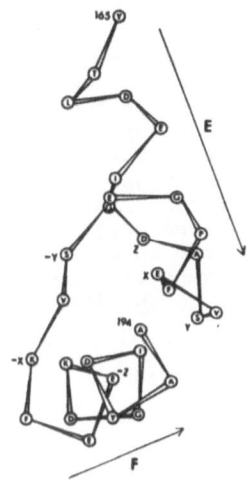

Figure 3. Possible EF hand tertiary structure for the potential calcium binding region of the 33kDa polypeptide.

```
                  10          20          30
                   .           .           .
Petunia   MATSAIQQSAFAGQTALKSQNELVRKIGSFGGGR
Tomato    MATCAIQQSAFVGQAVGKSQNEFIRKVGNFGEGR
Lemna     MAASAIQSSAFAGQTALKQREELVRKVGVS-DGR

                  40          50          60
                   .           .           .
Petunia   ATMRRTVKSAPQSIWYGEDRPKYLGPFSEQTPSY
Tomato    ITMRRTVKSAPQSIWYGEDRPKYLGPFSEQTPSY
Lemna     FSMRRTVKAVPQSIWYGADRPKFLGPFSEQTPSY
Wheat             GNDLWYGPDXVKYLGPFSAQTPS
```

Figure 4. Comparison of the *N*-terminal 68 amino acids derived from the LHC-II type II cab genes of petunia[17], tomato[18] and lemna[19] with the *N*-terminal amino acid sequence of wheat LHCII-24.

The 26 and 24kDa calcium binding polypeptides comigrate with the chlorphyll *a/b* binding LHC polypeptides. The isolated LHC II complex has been previously shown to bind calcium[20], but had not been resolved into its individual polypeptide components. To confirm their identity we have attempted to obtain the *N*-terminal sequence of these two polypeptides. The *N*-terminal sequence obtained from the 24kDa polypeptide is presented in

Figure 3.The 26kDa polpeptide was found to be blocked a the *N*-terminus. The 24kDa polypeptide *N*-terminal sequence shows very high homology to the amino acid sequence of LHC II polypeptides derived from type II *cab* genes, Fig 3. The N-terminal sequence of LHC II-24 predicts a processing site further into the mature polypeptide than predicted by analogy to the type I *cab* gene product. This would result in the type II *cab* gene product being approximately 1kDa smaller than the type I *cab* gene product and explain the increased mobility of LHC II-24 during SDS-PAGE.

Inspection of the amino acid sequences of the LHC II polypeptides does not provide any information on the location of the calcium binding site within the polypeptide. We have found that dicyclohexylcarbodiimide (DCCD) treatment of thylakoid membranes will inhibit the ability of the polypeptides to bind ^{45}Ca when immobilised to nitrocellulose. This would suggest that the Calcium binding domain is a carboxyl residue(s) normally within a hydrophobic domain of the polypeptide. It is not known whether calcium binding by the LHC II polypeptides is related to any of the observed effects of calcium depletion on photosystem II electron flow and oxygen evolution. A polypeptide of 24kDa was co-precipitated with hydrophilic polypeptides of the oxygen-evolving complex when partially solubilised photosystem II membranes were incubated with antibodies against the 23 or 33kDa polpeptides of the oxygen-evolving complex. This would indicate a close proximity of the 24kDa polypeptide with the oxygen-evolving complex. It will be important to establish if the 24kDa calcium binding polypeptide is related to the 24kDa polypeptide associated with the oxygen-evolving complex.

References

1. A. Boussac, B. Maison-Peteri, A-L Etienne and C. Vernotte, Biochim. Biophys. Acta 808: 231-260 (1985).
2. T. Ono and Y. Inoue, Biochim. Biophys. Acta 850: 380-389 (1986).
3. K. Satoh and S. Katoh, FEBS Lett. 190: 199-203.
4. K. V. Cammarata and G. M. Cheniae, Plant Physiol. 84: 587-598 (1987).
5. T. Ono and Y. Inoue, FEBS Lett. 227:147-152.
6. D. F. Ghanotakis, , and C. F.Yocum, FEBS Lett. 167: 127-130 (1984)
7. M. Miyao and N. Murata, FEBS Lett. 168: 118-120 (1984).
8. T. Ono and Y. Inoue, FEBS Lett. 168: 281-286 (1984).
9. J. P. Dekker, D. F. Ghanotakis, J. J. Plijter, H. J. VanGorkom and G.T. Babcock, Biochim. Biophys. Acta 767: 515-523 (1984).
10. D. A. Berthold, G. T. Babcock and C. F. Yocum, FEBS Lett.134: 231-234 (1981).
11. K. Maruyama, T. Mikawa and S. Ebashi, J. Biochem. 95: 511-519 (1984).
12. R. Wales, B. J. Newman, D. Pappin and J. C. Gray, Mol. Gen. Genet. in press.
13. C. S. Fullmer and R. H. Wasserman, J. Biol. Chem. 256: 5669-5674 (1981).
14. T. Hofmann, M. Kawakami, A. J. W. Hitchman, J. E. Harrison and K. J. Dorrington Can. J. Biochem. 57: 737-748 (1979).
15. C. Desplan, O. Heidmann, J. W. Lille, C. Auffray and M. Thomasset, J. Biol. Chem. 258 13502-13505 (1983).
16. R. M. Tufty and R. H. Kretsinger, Science 187: 167-169 (1975).
17. M. M. Stayton, M. Black, J. Bedbrook and P. Dunsmuir, Nucl. Acids Res. 14: 9781-9796 (1986).
18. E. Pichersky, N. E. Hoffman, V. S. Malik, R. Bernatzky, S. D. Tanksley, L. Szabo and A. R. Cashmore, Plant Mol. Biol. 9: 109-120 (1987).
19. G. A. Karlin-Neumann, B. D. Kohorn, J. P. Thornber and E. M. Tobin J. Mol. Appl. Genet. 3: 45-61 (1985).
20. D. J. Davis and E. L. Gross, Biochim. Biophys. Acta 387:557-567 (1975).

ISOLATION OF THE CHLOROPHYLL a/b PROTEIN COMPLEX CP29

Wolfgang P. Schröder, Michael Spangfort, Tomas Henrysson and
Hans-Erik Åkerlund

Department of Biochemistry, University of Lund, P.O.Box 124
S-221 00 LUND, Sweden

INTRODUCTION

The light harvesting apparatus of Photosystem II (PSII) has at least three
chlorophyll a/b binding protein complexes. Apart from the most abundant
complex , LHCII, there are the minor CP29 and CP24. The polypeptide composition and structure of CP24 have just recently been extensively characterized (Spangfort et al. these proceedings) but the precise structure and
function of CP29 is still unknown. CP29 is normally isolated by partly
denaturating sodium dodecyl sulphate polyacrylamide gel electrophoresis
(SDS-PAGE) and consists of one polypeptide with an apparent molecular
weight of 29 kDa which is unrelated to other chlorophyll-binding proteins
(1). In this work we report a new procedure for the isolation of CP29 by
mild detergent treatement of washed thylakoid membranes followed by a
single HPLC step.

MATERIALS AND METHODS

Photosystem II particles were prepared from spinach leaves according to
the detergent metod of (2). The excess of detergent was removed by washing
the PSII particles two times as in (3). To purify the CP29 protein, the
PSII particles were treated with 1 M NaCl, 0.06 % Triton X-100 and 10 mM
Mes pH 6.5 and with 1 M CaCl$_2$ and 10 mM Mes pH 6.5 as in (3). The PSII
particles (20 mg Chl) were then solubilized in 10 ml of a mixture of 2 %
Zwittergent TM-314, 2 % digitonin and 10 mM Mes pH 6.0. The solubilized
material was then loaded on a HPLC cation exchange column (TSK CM-3SW,
LKB, Sweden), preequilibrated with 10 mM Mes pH 6.0 and 0.1 % Zwittergent
TM-314. The bound material was then eluted with a NaCl gradient from 0 to
0.5 M, in 10 mM Mes pH 6.0 and 0.1 % Zwittergent TM-314.
Sodium dodecyl sulphate polyacrylamide gel electrophoresis with a 12 -
22.5 % polyacrylamide gradient was performed at room temperature using the
buffer system of Laemmli (4). The gels were stained with coomassie brilli-

ant blue R-250. The proteins were transfered from the gels to a polyvinylidene difluoride (PVDF) membrane using a JKA-Biotech semidry electroblotter essentially as in (5) except that the gels were not soaked in transfer buffer. The CP29 protein band were cut out and loaded on an Applied Biosystem model 470 sequenator (5) or analysed on a Beckman system 6300 high performance amino acid analyzer according to Beckman standard program.

RESULTS AND DISCUSSION

After removal of extrinsic proteins from PSII enriched membranes followed by solubilization with non-ionic detergents, a single polypeptide was purified by HPLC column chromatography (Fig.1). This polypeptide was found to be chlorophyll-binding with an apparent molecular weight of 29 kDa (Fig.2).

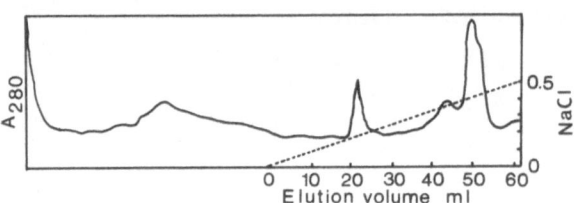

Fig. 1. Elution diagram from HPLC of the solubilized photosystem II particles. The proteins were eluted with a gradient from 0 to 500 mM NaCl in 10 mM Mes pH 6.0 and 0.1 % Zwittergent TM-314.

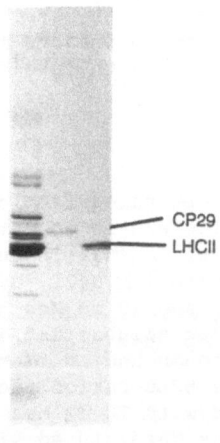

A B C

Fig. 2. SDS polyacrylamide gel electrophoresis of A. photosystem II particles B. purified CP29 and C. LHC II.

The identity of this protein was established by SDS-PAGE. Partly denaturating SDS-PAGE of PSII membranes resolves several minor chlorophyll a/b binding complexes, among them the CP29. When analysed by denaturating SDS-PAGE performed at room temperature, CP29 show one and sometimes two apo-polypeptides with apparent molecular weights of 29 kDa and 28 kDa. The present chlorophyll-binding polypeptide isolated by HPLC was found to comigrate with the 29 kDa apo-polypeptide of CP29. The chlorophyll a/b ratio of the purified 29 kDa polypeptide was about 2.0 and the absorbance spectra showed a maximun at 673.0 nm (Fig.3). These data are in agreement

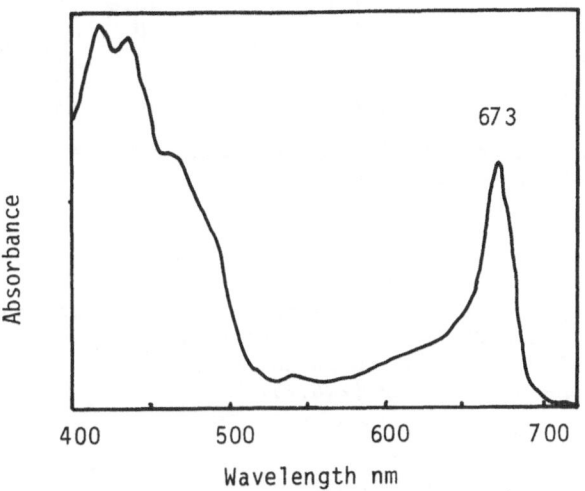

Fig. 3. Spectra of the HPLC fraction containing the CP29. The spectra were recorded with an Shimadzu UV-3000

with previously reported data of the CP29 isolated by partly denaturating SDS-PAGE (2). Polyclonal antibodies raised against the isolated 29 kDa polypeptide where found to crossreact with several other polypeptides of the thylakoid membrane. Among them where the polypeptides of LHCII but also two polypeptides associated with the light-harvesting antenna of PSI. This cross-reactivity between chlorophyll-binding polypepties is a common feature (2). The isolated 29 kDa polypeptide was found to be N-terminal blocked. The overall amino acid composition (Fig.4) reveal the presence of histidines, known as potential chlorophyll binding sites.

Amino acid	Mole %	Residues per protein (mol/mol)
Asp/Asn	8.8	21.1 (21)
Thr	6.2	14.9 (15)
Ser	5.4	13.0 (13)
Glu/Gln	11.4	27.4 (27)
Pro	5.9	14.2 (14)
Gly	13.9	33.4 (33)
Ala	8.6	20.6 (21)
Val	5.2	12.5 (13)
Met	0	0
Ile	4.6	11.0 (11)
Leu	13.4	32.2 (32)
Tyr	0	0
Phe	5.4	13.0 (13)
His	1.6	3.8 (4)
Lys	6.6	15.8 (16)
Trp	0	0
Arg	3.1	7.4 (7)
Total	100	240.3 (240)

Fig. 4. Amino acid composition of the CP29. Figures within parentheses denote the nearest integer.

The presented data strongly suggests that the isolated 29 kDa polypeptide is an apo-polypeptide of CP29. However, a more careful studie concerning the identity of the HPLC purified 29 kDa protein is in progress since the electrophoretical mobility of these polypeptides strongly depends on the running conditions of SDS-PAGE (2,6,7).

REFERENCES

1. Green, B.R. (1988) Photosynthesis Research 15,3-32
2. Ford, R.C. and Evans, M.C.W. (1983) FEBS Lett 160, 159-164
3. Ljungberg, U.,Åkerlund, H.-E. and Andersson, B. (1986) Eur. J. Biochem. 158, 477-482
4. Laemmli, U.K. (1970) Nature 227, 680-685
5. Matsudaira, P. (1987) J. Biol. Chem. 262, 10035-10038
6. Dunahay, T.G., Schuster, G. and Staehelin, L.A. (1987) FEBS Lett. 215, 25-30
7. Bassi, R. and Simpson, D. (1987) Progress in Photosynthesis Research (Biggens, J. ed.), Vol II, 81-88

SPECTROSCOPIC CHARACTERIZATION OF PURIFIED CHLOROPHYLL a/b

PROTIENS CP29, CP26, CP24 FROM MAIZE PSII ANTENNA COMPLEX

Fernanda Rigoni, Roberto Bassi and Giorgio M.Giacometti

Dipartimento di Biologia - Università di Padova

Via Trieste, 75 - 35121 Padova- Italy

INTRODUCTION

The antenna of Photosystem II (PSII) has been the subject of extensive investigation and recently has been proved to be a complex system made up of different populations of chlorophyll-protein complexes [1,2,3]. Besides the main chlorophyll a/b antenna component (LHCII) other CPs have been isolated such as CP29,CP26 and CP24. Scarce information exists on the structural, spectroscopic and functional properties of these antenna components also due to the difficulty of having them in a pure and native form.

Thanks to a newly developed purification method, based on isoelectrophocusing and sucrose gradient ultracentrifugation, we recently succeeded in isolating, in a very pure form, all the PSII antenna components including, at least three different populations of LHCII which can be distinguished in relation to their behaviour during the State I-State II transition [3]. One of these populations, which was proved to be phosphorylatable and yet to permanently stay in the grana partitions tightly associated to the PSII reaction centers, was further subjected to structural analysis. Two quaternary stable conformations were found corresponding to the so called "monomeric" and "oligomeric" forms of LHCII (also indicated as CPII and CPII'). Contrary to what previously proposed [4], we found a trimeric quaternary state for the former (CPII) and an ennameric (trimer of trimers) quaternary state for the latter (CPII').

A possible model for the "in vivo" relevance of the different quaternary states of LHCII has been proposed [5]. We present here a very first spectral characterization of the minor chlorophyll a/b protein complexes (CP24,CP26 and CP29) whose structure-function relationship is still to be clarified.

RESULTS AND DISCUSSION

The general characteristics of the three minor chl a/b complexes CP29,CP26 and CP24 are summarized in Table 1 and compared to those of LHCII. We may observe a systematic increase in the chla/chlb ratio in going from LHCII to CP29 suggesting different roles for these complexes. A small but systematic trend is also observed in the chla absorption maxima whose red shift with respect to LHCII increases in the order LHCII $\lesssim$ CP24 $<$ CP26 $<$ CP29. A relatively large shift in the adsorption maximum of chlb for CP24 and CP29 is indicative of a peculiar chl-protein and/or chl-pigment interaction in these complexes.

Apart from the polypeptide composition shown in Table 1, no information is available on the quaternary structure of the minor a/b chl-proteins, however sucrose gradient centrifugation shows that their mobility is comparable to that of CPII suggesting also for these antenna components a possible trimeric configuration.

Fig. 1 shows the absorption spectra of CP29, CP26, CP24 and LHCII for comparison. Besides the shift of the absorption maxima in the Soret region, discussed above, we may underline how the absorption band for chlb, normally observed at 650 nm, is much less intense and blue-shifted at 641 nm in CP26 and CP29. These data, taken together, seem to indicate that chlorophyll b in these minor proteins is not only present in lower amount but also is embedded in a special kind of surroundings, quite different from that of LHCII.

The CD spectra reported in Fig. 2, are rather complex, however some characteristic features can be identified in the spectrum of LHCII among which the most significant are: i) a chlb/chlb exciton splitting signal centred at 470 nm with extrema at 445 nm (+) and 495 nm (-). ii) a chla/chlb exciton splitting signal centred at 453 nm with extrema at 433 nm (+) and 473 nm (-). iii) a negative signal at 650 nm attributed to chlb absorption. iv) a chla/chla exciton splitting signal centred at 671 nm with extrema at 667 nm (+) and 679 nm (-).

The b/b exciton feature appears strongly reduced in CP26 and CP29 as well as the chlb negative band at 650 nm. This finding correlates well with the different absorption characteristics of chlb contained in these two complexes. CP24 also shows this kind of difference with respect to LHCII although to a lesser extent. For this complex a negative band at 480 nm is particularly evident for which no simple explanation is available.

Fig. 3 shows the excitation wavelength dependence of polarization of CP24, CP26 and CP29. To each of them the polarization spectrum of LHCII is superimposed for comparison. The emission was viewed at 719 nm through an interference filter. Similar spectra were obtained when the emission was viewed at 700 nm. Very low polarization values were obtained at all the excitation wavelengths with a polarization value at the red maximum of chla (675 nm) of about 0.17. The apparent increase of polarization at wavelengths higher than 675 nm is due to light scattering.

Similar results were obtained by Van Metter[6] on CPII (i.e. the trimeric form of LHCII) as to the limiting polarization value. However he could observe a clear plateau at 700 nm that in our case cannot be observed owing to the scattering artifacts.

A minimum of polarization was observed in correspondence of the red absorption maximum of chlb and this was particularly pronounced in CP29 at 640 nm. These findings indicate that also in the minor chl a/b protein complexes the high level of simmetry observed in LHCII for the mutual orientation of the pigments, is present. This is particularly true for the chlb organization in CP29. A trimeric quaternary structure of the same kind of that found for LHCII is highly probable also for these minor proteins.

Table 1 Characteristics of the chl a/b protein complexes of PSII membranes.

Name	Molecular mass of polypeptides	pI	a/b	chla red mx (nm)	chlb blue mx (nm)
LHCII	26, 28.5, 28.8, 29.5, 30	4.22	1.4	674.5	472
CP24	20, 25	4.43	1.6	674.5	464
CP26	28, 29	4.38	2.7	676	470.5
CP29	31	4.67	2.8	677	464

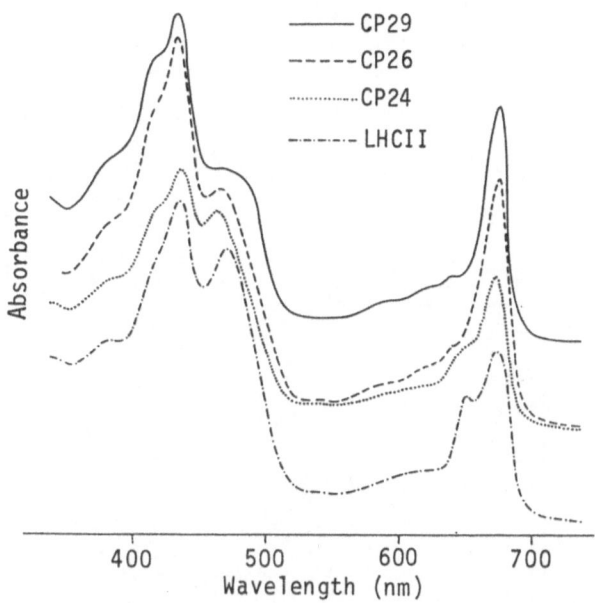

Fig.1 Absorption spectra of CP29, CP26, CP24 and LHCII.

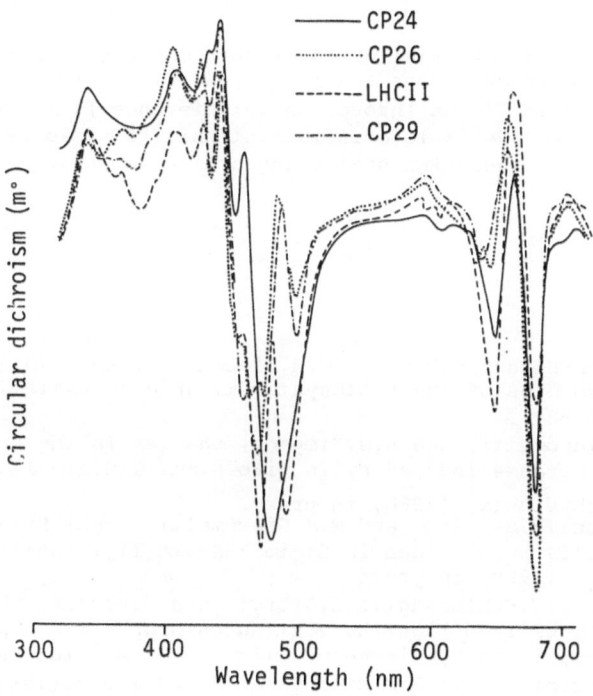

Fig.2 CD spectra of CP29, CP26, CP24 and LHCII.

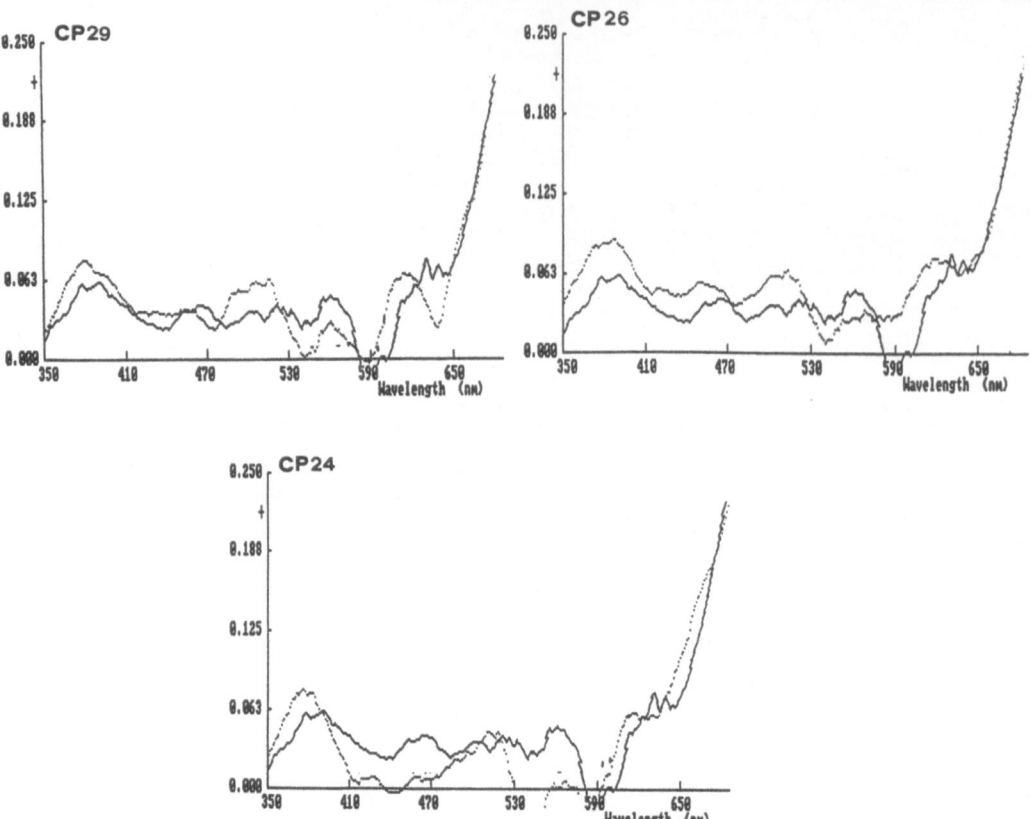

Fig.3 Comparison of the excitation wavelength dependence for the fluorescence polarization of CP29,CP26 and CP24 with that of LHCII (solid line). Emission was viewed at 719 nm through an interference filter having a band width of ± 15 nm. Reliable emission intensities could be measured up to 675 nm. At higher wavelengths scattering artifacts dominate.

REFERENCES

1. R.Bassi,G.Hoyer-Hansen,R.Barbato,G.M.Giacometti, and D.J.Simpson,Chlorophyll-proteins of the Photosystem II antenna system, J.Biol.Chem. 262: 13333 (1987).
2. R.Bassi,G.M.Giacometti, and D.J.Simpson, Changes in the organization of stroma membranes induced by in vivo State I-State II transition, Biochim.Biophys.Acta, (1988) in press.
3. R.Bassi,F.Rigoni,R.Barbato, and G.M.Giacometti, Light Harvesting Complex (LHCII) populations in State I-State II transitions, Biochim. Biophys.Acta, (1988) in press.
4. J.P.Ide,D.R.Klug,W.Kuhlbrandt,L.B.Giorgi, and G.Porter, The state of detergent solubilized light harvesting chlorophyll a/b protein complex as monitored by picosecond time resolved fluorescence and circular dichroism, Biochim.Biophys.Acta, 893: 349 (1987).
5. R.Bassi,M.Silvestri,P.Dainese,G.M.Giacometti, and I.Moya, Detergent effect on spectral properties and aggregation state of Light Harvesting Chlorophyll a/b protein complex (LHCII), (1988), submitted.
6. R.L.Van Metter, Excitation energy transfer in the light harvesting chlorophyll a/b protein, Biochim.Biophys.Acta, 462: 642 (1977).

THE 20 kDa APO-POLYPEPTIDE OF THE CHLOROPHYLL a/b PROTEIN COMPLEX CP24

M. Spangfort[1], U.K. Larsson[1], U. Ljungberg[1,4],
M. Ryberg[3], B. Andersson[2], R. Klein[4], N. Wedel[4],
and R.G. Herrmann[4]

1 Department of Biochemistry
  University of Lund, Sweden
2 Department of Biochemistry
  University of Stockholm, Sweden
3 Department of Plant Physiology
  University of Göteborg, Sweden
4 Botanisches Institut der Ludwig-Maximilians
  Universität, München, FRG

SUMMARY

A 20 kDa polypeptide is shown to be the only chlorophyll-binding protein of CP24. Immunoblotting of thylakoid subfractions and immunogold electron microscopy of spinach leaf sections demonstrate that CP24 is located in the PSII-rich, appressed grana regions. A cDNA clone encoding the entire precursor protein (261 aa) was isolated. The CP24 apo-polypeptide is nuclear encoded and is predicted to have two membrane spans with putative chlorophyll binding sites in regions which shows significant homology with LHC-II.

INTRODUCTION

The light-harvesting apparatus of higher plants is composed of several chlorophyll a/b binding complexes. Apart from LHC-II there are the minor CP29, CP27, CP24 and LHC-I.

The CP24 chlorophyll a/b complex has recently been identified (1) but its composition, structure and function need to be characterized. CP24 has been suggested to have a multisubunit composition and act as a linker between the light-harvesting antenna and the core of PSII (1) while others have associated it with the light-harvesting apparatus of PSI (2).

In this study we have characterized a 20 kDa polypeptide by a combination of biochemical and molecular genetical approaches and demonstrated that it is the only chlorophyll-binding polypeptide of the CP24. A full account of the data will be presented elsewhere (3).

145

MATERIAL AND METHODS

PSII enriched thylakoid membranes were isolated from spinach according to (4). To remove extrinsic proteins the PSII enriched membranes were treated with 3M NaSCN, 0.01% Triton X-100 (5). Stroma thylakoid vesicles were isolated as in (6). Partly denaturing SDS-PAGE was run as in (1) using octylglycoside and small amounts of SDS to solubilize PSII membranes in the presence of 40% glycerol. Denaturing 12-23% polyacrylamide gels containing 4M urea were run according to (7). The 20 kDa polypeptide was eluted from preparative polyacrylamide gels and injected into rabbit to obtain a monospecific antiserum. Immunoblotting was performed using standard procedures. Immunogold electron microscopy of thin sections of spinach leaves were performed as in (8). Isolation and sequencing of cDNA clones were performed essentially as in (9). Hydropathy plotting was performed according to (10) with an 11-point moving average.

RESULTS AND DISCUSSION

When PSII enriched thylakoid membranes are fractionated by denaturing SDS-PAGE in the presence of 4M urea, a distinct and heavily Coommassie-blue stained polypeptide with an apparent molecular weight of 20 kDa can be resolved (Fig. 1). Antibodies raised against this polypeptide were found to be highly monospecific. Moreover, when immunoblotting was performed after mild SDS-PAGE, which resolves the various chlorophyll-proteins, the antibody did only react with the green CP24 band. When CP24 was re-electrophoresed under denaturing conditions several polypeptides including the 20 kDa polypeptide were resolved. However, when the PSII enriched membranes were washed with 3M NaSCN, 0.1% Triton X-100 to remove loosely bound hydrophobic proteins prior to the two-step electrophoretic analyses of CP24 only the 20 kDa polypeptide was obtained (Fig. 1). Since no chlorophyll was released during the washing procedure, we conclude that the 20 kDa polypeptide is the only chlorophyll binding polypeptide of CP24.

Fig. 1. SDS-PAGE
a. PSII membranes,
b. excised CP24
(narrowed lane),
c. PSII membranes,
NaSCN treated

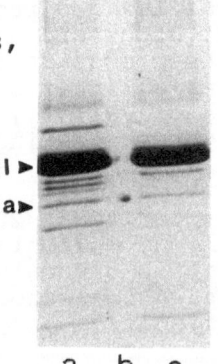

LHC-II▶
20 kDa▶

a b c

Fig. 2. Immunoblotting of a. stroma lamellae vesicles, b. grana thylakoid membranes using antibodies against the 20 kDa protein

a b

The antibody raised against the 20 kDa polypeptide was used in an immunoblot analyses of membrane fractions derived from the appressed or stroma exposed thylakoid regions (Fig. 2). The result shows that CP24 is heavily enriched in the appressed PSII rich region while stroma thylakoids contain only very small amounts. This connection of CP24 with PSII was further substansiated by immunogold electron-microscopy (Fig. 3). In thin sections of spinach leaves treated with the antibodies the labeling was almost exclusively found in the appressed regions of the thylakoid membrane.

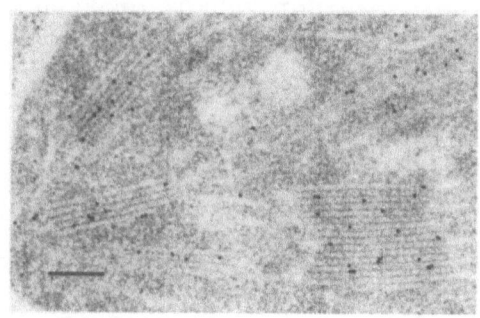

Fig. 3. Immunogold labeling of thin sections of spinach leaves using antibodies against the 20 kDa protein. Bar represents 1μm.

The antibody preparation was also used to isolate cDNA clones encoding the entire precursor protein of 261 amino acids. The 20 kDa polypeptide is nuclear encoded and contains 210 amino acids in its mature form. The transit sequence is relatively long (65 aa) and secondary structure predictions indicate a typical hydrophobic domain and an amphipathic β-sheet, typical for lumenal proteins (11), which preceed the terminal cleavage site. It should be noted that the determined amino acid sequence does not correspond to a previously reported sequence which has been ascribed to CP24 (2).

The overall homology between the chlorophyll-binding 20 kDa polypeptide and LHC-II is only about 40% but there are clearly conserved domains (60-80%) which are predominantly located in the two predicted transmembrane segments and their flanking regions of the 20 kDa protein (Fig. 4). These segments contain the sequence Ala-X-X-X-His typical for bacterial light-harvesting chlorophyll a/b proteins (12). Interestingly, the two predicted membrane spans of the 20 kDa polypeptide can only be aligned with the first and third α-helix of LHC-II predicted in (13). If correct, the surprising consequence would be that almost identical helices might be arranged either antiparallell as in CP24, or parallell as in LHC-II. Alternatively, the model needs to be revised. One alternative model for LHC-II proposed by Anderson and Goodchild (14) predicts four transmembrane helices, out of which 1 and 4 correspond to the spans predicted for CP24 in this work.

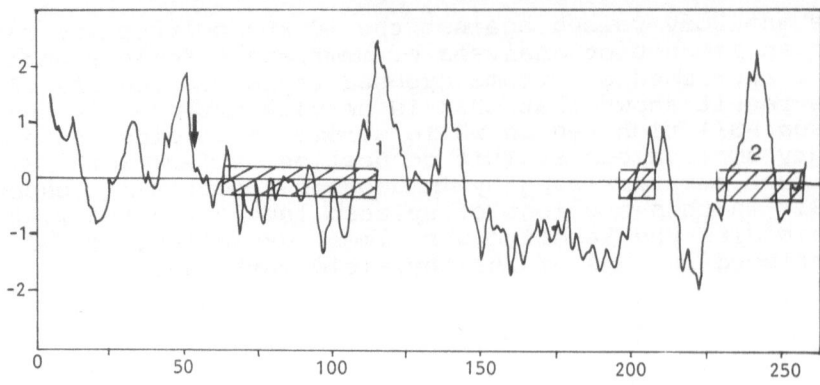

Fig. 4. Hydropathy analysis of the precursor 20 kDa protein. Boxed areas and arabic numbers indicates regions of homology with LHCII and putative membrane spanning segments. Arrow indicates cleavedge site of transit peptide.

ACKNOWLEDGEMENTS

This work was supported by the Swedish Natural Science Research Council, the German Science Foundation and the Fonds der Chemische Industrie.

REFERECENCES

1. T. G. Dunahay and L. A. Staehelin, Plant Physiol. 80:429 (1986)
2. M. M. Stayton, P. Brosio and P. Dunsmuir, Plant. Mol. Biol. 10:127 (1987)
3. N. Wedel, R. Klein, D. Bartling, R. G. Herrmann, M. Spangfort, U. K. Larsson, U. Ljungberg and B. Andersson, (in preparation).
4. D. A. Berthold, G. T. Babcock and C. F. Yocum, FEBS Lett. 134:231 (1981).
5. U. Ljungberg, H. -E. Åkerlund and B. Andersson, Eur. J. Biochem. 158:477 (1986).
6. U. K. Larsson, B. Jergil and B. Andersson, Eur. J. Biochem. 136:25 (1983).
7. U. K. Laemmli, Nature (Lond.) 227:680 (1970).
8. M. Ryberg and K. Dehesh, K., Physiol. Plant. 66:616 (1985).
9. T. Jansen, H. Reiländer, J. Steppuhn and R. G. Herrmann, Curr. Genet, 13:517 (1988).
10. J. Kyle and R. F. Doolittle, J. Mol. Biol. 157:105 (1982).
11. A. Tyagi, J. Hermans, J. Steppuhn, C. Jansson, F. Vater and R. G. Herrmann, Mol. Genet. 207:289 (1987).
12. H. Zuber in: "Encyclopedia of Plant Physiology, (L.A. Staehelin, C. J. Arntzen, eds., Springer-Verlag, Berlin) vol. 19, pp. 238-251 (1986).
13. G. A. Karlin-Neumann, B. D. Kohorn, J. P. Thornber and E. M. Tobin, J. Mol. Appl. Genet. 3:45 (1985).
14. J. M. Anderson and D. J. Goodchild, FEBS Lett. 213:29 (1987).

LHC II-APOPROTEINS OF <u>CHLAMYDOMONAS REINHARDII</u>:

ISOLATION, ISOELECTRIC FOCUSING, ASSOCIATION WITH LIPIDS

M. Sigrist, C. Zwillenberg-Fridman, Ch. Giroud,
W. Eichenberger and A. Boschetti

Institut für Biochemie, Universität Bern
Freiestrasse 3, CH-3012 Bern, Switzerland

INTRODUCTION

Light-harvesting chlorophyll a/b-protein complexes of the photosystem II (LHC II) are major components of thylakoids, and they contain chlorophyll a and b, xantophylls and 2-5 hydrophobic proteins, the LHC II-apoproteins.[1] In the green alga <u>Chlamydomonas reinhardii</u>, there exist three immunologically cross-reacting LHC II-apoproteins of 24, 25 and 29 kD.[2] Up to now, only phosphorylation and the blocked N-terminus are known as secondary modifications. We developed a method for the isolation of the LHC II-apoproteins on a preparative scale and characterized them by isoelectric focusing (IEF) and analysis of secondary modifications, such as binding of lipids or fatty acids. Furthermore, the apoproteins of a chlorophyll b-deficient mutant (pg 113),[2] from which the LHC II-complex can not be isolated, have been studied.

METHODS

Pigment-free LHC II-apoproteins were isolated by extraction of thylakoids with chloroform, methanol and trifluoracetic acid according to a modified procedure of Brunisholz.[3] For fatty acid analysis, we used the common methods of transesterification[4] and gas liquid chromatography.[5] Analysis of carbohydrates was described by Alpenfels.[6] IEF was performed in tubes containing 2% Servalyt 2-11, 3% polyacrylamide, 8 M urea and 2 % Nonindet NP 40.

RESULTS AND DISCUSSION

From 20 l of a culture of <u>Chlamydomonas reinhardii</u> containing 2×10^6 cells/ml, the isolation of the LHC II-apoproteins yielded about 10 mg of colorless protein, from both the parent strain and the mutant pg 113. Electrophoretic analysis of the product showed a mixture of three polypeptides with apparent molecular weights of about 24, 25 and 29 kD (Fig.1,a-c). They were also identified by immunostaining (Fig.1,d+e). Isoelectric focusing of the mixture of the LHC II-apoproteins resulted in more than 10 bands (Fig.2). The proteolytic fragmentation

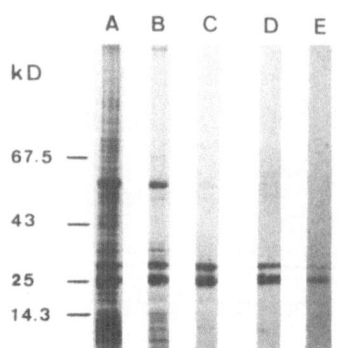

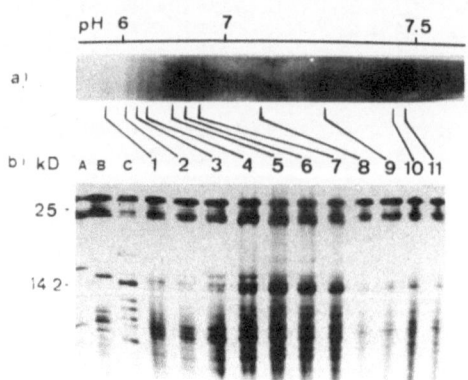

Fig.1: Purification of
the LHC II-apoproteins
from C. reinh. (parent
strain). SDS-PAGE on a
7.5-15% gel, stained
with Coomassie.
A: Cells, B: Washed
thylakoids, C: Final
product, D and E:
Immunostaining of the
proteins from lane C
with antibodies against
the 25 and 23 kD LHC II-
apoproteins.[2]

Fig.2: IEF of the 3 isolated LHC
II-apoproteins and proteolytic
fragmentation of the separated
bands. a) Isolated LHC II-apoproteins as characterized in
Fig.1c were subjected to IEF on
rod gels. b) The bands separated by IEF were cut out and the
proteins digested with S.aureus
V8-protease.[2] Shown are the
silver-stained fragmentation patterns on SDS-PAGE (17% acrylamide). A-C: Fragmentation of 24,
25 and 29 kD-proteins. 1-11:
Fragmentation of bands from IEF.

patterns of each of these bands showed that (a) each LHC II-
apoprotein occured in several IEF-bands and (b) most IEF bands
contained at least 2 different apoproteins. We don't know
whether these bands reflect merely stoichiometrically different
aggregates of the 3 LHC II-apoproteins or different chemical
modifications (e.g. phosphorylation) of the polypeptides.

The isolation procedure for the LHC II-apoproteins includes
a gel filtration in a very strong lipid solvent. Therefore, no
lipid would be expected to be associated anymore to the proteins
present in the exclusion volume after gel filtration. Surprisingly, the analysis of the isolated proteins revealed a considerable amount of fatty acid, predominantly palmitic acid, associated with the proteins (Table 1). Assuming a mean molecular
weight of 26 kD for each of the three LHC II-apoproteins, about
4 molecules of fatty acids per molecule of protein were found.
In order to decide, whether these fatty acids were linked to the
polypeptides themselves or to the protein-associated membrane

Table 1. Composition of fatty acids found in the LHC II-
apoproteins. The molecular weight was assumed as 26 kD.

| | mol fatty acid per mol protein | |
	parent strain	pg 113
16:0	3.0 (79 %)	2.7 (69 %)
18:0	0.05 (1 %)	0.15 (4 %)
18:1	0.3 (9 %)	0.34 (9 %)
18:3	0.1 (3 %)	not determinated
Total	3.8	4.0

Table 2. Fatty acid composition of the lipids separated by TLC. Fraction a: Material left at the starting point. Fraction f: Material migrating with the front (average of 4 determinations):

Fraction	amount of 16:0 in the separated lipids		amount of all fatty acids in the separated lipids	
	p.s.	pg 113	p.s.	pg 113
a (protein)	3.9 %	7.2 %	5.2 %	10.2 %
b (DGDG,PI)	7.0 %	14.4 %	6.1 %	14.8 %
c (SQDG)	55.2 %	60.8 %	44.8 %	36.5 %
d (PE,PG)	24.8 %	5.2 %	22.5 %	7.4 %
e (MGDG, DGTS)	0.4 %	2.1 %	5.2 %	6.3 %
f (pigments, free fatty acids)	8.7 %	10.3 %	16.2 %	24.4 %
Total	100 %	100 %	100 %	100 %

lipids, extracted proteins were chromatographed on a TLC-plate together with a total lipid extract from Chlamydomonas reinhardii as a standard. The silica gel containing the presumptive LHC II-associated lipids was scraped off. The fatty acids in these samples were analysed by transesterification and gas liquid chromatography (Table 2). The starting point (fraction a), where the protein has been retained, contained less than 10 % of the total fatty acids originally associated with the proteins. The spot with the greatest amount of fatty acid was that corresponding to SQDG, the sulfoquinovosyldiacylglycerol. It was shown to contain more than 55 % of the palmitic acid. When analyzing the fractions d, e and f (Table 2) for free fatty acids using diazomethane for esterification, we did not find significant quantities of free fatty acids in those regions in which they are normally found in TLC. Moreover, no fatty acids could be detached from the delipidated protein (fraction a, Table 2) by transesterification with methanolic HCl, showing that no amide-bound acyl groups were present. Obviously, lipids consisting mainly of SQDG are bound to the LHC II-apoproteins exclusively by hydrophobic interactions, and this in both the parent strain and the chlorophyll b-deficient mutant.

To ascertain the identity of the bound lipids, the carbohydrates were released from the isolated proteins and analyzed as dansyl hydrazones by HPLC (Fig. 3 and 4). Obviously, the dansyl hydrazone derived from the LHC II-apoproteins is a derivative of sulfoquinovose, the carbohydrate moiety of SQDG. By use of mannose as an external standard, we calculated 1.8 moles of SQDG per mol of protein, assuming a mean molecular weight of 26 kD for each of the three LHC II-apoproteins. This corresponds well to the content of SQDG calculated from the fatty acid analysis (Table 1). The carbohydrate analysis indicates also that there are no carbohydrates covalently attached to the LHC II-apoproteins.

From our experiments we can conclude that the three LHC II-apoproteins are not covalently modified with fatty acids, lipids or carbohydrates, but that there are 1-2 moles of sulfolipid tightly associated to one mole of them.

Much has been speculated on the function of SQDG.[7] The fact that all lipids in the membranes, except SQDG, are accessible to lipase attack, was explained by a tight binding and shedding of SQDG by proteins.[8] More recently, it has been re-

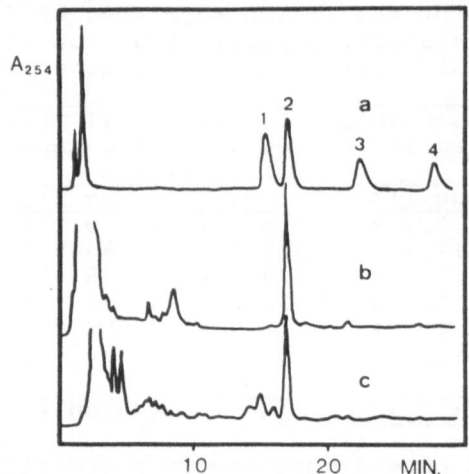
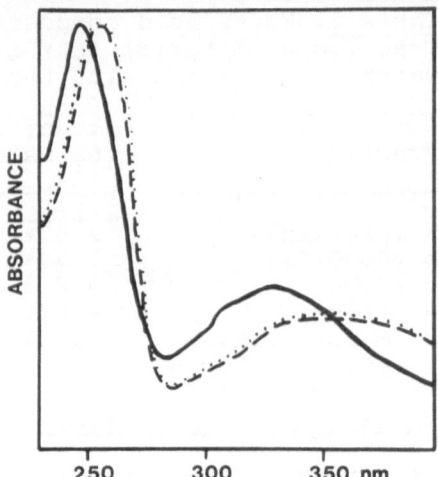

Fig.3. Separation of the dansyl hydrazones by HPLC on a Spherisorb S5-ODS column with 22% acetonitril as eluent.[6] a) Dansyl hydrazones of galactose (1), mannose (2), xylose (3) and fucose (4). b) SQDG derivated with dansyl hydrazine. c) LHC II-apoproteins derivated with dansyl hydrazine.

Fig.4. UV/VIS absorption spectra of the eluted dansyl hydrazones.
a) Dansyl hydrazone of mannose (———)
b) Main peak of Fig.3b (— — — —)
c) Main peak of Fig.3c (• • • • • • •)

ported that the CF_o/CF_1-ATPase contains almost exclusively sulfolipids, which are required for ATPase activity.[9] The reaction centre of PS II was also shown to be enriched in SQDG.[10] However, addition of an excess of SQDG together with PG to this preparation resulted in an inhibition of its activity.[11] In higher plants, SQDG has a concentration of about 12% of the lipids. It is, together with PG and PI, the only lipid with a net negative charge, and for this reason, it may play a specific role in the structural organisation of the membrane.

AKNOWLEDGEMENT

This work has been supported by the Swiss National Foundation for Scientific Research.

REFERENCES

1. H. Zuber, Photochem. Photobiol. 42:821 (1985)
2. H. Michel, M. Tellenbach and A. Boschetti, BBA 725:417 (1983)
3. R. Brunisholz et al., FEBS Lett. 129:150 (1981)
4. M. Kates, "Techniques of lipidology," Elsevier, Amsterdam (1986)
5. C. Giroud, A. Gerber and W. Eichenberger, Plant and Cell Physiol. 29 (1988), in press.
6. W.F. Alpenfels, Anal. Biochem. 114:153 (1981)
7. J. Barber and K. Gounaris, Photosynth. Res. 9:239 (1986)
8. A. Shaw, M. Anderson and R. Mc Carty, Plant. Physiol. 57:724 (1976)
9. U. Pick et al., BBA 808:415 (1985)
10. K. Gounaris and J. Barber, FEBS Lett. 188:68 (1985)
11. K. Gounaris, D. Whitford and J. Barber, FEBS Lett. 163:230 (1983)

EXCITATION ENERGY DISTRIBUTION IN AN ORGANISM WITH A

CHL A/C/CAROTENOID LIGHT HARVESTING COMPLEX

Pamela B. Gibbs and John Biggins

Division of Biology and Medicine

Brown University, Providence, RI, USA

INTRODUCTION

Photosynthetic organisms have the capacity to regulate the distribution of excitation energy between PS2 and PS1 via a mechanism termed the light state transition (4, 11). It has been argued that this phenomenon serves to maintain equal turnover of PS2 and PS1 ensuring efficient linear electron transport . To date, state transitions have been described in organisms that use either a Chl a/b LHC or a phycobilisome as light harvesting antenna and models have been proposed to describe the state transition with regards to the particular type of antenna complex employed (see 1, 2, 7 and 13). However, not all photosynthetic algae use these types of light harvesting complex. A Chl a, Chl c and carotenoid antenna complex is used by the various members of the Chromophyta. These algae are taxanomically diverse (Bacillariophyta, Dinoflagellata, Chrysophyta, Phaeophyta, Prymnesiophyta) and contribute a major fraction to the primary productivity of the oceans. However, surprisingly few studies have been concerned with the details of photosynthesis in these organisms (for review see 10).

The unicellular Chrysophyte, *Ochromonas danica* uses Chl a, Chl c and the carotenoid fucoxanthin as light harvesting pigments. These pigments are most likely complexed to proteins which are located within the thylakoid membranes although the details of their arrangement are unknown (5, 6). The thylakoids themselves have a unique organization common to all members of the Chromophyta in that thylakoids are invariably arranged in bands of three and a girdle lamellae is present (8). The extent and nature of thylakoid interaction and thylakoid heterogeneity is also currently unknown.

In a previous report conducted on the diatom *Phaeodactylum tricornatum*, Owens (12) concluded that excitation energy distribution might be dependent on light intensity rather than wavelength. However, in this report we demonstrate that preferential illumination of PS2 and PS1 leads to changes in both the room temperature PS2 fluorescence and the low temperature (77K) chlorophyll fluorescence emission spectra (8).

METHODS

O. danica (UTEX) was grown photoheterotrophically on a modified Ochromonas medium (8) at 23 $^\circ$. These conditions led to growth and light dependent oxygen evolution as assayed polarographically. Cells were provided 35 uE m^{-2} s^{-1} light and grown on a 12 hour light/dark cycle. Whole cells were pre-illuminated as indicated in the figure legends and 77K fluorescence emission was measured as previously described (3). Room temperature fluorescence was measured using a system similar to that described by Bonaventura and Myers (4).

RESULTS AND DISCUSSION

In Fig. 1, 77K fluorescence emission spectra of *O. danica* cells are shown. We have tentatively concluded that the 2 peaks in the emission spectra represent fluorescence emission from PS2 (~690 nm) and PS1 (~720 nm). An excitation wavelength at 77K of 500 nm was selected as it is strongly absorbed by the carotenoid, fucoxanthin. Excitation spectra for peaks at 690.5 nm and 720.5 nm indicate that Chl a, Chl c and fucoxanthin transfer absorbed light energy to species fluorescing at these wavelengths (Fig. 2).

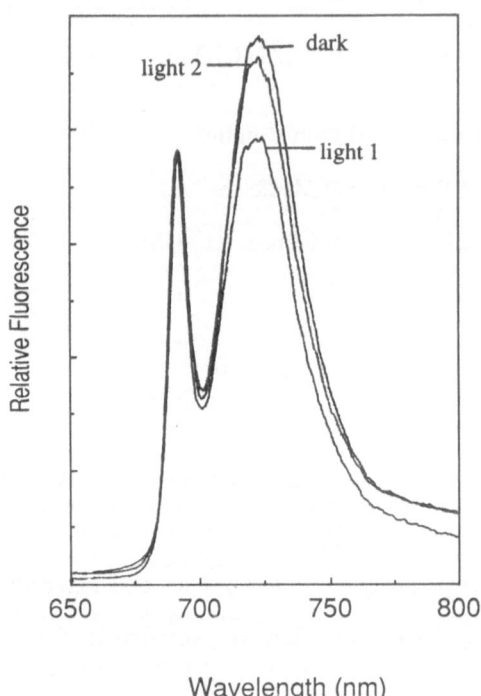

Figure 2. Fluorescence emission spectra (77K) of *O. danica* . Cells were pre-illuminated for 10 minutes with light 2 (640 nm) or light 1 (700 nm) (55 uE m^{-2} sec^{-1}) or dark adapted before being rapidly frozen for 77K fluorescence analysis. Excitation wavelength was 500 nm. Fluorescence emission spectra were normalized at the 690 nm maximum.

The relative emission from the 690 nm and 720 nm peaks was dependent on pre-illumination conditions in a manner indicative of a light state transition. When cells were pre-illuminated with light 2 (640 nm), there was a relative increase in fluorescence eminating from the 720 nm peak. In the converse situation, pre-illumination with light 1 (700 nm) led to a relative increase in the fluorescence yield of the 690 nm peak. It should also be noted that the spectra of dark adapted cells was similar, but not always equivalent, to that of light 2 adapted cells. These changes were reversible and were consistent with the state 1/state 2 transitions previously observed in organisms which use phycobilins or Chl b as accessory pigment following preferential illumination of PS2 and PS1 (3, 7, 11).

Conditions which led to reversible changes in the 77K fluorescence spectra also induced changes in the room temperature fluorescence yield of PS2. A room temperature fluorescence signal (680 nm) was induced by modulated light 2 (500 nm) and the yield of this fluorescence signal was quenched by the addition of background light 1. The fluorescence traces from a typical experiment are shown in Fig. 3 in which cells are initially brought to state

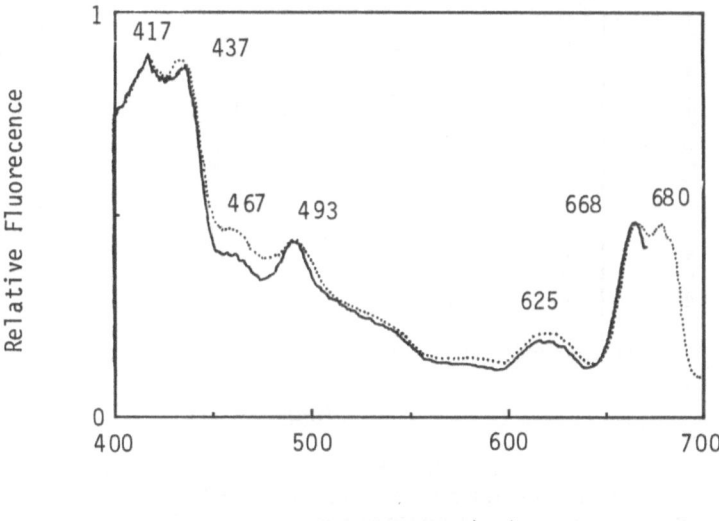

Wavelength (nm)

Figure 2. Fluorescence excitation spectra (77K) of *O. danica* cells. Cells were removed from the growth changer and rapidly frozen in liquid nitrogen. PS1 emission (__), was measured at 690.5 nm and PS2 emission (.....), measured at 720.5 nm. Spectra were normalized at the 417 nm maximum. Maxima have been tentatively assigned to the following species: Chl a (417 nm, 437 nm, 668 nm, 680 nm), Chl c (467 nm, 625 nm) and fucoxanthin (493 nm).

1 by illumination with modulated light 2 and background light 1. When light 1 was removed, there was a rapid increase in the fluorescence yield which was followed by a slow quenching. The restoration of background light 1 caused a rapid quenching in the fluorescence signal.

To confirm that these room temperature changes represented a redistribution of excitation energy, samples were removed at the indicated times for 77K analysis and the results of this experiment are presented in Tab. 1. Initially, the F690/F720 ratios were typical of state 1. The removal of background light 1 led to the conversion to state 2 (low F690/F720 ratio). Upon restoration of light 1, both the room temperature changes and the changes in the 77K emission spectra were reversed.

Detailed kinetic studies performed at room temperature and analyzed at 77K indicated that the half times for the state 1 to state 2 conversion was 45 sec - 2.0 min and that for the state 2 to state 1 transtion was 20 sec to 1.5 min (data not shown). Although the times required for the state transition were slower than those documented for red algae (2), they were similar to those we previously observed in green algae (unpublished data). The times required for the changes in the 77K emission spectra (~ 5 min) were kinetically consistent with the slow quenching phase of the 680 nm signal at room temperature. Therefore, rapid changes in the room temperature fluorescence signal following the removal or restoration of light 1 may be indicative of the redox status of Qa as has been observed for other systems. This proposal is supported by the observation that 1 uM DCMU prevented the state 2 and led to a rapid rise in the 680 nm fluorescence signal at room temperature (data not shown).

In conclusion, preferential excitation of PS2 or PS1 does lead to changes in both the PS2 room temperature fluorescence yield and the corresponding low temperature fluorescence emission spectra of *O. danica* cells. Confirmation of the photosynthetic significance of these

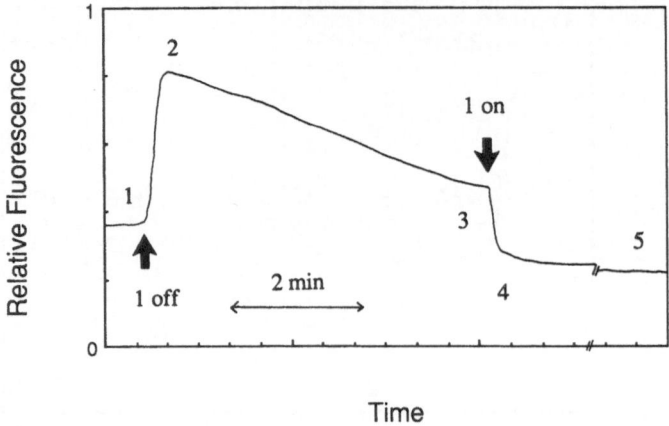

Time

Figure 3. Room temperature fluorescence trace of *O. danica* cells showing the fast and slow fluorescence changes at 680 nm. The cells were initially brought to state 1 in response to modulated light 2 (500 nm, 15 uE m-2 sec-1) and continuous background light 1 (720 nm, 650 uE m-2 sec -1). Removal and restoration of light 1 are indicated by the arrows. At the times indicated, samples were withdrawn for 77K analysis.

Table 1. The normalized F690/F720 ratios for the 77K analysis of the room temperature fluorescence trace shown in Fig. 4.

Sample	F690/F720
1. light 1 + 2, 5 min	0.95
2. light 2, 15 s	0.89
3. light 2, 5 min	0.80
4. light 1 + 2, 15 s	0.83
5. light 1 + 2, 5 min	0.99

observations will require analysis of oxygen evolution rates and changes in the turnover of PS2 and PS1. It remains to be determined whether wavelength dependent state transtions in the chromophytes will occur via a mechanism similar to that used by organisms with phycobilisomes, the Chla/b LHC or one unique to organisms with Chla/c/carotenoid antenna.

References

1. Allen JF, Bennett J, Steinback KE and Arntzen CJ (1980), Nature, 291:25-29.
2. Biggins J and Bruce D (in press), Photosyn Res.
3. Biggins J (1983), Biochim Biophys Acta, 724:111-117.
4. Bonaventura C and Myers J (1969), Biochim Biophys Acta, 189:366-383.
5. Brown JS, Alberte RS and Thornber JP (1974), In: Avron M, ed., Third International Congress on Photosynthesis, pp. 1951-1962, Amsterdam, Elsevier Scientific Publishing Company.
6. Caron L and Brown J (1987), Plant Cell Physiol 28,:775-785.
7. Fork DC and Satoh K (1986), Ann Rev Plant Physiol, 37:335-366.
8. Gibbs PB and Biggins J (submitted).
9. Gibbs SP (1970), Ann N.Y. Acad Sci, 175:454-473.
10. Larkum AWD and Barrett J (1985), Adv in Bot Res, 10:1-210.
11. Murata N (1969), Biochim Biophys Acta, 172:242-251.
12. Owens TG (1986), Plant Physiol, 80:739-746.
13. Williams WP and Allen JF (1987), Photosyn Res, 13:19-45.

SECTION II

MEMBRANES AND ELECTRON TRANSPORT

REGULATION OF THYLAKOID MEMBRANE STRUCTURE AND FUNCTION BY SURFACE ELECTRICAL CHARGE

J. Barber

AFRC Photosynthesis Research Group
Biology Department, Imperial College
London SW7 2BB, UK

INTRODUCTION

The thylakoid membrane of chloroplasts has many features which distinguish it from other biological membranes. It contains the molecular machinery, in the form of protein complexes, which is able to intercept light energy and efficiently convert it into chemical potential. It is also characterised by its unusual lipid composition (1). The dominating lipids of this membrane system are the electroneutral species, monogalactosyldiacylglycerol (MGDG) and digalactosyldiacylglycerol (DGDG) with minor lipids being the negatively charged phosphatidylglycerol and sulpholipid, sulphoquinovosyldiacylgycerol. Despite these difference, however, this chloroplast membrane shares the same common property as other biological membranes in that at physiological pH its surfaces carry excess negative electrical charge. This negative charge seems to be due mainly to carboxyl groups of glutamic and aspartic acid residues of exposed segments of integral membrane proteins with little or no contribution from the head groups of the acidic lipids mentioned above (2). Below pH 4.3 the surface becomes positively charged. The origin of this net positive charge seems to result from the guanidine group of exposed arginine residues (3).

In this article I wish to focus attention on the surface electrical properties of the thylakoid membrane in order to present basic concepts and at the same time emphasise how these concepts have been used in my laboratory to interpret experimental results, especially in terms of membrane organisation and dynamics.

THE ELECTRICAL DOUBLE LAYER

When an electrically charged surface is in contact with a liquid medium it attracts ions of opposite charge (counterions) and repels ions of the same charge (coions). Therefore the counterion concentration progressively increases towards the surface and the coion concentration progressively decreases (see Fig.1A and C). This region of unequal positive and negative ion concentration near a charged surface is known as the diffuse electrical double layer. The asymmetric charge distribution is created in this layer because of the development of electrical potential profile which falls off with distance and is a consequence of fixed surface charges (see Fig 1B.). The idea that counterions form a diffuse double layer, rather than a non-diffuse single plane immediately adjacent to the surface (i.e. a Helmholtz double layer), was proposed by Gouy in 1910 (ref. 4), while the appropriate mathematical formulation, using the Poisson equation, was developed by Chapman in a paper on electrocapillarity in 1913 (ref. 5). Therefore, the following theoretical treatment of the diffuse double layer is attributed to these two workers and is often known as the Gouy-Chapman Theory.

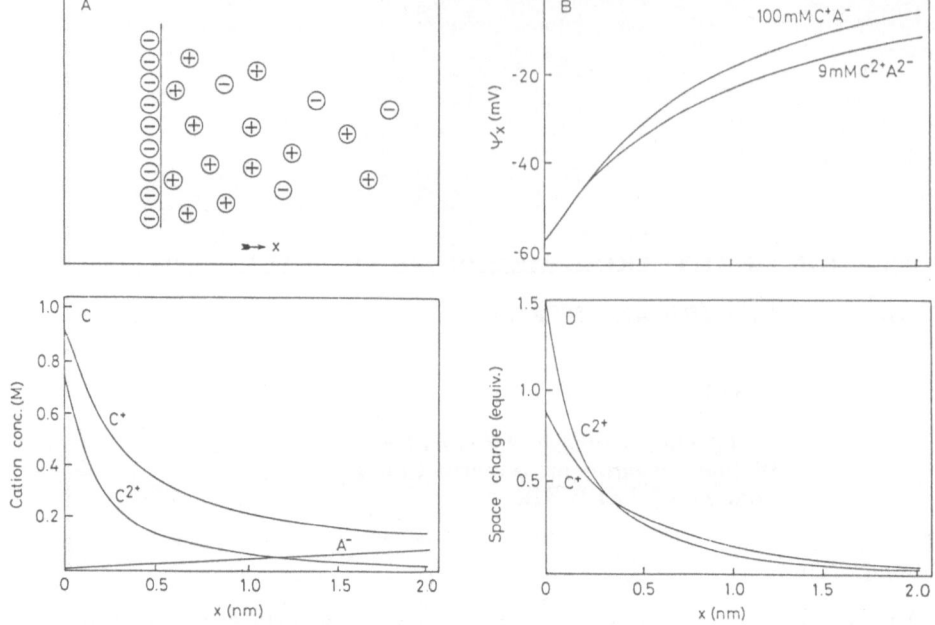

Fig. I   (A) Schematic diagram of the distribution of ions near a negatively charged surface. (B)$_2$ Electrical potential profiles at a negatively charged surface of -0.05 C/m^2 immersed either in 100 mM C$^+$A$^-$ or 9 mM C^{2+}A^{2-}. The curves were calculated using Eqn(11). (C) Curves showing the distribution of mono- and divalent ions in the diffuse layers generated in the presence of the two salt conditions. The curves were computed from Eqn.(4) and the values of A$^-$ are too low to be included. (D) The two curves express the amount of positive space charge density at various distances from the surface for the two salt conditions calculated using Eqn.(2) and the data in (C). The curves demonstrate the effectiveness of divalent cations to increase the space charge density near the surface relative to the effectiveness of mono-valent cations even when the surface potential is the same for both salt conditions.

The G-C theory is derived by combining the equations of Poisson and Boltzmann. For a planar electrically charged surface the Poisson equation is given by

$$d^2\psi_x/dx^2 = -\rho_x/\varepsilon_r\varepsilon_o \qquad (1)$$

where ψ_x is the electrostatic potential at distance x from the surface relative to the potential of the bulk medium (i.e. $\psi_x = 0$), ε_r is the relative permittivity of the solution and ε_o is the permittivity of a vacuum and has the value 8.854 x 10^{-12} C^{2}J^{-1} m^{-1}. It is usually assumed that the dielectric constant or permittivity, ε_r, remains constant throughout the aqueous phase including the diffuse layer. ρ_x is the net space charge density of ions in solution in a plane parallel to the membrane surface at distance x. Clearly this latter quantity is given by

$$\rho_x = \Sigma_i Z_i FC_{ix} \qquad (2)$$

where Z_i is the charge carried by ion i, F is the Faraday constant and C_{ix} is the concentration of ion i in the diffuse layer.

The Boltzmann equation simply stems from the fact that any ion i in the diffuse layer must be in thermodynamic equilibrium with the same species in the bulk solution.

$$\mu_{ix} = \mu_{i\alpha} \qquad (3)$$

where $\mu_{i\,x}$ is the electrochemical potential of species i at a point x near the surface while $\mu_{i\,\alpha}$ is the electrochemical potential of same species at an infinite distance from the membrane (i.e. in the bulk). Assuming that μ_i is completely described by concentration and electrical terms and neglects pressure terms and differences in activity coefficients, the distribution of ion i at any point x from the surface is given by

$$C_{i\,x} = C_{i\,\alpha} \exp(-Z_i F\psi_x /RT) \tag{4}$$

Combining equations 1,2 and 4 yields the non-linear Poisson-Boltzmann relationship

$$\frac{d^2\psi_x}{dx^2} = -1/\varepsilon_r \varepsilon_o \, \Sigma_i \, ZFC_{i\,\alpha} \, \exp(-Z_i F\psi_x /RT) \tag{5}$$

By recognising that $d\psi/dx \rightarrow O$ as $x \rightarrow \alpha$ and $\psi_x \rightarrow O$, this equation can be integrated to give

$$\frac{d\psi_x}{dx} = \pm \, [2RT/\varepsilon_o \varepsilon_r \, \Sigma_i \, C_{i\,\alpha} \, (\exp(-Z_i F\psi_x /RT) -1)]^{1/2} \tag{6}$$

$d\psi_x/dx$ is the field strength and can be related by the Gauss equation to the electrical charge density (σ) on the surface.

$$(d\psi_x/dx)_{x=0} = -\sigma/\varepsilon_r \varepsilon_o \tag{7}$$

Therefore combining equations (6) and (7) yields

$$\sigma = [2RT\varepsilon_r \varepsilon_o \Sigma_i \, C_{i\,\alpha} \, (\exp(-Z_i F\psi_o /RT) -1]^{1/2} \tag{8}$$

where ψ_o is the electrical potential at the surface (i.e. $x=0$).

When the surface is bathed in a medium containing a Z-Z electrolyte such as KCl or $MgSO_4$ equation (8) reduces to

$$\sigma = 2A \, (C_{i\,\alpha})^{1/2} \, \sinh(ZF\psi_o /2RT) \tag{9}$$

where $A = (2RT\varepsilon_r \varepsilon_o)^{1/2}$

Numerical substitution assuming $T = 298$ K gives

$$\sigma = 0.1174 \, (C_{i\,\alpha})^{1/2} \, \sinh(ZF\psi_o /51.7) \tag{10}$$

where σ is in C m^{-2}, $C_{i\,\alpha}$ is in mol dm^{-3} and ψ_o is in mV.

ψ_x decays to zero with increasing distance perpendicular to the plane of the surface and its value can be determined at any point x by integrating equation 6.

For a Z-Z electrolyte the integration gives

$$\kappa x = \ln[\tanh(ZF\psi_o /4RT)] - \ln[\tanh(ZF\psi_x /4RT)] \tag{11}$$

where $\kappa = [2Z^2 F^2 C_{i\,\alpha}/RT\varepsilon_o \varepsilon_i]^{1/2}$

Equation 11 has been used to calculate the curves shown in Fig. 1(B) where σ has been taken to be -0.05 C m^{-2} and the electrolyte level as either 100 mM $C^+ A^-$ or 9 mM $C^+ A^{2-}$. Fig. 1(C) is the result of substituting values of ψ_x into the Boltzmann equation, (Equ.(4)), to calculate the distribution of cations and anions within the double layer while Fig. 1(D) is a replot of Fig. 1(C) but in terms of electrical equivalents (i.e. net space charge ρ_x) using Equ.(2).

These simple calculations emphasise three important features:

i) A significantly lower concentration of divalent electrolyte is required, compared to monovalent, to give the same surface potential, ψ_o (this is due to the valency terms appearing in the exponential factors of Equ.(10)).

ii) The concentration of anions in the diffuse layer is very low.

iii) Despite the fact that the electrical potential profiles are similar for the two electrolyte conditions, the distribution of net positive space charge, ρ_x, within the diffuse layer is quite different. This difference is particularly evident close to the surface especially if the integrated space charge density ρ_x' is considered, where ρ_x' is defined by:

$$\rho_x' = \int_{x=o}^{x} \rho_x \, dx \qquad (12)$$

FUNCTIONAL SIGNIFICANCE OF SURFACE POTENTIAL AND SPACE CHARGE DENSITY

From studies with isolated thylakoid membranes it has become clear that it is important to distinguish between those phenomena which relate directly to the electrical potential profile and those which are governed by the space charge density. This differentiation is aided by the fact that under certain mixed electrolyte conditions (with both mono- and di-valent salts) there is a complicated relationship between ρ_x' and ψ_x (ref.6).

The electrical potential governs the concentration of ions at the surface including H^+ and charged metabolites which may effect the conductance of the membrane as well as its specific functions. Moreover, the surface potentials on each side of the membrane are likely to be different, therefore giving rise to a trans-membrane electrical gradient. These effects are relatively easy to understand and predict (7,8). In contrast, the significance of space charge density on membrane organisation and functioning is less obvious. It turns out that it is this quantity which governs the electrostatic screening of the fixed charges on the surface. Such screening is vital if another electrically charged surface with the same polarity wishes to form a close spatial association with the membrane. The distance of approach will be a balance between electrostatic repulsive and attractive van der Waals forces (7). General expressions to estimate the van der Waals force between two macroscopic surfaces have been formulated based on the Lifshitz theory and has been considered for the specific case of biological membranes (9).

Normally the van der Waals force would remain constant under different electrolyte conditions while electrostatic screening would be variable thus changing the coulombic repulsion between the two adjacent surfaces. It is this concept which forms the basis of the Derjaguin-Landau-Verwey-Overbeek (DLVO) theory to describe the aggregation of lipophobic colloids (10). Thus it is in this way that the space charge density ρ_x' controls processes on and between biological membranes. At least three different organisational changes can be visualised as occurring when the coulombic repulsion is decreased by increasing the value of ρ_x':

i) **Membrane stacking.** Two membrane surfaces come close to each other with the minimum distance determined by the existence of structured water layers and whether any short-range specific interactions occur.

ii) **Clustering of intrinsic protein complexes**. Such an aggregation may be functionally important and is dependent on the fluidity of the lipid matrix.

iii) **Interactions between membrane surfaces and large extrinsic proteins or protein complexes.** In this case the initial approach towards the membrane surface is controlled by the long range coulombic force but additional short range forces (e.g. coulombic attraction) will be necessary to facilitate an intimate interaction or binding with a specific intrinsic membrane component.

SALT INDUCED CHANGES IN THYLAKOID ORGANISATION

When unwashed, freshly isolated thylakoid membranes are suspended in very low salt media they maintain their normal configuration of stacked and unstacked regions and have a high F_m level of chlorophyll fluorescence (F_m is the level of fluorescence when all photosystem two (PS2) traps are closed) and in this type of experiment 3-(3,4-dichlorophenyl)-1,1-dimethylurea (DCMU) is added to block electron flow. When low levels of monovalent electrolyte (e.g. 1 to 10 mM KCl) are introduced into the suspension the membranes totally unstack and the F_m level is lowered (11,12).

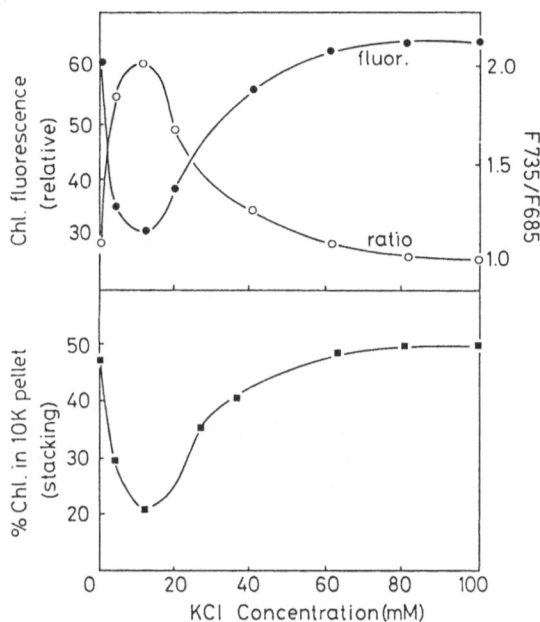

Fig.2   Effect of KCl concentration on the chlorophyll fluorescence at room temperature (closed circles) the F735/F685 at 77 K (open circles) and the degree of thylakoid stacking (closed squares).

For the experiment shown in Fig.2 (see ref.13) the degree of stacking was monitored by the digitonin method. The experiment also shows that these changes are accompanied by an increase in the level of 735 relative to 685 nm emission measured at 77K. From this condition the restacking and associated changes in the F_m level and F735/F685 ratio can be accomplished by the further addition of KCl. The changes in the F735/F685 have been shown to be due to changes in energy transfer from PS2 to photosystem one (PS1) since F685 originates from PS2 and F735 from PS1. Restacking and chlorophyll fluorescence changes can also be induced by adding divalent or trivalent cations as shown in Fig.3, but much lower concentrations are required as compared with the levels necessary with monovalent salts (15). These changes were found not to be influenced by the nature of the anion or associated changes in osmotic strength (14). In fact the order of effectiveness of different cations ($C^{3+} > C^{2+} > C^+$) and the independence of the chemical species within a valency group (all alkali metal cations gave essentially the same response as did the alkaline earth cations) are clear indications that the rather complicated changes in organisation and fluorescence properties of the thylakoid membrane are governed by electrostatic properties of the system. The dependency on valency could, however, imply that either ψ_o or ρ_x' is the controlling factor but there are two observations which suggest that the latter parameter underlies the salt induced effects. Firstly the antagonistic action between low levels of C^{2+} and C^+ can be explained theoretically in terms of ρ_x' if at the beginning of the experiment described above the thylakoid membrane had divalent cations within its diffuse layer. Analysis of the ionic composition of intact chloroplasts does indeed suggest that this is the case (16). Therefore as discussed above, the divalent cations within the diffuse double layer tend to distribute themselves closer to the membrane surface than monovalent cations and therefore are better at electrostatic screening. Addition of low levels of monovalent cations displaces divalent cations from the double layer and decrease the space charge density close to the membrane surface. Only with higher levels of monovalent

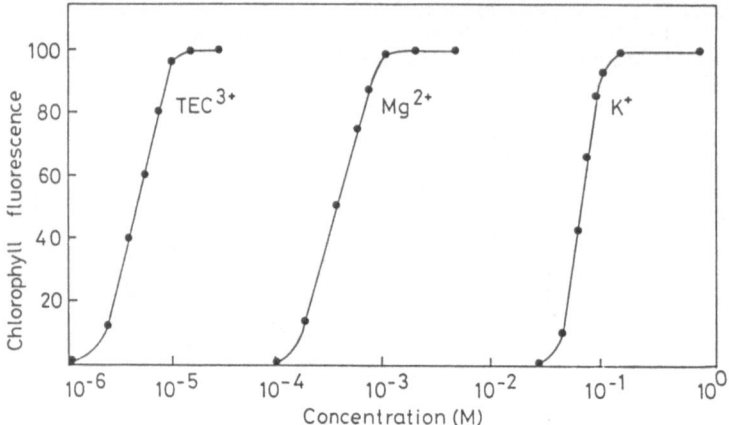

Fig.3   Concentration requirements for the cation induced increase in room temperature chlorophyll fluorescence yield from isolated pea thylakoids showing the differential effect of Tris (ethylenediamine)-cobaltic cation (TEC)$^{3+}$, Mg^{2+} and K^+. All salts were added as their chlorides (from ref.15).

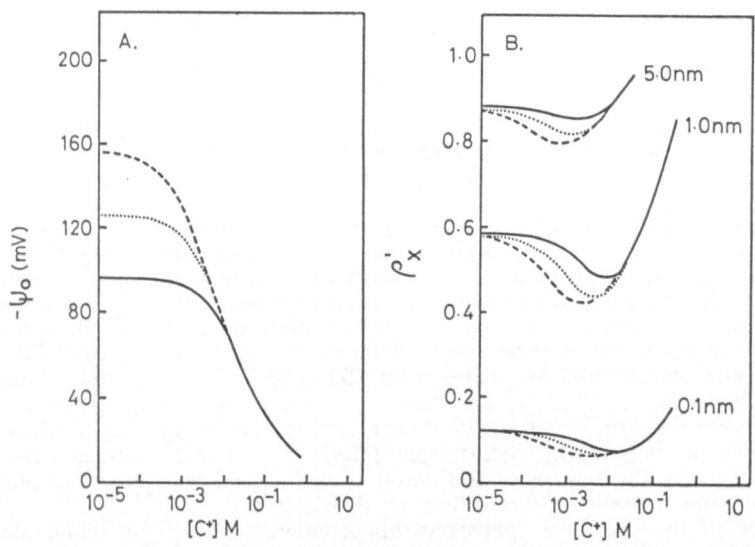

Fig.4   Computer derived curves showing how changes of the surface potential (ψ_o) and the integrated space charge density (ρ'_x) varies under different electrolyte conditions. For the determination of ρ'_x three volumes have been considered per unit area between x = 0.1 nm, 1.0 nm and 5.0 nm and the calculations have used expressions given in ref. (6). Three background divalent levels have been considered (dashed 10^{-6}m, dotted 10^{-5} and solid 10^{-5} m) and a surface charge density of -0.025 c m^{-2} used.

164

electrolytes present, when the value of ψ_o significantly decreases, does the value of ρ'_x increase to the original level. This "dip" in the value of ρ'_x with increasing monovalent cations levels is shown in Fig.4 for three different background levels of divalent salts.

The calculated curves of Fig.4 also indicate the surprising fact that the "dip" phenomenon stretches well out into the diffuse layer. If similar calculations are conducted but with only monovalent cations in the diffuse layer, no such minimum in ρ'_x is observed.

A second and more direct indication that the salt induced phenomena follow changes in ρ'_x and not ψ_x comes from measuring the fluorescence from chlorophyll and 9-aminoacridine (9-AA). We have shown that the level of fluorescence from 9-AA can be used as an indirect monitor of ψ_o as long as the membranes are treated with DCMU or uncouplers to prevent energisation (17). The technique relies on the fact that 9-AA is a monovalent cation which becomes non-fluorescent when it is attracted into the diffuse double layer. Therefore addition of other monovalent cations gives rise to an increase in 9-AA fluorescence whereas the simultaneous measurement of chlorophyll fluorescence shows the usual "dip" effect (18).

Clearly the salt induced effects relate to changes in ρ'_x, but there are three important characteristics which suggest that the coming together of adjacent thylakoid membrane surfaces on addition of electrolytes is not simply due to the screening of fixed surface electrical charges as depicted in the DLVO theory . Firstly not all the membrane surfaces become appressed (i.e. stacked and unstacked regions form) indicative of heterogeneity in surface properties. Secondly where appression does occur the distance between adjacent surfaces is very short, being less than 4 nm. If the simplest form of DLVO theory was applicable then, because the negative charge density on the thylakoid surface is relatively large (0.025 C m^{-2} or more, see ref. 17), the expected inter-membrane distance would probably be greater than 10 nm. In fact to account for a 4 nm inter-membrane distance requires a very strong van der Waals interaction with a low coulombic repulsion (19,20). The third factor which must be explained is the variation in yield of chlorophyll fluorescence due to changes in energy transfer between PS2 and PS1. In order to account for these characteristics I proposed that as the two membrane surfaces approach each other there is a lateral redistribution of components in accordance with Le Chatelier's principle (7,21). It was suggested that components carrying significant levels of electrical charges on their exposed surfaces would migrate away from the area where membrane appression occurs while electroneutral components or components with low surface charge densities would be preferentially located in the stacked region. Because the formation of appressed and non-appressed membrane regions is paralleled by changes in energy transfer between PS2 and PS1, it was suggested that the two photosystems were located in separate domains, PS2 in appressed and PS1 in non-appressed regions. Evidence for this physical separation of PS2 and PS1 protein complexes has been in the literature since the pioneering work of Boardman and Anderson (22) but was most thoroughly studied by Andersson and colleagues (see 23). It is now quite clear that it is the light harvesting chlorophyll a/b complex (LHC-2) associated with PS2, which is the major component causing membrane stacking. Its presence increases the protein to lipid ratio of the appressed over that of the non-appressed membranes (24) and gives rise to a significant level of exposed protein surface as judged from structural studies on the isolated complex (25). Both features will tend to increase van der Waals attractive forces.

There are a number of observations which support the charge displacement model presented above. It has been shown that, as expected, the restacking of thylakoids by adding salts requires the membrane to be fluid and is therefore very sensitive to temperature (26). The postulated lateral migration of complexes has been observed by freeze-fracture microscopy (27). But perhaps more striking is the fact that stacking of thylakoids by increasing ρ'_x contrasts with the trivial mechanism involving neutralization of the surface charges by binding of counterions or by lowering pH. In this case more extensive stacking occurs which is independent of fluidity and which maintains a randomisation of complexes (13,26). As would be expected, with this type of stacking there are no changes in chlorophyll fluorescence indicative of a decrease in energy transfer from PS2 to PS1.

In conclusion it can be stated with confidence that despite the many assumptions implicit in the Gouy-Chapman theory (see ref.7) it can be used to relate rather complicated experimental data to relatively simple electrostatic parameters. Such a correlation means that at the first approximation the organisation of the thylakoid membrane can be viewed in terms of physico-chemical principles. Because of this we can attempt to predict what conformational changes are likely to occur in this membrane system when its electrical properties are perturbed, for example, by protein phosphorylation or by specific proteolytic digestion.

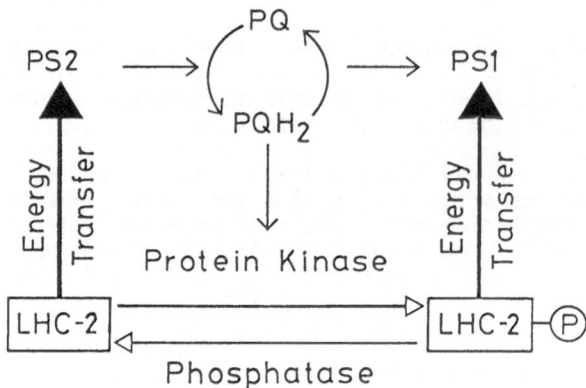

Fig.5   A model of the mechanism by which reversible phosphorylation of LHC-2 is controlled by the redox state of the plastoquinone (PQ) pool.

CONSEQUENCES OF SURFACE ELECTRICAL PROPERTIES

The above considerations indicate that the organisation of the thylakoid membrane in higher plants, and probably in green algae, including the spatial relationships between various intrinsic protein complexes, is governed by the interplay of electrostatic and electrodynamic forces which can be easily disturbed. Other factors must also come into play such as steric hinderence as in the case of the coupling factor complex with its large CF_1 component extruding well into the aqueous phase. It is probably this reason why this complex is located specifically in non-appressed regions. The rather even distribution of the cytochrome b-f complex between appressed and non-appressed membranes is not easily understood (28) although the explanation may simply reside in the fact that its "partition coefficient" for the two phases is close to 0.5 which also seems to be the case for most of the acyl lipids (29). In the case of the latter, the only consistent difference detected is that the appressed, compared with the unappressed lamellae has a higher MGDG to DGDG ratio. Also fatty acid analyses have not revealed any major differences between the two regions, although the appressed membranes seem to be slightly more unsaturated (24).

The dramatic changes in thylakoid membrane organisation induced by varying the ionic composition of the bathing medium is an extreme example of perturbing the electrostatic forces. Another possibility is to alter the nature or extent of the electrical charge on the surface of a particular membrane complex. It is the latter possibility which seems to underlie the molecular mechanism by which protein phosphorylation can regulate energy distribution between PS1 and PS2 via a phenomenon known as State transitions (30-33).

The phosphorylation occurs at the threonyl residue close to the N-terminus of the LHC-2 polypeptide. The phosphorylation is catalysed by a membrane-bound kinase activated when PS2 is excited more than PS1 so that the electron transport system which functionally links the two photosystems becomes over reduced. The trigger for the kinase activation seems to involve redox components within the cytochrome b-f complex. When the kinase is not activated, for example, in the dark under oxidising conditions, or in excess PS1 light, a membrane-bound phosphatase brings about the dephosphorylation of LHC-2. The processes have been shown to give rise to changes in energy distribution between PS2 and PS1 (34,35) in a way similar to that observed during State transitions (36-38). The concept that reversible phosphorylation of the LHC-2 can regulate energy distribution between PS2 and PS1 is shown in Fig.5. But how can such a process occur? Bearing in mind the surface electrical properties of thylakoid membranes as described above, it could be predicted that the phosphory-

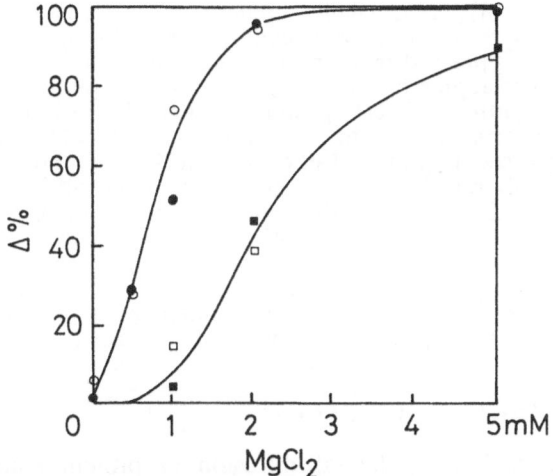

Fig.6   Comparison of the percentage changes in stacking measured from electron micrographs (open symbols) and from the 90° light scattering data (closed symbols), as a function of Mg^{2+} concentration, with (squares) and without (circles) LHC-2 phosphorylation.

lation/dephosphorylation of LHC-2 would alter its surface charge properties and thus dictate its preference to partition into appressed and non-appressed regions of the membrane (39). There is considerable experimental support for this concept ranging from freeze fracture (40) and biochemical analyses (41,42) to energy transfer and quantum-yield studies (43,44). As yet however, quantitative details of this regulatory process are not firmly established. It seems that only a proportion of the LHC-2 complexes are normally involved in this process and are therefore mobile. But is is unclear whether all of this mobile form becomes functionally linked to PS1. Neither do we know whether LHC-2 always moves alone or whether some LHC-2–PS2 complexes also laterally diffuse into the non-appressed regions. If the latter occurs then there will be direct changes in energy transfer between PS2 and PS1 (spillover) as well as alterations in the absorption cross section of the two photosystems. Indeed, as shown in our work with isolated thylakoids (45,46) the degree of inter-mixing of complexes in response to LHC-2 phosphorylation is, as would be expected for the electrostatic model of the thylakoid membrane, highly dependent on the level of cations in the bathing medium. For example, Fig.6 shows that higher levels of Mg^{2+} are required to induce stacking of thylakoids (and therefore segregation of complexes) after they have been subjected to phosphorylation.

Reshuffling of protein complexes within the thylakoid membrane can also occur in response to heating to about 40°C (47,48). Again this reorganisation emphasises the delicate balance of interactions which exists between the various complexes. In this case there seems to be a weakening of the attractive forces between the LHC-2 and PS2 complexes so that the latter (which probably carry net negative charge) repartition into the non-appressed regions while the LHC-2 rich domains continue to maintain appropriate physico-chemical properties necessary for tight association between adjacent membrane surfaces. This lateral separation of LHC-2 and PS2 is reversible and may play a physiological role in helping to minimise damage to PS2 reaction centres when there is excess light and when local heating occurs in the membrane.

The inhibitory effect of high light is known as photoinhibition. It is now quite clear that the PS2 reaction centre is composed of a heterodimer of the D1 and D2 polypeptides (see 49). The D1 polypeptide is a remarkable protein in that it turns over very rapidly as compared with other chloroplast proteins, including the D2 polypeptide. There is evidence that this turnover of D1 is related to the vulnerability of the PS2 reaction centre to high light intensities and it is the re-synthesis of the D1 polypeptide which underlies the recovery after photoinhibitory treatment (50). Clearly there is a spatial problem to be overcome since PS2 is located mainly in the appressed membranes while chloroplast ribosomes are located on the surface of the non-appressed membranes. Using a pulse-chase procedure it was shown that newly synthesised D1 protein is initially inserted into non-appressed lamellae (51). It is processed from a 33.5 kD to a 32 kD form and then laterally diffuses to the appressed regions. Little more is known about the disassembly and reassembly of the PS2 reaction centre and where it occurs. Presumable there are processes involved which dictate the movement of the complex and in this regard it is of interest to note that both the D1 and D2 polypeptides can be reversible phosphorylated at the N-terminus (52,53). The function of this protein phosphorylation is unclear but must disturb the electrostatic properties of the PS2 environment. It is also of interest to note that the D1 protein undergoes a post-translational palmitoylation which could act as a trigger for the assembly process (51).

THYLAKOID MEMBRANE FLUIDITY

The above discussions of lateral diffusion of protein complexes indicate that the thylakoid membrane is normally a highly fluid system. Such a conclusion is also supported by the high degree of unsaturation of the acyl lipid which typically has an average double bond index greater than five per molecule (the dominating fatty acid being linolenic acid). Direct measurement of thylakoid membrane fluidity has come from steady-state and time-resolved fluorescence polarization studies using the probe 1,6-diphenyl-1,3,5-hexatriene (DPH). Steady-state measurements gave polarisation (p) values in the range of 0.2 to 0.25 at 25°C (ref.54) which contrasts, for example, with the "stiff" purple membrane of Halobacterium halobium or human eye cortex, which have p values of about 0.4, and with a very fluid system like human milk globules where p = 0.12. The steady-state fluorescence measurements using DPH and EPR spin probe studies, also revealed that the appressed thylakoid membranes were more rigid than the unappressed (24).

Time-resolved fluorescence studies using DPH can give additional information by allowing the properties of the emission to be divided into dynamic and static components (55,56). The former relates to the time of rotation of the probe molecule within the bilayer while the latter gives information about acyl chain ordering. This is to say that the final distribution of emitting dipoles is anisotropic and the time dependent anisotropy r(t) follows an equation of the following form:

$$r(t) = (r_o - r_\infty) \exp(-t/\phi) + r_\infty \tag{13}$$

where r_o is the initial fluorescence anisotropy, r_∞ is the time-limited anisotropy, t is time and ϕ is the rotational correlation time. The latter quantity is inversely related to the wobbling diffusion coefficient D_w which can be used to calculate a viscosity of the bilayer using the equation

$$\eta_m = \frac{kT}{6D_w V_e f} \tag{14}$$

where k is the Boltzmann constant, T is absolute temperature, V_e is the volume of the probe and f is its shape factor. Using this equation a viscosity of 0.34 P was estimated for thylakoid membranes isolated from peas when the measuring temperature was 25°C (55). This value contrasts with 0.42 P at 35°C for the cell membranes of human erythrocytes and 0.82 at 35°C for the purple halobacteria.

According to our DPH measurements with thylakoids isolated from peas there are no obvious gross phase changes of the bilayer over the temperature range 55 to -20°C (57) but interestingly there are changes in the degree of fluidity as a result of growth temperature. Low growth temperatures (e.g. 7°C) give rise to a more fluid thylakoid membrane than when plants are grown at higher temperatures (17°C). Therefore it seems that a homeostatic mechanism exists to maintain the thylakoid at an optimal level of fluidity (57).

As discussed in detail above the restacking of isolated thylakoids and the associated changes in distance between PS2 and PS1 can be monitored by an increase in chlorophyll fluorescence. By following the time course of the chlorophyll fluorescence yield changes as a function of temperature we were able to estimate diffusion coefficients for the lateral movement of the complexes (58). Values of 1.85×10^{-12} to 3.08×10^{-11} cm^2 s^{-1} were obtained for the temperature range of 10 to 30°C. At 25°C the lateral diffusion coefficient was estimated to be about 2×10^{-11} cm^2 s^{-1}. It is interesting to check whether such values are consistent with the fluidity measurements using DPH. This can be done by applying the Saffman-Delbruck equation which relates membrane viscosity with the lateral diffusion coefficient (D_L) for a protein of cylindrical shape:

$$D_L = \frac{kT}{4\Pi h \eta_m} \left(\ln \frac{h\eta_w}{a\eta_m} \right) - 0.5772 \tag{15}$$

where k is the Boltzmann factor, T is absolute temperature, η_m and η_w are the viscosity coefficients of the membrane and the aqueous medium surrounding the membrane (essentially water) respectively, h is the height of the protein and a is its radius. Taking a = 4 nm and h = 14 nm (as deduced for LHC-2 by Kuehlbrandt (25) and $\eta_w = 0.01$ P and $\eta_m = 0.34$ then the value for D_L at 25°C is approximately 2×10^{-11} cm^2 s^{-1} which matches that estimated from the chlorophyll fluorescence analysis.

CONCLUSION

Electrical charges on the surface of chloroplast thylakoid membranes seem to dictate its conformational state. Any perturbation of the surface electrical properties therefore bring about a reorganisation of the membrane (see Fig.7). The phosphorylation of LHC-2 seems to be a beautiful demonstration of how such perturbations are used in order to regulate and optimise the efficiency of photosynthesis. Other proteins of the thylakoid membrane can be reversibly phosphorylated but the precise reason why is as yet unclear. It could be that in a similar way to LHC-2 these phosphorylations bring about key changes in surface electrical properties of the exposed portions of the proteins and therefore effect their interaction with neighbouring complexes by changes in coulombic forces.

REFERENCES

1. K. Gounaris and J. Barber, Trends in Biochem. Sci. 8:378-381 (1983)
2. H.Y. Nakatani, J. Barber and J.A. Forrester, Biochim. Biophys. Acta 504:215-225 (1978)

3. H.Y. Nakatani and J. Barber, Biochim. Biophys. Acta 591:82-91 (1980)
4. M. Gouy, J. Phys. (Paris) 9:457-468 (1910)
5. D.L. Chapman, Phil. Mag. 25:475-481 (1913)
6. B.T. Rubin and J. Barber, Biochim. Biophys. Acta 592:87-102 (1980)
7. J. Barber, Biochim. Biophys. Acta 594:253-308 (1980)
8. J. Barber, Ann. Rev. Plant Physiol. 33:261-295 (1982)
9. V.A. Parsegian and B.W. Ninham, J. Theor. Biol. 38:101-109 (1973)
10. J.Th.G. Overbeek, J. Colloid Interface Sci. 58:408-422 (1978)
11. E.L. Gross and S.C. Hess, Biochim. Biophys. Acta 339:334-346 (1974)
12. E.L. Gross and S.H. Prasher, Arch. Biochem. Biophys. 164:460-468 (1974)
13. C. Scoufflaire, R. Lannoye and J. Barber, Photosyn. Res. 6:133-146 (1985)
14. J.D. Mills and J. Barber, Biophys. J. 21:257-272 (1978)
15. J. Barber and G.F.W. Searle, FEBS Lett. 92:5-8 (1978)
16. H.Y. Nakatani, J. Barber and M. Minski, Biochim. Biophys. Acta 545:24-35 (1979)
17. W.S. Chow and J. Barber, J. Biochem. Biophys. Meth. 3:173-185 (1980)
18. J. Barber and G.F.W. Searle, FEBS Lett. 103:241-245 (1979)
19. M.J. Sculley, J.T. Duniec, S.W. Thorne, W.S. Chow and N.K. Boardman, N.K. Arch. Biochem. Biophys. 201:339-346 (1980)
20. B.T. Rubin, W.S. Chow and J. Barber, Biochim. Biophys. Acta 634:174-190 (1981)
21. J. Barber, FEBS Lett. 118:1-10 (1980)
22. N.K. Boardman and J. Anderson, Nature (London) 203:166-167 (1964)
23. B. Andersson, in: Methods in Enzymology, A. Weisbach and H. Weisbach, eds., Academic Press, New York (1986)
24. R.C. Ford, D.J. Chapman, J. Barber, J.Z. Pedersen and R.P. Cox, Biochim. Biophys. Acta 681:145-151 (1982)
25. W. Kuehlbrandt, Nature (London) 307:478-480 (1984)
26. J. Barber, W.S. Chow, C. Scoufflaire and R. Lannoye, Biochim. Biophys. Acta 591:92-103 (1980)
27. L.A. Staehelin, J. Cell Biol. 71:136-158 (1976)
28. J. Barber, Plant, Cell and Environment 6:311-312 (1983)
29. D.J. Chapman, J. DeFelice and J. Barber, Photosyn. Res. 9:239-249 (1986)
30. J. Barber, Photobiochem. Photobiophys. 5:181-190 (1983)
31. J. Bennett, Biochem. J. 212:1-13 (1983)
32. P. Horton, FEBS Lett. 152:47-52 (1983)
33. J.F. Allen, Trends Biochem. Sci. 8:369-373 (1983)
34. J.F. Allen, J. Bennett, K.E. Steinback and C.J. Arntzen, Nature 291:21-25 (1981)
35. P. Horton and M.T. Black, FEBS Lett. 119:141-144 (1980)
36. A. Telfer and J. Barber, FEBS Lett. 129:161-165 (1981)
37. A. Telfer, J.F. Allen, J. Barber and J. Bennett, Biochim. Biophys. Acta 722:176-181 (1983)
38. O. Canaani, J. Barber and S. Malkin, Proc. Natl. Acad. Sci. USA, 81:1614-1618 (1984)
39. J. Barber In: Encyclopedia of Plant Physiology, New Series, L.A. Staehelin and C.J. Arntzen, eds., Springer Verlag, Berlin, 19:653-664 (1986)
40 D.J. Kyle, L.A. Staehelin and C.J. Arntzen, Arch. Biochem. Biophys. 222:527-541 (1983)
41. W.S. Chow, A. Telfer, D.J. Chapman and J. Barber, Biochim. Biophys. Acta 638:60-68 (1981)
42. B. Andersson, H.-E. Akerlund, B. Jergil and C. Larsson, FEBS Lett. 149:181-185 (1982)
43. J.N. Farchaus, W.R. Widger, W.A. Cramer and R.A. Dilley, Arch. Biochem. Biophys. 217:362-367 (1982)
44. A. Telfer, J. Whitelegge, H. Bottin and J. Barber, J. Chem. Soc. Faraday Trans. 2 (Special Edition) 82:2207-2215 (1986)
45. A. Telfer, M. Hodges and J. Barber, Biochim. Biophys. Acta 724:167-175 (1983)
46. A. Telfer, M. Hodges,, P.A. Millner and J. Barber, Biochim. Biophys. Acta 766:554-562 (1984)
47. K. Gounaris, A.P.R. Brain, P.J. Quinn and W.P. Williams, Biochim. Biophys. Acta 766:198-208 (1984)
48. C. Sundby and B. Andersson, FEBS Lett. 191:24-28 (1985)
49. J. Barber, Trends Biochem. Sci. 12:321-326 (1987)
50. D.J. Kyle and I. Ohad, in: Encyclopedia of Plant Physiol. (New Series) L.A. Staehelin and C.J. Arntzen, eds., Springer Verlag, Berlin 19, 468-476 (1986).

51. A.K. Mattoo and M. Edelman, Proc. Natl. Acad. Sci. USA, 84:1497-1501 (1987)
52. A. Telfer, J.B. Marder and J. Barber, Biochim. Biophys. Acta 893:557-563 (1987)
53. J.B. Marder, A. Telfer and J. Barber, Biochim. Biophys. Acta 932:362-365 (1988)
54. R.C. Ford and J. Barber, Photobiochem. Photobiophys. 1:263-270 (1981)
55. R.C. Ford and J. Barber, Biochim. Biophys. Acta 722:341-348 (1983)
56. P.A. Millner, R.A.C. Mitchell, D.J. Chapman and J. Barber, Photosyn. Res. 5:63-76 (1984)
57. J. Barber, R.C. Ford, R.A.C Mitchell and P.A. Millner, Planta 161:375-380 (1984)
58. B.T. Rubin, J. Barber, G. Paillotin, W.S. Chow and Y. Yamamoto, Biochim. Biophys. Acta 683:69-74 (1981)

THYLAKOID MEMBRANE POLAR LIPIDS

Kleoniki Gounaris

AFRC Photosynthesis Research Group, Department of Pure & Applied Biology
Imperial College, London SW7 2BB UK

INTRODUCTION

Oxygenic photosynthetic organisms have maintained a highly conserved polar lipid composition of the thylakoid membranes through their evolution. Essentially four, glycerol based lipid classes comprise the total polar lipid content of the thylakoids of higher plant chloroplasts, eukaryotic algae and cyanobacteria. Of these lipid classes three are glycosylated, to either galactose or glucose, and the glycolipids account for approximately 90% of the total polar lipid fraction. Phospholipid, a major component of other membrane systems, comprises only about 10% of the total.

Fatty acyl residues associated with the above lipids are either 16 or 18 carbon atoms long. The degree of unsaturation, however, has been shown to be variable when comparing different species, different strains of the same genus or when, particularly in the case of algae and cyanobacteria, the growth temperature is varied.

The ubiquitous distribution of these lipids in photosynthetic membranes of oxygenic organisms as well as their properties in aqueous environments have stimulated extended investigations and interesting postulations have been put forward concerning the involvement of these lipids in the organisation and function of the photosynthetic membranes.

Here the properties of polar lipids isolated from higher plant thylakoid membranes will be briefly reviewed followed by an account of the factors that affect these properties. Finally, possible roles of these lipids in the functioning of the membrane will be outlined.

1. POLAR LIPID COMPOSITION

The polar lipids of the thylakoids of higher plant chloroplasts are characterised by a very high degree of unsaturation. Although the precise fatty acyl composition can vary with the plant species, the main fatty acid found esterified to glycerol is linolenic acid (C18:3) which can account for 80% of the total acyl chains. Other fatty acyl residues occurring in thylakoid polar lipids include: palmitic (C16:0), stearic (C18:0), oleic (C18:1) and linoleic (C18:2) acids. Glycerol has the same conformation in both glycolipids and phospholipids, in the sense that the hydrocarbon chains are attached to the 1- and 2- (sn)- atoms of the glycerol moiety. Carbohydrate is glycosidically attached to the 3-(sn)-carbon of glycerol in the case of glycolipids while a phosphate ester replaces the carbohydrate residue in the phospholipid.

The glycolipids of thylakoid membranes are in order of abundance, monogalactosyldiacylglycerol (MGDG), digalactosyldiacylglycerol (DGDG) and sulphoquinovosyldiacylglycerol (SQDG) or plant sulpholipid (SL). The only phospholipid unequivocally attributed to the thylakoids is phosphatidylglycerol (PG). It is worth noting that although galactose represents the carbohydrate moiely of mono- and digalactosyl diacylglycerol lipids, the carbohydrate residue in SQDG is glucose. In DGDG, the two galactose molecules are linked with a $1 \rightarrow 6$ bond. The two electroneutral galactolipids account for more than two-thirds of the total polar lipid content of thylakoids from all higher plant chloroplasts and they occur in a molar ratio of 2:1 (MGDG:DGDG).

On the basis of the fatty acyl content, two types of photosynthetic tissue have been desribed. Plant species such as Solanaceae and Brassicaceae are characterised by the occurrence of C16:3 together with C18:3 fatty acids in their galactolipids (16:3 plants). In other species, the polyunsaturated fatty acid components of galactolipids are almost exclusively C18:3 (18:3 plants) (1). The poly-cis double bond system is extended towards the methyl end, leaving only two saturated terminal carbons. In 16:3 plants, the C18:3/C18:3 species occurs almost exclusively in MGDG with the C16:3 fatty acid esterified to the C-2 position (2). The glucolipid SQDG usually accounts for 10-15% of the total polar lipid content and is strongly acidic. The prefix sulpho-denotes a sulphonic acid ($-C-SO_3 H$) group which is attached to 6-deoxy-D-glucose (D-quinovose). SQDG contains high amounts of the saturated fatty acid palmitate ($\sim 35\%$) and in most plants examined this lipid is mainly composed of the palmitoyl/linolenoyl and palmitoyl/linoleoyl species. (2,3).

Phosphatidylglycerol, PG, the other negatively charged lipid of thylakoids contains a characteristic fatty acid, trans-Δ^3-hexadecenoic acid, esterified to the C-2 position. The precise reasons for the specific association of this fatty acid with phosphatidylglycerol of thylakoid membranes remain unclear although some proposals have been made concerning, for example, its direct involvement in the process of chilling-sensitivity (4).

Finally, the majority of the reactions required for thylakoid polar lipid biosynthesis takes place in the chloroplast but co-operation with other cellular compartments also occurs (5,6).

2. PROPERTIES OF THYLAKOID POLAR LIPIDS IN AQUEOUS SYSTEMS

A variety of physical techniques has been employed to study the properties of isolated polar lipids extracted from thylakoid membranes. Investigations have, however, been concentrated on the two galactolipids, MGDG and DGDG, and information on SQDG and PG is scarce. An account will be presented on the properties of pure lipid classes and of their mixtures and the effect various factors exert on such properties in aqueous environments.

2.1 PURE POLAR LIPID CLASSES

The structures adopted by the two thylakoid galactolipids as examined by x-ray diffraction techniques were first reported in 1969 (7). It was observed that the galactolipids can incorporate minimal amounts of water between molecules with excess water remaining as a separate phase. It was further reported that MGDG adopts a hexagonal type II configuration (H_{II}) while DGDG formed a lamellar (Lα) structure. Subsequent reports (8,9) confirmed and expanded the above. Thus it was established that DGDG formed a liquid crystalline bilayer structure over the temperature range of -10 to 80°C and concentration range C=1.0-0.6 (C=g lipid/g lipid + water). The lamellae, parallel and equidistant, had a maximum repeat distance of 5.42nm and a bilayer thickness of 4.2nm at C=0.78 when maximum hydration occured. MGDG also exhibited maximum hydration at C=0.78 and a hexagonal II phase was observed over a concentration range C=1.0-0.5 and temperature range -15 to 80°C. The hexagonal type II phase is periodic in two dimensions, the structure being formed by cylinders in a two dimensional array. The interior of the cylinder contains water. It was found that the distance between the cylinders increased with temperature reaching a limiting value of 6.25nm at 0°C and C=0.78 (9).

Thermal measurements also cited in (9) indicated that the gel to liquid crystalline transition for DGDG, extracted from pelargonium leaves, occurs at -50°C. MGDG was also found in a liquid crystalline phase down to -30°C when a single thermal transition took place.

Wide-angle x-ray diffraction studies showed that all four polar lipid classes were in the liquid crystalline phase at 25°C by exhibiting a diffuse spacing centred at 0.46nm indicative of a disordered organisation of the hydrocarbon chains. (10)

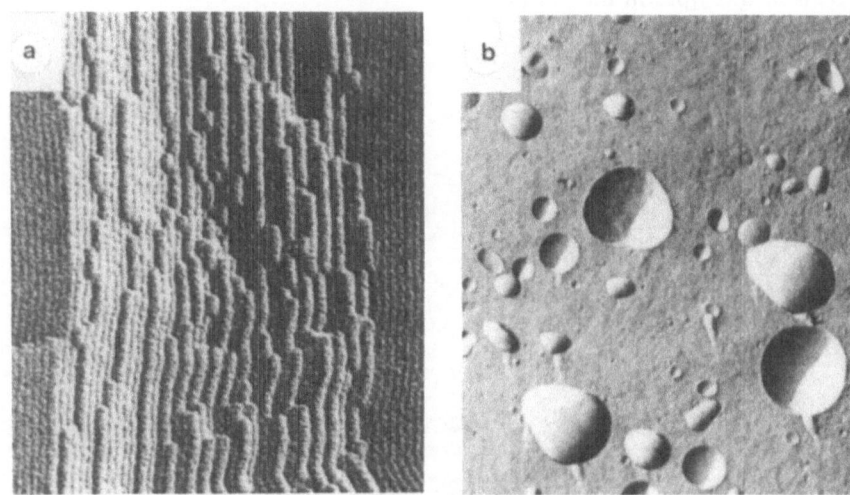

Fig. 1   Typical electronmicrographs of freeze-fracture replicas of (a) the hexagonal type II configuration adopted by monogalactosyldiacylglycerol and (b) the liposomes formed by the other lipid classes of the thylakoids, dispersed in distilled water. (a) x 200,000; (b) x 54,000.

Freeze-fracture electronmicroscopy demonstrated that DGDG, SQDG and PG form closed bilayers on hydration while MGDG molecules are arranged in the hexagonal type II configuration (Fig. 1).

Measurements of force-area isotherms at the air-water interface have indicated that all four polar lipid classes form stable monolayers at 20°C on a water sub-phase (10,11) No hysterisis effects were observed if the films were not compressed beyond their collapse point. SQDG and PG form more condensed monolayers than the two galactolipids an observation accounted for by the presence of saturated fatty acyl residues in SQDG and the presence of a trans double bond in PG molecules. It thus, appears that the condensing effect exerted by the hydrocarbon chains overcome the repulsion expected to occur between the head groups of the two charged lipids, SQDG and PG.

2.1.1 FACTORS AFFECTING THE PROPERTIES OF PURE LIPID CLASSES

A number of studies have indicated that parameters such as temperature, fatty acyl chain unsaturation or the presence of cations in the suspending medium can affect the physical state of biomembrane lipids. The effect exerted by such parameters on the structure and properties of isolated polar lipids of thylakoid membranes has also been examined. Freeze-fracture electronmicroscopy showed that there was no apparent effect on the structures adopted by DGDG, PG or SQDG between 40°C and -28°C. The structures observed for samples quenched from these temperatures were similar to those observed at 25°C. MGDG remained in the hexagonal type II configuration when heated above 25°C but there were marked changes when the temperature was lowered to -28°C, as shown in Fig. 2. At this temperature there is no evidence of hexagonal structures and the molecular arrangement is that of a bilayer. It is, however, apparent that these bilayers are not arranged in a closed liposomal structure, but instead what appears to be flat open sheets of lamellae are observed. It is probable that these sheets of lamellae represent the structural arrangement of this lipid in the gel phase.

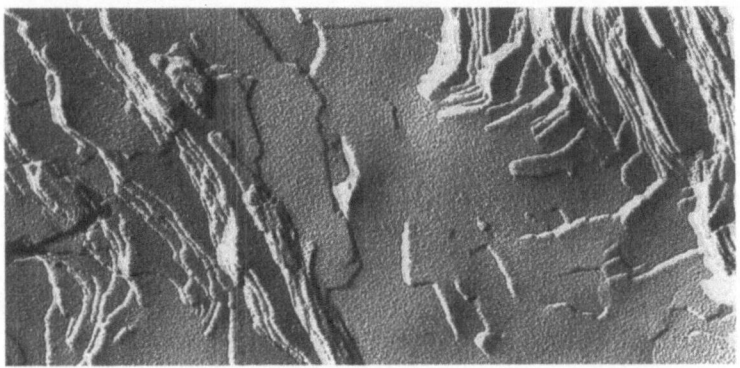

Fig. 2   Freeze-fracture electronmicrograph of replica obtained from MGDG samples quenched from -28°C (x120,000).

Fully saturated molecular species of DGDG and MGDG have been examined by Sen et al (12). Thus, the 1,2 distearoyl derivatives of the galactolipids have been studied by means of electronmicroscopy, x-ray diffraction and calorimetry and have been reviewed in (13). It was shown that the fully saturated MGDG formed flat sheets of lamellae, similar to those shown in Fig 2., and no evidence of hexagonal II configuration was obtained. A number of inter-related thermotropic phases for this lipid species were also described, along with the thermal behaviour of 1,2 distearoyl derivative of DGDG.

It appears that the ability of MGDG to adopt a hexangonal type II configuration on hydration strongly depends on the degree of unsaturation of its fatty acyl chains and on the temperature. A detailed examination of the effect of the number of double bonds per lipid molecule on the structural configuration of this lipid has been carried out (14). Freeze-fracture electronmicrographs of MGDG samples with different degrees of unsaturation obtained by catalytic hydrogenation of the double bonds are shown in Fig. 3. It appears that the transformation of the hexagonal type II configuration to the open sheets of lamellae proceeds through on organisation where both structures are present indicating that an equilibrium exists between these two

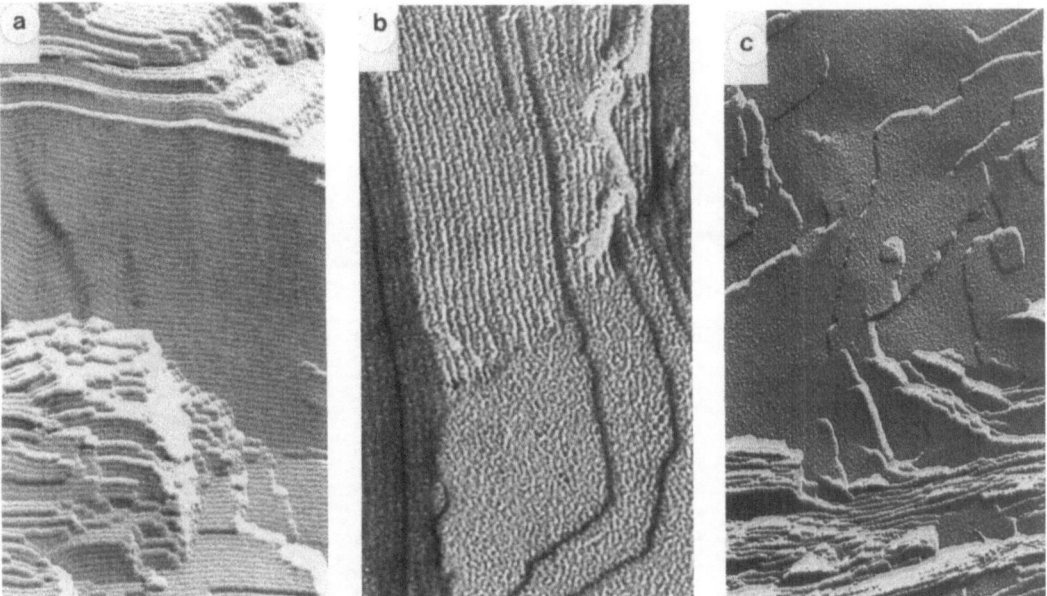

Fig. 3   Electronmicrographs of freeze-fracture replicas prepared from sonicated
dispersions of MGDG samples with different degrees of unsaturation. (a)
double bonds per lipid molecule 5.4 (x 110,000) (b) double bonds per lipid
molecule 5.1 (x 260,000) (c) double bonds per lipid molecule 3.9
(x 120,000).

arrangements in MGDG.   Wide angle x-ray diffraction studies indicated that all
samples with a double bond index value of about 4.5 or below were in the gel state at
20°C.

Lowering the temperature had a similar net effect on the structural
arrangements of MGDG as the decrease in the unsaturation level. Thus, flat sheets of
lamellae were observed at -28°C but examination of the structures adopted at
temperatures between 25°C and -28°C showed that the transformation to the lamellae
probably proceeds via a series of intermediate structural arrangements.   The
characteristic intermediate phase observed between the hexagonal type II and the
lamellae arrangements was the occurrence of lipidic particles corresponding to
inverted lipid micelles (Fig. 4).

MGDG accounts for about 50% of the total polar lipid content of thylakoids and
is the only lipid present in these membranes which is capable of forming non-bilayer
structures in aqueous dispersion. It would therefore be expected that the behaviour
of this lipid would typify that of the whole membrane lipid extract. Studies on
thylakoid membrane polar lipid mixtures are described below.

2.2 POLAR LIPID MIXTURES

Mixtures of the two galactolipids MGDG and DGDG in their naturally occuring
molar ratio of 2:1 (MGDG:DGDG) have been examined by freeze-fracture
electronmicroscopy. A dominating feature of replicas obtained from such mixtures is
the presence of freeze-fracture particles (Fig. 5). This molecular organisation has
been termed the "lipidic particle" and and has been shown to correspond to inverted
lipid micelles sandwiched within a bilayer (15,13).   Other molecular arrangements
observed in galactolipid mixtures are believed to be related and, under certain
conditions, interconvertible (16).

Fig. 4   Freeze-fracture electronmicrographs obtained from MGDG samples quenched from different temperatures (a) 2°C; x 160,000 (b) -10°C; x 200,000.

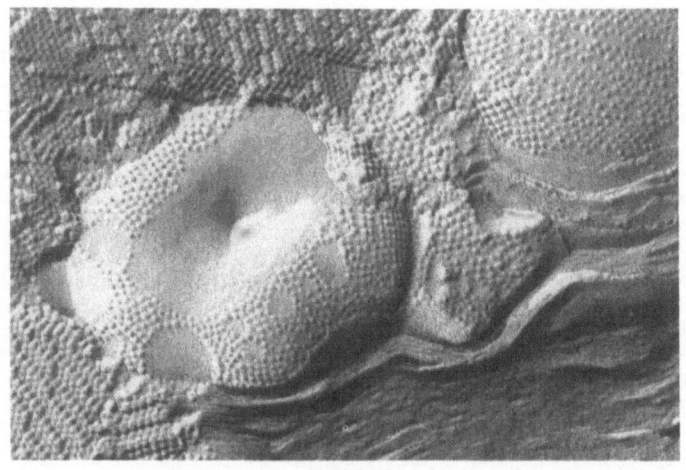

Fig. 5   Freeze-fracture electronmicrograph obtained from aqueous dispersions of mixtures of MGDG and DGDG (2:1 molar ratio respectively) x 110,000.

178

The structures adopted by total polar lipid extract of thylakoid membranes in aqueous dispersions are shown in Fig. 6.

It is apparent that small liposomes are formed and there is no evidence of inverted lipid micelles or other lipidic organisations similar to those observed in the galactolipid mixtures. The only difference in the composition of the two types of mixtures is the presence of the anionic lipids SQDG and PG in the total lipid extract. The only lipid class in the mixture capable of adopting non-bilayer configurations is MGDG; it therefore appears that the presence of the charged lipids SQDG and PG in the mixture prevents MGDG from adopting such structures. It is worth noting that in total polar lipid extract very small liposomes are formed probably due to electrostatic repulsion between the negatively charged lipid head-groups and it is possible that the absence of non-bilayer configuration is due to the fact that such vesicles are too small to accomodate such structures. Neutralisation of the anionic lipids by lowering the pH of the dispersions or the addition of cations leads to fusion of the vesicles to form larger liposomes. In such vesicles geometric constraints are removed and non bilayer configurations become apparent (17).

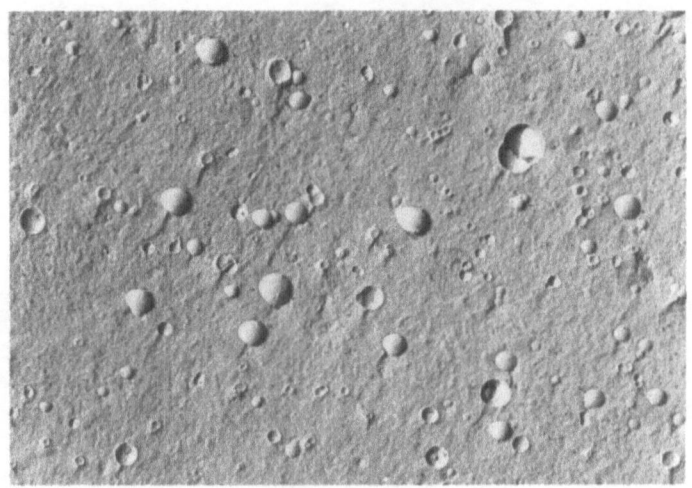

Fig. 6   Electronmicrographs of freeze-fracture replicas obtained from total polar lipid extract dispersed in distilled water (x 54,000).

Addition of inorganic salts to the thylakoid total polar lipid extracts led to aggregation effects in a distinct manner. Low concentrations of cations (<2mm) resulted in the fusion of the small unilamelar liposomes. Larger vesicles thus appearing contained a few inverted lipid micelles (Fig. 7). This effect was found to be triggered with increasing efficiency by monovalent, divalent and trivalent cations and it is possible that either electrostatic shielding of the negative charges of the acidic lipids or direct binding of the cations to the lipid headgroups is involved (17). A different process appears to be implicated in the aggregation of the thylakoid lipids induced by higher concentrations of polyvalent but not monovalent cations. On increasing the cation concentration above 10 mM large lipid aggregates are formed containing extensive non-bilayer configurations (Fig. 8).

This aggregation appeared to depend, at least in the case of the divalent cations, on their ionic radius (17). In conclusion, the formation of non-bilayer

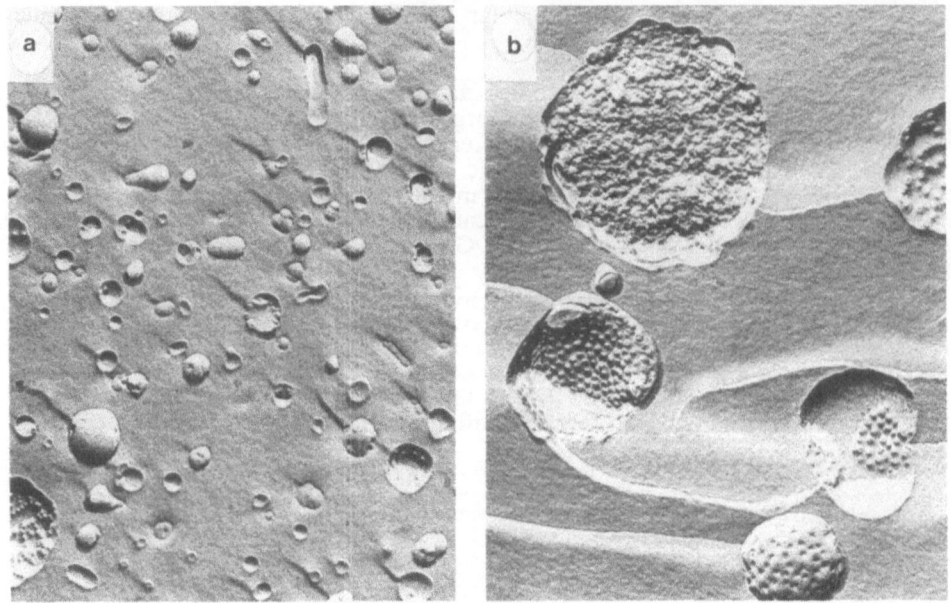

Fig. 7   Electronmicrographs of freeze-fracture replicas prepared from dispersions of total polar lipid extracts suspended in (a) 1mM $MgCl_2$ (x 64,000); (b) 7.5mM $MgCl_2$ (x 64,000).

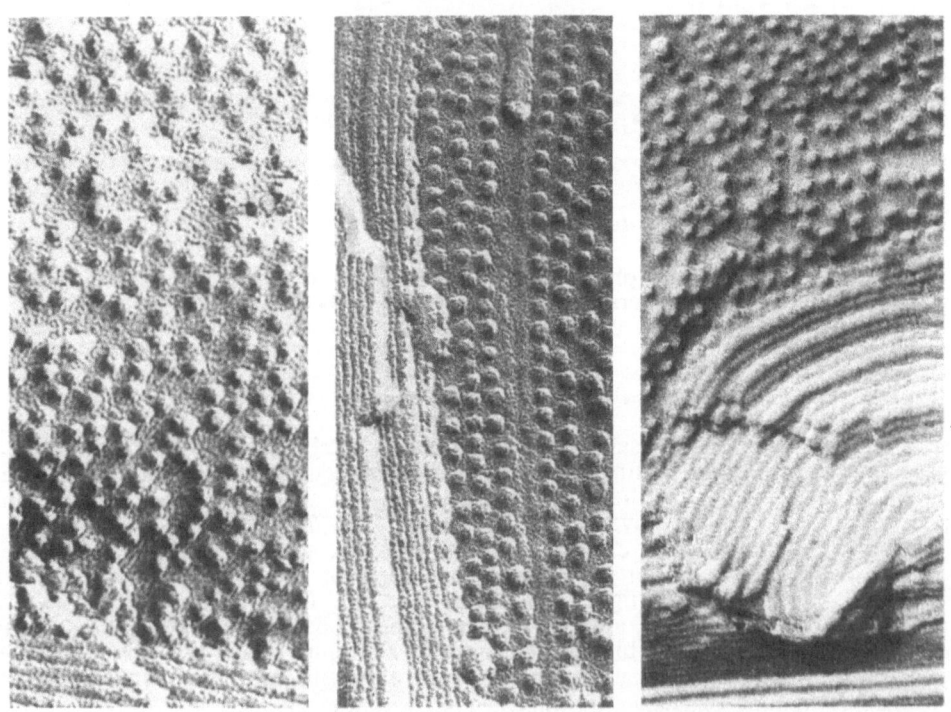

Fig. 8   Electronmicrographs showing the main features of freeze-fracture replicas prepared from total polar lipid extracts in the presence of 30mM $MgCl_2$ (x 200,000).

structures in total polar lipid extracts seem to depend on the neutralisation of the charges of the acidic lipid fraction.

It is of interest to note however, that even in the presence of high concentrations of cations the formation of non-bilayr configurations was found to depend on two main factors: temperature and the degree of unsaturation of the fatty acyl chains of the lipids, particularly those of MGDG. Thus, decreasing the temperature of the lipid suspension or decreasing the degree of unsaturation of the fatty acyl chains leads to the formation of bilayer configurations in total lipid extracts even in the presence of cations. Bilayer arrangements induced in this way are quite different to the conventional liposome type. Open sheets of lamellae are observed similar to those shown in Fig. 2.

Given that MGDG is the only lipid present in the mixture that can adopt non-bilayer structures, when dispersed alone in water, and the fact that its ability to form such structures, as previously discussed, depends markedly on both temperature and the extent of saturation it is not surprising that the behaviour of this lipid typifies that of the total polar lipid extract. The physical state of the lipid extracts under such conditions has been studied in addition to electronmicroscopy, by fluorescence polarisation and wide-angle x-ray diffraction measurements (14). Although the appearance of flat sheets of lamellae could be correlated with high fluorescence polarisation values (typical of gel phase lipids) no evidence of gel phase was obtained by x-ray diffraction measurements.

Fig. 9   Electronmicrographs of freeze-fracture replica prepared from thylakoid membranes that had been subjected to heating at 50°C for 5 min (x 190,000).

CONCLUSIONS

Despite the tendency of thylakoid polar lipid extracts to adopt non-bilayer structures, there is no evidence to suggest that analogous configurations are present in the thylakiod membrane under physiological conditions. This suggests that the formation of such structures is suppressed by other membrane components. If this is the case then non-bilyar forming lipids are directly interacting with the membranes' protein complexes.

It has been suggested (18) that non-bilayer lipids are required in order to satisfy the packing requirments of the protein components; this structural coupling results in non-bilayer lipid-protein interactions which modify the structures formed by the lipids on isolation. Disturbance of these interactions would therefore lead in non-bilayer lipid phase separations detrimental to the integrity of the thylakoids. Indeed, it has been shown that treatments which favour non-bilayer configurations can lead to the formation of such structures in the membrane. For example, dehydration, freeze thawing cycles or increase in temperature (Fig. 9) would induce the appearance of non-bilayer structures.

Apart from this structural role, thylakoid polar lipids appear to be involved directly in the functional processes of the thylakoid membranes. In a number of experiments (reviewed in 5) it has been demonstrated that lipids are essential for the maintenance of enzymatic activities. In a detailed study (19,20,21) it was shown that specific polar lipids, in defined proportions and with precise degree of unsaturation are required for the optimal activities of the CF_o-CF_1 ATP synthase complex. It therefore appears that thylakoid polar lipids may fulfil other functions besides the provision of a semipermeable matrix. Their involvement in the structural organisation and functional activities of the membrane system could account for their unique and conserved composition among oxygenic photosynthetic organisms.

REFERENCES

1. Roughan, P.G. and Slack, C.R. Annu. Rev. Plant Physiol. 33:97-132 (1982)
2. Nishihara, M., Yokota, K. and Kito, M. Biochim. Biophys. Acta. 617:12-19 (1980)
3. Harwood, J.L. in Biogenesis and Function of Plant Lipids (Mazliak, P., Benveniste, P., Costes, C. and Douce, R. eds.) pp. 143-152, Elsevier, Amsterdam (1980)
4. Murata N. and Yamada, J. Plant Physiol. 74:1016-1024 (1984)
5. Gounaris, K., Barber, J. and Harwood, J.L. Biochem. J. 237:313-326 (1986)
6. Joyard J. and Douce, R. The Biochemistry of Plants, (Stumpf, P.K. ed) vol. 9, pp 215-274, Academic Press, New York (1987)
7. Rivas, E. and Luzzati, V.C. J. Mol. Biol. 41:261-275 (1969)
8. Kreutz, W. Adv. Bot. Res. 3:54-165 (1970)
9. Shipley, G.G., Green, J.P. and Nichols, B.W. Biochim. Biophys. Acta 311:531-544 (1973)
10. Gounaris, K. PhD Thesis University of London (1983)
11. Bishop, D.G., Kenrick, H.R., Baytson, J.H., MacPherson, A.S. and Johns, S.R. Biochim. Biophys. Acta 602:248-259 (1980)
12. Sen, A., Williams, W.P. and Quinn, P.J. Biochim. Biophys. Acta 663:380-389 (1981)
13. Quinn, P.J. and Williams, W.P. Biochim. Biophys. Acta 737:223-266 (1983)
14. Gounaris, K., Mannock, D.A., Sen, A., Brain, A.P.R., Williams, W.P. and Quinn, P.J. Biochim. Biophys. Acta 732:229-242 (1983)
15. Sen, A., Williams, W.P., Brain, A.P.R. and Quinn, P.J. Biochim. Biophys. Acta 685:297-306 (1982)
16. Sen, A., Brain, A.P.R., Quinn, P.J. and Williams, W.P. Biochim. Biophys. Acta 686:215-224 (1982)
17. Gounaris, K., Sen, A., Brain, A.P.R., Quinn, P.J. and Williams, W.P. Biochim. Biphys. Acta 728:129-139 (1983)
18. Williams, W.P., Gounaris, K. and Quinn, P.J. In Advances in Photosynthesis Research (Sybesma, C. ed) vol. 3, pp 123-130. Martinus Nijhoff/Dr. W. Junck, The Hague (1984)
19. Pick, U., Gounaris, K., Admon, A. and Barber, J. Biochim. Biophys. Acta 765:12-20 (1984)
20. Pick, U., Gounaris, K., Weiss, M. and Barber, J. Biochim. Biophys. Acta 808:415-420 (1985)
21. Pick, U. Weiss, M., Gounaris, K. and Barber, J. Biochim. Biophys. Acta 891:28-39 (1987)

QUINONE BINDING AND HERBICIDE ACTIVITY IN THE ACCEPTOR QUINONE COMPLEX OF BACTERIAL REACTION CENTRES

Colin A. Wraight and Robert J. Shopes

Department of Physiology and Biophysics
University of Illinois
Urbana, IL 61801

INTRODUCTION

The electron acceptor function of bacterial reaction centres is carried out by two quinones, Q_A and Q_B, which function in series to provide a unique, two-electron gate that releases reducing equivalents only in pairs. In a series of flashes, the quinones undergo reduction and oxidation as follows [1]:

$$\text{Odd flashes:} \quad Q_A Q_B \xrightarrow{\ h\nu\ } Q_A^- Q_B \underset{(H^+)}{\overset{K_2}{\longleftrightarrow}} Q_A Q_B^- (H^+) \text{ (stable)}$$

$$\text{Even flashes:} \quad Q_A Q_B^-(H^+) \xrightarrow{\ h\nu\ } Q_A^- Q_B^-(H^+) \underset{(H^+)}{\overset{K_3}{\longleftrightarrow}} Q_A Q_B H_2 \overset{Q}{\underset{QH_2}{\longleftrightarrow}} Q_A Q_B$$

(The photooxidized donor, P^+, is rereduced after each flash, by a secondary donor - a c-type cytochrome, in vivo - and is omitted). A very similar electron acceptor activity is observed in PS II of plants and algae.

Significant functional differences between the two quinones are implicit in this scheme - Q_A is a tightly bound, one-electron redox couple, operating between the quinone and semiquinone forms, while Q_B is a reversibly bound, two-electron couple with quinone, semiquinone and quinol all involved in turnover. Since in many species Q_A and Q_B are chemically identical, these radically different properties are imparted to the quinones by interactions with the protein of their respective binding sites. Current work on RCs from Rhodobacter sphaeroides is aimed at finding out exactly what this means. This involves mutagenesis studies on the protein domain of the quinone binding sites, in addition to the extensive studies using non-native quinones to reveal details of the structural requirements of the quinone for Q_A and Q_B function [2,3,9].

BINDING AND REDOX PROPERTIES OF THE SECONDARY QUINONE

The redox properties of Q_B can be viewed as a direct consequence of the differential affinity that the Q_B site has for the oxidized, semireduced and fully reduced species, Q, $Q^-(H^+)$ and QH_2 [4]. Each one-electron couple can be viewed separately, viz. Q/Q^- and Q^-/QH_2. For the first reduction step, for example, the characteristic feature is its unusually high midpoint potential, i.e., the ease with which the semiquinone can be generated. This is achieved by a strong preferential binding (equivalent to "stabilization") of the reduced form, the semiquinone, relative to the oxidized quinone. Quantitatively, this can be expressed as:

$$\Delta E_m = 60/n \cdot \log K_r / K_o \quad mV \qquad \text{Eqn. 1}$$

where ΔE_m is the difference between the midpoint potential in the absence of any binding interactions and that observed with binding, E_m (free)-minus-E_m (bound), n is the number of electrons involved in the oxidation-reduction, and K_r and K_o are the dissociation constants for the semiquinone and quinone states, respectively. 60 mV is a convenient approximation for the term $2.303 \cdot RT/F$.

The E_m for Q_B/Q_B^- has been measured and/or estimated to be in the range 0 to + 100 mV in various species [5,6]. Since the Q/Q^- midpoint, in vitro, is not likely to be higher than -300mV [7], the shift in E_m induced by binding to the Q_B site is considerable (>300 mV) and implies a preferential binding of Q^- by a factor of at least 10^5 [1]. For RCs in detergent, measured values of K_o for ubiquinone are on the order of 1 μM [3,8], and K_s must therefore be no larger than a few picomolar, possibly much less.

Similar considerations apply to the net, two-electron redox activity of Q_B ($Q \longrightarrow QH_2$). In chromatophores of <u>Rb. sphaeroides</u>, the effective midpoint potential (the average of the two one electron couples) has been determined to be about 30 mV more negative than the E_m of the quinone pool (i.e., unbound) in the membrane [1,6]. This implies that the oxidized quinone is bound about ten times more strongly than the fully reduced quinol.

We can therefore summarize the properties of the Q_B site as binding semiquinone very tightly, quinone and quinol quite weakly, and quinol more weakly than quinone. A consistent picture for this would be of a quinone pocket with hydrogen bond donating groups available for interaction with the carbonyl oxygens [2,3,9-11]. Other binding interactions, especially van der Waals type, are also very important, but with respect to specific hydrogen bond formation quinol would be weakly bound compared to quinone. Since the semiquinone is maintained in its charged, anionic form, much of the additional binding energy must come directly from electrostatic sources through, e.g., increased strength of charge-dipole interactions and hydrogen bonds. Although the semiquinone is anionic, the net gain in charge of the RC-Q^- complex is largely neutralized by protonation of the protein [12] and the charge compensation mechanism involves pK shifts of several amino acid groups [13-15]. How this influence is relayed from the quinone site to solvent accessible groups is unknown at the present time.

The weakness of binding of the quinone and quinol to the Q_B site of the RC opens the door to tight-binding competitors and many commercially important herbicides, such as atrazine, appear to act by displacing the secondary quinone from the RCs of Photosystem II and photosynthetic bacteria [16,17]. The binding equilibrium for quinone and inhibitor can be added to the two-electron gate by modifying the pre-first flash states, as follows [8,17,18]:

First Flash

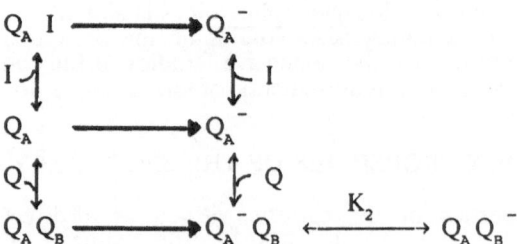

Direct competition between the binding of quinone and herbicides is now the accepted mode of action for these inhibitors of PS II-type RCs: inhibition of Q_A oxidation is brought about by competitive displacement of Q_B [17]. This is confirmed by the recent X-ray structures of the <u>Rhodopseudomonas</u> viridis RC, which show terbutryn, a well known triazine herbicide, or o-phenanthroline, a widely used research inhibitor, bound in the Q_B pocket of the L subunit [11]. The binding locations of these inhibitors and of ubiquinone (Q_B) are not identical, however, and understanding the functional properties of the secondary quinone and the molecular mechanism of herbicide action, can now be redefined as interpreting the implications of the binding structures. This is not a trivial task even with the detailed structures available, and we have begun to approach it through the generation and characterization of herbicide resistant strains of Rb. sphaeroides [18] and Rps. viridis [19]. The responses of quinone and herbicide function to mutations in the binding site structure provide a good example of the fundamental nature of the relationship between binding and redox properties [4].

KINETIC METHODS OF STUDYING THE ACCEPTOR QUINONE COMPLEX

A simple and convenient method of studying the functional properties of the acceptor quinone complex, is through the kinetics of the back reaction or charge recombination seen in the absence of an electron donor to the RC:

$$P(Q_A Q_B) \xrightarrow{\text{hv}} P^+(Q_A Q_B)^-$$

It is now well established that the only significant recombination process is between P^+ and Q_A^-. Thus, when Q_B is active, the recombination occurs via $P^+ Q_A^-$ which is repopulated by back transfer of the electron from Q_B, according to the electron transfer equilibrium constant, K_2:

$$P Q_A Q_B \underset{k_{QA}}{\xrightarrow{\text{hv}}} P^+ Q_A^- Q_B \xleftrightarrow{\quad K_2 \quad} P^+ Q_A Q_B^-$$

Thus, the decay of P^+ when Q_B is functional can be described as follows [8,12]:

$$k_{QB} = k_{QA} \{1 + K_2\}^{-1} \qquad\qquad \text{Eqn. 2}$$

where k_{QA} is the rate constant for $P^+ Q_A^-$ recombination, observed when Q_A^- reoxidation by Q_B is blocked, and k_{QB} is the apparent rate constant for $P^+ Q_B^-$ recombination.

In fact, since Q_B is weakly bound, the above expression should include the dependence of Q_B function on the quinone concentration. Furthermore, since the binding of quinone (as Q_B) is affected by competition from agents such as herbicides, the complete expression becomes [8,18,20]:

$$k_{QB} = k_{QA} \{1 + K_2^{app}\}^{-1} \qquad\qquad \text{Eqn. 3}$$

where: $$K_2^{app} = K_2 \frac{[q]/K_q^L}{1 + [q]/K_q^L + [i]/K_i^L} \qquad\qquad \text{Eqn. 4}$$

Equations 3 and 4 describe the expected kinetics of recombination (actually P^+ decay) when quinone and inhibitor are all in rapid binding equilibrium with the Q_B binding site - rapid, that is, compared to the fast recombination from $P^+ Q_A^-$. Thus, the back reaction rate can be continuously dependent on the concentrations of both quinone and inhibitor, and tends towards the value given by Equation 2 as $[q] \longrightarrow \infty$ and $[i] \longrightarrow 0$. In Rb. sphaeroides, $P^+ Q_A^-$ recombines in about 100 ms; in Rps. viridis, it occurs in 1 ms. These differences are not fundamentally important, but they do lead to some distinctive behaviour which is of practical importance. In Rb. sphaeroides, the quinone binding equilibrium occurs on the same timescale as k_{QA}, so that Equations 3 and 4 are not properly valid. Instead, the observed back reaction kinetics of $P^+ Q_B^-$ are quinone concentration dependent in a manner that is not analytically soluble. In Rps. viridis, on the hand, k_{QA} is much faster than the quinone binding equilibrium and the most notable feature of the recombination kinetics is that they are biphasic, with a fast phase arising from RCs which do not have Q_B bound at the time of the flash

$(P^+_+ Q^-_{A-})$, and a slow phase from RCs that do have Q_B bound at the time of the flash $(P^+_+ Q_B)$. The relative amplitude of the slow phase (ΔS) is a direct measure of the pre-flash binding equilibrium [8,18,20]:

$$\Delta S = \frac{[q]/K_q}{1 + [q]/Kq^D + [i]/K_i^{\ D}} \qquad \text{Eqn. 5}$$

Taking the analogy further suggests the possibility that herbicides might be semiquinone analogues. In Photosystem II, many herbicides bind very tightly, with dissociation constants on the order of 1 nM. In bacterial RCs, the same herbicides are not so efficacious and exhibit binding affinities of about 1 μM, quite similar to measured values for ubiquinones. However, for a homologous series of herbicides, the same order of activity is observed in PS II and in bacterial RCs [18]. We have tried to address the structural basis of herbicide activity by characterizing the behavior of herbicide-resistant bacterial RCs, derived either by chemical mutagenesis and selection [18,19] or by oligonucleotide-directed methods. Our initial studies on RCs from Rb. sphaeroides seemed to support the notion of herbicides as semiquinone analogues, but these studies may have been confounded by the mixed kinetic behaviour of Q_B as a function of quinone concentration (see above). The more clear cut kinetic behaviour of RCs from Rps. viridis has revealed a different conclusion [19].

Table 1   Quinone and Herbicide Dissociation Constants for Reaction Centres of Wild Centres of Wild Type and Mutant Rps. viridis

		K_q or K_i $(\mu M)^a$			K_i/K_q	
		WT	RV4	RV1	RV4/WT[b]	RV1/WT
Ubiquinone (Q-10)		4.5	100 (22)[c]	1600* (355)		
O-phen		10	1500 (150)	10000(1000)	7.5	3
	R4					
Simetryn[d]	(ethyl)	1	100 (100)	900 (900)	4	2
Ametryn	(i-prop.)	0.4	75 (200)	1250 (3000)	7.5	7.5
Terbutryn	(t-but.)	0.18	500 (3000)	250 (1300)	125	4
	X					
Atrazine	(Cl-)	2.5	4000 (1600)	1500 (600)	65	1.7
Atraton[d]	(MeO-)	2.5	4000 (1600)	8000 (3200)	65	8
Ametryn	(MeS-)	0.4	75 (200)	1250 (3000)	7.5	7.5

All data were obtained in 10 mM Tris, pH 9.0, 0.1% Triton X-100, except* which was estimated from measurements in 0.1, 0.025 and 0.006% Triton. The inhibitor data were obtained in the presence of Q-10 at a concentration close to k_a for each strain.

a. The inhibitor data were originally reported in ref. 19 as I_{50} values but here are given as k_i, calculated according to Eqn. 5, using the known [q] and k_q.

b. These columns are double ratios of K_i/K_q normalized to the wild type ratio, as described in the text.

c. Values in parentheses are ratios of the mutant to wild type, viz. K_{RV}/K_{WT}

d. The ametryn data are shown twice, as representative of both herbicide groups.

RESULTS AND DISCUSSION

Table 1 shows quinone and inhibitor binding data for isolated RCs from two herbicide resistant mutants of Rps. viridis, (provided by I. Sinning and H. Michel). The data, which were presented earlier [19], can be viewed in several ways. It is quite clear from the raw data that, compared to the wild type, both mutants are much less sensitive to all inhibitors tested, including o-phenanthroline which binds in a distinct fashion from the s-triazines and sits deeper in the Q_B pocket. The two Rps. viridis mutants also bind quinone (ubiquinone-10) much less avidly than the wild type, RV1 being especially impaired. This is reasonable in view of the overlapping nature of the herbicide and quinone binding domains, but may not be a necessary consequence.

The concomitant effects on quinone and herbicide binding somewhat obscures any discrimination that the mutations may have between the two moieties. This is reavealed more readily by a comparison of the mutant (RV) and wild-type (WT) herbicide affinities, normalized to the respective quinone affinities: $(K_i/K_q)_{RV}/(K_i/K_q)_{WT}$. This procedure clearly shows that RV4 exhibits greater discrimination than RV1, especially with respect to terbutryn, against which the two mutants were selected. The lesser discrimination and the huge change in quinone affinity in RV1 suggests an extensive disruption of the Q_B site.

Further aspects of the dysfunction of the Q_B site in the mutants are revealed by alterations in the redox properties of Q_B which, as described above, are sensitive to the differential binding of the oxidized and reduced forms. In marked contrast to the reported behaviour of herbicide resistant mutants of Rb. sphaeroides [18], both RV1 and RV4 exhibit slower $P^+Q_B^-$ recombination kinetics than the wild type (Table 2). This implies that the $Q_A^-Q_B-->Q_AQ_B^-$ electron transfer equilibrium, K_2, lies further to the right in the mutants. Since the primary effect of the mutation is likely to be on Q_B rather that Q_A, this implies that the midpoint potential of Q_B/Q_B^- is raised. This, in turn, means that the semiquinone species is more stabilized in the mutants or, equivalently, that the quinone is destabilized. The data of Table 1 show that, compared to the wild type, quinone is bound 20 times more weakly in RV4 and 360 times more weakly in RV1.

Table 2	Kinetic and Electron Transfer Data for RCs from Wild Type and Mutant Rps. viridis			
	$k_{QA}(s^{-1})$	$k_{QB}(s^{-1})$	K_2	$\Delta E_m (mV)$
Wild Type	700	9.5	75	110
RV4	700	1.4	500	160
RV1	900	1.3	690	170

Conditions as for Table 2.

In view of the possibly extensive alterations in the Q_B site of RV1, indicated by the massive loss of binding affinity for herbicides and quinone alike, it is noteworthy that this mutant also exhibits altered $P^+Q_A^-$ recombination kinetics. In Rps. viridis, unlike Rb. sphaeroides recombination from $P^+Q_A^-$ occurs substantially by thermal excitation to a state closely related to P^+I^-, the intermediate of the charge separation pathway, which then decays to the ground state [21]. Consequently, the kinetics of recombination from $P^+Q_A^-$ are very sensitive to the energy gap between $P^+Q_A^-$ and P^+I^-, i.e., the redox midpoint potential span between I/I^- and Q_A/Q_A^-. The larger value of k_{QA} in RV1 (900 s^{-1} in RV1 vs. 700 s^{-1} in WT) can be accounted for by a 5-6 mV decrease in the $E_m(Q_A/Q_A^-)$ relative to $E_m(I/I^-)$.

Comparison of k_{QA} and k_{QB}, according to Equation 2, yields K_2, from which we find the midpoint potential spans between Q_A/Q_A^- and Q_B/Q_B^- to be 110, 160 and 170 mV, in WT, RV4 and RV1, respectively (Table 2). Taking into account the small shift in the E_m

of Q_A in RV1, these values indicate positive shifts in E_m (Q_B/Q_B^-) of 50 and 55 mv, in RV4 and RV1 respectively, equivalent to 7 and 9-fold changes (decreases) in the binding affinity of quinone relative to semiquinone (see Eqn. 1). The binding measurements of Table 1 show decreases in quinone affinity of 20-fold and 360-fold, in these mutants. Thus, in RV4 the semiquinone affinity need change by only a factor of 3, while in RV1 the semiquinone affinity must decrease by at least 40. In either case, however, the major effect is on the quinone affinity. Clearly, therefore, the mutational events "view" the herbicides more as quinone analogues than as semiquinone analogues. However, this may not reflect significant structural distinctions between the nature of the quinone and semiquinone interactions with the protein, but more the fact that the energetics of semiquinone binding are dominated by ion-dipole interactions that are not present for the neutral quinone and herbicide molecules.

The Q_B site, shown schematically in Figure 1, is defined by a relatively small number of residues from the D and E helices of the L subunit and from the interhelix loop (including a short segment of α-helix, "de") joining them [10,11]. The principal residues are His190 and Ser223 which form hydrogen bonds with the quinone carbonyls, and Ile229, Leu193, Glu212 and Phe216, which form close contacts over the surface of the head group. The Q_A site is situated symmetrically with respect to the Q_B site, related to it by a C_2 rotation axis through the iron atom (normal to the page). The Q_A and Q_B binding domains are similar in nature with the marked exception of the Glu212 residue, for which there is no polar equivalent in the Q_A site. Although there are significant binding interactions with the isoprene side chains of the ubiquinone series, these do not contribute to the functional specificity which resides in the head group alone [2,3].

Some residues of less obvious role are also shown in Figure 1. Tyrosine 222 is known to be a sensitive site in Rb. sphaeroides for the production of herbicide resitant mutants [22, and E Takahashi and C A Wraight, unpublished observations). It is not directly involved in the Q_B and herbicide binding domain, but does hydrogen bond to the peptide backbone of the M subunit. Loss of the H-bond might allow some collapse of the Q_B pocket. Arginine 217 is shown for similar reasons. It is apparently active as a counter ion to the net dipole of the neighbouring α-helix, and loss may induce relatively large structural effects. It, too, has been implicated in some reported herbicide resistances [23] and is found as one of a double mutation in cases where Ser223 has been altered to alanine. In Rb. capsulatus the single mutation Ser223-->Ala has been reported to be an assembly mutation, with no viable reaction centres present [24]. Thus, the occurrence of Arg217-->His, in the otherwise unlikely event of a double mutation, may arise because of the intense selective pressure to suppress the assembly mutation and may not indicate any direct role of this residue in herbicide resistance. This can be tested by site directed mutation studies on this position.

The X-ray structural work on the Rps. viridis RC shows the binding of two Q_B-inhibitors - terbutryn (a triazine) and o-phenanthroline [11]. The former is located more towards the Ser223 end of the Q_B pocket. It forms a hydrogen bond between the 2-ethyl-amino hydrogen and the Ser-OH oxygen, and another between the 1-aza nitrogen of the triazine ring and the peptide NH of Ile224. The tertiary butyl group of terbutryn is located close to Phe216 and Val220, and the 6-thiomethyl group points towards the D helix. It is noteworthy that the role of Ser223 in hydrogen bonding to terbutryn, where it is a H-bond acceptor, is inverted relative to its contribution to Q_B binding, where it is a H-bond donor. When binding terbutryn, the hydrogen of the Ser223-OH is H-bonded to the amide oxygen of AsN213 (Asp213 in Rb. sphaeroides [25].

O-phenanthroline is found to bind deeper in the Q_B pocket, the two aza-nitrogens sharing H-bonds with the NH of His190. The molecule forms good contacts with Ile229, Leu193 and Glu212 [11].

The selectivity of the two RV mutants towards inhibitors of different structure should reveal some information on the structure of the binding site. At present, however, the principles of structure/function determination are poorly understood. Furthermore, the protein sequence of the two mutants are not reliably known as genotypic changes may have occurred between the functional characterizations described here and the sequence determinations [I. Sinning and H. Michel, personal

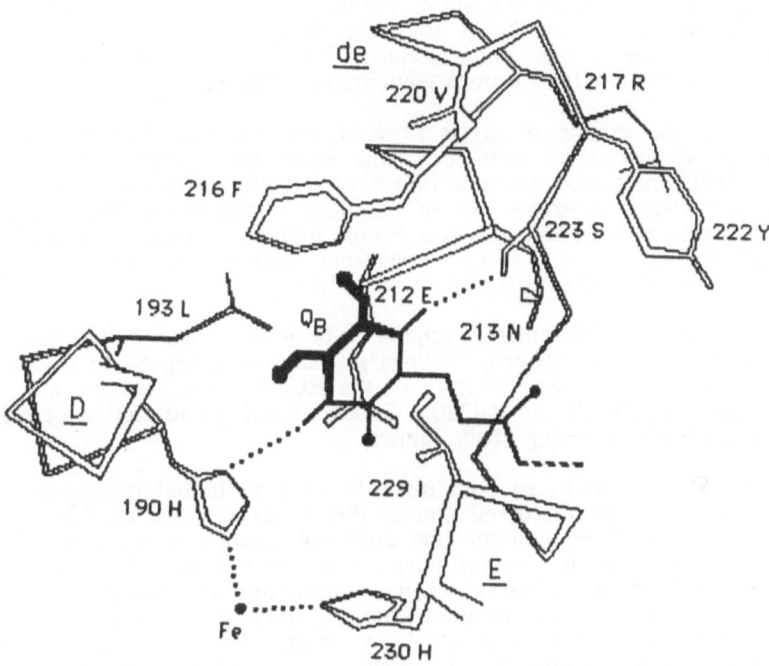

Figure 1.   The Q_B binding site of reaction centres from Rps. viridis (adapted from ref. 11). Principal residues are indicated by the standard one letter code and numbered according to their position in the L-subunit sequence. The two transmembrane helices are denoted D and E, and de is the connecting helix.

communication]. We may, however, use the available data in a speculative exercise in which the triazines are best compared within two classes represented here. Simetryn, ametryn and terbutryn represent a homologous series : 2-ethylamino-4-(R_4)-amino-6-methylthio-s-triazines, where R4 is ethyl (simetryn), isopropyl (ametryn) or tertiary butyl (terbutryn) as given in Table 1. It is clear that the larger the substituent in position 4, the more dramatic the resistance of mutant RV4, compared to the wild type affinity, and the greater the discrimination between herbicide and quinone. This is particularly true for terbutryn which was the herbicide used in the selection for these mutants. For mutant RV1, resistance to all three methylthio-triazines is comparable, as is the discrimination factor.

The other herbicide group includes 2-ethylamino-4-isopropylamino-6-X-s-triazines, differing only in the 6-position: chloro (atrazine), methoxy (atraton) and methylthio (ametryn). Here the response of RV4 is of relatively greater resistance to atrazine and atraton, than to ametryn. This may arise from the greater bulk and bond lengths of the methylthio group compared to the chloro or methoxy substituent, or it may reflect the lesser electronegativity of the sulfur atom. In RV1, the difference in response to the three triazines is less marked - they are all quite strongly affected - but atraton and ametryn are more discriminated against. Possibly this may reflect the greater bulk of the methoxy and methylthio groups compared to chlorine.

In general, the effects observed in RV1 are dramatic but do not show strong discrimination against any herbicide, and even the quinone affinity is strongly depressed. This may arise from a rather global alteration in the Q_B site, affecting all regions - perhaps by some sort of collapse. This is consistent with preliminary sequence data which shows that this is a double mutant - a phenomenon also reported by Sinning and Michel [23]. It is also consistent with the fact that even the properties of Q_A appear to be altered in this mutant, viz. the E_m.

The effects of the RV4 mutation appear to be more selective than those of RV1, and significant discrimination between triazines is made with regard to the bulk of the R_4 substituent. From the known structure of the RC, this suggests the area affected may be in the region of Phe216 or Val220. Substantiation or refutation of this speculation will have to await resequencing of the mutant.

The effect on Q_B itself, in both mutants, is a subtantial raising of the E_m of the first reduction step. As discussed above, this is largely accountable by the decrease in affinity for the oxidized quinone and does not require comparably large changes in the affinity for semiquinone. Nevertheless, it is worth remarking, as a more molecular view of the origin of the redox properties of ubiquinones, that they are very sensitive to the angular displacement of the two methoxy groups (positions 2 and 3 on the quinone ring). This is because steric hindrance between the two prevents them both aligning in the ring plane, which would be the position of maximum effect of the mesomeric resonance structures involving the lone p-orbitals of the methoxy oxygens [7,26]. Thus, the magnitude of this electron donating behaviour of the methoxy groups is sensitive to how close they can come into the ring plane. Relatively minor perturbations of the Q_B pocket can, therefore, be expected to have significant effects on the in situ redox potential of the quinone couples and their binding affinities.

ACKNOWLEDGMENTS

This work was supported by grants from the Competitive Research Grants Office of the USDA (AG-86-CRCR-1-2149) and from the NSF (DMB 86-17144).

REFERENCES

1. Crofts AR and Wraight CA (1983) Biochim Biophys Acta 726: 149-185
2. Gunner MR, Robertson DE and Dutton PL (1986) J. Phys Chem 90: 3783-3795
3. McComb JC (1987) PhD Thesis, University of Illinois, Urbana IL
4. Wraight CA (1982) in Function of Quinones in Energy Conserving Systems (Trumpower BL, ed) pp 181-197, Academic Press

5. Rutherford AW, Heathcote P and Evans MCW (1979) Biochem J 182: 515-523
6. Rutherford AW and Evans MCW (1980) FEBS Lett 110: 257-261
7. Prince RC, Dutton PL and Bruce JM (1983) FEBS Lett 160: 273-276
8. Stein RR (1985) PhD Thesis, University of Illinois, Urbana IL
9. Gunner MR and Dutton PL (1986) Biochim Biophys Acta. In press
10. Deisenhofer J, Epp O, Miki K, Huner R and Michel H (1985) Nature 318: 681-624
11. Michel H, Epp O and Deisenhofer J (1986) EMBO J 5:2445-2451
12. Wraight CA (1979) Biochim Biophys Acta 548: 309-327
13. Maroti P and Wraight CA (1988) Biochim Biophys Acta 934: 314-328
14. Maroti P and Wraight CA (1988) Biochim Biophys Acta 934: 329-347
15. McPherson PA, Okamura MY and Feher G (1988) Biochim Biophys Acta 934: 348-368
16. Pfister K and Arntzen CJ (1979) Z Naturforsch 34c: 996-1009
17. Wraight CA (1981) Israel J Chem 24: 348-354
18. Stein RR, Castellvi AL, Bogacz JP and Wraight CA (1984) J Cell Biochem 24: 243-259
19. Shopes RJ and Wraight CA (1987) in Progress in Photosynthesis Research (Biggins J, ed) Vol 2, pp 397-400, Martinus Nijhoff, Dordrecht
20. Wraight CA and Stein RR (1983) in Oxygen Evolution System of Plant Photosynthesis (Inoue Y, et al, eds) pp 383-392, Academic Press
21. Shopes RJ and Wraight CA (1987) Biochim Biophys Acta 893: 409-425
22. Paddock ML, Rongey SH, Abresch EC, Feher G and Okamura MY (1988) Photosynth Res 17: 75-96
23. Sinning I and Michel H (1987) Z Naturforsch 42c: 751-754
24. Bylina EJ and Youvan DC (1988) Gene. In press.
25. Williams JC, Steiner LA and Feher G (1986) Proteins 1: 312-325
26. Prince RC, Halbert TR and Upton TH (1988) in Advances in Membrane Biochemistry and Bioenergetics (Kim CH, Tedeschi H, Diwan JJ and Salerno JC, eds) pp 469-478, Plenum Press

STRUCTURAL AND FUNCTIONAL ASPECTS OF THE PHOTOSYNTHETIC ELECTRON TRANSPORT CHAIN OF *RHODOBACTER CAPSULATUS*

Giovanni Venturoli and B. Andrea Melandri

Department of Biology, University of Bologna

Bologna, Italy

INTRODUCTION

The electron transfer chain in membrane fragments of non-sulphur purple bacteria (*Rhodospirillaceae*) such as *Rhodobacter sphaeroides* and *Rb. capsulatus* is the simplest and best characterized photosynthetic system under study. It consists of only two electron transfer complexes, the photosynthetic reaction center (RC) and the ubiquinol-cyt. c_2 oxidoreductase, and of two mobile electron carriers, of which one, ubiquinone-10 (UQ) diffuses in the lipid phase of the membrane, and the second, cyt. c_2, interacts with the complexes at the water- membrane interface. A cyclic electron transfer takes place in this system, where the two complexes can exchange electrons through the action of the mobile electron carriers (Crofts and Wraight, 1983; Melandri and Venturoli, 1984; Dutton, 1986).

In the following we will review some structural and functional aspects of this bacterial system. Due to the limitation in space we will not be able to quote all the relevant work done in this area. For this we apologize in advance. We have also, somewhat arbitrarily, considered chromatophores of *Rb. capsulatus* and *Rb. sphaeroides*, the two species most extensively studied, as equivalent systems. Although this is amply justified by the strict similarity observed between these two bacteria, one should always bear in mind that some structural or functional differences certainly exist.

THE PHOTOSYNTHETIC REACTION CENTER

The structural organization of the reaction center (RC) is discussed in detail in another chapter of this book. Some aspects of the organization of the RC will be reviewed here in order to discuss the functional interactions with the other electron transfer components present in the membrane. The RC proteinaceous structure is formed by three subunits, indicated as L, M and H; in *Rps. viridis* a permanently bound cytochrome c can be considered as a fourth subunit. The functional cofactors present in one reaction center are: 4 molecules of bacteriochlorophyll, 2 molecules of bacteriopheophytine, 2 molecules of quinone and one atom of Fe (Okamura et al., 1982). The precise nature of these cofactors is dependent on the bacterial species studied. Of the two RCs so far crystallized and structurally characterized, the one from *Rps. viridis* contains bacteriochlorophyll b's, bacteriopheophytine b's and one menaquinone (Q_A, being the second quinone (Q_B) lost during crystallization); the RC from *Rb. sphaeroides* contains bacteriochlorophyll and bacteriopheophytine a's and ubiquinone-10. These cofactors are arranged across the membrane, bound to a bundle of ten membrane-spanning helices, which in turn are part of the structure of the L and M apoprotein subunits of the complex. A transmembrane helix formed by the third subunit of the complex (the H subunit) does not interact directly with the cofactors (Deisenhofer et al., 1985; Michel et al., 1986; Allen et al., 1987a).

The cofactors are arranged in the RC in such a way to form a transmembrane sequence; they are positioned with an approximate twofold rotational symmetry axis, which is nearly perpendicular to the membrane plane. Thus, two nearly identical branches of cofactors are formed, each containing a sequence [BChl]$_z$ → [BChl] → [BPheo] → [Q], where [BChl]$_z$ is the doublet acting as primary electron donor, and [BChl] is a second monomeric bacteriochlorophyll molecule. In each chain the cofactors are at a distance such to assure an efficient electron exchange, although some aromatic residues of the apoproteins might be also involved in electron transfer (Michel et al., 1986; Allen et al., 1987b). In spite of the apparent similarity, the two branches are not functionally equivalent in the photochemical charge separating process; only branch A (i.e. the one including the Q_A quinone) is actually active in photosynthesis. The reasons of this functional asymmetry are practically unknown, and are possibly related to subtle differences in the positioning of the cofactors on the apoprotein.

Relevant to the functional studies of the RC as an electrogenic device in the membrane, are the position of the whole complex in the membrane lipid layer, and the positions of the cofactors within the complex. Minima of free energy for protein-lipid interaction were found in correspondence with a position in which the twofold axis is perpendicular to the membrane plane and the complex is located inside the hydrophobic core of the membrane for a depth of 40 - 45 A. With this positioning, which is dictated by the length and angles of the hydrophobic helices of the apoproteins the [BChl]$_z$ doublet is placed at about 26 - 28 A from the cytoplasmic face of the membrane dielectric and 12 - 14 A from the periplasmic face (for these calculations a total thickness of 42 A was utilized) (Yeats et al., 1987). The two quinones and the Fe atom are coplanar and situated between 0 and 3 A from the cytoplasmic face.

The positioning of [BChl]$_z$ some 14 A inside the membrane dielectric implies that part of the charge separating process occurs during the reduction of [BChl]$_z^+$ by the cytochrome c_2 acting as secondary electron donor. In *Rb. sphaeroides*, as opposed to *Rps. viridis* which possesses a four-heme cytochrome c permanently bound to the RC, this electron carrier is a hydrophylic cytochrome c, which diffuses along the membrane surface, shuttling electrons from the UQH_2- cyt. c_2

oxidureductase complex to the RC (and, in the respiratory metabolism, to a cyt. c_2 oxidase (Zannoni and Baccarini-Melandri, 1980). In the X-ray studies of the *Rb. sphaeroides* RC the docking groups for cyt. c_2 have been identified as a few Asp and Glu residues on the non-helical portion of the L and M subunits protruding from the membrane on the periplasmic side. They are interacting electrostatically with corresponding Lys residues of the cyt. c_2 protein. In this construct the heme -[BChl]$_2$ distance (center to center) was 19 A (Allen et al., 1987a).

Electron transfer from reduced cyt. c_2 to [BChl]$_2^+$, therefore, induces a considerable charge displacement across the membrane. This structural conclusion is in agreement with evidence for a charge separating process coupled to the reduction of [BChl]$_2^+$ and evaluated from the electrochromic signal of carotenoids associated with the redox reactions within the RC. The phase of this electrochromic signal (whose amplitude for one turnover of the RC is believed to be coupled to the transfer of one electron per RC across the membrane) is composed of two subphases (phase I and II), respectively coincident with the [BChl$_2$] $\rightarrow$ [Q_A] and the cyt.c_2 $\rightarrow$ [BChl]$_2^+$ electron transfer processes (Jackson and Crofts, 1971; Dutton and Prince, 1978) The relative amplitudes of phase I vs. II is 6:4, and compares well with the atomic distances 28: 19 A as revealed by the X-ray studies. A corollary of the above conclusions is that the onset of a transmembrane potential difference will displace the redox equilibrium of the cyt.c red/ cyt.c ox vs. [BChl]$_2$/[BChl]$_2^+$ couples, and have kinetic influence on the rate constants for the forward and reverse reaction. The former of these expectations has been demonstrated experimentally (Takamiya and Dutton, 1977).

In the resolved X-ray structure of the *Rb. sphaeroides* RC the two ubiquinones Q_A and Q_B and the Fe atom are coplanar with each other and with the surface of the membrane. This also is in agreement with electrochromic measurements, which always failed in demonstrating any electrogenic event coupled to the inter-quinone electron transfer (see however Semenov et al. (1986)). The quinones and Fe are not directly coordinated; the non-heme Fe is coordinated by four His and one Glu residues. Two of these His (L190 and M217) are located between Q_A and Q_B and could contribute to the mechanism of electron transfer between Q_A and Q_B ($t_{1/2}$= 100 μs); The role of the Fe atom in this process is still obscure (Allen et al., 1987b). Direct evidences for magnetic coupling between semiquinone radicals of Q_A and Q_B with the Fe have been obtained by EPR spectroscopy (Okamura et al., 1982).

The two-quinone/Fe complex is involved with the two-electron gate mechanism, whereby the one-electron photooxidation of [BChl]$_2$ is coupled to the two-electron process of quinone reduction (Crofts and Wraight, 1983; Okamura et al., 1982). This mechanism originates from the stabilization, through binding to the apoprotein, of the semiquinone radical at the Q_B site. This site can therefore exchange quinones with the pool in the membrane only when the quinone is either fully reduced or fully oxidized. The structural mechanism of this Q_B^- stabilization is unresolved. Q_A, on the contrary, appears to be permanently bound to the RC, and can be reduced only with one electron giving rise to the Q_A^- anionic radical.

The physiological consequence of the stabilization of Q_B^-, is that the quinol can be released from the RC into the membrane lipids only after two subsequent turnovers of the RC (the two electron gate). Another thermodynamic consequence of the stabilization of Q_B^- is that the difference in the midpoint potentials of the Q_B/Q_B^- and the Q_B^-/Q_BH_2 couples is narrowed, and the two subsequent electron donation steps of quinone reduction can occur at comparable redox potentials (Kleinfeld et al., 1984; Kleinfeld et al., 1985).

THE UBIQUINONE POOL

The chromatophores of *Rb. capsulatus* contain a large pool of ubiquinone-10, in an excess of about 25-fold the RC (Baccarini-Melandri and Melandri, 1977). This pool is believed to be freely diffusable in the lipid bilayer, and to distribute the reducing equivalents generated by different photosynthetic complexes to quinol-oxidizing complexes. To the best of the present knowledge, in *Rb. capsulatus*, two of such complexes exist: a UQH_2- cyt.c_2 oxidoreductase (or bc_1 complex), involved both in photosynthesis and respiration, and a quinol oxidase (alternative oxidase or cyt.$b_{2\infty}$), involved in one branch of the respiratory chain (Zannoni and Baccarini-Melandri, 1980). The value of the diffusion coefficient of UQ in the membrane lipids is still a matter of debate, since a value of $3\ 10^{-9}\ cm^2s^{-1}$ has been evaluate by FRAP (Fluorescence Recovery After Photobleaching) (Gupte et al., 1984), while a 1000 fold faster diffusion has been extimated by a fluorescence quenching approach (Fato et al., 1986).

The quinone pool in *Rb. sphaeroides* appears to be thermodynamically homogeneous: its midpoint potential at pH 7.0 is 90 mV, with a pH dependence of -60 mV / pH unit (Takamiya and Dutton, 1979). In the presence of the redox mediators utilized for equilibrium titrations, therefore, UQ behaves as a two-electron/two-proton carrier between at least pH 5 and 9. A fraction of the quinone complement, about 20% of the total, cannot be reduced at E_h as low as -10 mV, pH 7; it was proposed that this quinone is not operative in the pool, but is either inactivated or bound to protein complexes (as are Q_A and Q_B in the RC).

Another property of the UQ pool in agreement with its free diffusion in the membrane lipids is its ready extractability with apolar solvents from dehydrated membranes. Repetitive washings of lyophilized chromatophores can extract the total UQ pool. This quinone can be reincorporated into the lipid layer if pure quinone is added to the dehydrated membrane; the maximal concentration that can be reached by this procedure can be up to threefold that of the native unextracted chromatophores. Rehydrated chromatophores perfectly functioning, except for an altered size of the UQ pool, can be obtained with this extraction and/or enrichment procedure. After extensive extraction also the amount of Q_B in the RC can be decreased, in agreement with the ability of Q_B in exchanging with the UQ pool (Baccarini-Melandri and Melandri, 1977; Baccarini-Melandri et al., 1982).

THE UQH$_2$-CYT.C$_2$ OXIDOREDUCTASE

The UQH$_2$-cyt c$_2$ oxidoreductase (or bc$_1$ complex) is the second oligomeric electron carrier forming the cyclic photosynthetic chain of *Rb. capsulatus* (Hauska et al, 1983).

The structure of the bc$_1$ complex in some bacteria is known at the level of the primary sequence of the apoproteins (based on the sequencing of the operon containing the structural genes), and the stoichiometry and the putative location of the prostetic groups (Gabellini and Sebald, 1986; Davidson and Daldal, 1987). The complex is formed by three subunits, respectively apoproteins of a cytochrome of b type, a cytochrome of c type (or cyt.c$_1$), and a FeS protein. Of these three electron carriers, only cyt. b presents a sequence typical of intrinsic membrane proteins and can in principle catalyze a transmembrane electron transfer (Widger et al., 1984)).In most of the structural models based on DNA sequences this cytochrome has been considered to have nine transmembrane helices, forming a structural bundle across the lipid dielectric. The two heme groups were placed near the opposite faces of the membrane, coordinated by two couples of conserved His, present in helices II and V. This arrangement is somewhat reminiscent of the structure of the RC, and could be very apt to electron conduction across the membrane. In this model the distance of the first heme (cyt.b$_L$ or cyt. b$_{566}$) is about 15 A from the membrane surface, and the inter-heme distance (center to cente) is 20 A. The heme b doublet is often referred to as the low potential chain of the bc$_1$ complex. Recently, after comparing several sequences of bc$_1$ complexes, Crofts et al (1987) proposed that one of the helices (helix IV) has not a sufficiently conserved hydrophobic structure to be considered as a transmembrane element; these authors have, therefore, suggested an eight helix strucure. This alteration of the model has no bearing on the location of the two heme groups (although helices II and V from antiparallel become parallel), but are important for the topological location of the binding sites of the inhibitors of the Q$_o$ and Q$_i$ sites on the two faces of the membrane (see below).

The major portion of the other two apoproteins is hydrophilic; these portions contain the binding sites for the prostetic groups and are attached to the membrane only through a hydrophobic segment. This is particularly true for cyt.c$_1$, while part of the FeS apoprotein, containing the putative ligand groups of the Fe$_2$S$_2$ cluster presents a limitately hydrophobic portion, compatible with a partial insertion in the lipid bilayer (Link et al., 1987). This is important in view of the possibility that the FeS protein contributes, together with cyt. b$_{566}$, to the formation of the UQH$_2$ oxidizing site (site Q$_o$),located on the periplasmic face of the membrane. Since cyt. c$_1$ and the FeS protein form the so called high potential chain of the complex, the electron transfer in this segment should be equipotential and should not perform electrical work.

Any possible mechanism of electron transfer within the bc$_1$ complex must consider the thermodynamic constrains of the

redox potential of electron carriers. Besides the midpoint
potentials of the electron donor and acceptor (E^7_m= 90 mV for
UQ and E^7_m=330 mV for cyt.c$_2$), the intracomplex carriers have
the following redox potentials (in *Rb. capsulatus* and at pH
7):FeS protein 310 mV, cyt.c$_1$ 285 mV, cyt. b$_{561}$ 50mV and
cyt.b$_{566}$ -120 mV (Garcia et al., 1987). The two b-type hemes,
although titrating at two distinct redox potentials, are, in
fact, the prostetic groups of a single di-heme cytochrome.

A   B

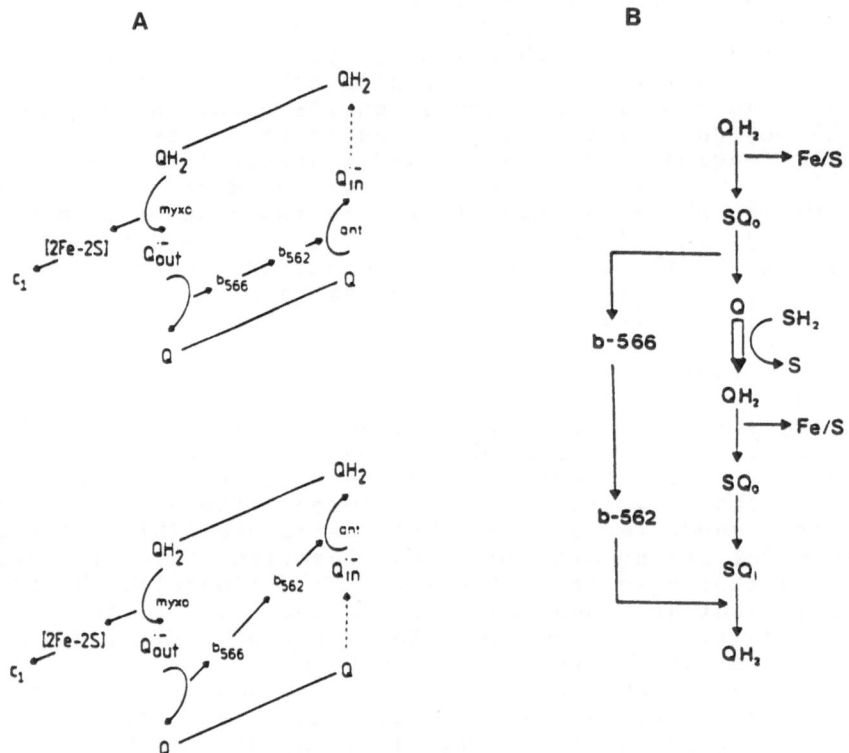

Fig.1. Schemes showing the pathway of electron transfer in the
        Q-cycle (A) and in the semiquinone-cycle (B). In B, the
        reaction between Q and SH$_2$ represents hydrogen transfer
        from hydrogenated substrates to the ubiquinone pool.
        Scheme A is taken from de Vries (1986) and scheme B
        from Wikstrom and Krab (1986).

     Two models are presently under discussion for the redox
mechanism within the bc$_1$ complex. Both models suggest a
non-linear electron transfer pathway in order to account for
the involvement of the low potential cyt.b$_{566}$. In both model
(see Fig.1 A and B) this cytochrome is thought to act as
electron acceptor from the strong reductant UQ$^-$, produced upon

the one-electron oxidation of UQH_2 by the high potential chain, acting in the sequence: $Fe_2S_2 \rightarrow cyt.c_1 \rightarrow (cyt.c_2)_{sol}$. The two one-electron oxidation steps are thought to occur in a concerted reaction at the QH_2 oxidizing site Q_o. The sequence of the electron transfer should include the formation of a Q^- radical, stable enough to deliver efficiently one electron to cyt. b_{566}, but unstable enough to satisfy the thermodynamic constrains dictated by the large redox potential gap existing between cyt.b_{566} and the FeS protein(about 400 mV, corresponding to a stability constant for the dismutation reaction of the semiquinone equal to $10^{-6.5}$ (see Appendix)). Such value of the stability constant requires a very fast electron transfer to cyt. b_{566}, or a conformational change in the Q_o site such to shield Q^- from any exchange with the UQ pool (Rieske, 1986). The two protons of the UQH_2 oxidation are proposed to be released in the periplasmic space, in agreement with the location of the high potential chain components. Since the Q_B site of the RC is located on the opposite face of the membrane, the transfer of two electrons from the RC to the bc_1 complex forms the H^+-translocating arm of a protonic loop, utilizing UQH_2 as a transmembrane proton carrier.

According to all models the production of a strong reductant in the cyt. b doublet is utilized for transferring the electron on the opposite side of the membrane, to cyt.b_{561} and, thus, for performing electrical work. The distinctive features between the two models resides in the nature of the electron acceptor from cyt b_{561}. In model A (the Q cycle of Mitchell (1976)) this acceptor is a fully oxidized UQ in the lipid phase, encountering collisionally the bc_1 complex in a second UQ reductase site (site Q_i); in the alternative scheme B (the semiquinone cycle proposed by Wikstrom and Krab (1986)) the acceptor is a semiquinone radical, translocated within the complex from the UQH_2 oxidase site (site Q_o) to the opposite face of the membrane. In this latter case the electrogenic step will coincide also with the traslocation of the negative charge of the semiquinone anion. The basic difference of the two mechanisms is therefore that in the latter the oxidant of cyt. b_{561} is produced inside the bc_1 complex, while in the Q cycle this oxidant is already present in the membrane as an independent species. The existence of either one of the two mechanisms can therefore be demonstrated by studying the kinetics of interaction of the complex with the UQ pool both at the Q_o and at the Q_i sites.

Only in the semiquinone cycle the production of quinol at the Q_i site is a one-electron reaction. In the Q cycle scheme two sequential electron donations to UQ are needed in a sort of two-electron gate mechanism. In order to avoid insuperable endoergonic conditions for the formation of Q^- from UQ by an electron transfer from cyt.b_{561} (Em = 50 mV at pH ?), the semiquinone must be largely stabilized at the Q_i site. This requirement is analogous to that already resolved structurally at the Q_B site of the RC. Alternatively, UQ could be directly reduced in a concerted two-electron reduction involving the cooperation of two Q_i sites in a dimeric structure of the bc_1 complex (de Vries, 1986). Although there is some evidence for the formation of dimers in the isolated bc_1 complex, no clear functional requirement for dimerization has been so far

demonstrated. On the other hand, the existence of a stable
semiquinone at the Q_i site in bacterial chromatophores has
been demonstrated by EPR spectroscopy (Robertson et al.,
1984). The topological location of the Q_i site in the bc_1
complex is still debated. The Q cycle mechanism proposes a
location opposite to that of the Q_o site, i.e. on the
cytoplasmic face of the membrane (cf. the schemes in Fig.2).
This is in agreement with the topological location of the
mutations conferring resistence to Q_i-specific inhibitors,
like antimycin, in the 8-helix structure proposed by Crofts et
al. (1987). Also, electrochromic signals coupled to the
activity of the bc_1 complex (i.e. sensitive to Q_o specific
inhibitors) have been identified: the extent of these signals

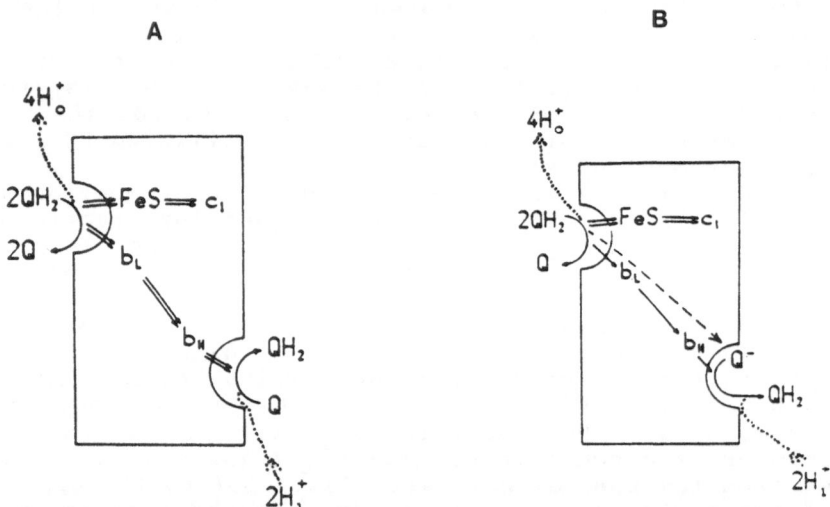

Fig.2. Schemes of the Q-cycle (A) and of the semiquinone-cycle
        (B), emphasizing the topology of electron transfer with
        respect to the Q_o and Q_i sites of the bc_1 complex.
        Solid lines indicate electron transfer; dotted lines,
        proton transfer; dashed lines, semiquinone transfer.
        Taken from Rich (1986).

(Phase III of the flash induced electrochromic shift)
indicates that the electron crosses the entire membrane
dielectric (Jackson and Dutton, 1973). The sensitivity of
Phase III to antimycin is only partial; since this inhibitor
is thought to displace UQ from the Q_i site, blocking in this
way the oxidation of cyt.b_{561}, this is considered as evidence
for an electrogenic step during the electron transfer from
cyt.b_{561} to the Q_i site (Glaser and Crofts, 1984). This
conclusion is also in agreement with the redox equilibrium
between cyt. b_{561} and this site in the presence of a membrane
potential (Venturoli et al., 1987; Robertson and Dutton,
1987). Whether the electrogenic step coincides with the
translocation of one electron from heme b_{561} inside the
membrane to the Q_i site on the surface, or rather of one

proton to an intramembrane quinone site is an unsolved problem which requires further structural resolution (Robertson and Dutton, 1987).

Whatever the case may be, the production of QH₂ at the Qi site occurs at a side of the membrane opposite to that where the oxidation of QH₂ takes place (cf. the schemes in Fig.2). The diffusion of QH₂ from Qi to Qo, together with the transfer of electrons through the cyt.b doublet, produces an additional protonmotive loop. The entire cyclic system of *Rb. capsulatus* (or of other similar non-sulphur purple bacteria like *Rb. spheroides*) includes therefore two protonmotive loops, whose electron transferring arms are the RC (from cyt.c₂ to Qв) and the cyt.b₅₆₆-cyt.b₅₆₁ doublet (from Qo to Qi), and the proton-trasferring arms are coincident with two eletrons transferred electroneutrally by the transmembrane diffusion of one quinol molecule.

THE KINETICS OF THE QUINOL OXIDIZING SITE (Qo)

The reaction of ubiquinol at the Qo site can be studied in several ways, following the reduction rate of components of the bc₁ complex; the most convenient experimental procedure in bacterial chromatophores is to monitor the reduction of cyt. b₅₆₁, after inhibition of its oxidation at the Qi site with antimycin. This experiment can be performed in anaerobiosis, under controlled redox potential, inducing the reaction with one flash of actinic light; this, photooxidizing the RC, produces a corresponding number of electron holes in the high potential chain of the bc₁ complex. The kinetic traces of cyt. b₅₆₁ reduction present two characteristic features: a) a short lag period before the onset of the reduction; b) a steady rate of reduction untill the maximal level of reduction is reached. Both features have been interpreted on the basis of the interaction of the Qo site with the UQ pool in the membrane (Crofts et al., 1983).

If the rate of reduction, after the lag , is measured as a function of Eh between 200 and 50 mV at pH 7, it can be demonstrated that the reduction is progressively accelerated at Eh< 150 mV, untill, after reaching a maximum value, again declines. This kinetic behaviour has been attributed to the prereduction of the UQ pool upon lowering the Eh below 150 mV, and, consequently, to the increasing availability of UQH₂ as substrate at the Qo site (Crofts et al., 1987). This interpretation has been verified accurately in our laboratory by exploiting the possibility of increasing or decreasing the size of the UQ pool by extraction and enrichment procedures (Venturoli et al., 1986). It was observed, as expected from the model, that when the size of the pool is decreased, the titration curve of the rate vs. Eh is shifted towards values more negative than the controls, while the contrary occurs following an enrichment (Fig.3). It is clear, therefore, that the turnover at the Qo site depends both on the size of the UQ pool and on its degree of reduction.

In a more quantitative approach, the rate of reaction at the Qo site has been expressed as a Michaelis- Menten equation, in which the concentration of the substrate UQH₂ is a function of the Eh and of the total concentration of UQ;

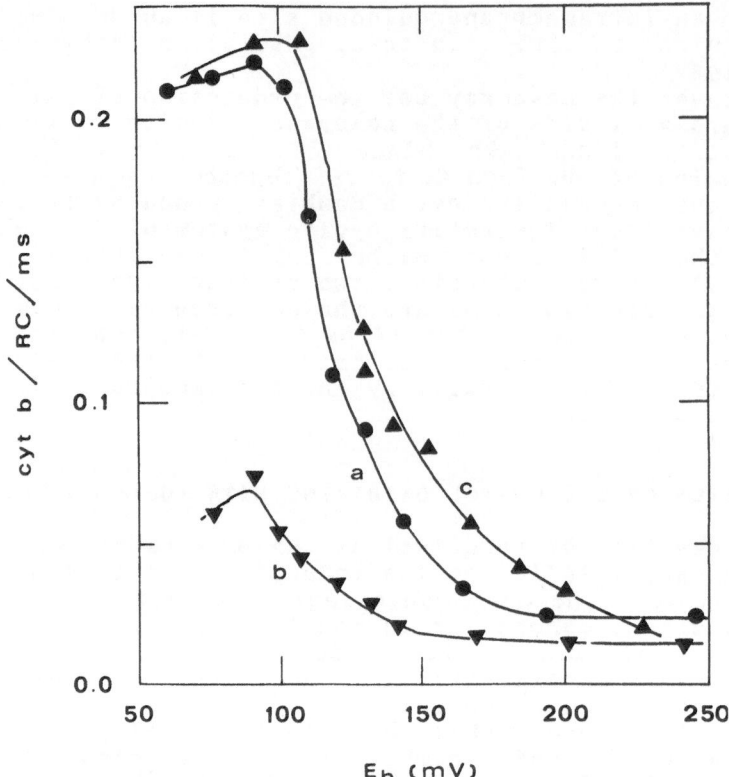

Fig.3. Titration curves of the initial reduction rate of cyt
b_{561} following a single turnover flash in UQ-extracted
and reconstituted/enriched chromatophores. a: control
lyophilized chromatophores (24 UQ/RC); b: partially
UQ-extracted chromatophores (8 UQ/RC); c:
UQ-reconstituted/enriched chromatophores (53 UQ/RC)
obtained by reincorporation of UQ-10 in the extracted
preparation (a) characterized by 8 UQ/RC. Taken from
Venturoli et al. (1986).

analogously, the V_{max} of the reaction was put as a function of
the E_h, since in our experimental conditions, the rate can be
affected by the prereduction of cyt. b_{561} before the flash.
    Formally:

$$V_o (E_h, Q_T) = V_{max}(E_h) [s(E_h, Q_T)/ (K_m + s(E_h, Q_T)] \qquad (1)$$

The dependence of the UQH_2 concentration on E_h
and Q_T has been expressed by means of the Nernstian behaviour
of an UQ pool of size Q_T, as follows:

$$s(E_h, Q_T) = 0.5 + Q_T/ [1 + \exp((E_h - E_{m,Q}) 2F/ RT)] \qquad (2)$$

where $E_{m,Q}$ is the midpoint potential of the UQ pool and 0.5
represents the number of UQH_2 molecules produced per flash
per RC.

The dependence of the V_{max} on E_h can be expressed considering the prereduction of cyt. b_{561}. In this case $V_{max}(E_h)$ would correspond to the maximal rate of cyt. b_{561} reduction normalized to the fraction of bc1 complex left oxidized at a given E_h:

$$V_{max}(E_h) = V_{max} \frac{\exp((E_h - E_{m,b})F/RT)}{1 + \exp((E_h - E_{m,b})F/RT)} \qquad (3)$$

The insertion of eq. 2 and 3 into eq. 1 yields the simulation shown in Fig.4, which is significantly similar to the experimental results. As an additional test, the experimental rates, both for the native and extracted chromatophores, were grouped in a single double reciprocal plot as a function of $[UQH_2]^{-1}$ evaluated from eq. 2). All experimental points could be fitted with a single Michaelis Menten equation, with an average K_m of about 10 mM in the

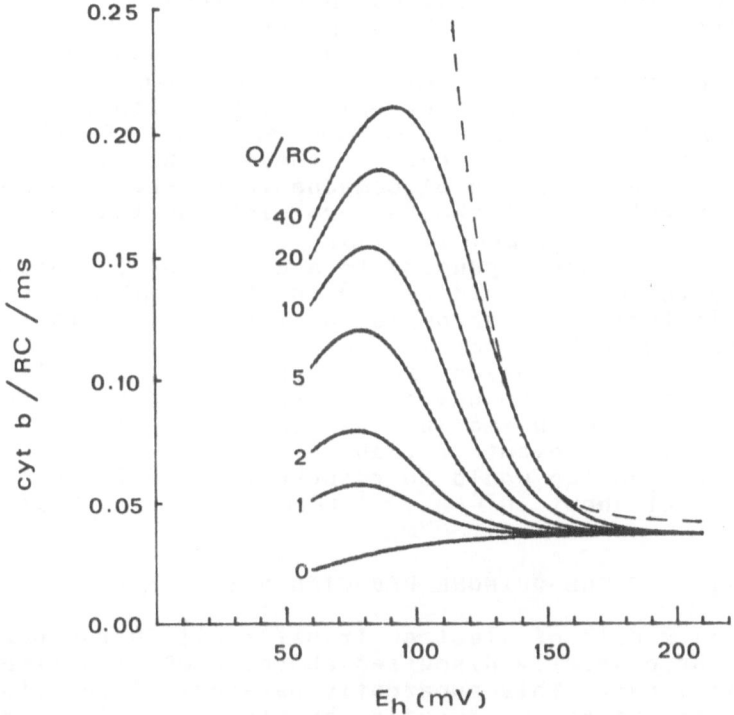

Fig.4. Theoretical titration curves of cyt b_{561} reduction rate as a function of the size of the UQ pool, calculated according to eqns.1, 2 and 3. V_{max}=0.3 cyt b_{561}/RC/ms; K_m=3.5 UQH_2/RC. The dotted line was calculated assuming a second-order kinetics, first-order in ubiquinol, for a pool of 20 UQ/RC. Taken from Venturoli et al. (1986)

lipid phase. The take home message from these experiments is that the bc1 complex interacts collisionally with the UQ pool, and obeys, for the Q_o site reaction, to a saturation kinetic law as it were a soluble enzyme interacting with a soluble substrate. This implies a fast binding and exchange of quinone and quinol at the Q_o site and a fast diffusion of quinone in the lipid phase, so that the average concentration of UQH_2 near the Q_o site can be considered in nernstian equilibrium with the UQ pool.

As mentioned above, the reduction of cyt. b_{561} starts after a lag period; this lag, also, is a function of the E_h, varying in unextracted chromatophores, from about 1 ms at $E_h > 180$ mV to about 200 μs at $E_h < 80$ mV at pH 7.0. Also the titration of the lag duration vs. E_h depends on the size of the UQ pool in a fashion suggesting that the absolute concentration of UQH_2 in the membrane is the decisive factor. A comparison of the duration of the lag with the actual concentration of UQH_2 suggests that the decrease in the lag duration is related to the availability of ubiquinol at the Q_o site before the flash. At $E_h > 180$ mV in fact, the whole UQ pool is oxidized, and the only UQH_2 available is the one produced at the Q_B site of the RC. The lag preceding the oxidation of this UQH_2 molecule at the Q_o sites (lasting about 1 ms), therefore, includes all the steps of UQH_2 dissociation from the Q_B site, the diffusion in the lipid phase, the (repetitive) collisions with the Q_o site, and the formation of the substrate-enzyme complex at this site. When the prereduced UQH_2 is already present at the Q_o site at lower E_h, the much shorter lag (200 μs) includes only the reaction steps within the bc1 complex, preliminary to the reduction of cyt. b_{561}. As discussed in the earlier section, these steps must include necessarily the oxidation of components of the high potential chain (in particular of the FeS protein) and the one-electron transfer from UQH_2 to produce UQ_o.
The lag of 1 ms, corresponding to a condition in which UQ is reduced by the RC and oxidized by the bc1 complex, sets an upper limit for the intercomplex diffusion time: this limit must be considered as grossly overextimated, since the lag includes several diffusion-independent steps, and also because it is expected that the number of useful encounters for the reaction will be much smaller than the number of random collisions. In any event, for an intercomplex average distance of 10-20 nm, the lag would correspond to a diffusion coefficient of about 10^{-9} cm^2 s^{-1} (Crofts et al., 1983).

THE KINETICS OF THE QUINONE REDUCING SITE (Q_i)

In all models of electron transfer within the bc1 complex that have been briefly discussed above, a UQ reduction step is an essential part. This apparently paradoxical function was suggested in the early seventies by Wikstrom and Berden (1972), and subsequently developped by Mitchell in the Q cycle reaction scheme. The non-linear mechanism, by which ubiquinone is simultaneously oxidized and reduced by the same enzyme complex, explains clearly the non-linear behaviour of electron transfer in the so called oxidant induced reduction of cyt.b_{561}. From an evolutionary point of view, such an unusual mechanism exploits at its best the bioenergetics of the two-electron oxidation of ubiquinol to ubiquinone, utilizing the large reducing power of the UQ^-/UQ couple for increasing the H^+/e^- stoichiometry of the H^+-pumping electron transfer reactions.

The nature of the quinone reducing site and its mechanism is the heart of the different models presently under discussion. In the classic Q cycle scheme, a first one-electron reduction at the Q_i site was completed by the donation of a second electron from a different membrane complex (a respiratory dehydrogenase in the original respiratory scheme, the RC in photosynthetic systems). Such a possibility seems quite unlikely to us, given the two electron gate mechanism operative in the RC, or, more in general, the functional independence and the lateral mobility of intramembrane complexes (Gupte et al, 1984). This mobility seems incompatible with a one-electron exchange between two complexes, involving necessarily the diffusion of a semiquinone, a species quite unstable in the lipid environment (Mitchell, 1976). Any other model, coupling the two one-electron steps of UQH_2 formation within the bc_1 complex, has to consider: i) either variation of Mitchell's suggestion (e.g. double turnover of the complex with a two electron gate mechanism at the Q_i site); ii) a dimeric scheme in which a two-electron reduction of UQ to UQH_2 occurs in a dimeric Q_i site (de Vries, 1986); iii) the mobility within the complex of the semiquinone, generated by the high potential chain within a single hydrophobic pocket in a region of the complex, and moving to the "Q_i region" for accepting one electron from the reduced cyt. b_{561} (Wikstrom and Krab, 1986). This last mechanism (essentially the semiquinone cycle) would require structural rearrangements of the Q_o and Q_i sites, for which there is so far no experimental evidence. The distinctive feature between mechanisms i) and ii) as opposed to mechanism iii), is that only in the first two schemes the quinone accepting electrons at the Q_i site originates from the UQ pool in the membrane. Studies of the kinetics of the Q_i site as a function of the size and the degree of reduction of the UQ pool, analogous to those previously discussed for the Q_o site, could possibly bring elements useful for discriminating between these two possibilities.

An antimycin sensitive reduction of cyt. b_{561} induced by a single turnover flash can be observed in chromatophores of *Rb. sphaeroides*, when the Q_o site is inhibited by myxothiazol; this reduction is however only detectable at very alkaline pH's (>9 (Glaser et al., 1984; Robertson et al., 1984)) when, owing to the pH dependence of the midpoint potentials of cyt. b_{561} and of the UQH_2/UQ couples, this redox step is thermodynamically more favoured; alternatively, the reduction can be observed in UQ extracted chromatophores, where the small size of the pool allows the reaching of a high degree of reduction of the pool in a single flash (i.e. a sufficiently negative potential of the UQ pool). This antimycin sensitive reduction has been considered as the reversal of the UQ reductase activity at the Q_i site, and it is supposed to occur with a mechanism completely independent of the normal reduction pathway of cyt. b_{561} through the Q_o site. In addition to the sensitivity of this type of reaction to antimycin, as opposed to its insensitivity to myxothiazol, the structural independence of the Q_o and the Q_i sites is also supported, in *Rb. capsulatus* by the isolation of the mutant strain R126. This mutant is lesioned in the bc_1 complex, and, therefore, is photosynthetically incompetent. It presents however, a normal functionality of the Q_i site, as can be judged from the antimycin sensitive cyt.b_{561} reduction, the detection of an EPR signal attributed to the UQ^- bound to the Q_i site and from the ability of antimycin to affect the spectrum of cyt. b_{561} (Robertson et al., 1986). On the contrary all Q_o dependent phenomena, such as the oxidant-induced reduction of cyt. b_{561}, or the effect of UQ in the EPR spectrum of the FeS protein, are totally impaired. The genetic lesion, corresponding to the substitution of one aminoacid on the cyt. b apoprotein (Daldal, personal communication), seems therefore to affect specifically only the Q_o site. This conclusion is fully supported by the behaviour of the purified bc_1 complex isolated from the R126 strain, in which cyt.b can only be reduced via the antimycin sensitive pathway, being the Q_o dependent, oxidant-induced reduction totally inactivated (Fernandez-Velasco et al., in preparation).

Although the functional independence of the Q_o and Q_i sites is now firmly established,the rather exotic conditions, under which the electron transfer through the Q_i site can be demonstrated, are not suitable for a detailed kinetic study of this reaction. In addition this electron transfer step can be studied only in a direction opposite to the physiological one, since, under normal conditions in uninhibited chromatophores, the rate of reduction of cyt. b_{561} via the Q_o site is nearly perfectly matched by its rate of oxidation; this prevents the accumulation of any significant level of reduced cytochromes, either in the steady state, or transiently after a flash, and consequently does not allow any measurement of cyt. b_{561} oxidation through the Q_i site (Meinhardt and Crofts, 1983).

For these reasons in our laboratory we have studied the rate of electron trasfer induced by a flash through the entire bc_1 complex in an indirect way, by monitoring the myxothiazol sensitive phase III of the flash-induced electrochromic signal of carotenoids. As illustrated above, this phase has been associated to the electron transfer along the cyt. b doublet. If the proposed topology of the Q_o and the Q_i sites across the membrane is correct, this electron transfer reaction should span completely the membrane dielectric and its kinetics should include all electrogenic and non-electrogenic steps leading from Q_o to Q_i. Owing to the analogy between the topology of the bc_1 complex and the RC, moreover, the amplitudes and rates of phase III can be calibrated against those of phase I+II, which is associated with a measurable concentration of RC in the membrane. We have sistematically studied the kinetics of phase III, as a function of the degree of reduction of the UQ pool and of its size, altered with the extraction-enrichment procedure (Fig.5) (Venturoli et al., 1988).

In order to discriminate between the semiquinone- or the Q- cycles, the kinetic data are particularly relevant at low E_h's, when the concentration of the oxidized UQ at the Q_i site could become limiting. At physiological pH's, however, a marked reduction of the pool can be obtained only at $E_h < 50$

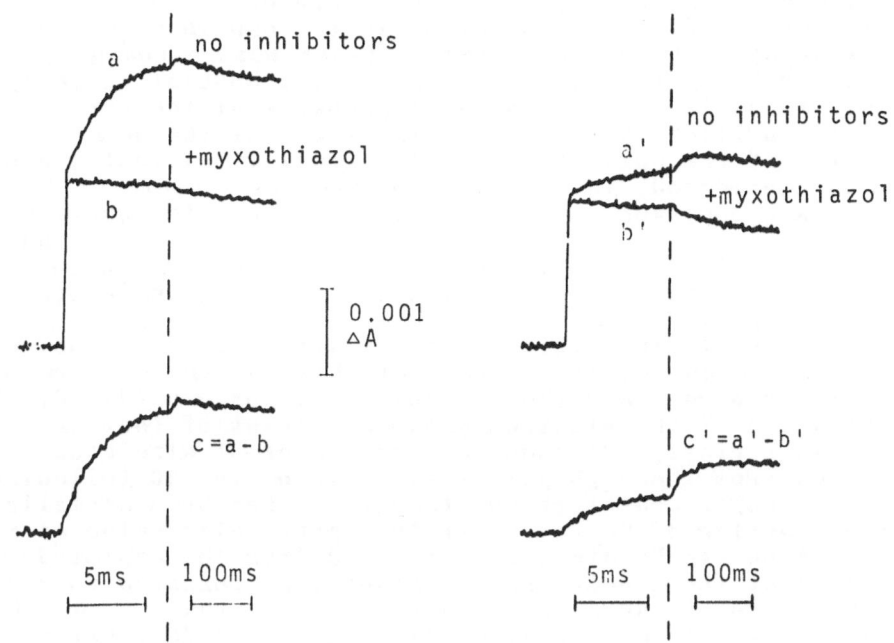

Fig.5. Kinetics of the carotenoid electrochromic change
induced by a single turnover flash, measured at 503 nm.
(A) Control-lyophilized chromatophores, characterized
by 16 UQ/RC. (B) Partially UQ-extracted chromatophores
(6 residual UQ/RC). E_h = 120 mV. Each signal was
recorded at two different sampling rates: the dashed
lines indicate the sweep changeover point. Traces c and
c' are the differences between traces without inhibitor
(a,a') and traces in the presence of myxothiazol
(b,b'). Taken from Venturoli et al. (1988).

mV, sufficiently near to the midpoint potential of Q_A, to
cause the reduction before the flash of a fraction of this
primary acceptor in the membrane. This causes a partial
inactivation of the RC and requires a correction in the
kinetic treatment of the data.

A generic reduction of UQ at the Q_i site can be
represented by the following rate equation:

$$v_i = k_i\,[UQ]_{pool}\,[UQ_iH_2\ b_{566}^-\ b_{561}^-] \qquad (4)$$

in which k_i is the second order rate constant of the reaction,
$[UQ]_{pool}$ the concentration of oxidized UQ in the membrane, and

[UQiH$_2$ b$_{566}$ b$_{561}$] is the concentration of any state of the bc$_1$
complex competent for the reduction of UQ at the Qi site. The
formalism used in eq. 4), in order to represent an electron
donating state of the bc1 complex, is the one in which the Qi
site is occupied by a UQH$_2$ molecule (that must exchange with
an oxidized UQ in the pool) and the cyt. b doublet is fully
reduced. This state seems the most probable at the very
reducing conditions suitable for the study of the Qi kinetics,
but it is not the only one possible. Equally relevant states
could be considered, in which the Qi site is vacant, or the
cyt. b chain contains only one electron. The rate equation, as
it is written above , is meant only to indicate the second
order nature of the reaction at the Qi site in any scheme
(such as the Q cycle or its variants) involving an interaction
with the UQ pool.

In the bacterial cyclic electron transfer chain, poised
at an Eh at which the high potential chain of the bc$_1$ complex
is totally reduced, and the low potential chain oxidized, the
concentration of the electron donating states of the bc$_1$
complex is strictly dependent on the number of electrons
subtracted from the high potential chain by the RC following a
flash. Any experimental trace, therefore, has been normalized
for the fraction of RC active at that particular value of Eh;
this fraction can be directly evaluated from the amplitude of
phase I+II of the electrochromic signal, as compared to the
amplitude when all RC's are open (i.e. at Eh>100 mV). By this
normalization the rate equation 4) is made pseudo-first order
for [UQ]pool, and the effect of the concentration of UQ in the
membrane lipids on the reaction rate can be evaluated.

The rate and the amplitude of phase III after a flash
result in fact from the convolution of more than one
electrogenic step within the bc$_1$ complex. A general consensus
exists that the charge transferring steps include both the
cyt.b$_{566}$ → cyt.b$_{561}$ and the cyt.b$_{561}$ → Qi electron transfers
(Glaser and Crofts, 1984; Robertson and Dutton, 1987;
Venturoli et al., 1987). In addition, since generally the RCs
are in excess over the bc$_1$ complexes in the membrane (Crofts
et al., 1983), phase III includes more than one turnover of
the complex, during which more than one molecule of UQH$_2$ is
oxidized at the Qo site, untill all electron holes in the high
potential chain are filled. The initial rate of phase III,
however, does not include this latter aspect, and comprises
only the sequence of the two consecutive electrogenic steps in
the low potential chain. It contains therefore information on
the kinetics of the Qo and Qi sites.

The experimental rates of charge transfer measured by the
myxothiazol sensitive phase III are plotted in Fig.6 as a
function of Eh. The rates have been corrected for the
fraction of closed RCs. The rates, moreover, are directly
compared with rates of reduction of cyt. b$_{561}$, measured in the
same chromatophores preparation in the presence of antimycin:
under this latter experimental conditions, only the rate
through the Qo site is measured. The data clearly indicate
that for Eh > 100 mV the initial rate of charge transfer and
that of cyt. b$_{561}$ reduction perfectly coincide, with a ratio
of one electron per cytochrome reduced. This is interpreted to
mean that under these conditions the overall turnover of the
bc$_1$ complex is totally controlled by the rate at the Qo site.
This result is expected since at Eh > 100 mV, the UQ pool is
markedly oxidized and the [UQH$_2$] in the membrane is limiting,

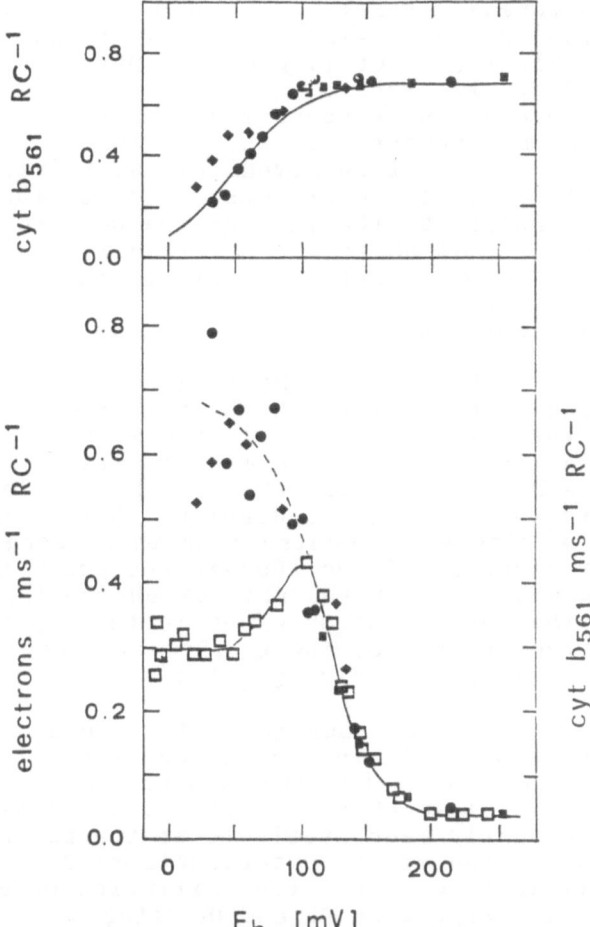

Fig.6. Direct comparison between the initial rates of phase III of the carotenoid signal (-□-) and of the reduction of cyt b_{561} (-●-, -■-, -◆-), as a function of E_h (lower panel). The upper panel shows the maximum extent of cyt b_{561} reduction induced by a single flash. Antimycin and valinomycin were added to the sample for the measurement of cyt b_{561} reduction rates and extents. These data were used for a normalization of the rate of reduction of cyt b_{561} at E_h < 90 mV. The rates of the carotenoid signal were corrected for the fraction of Q_A pre-reduced before the flash.

as compared to the K_m value of the site for ubiquinol. At E_h of about 100 mV, however, the rate of charge transfer reaches a maximum and declines at more negative E_h's. At this point the overall rate of charge transfer diverges from the potential rate at the Q_o site (indicated by the rate of reduction of cyt. b_{561}). It is evident, therefore, that at E_h < 100 mV the rate at the Q_i site begins to limit the overall reaction, in parallel with the decrease of the concentration

of UQ in the membrane. At Eh< 50 mV the rate of charge
transfer remains constant, since a minimal fixed concentration
of one UQ molecule for 2 RCs is produced per flash at the Q_o
site (Venturoli et al., 1988).

This interpretation is supported by the data obtained in
UQ-extracted chromatophores (not shown). The increase in the
rate of charge transfer is observed only at Eh values more
negative than in controls, confirming that, at these Eh's the
turnover is controlled by the rate at the Q_o site. At more
negative Eh the rate of phase III levels off, since, when the
size of the pool is decreased, the rate at the Q_i site can
never become limiting over that at the Q_o site.

The conclusions that can be drawn from these results are
the following:
i) at very positive Eh's, when the [UQH₂] is limiting, the
rate of electron transfer through the bc_1 complex is totally
controlled by the rate of UQH₂ oxidation at the Q_o site;
ii) at more negative Eh's, the potential rate of electron
transfer at the Q_o site exceeds that of the complete charge
transfer across the membrane, indicating that reaction steps
involving the Q_i site start controlling this process;
iii) a further decrease of the [UQ] in the pool, by lowering
the Eh, causes a decline of the rate of charge transfer. This
suggests that the concentration of UQ in the lipid phase,
affects the turnover rate at the Q_i site as expected for a
collisional interaction involving extra-complex diffusable UQ
molecules;
iv) at very negative Eh's, when the only oxidized quinone
present in the membrane is that produced by the oxidation of
UQH₂ at the Q_o site, no marked lag in phase III of the
electrochromic signal is observed, indicating that these UQ
molecules are promptly made available at the Q_i site;
v) the time resolution of our experiments renders the above
results compatible also with a true diffusion of UQ through
the two physically separated Q_o and Q_i sites.

As a whole, therefore, our data support clearly the
Q-cycle reaction scheme rather than the semiquinone cycle.

APPENDIX. THE THERMODYNAMICS OF THE REDOX REACTIONS OF
UBIQUINONE

Formally the two-electron reduction of ubiquinone to
ubiquinol can be separated into two consecutive one-electron
steps, as follows:

step 1 $\qquad$ UQ + e⁻ $\longrightarrow$ UQ⁻

step 2 $\qquad$ UQ⁻ + e⁻ + 2H⁺ $\longrightarrow$ UQH₂

From the additivity of the free energies of the two partial
steps 1 and 2 (ΔG_1 and ΔG_2) to yield the total free energy
change for the two-electron reduction (ΔG_T), it can be
obtained:

$$\Delta G_1 + \Delta G_2 = -F (\Delta E_1 + \Delta E_2) = \Delta G_T = -2 F \Delta E_T \qquad (A1)$$

Under standard conditions, remembering that every redox couple
is referred to the standard hydrogen electrode, relation (A1)

yields:

$$E_{m,Q^-/Q} + E_{m,QH_2/Q^-} = 2\, E_{m,QH_2/Q} \qquad (A2)$$

i.e. the midpoint potential of the dielectronic reaction is the average of those of the two monoelectronic steps.

The redox span separating the values of $E_{m,Q^-/Q}$ and $E_{m,QH_2/Q}$ is related to the stability constant of the semiquinone radical for the dismutation reaction defined as

$$K_{Q^-} = [Q^-]^2/([Q]\,[QH_2]) \qquad (A3)$$

In fact, at equilibrium, it must be:

$$E_{m,Q^-/Q} + RT/F \ln (Q/Q^-) = E_{m,QH_2/Q^-} + RT/F \ln (Q^-/QH_2) \quad (A4)$$

Combining eqn.A3 and eqn.A4 it can be easily obtained:

$$E_{m,Q^-/Q} - E_{m,QH_2/Q^-} = RT/F \ln K_{Q^-} \qquad (A5)$$

Eqn.A5 means that the more the semiquinone is unstable in its equilibrium with the fully oxidized and reduced forms, the wider is the difference between the midpoint potentials of the two redox couples (60 mV every tenfold decrease in the value of the stability constant). Any situation stabilizing the semiquinone, as compared to its equilibrium concentrations when free in the membrane lipids, will increase the stability constant and therefore will reduce the gap between the two midpoint potentials.

AKNOWLEDGEMENTS

The support of the C.N.R. and of the Ministry of the Public Instruction of Italy are gratefully acknowledged.

REFERENCES

Allen,J.P., Feher,G., Yeates,T.O., Komiya,H. and Rees,D.C., 1987a, Structure of the reaction center from *Rhodobacter sphaeroides* R-26: The protein subunits, Proc. Natl. Acad. Sci. USA, 84:6162.

Allen,J.P., Feher,G., Yeates,T.O., Komiya,H. and Rees,D.C., 1987b, Structure of the reaction center from *Rhodobacter sphaeroides* R-26: The cofactors, Proc. Natl. Acad. Sci. USA, 84:5730.

Baccarini-Melandri,A. and Melandri,B.A., 1977, A role for ubiquinone-10 in the b-c₂ segment of the photosynthetic bacterial electron transport chain, FEBS Lett., 80:459.

Baccarini-Melandri,A., Gabellini,N., Melandri,B.A., Jones,K.R., Rutherford,A.W., Crofts,A.R. and Hurt,E., 1982, Differential extraction and structural specificity of specialized ubiquinone molecules in secondary electron transfer in chromatophores from *Rhodopseudomonas sphaeroides*, Ga, Arch. Biochem. Biophys., 216:566.

Crofts,A.R. and Wraight,C.A., 1983, The electrochemical domain of photosynthesis, <u>Biochim. Biophys. Acta</u>, 726:149.

Crofts,A.R., Meinhardt,S.W., Jones,K.R. and Snozzi,M., 1983, The role of the quinone pool in the cyclic electron-transfer chain of *Rhodopseudomonas sphaeroides*. A modified Q-cycle mechanism, <u>Biochim. Biophys. Acta</u>, 723:202.

Crofts,A.R., Robinson,H., Andrews,K., Van Doren,S. and Berry,E., 1987, Catalytic sites for reduction and oxidation of quinones, <u>in</u>: "Cytochrome Systems. Molecular Biology and Bioenergetics", S.Papa, B.Chance and L.Ernster, eds., Plenum Press, New York and London.

Davidson,E. and Daldal,F., 1987, Primary structure of the bc_1 complex of *Rhodopseudomonas capsulata*, <u>J. Mol. Biol.</u>, 195:13.

Deisenhofer,J., Epp,O., Miki,K., Huber,R. and Michel,H., 1985, Structure of the protein subunits in the photosynthetic reaction center of *Rhodopseudomonas viridis* at 3Å resolution, <u>Nature</u>, 318:618.

Dutton,P.L., 1986, Energy transduction in anoxygenic photosynthesis, <u>in</u>: "Encyclopedia of Plant Physiology, vol.19. Photosynthesis III", L.A.Staehelin and C.J.Arntzen, eds., Springer-Verlag, Berlin Heidelberg.

Dutton,P.L. and Prince,R.C., 1978, Reaction-center-driven cytochrome interactions in electron and proton translocation and energy coupling, <u>in</u>: "The Photosynthetic Bacteria", R.K.Cayton and W.R.Sistrom, eds., Plenum Press, New York and London.

Fato,R.,Battino,M., Degli Esposti,M., Parenti Castelli,G. and Lenaz,G., 1986, Determination of partition and lateral diffusion coefficients of ubiquinone by fluorescence quenching of n-(9-anthroyloxy)stearic acids in phospholipid vesicles and mitochondrial membranes, <u>Biochemistry</u>, 25:3378.

Gabellini,N. and Sebald,W., 1986, Nucleotide sequence and transcription of the fbc operon from *Rhodopseudomonas sphaeroides*, <u>Eur. J. Biochem.</u>, 154:569.

Garcia,A.F., Venturoli,G., Gad'on,N., Fernandez-Velasco,J.G., Melandri,B.A. and Drews,G., 1987, The adaptation of the electron transfer chain of *Rhodopseudomonas capsulata* to different light intensities, <u>Biochim. Biophys. Acta</u>, 890:335.

Glaser,E.G. and Crofts,A.R., 1984, A new electrogenic step in the ubiquinol:cyt c_2 oxidoreductase complex of *Rhodopseudomonas sphaeroides*, <u>Biochim. Biophys. Acta</u>, 766:322.

Glaser,E.G., Meinhardt,S.W. and Crofts,A.R., 1984, Reduction of cyt b_{561} through the antimycin-sensitive site of the ubiquinol-cyt c_2 oxidoreductase complex of *Rhodopseudomonas spaeroides*, <u>FEBS Lett.</u>, 178:336.

Gupte,S., Wu,E., Hoechli,M., Jacobson,K., Sowers,A.E. and Hackenbrock,C.R., 1984,Relationship between lateral diffusion, collisional frequency and electron transfer of mitochondrial inner membrane oxidation-reduction components, <u>Proc. Natl. Acad. Sci. USA</u>, 81:2606.

Hauska,G., Hurt,E., Gabellini,N. and Lokau,W., 1983.Comparative aspects of quinol-cytochrome c/plasocyanin oxidoreductases, <u>Biochim. Biophys. Acta</u>, 726:97.

Jackson,J.B. and Crofts,A.R., 1971, The kinetics of light induced carotenoid changes in *Rhodopseudomonas sphaeroides* and their relation to electrical field generation across the chromatophore membrane, Eur. J. Biochem., 18:120.

Jackson,J.B. and Dutton,P.L., 1973, The kinetic and redox potentiometric resolution of the carotenoid shifts in *Rhodopseudomonas sphaeroides* chromatophores: their relationship to electric field alterations in electron transport and energy coupling, Biochim. Biophys. Acta, 325:102.

Kleinfeld,D., Okamura,M.Y. and Feher,G., 1984, Electron transfer in reaction centers of *Rhodopseudomonas sphaeroides*. I. Determination of the charge recombination pathway of $D^+Q_AQ_B^-$ and free energy and kinetic relations between $Q_A^-Q_B$ and $Q_AQ_B^-$, Biochim. Biophys. Acta, 766:126.

Kleinfeld,D., Okamura,M.Y. and Feher,G., 1985, Electron transfer in reaction centers of *Rhodopseudomonas sphaeroides*. II. Free energy and kinetic relations between the acceptor states $Q_AQ_B^-$ aand $Q_AQ_B^-$, Biochim. Biophys. Acta, 809:291.

Link,T.A., Schaegger,H. and Von Jagow,G., 1987, Structural analysis of the bc_1 complex from beef heart mitochondria by sisded hydropathy plot and by comparison with other bc complexes, in: "Cytochrome Systems. Molecular Biology and Bioenergetics", S.Papa, B.Chance and L.Ernster, eds., Plenum Press, New York and London.

Meinhard,S.W. and Crofts,A.R., 1983, The role of cytochrome b_{566} in the electron transfer chain of *Rhodopseudomonas sphaeroides*, Biochem. Biophys. Acta, 723:219.

Melandri,B.A. and Venturoli,G., 1984, Photosynthetic electron transfer, in: "New Comprehensive Biochemistry. Bioenergetics", L.Ernster, ed., Elsevier, Amsterdam.

Michel,H., Epp,O. and Deisenhofer,J., 1986, Pigment-protein interactions in the photosynthetic reaction center from *Rhodopseudomonas viridis*, EMBO J., 5:2445.

Mitchell,P., 1976, Possible molecular mechanisms of the protonmotive function of cytochrome systems, J. Theor. Biol., 62:327.

Okamura,M.Y., Feher,G. and Nelson,N., 1982, Reaction centers, in: "Photosynthesys, vol.I. Energy Conversion by Plants and Bacteria", Govindjee, ed., Academic Press, New York.

Rich,P.R., 1986, A perspective on Q-cycles, J. Bioenerg. Biomembr., 18:145.

Rieske,J.S., 1986, Experimental observations on the structure and function of mitochondrial complex III that are unresolved by the protonmotive ubiquinone-cycle hypothesis, J. Bioenerg. Biomembr., 18:235.

Robertson,D.E., Giangiacomo,K.M., Moser,C.C. and Dutton,P.L., 1984, Two distinct quinone modulated modes of antimycin-sensitive cytochrome b reduction in the bc_1 complex, FEBS Lett., 178:343.

Robertson,D.E., Prince,R.C., Bowyer,J.R., Matsuura,K., Dutton,P.L. and Ohnishi,T., 1984, Thermodynamic properties of the semiquinone and its binding site in the ubiquinol-cytochrome c (c_2) oxidoreductase of respiratory and photosynthetic systems, J. Biol. Chem., 259:1758.

Robertson,D.E., Davidson,E., Prince,R.C., van den Berg,W.H., Marrs,B.L. and Dutton,P.L., 1986, Discrete catalytic sites for quinone in the ubiquinol-cytochrome c_2 oxidoreductase of *Rhodopseudomonas capsulata*. Evidence from a mutant defective in ubiquinol oxidation, J.

Biol. Chem., 261:584.

Robertson,D.E. and Dutton,P.L., 1987, The redox reaction
between cytochrome b$_H$ and the semiquinone-quinol couple
of Q$_c$ is electrogenic in ubiquinol cyt c$_2$ oxidoreductase,
in: "Cytochrome Systems. Molecular Biology and
Bioenergetics", S.Papa, B.Chance and L.Ernster, eds.,
Plenum Press, New York and London.

Semenov,A.Yu., Drachev,L.A., Kaminskaya,O.P. and
Konstantinov,A.A., 1986, Electrogenic protonation of the
secondary quinone acceptor in photosynthetic bacterial
chromatophores, in: "European Bioenergetic Conference
Reports, Vol.4", Congress Edition, Prague.

Takamiya,K. and Dutton,P.L., 1977, The influence of
transmembrane potentials on the redox equilibrium
between cytochrome c$_2$ and the reaction center in
Rhodopseudomonas sphaeroides chromatophores, FEBS Lett.,
80:279.

Venturoli,G., Fernandez-Velasco,J.G., Crofts,A.R. and
Melandri,B.A., 1986, Demonstration of a collisional
interaction of ubiquinol with the ubiquinol-cytochrome c$_2$
oxidoreductase complex in chromatophores from *Rhodobacter
sphaeroides*, Biochim. Biophys. Acta, 851:340.

Venturoli,G., Virgili,M., Melandri,B.A. and Crofts,A.R., 1987,
Kinetic measurements of electron transfer in coupled
chromatophores from photosynthetic bacteria, FEBS Lett.,
219:477.

Venturoli,G., Fernandez-Velasco,J.G., Crofts,A.R. and
Melandri,B.A., 1988, The effect of the size of the
quinone pool on the electrogenic reactions in the
ubiquinol-cytochrome c$_2$ oxidoreductase of *Rhodobacter
capsulatus*. Pool behaviour at the quinone reductase site,
Biochim. Biophys. Acta, in press.

de Vries,S., 1986, The pathway of electron transfer in the
dimeric QH$_2$:cytochrome c oxidoreductase, J. Bioenerg.
Biomembr., 18:195.

Widger,W.R., Cramer,W.A., Herrmann,R.G. and Trebst,A., 1984,
Sequence homology and structural similarity between
cytochrome b of mitochondrial complex III and the
chloroplast b$_6$-f complex: position of the cytochrome b
hemes in the membrane, Proc. Natl. Acad. Sci. USA,
81:674.

Wikstrom,M.K.F. and Berden,J.A., 1972, Oxidoreduction of
cytochrome b in the presence of antimycin, Biochim.
Biophys. Acta, 283:403.

Wikstrom,M.K.F. and Krab,K., 1986, The semiquinone cycle. A
hypothesis of electron transfer and proton translocation
in cytochrome bc-type complexes, J. Bioenerg. Biomembr.,
18:181.

Yeates,T.O., Komiya,H., Rees,D.C., Allen,J.P. and Feher,G.,
1987, Structure of the reaction center from *Rhodobacter
sphaeroides* R-26: Membrane-protein interactions, Proc.
Natl. Acad. Sci. USA, 84:6438.

Zannoni,D. and Baccarini-Melandri,A., 1980, Respiratory
electron flow in facultative photosynthetic bacteria, in:
"Diversity of Bacterial Respiratory Systems, vol.II",
C.J.Knowles, ed., CRC Press, Boca Raton, Florida.

ARE ENVELOPE MEMBRANES FROM CHLOROPLASTS

AND NON-GREEN PLASTIDS IDENTICAL ?

Roland Douce, Claude Alban and Jaques Joyard

Laboratoire de Physiologie Cellulaire Végétale
UA CNRS n° 576, Département de Recherche Fondamentale
Centre d'Etudes Nucléaires et Université Joseph Fourier
85X, F-38041 Grenoble-cédex, France

INTRODUCTION

The feature shared by all plastids (proplastids, leucoplasts, amyloplasts, etioplasts and chloroplasts) is a pair of outer membranes, known as the envelope (Douce et al, 1984). Together, these membranes provide a flexible boundary between the plastid and the surrounding cytosol. Unfortunately, work on the envelope membranes from non-green plastids has lagged far behind that conducted on their chloroplast counterpart mostly because it is very difficult to prepare large amounts of intact non-green plastids free from contaminating membranes and mitochondria. Among non-green plastids, chromoplasts and etioplasts have been the main sources for investigations on the structure, chemical composition and functions of envelope membranes (for a review, see Douce et al, 1984). However, except for the work of Fishwick and Wright (1980) on envelope membranes from potato tuber amyloplasts, no studies on starch-containing plastids are available, mainly because the presence of starch grains makes the preparation of intact plastids, and consequently of envelope membranes, difficult. Having developed in our laboratory procedures for the preparation of large amounts of non-green plastids (Figure 1) from cauliflower buds (Journet and Douce, 1985) and amyloplasts from sycamore cells (Journet et al, 1986), we have been able recently to prepare and characterize envelope membranes from these plastids (Alban et al, 1988). The purpose of this article is to review some of the major properties of envelope membranes from chloroplasts and non-green plastids in order to determine whether a general picture can emerge.

ENVELOPE PURIFICATION

The key step in the purification of envelope membranes from chloroplasts or non-green plastids is actually the preparation of large amounts of intact and pure organelles. For many years, it has been possible to obtain excellent chloroplast preparations, devoid of extra-plastidial contaminations, owing to purification using Percoll gradients. Such procedures are now available for non-green plastids, although a double purification using Percoll gradients is necessary to obtain a minimal contamination by mitochondria. Cauliflower buds are a material of choice to prepare large amounts of non-green plastids almost devoid of mitochon-

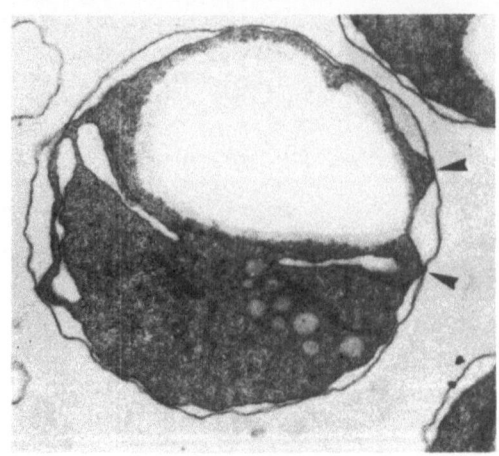

<u>Figure 1.</u> **Purified cauliflower bud plastid (Journet and Douce, 1985)**

This plastid exhibit intact envelope membranes and a dense stroma. The internal membrane system is poorly developed and is connected to the inner envelope membrane. Starch grains are of limited size. See the contact points between the two envelope membranes (arrows).

dria by classical grinding procedures (Journet and Douce, 1985). We also used sycamore cells for the preparation of amyloplasts. This can only be achieved using protoplasts, and the yields are therefore much lower than with cauliflower.

From chloroplasts, it is possible to prepare total envelope, containing both the outer and the inner envelope membrane, and membrane fractions enriched in outer or inner envelope membranes. The procedures have been extensively described. To prepare total envelope membranes, intact chloroplasts are broken by a gentle osmotic shock in a hypotonic medium (Douce et al, 1973). At this stage, the two envelope membranes fuse along their breaking edges and it is no longer possible to separate the outer from inner membrane. This can only be achieved by using an hypertonic incubation of the plastids to enlarge the space between the two membranes, followed by rupture of the chloroplasts using freeze-thawing (Cline et al, 1981, Keegstra and Yousif, 1986) or mechanical (Block et al, 1983) procedures.

The fractionation of purified amyloplasts from sycamore cells was done essentially as for spinach chloroplasts (Alban et al, 1988). However, fractionation of non-green plastids from cauliflower buds is not easy: a gentle osmotic shock, which is the most efficient method for the rupture of chloroplast envelope membranes, broke 70 % of the cauliflower plastids. Freeze-thawing alone is even less efficient: only 20 % of cauliflower plastids are broken. However, when intact cauliflower plastids are frozen (for 30 min at -80°C), then thawned to room temperature, treated with hypotonic medium and homogenized with a Potter-Elvehjem apparatus having a loose fitting pestle, the envelope membranes of almost 100 % of the plastids are ruptured, as judged by the lack of latency of gluconate 6-phosphate dehydrogenase, a stromal enzyme (Journet and Douce, 1985).

216

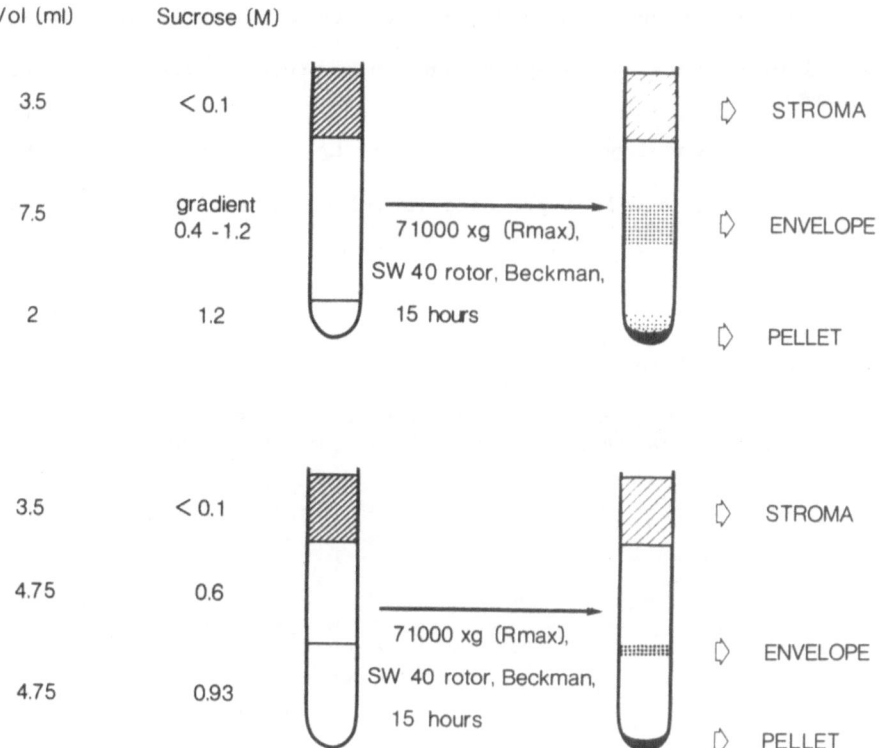

Figure 2. **Purification of envelope membranes from cauliflower bud plastids.**

Schematic representation of sucrose gradient fractionations of cauliflower plastids after rupture of envelope membranes.
The experimental conditions are described by Alban et al (1988).
Top : purification of envelope membranes using a continuous sucrose gradient.
Bottom : purification of envelope membranes using a discontinuous sucrose gradient.

The preparation of amyloplasts envelope membranes from sycamore cells presents few differences with that of cauliflower bud plastids: (a) a simple osmotic shock was sufficient for disruption of the envelope membranes, (b) the swelling medium as well as the different layers of the sucrose gradient should be devoid of $MgCl_2$, (c) protease inhibitors should be added in all the medium used and at all the steps of the purification.

The broken plastids were then layered on top of a sucrose gradient, either a linear or a step gradient can be used. In addition, the gradients can be used for chloroplasts as well as for non-green plastids (Figure 2), since envelope membranes from all these plastids have the same density (1.12 g/cm^3). The detailed procedures for the preparation of envelope membranes are described by Douce and Joyard (1980) and Block et al (1983) for chloroplasts, and by Alban et al (1988) for non-green plastids. It is not yet possible to separate the inner and the outer envelope membrane from non-green plastids.

The yields for envelope membrane purification are the following:

- about 8-10 mg total chloroplast envelope proteins from 2 kg spinach leaves ;

- about 1-1.5 mg envelope proteins from 4-5 kg cauliflower bud plastids ;

- about 0.2-0.3 mg envelope proteins from 150-200 g of isolated sycamore cells.

CHEMICAL COMPOSITION

Polypeptide composition

In chloroplasts as well as in cauliflower bud plastids, the polypeptides have Mr values less than 15,000 to more than 100,000. The poly-

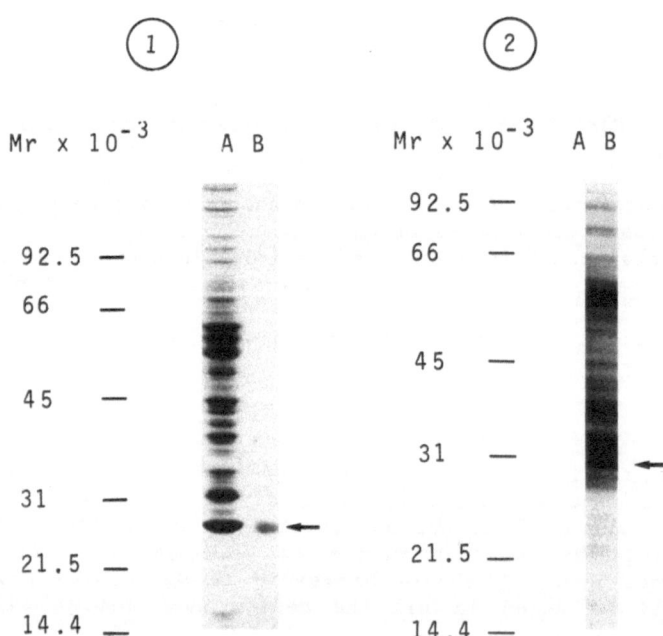

<u>Figure 3</u>. **Electrophoretic and immunochemical analyses of purified envelope membranes from cauliflower bud plastids (1) and sycamore amyloplasts (2)**

The experimental conditions are described by Alban et al (1988).
A : coomassie blue staining of envelope polypeptides separated by LDS-PAGE at 4°C with 7.5-15 % acrylamide gradient.
B : autoradiography of antigen-antibody-^{125}I protein A complexes after electrophoretic transfer of envelope polypeptides from a polyacrylamide gel (A) to a nitrocellulose sheet. Antibody to the phosphate translocator from spinach chloroplast envelope was used.

peptide patterns are extremely complex, each fraction analyzed containing more than 100 distinct bands after staining with coomassie blue. Some polypeptides have apparently the same electrophoretic mobility when all envelope membrane profiles are analyzed. But more precise analyses should be done. To determine whether the same polypeptides were present in envelope membranes from chloroplasts and non-green plastids, we have used antibodies raised against several chloroplast envelope polypeptides: E10, E24, E30 and E37 (Joyard et al, 1982, 1983). By immunoblotting experiments, we have demonstrated that all of them led to a reaction with envelope proteins from non-green plastids (cauliflower and sycamore). The most interesting result was obtained with the antibody to E30, since this polypeptide is involved in the phosphate/triose phosphate transport across the inner envelope membrane (Flugge and Heldt, 1976). Western blotting experiments with our antibody to spinach E30 clearly demonstrate that the major 28,000-dalton polypeptide of the cauliflower envelope fraction and the phosphate translocator from spinach chloroplasts have closely related antigenic sites (Figure 3). The same experiment made using envelope membranes from sycamore amyloplasts demonstrates that the major 30,000-dalton polypeptide also reacts with antibody against the spinach phosphate translocator (Figure 3). These observations demonstrate that there are probably only limited differences between the sequence of the phosphate translocator in spinach, cauliflower and sycamore, since good cross-reactivity of the antibody was observed between all these species, despite the differences in Mr.

Glycerolipid composition

Galactolipids in thylakoid and envelope membranes from chloroplasts contain a high amount of polyunsaturated fatty acids: up to 95 % (in some species) of the total fatty acids is linolenic acid. In non-green plastids, 18:3 is still a major component although appreciable amounts of 18:1 and 18:2 are present. Therefore, the most abundant molecular species of MGDG and DGDG in plants have 18:3 at both sn-1 and sn-2 positions of the glycerol backbone. Some plants, such as pea, having almost only 18:3 in MGDG are called "18:3 plants". Other plants, such as spinach, contain important amounts of 16:3 in MGDG, they are called "16:3 plants" (Heinz, 1977). These two types of plants are also found when non-green plastids are analyzed: sycamore is a 18:3 plant whereas cauliflower is a typical 16:3 plant. The positional distribution of 16:3 in MGDG is highly specific: this fatty acid is only present at the sn-2 position of glycerol, and is almost excluded from sn-1 position. Therefore, two major structures are found in galactolipids, one with C18 fatty acids at both sn position and one with C18 and C16 fatty acids respectively at sn-1 and sn-2 position. The first one is typical of "eukaryotic" lipids (such as PC) and the second one corresponds to a "prokaryotic" structure. These differences are probably due to galactolipid biosynthetic pathways (see below).

Plastid membranes from isolated chloroplasts or non-green plastids sometimes contain galactolipids with 3 (tri-GDG) and 4 (tetra-GDG) galactoses. They are formed by an enzymatic galactose exchange between MGDG and DGDG, owing to a galactolipid:galactolipid galactosyltransferase shown in chloroplasts by Van Besouw and Wintermans (1978) and which catalyzes the following reactions:

$$2 \text{ MGDG} \longrightarrow \text{DGDG} + \text{diacylglycerol}$$

$$\text{MGDG} + \text{DGDG} \longrightarrow \text{tri-GDG} + \text{diacylglycerol}$$

$$2 \text{ DGDG} \longrightarrow \text{tetra-GDG} + \text{diacylglycerol}$$

Table I. Glycerolipid composition of membranes from mitochondria and plastids isolated from various plants

Except for envelope membranes from sycamore cells, all these analyses have been done after thermolysin treatment of intact plastids prior to fractionation. Experiments on pea etioplasts were done in collaboration with Dr. J. Soll (Munich, RFA).

ORGANELLE	MGDG	DGDG	SL	PC	PG	PI	PE	DPG
MITOCHONDRIA								
- cauliflower buds	tr	tr	0	37	2	8	38	13
- mung bean hypocotyls								
total	0	0	0	36	1	2.5	46	14
inner membrane	*0*	*0*	*0*	*29*	*1*	*2*	*50*	*17*
outer membrane	*0*	*0*	*0*	*68*	*2*	*5*	*24*	*0*
- sycamore cells								
total	0	0	0	43	3	6	35	13
inner membrane	*0*	*0*	*0*	*41*	*2.5*	*5*	*37*	*14.5*
outer membrane	*0*	*0*	*0*	*54*	*4.5*	*11*	*30*	*0*
PLASTIDS								
- spinach chloroplasts								
thylakoids	52	26	6.5	4.5	9.5	1.5	0	0
inner envelope								
membrane	*49*	*30*	*5*	*6*	*8*	*1*	*0*	*0*
outer envelope								
membrane	*17*	*29*	*6*	*32*	*10*	*5*	*0*	*0*
total envelope	32	30	6	20	9	4	0	0
- pea etioplasts								
prothylakoids	42	35	6	9	5	2	0	0
envelope	34	31	6	17	5	4	0	0
- cauliflower bud								
non-green plastids								
envelope	31.5	27.5	6	20	9	4.5	1	0
- sycamore cells								
envelope	24	21	4	21,5	9	5.5	0.8	0

Abbreviations: MGDG, monogalactosyl diacylglycerol ; DGDG, digalactosyl-diacylglycerol ; SL, sulfolipid ; PC, phosphatidylcholine ; PG, phospha-tidylglycerol ; PI, phosphatidylinositol ; PE, phosphatidylethanolamine ; DPG, diphosphatidylglycerol.

We have demonstrated that this enzyme (a) is located on the cyto-solic side of the outer envelope membrane (Dorne et al, 1982), (b) is susceptible to proteolytic digestion by thermolysin, a non-penetrant protease and (c) is present in all plastids analyzed so far (chloro-plasts, non-green plastids). Consequently, in order to obtain a glycero-lipid composition which could represent the *in vivo* situation within plastid membranes, the galactosyltransferase should be destroyed prior to fractionation of plastids. Not all plastids can survive this treat-ment, for instance, amyloplasts from sycamore cells are too fragile and are rapidly destroyed during the proteolytic digestion, in contrast, plastids from cauliflower buds can be treated like chloroplasts.

The glycerolipid composition of envelope membranes from chloroplasts and non-green plastids is given in table 1. As expected after thermolysin treatment, no diacylglycerol was detected (whereas it represents about 10 % of the total envelope glycerolipids from non-treated plastids). The most striking feature is that the glycerolipid pattern is almost identical in envelope membranes from chloroplasts, etioplasts or other non-green plastids. Although separation of outer from inner membrane cannot be achieved by the available procedures, it is likely that the glycerolipid composition of the two envelope membranes is almost identical in chloroplasts and non-green plastids. The high amount of MGDG

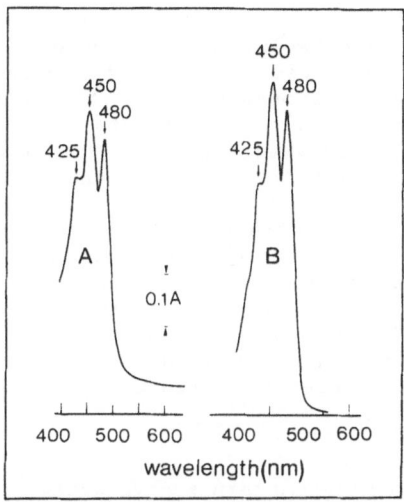

Figure 4. **Absorption spectra of envelope membranes from spinach chloroplasts (A) and cauliflower bud plastids (B)**

The experimental conditions are described by Neuburger et al (1982).

reflects the presence of the inner envelope membrane which has a glycerolipid composition close to that of thylakoids as shown in pea (Cline et al, 1981) or spinach (Block et al, 1983): in the chloroplast inner envelope membrane and in thylakoids, MGDG is the major component. The presence of PC in envelope membranes reflects the presence of the outer envelope membrane: using phospholipase C digestion of intact chloroplasts, we have demonstrated that no PC is present in thylakoids. In contrast, PC represents about 30-35 % of the outer envelope glycerolipids, where it is concentrated in the outer leaflet of the membrane (Dorne et al, 1985), as shown in Figure 1. In chloroplasts, the major phospholipid in the inner envelope membrane and in thylakoids is PG,

which is unique because of a $16:1_{trans}$ fatty acid at sn-2 position
of the glycerol backbone. This phospholipid is therefore different
from that found in extra-plastidial membranes and in PG from non-green
plastids. Finally, table 1 also confirms the observation that plastid
membranes from chloroplasts or non-green plastids are devoid of PE,
which is a major component (together with PC) of mitochondrial, endoplasmic reticulum or other extra-plastidial membranes.

Pigments

In contrast to thylakoids, envelope membranes from chloroplasts
or non-green plastids are yellow, due to the presence of carotenoids
and the absence of chlorophyll. Envelope membranes from chloroplasts
and from amyloplasts present the same absorption spectrum (Figure 4),
thus demonstrating that they have a very close pigment composition.
The precise carotenoid composition of envelope membranes from non-green
plastids is not yet available. It is probably very similar to that
of chloroplasts as suggested by the absorption spectra (Figure 4).
Therefore, violaxantin is probably as in chloroplast envelope membranes
a major component. In fact, chloroplasts envelope membranes contain
a higher proportion of violaxanthin (even the outer membrane), compared
to thylakoids which are rich in β-carotene and xanthophyll.

Together, these observations demonstrate that envelope membranes
from the different types of plastids analyzed contain carotenoids.
These pigments are also present in thylakoids, but in different amounts,
probably due to their role within envelope membranes. Unfortunately,
if the functions of pigments are well established for thylakoids, nothing
is known for their function in envelope membranes. The use of non-green
plastids could be most interesting to determine the specific functions
of envelope pigments.

ENVELOPE MEMBRANES AND GLYCEROLIPID BIOSYNTHESIS

The first observation that envelope membranes could be involved
in the biosynthesis of plastid components was provided by Douce (1974):
isolated envelope membranes purified from spinach chloroplasts were
able to catalyze MGDG synthesis from diacylglycerol and UDP-galactose
(UDP-gal) at high rates. Following the observations of Douce and
Guillot-Salomon (1970) who established the presence of the enzymes
from the Kornberg-Pricer pathway in chloroplasts and non-green plastids,
a considerable body of data has been accumulated which clearly demonstrate that plastids are able to incorporate sn-glycerol 3-phosphate
into lysophosphatidic acid, phosphatidic acid and diacylglycerol and
then into MGDG (after addition of UDP-gal). Glycerolipid biosynthesis
requires the assembly of three parts (Joyard and Douce, 1987): fatty
acids, glycerol and a polar head group (galactose, for galactolipids ;
sulfoquinovose, for sulfolipid ; and phosphorylglycerol, for PG).

In the plant cell, the plastid Kornberg-Pricer pathway is unique
in the sense that it is closely associated with fatty acid synthesis,
which is localized within the plastid stroma. All the enzymes of this
pathway are closely associated with the envelope membranes (Joyard
and Douce, 1977), and more precisely with the inner membrane (Block
et al, 1983 ; Andrews et al, 1985).

The first enzyme of the pathway is a "soluble" enzyme which produces
lysophosphatidic acid (lyso-PA):

$$\text{\underline{sn}-glycerol 3-phosphate + acyl-ACP} \longrightarrow$$

$$\text{1-acyl-\underline{sn}-glycerol 3-phosphate + ACP}$$

In fact, this enzyme is closely associated with the inner envelope membrane and lyso-PA is released directly into the membrane (Joyard and Douce, 1977). This enzyme is most interesting since it is specific for acyl-ACP thioesters (Frentzen et al, 1983). In addition, regardless of the thioester used, a striking specificity for <u>sn</u>-1 position of

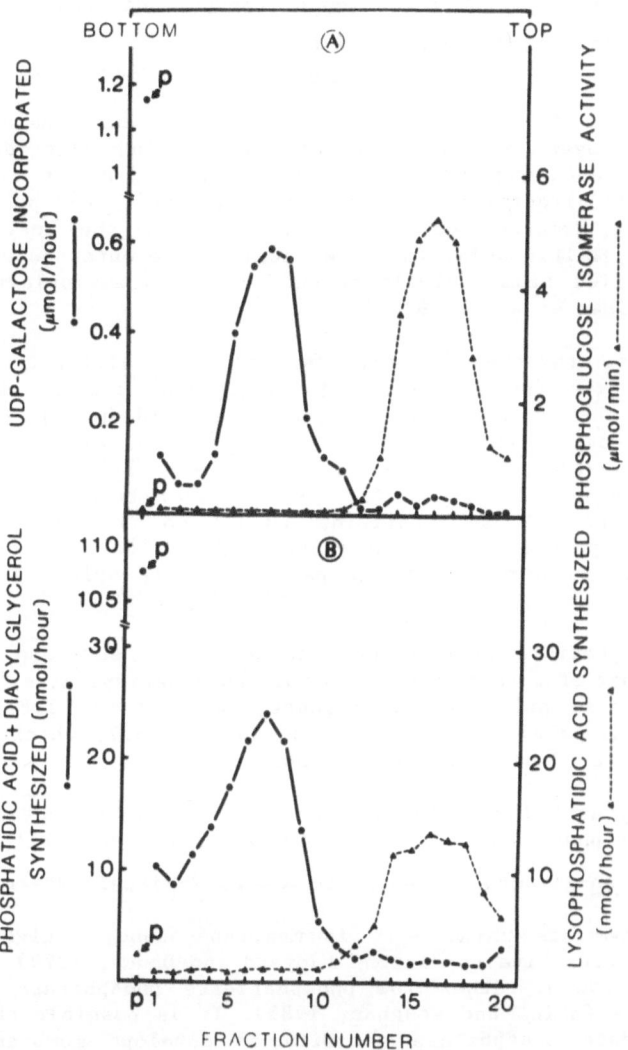

Figure 5. Distribution of marker enzymes (A) and acyltransferases (B) along the linear sucrose gradient used for fractionation of cauliflower bud plastids.

The experimental conditions are described by Alban et al (1988).
Phosphoglucose isomerase was used as a marker for stroma.
MGDG synthase (UDP-galactose incorporated) was used as a marker for envelope membranes.

the glycerol is observed (Joyard et al, 1979 ; Frentzen et al, 1983). Finally, when a mixture of 16:0 and 18:1 thioesters is offered, 18:1 is preferably used (Frentzen et al, 1983). We have found that in non-green plastids from cauliflower buds and amyloplasts from sycamore cells, this enzyme was also soluble (Figure 5) and presents the same specificities and selectivities as in mature chloroplasts (Alban et al, 1988).

The second enzyme, acyl-ACP:1-acyl-sn-glycerol 3-phosphate acyl-transferase, catalyses the formation of phosphatidic acid.

1-acyl-sn-glycerol 3-phosphate + acyl-ACP —————→
                                1,2-diacyl-sn-glycerol 3-phosphate + ACP

This enzyme is firmly bound to the inner envelope membrane from chloroplasts (Joyard and Douce, 1977) and non-green plastids (Figure 5). Since lyso-PA used for this reaction is esterified at sn-1 position, the enzyme will direct fatty acids to the available sn-2 position (Joyard et al, 1979 ; Frentzen et al, 1983). However, the enzyme is highly specific for palmitic acid. The same results were obtained with non-green plastids purified from cauliflower buds and with amyloplasts from syca-more cells (Alban et al, 1988).

Therefore, the two acyltransferases have distinct specificities and selectivities for acylation of sn-glycerol 3-phosphate: together, they led to the formation of phosphadicic acid having 18:1 and 16:0 fatty acids respectively at sn-1 and sn-2 position of the glycerol backbone. This structure is typical for the so-called prokaryotic glyce-rolipids which are found in 16:3 plants (see above). In contrast, in 18:3 plants, plastid glycerolipids (MGDG) do not contain C16 fatty acids at sn-2 position, and therefore probably do not derive directly from the envelope Kornberg-Pricer pathway, their origin is still under investigation.

The phosphatidic acid formed in the inner envelope membrane can be used either for the synthesis of phosphatidylglycerol (this has been demonstrated only in chloroplasts, Mudd et al, 1987) or for the formation of diacylglycerol (Joyard and Douce, 1977) which is the subs-trate for MGDG and sulfolipid biosynthesis.

Diacylglycerol biosynthesis occurs in the inner envelope membrane owing to a phosphatidate phosphatase (Joyard and Douce, 1977):

1,2-diacyl-sn-glycerol 3-phosphate ————→1,2-diacylglycerol + Pi

This enzyme is unique: it is membrane-bound, active at alkaline pH and highly sensitive to cations (Joyard and Douce, 1979). In addition, 18:3 plants have a rather low phosphatidate phosphatase, in contrast to 16:3 plants (Heinz and Roughan, 1983). It is possible that the level of phosphatidate phosphatase activity in envelope membranes could be responsible for the difference observed between 18:3 and 16:3 plants at the level of MGDG synthesis. We have demonstrated that this was also true for non-green plastids (Figure 6). In addition, we have shown that in envelope membranes from sycamore (18:3 plant), diacylglycerol formation was enzymatic and not artifactual as previously suggested. Envelope membranes isolated from 16:3 and 18:3 plants have almost the same capacity to form MGDG when supplied with diacylglycerol and UDP-gal. However, in contrast to 16:3 plants, very little MGDG is formed from sn-glycerol 3-phosphate in 18:3 plastid (chloroplasts as well as amylo-plasts) envelopes, although the biosynthesis of phosphatidic acid is very active, as shown in figure 6, thus suggesting that the limiting step is indeed the formation of diacylglycerol (Heinz and Roughan, 1983).

224

The last enzyme for MGDG synthesis is a UDP-galactose:diacylglycerol galactosyltransferase (or MGDG synthase) which transfers a galactose from a water soluble donor, UDP-gal, to an hydrophobic acceptor molecule, diacylglycerol, for synthesizing MGDG (Douce, 1974):

UDP-gal + 1,2-diacyl-<u>sn</u>-glycerol ⟶
        1,2-diacyl-3-O-β-D-galactopyranosyl-<u>sn</u>-glycerol + UDP

In spinach chloroplasts, this enzyme is localized on the inner envelope membrane (Block et al, 1983). In non-green plastids, this enzyme is very active too, despite the fact that the internal membrane

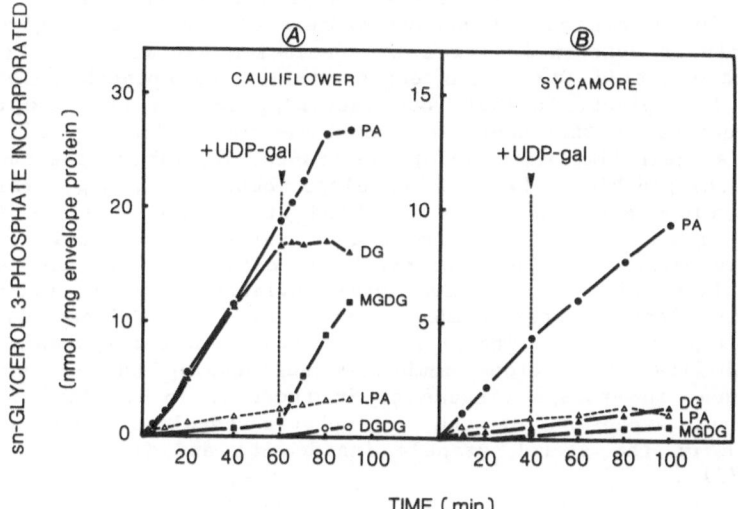

Figure 6. Effect of UDP-galactose on the incorporation of sn- [^{14}C]-glycerol 3-phosphate into envelope lipids from cauliflower plastids (A) and sycamore amyloplasts (B).

The experimental conditions are described by Alban et al (1988).
Note that the major difference between cauliflower and sycamore is the very low level of diacylglycerol (and therefore of MGDG) formed in sycamore.

system is far less developped than thylakoids. In fact, this enzyme is an excellent marker for envelope membranes isolated from chloroplasts as well as from other plastids (Figure 5).

The biosynthesis of DGDG is still a problem in chloroplasts, this is even worst in non-green plastids. Van Besouw and Wintermans (1978) have proposed that this galactolipid could be synthesized owing to the galactolipid:galactolipid galactosyltransferase, we have shown (see above) that this enzyme is present in all plastids.

Finally, other plastid glycerolipids (sulfolipid, PG) have been shown to be synthesized in intact chloroplasts (Joyard et al, 1986 ; Mudd et al, 1985, 1987). It is not yet known whether non-green plastids are also able to catalyze their synthesis, but preliminary data suggest that this is the case.

CONCLUSION

The results presented above demonstrate that envelope membranes from non-green, starch containing plastids have a chemical composition and enzymatic activities very close to that of mature chloroplasts. For instance, the presence of the phosphate translocator in chloroplast and in non-green plastids is extremely important to understand the flow of carbon through these different organelles. Furthermore, despite the strong structural differences, amyloplasts from sycamore cells and pea chloroplasts do not behave differently, as far as glycerolipid biosynthesis is concerned. The same conclusion can be drawn for cauli- flower bud plastids which are quite similar to spinach chloroplasts, a typical 16:3 plant. In fact, one must keep in mind that despite their major structural differences, a common feature shared by all kind of plastids is their limiting envelope membranes. Therefore, the observation that glycerolipid biosynthesis in envelope membranes is a general feature of all envelope membranes from any kind of plastids clearly indicates the major importance of this membrane system in the cell, at least for plastid biogenesis. These biosynthetic capacities are not restricted to chloroplasts with a well developped thylakoid network, but are also present in plastid whose membrane structures are almost limited to their envelope membranes. The presence of these enzymes probably reflects the flexibility of envelope membranes surrounding the starch grains. For instance, in storage tissues or in roots or in isolated cells such as sycamore, expansion of the amyloplasts envelope keeps pace with the growth of the starch granule inside it, and *vice versa* (Briarty et al, 1979).

Is it possible now to answer the initial question "Are envelope membranes from chloroplasts and non-green plastids identical ?". Our observations demonstrate that there is indeed a high degree of uniformity in the structure, chemical composition and functions of the plastid envelope membranes amongst the different plastid types. However, we must keep in mind that differentiation of all plastid types and their develop- mental transitions are associated with, or dependent on, marked changes in their specific enzymatic complement. It is obvious that this is also true for envelope proteins. Characterization of these tissue-spe- cific or plant-specific envelope components that are differentially expressed during plant development is a challenging goal for further studies.

REFERENCES

Alban, C., Joyard, J. & Douce, R. (1988) Plant Physiol. in press

Andrews, J., Ohlrogge, J.B. & Keegstra, K. (1985) Plant Physiol. **78**, 459-465

Block, M.A., Dorne, A.J., Joyard, J. & Douce, R. (1983) J. Biol. Chem. 258, 13280-13286

Block, M.A., Dorne, A.J., Joyard, J.   Douce, R. (1983) J. Biol. Chem.
    258, 13181-13286

Briarty, L.G., Hughes, C.E.   Evers, A.D. (1979) Ann. Bot. (London)
    44, 641-658

Cline, K., Andrews, J., Mersey, B., Newcomb, E.H. & Keegstra, K. (1981)
    Proc. Natl. Acad. Sci. U.S.A. 78, 3595-3599

Dorne, A.J., Joyard, J., Block, M.A. & Douce, R. (1982) FEBS Lett.
    145, 30-34

Dorne, A.J., Joyard, J., Block, M.A. & Douce, R. (1985) J. Cell Biol.
    100, 1690-1697

Douce, R. & Guillot-Salomon, T. (1970) FEBS Lett. 11, 121-126

Douce, R., Holtz, R.B. & Benson, A.A. (1973) J. Biol. Chem. 248, 7215-722

Douce, R. (1974) Science 183, 852-853

Douce, R. & Joyard, J. (1980) Methods Enzymol. 69, 290-301

Douce, R., Block, M.A., Dorne, A.J. & Joyard, J. (1984) In Subcellular
    Biochemistry (D.B. Roodyn, ed.), vol. 10, pp. 1-86, Plenum Press,
    New York

Fishwick, M.J. & Wright, A.J. (1980) Phytochemistry 19, 55-59

Flügge, U.I. & Heldt, H.W. (1976) FEBS Lett. 68, 259-262

Frentzen, M., Heinz, E., McKeon, T.A. & Stumpf, P.K. (1983) Eur. J.
    Biochem. 129, 629-639

Heinz, E. (1977) In Lipids and Lipid Polymers (M. Tevini and H.K.
    Lichtenthaler, eds.), pp. 102-120, Springer, Berlin

Heinz, E. & Roughan, P.G. (1983) Plant Physiol. 72, 273-279

Journet, E.P. & Douce, R. (1985) Plant Physiol. 79, 458-467

Journet, E.P., Bligny, R. & Douce, R. (1986) J. Biol. Chem. 261,
    3193-3199

Joyard, J. & Douce, R. (1977) Biochim. Biophys. Acta 486, 273-285

Joyard, J. & Douce, R. (1979) FEBS Lett. 102, 147-150

Joyard, J., Chuzel, M. & Douce, R. (1979) In Advances in Biochemistry
    and Physiology of Plant Lipids (Appelquist L.A and Liljenberg C.
    eds.), 181-186, Elsevier/North-Holland Amsterdam)

Joyard, J., Grossman, A.R., Bartlett, S.G., Douce, R. & Chua, N.H.
    (1982) J. Biol. Chem. 257, 1095-1101

Joyard, J., Billecocq, A., Bartlett, S.G., Block, M.A., Chua, N.H.
    & Douce, R. (1983) J. Biol. Chem. 258, 10000-10006

Joyard, J., Blée, E. & Douce, R. (1986) Biochim. Biophys. Acta 879,
    78-87

Joyard, J. & Douce, R. (1987) In The Biochemistry of Plants, volume 9, Lipids (P.K. Stumpf, ed.) pp. 215-274, Academic Press, New York

Keegstra, K. & Yousif, A.E. (1986) In Meth. Enzymol. **118**, 316-325

Mudd, J.B., Andrews, J. & Sparace, S.A. (1987a) Methods Enzymol. **148**, 338-345

Mudd, J.B., Kathryn, F. & Kleppinger-Sparace (1987b) In The Biochemistry of Plants, volume 9, Sulfolipids (P.K. Stumpf, ed.) pp. 275-289, Academic Press, New York

Neuburger, M., Journet, E.P., Bligny, R., Carde, J.P. & Douce, R. (1982) Arch. Biochem. Biophys. **217**, 312-323

Van Besouw, A. & Wintermans, J.F.G.M. (1978) Biochim. Biophys. Acta **529**, 44-53

SOME ZERO- AND HIGH-FIELD ELECTRON SPIN RESONANCE AND TIME-RESOLVED FLUORESCENCE STUDIES ON ISOLATED PHOTOSYSTEM II PARTICLES

Geoff Searle, Frans van Mieghem, and Tjeerd Schaafsma

Dept. of Mol. Physics, Wageningen Agricultural University

Dreijenlaan 3, Wageningen, 6703 HA   The Netherlands

INTRODUCTION

Primary photochemical processes in photosystem II (PSII) - energy transfer and charge separation - can be best studied free from possible interference from photosystem I (PSI), by separating the photosystems using biochemical techniques or by the use of mutants of plants which specifically lack PSI. The use of surfactants necessary for the preparation of isolated PSII can give rise to artefacts, whilst the effects of mutations are not always simple. A combined approach is therefore preferable.

The kinetics of singlet excitation transfer and trapping in PSII can be studied with chlorophyll (chl) fluorescence on a sub-ns timescale using a pulsed laser.

The radical ion states, which result from charge separation in PSII, and the triplet states, which arise from subsequent recombination, can be studied using steady-state electron spin resonance (ESR) techniques. The zero-field form of ESR coupled to optical detection is a very sensitive method of investigating chl triplet states[1].

We present an initial characterisation of primary processes in a standard isolated PSII particle preparation, free from PSI and light harvesting chl (LHC), but still containing the antenna chl (on the 47 kD and 43 kD polypeptides). Both the electron donor side (water oxidation), and the electron acceptor (quinone) remain intact, in contrast to the reaction centre (r.c.) preparations completely free of antenna chl.

MATERIALS AND METHODS

PSII Isolation. Broken thylakoids were isolated from greenhouse spinach. PSI was removed using Triton-X100, and the PSII-LHC complex was then treated with n-octyl-glucoside (OG), followed by a precipitation step to remove the LHC[2]. The isolated PSII was stored at 77 K.

Barley Mutants. Mutants of barley, Hordeum vulgare L., were grown at the Carlsberg Laboratory (Dr. D. Simpson) under continuous white light at 20 C, and leaves from the homozygous plants were harvested after 7-10 days. Broken thylakoids were isolated and stored at 77 K.

Time-resolved Fluorescence Kinetics. The samples of PSII particles at 4 C in a 10 mm cuvette, were excited by light pulses from a Rhodamine 6G dye laser at 590 nm. The dye laser was synchronously pumped by a mode-locked Ar ion laser. An electro-optical modulator reduced the pulse frequency to 330 kHz. Fluorescence was detected by a multi-channel plate photomultiplier, through a polariser placed at 54.7 deg. The decay kinetics were recorded using single photon counting, and analysed as a sum of exponentials. The lifetime of the reference (mimic) can also be variable during the fitting procedure. The quality of the fit was judged from the values of chi-square and Durbin-Watson, and from the weighted-residual and auto-correlation plots.

High Field ESR. ESR was measured on samples of PSII particles to which ethyleneglycol was added to 66% to obtain an optically transparent glass at 9 K. An optical cavity allowed illumination in situ.

Zero Field ESR - Fluorescence Detected Magnetic Resonance (FDMR). Samples were excited at 4.2 K by a CW Ar ion laser at 458 nm, 40-80 mW, and the fluorescence detected at the emission maximum. This is 692 nm (2 nm bandwidth) for PSII particles, and emission at wavelengths shorter than 680 nm is completely quenched (spectrum not shown). The small (1-10 ppm) changes in the chl fluorescence intensity induced by resonance of the spin-levels of the lowest triplet state with an isotropic microwave field of variable frequency (100 - 1124 MHz) were averaged over 1000-10 000 sweeps to achieve an acceptable S/N ratio. Sweep time 1 sec.

RESULTS AND DISCUSSION

Singlet Excitation Dynamics. Table 1 presents an analysis of the fluorescence decay kinetics of isolated PSII particles. The kinetics are adequately described by a sum of three exponential components. The 30 ps component is unusually short for PSII. PSI contamination can be ruled out both from the absence of 720-740 nm fluorescence at low temperature and from the absence of the 82/83 kD polypeptides on SDS-PAGE electrophoresis. We have also checked for the absence of pile-up distortion of the decay kinetics, which is known to introduce short lifetime artefacts. The use of the mimic procedure for deconvolution eliminates possible artefacts arising from a wavelength-dependence of the transit time in the photo-multiplier, or from a drift in the excitation pulse with time. The 30 ps component is unaffected by changing the redox state of the acceptors, which could indicate that the lifetime of this component is dominated by the rate constant for energy transfer from the antenna chl to a core of specialised antenna close to the r.c.

The second component has a lifetime, 300-400 ps, similar to that reported widely in the literature for PSII. We were not able to resolve more components in this lifetime region by using a 4 component fit. As our preparations are isolated from grana and are free of LHC they are indeed not expected to show any of the heterogeneity of PSII in thylakoids.

Table 1. Fluorescence Decay Kinetics of PSII Particles at 4 C

Addition	Amplitude(1-3),%			Lifetime(1-3),ns			Rel. Yield
None	78	20	2	0.03	0.40	5.1	0.186
10 mM ascorbate	68	24	8	0.02	0.33	4.1	0.394
2 mM dithionite	60	29	11	0.03	0.32	4.9	0.650

Concentration 5 microgram chl/ml in 0.4 M sucrose, 50 mM MES pH 6.0, 10 mM
NaCl, 5 mM $CaCl_2$. Detection 679 nm. St. dev. amplitudes : 3%(1), 1%(2,3).
St. dev. lifetimes : 2 ps(1), 10 ps(2), 0.1 ns(3). Rel. yield with 2 mM
ferricyanide added : 0.149.

The increase in amplitude of the 4-5 ns component accounts for most of
the increase in the relative fluorescence yield (Table 1) on reducing the
acceptors with ascorbate (dithionite gives a further increase). It is
consistent with a model for singlet energy distribution in which forward
and reverse transfer is possible between the bulk antenna, the r.c. with
postulated core antenna, and the r.c. radical pair.

High Field ESR. Fig. 1 shows the light minus dark ESR difference
spectrum found on illuminating PSII particles at 6 K. This is similar
to the Signal II spectrum of the Z or D neutral radical, and its induction
at 6 K indicates a high degree of intactness of electron donation to P680.
The experimental conditions were : microwave freq. 9.22 GHz, power 1.5
microwatt, mod. ampl. 4 G, gain 10 000, chl conc. 0.75 mg/ml. The PSII
particles show a low yield of triplets either on illumination or on
addition of dithionite. The cause of this low yield is not yet clear. A
triplet state is detected with 40 mM dithionite/1 mM EDTA, and
Fig. 2 shows that the characteristic AEEAAE polarisation pattern is found.
The experimental conditions were : microwave freq. 9.437 GHz, power 36 dB,
mod. ampl. 25 G, gain 400 000, chl conc. 1 mg/ml, temperature 4 K. From the
spectrum the zero field splitting parameters are found to be : D=0.0291 and
E=0.0045 /cm.

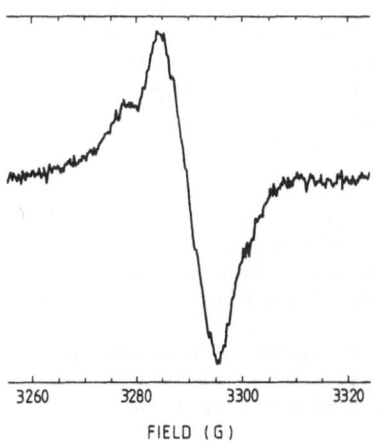

Fig. 1 ESR of PSII particles
        : Signal II induction.

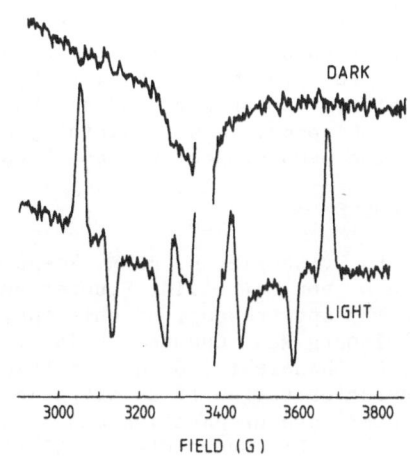

Fig. 2 ESR of PSII particles
        : polarised triplet

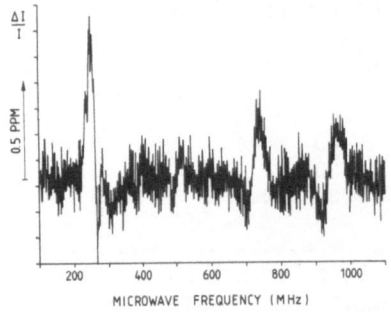

Fig. 3 FDMR of PSII particles

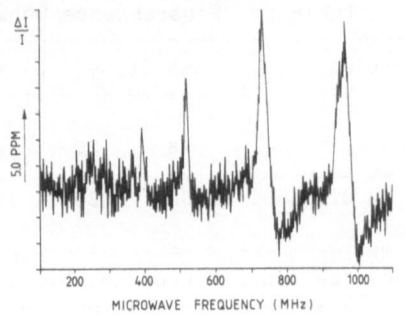

Fig. 4 FDMR of vir-zb63xMU2800

<u>Zero Field ESR</u>. PSII particles only gave an FDMR spectrum after the acceptor was reduced (dithionite 8 mg/ml, pH=7.4). The spectrum contains several resonances, all with positive amplitude. The linewidths appear not to be broadened by spin interactions with Fe or Mn. The resonances at 730 MHz and at 950 MHz (see Fig. 3) are assigned to the D−E and D+E transitions respectively of chl a, by comparison with those for chl in vitro. This triplet state has zero field splitting parameters D=0.0282, and E=0.0038 /cm, and we suggest it is that of P680. A third resonance at 510 MHz is not the 2E line, and is not yet assigned. Fig. 4 shows the FDMR spectrum of thylakoids of the crossed barley mutant viridis-zb63xMU2800, also treated with dithionite. This mutant lacks both PSI and chl b and therefore is expected to contain only PSII. The 600 −1000 MHz region resembles that of PSII particles, the 510 MHz line is virtually absent, but a pronounced 2E line is present. Other mutants of barley from the Carlsberg Laboratory are at present under investigation.

ACKNOWLEDGEMENTS

We thank A. van Hoek for his contribution to the time-resolved fluorescence measurements, and Dr. A. Holzwarth for interesting discussions on kinetic models for PSII. Some high-field ESR experiments were performed at the State University of Leiden, and we thank Prof. A. Hoff and M. Fischer for their co-operation. ESR measurements were also carried out at C. E. N. Saclay, in the laboratory of Dr. A. W. Rutherford. The mutant barley thylakoids were isolated at the Carlsberg Laboratory, Copenhagen, in collaboration with Prof. D. von Wettstein and Dr. D. Simpson. G.S. gratefully acknowledges the award of a Fellowship under the OECD Project on Food Production and Preservation.

REFERENCES

1. G. F. W. Searle, R. B. M. Koehorst, T. J. Schaafsma, B. L. Moller, and D. von Wettstein, Fluorescence detected magnetic resonance (FDMR) spectroscopy of chlorophyll-proteins from barley, Carlsberg Res. Commun. 46:183 (1981).
2. D. F. Ghanotakis, D. M. Demetriou and C. F. Yocum, Isolation and characterisation of an oxygen-evolving Photosystem II reaction center core preparation and a 28 kDa Chl-a-binding protein, Biochim. Biophys. Acta 891:15 (1987).

PHOTOCHEMISTRY IN THE ISOLATED PHOTOSYSTEM 2 REACTION CENTRE

Christalla Demetriou, Christopher J. Lockett
and Jonathan H.A. Nugent

Department of Biology, University College
London, Darwin Building, Gower Street
London WC1E 6BT, U.K.

INTRODUCTION

The recent isolation of a complex containing the polypeptides D1, D2 and cytochrome b_{559}, confirmed the proposal that D1 and D2 bind the reaction centre components of PS2[1,2]. The complex was shown to be photochemically active by the photoreduction of phaeophytin and a spin polarised triplet was also detected by epr[3]. The triplet species originates from a radical pair recombination between the oxidised primary donor chlorophyll P680[+] and the reduced phaeophytin acceptor I[-] [4]. Redox titrations of the reaction centre preparation using the spin polarised triplet signal confirmed the lack of a quinone electron acceptor but suggested that the non-haem iron still present may act as an electron acceptor when oxidised[5].

Experiments are described which examine the reaction centre for evidence of electron donation to P680.

MATERIALS AND METHODS

Photosystem 2 was prepared by the method of Ford and Evans[6] from pea _Pisum sativum_ var Feltham first. _Scenedesmus obliquus_ wild type and the mutant LF1 were grown heterotrophically in the dark and membranes prepared. PS2 particles prepared as for pea. PS2 from the LF1 mutant has a low manganese content and the water oxidation system is absent[7,8]. The PS2 reaction centre was isolated by the method based on that of Nanba and Satoh[1] as modified in[3]. Esr measurements were performed using a Jeol X-band spectrometer. Chlorophyll concentrations of reaction centres were 40ug/ml. Illumination in the cryostat was provided by a 150W light source directed by a light guide.

RESULTS AND DISCUSSION

Fig.1A shows the spin polarised reaction centre triplet

detected by esr in the reaction centre under continuous illumination at 4.2K[3,8]. The AEEAAE polarisation pattern of the spectrum is interpreted as originating from a radical pair recombination between P680[+] and I[-] [4]. The characteristic zero field splitting parameters were [D]=0.0297 cm[-1] and [E]=0.0041 cm[-1].

An additional light induced signal was seen in the reaction centre samples together with the spin polarised triplet (Fig.1A arrowed). The signal was symmetrical about the g=2 region with peaks observed above the high and low field peaks of the reaction centre triplet. The signal was only present under continuous illumination and the same conditions required to observe the major reaction centre triplet. The [D] value is larger than would be expected from a chlorophyll triplet[9] or PS1[10,11] but similar to that obtained for monomeric pheophytin a in solution although with the different polarisation pattern[12].The additional triplet may represent reaction centres where P680 is damaged i.e. has lost magnesium forming a pheophytin or perhaps centres where the excitation has migrated out of the reaction centre onto a nearby pheophytin molecule.

Fig.1B shows the spectrum in the dark at 4.2K following illumination. During illumination a photoinduced signal near g=2 was produced which was stable in the dark at 4.2K and consisted of two peaks split by approximately 12mT

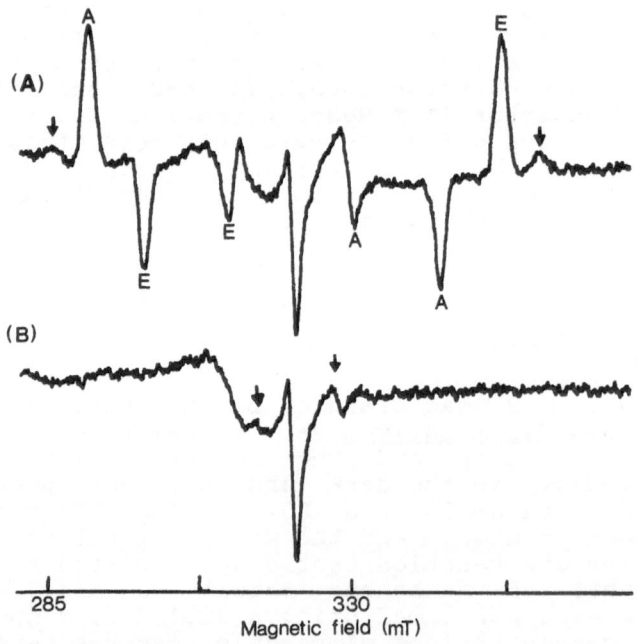

FIG.1 ESR SPECTRA OF TRIPLETS FORMED IN THE PS2 COMPLEX
Pea PS2 reaction centres were examined (A) under continuous illumination at 4.2K. The AEEAAE polarisation pattern is indicated and the additional peaks arrowed (B) at 4.2K in the dark following illumination for 8 min at 4.2K. The dark stable triplet is arrowed. Esr conditions Microwave power 25uW, modulation width 1mT and temperature 4.2K.

(Fig.1B arrowed). The shape and the magnitude of the splitting suggest that it arises through a magnetic interaction between two radicals. The signal is similar to one previously reported in sodium dithionite reduced pea PS2 particles[13] which was attributed to formation of Donor$^+$I$^-$ through electron donation to P680$^+$. This provides evidence for the presence of an electron donor in the reaction centre preparation.

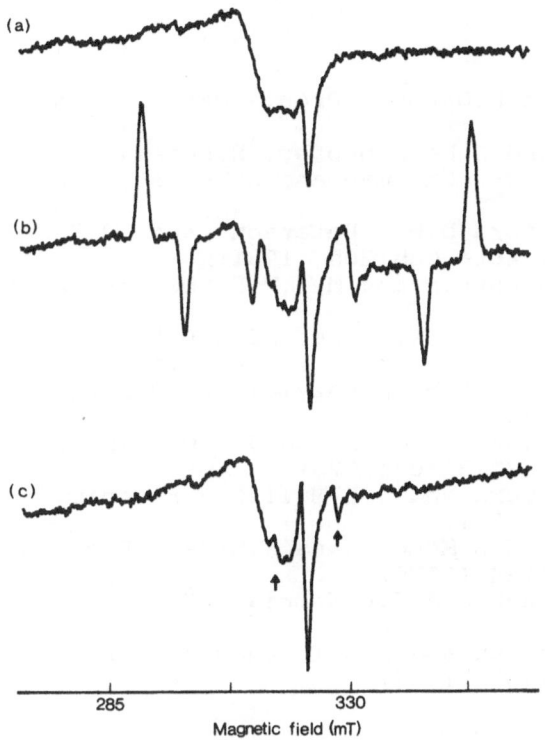

FIG. 2 ESR SPECTRA OF TRIPLETS IN A <u>SCENEDESMUS</u> REACTION CENTRE PREPARATION
(a) dark at 4.2K (b) continuous illumination at 4.2K (c) dark following 8 min illumination at 4.2K. The dark stable triplet is arrowed. Esr conditions as figure 1.

We also obtained a reaction centre preparation from both the wild type and the LF1 mutant of <u>Scenedesmus obliquus</u>. The LF1 reaction centre showed the same esr characteristics as the pea preparation (Fig.2). On continuous illumination the spin polarised triplet was formed (Fig.2b) and in the dark following illumination the dark stable radical pair was observed (Fig.2c arrowed). Warming the sample to 77K in the dark for 30s and remeasuring at 4.2K resulted in loss of the signal. This behaviour was noted in[13] for the signal in reduced PS2 membranes and can be attributed to recombination of the Donor[+]I[-] radical pair. The characteristics of electron donors D and Z which have a high redox potential in PS2 membranes[10,11] can be expected to change during purification due to the stripping away of polypeptides and greater exposure of the redox centre in the preparation.

We therefore conclude that the dark stable triplet arises from the Donor[+]I[-] radical pair and that this indicates that an electron donor to P680[+] was present in the PS2 reaction centre preparation.

REFERENCES

1. O.Nanba and K.Satoh, Proc. Natl. Acad. Sci. USA 84:109 (1987)
2. M.Michel and J.Deisenhofer, Biochemistry 27: 1 (1988)
3. J.Barber,D.J. Chapman and A.Telfer, FEBS Lett. 220:67 (1987)
4. A.W.Rutherford,D.R. Paterson and J.E.Mullet, Biochim. Biophys. Acta 635:205 (1981)
5. A.Telfer,J.Barber and M.C.W.Evans, FEBS Lett. 232:209 (1988)
6. R.C.Ford and M.C.W.Evans, FEBS Lett. 160:159 (1983)
7. J.G.Metz and M.Siebert, Plant Physiol. 76:829 (1984)
8. N.I.Bishop, Methods Enzymol. 23:372 Academic press, New York (1971)
9. M.C.Thurnauer,J.J.Katz and J.R.Norris, Proc. Natl. Acad. Sci. USA 72:3270 (1975)
10. A.W.Rutherford and J.E.Mullet, Biochim. Biophys. Acta 635:225 (1981)
11. H.A.Frank, M.B.McLean and K.Sauer, Proc. Natl. Acad. Sci. USA 76:5124 (1979)
12. M.C.Thurnauer and J.R.Norris, Chem. Phys. Lett. 47:100 (1977)
13. A.W.Rutherford and M.C.Thurnauer, Proc. Natl. Acad. Sci. USA 79:7283 (1982)

ACKNOWLEDGEMENTS

We thank the U.K. Science and Engineering Research Council for financial support. We also thank Prof. M.C.W. Evans for helpful discussion, Prof. N.I. Bishop for providing the LF1 mutant and Prof. J. Barber,Dr A. Telfer and Dr N. Gounaris for helpful comments about the preparative techniques.

236

ELECTRON ACCEPTORS IN PHOTOSYSTEM II

Julia A.M. Hubbard and Michael C.W. Evans

Department of Biology, University College London
Gower Street,
London WC1E 6BT

INTRODUCTION

Several components have been identified on the acceptor side of PSII including I (pheophytin), Q_A and Q_B (plastoquinones). However a number of experiments including fluorescence spectroscopy[1] and E.P.R.[2] suggest that the acceptor side is very complex and that additional electron acceptors may exist. We have used absorption spectroscopy to study the acceptor side of photosystem II by following the reduction of the P680 cation formed by a laser flash. In samples where oxygen evolution is active almost all P680 is reduced by various components on the donor side, this takes place in <50ns, faster than the response time of our spectrophotometer. Treatment with hydroxylamine plus illumination inhibits oxygen evolution and electron donation from the donor D. $P680^+$ is now reduced more slowly (μs & ms) by back reactions from reduced electron acceptors formed by the reaction $P680^+$ hv> $P680^+$ + e-.

METHODS

Spinach or pea Photosystem II particles with high rates of oxygen evolution were prepared as in Ford and Evans (1983)[3], and resuspended in 20mM glycine, 5mM MgCl, 10% glycerol, pH10.0. E.P.R. analysis indicated essentially no contamination with photosystem I.

Optical measurements of $P680^+$ reduction at 820nm were made using a laser flash spectrophotometer as described by Mansfield et al, (1987)[4] except that the measuring device was a photodiode type UDT 10D. A N_2 laser (LN 1000, Photochemical Research Associate Inc) was used to excite the sample (λ337nm, 800 ps flash) at either 2 or 5 H_2. Redox titrations carried out according to Dutton (1978)[5] employed redox mediators as in Evans (1987)[6].

RESULTS AND DISCUSSION

The reduction of P680+ show at least 2-3 phases with different kinetics. The presence of the different phases were found to depend upon the ambient redox potential (Fig.1). At potentials more oxidised than about -110 mV (at pH 10.0) the decay consists of 2 phases with half times of 300-400 μs and several ms. Below -110mV and above -300 mV 2 phases are still observed with half times of 100-200 μs and several ms. At potentials below –300mV but above –430 mV the decay is monophasic with half times of 100–200 μs.   Below –430 mV absorbance changes due to P680 cannot be observed in the response time of our instrument.

We suggest that the μs phases reflect back reactions from acceptors, while the ms phases reflect reduction by exogenous donors in centres in which the electron has been transferred onwards from the bound acceptors.

Changes in the decay kinetics of P680 occur at similar potentials to the steps seen in the titration of the electron acceptors of Photosystem II studied by E.P.R.   That is the 300 to 400 μs component is lost around –50mV, the ms component at –300 to –350 mV and the 100 to 200 μs component at about –420 mV.

These results indicated that multiple electron acceptors may exist on the acceptor side.   Thus although the molecular nature of the components giving rise to the different kinetics remains to be described, the laser–flash optical studies of photosystem II appear to confirm the E.P.R. studies, confirming that the initial stable acceptor has Em below –400 mV.

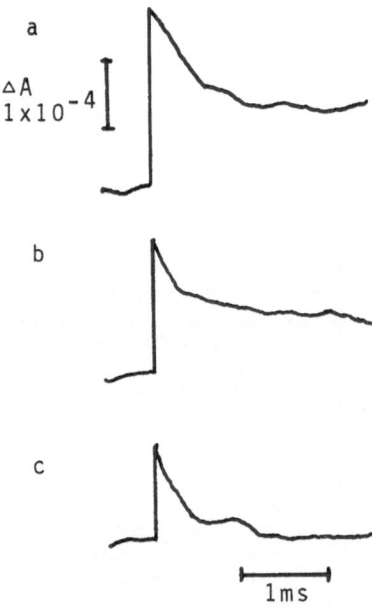

Fig.1.   Flash–induced absorbance changes at 820nm in hydroxylamine–inhibited PSII at pH 10 at several different redox potentials: (a) +110mV; (b) –110mV;   (c) –395mV.   In each case 64 measurements were averaged at a flash rate of 2 Hz.

REFERENCES

1.   W.A. Cramer and W.L. Butler, Potentiometric titration of the fluorescence yield of spinach chloroplasts, Biochim. Biophys. Actu. 172:503 (1969).

2.   M.C.W. Evans, Y.E. Atkinson, and R.C. Ford, Redox characterisation of the Photosystem II electron acceptors.   Evidence for two electron carriers between pheophytin and Q, Biochim. Biophys. Acta. 806:247 (1985).

3. R.C. Ford and M.C.W. Evans, Isolation of a photosystem 2 preparation from higher plants with highly enriched oxygen evolution activity FEBS Lett 160:159 (1983).

4. R.W. Mansfield, J.A.M. Hubbard, J.H.A. Nugent, and M.C.W. Evans, Extraction of electron acceptor A, from pea photosystem I, FEBS Lett. 220:74 (1987).

5. P.L. Dutton, Redox Potentiometry, in: Methods in Enzymology, Vol.54, Academic Press, 411 (1978).

6. M.C.W. Evans, Evidence for two Q_a-like quinone binding sites in the reaction centre of Rhodospeudomonas viridis, Biochim. Biophys. Acta. 894:524 (1987).

4. McLeod, L., R.J.H. Morey, MacLeod, J. Machine Co. Inc., 1975.

5. Ig., Igor, Igor, Piumbia, Piumbia, original products bulletin, No. 40-PQR, 1975.

6. G.W., Woodhall, L.H.W. Lucht, R.L. Struck, and R.L. Struggle, Bark Piump determination method, in Program organising 5 1975, 1975.

7. Jin, Hutch, R.L. L'encyclopédie, Features of chemica Jin, in Programming 5 (122), 1975.

8. Wild, Wood-compounding for the microcommputer programms bulletin, Bulletin 5 compounding 5 1975, Pirumbia, Piumbia, 1975.

CYT. b-559 IN PHOTOSYSTEM II

Ranjana Paliwal and G.S. Singhal

School of Life Sciences
Jawaharlal Nehru University
New Delhi-110067, India

INTRODUCTION

In chloroplasts, cytochrome b-559 is known to be associated with photosystem II. Though its role is not precisely understood, cyt b-559 functions probably on both the reducing and the oxidising sides of photosystem II.[1] A high as well as a low potential form of cyt b-559 with mid-point potentials respectively from +350 mV to +400 mV and from +50 mV to +100 mV at pH 7.8 have been reported.[2] A range of redox forms may be observed in the intermediate region.[3] The high potential cyt b-559 requires the structural integrity of the thylakoid membrane, and its disruption causes cyt b-559 to be modified to lower potential forms.[4] A heterogenous population of the high potential cyt b-559 has been reported.[5] Based on biphasic chemical-induced oxidation kinetics, they have suggested that the high potential cytochrome b-559 exists in the thylakoid membrane in two different environments.

Most of the cyt b-559 in the low potential form can be restored to the high potential form on incorporation of PS II particles into liposomes. The liposomal environment restores the normal variable fluorescence and increases the rate of oxygen evolution substantially.[6] The purpose of the present work was to investigate further the function of different potential forms of cyt b-559 in the electron transport process. We evaluated the amount of photo-reduced cyt b-559 and the rate constant of its photo-oxidation and photo-reduction in isolated PS II preparation suspended in buffer (for low potential form of cyt b-559) and in PS II incorporated into liposomes (for relatively higher or intermediate potential form of cyt b-559). We have used (1) atrazine which blocks electron transfer from Q_B and (2) TMPD and ADRY like reagent, to analyze the process of photo-oxidation and photo-reduction of cyt b-559 in PS II incorporated in liposomes as well as in isolated PS II particles.

MATERIALS AND METHODS

Spinach leaves obtained from local market were used to isolate PS II particles according to the method of Ghanotakis et al.[7] Liposomes were prepared by forming a thin layer of either Dimyrisitoyl phosphatidyl choline, DMPC, or Dipalmitoyl phosphatidyl choline DPPC on evaporation of the organic solvents under nitrogen in a rotary evaporator. The lipid was then dispersed in 20 mM Mes-NaOH pH = 6.0 containing 15 mM NaCl and 5 mM $MgCl_2$. Lipid and PS II particles were sonicated separately before mixing in a lipid to Chl ratio of 20:1. Immediately after mixing, the suspension was frozen in liquid nitrogen and thawed in H_2O at 25°C. TMPD and atrazine were added in buffer before freez-thawing, thus ensuring that it was trapped or incorporated inside the liposomes along with the PS II particles. Absorbance spectra and light-minus-dark difference spectra of cyt b-559 were scanned in the double wavelength mode of Shimadzu UV 3000 Spectrophotometer, using $\lambda_{ref} = 570$ nm and $\epsilon_{570-559} = 15$ mM^{-1} cm^{-1}. Actinic white light was provided by the cross-illumination attachment of the instrument (light intensity 30,000 lux, light source-tungsten halogen lamp, 50 W).

In the experiments with TMPD, the changes in ΔA_{559} were corrected for ΔA_{570} according to Agalidis and Velthuys,[8] by substracting the change at λ_{559} due to TMPD oxidation.

RESULTS

The absorbance spectrum of cyt b-559 (Fig. 1) in PS II in buffer and in PS II incorporated in DMPC liposomes was identical. Chemically induced difference redox spectra (Fig.2) show that most of the cyt b-559 is in the dithionite reducible (i.e. the low potential) form in PS II in buffer and in the ascorbate reducible (intermediate potential or relatively higher potential) form in PS II incorporated in DMPC liposomes. We have compared the behaviour of different potential forms of cyt b-559 by using these two preparations (a) PS II in buffer and (b) PS II in liposomes.

The rate constant (k) of light induced absorbance change at $\lambda = 559$ nm (Table 1) do not give the absolute rate constants, but can be used for comparing the effect of different concentrations of atrazine and TMPD. In PS II in buffer we observed an inhibition of the rate of photo-reduction of cyt b-559 with increasing concentrations of atrazine. However, increasing concentrations of TMPD enhanced the rate of photo-reduction, although compared to control (i.e. without TMPD) there was an inhibition (see Table 1).

In liposomes, the rate of photoreduction is only 12% of that in PS II in buffer. With addition of atrazine or TMPD the cyt b-559 in liposomes becomes photo-oxidised (in contrast to its photo-reduction in PS II in buffer). With increasing concentrations of atrazine or TMPD the rate of photo-oxidation increases.

DISCUSSION

Our observations show that the presence of atrazine inhibits the photo-reduction of cyt b-559 by blocking the electron transport at Q_B. This would indicate that cyt b-559 is reduced by the acceptor side of PS II (by Q_B). A similar conclusion has been earlier reported in thylakoids, that cyt b-559 is reduced by photosystem II through the plastoquinone pool.[9] However, it has also been reported that the high potential cyt b-559 does not function as an obligatory redox component in the main electron transport chain.[10]

In PS II incorporated in liposomes, illumination in presence of atrazine results in oxidation of cyt b-559 (Table 1). This may be due to the lipid environment provided by the DMPC liposomes which might be restoring electron transfer from reduced cyt b-559 to the donor side (probably Z). The redox state of cyt b-559 under continuous illumination would depend on the competition between its reduction by acceptor side and the oxidation by donor side of PS II.

TMPD affects the photo-reduction of cyt b-559. TMPD is known to accept electrons from Q_A and has also been shown to destabilise the donor side of PS II in a manner analogous to other ADRY reagents.[11]

Table 1. Per cent content of reduced cyt b-559 on continuous illumination and the rate of photoinduced A_{559} in presence of various concentrations of atrazine and TMPD

		PS II in buffer (cyt.b-559 LP)		PS II in liposomes (cyt. b-559 IP)	
		Reduced cyt.b-559 (%)	$K_{red} \cdot 10^{-3}$ (sec^{-1})	Reduced cyt.b-559 (%)	$K_{ox} \cdot 10^{-3}$ (sec^{-1})
No addition		67.5	16.0	51.9	$2.0(K_{red})$
Atrazine	0.45 µM	64.4	16.5	41.0	3.7
	1.05 µM	58.2	11.6	29.2	9.6
	1.5 µM	54.2	7.4	21.0	12.4
	2.1 µM	51.0	3.5	-	-
TMPD	4.5 µM	49.6	1.1	48.0	0.8
	7.5 µM	49.6	1.5	43.0	2.8
	10.5 µM	52.7	8.8	29.2	8.2
	13.5 µM	58.2	11.6	-	-
	15.0 uM	59.7	13.5	26.1	10.4

The content of reduced cyt. b-559 was 49.2% in dark. Assay mixture was same as in Fig.3. (The amount of chemically reduced cyt. b-559 by dithionite was taken as 100% reduced cyt. b-559).
LP - Low potential, IP - Intermediate potential.

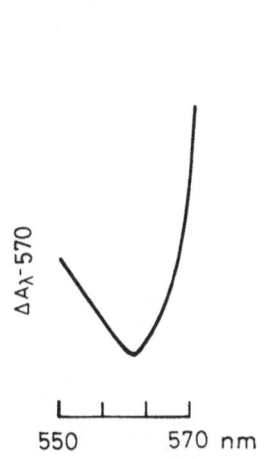

Fig.1. Absorbance spectrum of cyt. b-559 in dark adapted PS II particles without any additions. Spectrum was taken in the dual wavelength mode with λ_{ref}=570 nm. Assay mixture contained 20 mM Mes-NaOH pH = 6, 15 mM NaCl and 5 mM MgCl$_2$.

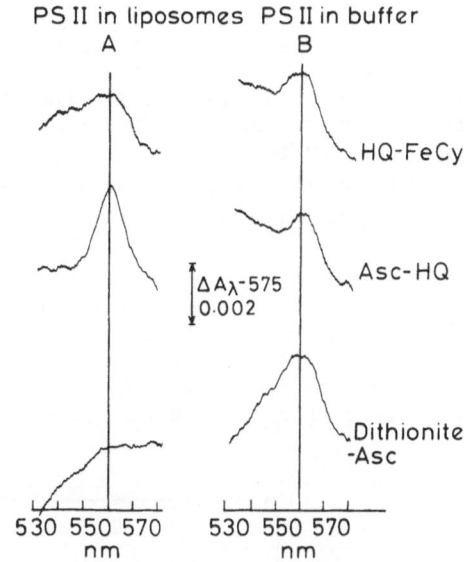

PS II in liposomes   PS II in buffer

A   B

HQ-FeCy

ΔA_λ-575 0.002

Asc-HQ

Dithionite -Asc

530 550 570 nm   530 550 570 nm

Fig.2. Redox difference spectra of cyt. b-559 in PS II in buffer and PS II incorporated into liposomes. Assay mixture same as in Fig.1. Chl conc. 50 µg/ml. Additions were made sequentially of 0.2 mM FeCy, 20 mM hydroquinone, 20 mM sodium ascorbate and a few grains of sodium dithionite.

In the presence of a mixture of TMPD and $TMPD^+$, Z^+ oxidises TMPD and Q_A^- reduces $TMPD^+$.[8] Like other ADRY reagents, TMPD is also oxidised by S_2 state and may be responsible for the oxidation of reduced cyt b-559.[12] The photo-oxidation of TMPD and its subsequent rereduction of cyt b-559 has been demonstrated by Yerkes and Crofts.[13] A charge recombination pathway between Q_A^- and Z^+ through a component C (likely to be cyt b-559) has been suggested.[14,15]

Under continuous illumination we presume that a steady state involving the photo-reduction of $TMPD^+$ by Q_A^- and cyt b-559, and photo-oxidation of TMPD by Z^+ and S_2 is reached depending on the concentration of TMPD.

Photo-oxidation of cyt b-559 with increasing concentrations of TMPD, only when PS II is incorporated into liposomes would suggest that only the intermediate potential form of cyt b-559 (which is restored in the liposomes) can reduce the donor side.

Our results may be best interpreted according to the following scheme (Fig.3), which represents the oxidation and reduction of cyt b-559 and the action of TMPD and atrazine in shifting the reaction (photo-reduction of cyt b-559) to the right or to the left. This involves the function of cyt b-559 on both the reducing and the oxidising sides of PS II. The conversion of low potential form of cyt b-559 to high potential form by protonation (H^+ taken from splitting of H_2O) has been suggested by Butler[16,1] and may be affecting the electron transfer.

Liposomes made from DMPC were able to reconstitute an intermediate potential (ascorbate reducible) form of cyt b-559. The lipid environment might be a factor in evaluating the proportion of photo-reduction and photo-oxidation of cyt b-559 by (1) restoring the integrity of OEC and PS II complex and/or (2) by modifying the environment in some way which facilitates e^- transfer from cyt b-559 to OEC component Z. This is also supported by the fact that the rate constants of cyt b-559 photo-reduction/oxidation in liposomes are considerably different from those in PS II in buffer.

The PS II particles incorporated into liposomes provide a method of distinguishing absorbance changes due to photo-oxidation and photo-reduction of different potential forms of cyt b-559.

On the basis of our results we conclude that the dithionite reducible form (low potential) and the ascorbate reducible form (intermediate potential) behave differently. The photo-oxidation and photo-reduction of cyt b-559 depends on its redox potential, on the ratio of reduced to oxidised cyt b-559 and on the concentration of electron transport perturbants (atrazine and TMPD).

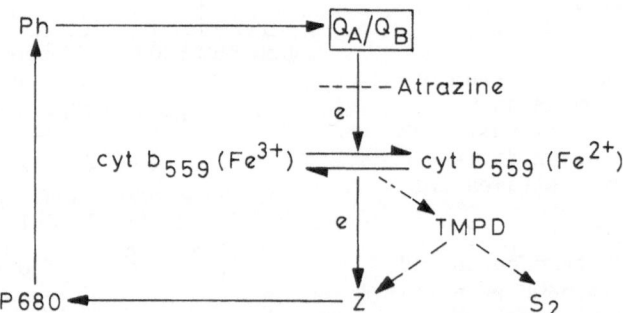

Fig.3. Schematic representation of electron transfer in PS II and the involvement of cyt. b-559 with other components.

REFERENCES

1. Butler, W.L. and Matsuda, H. (1983) Possible role of cytochrome b-559 in Photosystem II in The Oxygen Evolving System of Photosynthesis (Inoue, Y., Crofts, A.R., Govindjee, Murata, N., Renger, G. and Satoh, K. eds) pp 113-122 Acad. Press, Japan.

2. Cramer, W.A. and Whitmarsh, J. (1977) Photosynthetic cytochromes. Ann. Rev. Plant Phsyiol. 28, 133-172.

3. Cramer, W.A. and Crofts, A.R. (1982) Electron and Proton Transport in: Photosynthesis. Vol.I (Govindjee, ed.) pp 389-467 Acad. Press, New York.

4. Erixon, K., Lozier, R. and Butler, W.L. (1972) The redox state of cytochrome b-559 in spinach chloroplasts. Biochim. Biophys. Acta. 267. 375-382.

5. Selak, M.A., Koch-Whitmarsh, B.E. and Whitmarsh, J. (1984) Evidence for a heterogenous population of high potential cytochrome b-559 in the thylakoid membrane in: Advances in Photosynthesis Research (Sybesma, C. ed.) Vol.I. pp 493-496 Martinus Nijhoff/Dr. W. Junk Publishers. The Hague, Netherlands.

6. Matsuda, H. and Butler, W.L. (1983) Restoration of high potential cytochrome b-559 in photosystem II particles in liposomes. Biochim. Biophys. Acta. 725, 320-324.

7. Ghanotakis, D.F., Babcock, G.T. and Yocum, C.F. (1984) Structural and catalytic properties of the oxygen-evolving complex. Correlation of polypeptide and manganese release with the behaviour of Z^+ in chloroplasts and a highly resolved preparation of the PS II complex. Biochim. Biophys. Acta. 765, 388-398.

8. Agalidis, I. and Velthuys, B.R. (1986) Oxidation of Q_A and of Q_B of photosynthetic reaction centres by an artificial acceptor. FEBS Lett. 197, 263-266.

9. Whitmarsh, J. and Cramer, W.A. (1978) A pathway for the reduction of cytochrome b-559 by photosystem II in chloroplasts. Biochim. Biophys. Acta. 501, 83-93.

10. Whitmarsh, J. and Cramer, W.A. (1977) Kinetics of the photoreduction of cytochrome b-559 by photosystem II in chloroplasts. Biochim. Biophys. Acta. 460, 280-289.

11. Velthuys, B.R. (1983) Spectrophotometric methods of probing the donor side of photosystem II in: The Oxygen Evolving System of Photosynthesis (Inoue, Y., Crofts, A.R., Govindjee, Murata, N., Renger G. and Satoh, K. eds) pp 83-90. Acad. Press, Japan.

12. Ghanotakis, D.F., Yerkes, C.T. and Babcock, G.T. (1982) The role of reagents accelerating the deactivation reactions of water-splitting enzyme system Y (ADRY reagents) in destabilizing high potential oxidizing equivalents generated in chloroplast photosystem II. Biochim. Biophys. Acta.682, 21-31.

13. Yerkes, C.T. and Crofts, A.R. (1984) A mechanism for the ADRY induced photo-oxidation of cytochrome b-559 in: Advances in Photosynthesis Research (Sybesma, C. ed.) Vol.I. pp 489-492. Martinus Nijhoff/Dr. W. Junk Publishers, The Hague, Netherlands.

14. Tamura, N., Radmer, R., Lantz, S., Cammarata, K. and Cheniae, G. (1986) Depletion of photosystem II-extrinsic proteins. II. Analysis of the PS II/water oxidizing complex by measurements of N,N,N',N'-tetramethyl-p-phenylenediamine oxidation following an actinic flash. Biochim. Biophys. Acta. 850, 369-379.

15. Radmer, R., Cammarata, K., Tamura, N., Ollinger, O. and Cheniae, G. (1986) Depletion of photosystem II-extrinsic proteins. I. Effects on O_2 -and N_2 -flash yields and steady-state O_2 evolution. Biochim. Biophys. Acta. 850, 21-32.

16. Butler, W.L. (1978) On the role of cytochrome b-559 in oxygen evolution in photosynthesis. FEBS Lett. 95, 19-25.

A NEW CLASS OF POTENTIAL MECHANISM-BASED SUICIDE INHIBITORS OF PHOTOSYNTHETIC ACTIVITY

Jacqueline Tso and G. Charles Dismukes

Department of Chemistry
Princeton University
Princeton, NJ 08544 U.S.A.

INTRODUCTION

Inhibitors on the donor side of Photosystem II (PSII) in higher plants have been limited to an assortment of small amines (NH_2OH, NH_2NH_2, NH_3, etc.) and various derivatives of 2-anilino-thiophenes and phenylhydrazone derivatives which together form a group called the ADRY reagents. These compounds have all been shown to effectively disrupt the normal cyclization through the S-states of the Water-Oxidizing Complex (WOC), which is known to contain a special arrangement of four manganese ions. Identification of the proteins which bind these manganese ions and the specific amino acids which surround the site is still unknown. The reactivity of the first class of inhibitors at the manganese site is well documented, however, their usefulness in elucidating the protein environment surrounding the metal site has been limited. To approach this goal, we have developed a new class of inhibitors of the WOC which could react specifically with the manganese site to form a reactive species that is capable of covalently modifying the surrounding protein matrix in the active site. A similar approach has proven to be successful in identifying the binding site for Q_b, the secondary electron acceptor, to be on the D1 subunit using photoaffinity labelling with azido[^{14}C]atrazine, a herbicide which binds to the Q_b site (1). By an analogous method, the location of the PSII electron donor, Z^+, has also been proposed to be located on the D1 subunit, as shown by its photooxidation of [125]I^- (2).

In this report, we present the rationale for use of the hydrazones, in particular dimethyl-hydrazone, as potential covalent labelling reagents for the WOC, and the initial results on their effectiveness as inhibitors of photosynthetic water oxidation. The application of dimethylhydrazone for the covalent modification of proteins in the WOC is illustrated in Scheme 1.

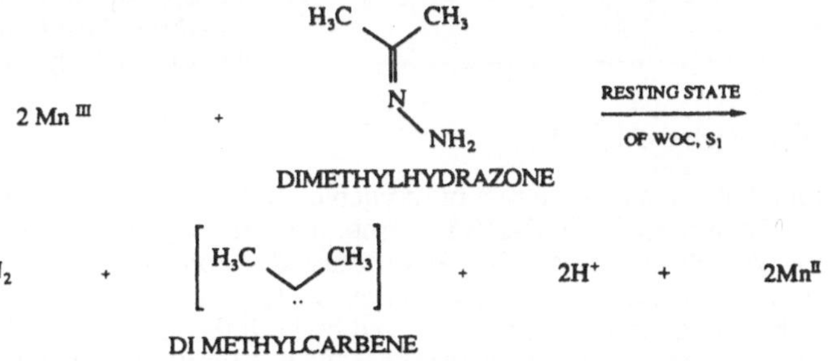

Scheme 1- Proposed reaction of Dimethyl-hydrazone with the WOC.

This shows that oxidation of dimethylhydrazone by a pair of the endogenous Mn(III) ions of the WOC can afford a potential route for the covalent modification of the active site. This reaction is based on oxidation chemistry that occurs with free Mn[III] in solution.

METHODS AND MATERIALS

Spinach BBY-PSII membranes were prepared according to the procedure outlined by Berthold, et al. (3) and stored at -80°C in Buffer A containing 0.4 M sucrose, 40 mM MES, 25 mM NaCl, pH=6.5 and 30% glycerol until further use. Oxygen-evolving OGP-core membranes were prepared according to the method of Ghanotakis, et al. (4) and also stored at -80°C until further use. Oxygen evolution measurements were taken with a Clark-type oxygen electrode using 2,5 - dichloro-p-benzoquinone (DCBQ) as the exogenous electron acceptor. The S_2-state of the WOC was formed by continuous illumination at 200 K in a solid CO_2/methanol bath. EPR measurements were taken with a Varian E-12 EPR spectrometer equipped with an Oxford Instruments helium cryostat. Typical EPR samples for multiline measurements contain 4-5 mg Chl/ml and 0.5 mM DCBQ. Samples were treated with dimethylhydrazone at a [Chl]= 0.5 mg/ml, centrifuged, resuspended in Buffer A and loaded into calibrated quartz EPR tubes. The Hill reaction (H_2O to DCIP) and DPC to DCIP photoactivation kinetics were recorded with a Hewlett Packard UV-VIS spectrophotometer by monitoring the rate of change in the 600 nm absorption band of DCIP. NaCl-treated PSII membranes lacking the 23 and 17 kDa polypeptides were prepared according to the procedure of Miyao, et al. (5). The Tris-treated membranes which lacks all three extrinsic polypeptides and manganese were prepared according to the procedure of Kuwabara, et al. (6).

RESULTS

Dimethylhydrazone can readily undergo a two-electron oxidation even under mild oxidizing conditions (7), forming primarily a very reactive dimethylcarbene intermediate (8). The chemistry of carbenes are well-known in organic chemistry and can undergo a variety of reactions, the common result of which is incorporation into a nearby substrate (8). Evidence supporting the reaction stoichiometry in Scheme 1 was obtained by analysis of the products formed upon treatment of dimethylhydrazone with a dinuclear Mn(III) complex, $(HBPz_3)Mn^{III}(O)(CH_3CO_2)_2Mn^{III}(Pz_3BH)$ (9). Reaction of this dinuclear complex with dimethylhydrazone resulted in the formation of a broad (>5000 Gauss) 6-line EPR (Electron Paramagnetic Resonance) pattern that was indicative of $Mn^{II}(Pz_3BH)_2$. To test for the formation of a carbene intermediate, a similar experiment was carried out using 2,3-dichloro-5,6-dicyano-1,4- benzoquinone as the oxidant. In the presence of benzaldehyde, the formation of a covalent adduct was established by thin-layer chromatography. The use of dipolarophiles, such as benzaldehyde, has been shown to trap reactive 1,3-dipolar intermediates that are formed from the carbene precursors (10).

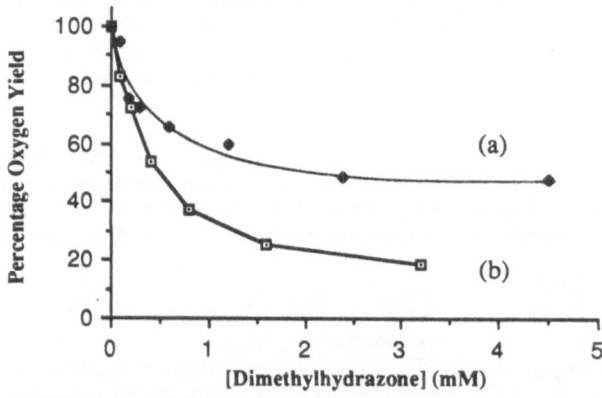

Figure 2a & 2b-Inhibition curves of oxygen activity with PSII membranes (a) and PSII core complexes (b) , respectively. The acceptor, 1 mM DCBQ, was added after treatment with the hydrazone. Typical oxygen rates for PSII membranes and PSII core complexes were 350-450 μmol O_2/mg Chl-hr and 600-800 μmol O_2/mg Chl-hr, respectively.

Dimethylhydrazone inhibited the steady-state rate of oxygen evolution with a concentration dependence shown in Figures 2a and 2b for PSII-BBY membranes and oxygen-evolving PSII core complexes, respectively. A 50% inhibition, I_{50}, was observed at concentrations of 2 mM for PSII-BBY membranes and 0.5 mM for PSII core complexes. The site of inhibition was observed to be at or close to the manganese center as detected by both the loss of the EPR S_2-multiline signal (Figure 3a) and the release of manganese. To test for photoreversibility of the centers inhibited by dimethylhydrazone, the same samples were thawed at 273K and illuminated under dim light, dark-adapted for 30 minutes at 273K and re-illuminated at 200K to produce the S_2-state. The partial recovery of the S_2-state signal is shown in Figure 3b. Evidence also exists which supports non-photoreversible damage to the site, as evident from the partial release of the three extrinsic polypeptides (17, 23 and 33 kDa) and manganese.

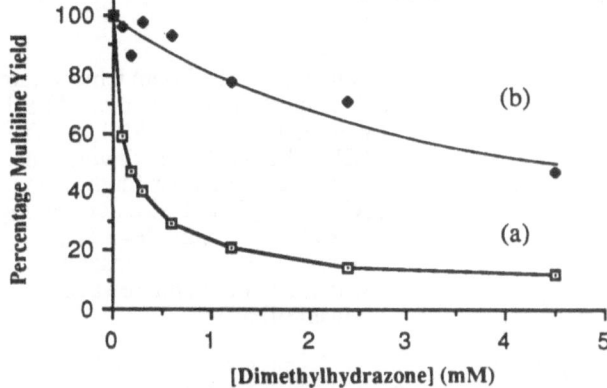

Figure 3a & 3b- Inhibition by dimethylhydrazone of the S_2-state multiline signal in PSII membranes (a) and its subsequent recovery by illumination (b). Each EPR sample contained 4-5 mg Chl/ml and 500 μM DCBQ.

The effects of dimethylhydrazone on the photoreduction kinetics of DCIP in PSII membranes, membranes depleted of the 17 and 23 kDa polypeptides and of all three extrinsic proteins and manganese are shown in Figure 4. The PSII-mediated electron transport rates in control PSII membranes (curve a) and in protein-depleted membranes (lacking the 17 and 23 kDa polypeptides, curve b) remain much higher (x10) than oxgen evolution rates at comparable concentrations of dimethylhydrazone. This implies that the PSII reaction center (P_{680}) remains functional and that the inhibitor can also act as an electron donor. To confirm this observation, dimethylhydrazone was found to restore PSII-mediated electron transport in manganese-depleted PSII-membranes (0.8M Tris, pH=9.0) as shown in curve c. This reaction is completely inhibited by addition of 0.5 mM DCMU. This result seems to indicate that the observed reactivity of dimethylhydrazone is compatible with the electron transport properties of PSII. These findings support the idea that dimethylhydrazone is also capable of reacting with other strong oxidants on the donor side of PSII in the absence of manganese, possibly the species Z.

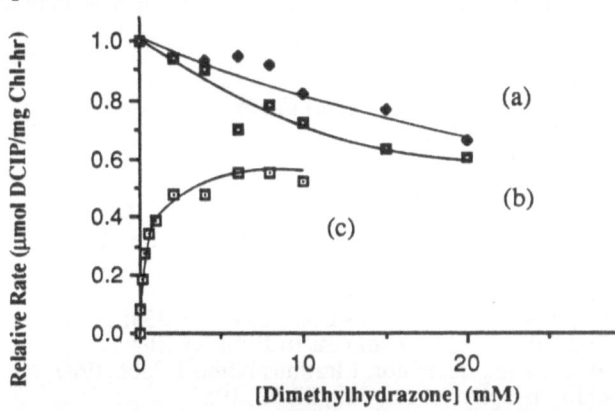

Figure 4- Effect of Dimethylhydrazone on the photoreduction rate of DCIP with PSII (a), NaCl-treated PSII (b) and Tris-treated PSII (c) membranes. Each sample cuvette contained [Chl]=20 μg/ml and [DCIP]=100 μM. Control rates were typically between 200-250 μmol DCIP/mg Chl-hr.

EPR evidence also supports the reactivity of dimethylhydrazone with the species Z in PSII membranes which have been depleted of manganese by Tris treatment. As previously reported by Babcock and coworkers a doubling of the intensity for EPR Signal II_s (D^+) arising from a contribution from EPR Signal II_f (Z^+) in Tris-treated PSII membranes can be observed under the condition of continuous illumination in the absence of an added artifical electron acceptor (11). As shown in Figure 5, the region showing the stable free radical signal attributed to the D^+ species which gives rise to the dark Signal II_s, increased by at least a factor of two upon continuous illumination, revealing the photo-oxidizable Z^+ species. However, when dimethylhydrazone was present, the light-induced contribution to the signal was not observed, presumably due to electron donation by dimethylhydrazone.

DISCUSSION

The mechanism of action of a new category of inhibitors has been studied for its effects on the various electron carriers on the donor side of PSII in spinach. One such compound, dimethylhydrazone, has been shown to inhibit formation of the S_2-state multiline EPR signal

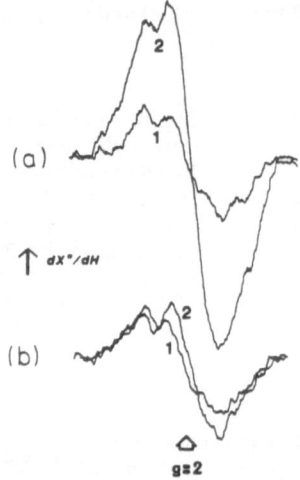

(a)

↑ dX°/dH

(b)

△
g≈2

Figure 5a & 5b- Room temperature EPR spectrum of Tris-treated PSII membranes. (a) Tris PSII, (b) Tris PSII + 1 mM Dimethylhydrazone. 1)dark, 2)continuous illumination. [Chl]=4 mg/ml. Instrument settings: Modulation amplitude=3.2G, Gain=10 x10^3, time constant=1, scan time=4 min, microwave power=30 mW.

and oxygen evolution activity in PSII-BBY membranes and PSII core complexes by a photoreversible mechanism when manganese is not released and at higher concentrations, by an irreversible process which results in release of manganese. This mechanism seems to be analogous to that found for NH_2OH (12) and H_2S (13). The site of inhibition seems to be located within the WOC, probably at the manganese site. In PSII membranes which have been depleted of manganese (i.e. Tris treatment), the primary interaction of the donor side with dimethylhydrazone seems to lie at the site of the Z species. This result is compatible with the observed accessibility of this site from the aqueous phase only after release of manganese (14). Recently, a location has been suggested for the Z species , its identity linked to the [161]Tyr residue on the D1 subunit (15).

Work is currently in progress to determine the fate of the anticipated carbene intermediate with the proteins in PSII.

ACKNOWLEDGEMENTS
    Research was funded from the following grants- DOE Soleras Grant #DE-FG02-84CH10199 and NIH Grant #GM39932.

REFERENCES

(1) Pfister, K., Steinback, K., Gardner, G. and Arntzen, C., Proc. Nat. Acad. Sci. U.S.A., **78**, 981-985, 1981.
(2) Ikeuchi, M. and Inoue, Y., FEBS Lett., **210**, 71-76, 1987.
(3) Berthold, D.A., Babcock, G.T. and Yocum, C.F., FEBS Lett., **134**, 231-234, 1981.
(4) Ghanotakis, D., Demetriou, D. and Yocum, C.F., **Progress in Photosynthesis Research, Vol. 1**, 1.5.681-1.5.684, J. Biggins, editor, Martinus Nijhoff Publ.,1987.
(5) Miyao, M. and Murata, N., Biochim. Biophys. Acta, **725**, 87-93, 1983.
(6) Kuwabara, T. and Murata, N., Plant Cell Physiol., **23**, 533-539, 1982.
(7) **Encyclopedia of Electrochemistry of the Elements, XIII**; Chp. 6, Bard, A.J., editor. Marcel Dekker, Inc., 1979.
(8) Gilchrist, T. and Rees, C., **Carbenes, Nitrenes, and Arynes**, Appleton-Century-Crofts, 1969.
(9) Sheats, J., Czernuszewicz, R., Dismukes, G., Rheingold, A., Petrouleas, V., Stubbe, J., Armstrong, W., Beer, R. and Lippard, S., JACS, **109**, 1435-1444, 1987.
(10) Padwa, A., Gasdaska, J., Tomas, M., Turro, N., Cha, Y. and Gould, I., JACS, **108**, 6739-6746, 1986.
(11) Babcock, G., Ghanotakis, D., Ke, B. and Diner, B., Biochim. Biophys. Acta, **723**, 276-286, 1986.
(12) Sivaraja, M. and Dismukes, G. C., Biochemistry, **27**, 3467-3475, 1988.
(13) Sivaraja, M. and Dismukes, G. C., Biochim. Biophys. Acta, in press.
(14) Ghanotakis, D. and Babcock, G., FEBS Lett., **153**, 231-234, 1983.
(15) Debus, R., Barry, B., Babcock, G. and McIntosh, L., Proc. Nat. Acad. Sci. U.S.A., **85**, 427-430, 1988.

MODIFICATION OF THE MODEL EXPLAINING FLASH-INDUCED OXYGEN EVOLUTION PATTERNS IN ISOLATED THYLAKOIDS

J. Dirk Naber and Jack J.S. van Rensen

Department of Plant Physiological Research
Agricultural University, Gen. Foulkesweg 72
6703 BW Wageningen, The Netherlands

Introduction

The photosynthetic oxygen evolution is schematically described by the model of Kok et al. [1]. An excited PS II reaction center is able to extract electrons from the oxygen evolving complex (OEC) in a 4-step sequence. The evolution of 1 molecule of O_2 requires cooperation of 4 oxidizing equivalents. The OEC contains 4 Mn atoms, which can successively be oxidized and thus accommodate positive charges.

At the acceptor side of PS II the primary one-electron acceptor Q_A is reduced on each flash, transferring the electron to the secondary acceptor Q_B on a 100-200 μs timescale. Q_B can be reduced in two steps to form Q_B^{2-}, which is protonated twice to Q_BH_2 within 10 ms. In the electron transfer the redox component Q_{400} is involved. This component was shown to be a high-spin Fe atom [2], located between 4 histidine residues of the D1 and D2 protein of the reaction center [3]. In PS II particles the Fe has been shown to oscillate between the Fe^{2+}- and Fe^{3+}-states, depending on the redox state of the quinone acceptors [4]. The Fe^{2+}-form is proposed to be able to donate an electron to the semi-reduced quinone Q_B^-, thereby inducing the formation of Q_BH_2 after a single flash [5]. This mechanism, demonstrated so far only with artificial quinones, is indicated as reductant-induced oxidation [6].

The oscillation with period 4 in oxygen evolution patterns can be explained with the Kok model by introducing miss and double hit parameters, indicated respectively with α and β. An explanation for the occurrence of misses is the presence of an electron on the reaction center, which may decrease the probability of the center trapping a light quantum [7]. The α- and β- parameters were assumed by most authors to be independent of the flash number, determined only by the chloroplast preparation and the equipment used.

Another factor which influences the oxygen evolution pattern is the action of an electron donor D, which is able to reduce the S_2- or S_3-state of the OEC [7]. It is suggested that both electron donors D and Z, the regular intermediate between the OEC and the reaction center, are tyrosine residues located on the D1 and D2 protein [8]. In our model the donor D is the only time-dependent factor influencing the damping of the pattern.

We propose a refinement of the Kok model, assuming a correlation between the miss parameters α_n ($\alpha_0 \leq n \leq 3$) and the redox state of the quinone acceptor complex. After a dark period longer than about 10-20 ms, the time needed for the exchange of Q_BH_2 with Q_B, this complex can exist only in the oxidized or the semi-reduced form. The probability for any reaction center to be closed, and thus unable to receive another light quantum, is higher when Q_A or Q_B is reduced. This results in an adaptation of the S-state model, predicting alternately a relatively high and a very low value for α on successive flashes.

Materials and methods

Thylakoids were isolated from Pisum sativum L. or Chenopodium album L. (wildtype or triazine resistant mutant). The thylakoids were dark adapted for at least 1 h and used for measurements not later than 4 h after thawing. The chlorophyll concentration was 0.5 mg/ml. The flash induced oxygen production was measured with a Joliot-type of electrode.

The flash parameters were calculated from the oxygen yields in a series of 10 saturating flashes. The data were normalized to a steady state production of 1 a.u. per flash. The flash frequency was varied from 0.25 to 10 Hz. The reaction constant for reduction by donor D, and the fraction of reaction centers in which it could take place, were determined from experiments in which the chloroplasts were preilluminated with 2 flashes. The time interval (Δt) between preflashes and flash train was varied from 0.2 to 10 s. The oxygen evolution at the third flash was used to calculate the amount and the velocity of back reactions that had occurred.

Results

For the reaction of donor D with the S-states an exponential decay mechanism is proposed, resulting in the following equation :

$$[D_{ox}]_t = [D_{red}]*(1-e^{-k_D*t})$$ (1)

Using Eqn. 1, the fraction D_{red} of centers containing reduced D, the reaction constant k_D and the half time were calculated. The results are presented in Table I. The fraction of oxygen evolving centers in state S_0 after thorough dark adaptation and the parameters α_n and β were calculated by fitting data using a least-squares fitting method (see Table II).

The S_0-fraction was found to be 0 in all experiments. This means that in the dark all centers are in state S_1, coupled to a semi-reduced acceptor complex. For α_n the best fits were obtained by choosing α_0 and α_2 to be 0, while the sum of α_1 and α_3 was about 0.5. The value for β is calculated to be 0.05 for pea chloroplasts, and a little higher for the Chenopodium biotypes.

Table I. Determination of fraction D and reaction constants.

		D_{red}	k_D (s^{-1})	$t_{1/2}$ (s)
Pea	a	0.19	0.41	1.7
	b	0.45	0.32	2.2
Chenopodium S	a	0.26	0.28	2.5
	b	0.32	0.34	2.1
Chenopodium R	a	0.26	0.27	2.5
	b	0.34	0.33	2.1

a: for S_2 to S_1 reduction; b: for S_3 to S_2 reduction.

Table II. Parameters used to describe flash-induced oxygen evolution.

	Pea	Chenopodium-S	Chenopodium-R
S_0	0	0	0
β	0.05	0.08	0.09
$\alpha 0$	0	0	0
α_1	0.19	0.17	0.21
α_2	0	0	0
α_3	0.35	0.37	0.50

Parameters were calculated from mean values of at least 7 series of experiments with flash frequencies 0.25, 0.5, 1, 2, 4 and 10 Hz. Sum of quadratical differences between experimental and theoretical data ($S = \Sigma_{1-10}(Y_{EXP} - Y_{TH})^2$) varied from 0.01 to 0.025.

When chloroplasts were incubated with dithionite the best fitting values were found for α_0 and α_2 = 0, and $\alpha_3 > \alpha_1 > 0$. Treatment with ferricyanide reversed this situation: $\alpha_1 = \alpha_3 = 0$ and $\alpha_2 > \alpha_0 > 0$. The oxygen evolution patterns are shown in Fig. 1.

Discussion

The decay rates of the S_3-state were equal for all biotypes used (Table I). This indicates that activity of the donor D is not influenced by the shift in equilibrium from $Q_A.Q_B^-$ to $Q_A^-.Q_B$, which is proposed for triazine resistant mutants [7]. In pea chloroplasts the reduction of the S_2-state is somewhat faster. An explanation for this observation could be a charge recombination between the reduced acceptor complex and the S_3-state. The measuring method does not allow a discrimination between a back reaction with donor D and this charge recombination, because both reaction constants are of the same order of magnitude.

The amount of reaction centers in the S_0-state after dark adaptation is found to approach 0. This might indicate that after dark adaptation the 4 Mn-ions of the water oxidizing complex are in the Mn^{3+}-state.

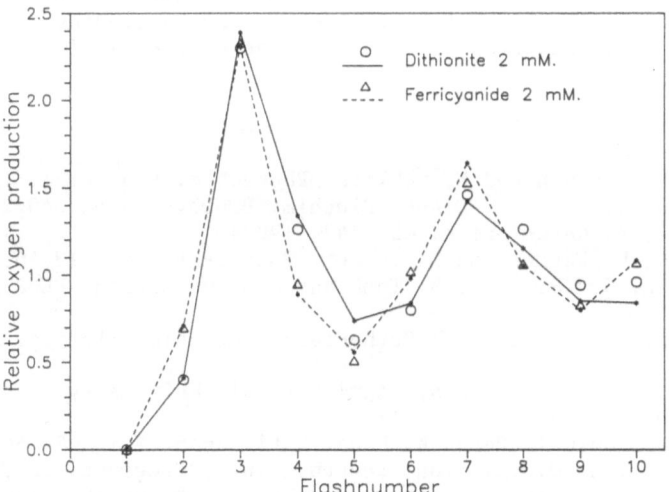

Figure I. Oxygen evolution in thylakoids treated with dithionite or ferricyanide. Lines represent experimental data, symbols show best fitted values.

For the sum of the α_n-parameters a value of about 0.5 is found. The α on even flashes is low, that on odd flashes high. This could be accounted for by assuming the redox situation of the quinone complex to play an important role in the occurrence of misses, thus introducing a 2-step oscillation. This suggestion is supported by the fact that incubation with ferricyanide shifts the high values for α from α_3 to α_2 and from α_1 to α_0, compared to an incubation with dithionite.

On account of the number of flashes fired, a semi-reduced quinone complex, and a high miss parameter, would be expected after an odd number of flashes. However, a high α is found after even flash numbers: α_1 and α_3 are high. The mechanism of reductant-induced oxidation [2,6] possibly can account for a dark one-step advancement of the redox state of the quinone complex. Compared to the oxidation state that is expected on account of the number of flashes fired, or the corresponding S-states of the oxygen evolving complex, the redox state of the $Q_A.Fe.Q_B$ complex has been advanced once more. The proposed scheme is represented as follows:

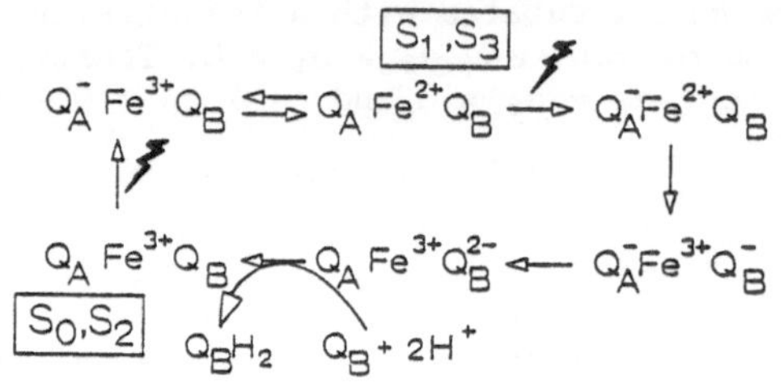

Figure II. Proposed scheme for the acceptor side reaction sequence.

The observation that α_3 is a little higher than α_1 could be caused by the fact that the S_3 to S_0 transition is somewhat slower than the other S-state transitions. Another explanation is that a charge recombination between the acceptor complex and the S_3-state of the OEC can occur, thus inducing an extra back reaction. In the triazine-resistant Chenopodium, the sum of the miss parameters is somewhat higher than in pea. This might be due to the shifted redox equilibrium in the quinone complex.

In conclusion, with this model parameters involved in describing the flash induced oxygen evolution can adequately be determined. They must be determined as good as possible, in order to use them in future calculations of the exchange parameters of herbicides on the Q_B binding site. The method for calculating these parameters has been described [7,9].

References

1) Kok, B., B. Forbush and M. McGloin, Photochem. Photobiol. 11:457 (1970).
2) Petrouleas, V. and B.A. Diner, Biochim. Biophys. Acta 849:264 (1986).
3) Trebst, A., Z. Naturforsch. 42c:742 (1987).
4) Renger, G., U. Wacker and M. Völker, Photosynt. Res. 13:166 (1987).
5) Van Rensen, J.J.S., W.J.M. Tonk and S.M. de Bruijn, FEBS Lett. 226:347 (1988).
6) Zimmermann, J.-L. and A.W. Rutherford, Biochim. Biophys. Acta 851:416 (1986).
7) Vermaas, W.F.J., Thesis, Agricultural University Wageningen, The Netherlands (1984).
8) Barry, B.A. and G.T. Babcock, Proc. Natl. Acad. Sci. USA 84:7099 (1987).
9) Naber, J.D. and J.J.S. van Rensen, in: "Progress in Photosynthesis Research," J. Biggins, ed., Martinus Nyhoff, Dordrecht, The Netherlands, Vol. III:767 (1987).

INVESTIGATION OF THE WATER OXIDISING SYSTEM OF PS2

THE ROLE OF CALCIUM IN WATER OXIDATION

Christopher J. Lockett, Christalla Demetriou
and Jonathan H.A. Nugent

Department of Biology, University College
Darwin Building, Gower Street
London WC1E 6BT, U.K.

INTRODUCTION

Oxygen evolution from photosystem 2 of plants, algae and cyanobacteria requires the storage of four positive charges by a complex thought to contain four manganese ions. A model for oxygen evolution, based on oxygen release after single turnover laser flashes, was devised with five oxidation states S_0 to S_4[1]. Oxygen evolution occurs when S_3 is oxidised to S_0. Calcium and chloride ions have been shown to be important for normal oxygen evolution, along with three extrinsic polypeptides of masses 18, 24 & 33 kDa. Several different washes have been developed to remove some or all of the extrinsic polypeptides from the reaction centre core. The absence of these polypeptides increases the concentration of calcium and chloride ions required for maximum oxygen evolution. Washing PS2 membranes with high salt concentrations removes the 18 and 24 KDa polypeptides[2]. The removal of calcium ions along with these polypeptides is a light dependent process[3]. The S_3 to S_0 transition is inhibited by the removal of calcium ions[4,5]. This suggests that calcium may be bound less firmly in certain 'S' states. EPR signals have only been identified for the S_2 state. These are termed the multiline signal and g = 4.1 signal[6,7]. The S_2 multiline signal has been shown not to be affected by the removal of calcium ions or the 18 & 24 kDa polypeptides[3]. Strontium has been shown to specifically replace calcium by both reactivation of oxygen evolution and by modification of the multiline EPR signal[3]. Various other metal ions have been used to compete with calcium ions for the binding sites, particularly lanthanum ions, but none show specific binding or stimulation of oxygen evolution[8].

In this paper we describe experiments where we have substituted calcium ions with vanadyl ions. Vanadyl ions have previously been used to replace calcium ions in several other proteins[9,10]. The advantage of replacement by vanadyl ions is that vanadyl is paramagnetic and may be used to probe the S states of oxygen evolution.

MATERIALS AND METHODS

Oxygen evolving PS2 was prepared from greenhouse grown peas (<u>Pisum sativum</u> var Feltham first) and from market spinach (<u>Spinacea oleracea</u>)[11]. It was resuspended and stored at 77K in 20mM Mes, 5mM $MgCl_2$, 15mM NaCl and 20% glycerol (pH6.3). Preparations with oxygen evolution rates of 400–1000 umol O_2 per mg Chl were used. Oxygen evolution measurements were made using a Clark type electrode with 25mM Mes, 20mM NaCl and other ions as described in the text. Removal of the 18 & 24 kDa polypeptides from oxygen evolving PS2 membranes was performed in room light[3]. The resulting particles were termed NaCl washed PS2. A low pH citrate wash for the extraction of calcium ions without loss of polypeptides was carried out[12]. The resulting particles were termed citrate washed PS2.

EPR spectrometry was performed at cryogenic temperatures using a JEOL X-band spectrometer with 100kHz field modulation and an Oxford instruments liquid helium cryostat. Samples of 0.3ml in 3mm diameter calibrated quartz tubes were used. EPR samples were prepared in room light, with any additions of metal ions in a solution of 20mM Mes pH 6.5, 20mM NaCl and 20% glycerol. Chlorophyll concentrations of 5–6 mg/ml were used for EPR. The EPR samples were placed in the dark for 4 hours to produce uniform S_1 states in the oxygen evolving complex[13]. The samples were frozen after this period of dark adaptation and illuminated at 200K with a 1000W projector for 8 minutes using an ethanol/solid CO_2 bath. EPR spectra of dark and illuminated samples were compared.

RESULTS AND DISCUSSION

Oxygen electrode measurements of NaCl washed PS2

Assay Conditions*	Rate (umoles O_2/mg Chl/hr)
Untreated PS2	
(1) No Additions	400
NaCl washed PS2	
(1) No Additions	24
(2) 20mM $CaCl_2$	182
(3) 20mM $SrCl_2$	121
(4) 20mM VO^{2+}	70

Measurements of PS2 washed with low pH citrate

Assay conditions*	Rate (umoles O_2/mg Chl/hr)
Untreated PS2	
(1) No Additions	583
Citrate washed PS2	
(1) No Additions	65
(2) 10mM $CaCl_2$	516
(3) 20mM $SrCl_2$	310
(4) 20mM VO^{2+}	260

*All assays carried out with 20mM Mes and 40mM Cl^-

Oxygen evolving activity is almost completely inactivated in both salt washed and citrate washed PS2. Table 1 shows the stimulation of oxygen evolution by calcium, strontium and vanadyl ions on samples following both treatments. These ions must therefore be replacing the function of calcium.

It has been suggested that there is a stoichiometry of 2 calcium atoms per reaction centre core[14]. The two calcium binding sites probably consist of a weak and a strong binding site. Citrate washed PS2 has been suggested to be depleted of one calcium ion[12]. Salt washed PS2 may be depleted of one calcium ion in the dark from the weak binding site and two calcium ions in room light with an electron acceptor present. The stronger binding site has been suggested to be depleted only when the S states of the OEC are advanced[3]. Calcium ions are probably bound more strongly in certain S states. Removal of calcium and replacement with other cations would therefore only be effective when the S states are cycled in the light. The presence of an electron acceptor such as PPBQ ensures that this cycling occurs. We have shown that the calcium binding sites can be reconstituted during an incubation in the light, with an electron acceptor, by calcium, strontium and vanadyl ions. The size and shape of the multiline formed on 200K illumination also supports the fact that the binding sites can be functionally reconstituted with these ions.

The high field section of the multiline spectrum has a superimposed iron-semiquinone signal which alters the shape of the multiline signal in this region. To avoid this interference, the low field section of the multiline was examined for changes in the spectrum caused by cation substitution. Salt washed PS2 reconstituted with calcium shows a normal multiline signal on 200K illumination. Strontium reconstitution however gives rise to a shifted multiline spectrum (Fig 1.). Strontium ions have a magnetic nucleus which affects the paramagnetic manganese complex. Only effects due to specific replacement of calcium ions in the oxygen evolving complex should be detected. Strontium causes splittings in the multiline signal that are not well resolved, but they result in a narrowing of the signal. This shifting of the multiline spectrum is good evidence that the strontium ions are entering the calcium binding sites close to the manganese complex and partially replacing the activity of calcium ions in oxygen evolution.

The interaction between two neighbouring paramagnetic components can also create changes in EPR signals. Vanadyl ions have previously been used to replace calcium ions in EPR studies of proteins. We have shown that the paramagnetic vanadyl ion reconstitutes the activity of calcium in salt washed and citrate washed PS2. We now hope to demonstrate changes in the multiline signal due to specific binding of vanadyl ions to the calcium binding sites of the oxygen evolving complex. This should provide a probe for the structure and relationship of the redox components of the oxygen evolving complex, particularly the S state cycle.

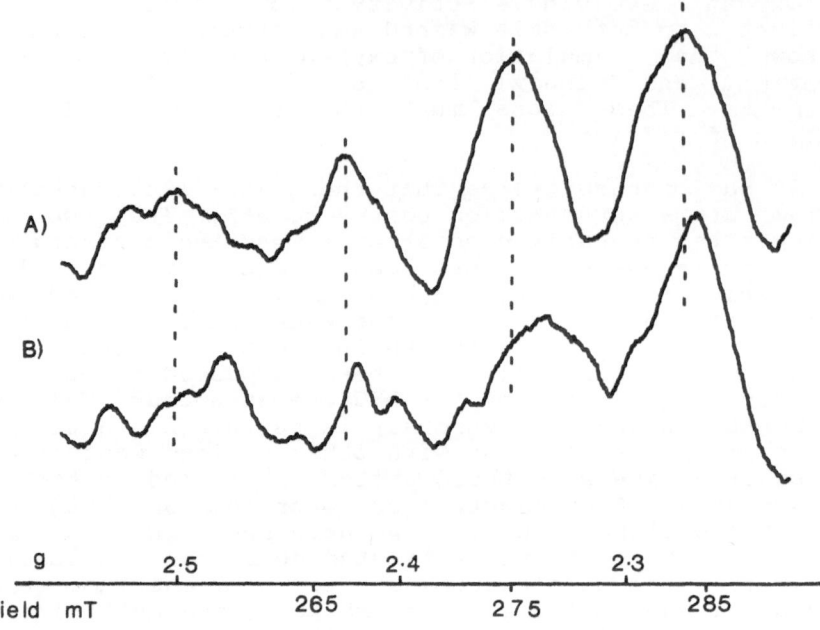

Fig 1. E.p.r spectrum of part of the low field of the S_2 state multiline, formed on 200K illumination in salt washed PS2 . (A) Reconstituted with calcium ions. (B) Reconstituted with strontium ions.

REFERENCES

1. B. Kok, B. Forbush and M. McGloin, Photochem. Photobiol 11:457 (1970)
2. M. Miyao and N. Murata, FEBS Lett. 168:118 (1984)
3. A. Boussac and A.W. Rutherford, Biochemistry U.S.A, 27:3476 (1988)
4. A. Boussac, B. Maison-Peteri, C. Vernotte and A.L. Etienne, Biochim. Biophys. Acta. 808:225 (1985)
5. T.A. Ono and Y. Inoue, Biochim. Biophys. Acta. 850:380 (1986)
6. G.C. Dismukes and Y. Siderer, FEBS lett. 121:78 (1980)
7. O. Hansson et al, Biophysics J. 51:825 (1987)
8. D.F. Ghanotakis, G.T. Babcock and C.F. Yocum, Biochim. Biophys. Acta. 809:173 (1985)
9. J. Nieves, L. Kim, D. Puett and L.Echegoyen. Biochemistry 26:4523 (1987)
10. N.D. Chasteen, in Biological Magnetic Resonance, 3:53 (1981)
11. R.C. Ford and M.C.W. Evans, FEBS lett. 195:285 (1983)
12. T.A. Ono and Y. Inoue, FEBS lett. 227:147 (1988)
13. W.F.J. Vermaas, G. Renger and G. Dohnt, Biochim. Biophys. Acta. 764:194 (1984)
14. N. Murata, M. Miyao, T. Omata, H. Matsunami and T. Kuwabara Biochim. Biophys. Acta. 765:363 (1984)

OXYGEN EVOLUTION FROM H_2O_2 AND H_2O IN RELATION TO Mn CONTENT

Johan Quensel, Wolfgang P. Schröder and
Hans-Erik Åkerlund

Department of Biochemistry, University of
Lund, P.O. Box 124, S-221 00 Lund, Sweden

INTRODUCTION

Manganese is required for the photosynthetic oxygen evolution, where it has a central role in the charge accumulation process. Important information about the stability of accumulated charges and kinetics of induvidual steps can be obtained from the oxygen yield pattern after flash illuminating the material on an oxygen electrode. Velthuys and Kok (1) found that the oxygen yield pattern changed upon H_2O_2 tretment, and that this was due to a two electron donation from H_2O_2 to the oxygen evolving complex. Photooxidation of H_2O_2 has been shown to occur in photosystem II particles both in the dark (2,3) and in the light (1,4,5).

Inoue has shown that Mn^{2+} is required for the H_2O_2 supported DCIP photoreduction in photosystem II reaction centre (6). Here we show that the oxygen evolution from H_2O_2 requires functional Mn. We also show that H_2O_2 extracts Mn from the photosystem II particles, and that the extraction is enhanced not only in particles treated with NaCl and $CaCl_2$ but also in reaction centre particles.

MATERIALS AND METHODS

Photosystem II enriched membranes (PS II membranes) were prepared according to Berthold et al.(7). NaCl- and $CaCl_2$- washed membranes were prepared by treating PS II membranes (500 ug chlorophyll/ml) for 30 min., 0°C, in 1 M NaCl, 10 mM Mes pH 6.5 and in 1 M $CaCl_2$, 10 mM Mes pH 6.5 respectivly. Reaction center complex particles (RC complex) were prepared according to Ghanotakis et al.(8). The material was either used directly or stored at -80°C.

H_2O_2 treatment was made by incubating PS II material (40 ug chlorophyll/ml, or 10 ug/ml when the RC complex was treated) for 5 min.,in the dark on ice, in 2 mM sucrose, 10 mM $CaCl_2$, 50mM Mes 6.0 and H_2O_2 concentrations as indicated, followed by centrifugation of the PS II material for 30 min. at 6000xg and resuspension in H_2O_2 free medium.

The oxygen evolution was measured in 2 mM sucrose, 10 mM NaCl, 50 mM Mes pH 6.0, with either 0.4 mM phenyl- p-bensoquinone or 120 mM H_2O_2 on a Clark-type electrode.

Bound Mn was analysed on PS II material in the presence of 1 mM HNO_3. The analysis was made on a Varian GTA-96 SpectrAA-10 atomic absorbtion spectrophotometer with graphite tube atomization.

RESULTS

1 mM H_2O_2 removed more than 75% of the the functional Mn in NaCl- and $CaCl_2$-washed PS II membranes and the RC complex (fig.1). The NaCl-washed material had originally 2.6 Mn/RC (1 RC=400 chlorophyll) and lost 2.2 Mn/RC. The $CaCl_2$-washed material lost 2.4 Mn/RC of the original 2.9 Mn/RC. The RC complex lost 2.8 Mn/RC (1 RC=65 chlorophyll) of 3.2 Mn/RC.

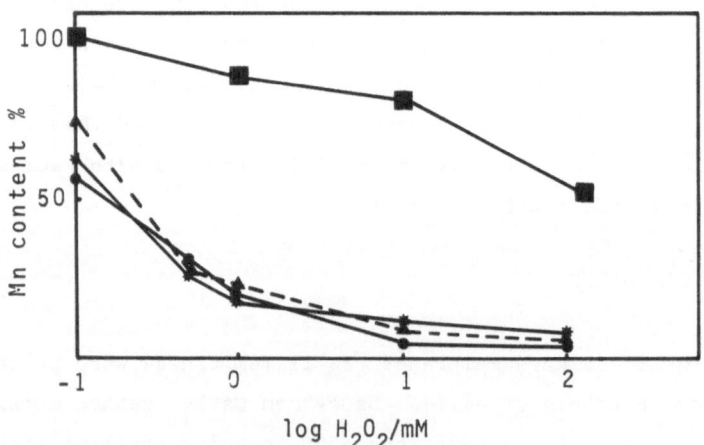

Fig. 1. Mn content in % of starting concentration, after treatment with H_2O_2 at different concentrations, (■) PS II membranes, (✳) PS II membranes NaCl washed and (●)$CaCl_2$ washed, and (▲)Reaction centre complex.

In contrast when untreated PS II membranes (PS II control, 5.4 Mn/RC.)
were subjected to 1 mM H_2O_2 , only 10% of the Mn was removed. With as
much as 120 mM H_2O_2 still only some 50% of the Mn was removed. At this
concentration of H_2O_2 more than 90% of the bound Mn in the salt treated
PS II membranes and the RC complex were removed.

The oxygen evolution from water, under light saturated conditions,
decreased upon H_2O_2 treatment and was roughly paralell to the loss of Mn
(fig. 2).

The oxygen evolution from H_2O_2 was, however, constant until the amount
of Mn in the PS II control material reached 70%. The oxygen evolution
then decreased from 2250 to 450 umol O_2/mg chlorophyll*h while the Mn
content only droped to 50%.

DISCUSSION

The results clearly show that, when PS II membranes is exposed to H_2O_2,
functional Mn is removed from its original site. The bound Mn is more
sensitive to H_2O_2 when the extrinsic 16, 23 and 33 kDa proteins is
removed by $CaCl_2$-washing. The 33 kDa protein in it self is unable to
protect functional Mn against H_2O_2 since the removal of the only 16 and
23 kDa proteins by NaCl-washing still enables H_2O_2 to extract Mn.

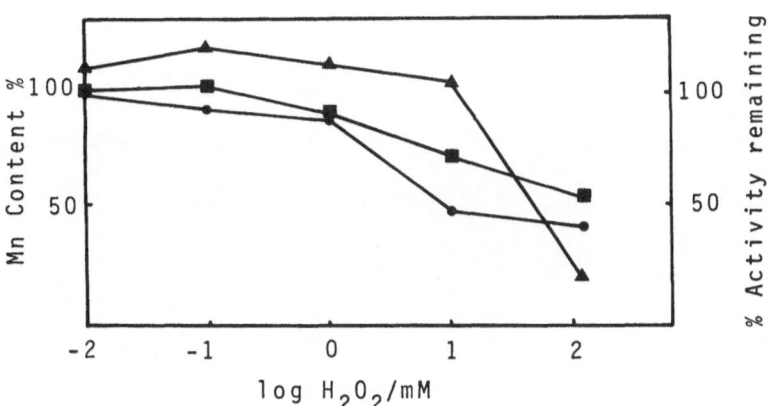

Fig. 2. Oxygen evolution from H_2O (●) and H_2O_2 (▲) and the Mn content (■)
in percent, from PS II membranes after treatment with different
H_2O_2 concentrations.

It has been suggested that oxygen evolution from H_2O_2 could be an effect of endogenous catalas (2). However, the results obtained here and those presented by Inoué (6) suggest that the O_2 evolution from H_2O_2 is dependent on the Mn associated with PS II. The difference in the Mn dependence between oxygen evolution from hydrogen peroxide and water (fig.2) may suggest that different parts of the PS II Mn pool is required for the two reactions.

REFERENCES

1. B. Velthuys and B. Kok, Biochim. et Biophys. Acta 502 (1978) 211-221
2. W.D. Frash and R. Mei, Biochim. et Biophys. Acta 891 (1987) 8-14
3. J. Mano, M. Takahashi, K. Asada, Biochemistry 26 (1987) 2495-2501
4. H. Inoué and M. Nishimura, Plant Cell Physiol. 12 (1971) 739-747 5.
5. W.P.K. Schröder and H-E. Åkerlund, Biochim. et Biophys. Acta 848 (1986) 359-363.
6. H. Inoué and T. Wada, Plant Cell Physiol. 28(5) (1987) 767-773
7. D.A. Berthold, G.T. Babcock and C.F. Yocum, FEBS Lett. 134 (1981) 231-234
8. D.F. Ghanotakis, D.M. Demetriou and C.F. Yocum, Biochim. et Biophys. Acta 891 (1987) 15-21.

STRUCTURE OF THE PRIMARY ELECTRON DONOR IN PHOTOSYSTEM I :

DIFFERENCE RESONANCE RAMAN SPECTROCOPY OF CP_1 PARTICLES

Pierre Moenne-Loccoz, Bruno Robert and Marc Lutz

Service de Biophysique, Département de Biologie, C.E.N.

Saclay, 91191 Gif/Yvette, France

Much interest has been devoted to the structure of the primary donors of electron in photosynthesis : indeed, the precise structure, as well as the ground state interactions involving its constitutive pigment(s) are of particular interest, because of their obvious relevance to the understanding of the mechanisms of the primary charge separation.

(RR) spectroscopy has been shown to be a powerful technique for studying (bacterio)chlorophyll-protein interactions. Indeed information can be obtained from this method about the conformations and intermolecular binding states of the dihydrophorbin ring of and its conjugated substituents (1). The highest information content of RR spectra is obtained when resonance is with the Soret electronic transition and includes data on the intermolecular binding states of both the conjugated carbonyls and the central Mg atom. The main problem in studying pigment-protein complexes in these conditions is that limited selectivity is obtained amid the contributions of the different chlorin pigments present in the sample. We recently developed (2) difference methods permitting selective observation of the primary donor in the neutral, ground state. These methods involve control of the steady-state equilibria built up during continuous illumination, and they have been successfully applied to bacterial reaction centers extracted from various strains (3,4). We report here the first selective observation of RR spectra of the primary donor in photosystem I (P_{700}).

Fig 1_1 displays RR spectrum of CPI particles (containing approximately 50 Chl a and 6 β-carotene molecules per P_{700}), at 20 K and excited at 441.6 nm. In this spectrum, the strong 1530 cm^{-1} band arises from the C=C stretching modes of the carotenoid molecules. Bands at 1555 and 1615 cm^{-1} arise from modes of the dihydrophorbin skeleton of Chl a. These bands are sensitive to the coordination number of the central Mg of Chl molecules, being respectively located at 1545 and 1600 cm^{-1} when this atom binds two external ligands, and at 1555 and 1615 cm^{-1} when it binds a single ligand only (5). Presence of a complex, unresolved, cluster in the 9 keto carbonyl stretching region (1640-1710 cm^{-1}) reveals existence of several unequivalent populations of Chl molecules in CP1 particles, having their keto carbonyl groups engaged in different intermolecular interactions. In our experimental conditions, the irradiance due to the laser beam reaching the frozen sample during recording of such spectrum is not high enough to induce steady-state accumulation of any excited state in CP1 particles (Moenne-Loccoz, work in preparation).

In order to selectively extract contributions of the P_{700} from RR spectra of CP1, we used two different methods: in the first one, we induced a sizable steady-state accumulation of the triplet state of P_{700}, by using suitably high irradiance from the Raman probe laser beam according to the reaction :

$$PA_0 \longrightarrow P^+A_O^- \longrightarrow P^R \ (6)$$

In the second method we chemically oxidized P in presence of ferricyanide. Both of these experiments induce reproducible modifications of RR spectra of CP1 particles (fig 1_2, 1_3). Difference spectra computed by subtracting RR spectra of CP1 obtained in high irradiance conditions or in presence of ferricyanide from RR spectra obtained in low irradiance conditions are presented in fig 2_1 and 2_2 respectively. For obtaining these differences, spectra have been normalized taking into account the intensity of the 1530 cm^{-1} band, which is expected not to vary during illumination at 20 K, the yield of formation of carotenoid triplet being low in these conditions (Sétif,P., personal communication).

These difference spectra essentially contain positive contributions from neutral, ground state Chl a and no negative contribution from the cation and triplet states. Because of the similarity of these two difference spectra obtained through different sequences of reactions, we conclude that they arise from neutral P_{700} only, the latter being the only molecular species common to the two types of experiments and the concentration of which is expected to vary in each of them. From these spectra, the following conclusions can be drawn about the structure of P_{700} :

1) The presence of two distinct C=O stretching bands at 1655 and 1675 cm^{-1} clearly indicates that P_{700} is constituted from two Chl a molecules. This result thus confirms previous models drawn from EPR and ENDOR measurements (for a review, see 7).
2) The presence of bands at 1555 and 1612 cm^{-1} unambiguously indicates that both of these Chl, each bind a single external ligand on their central Mg atom.

Fig 1.

Resonance Raman spectra
  of CP_1 particules :

1_1 Low illumination
1_2 High illumination
1_3 Ferricyanide-treated
     particules

Excitation wavelength :
441.6 nm
Sample temperature : 20 K

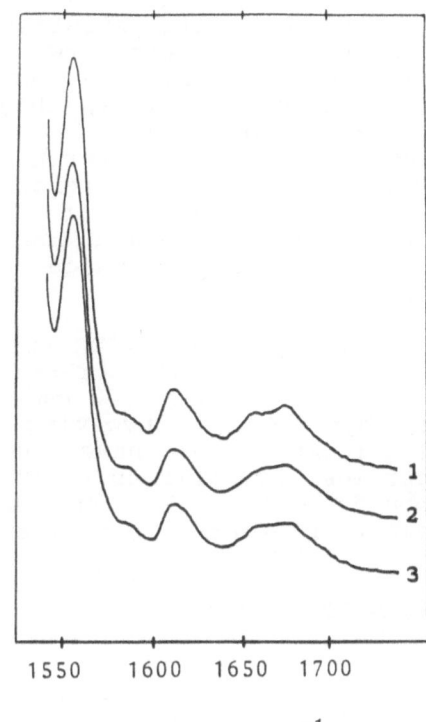

WAVENUMBER (cm^{-1})

Fig 2.

Difference resonance Raman
  spectra :

2_1 Low illumination
     -
     High illumination

2_2 Low illumination
     -
     Ferricyanide-treated
     particules

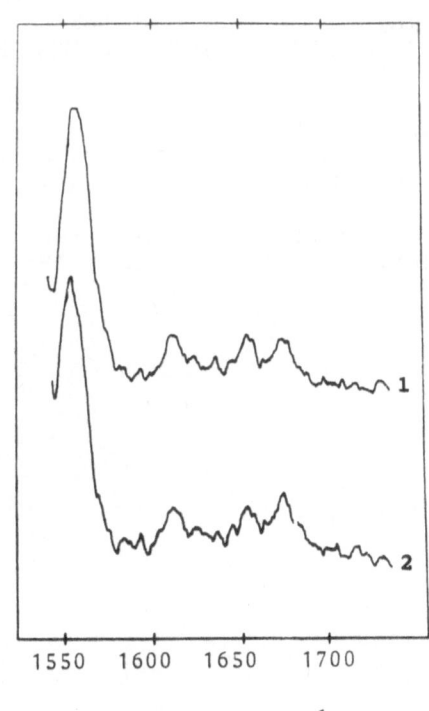

WAVENUMBER (cm^{-1})

3) However, the existence of two C=O stretching bands indicates that the two P_{700} Chl molecules are asymmetrically bound to the protein via their 9 keto carbonyl groups. Indeed, they appear to be both H-bonded (1), but the strengths of the H-bonds they are involved in are different : the difference in free energy corresponding to such a 20 cm^{-1} difference may be evaluated to 3 kCal/M (8). Such a difference in the interaction state of these groupings might explain the fact that the triplet state appears to be shared unequivalently between both of these molecules (7).

4) Moreover, these results exclude models based on water-bridged Chls, as RR spectra of such models compounds display both a 1618 cm^{-1}-located band and 9 keto C=O stretching bands at 1640 cm^{-1} in this spectral region (1). Other hypotheses, involving chloro-Chls, Chl epimers or enolic forms of Chla have been proposed for explaining P_{700} properties. Comparisons of RR spectra of P_{700} and of model compounds will allow the biological relevance of such models to be tested in a near future.

REFERENCES

1) LUTZ, M. and ROBERT, B. (1988) in : Biological Applications of Raman Spectroscopy (SPIRO, T. ed) Wiley, New York, Vol. 3, Chap. 8
2) ROBERT, B. (1987) Thesis, Université Pierre et Marie Curie, Paris
3) ROBERT, B. and LUTZ, M. (1986) Biochemistry 25, 2503-2509
4) ZHOU, Q., ROBERT, B. and LUTZ, M. (1987) Biochim. Biophys Acta 890, 368-376
5) FUJIWARA, M.and TASUMI, M. (1986) J. Phys. Chem. 90, 250-255
6) SETIF, P., HERVO, G. and MATHIS, P (1981) Biochim. Biophys. Acta 638, 257-267
7) MATHIS, P. and RUTHERFORD, W.A. (1987) in : Photosynthesis (Amesz, J. ed.) Elsevier, Amsterdam, pp 63-96
8) ZADOROZHNYI, B.A. and ISHCHENKO, I.K. (1965) Opt. Spectrosc. (eng.transl.) 19, 306-308

ANALYSIS OF THE SLOW COMPONENT OF THE FLASH-INDUCED P515 RESPONSE

IN CHLOROPLASTS

J.J.J. Ooms and W.J. Vredenberg

Laboratory of Plant Physiological Research
Agricultural University, Gen. Foulkesweg 72
6703 BW Wageningen, The Netherlands

Introduction

The flash-induced P515 absorbance change in dark adapted chloroplasts is known to consist of several components. Joliot and Delosme [1] described a fast (phase a) and a slow rising component (phase b). The fast rising component has a rise time less than 0.5 ms. This phase is generally interpreted as reflecting the primary charge separation in the reaction centers and is attributed to the rising part of reaction 1 [2]. The slow rising component (phase b) is referred to as the rising part of reaction 2 [2] and has been subject of numerous discussions.

A H^+-transporting Q-cycle or a modified version thereof has been suggested as an explanation for the slow rise in the flash-induced P515 signal [3,4]. In this interpretation one would expect equal decay rates for the fast and the slow component [5]. However, Schapendonk et al.[2] have presented ample evidence that the decay rate of reaction 2 in chloroplasts is substantially slower than the decay rate of reaction 1.

Recently it has been shown [6] that the slow phase in the flash-induced P515 signal in dark adapted intact chloroplasts is composed of two distinguishable components.

One component called reaction 1/Q has a rise time of about 10 ms and a decay time in the order of 50 -100 ms. Reaction 1/Q originates from an electrogenic Q-cycle and has approximately the same decay time as the response (reaction 1/RC) associated with the primary charge separation. For clarity we propose to call the electrogenic trans-membrane response associated with the primary charge separation reaction 1/RC. The other slow rising component is called reaction 2 as before [2] and has much slower rise and decay kinetics.

Here it is further demonstrated that reaction 1/Q reflects the trans-membrane charge separation as a result of a functional Q-cycle. The origin of reaction 2 still remains unclear, but it is shown that it is not a reflection of a Q-cycle.

Materials and Methods

Intact chloroplasts from spinach (Spinacea oleracea, c.v. Amsterdams breedblad) were isolated as described elsewhere [7]. Intactness and chlorophyll were determined according to adopted methods [8].

Flash-induced absorbance changes were measured using a modified Aminco

Chance spectrophotometer [7]. Signal processing was as described elsewhere [7].

Measurements were carried out at 5° C, unless otherwise indicated. Samples were taken from stock suspensions after dark adaptation of at least 2 hours. The reaction medium contained: 0.33 mol/liter sorbitol, 10 mmol/liter NaCl, 5 mmol/liter $MgCl_2$, 1 mmol/liter $MnCl_2$, 2 mmol/liter EDTA, 5 mmol/liter ascorbic acid, 75 mmol/liter Hepes/KOH pH 7.5 and chloroplasts equivalent to 25 μg chlorophyll per ml.

Results and Discussion

According to a current model [3] one might expect either an induction or a stimulation of a transmembrane Q-cycle after the addition of DQH_2 (reaction 1/Q). As a consequence this should lead to a complementary P515-signal. Fig.1 shows the P515 response in intact chloroplasts in the absence (trace A) and after addition of DQH_2 (trace B). The difference plot (fig.1C) shows the DQH_2 stimulated P515 response. Besides the DQH_2 stimulation an additional undershoot is shown. This undershoot might either be due to an inhibition of an other flash-induced P515 component, or to an acceleration of the decay of the P515 response. In the experiment shown in figure 1, we used a flash train with a frequency of 6.7 Hz.

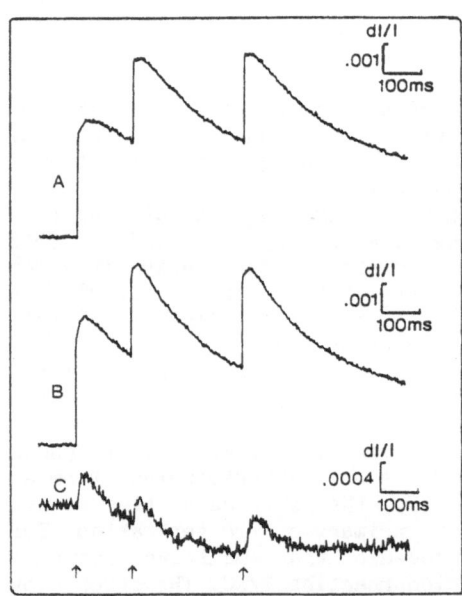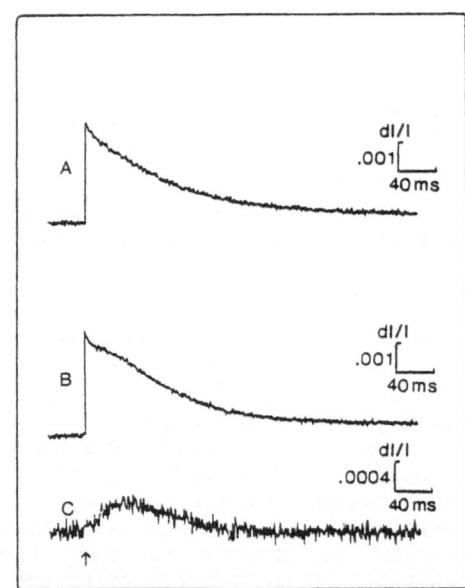

Figure 1. Flash-induced P515 absorbance change in dark adapted intact chloroplasts. Averages of five sweeps at a frequency of 0.05 Hz. Arrows indicate the firing time of the flash. A: No additions. B: As A with 300 μmol/liter DQH_2. Trace C, difference curve, B-A, shows the effect of DQH_2 on the P515 response on an expanded scale.

Figure 2. Flash induced P515 signal at 15°C in broken chloroplasts. The chloroplasts were preïlluminated (45 sec.) in the presence of 300 μmol/liter ATP, 1 mmol/liter DTE, 60 μmol methylviologen and 5000 units catalase. Curves are averages of 50 sweeps at a frequency of 0.5 Hz. Arrows indicate the firing time of the flash. Trace A: no further additions, trace B: with 450 μmol/liter DQH_2, dissolved in methanol. Trace C, difference curve B-A on an expanded scale, showing the effect of DQH_2 on the flash-induced P515-signal.

After the second flash, the undershoot reaches saturation. This is taken as evidence that DQH_2, besides being stimulatory, inhibits a saturable P515 component.

As documented in detail elsewhere [8,9] the decay of the overall P515 signal under ATP hydrolyzing conditions is relatively fast. It has been demonstrated that this faster decay is due to the absence of reaction 2, rather than to an enhanced membrane conductance. In figure 2 the flash-induced P515 responses after light activation of the ATP-ase is shown in the absence and presence of DQH_2 respectively. The difference plot in figure 2C shows that after suppression of reaction 2 by ATP hydrolyses, it is still possible to induce an extra P515 response with DQH_2. The DQH_2 stimulated P515 response is fully blocked by the PQ antagonist DBMIB (not shown). According to the model of Jones and Whitmarsh [3] it seems reasonable to ascribe this stimulated P515 response to the Q-cycle.

In a functional Q-cycle electrons are transported through the membrane with cytochrome b as an intermediate. As a consequence each turn-over of a Q-cycle giving rise to an associated P515 response (reaction 1/Q), should be accompanied by a turn-over of cytochrome b. In figure 3A the flash-induced turn-over of cytochrome b is shown in dark adapted chloroplasts, with two consecutive flashes, both showing a turnover of cytochrome b. Whereas it is known that reaction 2 is suppressed in a flash train of a few saturating flashes [2], reaction 1/Q is recurring in two or three consecutive flashes (fig.1).

The turn-over of cytochrome b after light activation of the ATP-ase is shown in figure 3B. The kinetics from the cytochrome b turn-over differ hardly if at all from those in fig. 3A. Thus the total suppression of reaction 2 by activation of the ATP-ase does not influence the Q-cycle dependent cytochrome b turn-over. These results confirm the earlier conclusion of Van Kooten et al. [10] that the slow phase of P515 caused by reaction 2 is not associated with a turnover of cytochrome b.

The origin of reaction 2 is still obscure. The suggestion that it finds its origin in the liberation and stabilization of protons in inner-membrane domains [11,12] needs still confirmation.

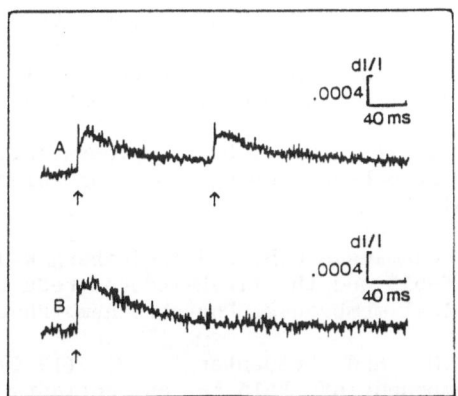

Figure 3. Turnover of cytochrome b, measured as the difference of absorbance changes at 563 nm and 575 nm. Arrows indicate the firing time of the flash. Trace A: intact dark adapted chloroplasts, with 450 μmol/liter DQH_2, 60 μmol/liter methylviologen, 5000 units catalase and 1.5 μmol/liter valinomycin. Average of 100 sweeps at a frequency of 0.2 Hz. Trace B: chloroplast treatment as in figure 2B i.e. with 45 sec. preïllumination and addition of 1.5 μmol/liter valinomycin. Average of 100 sweeps at a frequency of 0.5 Hz.

Acknowledgement   This research was supported in part by the Stichting Scheikundig Onderzoek Nederland (SON), financed by the Netherlands Organization for the Advancement of Pure Research (NWO).

REFERENCES

1. Joliot P. and Delosme R. (1974) Flash-induced 519nm Absorbance change in green algae. Biochim. Biophys. Acta 357: 267-284.

2. Schapendonk A.H.C.M., Vredenberg W.J. and Tonk W.J.M. (1979) Studies on the kinetics of the 515nm absorbance change in chloroplasts. Evidence for the induction of a fast and a slow P515 response upon saturating light flashes. FEBS Lett. 100: 325-330.

3. Jones R.W. and Whitmarsh J. (1985) Origin of the electrogenic reaction in the chloroplast cytochrome b/f complex. Photobiochem. Photobiophys. 9: 119-127.

4. Jones R.W. and Whitmarsh J. (1988) Inhibition of electron transfer and electrogenic reaction in the cytochrome b/f complex by 2-n-nonyl-4-hydroxyquinoline N-oxide(NQNO) and 2,5-dibromo-3-methyl-6-isopropyl-p-benzoquinone(DBMIB). Biochim. Biophys. Acta 933: 258-268

5. Vredenberg W.J. (1981) P515: A monitor of photosynthetic energization in chloroplast membranes. Physiol. Plant. 53: 598-602.

6. Vredenberg W.J., Ooms J.J.J. and Buurmeijer W.F. (1987) Electrogenesis in chloroplast membranes: Evidence for two slow components in the P515 signal after flash exitation. In: Stein WD (ed) Ion pumps: Structure, function and regulation. Alan Liss, Inc, Publ New York: 293-298.

7. Snel J. (1985) Regulation of photosynthetic electron flow in isolated chloroplasts by bicarbonate, formate and herbicides. Thesis, Agricultural University Wageningen.

8. Peters R.L.A., Bossen M., Van Kooten O. and Vredenberg W.J. (1983) On the correlation between the activity of ATP-hydrolase and the kinetics of flash-induced P515 electrochromic bandshift in spinach chloroplasts. J. Bioenerg. Biomemb. 15: 335-346.

9. Schreiber U. and Rienits K.G. (1982) Complementarity of ATP-induced and light-induced absorbance changes around 515nm. Biochim. Biophys. Acta 682: 115-123.

10. Van Kooten O., Gloudemans A.G.M. and Vredenberg W.J. (1983) On the slow component of P-515 and the flash-induced reduction of cytochrome b-563 in chloroplast membranes. Photobiochem. Photobiophys. 6: 9-14.

11. Schapendonk A.H.C.M. and Vredenberg W.J. (1979) Activation of the reaction 2 component of P515 in chloroplasts by pigment system 1. FEBS Lett. 106: 257-261.

12. Westerhoff H.V., Helgerson S.L., Theg S.M., Van Kooten O., Wikström M., Skulachev V.P. and Dancsházy Zs. (1983) The present state of the chemiosmotic coupling theory. Biochim. Biophys. Acad. Sci. Hung. 18:125-149.

THE SINGLE CHANNEL CONDUCTANCE OF CF_0

Gerd Althoff, Holger Lill and Wolfgang Junge
Biophysik, Universität Osnabrück, W. Germany

Introduction

The proton channel of the chloroplast ATP synthase, CF_0, was investigated by a flash spectrophotometric method, applied to a suspension of CF_1 depleted vesicles derived from thylakoid membranes (Lill, Engelbrecht, Schönknecht & Junge, 1986). Excitation of the vesicle suspension with short flashes of light generated an electric potential across the thylakoid membranes and a protonic charge pulse inside the vesicles. The relaxation of the voltage, due to ion fluxes across the membrane, was measured via electrochromic absorption changes of intrinsic pigments. It was compared with the proton fluxes which were determined by pH-indicating dyes in the lumen and in the medium.

Since there were only very few active conducting F_0 per vesicle (on the average between 1 and 2) the time averaged single channel conductance, G, of CF_0 has become accessible (Lill, Althoff & Junge, 1987). Evaluating the decay of electrochromic absorption changes, we determined G to be about 1 pS, equivalent to a turnover number of 2×10^5 $H^+/(CF_0 \cdot s)$ at 30 mV driving force. This was very high compared to reported turnover numbers for F_0-channels from mitochondria and bacteria, which have been inferred from reconstitution experiments (in the range of 10–100 $H^+/(F_0 \cdot s)$; for review see Schneider & Altendorf, 1987). But it was sufficient, and even higher, than supposed for the function as proton supply in the coupled enzyme, CF_0CF_1, in photophosphorylation (1200 $H^+/(CF_0CF_1 \cdot s)$). The latter was calculated with 400 $ATP/(CF_0CF_1 \cdot s)$ (Junesch & Gräber, 1985) and a stoichiometry of 3 H^+/ATP (Junge, Rumberg & Schröder, 1970)). Measurements of the single channel conductance of CF_0 under various conditions in the suspending medium should help to understand how the rapid proton supply to the coupling sites of the integral ATP synthase occurs. Are the results in agreement with the proposed conduction mechanism in CF_0 made of hydrogen bonded chains (Nagle & Morrowitz, 1978, Brünger, Schulten & Schulten, 1983)?

Results

We investigated the single channel conductance as function of the medium composition (ionic strenght, pH, pD (isotopic substitution), altered water and/or membrane structure and temperature) and we obtained the following results:

1) CF_0 is a proton specific channel. Complete tracking of proton flow (Schönknecht et al.,1986; Junge, 1987) showed, that the proton is the major charge translocated by

CF$_0$. Extending these studies we found that variation of the concentration of several monovalent cations in the range between 1 mM and 300 mM and of divalent cations from 100 μM to 30 mM did not influence the protonic single channel conductance. The selectivity of CF$_0$ for protons over e.g. Na$^+$ was greater than 10^7.

2) Between pH 5.6 and pH 8.0, the protonic single channel conductance of CF$_0$, did not vary as function of the medium pH.

3) Isotopic substitution of D$^+$ for H$^+$ caused a decrease of the time averaged single channel conductance by a constant factor of 1.7 over the whole pH/pD range mentioned under 2.

4) Addition of glycerol lowered the measured single channel conductance of CF$_0$. The isotope effect of D$^+$ over H$^+$ disappeared at about 50 % glycerol in solution.

5) The activation energy of proton translocation through CF$_0$ was 42 kJ/mol in H$_2$O and 47 kJ/mol in D$_2$O (compared with 30 kJ/mol for gramicidin, 65 kJ/mol for valinomycin, acting in the same membrane suspended in H$_2$O, and the activation energy for a diffusion limited process of about 20 kJ/mol).

Conclusion

1) Unlike suggested by the older literature CF$_0$ is kinetically competent to serve as a low impedance access ("proton well") to the protonic coupling site in the integral ATP synthase, CF$_0$CF$_1$.

2) CF$_0$ is distinguished by an extremely high selectivity for protons (H$^+$, OH$^-$, hydronium cation) over other cations and anions.

3) The absence of a pH–dependency suggests that the rate limiting step is not governed by a titratable group.

4) The H$^+$/D$^+$–isotope effect suggest rate limitation by the transit of a proton from one site to another one. The observed factor of 1.7, however, is small compared to published figures for such reactions between small acid–base molecules in solution. It is conceivable, though, that the full conduction cycle comprises at least two reactions in series, where the non–protonic ones becomes rate limiting only in D$^+$.

5) The activation energy of proton conduction by CF$_0$ is intermediate between the ones characteristic for diffusion and for chemical reactions.

6) If the environment of the channel mouth is disturbed (as in glycerol buffer) the supply of protons to the channel readily may become rate limiting.

From these results it is not clear if the proton pathway through CF$_0$ is made up from (one or more) hydrogen bonded chain(s). The very high specificity of CF$_0$ for protons and even the H$^+$/D$^+$–effect are in agreement with this model, but not the pH–independence in the neutral pH–range and the high turnover number (Brünger *et al.*, 1983).

References

A.Brünger, Z.Schulten & K.Schulten, 1983, *Z.Phys.Chem* **136**: 1–63

U.Junesch & P.Gräber, 1985, *Biochim.Biophys.Acta* **809**: 429–434

W.Junge, B.Rumberg & H.Schröder, 1970, *Eur.J.Biochem.* **14**: 575–581

W.Junge, 1987, *Proc.Natl.Acad.Sci. USA* **48**: 7084–7088

H.Lill, S.Engelbrecht, G.Schönknecht & W.Junge, 1986, *Eur.J.Biochem.* **160**: 627–634

H.Lill, G.Althoff, and W.Junge, 1987, *J.Membrane Biol.* **98**: 69–78

J.F.Nagle & H.J.Morrowitz, 1978, *Proc.Natl.Acad.Sci. USA* **75**: 298–302
E.Schneider & K.Altendorf, 1987, *Microbiological Reviews* **51**: 477–497
G.Schönknecht, W.Junge, H.Lill & S.Engelbrecht,1986, *FEBS Lett.* **203**: 289–294

RELATIONSHIP BETWEEN ENERGY DEPENDENT RELEASE OF ADENINE NUCLEOTIDES FROM CF_1 AND PHOTOPHOSPHORYLATION

Jai Parkash and G.S. Singhal

School of Life Sciences, Jawaharlal Nehru University
New Delhi-110 067, India

INTRODUCTION

On illumination of thylakoid membranes, nucleotides bind to CF_1 of CF_o–CF_1 complex. This reaction is an exchange of CF_1-bound adenine nucleotide for medium nucleotides.[1] On the other hand, in the de-energized membranes the nucleotides remain tightly bound to CF_1. Although the release of nucleotide is energy dependent, the rebinding in principle is energy independent. The rate of energy dependent, release of adenine nucleotide is too slow as compared to the rate of ATP synthesis,[2] however, the initial rate of exchange of nucleotide during 20-50 m sec of illumination are comparable to the initial rate of ATP synthesis[3].

The CF_o–CF_1 ATPase complex undergoes conformational changes when an electro-chemical gradient of H^+-ions ($\Delta\mu_H+$) is established across the thylakoid membrane[4]. It has been proposed that (a) the release of adenine nucleotide suggests a conformational change in CF_1, leading to activation of ATPase, (b) the fact that one ATPase molecule can exchange only one adenine nucleotide, the release of one nucleotide indicates the activation of one ATPase and (c) the fraction ($\emptyset$) of the activated ATPase is given by the magnitude of the fast phase of the adenine nucleotide release. The energy dependent activation of ATPase could arise through the opening of proton gate and release of adenine nucleotide from CF_1. The kinetic competence of energy dependent release of bound nucleotide and the development of ATPase activity suggests that CF_1 containing an empty nucleotide binding site is the species that is active in ATP synthesis/hydrolysis.

The chemical modifications of proteins, an important tool to study the structural-functional relationships in the enzyme molecule, can provide useful informations regarding the functional groups of CF_1 which are involved in regulating the energy dependent release of bound nucleotide and ATP synthesis. Different functional groups of CF_o–CF_1 complex e.g. $-COOH$, $-NH_2$, $-SH$ etc. have been modified by different[1] chemical modifiers and their role in the enzymatic activities have been studied[6-8].

In this article we present our findings on the role of ϵ–NH_2 group of lysine, phenolic hydroxyl group of tyrosine, $-SH$ group of cysteine and $-COOH$ group of CF_1 in energy dependent release of ADP and photophosphorylation.

MATERIALS AND METHODS

The salt-washed thylakoid membranes were prepared from spinach chloroplasts as described previously[9] and finally suspended in the washing medium to a chlorophyll (Chl) concentration of 2 mg/ml. The Chl concentrations were determined according to Arnon[10].

Chemical modification of thylakoid membranes

(1) With fluorescamine: The salt-washed thylakoid membranes were resuspended in 2 ml of <u>prelabelling medium</u> which contained 25 mM tricine pH 8.0, 50 mM NaCl, 1 mM MgCl$_2$, 0.5 mM methyl viologen, 2.5 μM [^{14}C]-ADP to a Chl concentration of 1 mg/ml and then illuminated at 25°C for 1 min with the help of 250 watt projector lamp (Kodak Co., USA). The light intensity at the center of reaction vessel was 290 Joules/m^2/sec. Just before putting off the illumination, 20 μl aliquot of fluorescamine solution in acetone was added to the suspension and then mixed thoroughly on a vortex mixutre. The samples were then transferred to 4°C and later used for prelabelling.

(ii) With 4-chloro-7-nitrobenzofurazan (NBD-Cl): The salt-washed thylakoid membranes were resuspended in the prelabelling medium in a final volume of 1 ml and Chl concentration of 1 mg/ml and then treated with 10 μl aliquots of NBD-Cl solutions of varying concentrations. The samples were incubated at 0°C for 4 hrs in dark. After the dark incubation, the thylakoid membranes were then prelabelled with [^{14}C]-ADP.

(iii) With dicyclohexyl carbodiimide (DCCD): The salt-washed thylakoid membranes were resuspended in prelabelling medium in a volume of 0.5 ml and Chl concentration of 0.5 mg/ml. 5 μl aliquots of DCCD solution of varying concentrations were added to the suspension and incubated for 20 min in dark. After the incubation period, the thylakoid membranes were prelabelled with [^{14}C]-ADP.

(iv) With n-ethylmaleiimde (NEM) in dark and then with iodosobenzoate (IBZ) in light: The salt-washed thylakoid membranes were resuspended in 3 ml of a medium containing 50 mM tricine, pH 8.0, 5 mM MgCl$_2$ and 50 mM NaCl to a Chl concentration of 0.33 mg/ml and then treated with 1.23 mM NEM in dark for 5 min at 25°C. The samples were then centrifuged at 5000xg for 5 min in SM-24 rotor of RC-5 Sorvall Centrifuge (Sorvall Instruments, Du Pont, USA). The pellets were washed twice with a buffer containing 2 mM tricine, pH 8.0, 0.4 M sucrose, 10 mM NaCl and 0.1% (w/v) bovine serum albumin and resuspended in the same medium to a Chl concentration of 2 mg/ml.

The control and NEM treated thylakoid membranes were centrifuged at 10000xg for 10 min at 0°C and the pellets were suspended in a reaction mixture containing 50 mM tricine, pH 8.0, 50 mM NaCl, 5 mM MgCl$_2$, 50 μM phenazine methosulfate and varying concentration of IBZ in a final volume of 3 ml and Chl concentration of 0.33 mg/ml. The samples were immediately illuminated with white actinic light (intensity=290 Joules/m^2/sec) for 1.5 min at 25°C. The samples were then centrifuged at 5000xg for 5 min at 0°C and the pellets were washed twice with ice-cold medium containing 0.4 M sucrose, 2 mM tricine, pH 8.0 and 10 mM NaCl and finally suspended in the same medium. For prelabelling of these thylakoid membranes with [^{14}C]-ADP the suspensions were again centrifuged at 10000xg for 10 min at 0°C and the pellets were suspended in the prelabelling medium.

Prelabelling of thylakoid membranes with [^{14}C]-ADP, the light-induced release of [^{14}C]-ADP bound to CF$_1$ and non-cyclic photophosphorylation coupled to electron transport from H$_2$O to methyl viologen were carried out as described previously[9].

RESULTS

As shown in Fig.1, the treatment of salt-washed thylakoid membranes with fluorescamine resulted in increase in the energy dependent release of $[^{14}C]$-ADP. The maximum increase in the activity was approximately 30 per cent at 1 μmole of fluorescamine. However, under the similar experimental condition, a significant decrease in the rates of non-cyclic photophosphorylation occurred when salt-washed thylakoid membranes were treated in their high energy state (i.e. in light) with fluorescamine. Even at 0.5 μmole fluorescamine, 55 per cent inhibition in the phosphorylation activity was observed.

The chemical modification of salt-washed thylakoid membranes with NBD-Cl did not affect significantly the energy dependent release of $[^{14}C]$-ADP bound to CF_1 (Fig.2). The maximum inhibition was only 12 per cent at 200 μM NBD-Cl. In contrast to this, a continual decline in the rate of photophosphorylation was observed on treatment of salt-washed thylakoid membranes with NBD-Cl. The maximum inhibition of photophosphorylation activity was 60 per cent at 400 μM NBD-Cl.

As shown in Fig.3, the incubation of salt-washed thylakoid membranes with increasing concentrations of DCCD resulted in increase in the energy dependent release of CF_1-bound $[^{14}C]$-ADP upto 30 per cent at 80 μM DCCD. However, at concentrations higher than 80 μM of this chemical modifier only a marginal increase in the energy dependent release of $[^{14}C]$-ADP could be seen. Under the similar experimental conditions addition of DCCD to the salt-washed thylakoid membranes caused inhibition of photophosphorylation activity. A maximum inhibition of 55 per cent in the phosphorylation activity occured at 80 μM DCCD.

On treatment of salt-washed thylakoid membranes with NEM alone in the dark, we observed a 16 per cent and 53 per cent inhibition of

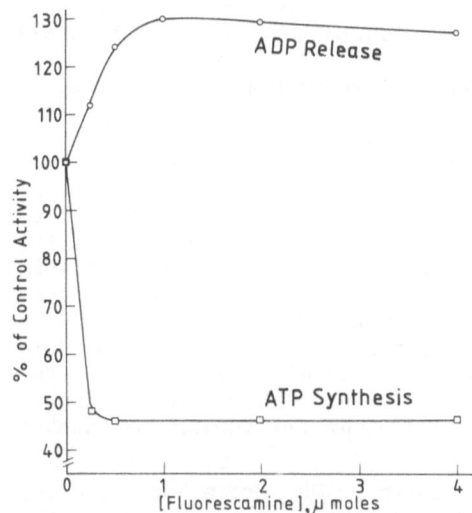

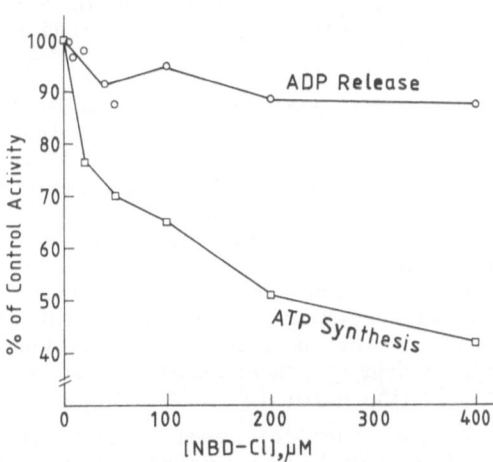

Fig.1. The energy dependent release of $[^{14}C]$-ADP and photophosphorylation as a function of fluorescamine concentration. Control activities were 0.69 nanomoles $[^{14}C]$-ADP released/mg Chl and 104 μmoles ATP formed/mg Chl/hr.

Fig.2. The effect of chemical modification of salt-washed thylakoid membrane with NBD-Cl on energy dependent release of $[^{14}C]$-ADP and photophosphorylation. Control activities were 0.5 nanomoles $[^{14}C]$-ADP released/mg Chl and 93 μmoles ATP formed/mg Chl/hr.

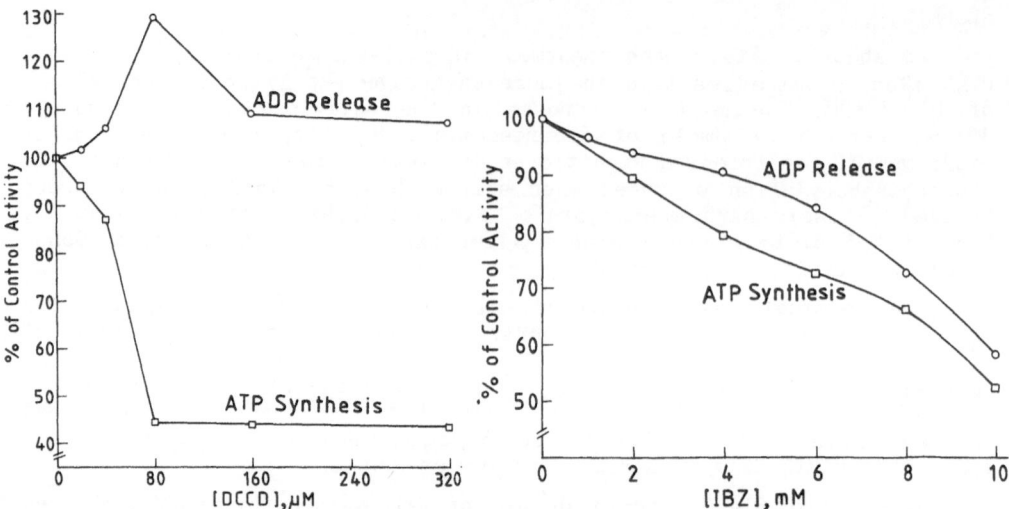

Fig.3. The energy dependent release of [^{14}C]-ADP and photophos- phorylation as a function of DCCD concentration. Control values were 0.5 nanomoles [^{14}C]-ADP released/mg Chl and 141 μmoles ATP formed/ mg Chl/hr.

Fig.4. Thiol modulation of salt-washed thylakoid membranes and its effect on energy dependent release of ADP and photophos- phorylation. Control values without any treatment with NEM were 0.52 nanomoles [^{14}C]-ADP released/mg Chl and 156 μmoles ATP formed/mg Chl/hr.

energy dependent release of [^{14}C]-ADP and photophosphorylation respec- tively (data not shown). Further modification of NEM treated thylakoid membranes with IBZ resulted in a continual decline in both the energy dependent release of [^{14}C]-ADP and the rate of photophosphorylation (Fig.4).

DISCUSSION

It has previously been shown by several workers that modification of ϵ-NH$_2$ group of lysine residues of CF$_1$ resulted in inhibition of ATP synthesis and hydrolysis[5,8,11]. The data as shown in Fig.1 indicate that ϵ-NH$_2$ group of lysine residues in CF$_1$ which get exposed during the high energy state are involved at the catalytic site of ATP synthesis. The inhibition of ATP synthesis by fluorescamine appears to be a sort of energy transfer inhibition since it has been shown with regard to certain energy transfer inhibitors e.g., DCCD, phloridzin as well as chemical modifiers that while they inhibit the rate of photo- phosphorylation (or ATPase activity) at the same time they increase the energy dependent release of ADP probably by increasing the number of binding sites for ADP on CF$_1$.

The chemical modification of phenolic -OH group of tyrosine residues by NBD-Cl caused inhibition of ATPase activity of spinach CF$_1$ and ATP synthesis/hydrolysis in <u>Rhodospirillum</u> <u>rubrum</u> chromato- phores[12,13,15]. Our observation as shown in Fig.2 indicates that modification of phenolic -OH group of tyrosine residues of CF$_1$ by NBD-Cl did not affect the energy dependent release of ADP whereas photophos- phorylation was inhibited. The inhibition of ATP synthesis by NBD-Cl seems to be an energy transfer type inhibition as has been observed with fluorescamine treatment (see above) thus suggesting that -OH group of tyrosine are also involved at the catalytic site of ATP synthesis.

DCCD, a well known energy transfer inhibitor of the membrane bound ATPases covalently reacts with -COOH group of proteolipid subunit e.g.

subunit III of CF_o, of F_o and blocks proton flux through F_o-F_1 complex[15]. It has also been observed that in isolated CF_1 or F_1 addition of DCCD caused inhibition of ATPase activity[16]. The inhibition of ATP synthesis by DCCD (Fig.3) may be explained as the decrease in the pool of ionizable groups due to modification of —COOH groups of CF_1 by DCCD resulting in decrease in the light-induced changes in buffering capacity of thylakoid membranes. However, it is also possible as studies with chemical modifications and spin-echo NMR have shown that —COOH group are involved at the catalytic site of ATP synthesis and therefore their modification with DCCD would cause inhibition of phosphorylation activity. However, the stimulation of energy dependent release of ADP (see Fig.3) with DCCD may arise because of exposure of more number of sites on CF_1 for binding of ADP.

The decrease in energy dependent release of ADP by 15 per cent and ATP synthesis by 53 per cent (data not shown) on treatment of salt-washed thylakoid membranes with NEM is due to energy transfer inhibition because dithiothreitol could not reverse the inhibitory effect of NEM. Our observations as shown in Fig.4 suggest that oxidation of —SH group, of γ-subunit of CF_1, which get exposed during energy dependent conformational changes in CF_1, with IBZ results in decline in the fraction of activated CF_1. It appears that thylakoid membrane becomes leaky to H^+-ions due to oxidation of —SH group of γ-subunit (also see ref.17). Since removal of tightly bound ADP from CF_1 requires certain threshold value of proton-motive force, the thiol modulation of —SH group of γ-subunit of CF_1 which participate directly in proton translocation, by IBZ, would result in decrease in pmf and thus inhibit both the activation of CF_1 (as shown by ADP release) and consequently the catalytic function (ATP synthesis) of CF_1[18].

In conclusion we would like to suggest that (i) ϵ-NH_2 group of lysine residues exposed during the high energy state, phenolic —OH group of tyrosine and —COOH group are involved at the active site of ATP synthesis and (ii) thiol modulation of —SH group of CF_1 which are exposed in high energy state regulates both the activation process and catalytic function of CF_1.

REFERENCES

1. H. Strotmann, and S. Bickel-Sandkötter, Biochim. Biophys. Acta 460: 126 (1977).
2. E. Schlodder, and H.T. Witt, Biochim. Biophys. Acta 635: 571 (1981).
3. E. Schlodder, P. Gräber, and H.T. Witt, in: "Topics in Photosynthesis", Vol.4, J. Barber, ed., Elsevier Biomedical, Amsterdam (1982).
4. H. Strotmann, S. Bickel-Sandkötter, U. Franek, and V. Gerke, in: "Energy Coupling in Photosynthesis", B.R. Selman and S. Selman-Reimer, eds, Elsevier, Amsterdam (1981).
5. D. Oliver, and A.T. Jagendorf, Fed. Proc. 34: 596 (1976).
6. S. Berg, S. Dodge, D.W. Krogmann, and R.E. Dilley, Plant Physiol. 53: 619 (1974).
7. H. Strotmann, and S. Bickel-Sandkötter, Ann. Rev. Plant Physiol. 35:97 (1984).
8. J. Parkash, and G.S. Singhal, in: "Adv. in Photosynth. Res.", C. Sybesma, ed., Martinus Nijhoff/Dr. Junk Publishers, The Hague (1984).
9. J. Parkash, and G.S. Singhal, Proc. Ind. Nat. Sci. Acad. B53:435 (1987).
10. D.I. Arnon, Plant Physiol. 24: 1 (1949).
11. Y. Sugiyama, and Y. Mukohata, FEBS Lett. 98: 276 (1979).
12. D.W. Deters, N. Nelson, H. Nelson and E. Racker, J. Biol. Chem. 250: 1041 (1975).
13. D. Khananshvili, and Z. Gromet-Elhanan, J.Biol. Chem. 258:3714 (1983).
14. L.P. Ting, and J.H. Wang, Biochemistry, 19: 5665 (1980).
15. K. Sigrist-Nelson, H. Sigrist, and A. Azzi, Eur. J. Biochem. 92:9(1978).
16. V. Shoshan, and B.R. Selman, J. Biol. Chem. 255: 384 (1980).
17. M.A. Weiss, and R.E. McCarty, J. Biol. Chem. 252: 8007 (1977).
18. P. Mitchell, FEBS Lett. 182: 1 (1985).

MATHEMATICAL MODEL OF THE MILLISECOND DELAYED FLUORESCENCE

Vasili Goltsev, Pavel Venedictov*, and Vladimir Shinkarev*

Department of Biophysics and Radiobiology, Biological
Faculty, Sofia University, Sofia, Bulgaria
*Department of Biophysics, Biological Faculty
Moscow State University, Moscow, USSR

INTRODUCTION

The delayed fluorescence (DF) of photosynthetizing organisms origina-tes as a result of a recombination of the primary separated charges in the reaction center of Photosystem 2 (PS2)[1,2]. The half-time for recombination of the radical pair $P680^{+}I^{-}$ is about 2-4 nsec[3], but the presence of slower components of luminescence is due to the equilibrium of this pair with other states including electron carriers both in the donor and in the ac-ceptor sides of PS2. The multitude of redox states, whose recombination le-ads to the delayed fluorescence, determines the big number of its kinetic components and considerably complicates the interpretation of data obtai-ned by this method. Mathematical description of the kinetics of transiti-ons between different states of the electron carriers with the rendering of changes in the functional parameters of the thylakoid membrane would increase the information obtaned by DF method. There exist mathematical models of the DF of simplified systems with a limited number of carriers (in the presence of diuron)[4,5].

In the present paper we attempted to describe the kinetics of the changes in the millisecond DF of chloroplasts by means of PS2 states taking part in the generation of this component of luminescence.

PS2 states model

For a description of the processes taking place in the chloroplasts, it is necessary to consider: the complex of electron carriers into the photosystems and between them; the presence of $\Delta\Psi$ and ΔpH gradients and their influence both on the electron flow rate and the quantum efficiency of luminescence; the redistribution of the exciting energy between the photosystems etc.

Even the effects pointed out make the supposed model very complicated and its analysis – difficult. Nevertheless, simplifying the object – thy-lakoid membranes without substrates of the phosphorylation and ignoring the changes in $\Delta\Psi$ and ion flows, we will consider only the noncyclic electron transfer reactions presented in the following scheeme:

$$Z_n \xrightarrow{} P680 \xrightarrow{k_1} Q_a \xrightarrow{k_2} Q_b \xrightarrow{k_3} PQ \xrightarrow{k_4} P700 \xrightarrow{k_5} A_1$$
$$\underset{k_6}{\xrightarrow{\hspace{1cm}}} A_2$$

where k_1 and k_4 are the light constants proportional to the number of light quanta reaching respectively one PS2 and PS1 reaction center per second; k_2 is the rate constant of the electron transfer between quinone carriers Q_a and Q_b; k_3 is the rate constant of Q_b^- oxidation by plastoquinone; k_5 and k_6 are the rate constants of PS1 and PS2 oxidation by the exogeneous acceptors A_1 and A_2 (or O_2).

Kinetic model

The kinetics of the reactions in this system could be considered on the basis of the concepts for the electron transport chain as a structural complex of molecules-carriers[6]. Then, the transitions between PS2 complex states could be presented in the following scheem:

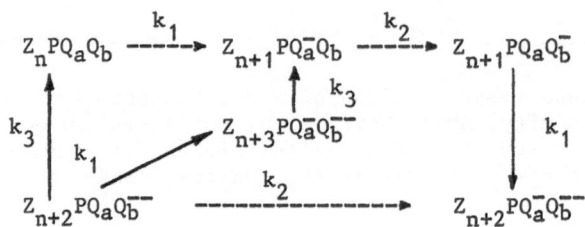

where Z_n, Z_{n+1}, Z_{n+2}, Z_{n+3} are different oxiding states of the O_2-evolving system and Q_a^-, Q_b^- and Q_b^{--} are reduced PS2 acceptors. Considering the millisecond DF, excited by continuous light, we ignore the short-lived states whose concentration is low at the moment of luminescence registration. We accept the S-states of the O_2-evolving system to be homogeneously "mixed" and do not account the transitions among them.

Designating the relative probability of the states $Z_{n+1}PQ_a^-Q_b$ as X_1, $Z_{n+1}PQ_aQ_b^-$ as X_2, $Z_{n+2}PQ_a^-Q_b^-$ as X_3, $Z_{n+2}PQ_aQ_b^{--}$ as X_4, $Z_{n+3}PQ_a^-Q_b^{--}$ as X_5, and relative PQH_2 concentration as X_6 at N actively functioning plastoquinone molecules per one electron transport chain, the changes in the relative concentrations of PS2 states could be expressed depending on time by a system of differential equations:

$$\frac{dX_1}{dt} = (1-X_1-X_2-X_3-X_4-X_5)k_1 + X_5(1-X_6)k_3 - X_1k_2$$

$$\frac{dX_2}{dt} = X_1k_2 - X_2k_1$$

$$\frac{dX_3}{dt} = X_2k_1 - X_3k_2$$

$$\frac{dX_4}{dt} = X_3k_2 - X_4(k_1+k_3)$$

$$\frac{dX_5}{dt} = X_4k_1 - X_5k_3$$

$$\frac{dX_6}{dt} = \frac{1}{2N}((X_4+X_5)(1-X_6)k_3 - X_6(1-\frac{X_7+H_0}{H_0 \circ H_m})(k_4k_5A_1 + k_6A_2))$$

The photoinduced electron flow is accompanied by the evolution of protons in the interthylakoid space at the two places of coupling - at the oxidation of water and at the oxidation of the reduced plastoquinone, what leads to a generation of the transmembrane electrochemical H^+ gradient. Taking into account the total buffer capacity (β) and the proton efflux driven by its concentration gradient, the changes in the internal proton concentration could be described by the following differential equation:

$$\frac{dX_7}{dt} = \frac{V_1 + V_2 + V_3 - V_{out}}{1 + \beta} \cdot \frac{S}{V}$$

$\frac{dX_7}{dt}$, the rate of increase in the proton concentration is proportional to the rates of electron transport in the sites of coupling without electron acceptors ($V_1 = (1-X_1-X_3-X_5)k_1$) and with them ($V_2 = X_6 k_4 k 5 A_1$ $\cdot(1-(X_7+H_0)/H_0 \cdot H_m)$ and $V_3 = X_6(1-(X_7+H_0)/H_0 \cdot H_m)k_6 A_2$). The expression ($1-(X_7+H_0)/H_0 \cdot H_m$) accounts for the decrease in the rate of PQH_2 oxidation at acidification of the internal phase (H_0 is the concentration of H^+ in the external phase and H_m is the proton concentration at which the direct and the back reactions are equilibrated: $PQH_2 \rightleftharpoons PQ + 2H^+$). V_{out} is the rate of the passive efflux of protons through th lipid bilayer equal to $p_H X_5$ (p_H is the permeability of the thylakoid membrane for protons). In this case, the proton efflux by the functioning of the ATP-synthase complex is not accounted. The ratio S/V is the area of one electron transport chain divided by the respective volume of the internal space of the thylakoid.

The buffer capacity β is determined as a sum of the contributions of the separate buffering membrane components and the internal thylakoid solution[7]:

$$\beta = \sum_i \frac{K_{ci} C_i}{(H_0 + X_7 + K_{ci})^2}$$

where C_i is the concentration of the i-th buffer and K_{ci} is its equilibrium constant for the dissociation.

Delayed fluorescence calculation

The considered system describes the changes in the concentrations of electron transport chain and pH components during continuous illumination of the thylakoid membranes. In this case, the probability (p) for recombination of the separated charges of the respective reduced states of the carrier Q^- and Z_n^+ will be proportional to the recombination rate constant (r):

$$p \sim r \left[Z_n^+ PQ^- \right]$$

At DF registration in the millisecond range with a quasi-steady-state illumination of the sample and with a duration of the cycle illumination/registration of the order of 10 msec, only the states formated during the illumination at the given cycle could be taken into account. In this case, $Z_n PQ_a Q_b$, $Z_{n+1} PQ_a Q_b^-$ and $Z_{n+2} PQ_a Q_b^{--}$ can be considered as active states for generation of DF, and the intensity of luminescence (L) will be given by:

$$L \sim k_1 (r_0(1-X_1-X_2-X_3-X_4-X_5) + r_2 X_2 + r_4 X_4$$

where r_0, r_2 and r_4 are recombination constants of the respective states.
Taking into account, that the photoinduced transmembrane proton gradient increases exponentially the probability for radiation recombination[8], we can write:

$$L \sim k_1 (r_0(1-X_1-X_2-X_3-X_4-X_5) + r_2 X_2 + r_4 X_4)(H_0+X_7)/H_0$$

If we solve numerically the system of differential equations regarding the concentrations of the states $X_1(t) \ldots X_6(t)$ and proton concentration $X_7(t)$, a description of the induction kinetics of the millisecond DF at the transitions of chloroplasts from "dark" to "light" state can be obtaned in the control, as well as in the presence of electron acceptors, inhibitors and uncouplers.

REFFERENCES

1. B.L.Strehler, W.Arnold, Light production by green plants,J.Gen.
   Physiol., 34:809 (1951)
2. J.Lavorel, Luminescence, in: "Bioenergetics of Photosynthesis",
   Govindjee, ed., Academic Press, London, New York (1975)
3. V.V.Klimov,S.I.Allakhverdiev, S.Demeter and A.A.Krasnovsky,
   Pheophetin photoreduction in the chloroplast photosystem 2 in
   dependence ofredox potential of the medium, Dokl.Acad.Sci.
   USSR, 249: 227 (1979)
4. E.M.Sorokin, Delayed luminescence kinetic of chlorophyll a in
   vivo in the absence of electron transport on acceptor side of
   photosystem II, Biofizika (Moscow), 21: 665 (1976)
5. P.Venedictov, V.Goltsev, V.Shinkarev, The influence of electric
   diffusion potential on delayed fluorescence light curves of
   chloroplasts treated with DCMU, Biocim.Biophys.Acta, 593 (1980)
6. A.B.Rubin, V.P.Shinkarev,"Electron transport in biological systems", Nauka, Moscow (1984)
7. J.Whitmarsh, Proton decay kinetics for vesicles containing buffers
   - an analytical solution, Photosynth.Res., 12:43 (1987)
8. A.R.Crofts, C.A.Wraight and D.E.Fleischmann, Energy conservation
   in the photochemical reactions of photosynthesis and its relation to delayed fluorescence, FEBS Lett., 15: 89 (1971)

ENERGY COUPLING TO PYRIDINE NUCLEOTIDE TRANSHYDROGENASE IN CHROMATOPHORES FROM PHOTOTROPHIC BACTERIA

B.F. Nore[1], T.M. Lever[2], N.P.J. Cotton[2], M.R. Jones[2] and J.B. Jackson[2]

Department of Biochemistry,[1] University of Stockholm, S-106 91 Stockholm and [2] University of Birmingham, POBox 363, Birmingham B15 2TT

INTRODUCTION

The membrane-bound pyridine nucleotide transhydrogenase of <u>Rhodobacter</u> <u>capsulatus</u> catalysing the reaction,

$$NADH + NADP^+ \rightleftharpoons NAD^+ + NADPH$$

like that from mitochondria and from <u>E. coli</u>, is a consumer of Δp in the direction from left to right. It offers several important advantages in the study of energy coupling (1).

RESULTS AND DISCUSSION

The rates of light-driven transhydrogenase (J_t) and ATP synthesis (J_p) were measured in chromatophores under similar experimental conditions during titrations with either FCCP, myxothiazol, antimycin or valinomycin/nigericin/K^+. In each case, for a given concentration of inhibitor, J_p was depressed more strongly than J_t. The nature of the relationship between J_t and J_p was independent of the way in which Δp was lowered (2). There are two possible conclusions.

Either there are no regulatory or "localised" interactions between the electron transport chain and the ATPsynthase and transhydrogenase. Or there are identical localised interactions between electron transport components and the ATP synthase and transhydrogenase. <u>A priori</u> the second possibility seems less likely.

The dependence of J_t on Δp (measured as $\Delta\Psi$ by electrochromism for $\Delta pH = 0$) in chromatophores treated with both venturicidin and nigericin was similar when $\Delta\Psi$ was reduced with either FCCP, myxothiazol, antimycin or valinomycin/K^+(3). This supports the conclution that there are no localised interactions between the electron transport components and transhydrogenase. The dependence was sigmoidal: at $\Delta\Psi = 0$, J_t was small but finite. There was an apparent threshold $\Delta\Psi$ below which J_t barely increased and at high $\Delta\Psi$ the rate tended towards saturation. These data are described by a model in which two first-order rate constants in transhydrogenase turnover exert control over the rate. K_1 is $\Delta\Psi$-

dependent. It domenates at low $\Delta\Psi$ and is respnsible for the threshold response. K_2 which is $\Delta\Psi$-independent is responsible for saturation at high $\Delta\Psi$.

$$V = \frac{K_2(E_{total})}{1 + K_2/K_1}$$

where $K_1 = K_1^0 \exp(zF\Delta\Psi/2RT)$. It is unnecessary according to this model to invoke conformationally active and relatively inactive states of the enzyme.

The effect of external pH on the dependence of J_t on $\Delta\Psi$ was examined. Decreasing pH led to a decrease in the threhold $\Delta\Psi$. Extremes of pH additionally led to a decrease in the maximum slope of $J_t/\Delta\Psi$. The former is attributed to a pKa on K_1 at neutral pH; the latter to pKa effects on K_2. Transhydrogenase activity in chromatophores is dependent on the presence of low concentrations of divalent cations or higher concentrations of monovalent cations, a situation somewhat different from that found in mitochondria (4). The effect of cation levels on the dependence of J_t on $\Delta\Psi$ in chromatophores suggests that only K_2 is affected.

The initial rate of decay of the electrochromic absorbance change after a period of illumination is a measure of the dissipative ionic flux (J_{dis}) across the chromatophores membrane driven by $\Delta\Psi$. J_{dis} is enhanced in the presence of transhydrogenase substrates (by approx. 10% in venturicidin-treated chromatophores) but $\Delta\Psi$ is not noticably depressed. A series of experiments was undertaken to check the calibration procedures for J_{dis} in Rb. capsulatus chromatophores. Using laser pulses to ensure single turn over conditions and after correcting for reaction centres"deficient" in electron donor, the contributions from the reactions, cytochromes (c_1+c_2) $\longrightarrow$ P and P $\longrightarrow$ Q_A to the generation of $\Delta\Psi$ were 46% and 54% respectively. Using these data J_{dis} was expressed as a flux of charge across the chromatophore membrane. The ratio of the difference in the extra charge flux in the presence of transhydrogenase substrates to the transhydrogenase rate ($H^+/H^{..}$) can be calculated.

REFERENCES

(1) Fisher, R.R. and Earle, S.R. (1982) in The Pyridine Nucleotide Coenzymes, pp.279-324, Academic Press.
(2) Cotton, N.P.J. and Jackson, J.B. (1988) FEBS Lett. 229, 303-307.
(3) Cotton, N.P.J., Myatt, J.F. and Jackson, J.B. (1987) FEBS Lett. 219, 88-92.
(4) Rydsröm, J., Teixeria da Cruz, A. and Ernster, L. (1970) Eur. J. Biochem. 17, 56-62.

CHARACTERISATION OF THE bc_1 COMPLEX

IN RHODOPSEUDEMONAS VIRIDIS

Maria Cully, Frances A.Jay, Nadia Gabellini
and Dieter Oesterhelt

Max-Planck-Institut für Biochemie
Martinsried
F.R.G.

The ubiquinol:cytochrome c - oxidoreductase (bc_1 complex) of the photosynthetic purple non-sulphur bacterium Rhodopseudomonas viridis (Rps. viridis, Fig.1) is thought to be involved in both cyclic electron transport and proton transport across the thylakoid membrane. Although the donor site of this electron cycle, the reaction centre, has been structurally and functionally well characterised (Fig.2) there is a little information available of the bc_1 complex.

For many years it was thought that bc_1 complex was absent from Rps. viridis membranes. However, in 1972 Jones and Saunders[2] reported that the electron transport chain in Rps. viridis is antimycin A - sensitive (an inhibitor of electron flow at sites involving b-type cytochromes). No evidence for b-type cytochrome was, however, found.

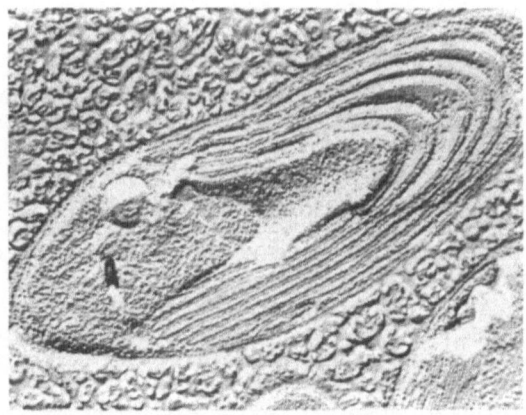

Fig.1   Freeze etch section of the photosynthetic bacterium Rhodopseudomonas viridis.

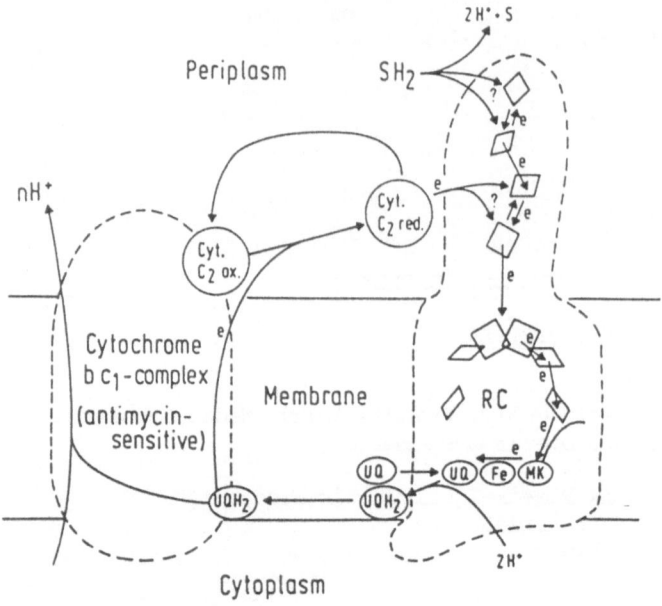

Fig.2 Postulated scheme of electron transport through the bc_1 complex in the bacterial photosynthetic apparatus[1].

Recent results from Wynn et al. 1986[3] showed that it was just present in exceptionally low amounts with a composition corresponding to that occuring in prokaryotes. It consists of three subunits - cytochrome b (high and low potential), cytochrome c_1 and Rieske FeS - protein. The gene coding for the bc_1 complex has provided the sequence of the bc_1 polypeptides from Rhodobacter capsulatus. This sequence shows significant amino acid homology to mitochondrial bc_1 and chloroplast b_6f complexes. The topology of the protein has been postulated on the basis of hydropathy plots, but little biochemical evidence regarding protein topography is available.

The isolation of the bc_1 complex as prepared by Wynn et al. used dodecylmaltoside solubilisation of Rps. viridis membranes. Although the complex thus obtained lacks ubiquinol:cytochrome c - oxidoreductase activity (probably associated with the loss of the FeS - protein) it was the first isolation procedure and characterisation of the bc_1 complex in Rps. viridis.

Modification of this isolation procedure by omitting the ammonium sulphate precipitation and by replacing cytochrome c affinity chromatography with ion exchange chromatography enabled us to obtain a complex which contained all three subunits and retained activity.

The isolated enzyme contains a minimum of three polypeptides (SDS-PAGE): cytochrome b with apparent molecular weight 36 kD, cytochrome c_1 (31 kD) and Rieske FeS - protein (25 kD). The cytochrome c_1 stains haem positive after SDS-PAGE. Difference, absolute and ESR spectra of the isolated bc_1 complex were monitored.

The midpoint potentials of the three redox components (cytochrome c_1, cytochromes b high and low potential) were estimated (Table I.).

Table I. Midpoint potential of cytochromes b high and low potential and cytochrome c_1.

Midpoint potential (mV)

	cyt b (high)	cyt b (low)	cyt c_1
pH 6	53	-57	253
pH 7	34	-106	266
pH 8	20	-155	258

The enzyme activity of the bc_1 complex is pH dependent, antimycin A - sensitive and affected by detergent concentration. The maximal achieved turnover is 80 s^{-1} with ubiquinol 10 and horse heart cytochrome c as substrates.

In order to investigate the structure of the bc_1 complex in Rps. viridis, the topology of the enzyme and the electron flow through the bc_1 complex, an immunological approach was taken. Monoclonal antibodies were raised against the bc_1 complex (native and denatured). Positive clones were selected using Dot Blot and Immunoblot assays. For the initial preliminary investigations, polyclonal antibodies against cytochrome c_1 were raised and used for immunolabelling on the thin cell sections.

SUMMARY

The bc_1 complex from Rhodopseudomonas viridis was isolated according to the method of Wynn et al. with some modifications. Our isolated enzyme has the advantage that it is functionally active, i.e. it exhibits ubiquinol:cytochrome c - oxidoreductase activity, which is sensitive to antimycin A. Monoclonal antibodies have been produced against the bc_1 complex of Rps. viridis, which will be applied to the investigation of the structure, function and topology of the enzyme.

REFERENCES

1. Michel,H. and Deisenhofer,J., Encycl. of Plant Physiol.19,1

2. Jones,O.T.G. and Saunders,V.A.,1972, Biochim.Biophys.Acta 275,417

3. Wynn,R.M.,Gaul,D.F.,Choi,W.K.,Shaw,R.W. and Knaff,D.B.,1986, Photosynt. Res.9,181

LIGHT DRIVEN AMINO ACID UPTAKE IN MEMBRANE VESICLES OF <u>STREPTOCOCCUS</u>

<u>CREMORIS</u> FUSED WITH LIPOSOMES CONTAINING BACTERIAL REACTION CENTERS

Wim Crielaard, Arnold J.M. Driessen, Douwe Molenaar, Klaas J. Hellingwerf and Wil N. Konings

Department of Microbiology, University of Groningen, Kerklaan 30, 9751 NN Haren, The Netherlands

INTRODUCTION

The incorporation of primary proton pumps in biological membranes has opened attractive possibilities for studies of proton motive force (Δp)-dependent processes in isolated membrane vesicles from bacterial[1], and eukaryotic[2] origin. Fused membranes obtained from liposomes containing a Δp-generating system and membrane vesicles of fermentative bacteria lacking an accessible proton pump have been shown to be excellent model systems for studies on the role of the Δp in solute transport[1,3,5,6]. Cytochrome <u>c</u> oxidase[1] and bacteriorhodopsin[3] have been used extensively as Δp-generating systems in these fused membranes.

Recently, we have shown that reaction centers (RC's) of phototrophic bacteria can successfully and functionally be reconstituted into liposomes[4]. At pH 7.0, in the presence of cytochrome <u>c</u>, ubiquinone-0 and ascorbic acid, a light induced Δp of approximately -170 mV can be generated in these liposomes[4]. Here we show that these reaction centers can also functionally be incorporated into bacterial membranes by fusing reaction center liposomes with bacterial membrane vesicles.

RESULTS AND DISCUSSION

Δp generation in Streptococcus cremoris membrane vesicles fused with RC-liposomes

In the light cyclic, electron transfer occurs in RC-proteoliposomes when ascorbate, cytochrome <u>c</u> and ubiquinone-0 are present as redox mediators. Electron transfer leads to the generation of a proton motive force (Δp) across the liposomal membrane[4]. It is possible to maintain a light induced Δp of around -170 mV for at least 30 min in RC-liposomes at pH 7.0. Under these conditions reaction centers could be used for the generation of a Δp in fused membranes. Therefore, it was tested whether the Δp-generating capacities of the reaction centers were affected by freeze/thaw/sonication fusion with <u>S.cremoris</u> membrane vesicles. More than 95% of the reaction center population in the fused membranes was shown to be orientated as <u>in vivo</u>, indicating that fusion does not reduce the amount of right-side out orientated RC's. Therefore also in the fused membranes almost all RC's are incorporated with the cytochrome c binding site exposed to the outer surface. Since cytochrome c is added to the outside of the fused membranes illumination should lead to an outward pumping of protons and the generation of a Δp. Under the experimental conditions (in the presence of nigericin) only an electrical potential (Δ ψ) could be generated which was recorded from the distribution of tetraphenyl-phosphonium ions (TPP[+]) between the external bulk phase and the inner compartment of the fused membrane vesicles (data not shown)· The Δψ generated in the light in these fused membranes was approximately -110 mV. This Δp is considerably lower than the Δp that can be generated in RCLH$_I$-liposomes (-170 mV)[4]. This can be due to a decline in the RC activity or to a higher proton or ion permeability of the fused membranes. However the Δp generated is still in a useful range.

Light driven leucine transport in the fused membranes

To investigate whether the membranes obtained by fusion of S.cremoris membrane vesicles and RC-liposomes are coupled, i.e. whether the Δp generated by the RC's can be used by one of the Δp consuming processes in the streptococcal membrane, the uptake of leucine in the fused membrane was followed (Fig. 1). In the dark the uptake of leucine was very slow and the level of uptake did not exceed equilibration. In the light, however, leucine was taken up rapidly. Initial rates of leucine transport of 0.54 nmol/(mg protein.min) were observed. A steady state level of leucine uptake was reached within 4 min after the addition of leucine. When the light was switched off or when the uncoupler carbonyl-cyanide-m-chlorophenylhydrazone (CCCP) was added, efflux of the accumulated leucine occurred immediately and within 4 min the internal leucine concentration equaled the external concentration under dark conditions.

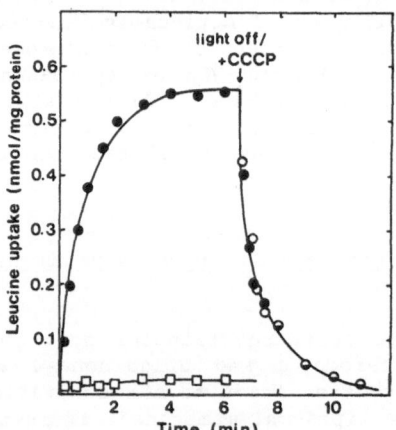

Figure 1.   The uptake of leucine by S.cremoris membrane vesicles fused with RC-liposomes. Experiment were performed at maximum light intensity (9.0 kW/m^2). The uptake was started by adding ^{14}C-leucine (4.5 µM). At the arrow the light was switched off (closed circles) or CCCP (2 µM) was added (open circles).Experiments were performed in the presence of ubiquinone-0 (UQ$_0$, 400 µM), cytochrome c (cyt c, 20 µM), ascorbate (500 µM) and nigericin (nig, 20 nM).

As was expected uptake of leucine could not be detected in non-fused liposomes or non-fused (and therefore non-energized) membrane vesicles. Uptake of leucine in the fused membranes supplies therefore strong evidence for a functional incorporation of the RC's in the S.cremoris membranes.

Relation between Δ p generation and leucine uptake

In RCLH$_I$-liposomes the Δp generated depends on the light intensity used for energization. This appeared also to be the case in the fused membranes. Figure 2 shows a simultaneous recording of both the Δ p and the uptake of leucine in the fused membranes. It is clear that at increasing light intensities both the Δp and accumulated level of leucine increased. This behaviour was investigated in more detail. Both Δ p and the accumulation level of leucine increased in with increasing light intensities up to 3.4 kW/m^2 (data not shown).

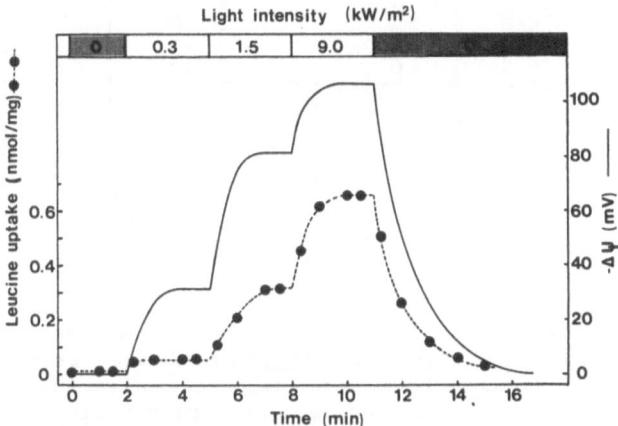

Figure 2.   The electrical potential (solid line) and the uptake of leucine (dashed line) in S.cremoris membrane vesicles fused with RC-liposomes at increasing light intensities. The $\Delta\psi$ and the uptake of leucine were measured simultaneously.

Since both the $\Delta\psi$ (= Δp, since a ΔpH is absent due to the presence of K$^+$ and nigericin) and the steady state level of leucine accumulation can be varied in these fused membranes by varying the light intensity the relation between Δp and $\Delta p_{leucine}$ (= (2.3 RT/F)log([leucine]$_{in}$/[leucine]$_{out}$), where R, T and F are the gas constant, the absolute temperature and the Faraday constant respectively) can easily be studied. The steady state level of accumulation of leucine depended linearly on the Δp, indicating a constant proton/leucine stoichiometry at different Δp values. An apparent stoichiometry of 0.8 can be calculated from the slope of the linear relationship (data not shown). This slope depends on the number of protons symported with one molecule of leucine by the leucine carrier and on the rate of non-carrier mediated efflux (passive diffusion) of leucine across the lipid membrane[5]. Considering the relatively high rate of passive diffusion of leucine across the membrane[5], the experimentally determined ratio of 0.8 indicates a mechanistic proton to leucine stoichiometry of one. The same stoichiometry was estimated in membranes of S.cremoris membrane vesicles fused with proteoliposomes containing beef heart cytochrome c oxidase[5].

Dependence of the initial rate of uptake on the Δp

It has been shown previously that the rate of amino acid uptake in membrane vesicles of S.cremoris depends sharply on the magnitude of Δp[6]. To investigate this phenomenon in more detail the initial rate of leucine uptake was measured at different light intensities, i.e. at different Δp levels. The initial rate was calculated from the uptake of leucine 30 s after the addition of leucine. The Δp was determined simultaneously. Figure 3 shows that the initial rate of leucine uptake varied exponentially with the Δp. A logarithmic replot of the data (Fig. 3, inset) yielded a linear relationship.

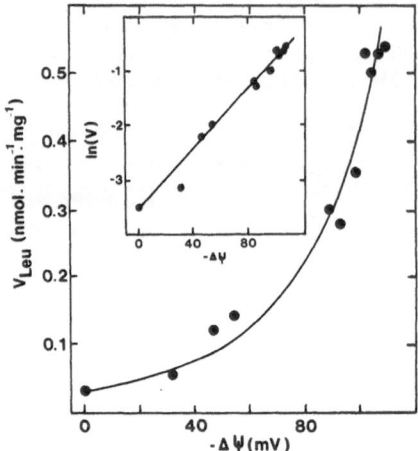

Figure 3. The relation between the initial rate of leucine uptake (V_{leu}) and Δp in S.cremoris membrane vesicles fused with RC-liposomes. Data were calculated from the uptake 30 s after the addition of leucine from experiments analogues to those described in Fig. 1. The inset shows a logarithmic replot of the data.

Since the K_t (the apparent K_m for transport) was found to be independent of Δp[11], it is to be expected that also the V_{max} increases exponentially with the Δp. An explanation for this phenomenon is not yet known.

CONCLUSIONS

By using reaction centers as a Δp generating system the attractive features of both bacteriorhodopsin and cytochrome c oxidase are combined. As for bacteriorhodopsin it is possible to control exactly the start and the stop of proton pumping activity by switching the light on and off and the rate of proton pumping by varying the light intensity. A Δp of any desired magnitude upto the maximum level can thus be generated and controlled at any time. As for cytochrome c oxidase the direction of proton pumping is always from the inner bulk phase to the outer bulk phase, since cytochrome c is added at the outside and interacts only with properly oriented RC's.

The major disadvantage of this pump is its narrow useful pH range. Because light driven cyclic electron flow is strictly dependent on the pH-sensitive chemical reaction between cytochrome c and ubiquinol-0, proton motive force generation by the reaction centers is only possible at relatively alkaline pH's (≥ 8.0). The addition of ascorbate as an extra electron donor extends this pH range, but at pH 6.0, even in the presence of ascorbate, transient Δp's are observed. One way to overcome this problem can be by co-reconstituting reaction centers with the photosynthetic bc_1-complex in liposomes. The direct pH-dependent chemical reaction between ubiquinol-0 and cytochrome c is than not needed for cyclic electron transport and proton pumping. Initial studies with proteoliposomes containing RC's and photosynthetic bc_1-complexes indicated that non-transient Δp's can indeed be generated at lower pH values (W. Crielaard and N. Gabellini, unpublished).

LITERATURE

1. A.J.M. Driessen, W. de Vrij, and W.N. Konings, Incorporation of beef-heart cytochrome c oxidase as a proton-motive-force-generating mechanism in bacterial membrane vesicles, Proc. Natl. Acad. Sci. USA. 82:7555 (1985).

2. M. Opakarova, A.J.M. Driessen, and W.N. Konings, Proton motive force driven leucine uptake in yeast plasma membrane vesicles, FEBS Lett. 213:45 (1987).

3. A.J.M. Driessen, K.J. Hellingwerf, and W.N. Konings, Light-induced generation of a proton-motive force and Ca^{2+} transport in membrane vesicles of Streptococcus cremoris fused with bacteriorhodopsin proteoliposomes, Biochim. Biophys. Acta 808:1 (1985).

4. D. Molenaar, W. Crielaard, and K.J. Hellingwerf, Characterization of protonmotive force generation in liposomes reconstituted from phosphatidylethanolamine, reaction centers with light-harvesting complexes isolated from Rhodopseudomonas palustris, Biochem. 27:2014 (1988).

5. A.J.M. Driessen, K.J. Hellingwerf, and W.N. Konings, Mechanism of energy coupling to entry and exit of neutral and branched chain amino acids in membrane vesicles of Streptococcus cremoris, J. Biol. Chem. 262:12438 (1987).

6. A.J.M. Driessen, S. de Jong, and W.N. Konings, Characterization of branched-chain amino acid transport in membrane vesicles of Streptococcus cremoris, J. Bacteriol. 169:5193 (1987).

ANALYSIS OF THE POLARIZED ELECTRON PARAMAGNETIC RESONANCE

SPECTRUM OF *RHODOBACTER SPHAEROIDES*

D.A. Hunter and P.J. Hore

Physical Chemistry Laboratory
Oxford University
Oxford. U.K.

INTRODUCTION

In the earliest steps of photosynthesis the energy of absorbed light is channelled to a reaction centre where the first chemical reactions occur. In intact systems these electron transfer processes take place in an essentially linear and irreversible fashion. However if the chain of molecules along which the electron travels is chemically or physically broken the electrons are constrained to remain within the reaction centre where they may be used as a probe of their environment. We show here that simulation of the Electron Paramagnetic Resonance (EPR) spectra of pairs of such electrons may be used to gain insight into the structure of the reaction centre.

We denote the first three components in the bacterial electron transfer chain by P, I and X. P is a bacteriochlorophyll dimer, I a bacterio-pheophytin and X an iron-quinone complex. The sequence of electron transfers may be represented as:

$$PIX \xrightarrow{h\upsilon} {}^1PIX \xrightarrow{3ps} P^+I^-X \xrightarrow{200ps} P^+IX^-$$

The intermediates, P^+I^- and P^+X^-, are known as radical pairs.

When the iron is uncoupled from the quinone, P^+X^- survives for a few milliseconds which is long enough to permit detection by EPR spectroscopy.[1] Its spectrum is observed to be more intense than expected and to contain features from both emission and absorption of radiation. These effects, known as Chemically Induced Magnetic Polarization (CIMP),[2] are well known in the EPR spectra of transient, liquid-phase radicals and arise from interactions within radical pairs. Previous attempts[3] to model the spectrum of P^+X^- have assumed that polarization is formed in P^+I^- and passes to P^+X^- by electron transfer. On this basis it proved possible to simulate the experimental spectrum but only with eccentric values of the various parameters involved. We have recently suggested that the lifetime of P^+I^- is too short to allow much polarization to develop and that the observed effects arise entirely from interactions in P^+X^-.[4] The results of this hypothesis are discussed below.

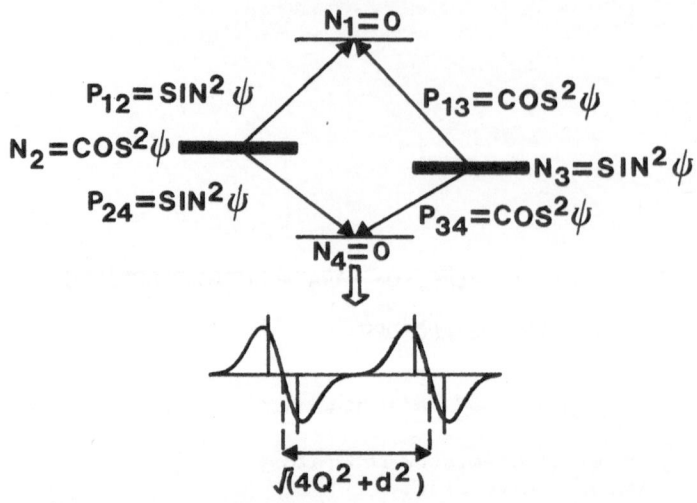

Figure 1. Top: The energy level diagram including
populations, N, and transition probabilities, P.
Bottom: The resulting spectrum.

METHODS

We shall discuss the simulation of the experimental spectrum by
reference to the energy level diagram of a prototype radical pair consisting
of two unpaired electrons (Figure 1). The time averaged populations, N, and
the transition probabilities for the allowed transitions, P, are shown as
functions of the variable ψ, defined by:

$$\tan 2\psi = 2Q/d$$

2Q is the separation of the EPR frequencies of P^+ and X^-. d is given by:

$$d = D(\cos^2\xi - \tfrac{1}{3})$$

where D is the parameter defining the strength of the electron–electron
dipolar coupling, taken to have axial symmetry, and ξ is the angle between
the P^+X^- axis and the magnetic field direction. These, non-equilibrium,
populations arise because the radical pair is formed from ^1P in a
spin–correlated singlet state.

The intensity of a transition is simply the product of the transition
probability and the difference in the populations of the two connected
levels. For the four allowed transitions, this intensity is $\pm\sin^2\psi \cos^2\psi$.
The pair of transitions ($3 \rightarrow 4$ and $2 \rightarrow 4$) in which there is a population
inversion are observed in emission, the other pair in absorption.
Unresolved nuclear hyperfine couplings cause a Gaussian linebroadening in
the spectrum of the uncoupled radicals. We incorporate this into the
simulation by making each of the resonances in the spectrum Gaussian in
shape with an adjustable width (figure 1).

The spectrum of randomly oriented reaction centres is obtained by summing the simulations for all values of ξ. d is given above explicitly in terms of ξ. Additionally the resonance frequency of X⁻ and hence the value of Q depends on the orientation of the semiquinone with respect to the field direction. This anisotropy is characterised by the three principal values of the g-tensor of X⁻: $g_X(X^-)$, $g_Y(X^-)$, and $g_Z(X^-)$.

RESULTS AND DISCUSSION

Experimental and "best-fit" simulated 9GHz EPR spectra of *Rb. sphaeroides* are shown in Figure 2. The parameter values so obtained are shown in Table I together with estimates based on the unrefined X-ray crystallographic co-ordinates for the related bacterium *Rps. viridis*. The agreement is encouraging though not perfect. There are several possible sources of discrepancy. First, the structures of *Rb. sphaeroides* and *Rps. viridis* would not be expected to be identical, not least because X is menaquinone in the former and ubiquinone in the latter. Second, it was assumed, for simplicity, that P⁺I⁻ is too short lived to give rise to significant polarization. Depending on the strengths of the exchange and dipolar couplings between P⁺ and I⁻ and the rate of electron transfer, this may or may not be a reasonable approximation. It would be a simple matter to extend the treatment described here to include polarization generated in P⁺I⁻ as described in reference 3. Third, unresolved hyperfine couplings were modelled as Gaussian linebroadenings. A much better, though time-consuming, approach would be to average over a range of hyperfine couplings.

A more challenging test of these ideas is presented by Stehlik *et al.*[9] who simulate the much better resolved spectra of deuterated photosystem I preparations at 24GHz EPR frequency. By adding a small g-tensor anisotropy for P⁺, they were able to account satisfactorily for the experimental results. The parameters so obtained were also in good agreement with the crystal structure of *Rps. viridis*.

In conclusion, it appears that the polarized EPR signals of bacterial reaction centres and plant photosystem I are due to pairs of interacting radicals. Spectral simulation provides information on the separation and orientation of the two radicals.

Table I. Best Fit Parameters for the simulated spectrum.

Experimental constants.[5][6]	
$g(P^+)$ = 2.0026 $g_X(X^-)$ = 2.0067 $\quad g_Y(X^-)$ = 2.0056 $\quad g_Z(X^-)$ = 2.0024	
Variables.[a] Value from this work.	Values from other work.[7][8]
D = -1.4G α = 59° β = 56°	D = -2.4G α = 72° β = 65°

a. α and β define the orientation of the P⁺X⁻ axis with respect to the pricipal axes of the semiquinone g-tensor.

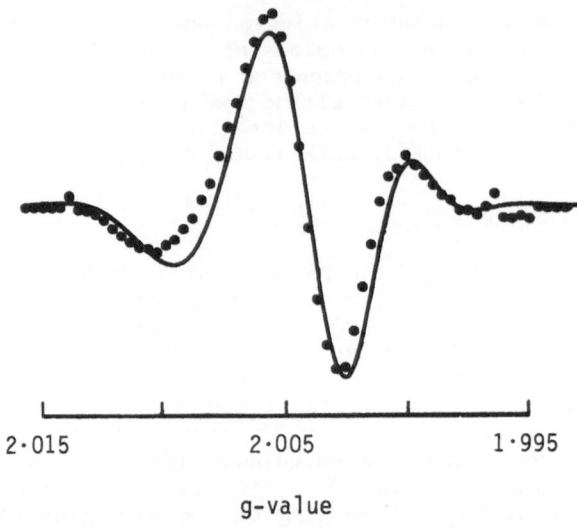

Figure 2. Experimental (dots) and simulated (solid line) EPR spectra of spin-correlated P⁺X⁻ radical pairs in *Rb. sphaeroides* The first derivative of the magnetic susceptibility is plotted as a function of magnetic field strength.

REFERENCES

1. A.J. Hoff, Electron Spin Polarization of Photosynthetic Reactants, Q. Rev. Biophys, 14:599 (1981).
2. L.T. Muus, P.W. Atkins, K.A. McLauchlan, and J.B. Pedersen, Eds, "Chemically Induced Magnetic Polarization," Reidel, Dordrecht (1977).
3. P.J. Hore, E.T. Watson, J.B. Pedersen and A.J. Hoff, Lineshape Analysis of Polarized Electron Paramagnetic Resonance Spectra of the Primary Reactants of Bacterial Photosynthesis, Biochim. Biophys. Acta, 852:106 (1986)
4. P.J. Hore, D.A. Hunter, C.D. McKie and A.J. Hoff, Electron Paramagnetic Resonance of Spin Correlated Radical Pairs in Photosynthetic Reactions, Chem. Phys. Lett, 137:495 (1987).
5. J.D. McElroy, G. Feher, and D.C. Mauzerell, Characterization of Primary Reactants in Bacterial Photsynthesis. I. Comparison of the Light Induced EPR Signal with that of a Bacteriochlorophyll Radical, Biochim. Biophys. Acta, 267:363 (1972).
6. P. Gast, A. de Groot, and A.J. Hoff, Evidence for an Anisotropic Magnetic Interaction between the Intermediary Acceptor and the first Quinone Acceptor in Bacterial Photosynthesis, Biochim. Biophys. Acta, 723:52 (1983).
7. J. Deisenhofer, O. Epp, K. Miki, R. Huber and H. Michel, X-ray Structure Analysis of a Membrane Protein Complex, J. Mol. Biol, 180:385 (1984)
8. A. Ogrodnik, W. Lersch, M.E. Michel-Beyerle, J. Deisenhofer, and H. Michel, Spin Dipolar Interactions of Radical Pairs in Photosynthetic Reaction Centres, in "Antennas and Reaction Centres of Photosynthetic Bacteria," M.E. Michel-Beyerle Ed., Springer, Berlin (1986).
9. D. Stehlik, C.H. Bock and J. Petersen, J. Phys. Chem, in press.

DIELECTRIC DISPERSION IN HYDRATED PURPLE MEMBRANE

Imre Kovács, and György Varo*

Central Research Institute for Physics, Budapest 114
P.O.Box 49. H—1525 Hungary
*Institute of Biophysics, Biological Research Center, Szeged
H-6701 Hungary

SUMMARY

Dielectric dispersion effects were studied in purple membrane of different hydration levels. The capacitance and conductivity were measured over the frequency range 10^2 Hz to 10^5 Hz. With increase in the hydration level, the conductivity increases sharply at the critical hydration h_c=0.06 gH_2O/gbR. This critical hydration is close to the extent of the first continuous strongly bound water layer and is interpreted as the threshold for percolative proton transfer. Above 0.1 gH_2O/gbR water content Maxwell—Wagner relaxation also appears, showing the presence of a bulk water phase.

INTRODUCTION

Bacteriorhodopsin is the main protein component of the purple membrane of Halobacterium halobium. After light excitation bR runs through a cycle and pumps proton across the plasma membrane creating an electrochemical gradient. The energy stored in this gradient is used by the cell for ATP synthesis and other energy recquiring processes[1,2].

Bacteriorhodopsin preserves some its activity in the dried form, but the photocycle strongly depends on the water content. On removal of the water, the light adaption gradually vanishes[3,4,5]. The first steps in the photocycle, leading to the M state are accelerated at low water content. The Arrhenius parameters of the photocycle show an abrupt change at a water content of around 0.06 gH_2O/gbR[6]. It has been demonstrated that the first continuous water layer on the purple membrane is completed at this water content and some structural change also occurs[7]. The M decay is hindered in the dried sample. At low water contents the protons from the M state return to their original sites. The 0 form and the proton pumping activity appear over 0.1 gH_2O/gbR, when some bulk water is present[7].

In the present work we have studied the contribution of the adsorbed water to the dielectric properties of the dried bR samples. The role of the adsorbed water in the functioning of bR is also discussed.

MATERIAL AND METHODS

The PM used in our measurements was isolated by the standard procedure from Halobacterium halobium strain ET 1001[8]. The sample consisted of about 2000 layers of oriented PM sheets electrophoretically deposited on a SnO_2-covered glass slide. Details of sample preparation can be found elsewhere[9]. A conducting rubber sheet pressed to the upper surface of the sample was the second electrode. The sample was placed in a closed chamber over saturated salt solution giving the desired humidity and left for at least 24 hours for equilibration[6]. For the quantity of the adsorbed water, we used the values determined earlier[4] by Váró and Keszthelyi[5].

The parallel capacitance (C_p) and conductivity (G_p) of the sample were measured with a Hewlett-Packard 4274A RCL bridge in the frequency range from 10^2 Hz to 10^5 Hz.

RESULTS AND DISCUSSION

Plots of capacitance against frequency at different humidities are shown in Fig.1. There is no detectable dielectric relaxation below a water content of 0.06 gH_2O/gbR.

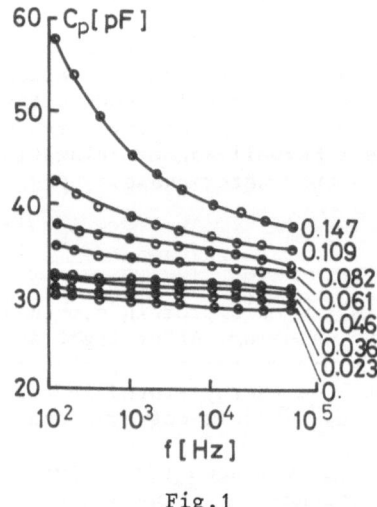

Fig.1

The parallel capacitance of the bacteriorhodopsin sample against frequency at different water content. The hydration levels ($h[gH_2O/gbR]$) are shown at the curves.

At a higher water contents, i.e. from o.1 gH_2O/gbR, there is a large dielectric relaxation at low frequencies (Fig.1). It is presumed that this is a Maxwell-Wagner relaxation process[10] caused by the appearance of a bulk water phase[1]. It has been observed[6] that at 0.1 gH_2O/gbR water content there is a change in the potential barrier system of the bR and at higher water contents these are similar to that of bR in suspension. The large increase in capacitance above a water content of 0.1 gH_2O/gbR shows that the dielectric screening increases. It can be supposed[2] that the increased dielectric screening plays an important role in determining the internal potential barriers in bR.

At a given frequency the G_p conductivity can be used to describe the hydration dependence of dielectric absorption. The conductivities at frequencies 10^2Hz, 10^3Hz are shown as function of the water content in Fig.2. These plots consist of two linear parts with a break-point, indicating the existence of critical hydration. Up to the sample water content of $h_c=0.06$ gH_2O/gbR the conductivity is nearly constant. Above h_c, the conductivity increases rapidly. The observed conductivity change is an accord with the percolation model of protonic conduction on partially hydrated protein[11]. It has been observed[6] that the Arrhenius parameters show an abrupt change at a water content of 0.06 gH_2O/gbR.

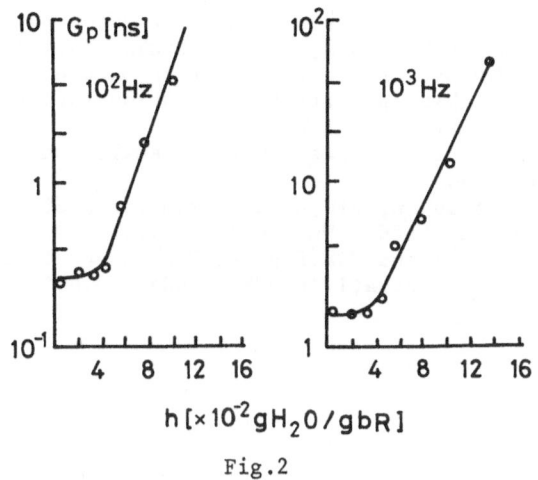

Fig.2

The conductivity as function of water content at frequencies 10^2Hz and 10^3Hz.

Our results confirm that this change is directly connected with the proton conduction inside the protein, as supposed earlier[7].

This study demonstrates that through its charge screening effect, water plays an important role in the functioning of bR. The screening effect in fact weakens the electrostatic interaction between the charges inside the protein making its fluctuations possible. These fluctuations as the function of water content had been studied by Váró and Eisenstein[7], and their observations are in agreement with our results. The appearance of the percolating protons, strongly correlated with the high water content of the sample, shows that the water fulfils a very important role for the normal functioning of bR.

ACKNOWLEDGEMENT

The authors are grateful to Professor L.Keszthelyi for helpful discussions.

REFERENCES

1. W.Stoeckenius, R.H.Lozier, R.A.Bogomoli,Bacteriorhodopsin and the Purple Membrane of Halobacteria, Biochimica et Biophysica Acta, 505;215(1979).

2. K.J.Lányi, Bacteriorhodopsin and related light-energy converters, in "Bioenergetics", L.Ernster, Elsevier Amsterdam, New York, Oxford (1984).

3. R.Korenstein and B.Hess, Hydration effects on CIS-TRANS isomerisation of bacteriorhodopsin, FEBS Lett., 82:7(1977).

4. R.Korenstein and B.Hess, Hydration effects on the photocycle of bacteriorhodopsin in thin layers of purple membrane, Natur,270:184 (1977).

5. G.Váró, L.Keszthelyi, Photoelectric signals from dried oriented purple membranes of Halobacterium halobium, Biophys. J., 43:47(1983).

6. G.Váró, L.Keszthelyi, Arrhenius parameters of the bacteriorhodopsin photocycle in dried oriented samples, Biophys.J.47:243(1985).

7. G.Váró, L.Eisenstein, Infrared studies of water induced conformation changes in bacteriorhodopsin, Eur.Biophys.J. 14:163(1987).

8. D.Oesterhelt, W.Stoeckenius, Isolation of purple membrane of Halobacterium halobium and its fractionation into red and purple samples, Methods Enzymol. 31:667 (1974).

9. G.Váró, Dried Oriented Purple Membrane Samples, Acta Biol. Acad. Sci. Hung. 32:301(1981).

10. R.Pethig, Dielectric and Electronic Properties of Biological Materials, John Wiley and Sons, Ltd., New York (1979).

11. G.Careri, A.Giasanti and J.A.Rupley, Proton percolation on hydrated lysozyme powders, Proc.Natl.Acad.Sci. USA, 33:6810 (1986).

SECTION III

CARBON FIXATION: MEASUREMENT AND REGULATION

RECENT DEVELOPMENTS IN RUBISCO RESEARCH: STRUCTURE, ASSEMBLY, ACTIVATION, AND GENETIC ENGINEERING

Robert T. Ramage[1] and Hans J. Bohnert[1,2]

[1]Department of Biochemistry and [2]Departments of Molecular and Cellular Biology and Plant Sciences
University of Arizona
Tucson, AZ 85721 U.S.A.

I. INTRODUCTION

Ribulose-1,5-bisphosphate carboxylase/oxygenase (Rubisco, E.C. 4.1.1.39) catalyzes the rate limiting steps in the diverging pathways of photosynthetic carbon assimilation and photorespiration. Rubisco is most likely the most abundant protein in the biosphere and it is definitely the most extensively studied plant enzyme. Recent developments in research on Rubisco have resulted in new information about several aspects of the enzyme. These include the determination of the structure of two Rubisco enzymes with different subunit stoichiometry, the identification of auxiliary systems for enzyme assembly and activation, and the application of molecular biology techniques for genetic engineering and analysis of the reactions catalyzed by Rubisco. This survey covers such recent developments while we refer to several other reviews[1,2,3] which deal with the intensively studied aspects of subunit biosynthesis, reaction chemistry and activity.

The most common form of Rubisco is composed of eight large subunits (LSU) and eight small subunits (SSU) which are assembled into the holoenzyme, L_8S_8. The LSU contains all known sites for activation and catalysis while the function of the SSU remains elusive. Other forms of Rubisco, known from photsynthetic bacteria, occur as L_2 and L_6, without any SSU. The holoenzyme is activated by covalent binding of activator CO_2 to a lysine residue in the LSU followed by the coordination of Mg^{2+} to yield the "activated ternary complex". This complex then binds the substrate, ribulose-1,5-bisphosphate (RuBP). After this step is completed the holoenzyme is competent to catalyze either the carboxylation reaction or the oxygenation reaction. Carboxylation of RuBP yields two molecules of 3-phophoglycerate and provides substrate for the Calvin cycle reactions. The competing oxygenation reaction produces one molecule each of 3-phosphoglycerate and 2-phosphogylcolate, the latter serves as substrate for the photorespiration pathway. Many of the enzymatic steps have been studied using the transition state analog of the carboxylation reaction, 2-carboxy arabinitol-1,5-bisphosphate (CABP). CABP binds very tightly to the activated enzyme. This tight binding has made CABP a formidable tool in Rubisco research. CABP has been used to assess the activation state, to lock-in the activator molecules, and to quantitate enzyme amount. It has also been used to arrest the enzyme into a structure that is believed to be equivalent to the structure of the enzyme during catalysis of the carboxylation reaction.

Rubisco occurs in all photosynthetic organisms; bacteria, cyanobacteria, algae and higher plants. The organization of the genes, rbcL and rbcS, coding for the LSU and SSU respectively, has been used to trace gene evolution. In photosynthetic prokaryotes the genes coding for the two subunits (when both are present) are encoded in an operon with rbcL preceding rbcS. This organization has been altered in most photosynthetic eukaryotes. In many algae and in higher plants Rubisco subunit genes are distributed in two subcellular compartments, rbcL in the chloroplast genome and rbcS in the nuclear genome. The presumptive migration of rbcS to the nucleus required several changes in Rubisco subunit biosynthesis and assembly of the holoenzyme. These changes involve control of coordinated gene expression, transport of the SSU into the chloroplast, helper proteins for assembly of the holoenzyme, and control of enzyme activation and activation status with regard to cytosolic control mechanisms. We will discuss new results in Rubisco research with focus on the crystallographic structure of holoenzymes, assembly and activation of the eukaryotic enzyme and genetic engineering of Rubisco.

II. STRUCTURE OF RUBISCO

Two types of Rubisco with different structures have been distinguished, form I and form II. Form I is the most common and consists of eight LSU and eight SSU assembled into the holoenzyme. Form I Rubisco is found in some photosynthetic bacteria and in cyanobacteria, algae and higher plants. Form II Rubisco is comprised of only LSU molecules, examples are enzymes from the photosynthetic bacteria *Rhodospirillum rubrum* (L_2) and *Rhodobacter sphaeroides* (L_6). Information has recently been obtained for the structure of the *R. rubrum* L_2 Rubisco and the tobacco L_8S_8 enzyme.

A. *R. rubrum* Rubisco Structure

R. rubrum Rubisco is a LSU dimer with a molecular weight of approximately 100 kDa. The *R. rubrum* Rubisco structure was determined using enzyme which was synthesized from the cloned gene in *E. coli*. The structure of the non-activated enzyme was determined to 2.9 Å resolution[4]. This L_2 enzyme has the shape of an elongated cylinder with the approximate dimensions of 50 x 72 x 105 Å (Fig 1). The two subunits

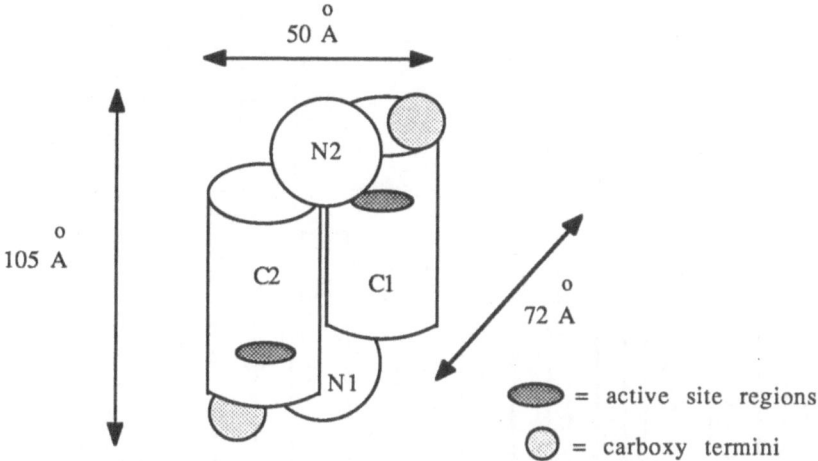

Figure 1. Schematic presentation of the *R. rubrum* L$_2$ Rubisco structure. Each subunit of the dimer contains two domains. The smaller N terminal domain is represented by an open circle and the larger C terminal domain is indicated by a cylinder. The C terminal extension is shown as a shaded circle. The active site is located at the C terminal ends of the β-strands of the α/β barrel.

are assembled into a 'head to tail' arrangement. Each subunit has two domains, a smaller amino-terminal (N terminal) domain and a larger carboxy-terminal (C terminal) domain. The C terminal domain contains the main structural feature of the protein, an eight stranded α/β barrel similar to barrels found in a number of functionally non-related proteins[5].

The *R. rubrum* Rubisco LSU is 466 amino acids (aa) long. The N terminal domain contains 137 aa and is a five stranded mixed β-sheet with three α-helices parallel to the β-sheet. This five stranded β-sheet has three parallel β-strands and two antiparallel β-strands (Fig. 2). The three α-helices occupy both sides of the β-sheet with two on one side and one on the opposite side. The larger C terminal domain consists of 329 aa and contains an eight stranded α/β barrel and three α-helices at the C terminal end (Fig. 2). The barrel can be imagined as a ring of eight parallel β-strands surrounded by, and alternating with, eight α-helices. Residues in the loops that make the strand to helix connections are important for intersubunit interactions and formation of the active site.

The previously identified active site residues are located at the carboxy-terminal ends of β-strands or in the loops which connect the β-strands to the α-helices of the barrel[4]. Lysine 166 is located in loop 1, lysine 191 is the last residue of β-strand 2 and lysine 329 is part of loop 6. The amino group of lysine 191 bonds the activator CO_2 molecule to form a carbamate during enzyme activation. This carbamate is stabilized by Mg^{2+}. Two conserved acidic rsidues, aspartate 193 and glutamate 194, are very close to the carbamated lysine. These residues may be involved in metal binding[6]. The side chain of histidine 287 is close to, and may

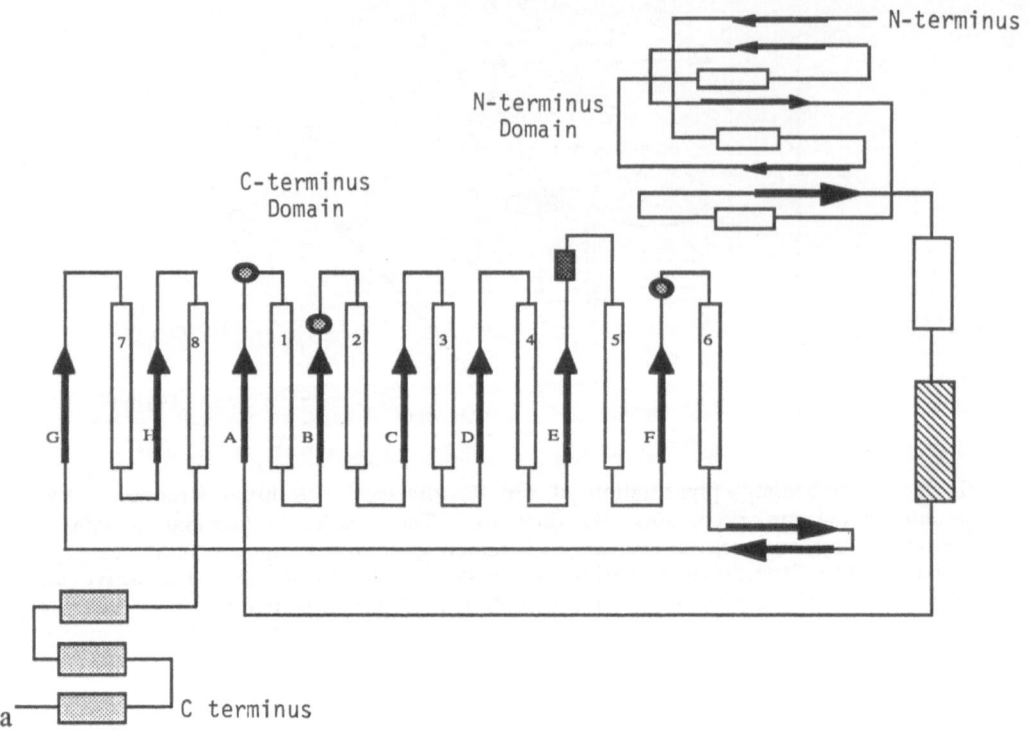

Figure 2. Connectivity diagram of the *R. rubrum* L$_2$ Rubisco LSU redrawn from Schneider et al. (ref. 8). β-strands are represented by arrows, α-helices by boxes, and active site lysine residues by circles. The 8 β-strands of the barrel are designated A to H and the 8 α-helices are numbered 1 to 8. The α-helix which closes off one end of the barrel is shown as the cross-hatched box. β-strand number 5 of the N terminal domain is located close to the pair of strands from loop 6, these are represented as large arrows. The α-helix that provides residues for the phosphate binding site of the substrate is indicated by the shaded box located in loop 5.

also be part of, the Mg-binding site. Loop 5 of the α/β barrel has an additional α-helix which provides residues that bind the phosphate group of the substrate (Fig. 2). An additional α-helix, providing residues to bind a phosphate group of the substrate, has been found in other enzymes which contain α/β barrels[4].

The electron density maps suggest that there are few interactions between residues within one subunit. Extensive interactions are, however, suggested at the interface between the two subunits. The two most important intrasubunit interactions are an α-helix which closes off one opening of the barrel and interactions with a pair of β-strands extending from loop 6 of the α/β barrel. The α-helix lies in the region connecting the N terminal and C terminal domains (Fig. 2) and is located at the N terminal end of the eight β-strands and closes off one end of the barrel. The position of a helix covering one end of the barrel has been observed in the structure of phospho-gluconate aldolase[4]. The second intrasubunit interaction occurs between the additional antiparallel pair of β-strands

310

from loop 6 (Fig. 2). This pair of β-strands are nearly parallel to the N terminal β-sheet and located close to β-strand 5 of the N terminal domain.

Interactions between the two subunits are found in two regions. One set of interactions occurs between the C terminal domains of the sub-units. The other set of interactions are formed from the N terminal domain of one subunit to the C terminal domain of the second subunit. Interactions between C terminal domains involve residues in loops 3, 4, and 5 of the α/β barrel and occur as homologous interactions across a local 2-fold axis with corresponding residues in the second subunit. Interactions between the N terminal domain of one subunit to the C ter-minal domain of the other subunit involves residues in loops 1, 2, and 3. The interdomain interaction contributes amino acids essential for the formation of the active site. The location of the active site at the interface of two subunits has been observed in other enzymes, for example glu-tamine synthetase[7].

B. Tobacco Rubisco Structure

The structure of a form I Rubisco at a resolution of 2.6 Å has recently been reported for the non-activated tobacco enzyme[8]. The L_8S_8 holoenzyme is composed of an octameric core of four L_2 units each of which resemble the *R. rubrum* L_2 Rubisco structure. The SSUs of the tobacco enzyme are arranged as 2 tetramers located at both poles of the L_8 core.

The structure of a single tobacco LSU (477 aa) is very similar to the *R. rubrum* LSU with a smaller N terminal domain and a larger C terminal domain. The main structural feature of the protein is, as in the L_2 form of the enzyme, an eight stranded α/β barrel. The nomenclature for the tobacco structure differs slightly from that of the bacterial enzyme. Chapman et al.[8] consider each of the two LSU domains to have C terminal extensions thereby yielding the assignment of 4 regions; N terminal (residues 5-134), connector (135-168), barrel (169-432), and C terminal (433-477). The N terminal domain is comprised of a four stranded anti-parallel β-sheet with flanking α-helices. Residues in the connector region between N terminal and barrel domains include a short helix which closes off one end of the barrel. The barrel domain consists of an eight stranded α/β barrel which is similar to that of the *R. rubrum* LSU. Additional structural elements are encoded in some of the loop regions. Some active site residues are located in the loops at the C terminal ends of the β-strands. Active site residues are also located in the N terminal domain of the LSU comprising the LSU pair. The C terminal domain contains an α-helix. The C terminal extension of LSU runs between and above the SSU to form the extreme top, and bottom, of the Rubisco holoenzyme.

In summary, the structure of the individual tobacco LSU and the L_2 units of the tobacco enzyme are similar to that of the *R. rubrum* Rubisco structure. However the tobacco enzyme has the added complexity that the four L_2 units are assembled into an L_8 core. The tobacco enzyme includes in addition the structures and interactions of the L_8 core with the eight SSU molecules. One might expect that differences in the LSU structures and interactions, comparing the L_2 and L_8S_8 enzymes, are involved in the more complex assembly and activation pathways of the L_8S_8-type Rubisco.

The structure of a single tobacco SSU (123 aa) consists of an anti-parallel β-domain of four β-strands with flanking α-helices. The SSU has a long C terminal extension which extends down the center of the holo-enzyme and contacts the back of the barrel domain of a LSU. Residues of SSU may contribute to the environment of the active site. The pattern of intersubunit interactions in complicated. Each LSU contacts four other LSUs with the most extensive interactions with the LSU comprising the L_2 pair. Each LSU contacts three different SSUs while each SSU is in contact with two other SSUs and three LSUs. Interactions from SSU to LSU occur to all four regions of neighboring LSUs. In the barrel domain of the LSU contact is made to two different SSUs. It is this complex interaction between subunits in the assembled enzyme which appears to confer long-range conformational stability to the active site that is the reason for the profound influence of SSU on enzyme activity (see below).

It has to be noted that for identification of contact regions Chapman et al.[8] have defined contact as a difference in the solvent accessible surface area between computations performed in the presence and absence of neighboring domains. The authors state that this operational definition of "contact" may be regarded as generous. Other Rubisco X-ray crystallographic structures are currently underway (G. Schneider, C.I. Branden, personal communication). Rubisco structures can be expected for activated tobacco enzyme with bound transition state analog (CABP), spinach Rubisco with and without CABP, and bacterial enzymes from *Alcaligenes* and *Chromatium* .

C. α/β Barrels in Protein Structures

The structure resembling a barrel composed of alternating β-strands and α-helices has meanwhile been reported for 14 enzymes[5]. Proteins which form α/β barrels have little similarity in amino acid sequence and are probably unrelated. The first demonstration came from the structure of triose phosphate isomerase (TIM)[9]. The α/β barrel structure consists of eight tilted, parallel β-strands alternating with eight α-helices. Residues in the loops that connect the strands to helices have been shown to be important in interactions between domains and formation of the active sites. The basic barrel structure is modified in several ways to allow different specificities in particular proteins[5]. In some examples, such as TIM, the subunits are formed of a single barrel domain. In other examples, such as Rubisco, there are other domains comprising the subunit. One example, N-(5'-phosphoribosyl) anthranilate isomerase-indole-3-glycerol-phosphate sythase, has two α/β barrels in one

312

monomeric subunit[10]. These two barrels catalyze consecutive reactions in the biosynthetic pathway of tryptophan. The cross section of the barrels varies in shape from circular to elliptical. In some enzymes additional structural elements are coded in the loops that make the connections from strand to helix and from helix to strand. A phosphate binding site located in an additional α-helix has been observed in several enzymes, for example Rubisco, glycolate oxidase and TIM[4]. In all cases the enzyme active site is located at the carboxy-terminal end of β-strands. Two enzymes with α/β barrels, enolase and pyruvate kinase which catalyze consecutive reactions in the glycolytic pathway, may share a common ancestor based on similarities in the geometry of the active sites[11]. However, sequence similarities of other proteins with barrels are probably due to structural requirements.

III. ASSEMBLY OF RUBISCO

Rubisco holoenzymes are multisubunit enzymes which must be assembled from subunit monomers. This assembly pathway is essentially unknown. Circumstantial evidence suggests that the pathway may be simple in prokaryotes, while it appears to be complex in eukaryotic organisms. In photosynthetic prokaryotes the Rubisco genes are coded in an operon and are co-transcribed[12,13]. The subunit proteins are sequentially synthesized and apparently assemble spontaneously. The production of active bacterial and cyanobacterial Rubisco holoenzyme, following the expression of cloned genes in *E. coli*, might imply that auxiliary proteins are not necessary for prokaryotic Rubisco assembly. Alternatively, it also is conceivable that *E. coli* cells contain proteins that may functionally replace the proteins necessary for Rubisco assembly in photosynthetic bacteria or cyanobacteria. In most photosynthetic eukaryotes the two Rubisco subunits are encoded in different subcellular compartments (Fig. 3). The LSU is encoded by chloroplast DNA and synthesized in the chloroplast. The SSU is nuclear encoded and synthesized as a precursor protein which has to be post-translationally imported into the chloroplast. During or after import the SSU precursor is proteolytically processed to remove its transit peptide and the mature portion of the SSU is released into the chloroplast. The subsequent assembly is known to be rapid and complete since there are no substantial pools of unassembled subunits[14]. It is not known whether modifications to the LSU, proteolytic processing and acetylation (see Activation section), occur before or after assembly of the holoenzyme. Studies of Rubisco assembly have included experiments involving dissociation of holoenzyme and subsequent reassociation of soluble subunits, examination of newly synthesized LSU in isolated organelles (which led to the identification of the Rubisco subunit binding protein), and examination of precursor SSU imported into isolated chloroplasts.

A. Dissociation-Reassociation of Cyanobacterial Rubisco

Cyanobacterial Rubisco can be dissociated into soluble L_8 cores and

SSU monomers[15,16]. The isolated L_8 cores have little carboxylase activity, which is due to 2-4% residual SSU that apparently cannot be removed. The L_8 cores can bind the transition state analog CABP[17,18]. Reassociation of the L_8 core and monomeric SSU readily occurs to reform active holoenzyme. The amount of added SSU, at sub-saturating molar ratios, corresponds to the increase in carboxylase activity thereby identifing the essential role of SSU in enzyme activity. The isolated L_8 cores can assemble with heterologous SSU, either from a different cyanobacteria or from higher plant Rubisco[19,20]. Heterologous Rubisco comprised of a cyanobacterial LSU and spinach SSU had a specific activity of about half the cyanobacterial enzyme and a $K_m(CO_2)$ about 2 times that of the native enzyme[20]. The specificity factor (a species-specific measure of the ratio by which carboxylation is preferred over oxygenation) of the heterologous enzyme was identical to the native enzyme. This is the best indication that the LSU contains the information for determining the specificity of the enzyme. One limitation of the reassociation system in the study of assembly is that conclusions are drawn about specific SSU aggregation with preassembled L_8 cores. LSU assembly can not be examined as the LSU is isolated and used as the L_8 core. In addition, this type of experiment can only be performed with cyanobacterial Rubisco. Dissociation of higher plant Rubisco inevitably yields an insoluble aggregate of LSU[21], presumably of varying aggregation status, while SSU is soluble and can be isolated free of LSU. Using expression systems for cyanobacterial LSU it is now possible to examine the properties of LSU in the complete absence of SSU[22]. The LSU, in the absence of SSU, aggregates to a prefered octameric stoichiometry. This LSU can bind the transition state analog CABP. LSU alone catalyzes the carboxylase reaction with a k_{cat} equivalent to 1% of the holoenzyme[22]. As carboxylation can take place in the absence of SSU it therefore cannot be directly involved in catalysis.

B. Rubisco Subunit Binding Protein

Experiments employing protein synthesis by isolated chloroplasts identified the LSU as an abundant product of biosynthesis. The LSU is found in three different multisubunit complexes of sedimentation coefficients (and approximate molecular weights) of 29S (720 kDa), 18S (550 kDa holoenzyme), and 7S (117 kDa). Assembly of newly synthesized LSU into holoenzyme was dependent on the osmoticum used in the experiment, use of sorbitol allowed assembly while use of KCl prevented assembly[23]. Addition of MgATP caused dissociation of the 720 kDa complex in dilute chloroplast extracts[24]. The dissociated subunits were not phosphorylated or adenylated and the dissociation was reversible when the concentration of ATP was lowered[25]. Time course experiments suggest that labeled LSU moves from the 720 kDa complex into the 117 kDa complex and subsequently into the holoenzyme.

The large subunit binding protein, BP (now simply called binding protein as it also binds SSU[26]), had been identified as a component of the 720 kDa complex and possibly the 117 kDa complex. The BP has been purified from pea[25] and barley[27]. The protein has been examined by

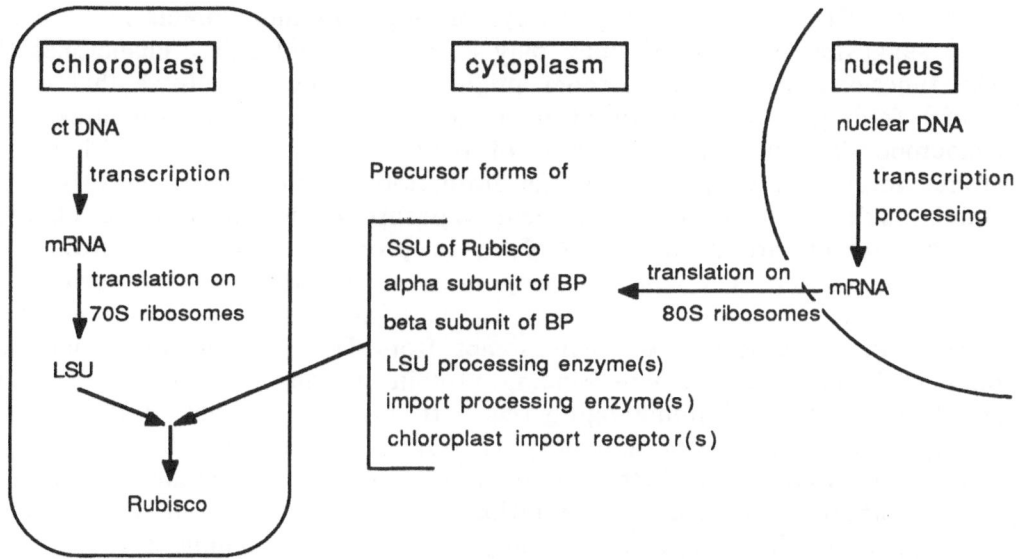

Figure 3. Diagram of the components of the Rubisco synthesis and assembly pathways.

partial proteolysis, N terminal peptide sequencing and antibody production (against the pea BP) and reactivity. These experiments demonstrated that BP is a distinct protein, not an aggregated form of Rubisco LSU. Two different types of BP were found and identified as the α-subunit (61 kDa) and β-subunit (60 kDa)[25]. These two subunits are immunologically distinct, have different partial protease digestion patterns and different N terminal sequences. These two BP subunits occur in equal amounts implying that the stoichiometry of the 720 kDa complex is α_6-β_6. More recently these authors have speculated that the stoichiometry of the BP complex may be α_7-β_7 based on an analogy to the bacterial *groEL* gene product[28]. The BP subunits are nuclear encoded and post-translationally imported into the chloroplast (Fig. 3). The α-subunits of BP from castor bean and wheat have been cloned and sequenced[28]. The BP subunits have homology to the bacterial *groEL* gene product, an essential *E. coli* protein initially identified because it is required for phage lambda and T4 head assembly. The *groEL* protein is a 14 subunit homo-oligomer composed of 2 stacked rings of 7 subunits each. The *groEL* product acts as an ATPase and the *groEL* and *groES* products form a complex with each other in the presence of MgATP but not in its absence.

Distribution of the BP has been examined immunologically, evidence for BP has been found in a number of higher plants and in some bacteria, even some bacteria that do not contain Rubisco. Extracts of the cyanobacterium *Anacystis* failed to show reaction to the pea BP antibodies[26]. Antibodies to the BP inhibit assembly of the LSU into Rubisco suggesting that assembly-competent LSU are associated with the BP[29]. The BP may be obligatory for Rubisco assembly but definite evidence is lacking. The BP does however appear to be a participant in the reactions of the LSU. One hypothesis that has been advanced is that the BP might function to keep the LSU soluble.   Alternatively, or probably more precisely, it has been suggested that the BP may play a role as molecular "chaperones"[30]. The functions of such chaperones might be defined as helpers that ensure proper folding and assembly of monomers into oligomeric structures.   The chaperone does not becoming part of the assembled structure.   Chaperones might act in preventing the formation of improper interactions which could arise from the transient exposure of hydrophobic or charged surfaces which are normally involved in protein-protein interactions. There are examples where auxiliary proteins are required for correct assembly of oligomeric structures[30].   These include the action of nucleoplasmin in the assembly of nucleosomes from DNA and histones and the immunoglobulin heavy chain binding protein in the assembly of the light and heavy chains of immunoglobulins.   It has been suggested that assisted post-translational assembly of oligimeric proteins is a general cellular phenomenon and that proteins that act as molecular chaperones may be ubiquitous.   This would reflect a general and essential function for chaperones in catalyzing assembly of oligomeric complexes.

C. Import Into Chloroplasts

Uptake experiments using labeled precursor SSU imported into isolated chloroplasts have been conducted to examine the post-translational transport process.   It has been shown that the transit peptide is necessary and sufficient for import and that import is an energy dependent reaction[31].   Several import experiments have yielded results that are pertinent to the study of the Rubisco assembly pathway.   Chua and Schmidt reported that newly transported pea SSU assembled with endogenous pea LSU into holoenzyme[32].   Formation of heterologous Rubisco was reported using transformed plant cells, pea SSU (and petunia SSU) assembled with petunia LSU to form mixed holoenzyme in transformed tissue[33].

We have used import assays to determine the function of a higher plant Rubisco SSU specific region[34].   When sequences of SSU (see Genetic Engineering section) are compared between cyanobacteria and higher plants only one feature distinguishes the SSU from the two groups.   This distinguishing character is the presence of a 12-16 aa insert present only in higher plant SSUs.   Precursor forms of chimeric SSUs that either contained or lacked this insert were constructed and expressed in *in vitro* systems.   Labeled precursor protein was examined in protein uptake assays into chloroplasts for characteristics of import, processing and assembly.   From the results it could be deduced that chimeric SSU lacking the higher plant SSU specific insert cannot assemble with endogenous pea LSU.   In contrast, a cyanobacterial SSU, to which the higher plant-specific insert was added, was made competent for assembly with higher plant LSU.   This region of higher plant SSU, which in some genes constitutes a separate exon[35], is thus defined as a SSU assembly domain, a necessary component of SSU for Rubisco assembly in organisms where the subunit proteins are encoded in different compartments.

IV. ACTIVATION AND ACTIVITY OF RUBISCO

Rubisco is activated by carbamation of a lysine residue (lysine 191 in the L_2 enzyme) in the LSU and the coordination of a divalent metal ion, preferably Mg^{2+}. The non-activated higher plant enzyme which binds the substrate RuBP is essentially removed from the activatible enzyme pool[36,37]. This simple statement, however, hides a very complex process. Rubisco is present in the chloroplasts of higher plants in large amounts, estimated to be 4-8 mM. A variable, often substantial, amount of holo-enzyme is inactive and is thus not participating in any productive reactions. This behaviour necessitates a regulatory mechanism of activation which is one of several mechanisms that must be important for higher plants in control of Rubisco activation state. Cyanobacterial Rubisco is not inhibited by substrate binding to non-activated enzyme[38]. Two regulatory processes have recently been identified that affect Rubisco enzyme activation. One of these is the metabolism of a natural inhibitor of Rubisco, the second involves an enzyme, Rubisco activase. Recent progress in the study of the activation process has revealed a picture that allows us to begin to merge many isolated observations into a unified hypothesis.

A. Natural Inhibitor of Rubisco

The natural Rubisco inhibitor 2-carboxy arabinitol-1-phosphate (CA1P) binds at the active site of the activated Rubisco enzyme during periods of reduced light and in the dark. The inhibitor was discovered by comparing the activity of Rubisco isolated from light and dark grown leaves[39,40]. The inhibitor was purified and its structure determined showing it to be a phosphorylated sugar (Fig. 4) similar in structure to the substrate RuBP and the transition state analog CABP[41,42]. The inhibitor binds activated Rubisco more tightly, $K_d(CA1P) = 3.2 \times 10^{-8}$ M[42], than the substrate, $K_m(RuBP) = 1$ to 13×10^{-5} M[43], but not as tightly as CABP, $K_d(CABP) = 10^{-11}$ M[44]. The value for $K_d(RuBP)$ to inactive enzyme is 2×10^{-8} M[36]. Inhibitors are rapidly released from the enzyme upon exposure to light[39]. The binding of the inhibitor presumably maintains the enzyme in the activated state without turnover. This suggests that the enzyme complexed with CA1P will be competent upon emergence from darkness. The biosynthetic pathways leading to the synthesis of the inhibitor, its regulation, release and degradation are not yet known, nor is the variable distribution of the inhibitor among different plant species[45]. In some plants the inhibitor is present in amounts greater than the Rubisco active site concentration, while in other species it is present in sub-saturating amounts[46]. A recent report shows that Rubisco activase can cause the release of the natural inhibitor from the dark-inactivated enzyme[47].

B. Rubisco Activase

Rubisco activase was originally discovered to be responsible for a mutation found in *Arabidopsis*[48] that contained inactive Rubisco holo-

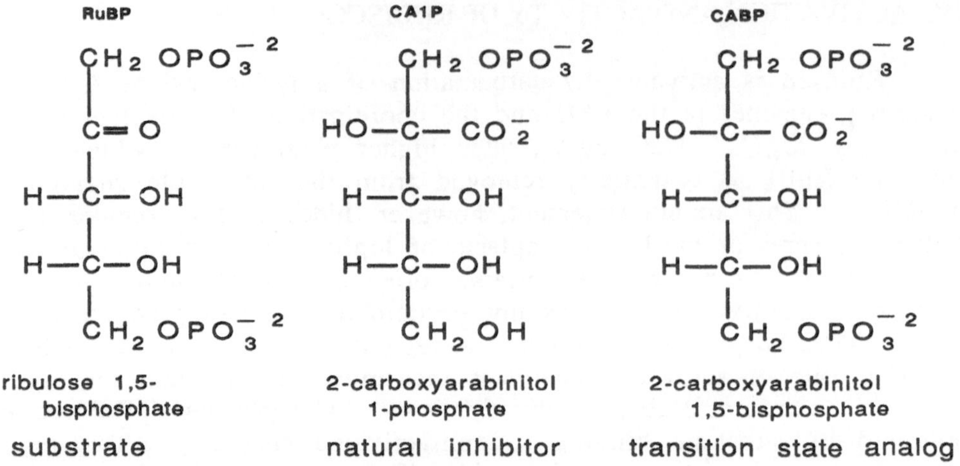

RuBP

$$CH_2\,OPO_3^{-2}$$
$$|$$
$$C=O$$
$$|$$
$$H-C-OH$$
$$|$$
$$H-C-OH$$
$$|$$
$$CH_2\,OPO_3^{-2}$$

ribulose 1,5-
bisphosphate

substrate

CA1P

$$CH_2\,OPO_3^{-2}$$
$$|$$
$$HO-C-CO_2^-$$
$$|$$
$$H-C-OH$$
$$|$$
$$H-C-OH$$
$$|$$
$$CH_2\,OH$$

2-carboxyarabinitol
1-phosphate

natural inhibitor

CABP

$$CH_2\,OPO_3^{-2}$$
$$|$$
$$HO-C-CO_2^-$$
$$|$$
$$H-C-OH$$
$$|$$
$$H-C-OH$$
$$|$$
$$CH_2\,OPO_3^{-2}$$

2-carboxyarabinitol
1,5-bisphosphate

transition state analog

Figure 4.   Representations of the structures of sugar phosphates which are important in the biology and chemistry of Rubisco.

enzyme. This protein has been purified and consists of two polypeptides of 44 and 41 kDa. Antibodies have been prepared[49] and it has been shown that the two proteins are cross-reactive[50]. One or both of these subunits have been found in all higher plant species examined[49]. Rubisco activase is a nuclear encoded chloroplast protein. The gene has been cloned and sequenced[50]. Expression of the cloned gene in *E. coli* produced two immunoreactive fusion proteins with apparent molecular masses nearly identical to Rubisco activase isolated from spinach leaves[50]. The 45 kDa polypeptide has been purified from *E. coli* extracts and shown to have Rubisco activase activity[51]. The activase protein has kinase activity in the presence of Rubisco, ATP, light, and thylakoids[52]. It has been suggested that the activase protein may act by displacing RuBP from non-activated Rubisco. A recent report showed that inactivated enzyme, to which substrate is bound, can be activated by Rubisco activase[53].

C. Modifications of Rubisco LSU

A point of controversy has been the exact molecular weight of LSU. Different size classes of LSU can be obtained and it has been shown that some of these isoforms are isolation artefacts or represent turnover of LSU dependent on leaf age, physiological status and developmental stage. There are, however, processing and modification steps of functional significance. Mulligan et al.[54] showed loss of enzymatic activity of trypsin treated Rubisco with little corresponding loss in CABP binding properties. Partial tryptic digestion resulted in cleavage of three fragments of the LSU originating from the amino-terminus (residues 3-8 and 9-14) and carboxy-terminus (467-475). Treatment of enzyme that had been activated and bound CABP yielded only the two N terminal peptides. Loss of susceptibilty of lysine 466 to trypsin reflects altered conformation of the LSU when CABP is bound to the enzyme. Analysis of the amino-terminus revealed post-translational modification occurs. In spinach Rubisco the LSU is modified by proteolytic processing, to remove methionine 1 and serine 2, and acetylation of the resulting N terminal residue, proline 3.

```
         1         10          20            30           40           50           60
Rr    M.DQSS.... ........RY  VNLALKEEDL   IAGGEHVLCA   YIMKPKAGYG   YVATAAHFAA   ESSTGTNVEV
An    M.PKT.QSA. AGTKAGVKDY  .KLTYYTPDY   TPKDTDLLAA   FPVSPQPGVP   ADEAGAAIAA   ESSTGTWTTV
Nt    MSPQTETKAS VGFKAGVKEY  .KLTYYTPEY   QTKDTDILAA   FRVTPQPGVP   PEEAGAAVAA   ESSTGTWTTV
con   =  -  -    -  ----- =  -=------ -         -=-   - - - -=---   ---=- ==  ========---=
2°                                       <--βN1-->     <--αN1>


        70          80          90           100          110          120          130
Rr    CTTDDFTR.G V.DALVYEVD  EARELTKIAY   PVAL.FDRNI   TDGKAMIASF   LTLTMGNNQG   MGDVEYAKMH
An    .WTDLLTDMD RYKGKCYHIE  PVQG.EENSY   FAFIAYPLDL   FEEGSVTNIL   TSI.VGNVFG   FKAIRSLRLE
Nt    .WTDGLTSLD RYKGRCYRIE  RVVG.EKDQY   IAYVAYPLDL   FEEGSVTNMF   TSI.VGNVFG   FKALRALRLE
con   -== -=  -  - ------=  --   - -  =  -  - ------- ---------  --- -==-=- --- - ----
2°              <-βN2>       <--βN3-->    <--αN2-->              <-βN4>


       140         150         160          170          180          190          200
Rr    DFYVPEAYRA LFDGPSVNIS  ALWKVLGRPE   VDGGLVVGTI   IKPKLGLRPK   PFAEACHAFW   LGG.DFIKND
An    DIRFPVALVK TFQGPPHGIQ  VERDLLTK..   .YGRPMLGCT   IKPKLGLSAK   NYGRAVYECL   RGGLDFTKDD
Nt    DLRIPPAYVK TFQGPPHGIQ  VERDKLNK..   .YGRPLLGCT   IKPKLGLSAK   NYGRAVYECL   RGGLDFTKDD
con   =  -  - =  -=-=-=--=--  ---- ⬅ -      -=-- -==- ========== ---=-=---- ---=-=--
2°              <-αN3->                   <-βB1>                 <----αB1--->   <βB2>


       210         220         230          240          250          260          270
Rr    EPQGNQPFAP LRDTIALVAD  AMRRAQDETG   EAKLFSANIT   ADDPFEIIAR   GEYVLETFGE   NASHVALLVD
An    ENINSQPFQR WRDRFTFVAD  AIHKSQAETG   EIKGHYLNVT   APTCEEMMKR   AEFAKE.LGM   PITMHDELTA
Nt    ENVNSQPFMR WRDRFLFCAE  ALYKAQAETG   EIKGHYLNAT   AGTCEEMIKR   AVFARE.LGV   PIVMHDYLTG
con   =-- --==== -   -==--- = =   - =-=== =-==---= = = ----=- -=  - -- --    -- --- =-
2°              <------αB2------->        <-βB3->     <--αB3--->    <-βB4>


       280         290         300          310          320          330
Rr    GYVAGAAAIT TARRRFPDNF  LHYHRAGHGA   VTSPQSKRGY   TAF.VHCKMA   RLQGASGIHT   GTMGFGKMEG
An    GFTANTT.LA KWCRD.NGVL  LHIHRAMH.A   VIDRQKNHG.   IHFRVLAKLL   RLSGGDHLHS   GTVV.GKLEG
Nt    GFTANTS.LA HYCRD.NGLL  LHIHRAMH.A   VIDRQKNHG.   IHFRVLAKAL   RMSGGDHIHS   GTVV.GKLEG
con   =--=-- -- -=- -- - ==-=-==-= = =---==- --=-=-= - = ---- =- ==-=- ==-=-
2°    <αB4>     <--βB5->                  <--αB5--->   <-βB6->       <


       340         350         360          370          380          390          400
Rr    ESSDRAIAY. .MLTQD..EA  Q...GPFYRQ   SWGGMKACTP   IISGGMNALR   MPGFFENLGN   ANVILTAGGG
An    DK.ASTLGFV DLMREDHIER  DRSRGVFFTQ   DWASMPGVLP   VASGGIHVWH   MPALVEIFGD   DSV.LQFGGG
Nt    ER.DITLGFV DLLRDDFVEQ  DRSRGIYFTQ   DWVSLPGVLP   EASGGIHVWH   MPALTEIFGD   DSV.LQFGGG
con   -----  -- -  --- ----  -=- --  = - ---=== -= --- ===-=-- ==- =-=--- -= ==-==≡
2°    ----αB6------>  <βB6a>            <-βB6b>        <βB7>        <αB7>  <-βB8>


       410         420         430          440          450          460          470
Rr    AFGHIDGPVA GARSLRQAWQ  A.WRDGVPVL   DYAREHKELA   RAFESFPGDA   DQIYPGWRKA   LGVEDTRSALPA
An    TLGHPWGNAP GATANRVALE  ACVQARNEGR   DLYREGGDIL   REAGKWSPEL   AAALDLWKEI   KFEFETMDKL
Nt    TLGHPWGNAP GAVANRVALE  ACVKARNEGR   DLAQEGNEII   REACKWSPEL   AAACEVWKEI   VFNFAAVDVLDK
con   --=≡--=-- == --=-=-= =-- ------ =- =- - =-- ----- --- =---
2°    <----αB8---->  <αC1>
```

Figure 5. Amino acid sequence comparison of Rubisco LSUs from *R. rubrum* (line 2, ref. 75), *Anacystis* (line 3, ref. 76), and tobacco (line 4, ref. 77). The numbers in line 1 refer to the tobacco sequence. Dashes have been added to optimize alignment. The symbols in line 5 indicate amino acid identity between all three sequences (=) or between *Anacystis* and tobacco (-). Regions of defined secondary structure are indicated in line 6 using the nomenclature of Chapman et al. (ref. 8).

V. GENETIC ENGINEERING OF RUBISCO

A number of Rubisco genes have been cloned and sequenced including genes from all classes of photosynthetic organisms; photosynthetic bacteria, cyanobacteria, algae, and higher plants. Sequence comparisons show the LSUs to be highly homologous; greater that 90% aa identity is found among higher plants and 80-90% between cyanobacterial and higher plant subunits (Fig. 5). The homology between R. $rubrum$ L_2 Rubisco and LSU from L_8S_8-type Rubisco is only about 25%. However, the homology is nearly absolute in regions identified to contain active site residues. In general, homology is high in the loop regions of the α/β barrel structure of the LSU[8] (Fig. 5). The SSU show considerable sequence identity, 70-90% among higher plants and 40-50% between cyanobacterial and higher plant proteins. Within the SSU three regions of particularly high homology can be distinguished. These regions in cyanobacterial as well as higher plant sequences are labeled CS1, CS2, and CS3 in Fig. 6. The mature SSU ranges in size from 109 to 140 aa; in most higher plants the SSUs are 123 aa long. A unique feature of higher plant SSU is a 12-16 aa insert, which was recently identified as an assembly domain[34] (see Assembly section).

Active Rubisco has been produced from bacterial and cyanobacterial genes expressed in $E. coli$. Form II Rubisco (comprised of LSU only) has been expressed from the cloned genes from $R. rubrum$[55] and $R. sphaeroides$[56,57]. Active form I Rubisco (L_8S_8 enzyme) has been expressed from the naturally occuring operons from $R. sphaeroides$[58], $Chromatium$[59], $Anacystis$[60-62], and $Anabaena$[63]. Enzymatically active heterologous enzyme has also been produced from expression of $Anacystis$ LSU and wheat SSU[64]. A current limitation for examination of the L_8S_8 Rubisco is the lack of X-ray crystallographic structure information for the cyanobacterial enzyme. The structure of the tobacco Rubisco is published[8] but higher plant LSU cannot be obtained in active form from expression in $E. coli$ systems. Although expression of higher plant LSU can be achieved, it yields insoluble material which cannot be solubilized and cannot assemble into active holoenzyme[65]. This problem is not circumvented when higher plant LSU is expressed in $E. coli$ in the presence of SSU[66].

A. Mutants of *R. rubrum* Rubisco

A number of oligonucleotide-directed mutations have been generated in the $rbcL$ gene of the $R. rubrum$ L_2 Rubisco (Fig. 7). These experiments were started before structure information was available. As far as the results are published, the residues initially targeted were those identified by chemical modification techniques. Mutagenesis of the active site residues yielded the predictable results of functional impairment of the enzyme. Molecular biology techniques can now be applied with the benefit of crystallographic structure information to identify functional residues and determine the action of the enzyme in much greater detail.

B. Mutants of *Anacystis* Rubisco

The mutagenesis of form I Rubisco requires the expression of two subunits. Initial reports of active L_8S_8 Rubisco from expression of cloned genes in $E. coli$ made use of naturally occuring Rubisco operons from cyanobacteria. Subsequent reports have shown that it is possible to

```
               1        10          20          30          40          50          60          70
                  <<---CS1---->>                                               <<--HPIN---->><<--
Ae    MR-------I TQGTFSFLPE LTDEQITKQL EYCLNQGWAV GLEYTDDPHP RNT------- -----YWEMF
An    MSMKTLPKER RFETFSYLPP LSDRQIAAQI EYMIEQGFHP LIEFNEHSNP EEF------- -----YWTMW
Av    M--QTLPKER RYETLSYLPP LTDVQIEKQV QYILSQGYIP AVEFNEVSEP TEL------- -----YWTLW
Cp    M--QTLAVER KFETLSYLPP LNDQQIARQL QYALSNGYSP AIEFSFTGKA EDL------- -----VWTLW
Eg    MKVWNPVNNK FWETFSYLPP LSDAEIAKQV DMIIAKGWIP CLEFSLREIS ERA---1-- YYDNRYWTMW
Cr    MMVWTPVNNK MFETFSYLPP LTDEQIAAQV DYIVANGWIP CLEFAEADKA YVS---2-- YYDNRYWTMW
Bn    MQVWPPVGKK KFETLSYLPD LTEVELGKEV DYLLRNKWIP CVEFELEHGF VYREHGSTPG YYDGRYWTMW
Zm    MQVWPAYGNK KFETLSYLPP LSTDDLLKQV DYL-RNGWIP CLEFSKV-GF VYRENSTSPC YYDGRYWTMW
Ps    MQVWPPIGKK KFETLSYLPP LTRDQLLKEV EYLLRKGWVP CLEFELLKGF VYREHNKSPR YYDGRYWTMW
Gm    MQVWPPEGKK KFETLSVLPD LDDAQLAKEV EYLLRKGWIP CLEFELEHGF VYREHNRSP- YYDGRYWTMW
Ha    MQVWPPLGLK KFETLSYLPP LTTEQLLAEV NYLLVKGWIP PLEFEVKDGF VYREHDKSPG YYDGRYWTMW
Mc    MQVWPPLGKK KFETLSYLPP LSEESLMKEV QYLLNNGWVP CLEFEPTHGF VYREHGNTPG YYDGRYWTMW
Ft    MKVWPPVGKK KYETLSYLPE LTEAQLAKEV DYLLRNKWVP CLEFELEHGF VYRENASSPG YYDGRYWTMW
Ph    MQVWPPYGKK KYETLSYLPE LTDEQLLKEI EYLLNKGWVP CLEFETEHGF VYREYHASPG YYDGRYWTMW
St    MQVWPPINMK KYETLSYLPD LSDEQLLKEV EYLLKNGWVP CLEFETEHGF VYREHNSSPG YYDGRYWTMW
Nt    MQVWPPINKK KYETLSYLPD LSQEQLLLEP DYLLKDGWVP CLEFETEHGF VYRENNKSPG YYDGRYWTMW
con   = ----    - - -=-=-==  -     --       - - - =-    --      ------ --- ----==---
2°        <-----αS1------>         <-βS1->         <--βS2-> <αS2 -->
```

```
               80         90         100         110         120
                 -CS2---->>                  <<--------CS3-------->>
Ae    GLPMFDLRDA AGILMEINNA RNTFPNHYIR VTAFDSTHTV ESVVMSFIVN RPADEPGFRL continued
An    KLPLFDCKSP QQVLDEVREC RSEYGDCYIR VAGFDNIKQC QTVSFIVHRP GRY
Av    KLPLFGAKTS REVLAEVQSC RSQYPGHYIR VVGFDNIKQC QILSFIVHKP SRY
Cp    KLPLFGTQSP EEVLSEIQAC KQQFPNAYIR VVAFDSIRQV QTLMFLVYKP L
Eg    KLPMFGCTDA SQVLKELSPL EFAAPENFVR LAAFDSVKQV QVISFVVQRP SGSSW
Cr    KLPMFGCRDP MQVLREIVAC TKAFPDAYVR LVAFDNQKQV QIMGFLVQRP KTARDFQPAN KRSV
Bn    KLPLFGCTDS AQVLKEVQEC KTEYPNAFIR IIGFDNNRQV QCISFIAYKP PSFTGA
Zm    KLPMFGCNDA TQVYKELQEA IKSYPDAFHR VIGFDNIKQT QCVSFIAYKP PGSD
Ps    KLPMFGTTDP AQVVKELVAV AAYPEAFVR VIGFNNVRQV QCISFIAHTP ESY
Gm    KLPMFGCTDA SQVLKELQEA KTAYPNGFIR IIGFDNVRQV QCISFIAYKP PGF
Ha    KLPMFGGTDP AQVVNEVEEV KKAYPDAFVR FIGFDNKREV QCISFIAYKP AGY
Mc    KLPMFGCTDP SQVVAELREA KKAYPEAFIR IIGFDNVRQV QCISFIAYKP ASYDA
Ft    KLPMFGCTDS AQVMKELQEC KKEYPQAVIR IIGFDNVRQV QCVSFIASKP TGF
Ph    KLPMFGCTDA TQVLGELQEA KKAYPNAWIR IIGFDNVRQV QCISFIAYKP PGF
St    KLPMFGCTDG TQVLAEVQEA KNAYPQAWIR IIGFDNVRQV QCISFIAYKP EGY
Nt    KLPMFGCTDA TQVLAEVGEA KKAYPEAWIR IIGFDNVRQV QCISFIAYKP EGY
con   -==-=----   --   =   -     --      =    ---=-- --- -- ---- --
2°    <--αS3->              <βS3> <βS4>
```

Ae   VRQEEPGRTL RYSIESYAVQ AGPK

Inserts to the algal SSU sequences;   1   YPCCYIANDNTVRFSGTAAG
                                       2   NESAIRFGSVSCL

Ae *Alcaligenes eutrophus* ATCC17707 (78)   Ps *Pisum sativum* (85)
An *Anacystis nidulans* PCC6301 (79)   Gm *Glycine max* (86)
Av *Anabaena variabilis* PCC7120 (12)   Ha *Helianthus annuus* (87)
Cp *Cyanophora paradoxa* (80)   Mc *Mesembryanthemum crystallinum* (88)
Eg *Euglena gracilis* (81)   Ft *Flaveria trinerva* (89)
Cr *Chlamydomonas reinhardtii* (82)   Ph *Petunia hybrid* (90)
Bn *Brassica napus* (83)   St *Solanum tuberosum* (91)
Zm *Zea mays* (84)   Nt *Nicotiana tabacum* (92)

Figure 6. Amino acid sequence comparison of Rubisco SSUs. The numbers in line 1 refer to the tobacco sequence. The symbols in line 2 identify 3 regions of conserved sequence, labeled CS1, 2 and 3, and the higher plant insert, labeled HPIN. Dashes have been inserted to optimize sequence alignment. The symbols in the line labeled con indicate identity among all sequences (=) or between most of the SSU sequences (-). Regions of defined secondary structure are noted in the last line of each set using the nomenclature of Chapman et al. (ref. 8).

produce active Rubisco when the two subunits are separately cloned into compatible plasmids or when the subunits are separately expressed in different strains.

residue and change	biology, result, and conclusion	reference
Asp198 to Glu	residue near metal binding site, altered environment of metal ion, residue contributes to metal binding	92
Lys191 to Glu	carbamated residue, no stabilization of analog binding by activating conditions, carbamate binds activator metal ion	93
His291 to Ala	test result from chemical modification study, carboxylase activity approximately 40% of wild type, residue is not catalytically essential	94
Met330 to Leu	residue near active site residue Lys329, altered Km's and analog binding, reduced stability of endiol intermediate	95
Lys166 to Gly,Ala,Ser,Gln,Arg,Cys,His	chemically identified active site residue, decreased carboxylase activity, but binds activator CO_2 and analog, residue not required for activation or substrate binding but essential residue	96
Lys166 to Asp	active site residue, no assembly of subunits (crosslinking data also), catalytic site at interface of subunits	97
Glu48 to Gln	chemically identified active site residue, assembly and analog binding but reduced carboxylase activity, catalytic site at interface of subunits	98
Lys166 to Gly Glu48 to Gln	active site residues, as compatible plasmids, 20% carboxylase activity domains from both subunits necessary	99
Lys166 to Cys, Gly Lys 329 to Cys, Gly	active site residues, chemical treatment of Cys mutant partially restores carboxlase activity, requirement of Lys 166 and 329	100
Lys166 to Gly	active site residue, examined in partial reactions, residue not required for hydolysis, maybe for enolization	101
Asp193 to Asn Glu 194 to Gln His 287 to Asn,Lys,Thr	residues near metal binding site, reported in abstract results not reported in abstract	102

Figure 7. List of oligonucleotide-directed mutations prepared in the *R. rubrum* L_2 Rubisco.

    Woodrouw et al.[67] reported the construction of altered genes to express protein which contained site-directed mutations in the *Anacystis* SSU. Two conserved tryptophan residues, tryptophan 54 and tryptophan 57, were replaced by phenylalanine residues to create two single-site mutants and a double-site mutation. These two tryptophan residues are located in the conserved region CS2. Both single site mutants had the same $K_m(CO_2)$ as the wild-type enzyme but showed a reduced V_{max}. This demonstrated that modification of residues in the SSU can have a dramatic effect on Rubisco kinetic parameters.

Mutations in the N terminal portion of the *Anacystis* LSU have been prepared and examined by Ketttleborough et al.[68]. The N terminal 12 aa of the *Anacystis* LSU were replaced by the analogous region (15 aa) of two higher plant LSU's or by an unrelated sequence (6 aa) from the cloning vector. The chimeric LSUs with the N terminus of maize or wheat had similar properties to the native enzyme. The chimera containing an unrelated sequence had much reduced carboxylase activity. This reduction was reflected in a ten fold increase in $K_m(RuBP)$. This study demonstrated that the N terminal portion of the LSU has a function in catalysis.

C. Components of Rubisco Systems

Experiments involving transgenic expression of genes in higher plants have employed components of the Rubisco system. Constructions for expression of genes in higher plants have made use of the promoter of the pSSU and the transit peptide (and portions of the mature sequence) of the pSSU. In one case, inclusion of 23 aa of the mature SSU, in constructions using the transit peptide, was optimal for transport of a foreign protein into the chloroplast[69]. A functional shuttle vector system for the introduction of foreign proteins to the chloroplast has been demonstrated[70,71].

VI. PROSPECTS FOR RUBISCO

We can expect extentions of current research efforts in all areas of research on Rubisco. It will be especially interesting to compare the structure of the activated enzyme with the structures that are presently available. Computer assisted analysis of the published structures has already had some impact on future experiments employing genetic engineering. While "pre-structure"-age sequence comparisons led to guesses about which elements might be important, an examination of the three-dimensional structure can now lead to insights that are more profitably converted into mutagenesis experiments. Genetic engineering of regions surrounding the active site can now be planned much more rigorously and we expect the results to be much more significant. In addition, genetic engineering techniques will be applied to the study of the auxiliary Rubisco proteins; BP, Rubisco activase, factors involved in inhibitor metabolism, LSU processing and modification, and SSU processing.

Rubisco catalyzes the rate limiting step in photosynthetic carbon assimilation; it is the most abundant protein in plants; and it is ubiquitous among photosynthetic organisms and is necessary for life. Rubisco, by its mass and by the reactions it catalyzes, is the most important single enzyme on this planet. As the components of the Rubisco system appear to be worked out by now --although details in every aspect still await in-depth analysis-- exploitation of the potential to engineer this enzyme will, and should, take center stage in the future. As Rubisco is the rate limiting step in photosynthetic carbon assimilation it should be attempted to improve Rubisco by increasing the enzyme turnover number or by decreasing the "non-productive" oxygenase reaction. Genetic engineering techniques may also be applied to investigate whether the competing oxygenase activity of Rubisco serves a useful or essential function.

Considering the specificity factors, which range from 50 to 150, it has been argued that it would be close to impossible to change this ratio by genetic engineering. The atomic structures of CO_2 and O_2 are, after all, very much alike. In fact O_2 is the smaller molecule and engineering of an enzyme to preferentially`prevent its access to the active site will be impossible unless the different charge distribution in the competing sub-strates provides a manipulative handle. Such fine tuned alternatives depend on a much higher resolution of the electron densities of the crystals and on a structure of the activated enzyme. Another route might be the modification of SSU which has a profound effect on activity. As all SSU residues are some distance from the LSU residues in catalytically important regions, SSU must have a long range conformational effect on stabilizing the active sites.

If Rubisco can be improved to yield more product, either by a faster or more selective enzyme, work would remain to optimize the altered metabolism of the plant. The problems are not trivial as the interactions of the Rubisco system affect many aspects of the plant. Will the photo-systems be able to provide more energy to assimilate the extra product? Can the downstream assimilation reactions proceed at an increased rate? Compensating gene expression or adjustments in metabolism may be a part of flexible responses of plants. Improvements in other metabolic pathways, however, may be required to make use of an improved Rubisco.

A recent report[72] from Spreitzer's lab provides evidence that the Rubisco CO_2/O_2 specificity factor can be changed. A strain of *Chlamydomonas reinhardtii* was obtained, after mutagenesis employing a selective screening procedure, that had a higher specificity factor. The change in specificity factor was due to a single mutation in the LSU gene which caused leucine 290 to be replaced by a phenylalanine residue in the LSU protein. This residue, according to the structure of a non-acti-vated enzyme from tobacco, is indeed located close to the active site of the enzyme. Any potential benefit of this mutation is, however, negated by a corresponding decrease in the reaction rate, although it is exactly information of this kind that will help us to obtain a better understanding of the reaction mechanism. The report amply demonstrated that genetic modification of the CO_2/O_2 specificity factor of Rubisco is feasible.

As the most abundant plant protein, Rubisco could be used as a protein source[73] or its activation state could be used as an indicator of plant condition or health. One might consider the recent experiments of Ranty et al.[74] as a step into this direction. These authors examined the synthesis of Rubisco subunits and activity of the enzyme following fungal infection showing that synthesis and acitivity of Rubisco decreased within three days after inoculation and preceded the appearance of other disease symptoms. Rubisco has already served, and will continue to serve, as a model for understanding subunit-subunit interactions (assembly and active site formation), coordinate expression of two genomes (chloroplast and nuclear), and for the transport of proteins accross organelle membranes. While the last three years have probably brought more new results about this enzyme than all years in the past, we are convinced that even this is only a step in the attempts to understand the biochemistry of a plant-specific, exceedingly complex and crucially important reaction for life on this planet.

Acknowledgements: Our work is supported by NSF (DCB-8812191) and by Arizona Agricultural Experiment Station (ARZT#174452).

References

1.   H. M. Miziorko and G. H. Lorimer, Ribulose-1,5-bisphosphate carboxylase oxygenase, Ann Rev Biochem 52:507-535 (1983).
2.   T. J. Andrews and G. H. Lorimer, Rubisco: Structure, mechanisms, and prospects for improvement, in: "The Biochemistry of Plants, Vol. 10," M. D. Hatch and N. K. Boardman, eds., Academic Press, San Diego, CA (1987).
3.   S. Gutteridge and A. A. Gatenby, The molecular analysis of the assembly, structure and function of rubisco, in: "Oxford Surveys of Plant Molecular Biology Vol. 4," B. J. Miflin, ed., Oxford University Press, London (1987).
4.   G. Schneider, Y. Lindqvist, C-I. Branden, and G. Lorimer, Three-dimensional structure of ribulose-1,5-bisphosphate carboxylase/oxygenase from *Rhodospirillum rubrum* at 2.9 Å resolution, EMBO J 5:3409-3415 (1986).
5.   C. Chothia, The 14th barrel rolls out, Nature 333:598-599 (1988).
6.   C.-I. Branden, G. Schneider, Y. Lindqvist, I. Andersson, S. Knight, and G. Lorimer, Structural and evolutionary aspects of the key enzymes in photorespiration; RuBisCO and glycolate oxidase, in: "Cold Spring Harbor Symposia on Quantitative Biology, Vol. LII," Cold Spring Harbor Laboratory (1987).
7.   R. J. Almassy, C. A. Janson, R. Hamlin, N-H. Xuong, and D. Eisenberg, Novel subunit-subunit interactions in the structure of glutamine synthetase, Nature 323:304-309 (1986).
8.   M. S. Chapman, S. W. Suh, P. M. G. Curmi, D. Cascio, W. W. Smith, and D. S. Eisenberg, Tertiary structure of plant RuBisCO: Domains and their contacts, Science 241:71-74 (1988).
9.   D. W. Banner, A. C. Bloomer, G. A. Petsko, D. C. Phillips, C. I. Pogson, I. A. Wilson, P.H. Priddle, A. J.Furth, J. D. MMilman, R. E. Offord, J. D. Priddle, and S. G. Waley, Structure of chicken muscle triose phosphate isomerase determined crystallographically at 2.5Å resolution using amino acid sequence data, Nature 255:609-614 (1975).
10.   J. P. Priestle, M. G. Gruetter, J. L. White, M. G. Vincent, M. Kania, E. Wilson, T. S. Jardetzky, K. Kirschner, and J. N. Jansonius, Three-dimensional structure of the bifunctional enzyme N-(5'-phosphoribosyl) anthranilate isomerase-indole-3-glycerol-phosphate synthase from *Escherichia coli*, Proc Natl Acad Sci USA 84:5690-5694 (1987).
11.   L. Lebioda and B Stec, Crystal structure of enolase indicates that enolase and pyruvate kinase evolved from a common ancestor, Nature 333:683-686 (1988).

12. S. A. Nierzwicki-Bauer, S. E. Curtis, and R. Haselkorn, Cotranscription of genes encoding the small and large subunits of ribulose-1,5-bisphosphate carboxylase in the cyanobacterium *Anabaena* 7120, Proc Natl Acad Sci USA 81:5961-5965 (1984).

13. K. Shinozaki and M. Sugiura, Genes for the large and small subunits of ribuose-1,5-bisphosphate carboxylase/ oxygenase constitute a single operon in a cyanobacterium *Anacystis nidulans* 6301, Mol Gen Genet 200:27-32 (1985).

14. G. W. Schmidt and M. L. Mishkind, Rapid degradation of unassembled ribulose 1,5-bisphosphate carboxylase small subunits in chloroplasts, Proc Natl Acad Sci USA 80:2632-2636 (1983).

15. T. J. Andrews and B. Ballment, The function of the small subunit of ribulose bisphosphate carboxylase-oxygenase, J Biol Chem 258:7514-7518 (1983).

16. S. Asami, T. Takabe, T. Akazawa, and G. A. Codd, Ribulose 1,5-bisphosphate carboxylase from the halophilic cyanobacterium *Aphanothece halophytica*, Arch Biocem Biophy 225:713-721 (1983).

17. T. J. Andrews and B. Ballment, Active-site carbamate formation and reaction-intermediate-analog binding by ribulosebisphosphate carboxylase/ oxygenase in the absence of its small subunit, Proc Natl Acad Sci USA 81:3660-3664 (1984).

18 T. Takabe, A. Incharoensakdi, and T. Akazawa, Essentiality of the small subunit (B) in the catalysis of RuBP carboxylase/oxygenase is not related to substrate-binding in the large subunit (A), Biochem and Biophy Res Com 122:763-769 (1984).

19. T. J. Andrews, D. M. Greenwood, and D. Yellowlees, Catalytically active hybrids formed *in vitro* between large and small subunits of different procaryotic ribulose bisphosphate carboxylases, Arch Biochem Biophy 234:313-317 (1984).

20. T. J. Andrews and G. H. Lorimer, Catalytic properties of a hybrid between cyanobacterial large subunits and higher plant small subunits of ribulose bisphosphate carboxylase-oxygenase, J Biol Chem 260:4632-4636 (1985).

21. D. B. Jordan and R. Chollet, Subunit dissociation and reconstitution of ribulose-1,5-bisphosphate carboxylase from *Chromatium vinosum*, Arch Biochem Biophy 236:487-496 (1985).

22. T. J. Andrews, Catalysis by cyanobacterial ribulose-bisphosphate carboxylase large subunits in the complete absence of small subunits, J Biol Chem 263:12213-12219 (1988).

23. R. Barraclough and R. J. Ellis, Protein synthesis in chloroplasts IX. Assembly of newly-synthesized large subunits into ribulose bisphosphate carboxylase in isolated intact pea chloroplasts, Biochim Biophys Acta 608:19-31 (1980).

24. M. V. Bloom, P. Milos, and H. Roy, Light-dependent assembly of ribulose-1,5-bisphosphate carboxylase, Proc Natl Acad Sci USA 80:1013-1017 (1983).

25. S. M. Hemmingsen and R. J. Ellis, Purification and properties of ribulose bisphosphate carboxylase large subunit binding protein, Plant Physiol 80:269-276 (1986).

26. R. J. Ellis and S. M. van der Vies, The rubisco subunit binding protein, Photosyn Res 16:101-115 (1988).

27. J. E. Musgrove, R. A. Johnson, and R. J. Ellis, Dissociation of the ribulose bisphosphate-carboxylase large-subunit binding protein into dissimilar subunits, Eur J Biochem 163:529-534 (1987).

28. S. M. Hemmingsen, C. Woolford, S. M. van der Vies, K. Tilly, D. T. Dennis, C. P. Georgopoulos, R. W. Hendrix, and R. J. Ellis, Homologous plant and bacterial proteins chaperone oligomeric protein assembly, Nature 333:330-334 (1988).

29. S. Cannon, P. Wang, and H. Roy, Inhibition of ribulose bisphosphate carboxylase assembly by antibody to a binding protein, J Cell Biol 103:1327-1335 (1986).

30. J. Ellis, Proteins as molecular chaperones, Nature 328:378-379 (1987).

31. G. W. Schmidt and M. L. Mishkin, The transport of proteins into chloroplasts, Ann Rev Biochem 55:879-912 (1986).

32. N-H. Chua and G. W. Schmidt, Post-translational transport into intact chloroplasts of a precursor to the small subunit of ribulose-1,5-bisphosphate carboxylase, Proc Natl Acad Sci USA 75:6110-6114 (1978).

33. R. Broglie, G. Coruzzi, R. T. Fraley, S. G. Rogers, R. B. Horsch, J. G. Niedermeyer, C. L. Fink, J. S. Flick, and N-H. Chua, Light-regulated expression of a pea ribulose-1,5-bisphosphate carboxylase small subunit gene in transformed plant cells, Science 224:838-843 (1984).

34. C. C. Wasmann, R. T. Ramage, H. J. Bohnert, and J. A. Ostrem, Identification of an assembly domain in the small subunit of ribulose-1,5-bisphosphate carboxylase, Proc Natl Acad Sci USA (in press).

35. F. P. Wolter, C. C. Fritz, L. Willmitzer, J. Schell, and P. H. Schreier, rbcS genes in Solanum tuberosum: Conservation of transit peptide and exon shuffling during evolution, Proc Natl Acad Sci USA 85:846-850 (1988).

36 D. B. Jordan and R. Chollet, Inhibition of ribulose bisphosphate carboxylase by substrate ribulose 1,5-bisphosphate, J Biol Chem 258:13752-13758 (1983).

37. A. Brooks and A. R. Portis, Jr., Protein-bound ribulose bisphosphate correlates with deactivation of ribulose bisphosphate carboxylase in leaves, Plant Physiol 87:244-249 (1988).

38. T. J. Andrews and K. M. Abel, Kinetics and subunit interactions of ribulose bisphosphate carboxylase-oxygenase from the cyanobacterium, Synechococcus sp., J Biol Chem 256:8445-8451 (1981).

39. J. R. Seeman, J. A. Berry, S. Freas, and M. A. Krump, Regulation of ribulose bisphosphate carboxylase activity in vivo by a light-modulated inhibitor of catalysis, Proc Natl Acad Sci USA 82:8024-8028 (1985).

40. J. C. Servaites, Binding of a phsyiological inhibitor to ribulose bisphosphate carboxylase/oxygenase during the night, Plant Physiol 78:839-843 (1985).

41. S. Gutteridge, M. A. J. Parry, S. Burton, A. J. Keys, A. Mudd, J. Feeney, J. C. Servaites, and J. Pierce, A nocturnal inhibitor of carboxylation in leaves, Nature 324:274-276 (1986).

42. J. A. Berry, G. H. Lorimer, J. Pierce, J. R. Seeman, J. Meek, and S. Freas, Isolation, identification and synthesis of 2'-carboxy arabinitol-1-phosphate, a diurnal regulator of ribulose bisphosphate carboxylase activity, Proc Natl Acad Sci USA 84:734-738 (1987).

43. H-H. Yeoh, M. R. Badger, and L. Watson, Variations in kinetic properties of ribulose-1,5-bisphosphate carboxylases among plants, Plant Physiol 67:1151-1155 (1981).

44. J. Pierce, N. E. Tolbert, and R. Barker, Interaction of ribulosebisphosphate carboxylase/oxygenase with transition-state analogues, Biochem 19:934-942 (1980).

45. J. C. Servaites, M. A. J. Parry, S. Gutteridge, and A. J. Keys, Species variation in the predawn inhibition of ribulose-1,5-bisphosphate carboxylase/oxygenase, Plant Phsiol 79:1161-1163 (1986).

46. J. Kobza and J. R. Seeman, Mechanisms for light-dependent regulation of ribulose-1,5-bisphosphate carboxylase activity and photosynthesis in intact leaves, Proc Natl Acad Sci USA 85:3815-3819 (1988).

47 S. P. Robinson and A. R. Portis, jr, Release of the nocturnal inhibitor carboxy arabinitol-1-phosphate, from ribulose bisphosphate carboxylase/oxygenase by rubisco activase, FEBS Lett 233:413-416 (1988).

48. C. R. Somerville, A. R. Portis, and W. L. Ogren, A mutant of Arabidopsis thaliana which lacks activation of RuBP carboxylase in vivo, Plant Physiol 70:381-387 (1982).

49. M. E. Salvucci, J. M. Werneke, W. L. Ogren and A. R. Portis, Jr., Purification and species distribiution of rubisco activase, Plant Physiol 84:930-936 (1987).

50. V. J. Streusand and A. R. Portis, Jr., Rubisco activase mediates ATP-dependent activation of ribulose bisphosphate carboxylase, Plant Physiol 85:152-154 (1987).

51. M. A. J. Parry, A. J. Keys, C. H. Foyer, R. T. Furbank, and D. A. Walker, Regulation of ribulose-1,5-bisphosphate carboxylase activity by the activase system in lysed spinach chloroplasts, Plant Physiol 87:558-561 (1988).

52. J. M. Werneke, R. E. Zielinski, and W. L. Ogren, Structure and expression of spinach leaf cDNA encoding ribulosebisphosphate carboxylase/oxygenase activase, Proc Natl Acad Sci USA 85:787-791 (1988).

53. J. M. Werneke, J. M. Chatfield, and W. L. Ogren, Catalysis of ribulose bisphosphate carboxylase/oxygenase activation by the product of a rubisco

activase cDNA clone expressed in *Escherichia coli*, <u>Plant Physiol</u> 87:917-920 (1988).

54. R. M. Mulligan, R. L. Houtz, and N. E. Tolbert, Reaction-intermediate analogue binding by ribulose bisphosphate carboxylase/ oxygenase causes specific changes in proteolytic sensitivity: The amino-terminal residue of the large subunit is acetylated proline, <u>Proc Natl Acad Sci USA</u> 85:1513-1517 (1988).

55. C. R. Somerville and S. C. Somerville, Cloning and expression of the *Rhodospirillum rubrum* ribulosebisphosphate carboxylase gene in *E. coli*, <u>Mol Gen Genet</u> 193:214-219 (1984).

56. R. G. Quivey Jr. and F. R. Tabita, Cloning and expression in *Escherichia coli* of the form II ribulose 1,5-bisphosphate carboxylase/oxygenase gene from *Rhodopseudomonas sphaeroides*, <u>Gene</u> 31:91-101 (1984).

57. J. Chory, E. D. Muller, and S. Kaplan, DNA-directed in vitro synthesis and assembly of the form II D-ribulose-1,5-bisphosphate carboxylase/ oxygenase from *Rhodopseudomonas sphaeroides*, <u>J Bacteriology</u> 161:307-313 (1985).

58. J. L. Gibson and F. R. Tabita, Isolation of the *Rhodopseudomonas sphaeroides* form I ribulose 1,5-bisphosphate carboxylase/ oxygenase large and small subunits genes and expression of the active hexadecameric enzyme in *Escherichia coli*, <u>Gene</u> 44:271-278 (1986).

59. A. M. Viale, J. Kobayashi, T. Takabe, and T. Akazawa, Expression of genes for subunits of plant-type RuBisCO from *Chromatium* and production of the enzymatically active molecule in *Escherichia coli*, <u>FEBS Lett</u> 192:283-288 (1985).

60. A. A. Gatenby, S. M. van der Vies, and D. Bradley, Assembly in *E. coli* of a functional multi-subunit ribulose bisphosphate carboxylase from a blue-green alga, <u>Nature</u> 314:617-620 (1985).

61. J. T. Christeller, B. E. Terzaghi, D. F. Hill, and W. A. Laing, Activity expressed from cloned *Anacystis nidulans* large and small subunit ribulose bisphosphate carboxylase genes, <u>Plant Mol Biol</u> 5:257-263 (1985).

62. F. R. Tabita and C. L. Small, Expression and assembly of active cyanobacterial ribulose-1,5-bisphosphate carboxylase/oxygenase in *Escherichia coli* containing stoichiometric amounts of large and small subunits, <u>Proc Natl Acad Sci USA</u> 82:6100-6103 (1985).

63. M Gurevitz, C. R. Somerville, and L. McIntosh, Pathway of assembly of ribulosebisphosphate carboxylase/oxygenase from *Anabaena* 7120 expressed in *Escherichia coli*, <u>Proc Natl Acad Sci USA</u> 82:6546-6550 (1985).

64. S. M. van der Vies, D. Bradley, and A. A. Gatenby, Assembly of cyanobacterial and higher plant ribulose bisphosphate carboxylase subunits into functional homologous and heterologous enzyme molecules in *Escherichia coli*, <u>EMBO J</u> 5:2439-2444 (1986).

65. A. A. Gatenby, The properties of the large subunit of maize ribulose bisphosphate carboxylase/oxygenase synthesized in *Escherichia coli*, <u>Eur J Biochem</u> 144:361-366 (1984).

66. A. A. Gatenby, S. M. van der Vies, S. J. Rothstein, Co-expression of both the maize large and wheat small subunit genes of ribulose-bisphosphate carboxylase in *Escherichia coli*, <u>Eur J Biochem</u> 168:227-231 (1987).

67. G. Voordouw, P. A. de Vries, W. A. M. van den Berg, and E. P. J. de Clerck, Site-directed mutagenesis of the small subunit of ribulose-1,5-bisphosphate carboxylase/ oxygenase from *Anacystis nidulans*, <u>Eur J Biochem</u> 163:591-598 (1987).

68. C. A. Kettleborough, M. A. J. Parry, S. Burton, S. Gutteridge, A. J. Keys, and A. L. Philips, The role of the N-terminus of the large subunit of ribulose-bisphosphate carboxylase investigated by construction and expression of chimaeric genes, <u>Eur J Biochem</u> 170:335-342 (1987).

69. C. C. Wasmann, B. Reiss, S. G. Bartlett, and H. J. Bohnert, The importance of the transit peptide and the transported protein for protein import into chloroplasts, <u>Mol Gen Genet</u> 205:446-453 (1986).

70. P. H. Schreier, E. A. Seftor, J. Schell, and H. J. Bohnert, The use of nuclear-encoded sequences to direct the light-regulated synthesis and transport of a foreign protein into plant chloroplasts, <u>EMBO J</u> 4:25-32 (1985).

71. G. van den Broeck, M. P. Timko, A. P. Kausch, A. R. Cashmore, M. van Montagu, and L. Herrera-Estrella, Targeting of a foreign protein to chlorplasts by fusion to the transit peptide from the small subunit of ribulose 1,5-bisphosphate carboxylase, Nature 313:358-363 (1985).

72. Z. Chen, C. J. Chastain, S. R. Al-Abed, R. Chollet, and R. J. Spreitzer, Reduced CO_2/O_2 specificity of ribulose-bisphosphate carboxylase/ oxygenase in a temperature-sensitive chloroplast mutant of *Chlamydomonas*, Proc Natl Acad Sci USA 85:4696-4699 (1988).

73. T. C. Tso and S. D. Kung, Soluble proteins in tobacco and their potential use, in: "Leaf Protein Concentrates," L. Telek and H. D. Graham, eds., AVI Publishing Co., Inc, Westport, CN (1983).

74. B. Ranty, D. Roby G. Cavelie, and M.-T. Esquerre-Tugaye, Ribulose-1,5 bisphosphate carboxylase/oxygenase expression in melon plants infected with *Colletotrichum lagenarium*, Planta 170:386-391 (1987).

75. F. Nargang, L. McIntosh, and C. Somerville, Nucleotide sequence of the ribulosebisphosphate carboxylase gene from *Rhodospirillum rubrum*, Mol Gen Genet 193:220-224 (1984).

76. K. Shinozaki, C. Yamada, N. Takahata, and M. Sugiura, Molecular cloning and sequence analysis of the cyanobacterial gene for the large subunit of ribulose-1,5-bisphosphate carboxylase/ oxygenase, Proc Natl Acad Sci USA 80:4050-4054 (1983).

77. K. Shinozaki and M. Sugiura, The nucleotide sequence of the tobacco chloroplast gene for the large subunit of ribulose-1,5-bisphosphate carboxylase/ oxygenase, Gene 20:91-102 (1982).

78. K. Andersen and J. Caton, Sequence analysis of the *Alcaligenes eutrophus* chromosomally encoded ribulose bisphosphate carboxylase large and small subunit genes and their gene products, J Bacteriol 169:4547-4558 (1987).

79. K Shinozaki and M Sugiura, The gene for the small subunit of ribulose-1,5 bisphosphate carboxylase/ oxygenase is located close to the gene for the large subunit in the cyanobacterium *Anacystis nidulans* 6301, Nucleic Acids Res 11:6957-6964 (1983).

80. C. C. Wasmann, Ph D thesis, Michigan State University, East Lancing Michigan, (1985).

81. A. Sailland, I. Amiri, and G. Freyssinet, Amino acid sequence of the ribulose-1,5-bisphosphate carboxylase/ oxygenase small subunit from Euglena, Plant Mol Biol 7:213-218 (1986).

82. M Goldschmidt-Clermont and M. Rahire, Sequence, evolution and differential expression of the two genes encoding varitant small subunits of ribulose bisphosphate carboxylase/ oxygenase in *Chlamydomonas reinhardtii*, J Mol Biol 191:421-432 (1986).

83. C. L. Baszczynski, L. Fallis and G. Bellemare, Nucleotide sequence of a full length cDNA clone of a *Brassica napus* ribulose bisphosphate carboxylase-oxygenase small subunit gene, Nucleic Acids Res 16:4732 (1988).

84. M. Matsuoka, Y. Kano-Murakami, Y. Tanaka, Y. Ozeki and N. Yamamoto, Nucleotide sequence of cDNA encoding the small subunit of ribulose-1,5 bisphosphate carboxylase from maize, J. Biochem 102:673-676 (1987).

85. G. Coruzzi, R. Broglie, A. Cashmore and N-H. Chua, Nucleotide sequences of two pea cDNA clones encoding the small subunit of ribulose-1,5-bisphosphate carboxylase and the major chlorophyll a/b-binding thylakoid polypeptide, J Biol Chem 258:1399-1402 (1983).

86. S. L. Berry-Lowe, T. D. McKnight, D. M. Shah and R. B. Meagher, The nucleotide sequence, expression, and evolution of one member of a multigene family encoding the small subunit of ribulose-1,5-bisphosphate carboxylase in soybean, J Mol Appl Gen 1:483-498 (1982).

87. G. Waksman and G. Freyssinet, Nucleotide sequence of a cDNA encoding the ribulose-1,5-bisphosphate carboxylase/ oxygenase from sunflower (*Helianthus annuus*), Nucleic Acids Res 15:1328 (1987).

88. E. J. DeRocher, R. T. Ramage, C. B. Michalowski and H. J. Bohnert, Nucleotide sequence of a cDNA encoding rbcS from the desert plant *Mesembryanthemum crystallinum*, Nucleic Acids Res 15:6301 (1987).

89. C. A. Adams, M. Babcock, F. Leung and S. M. Sun, Sequence of a ribulose 1,5-bisphosphate carboxylase/ oxygenase cDNA from the C$_4$ dicot *Flaveria rinervia*, Nucleic Acids Res 15:1875 (1987).

90. P. Dunsmuir, S. Smith and J. Bedbrook, A number of different nuclear genes for the small subunit of RuBPCase are transcribed in petunia, Nucleic Acids Res 11:4177-4183 (1983).

91. B. J. Mazur and C-F. Chui, Sequence of a genomic DNA clone for the small subunit of ribulose bis-phosphate carboxylase-oxygenase from tobacco, Nucleic Acids Res 13:2373-2386 (1985).

92. S. Gutteridge, I. Sigal, B. Thomas, R. Arentzen, A. Cordova, and G. Lorimer, A site-specific mutation within the active site of ribulose-1,5-bisphosphate carboxylase of *Rhodospirillum rubrum*, EMBO J 3:2737-2743 (1984).

93. M. Estelle, J. Hanks, L. McIntosh, and C. Somerville, Site-specific mutagenesis of ribulose-1,5-bisphosphate carboxylase/ oxygenase, J Biol Chem 260:9523-9526 (1985).

94. S. K. Niyogi, R. S. Foote, R. J. Mural, F. W. Larimer, S. Mitra, T. S. Soper, R. Machanoff, and F. C. Hartman, Nonessentiality of histidine 291 of *Rhodospirillum rubrum* ribulose-bisphosphate carboxylase/ oxygenase as determined by site-directed mutagenesis, J Biol Chem 261:10087-10092 (1986).

95. B. E. Terzaghi, W. A. Laing, J. T. Christeller, G. B. Petersen, and D. F. Hill, Ribulose 1,5-bisphosphate carboxylase: Effect on the catalytic properties of changing methionine-330 to leucine in the *Rhodospirillum rubrum* enzyme, Biochem J 235:839-846 (1986).

96. F. C. Hartman, T. S. Soper, S. K. Niyogi, R. J. Mural, R. S. Foote, S. Mitra, E. H. Lee, R. Machanoff, and F. W. Larimer, Function of Lys-166 of *Rhodospirillum rubrum* ribulosebisphosphate carboxylase/ oxygenase as examined by site-directed mutagenesis, J Biol Chem 262:3496-3501 (1987).

97. E. H. Lee, T. S. Soper, R. J. Mural, C. D. Stringer, and F. C. Hartman, An intersubunit interaction at the active site of D-ribulose-1,5-bisphosphate carboxylase/ oxgenase as revealed by cross-linking and site-directed mutagenesis, Biochem 26:4599-4604 (1987).

98. F. C. Hartman, F. W. Larimer, R. J. Mural, R. Machanoff, and T. S. Soper, Essentiality of Glu-48 of ribulose bisphosphate carboxylase/ oxygenase as demonstrated by site-directed mutagenesis, Biochem Biophy Res Com 145:1158-1163 (1987).

99. F. W. Larimer, E. H. Lee, R. J. Mural, T. S. Soper, and F. C. Hartman, Intersubunit location of the active site of ribulose-bisphosphate carboxylase/ oxygenase as determined by *in vivo* hybridization of site-directed mutants, J Biol Chem 262:15327-15329 (1987).

100. H. B. Smith and F. C. Hartman, Restoration of activity to catalytically deficient mutnats of ribulosebisphosphate carboxylase/ oxygenase by aminoethylation, J. Biol Chem. 263:4921-4925 (1988).

101. G. H. Lorimer and F. C. Hartman, Evidence supporting Lysine 166 of *Rhodospirillum rubrum* ribulosebisphosphate carboxylase as the essential base which initiates catalysis, J. Biol. Chem. 263:6468-6471 (1988).

102. G. H. Lorimer, S. Gutteridge, and M. Madden, Exploring the active-site of Rubisco using partial reactions and site-direct mutants, with guidance from crysallography, abstract 1-01-2 from XIV International Botanical Congress, Berlin, July 24-Aug 1, 1987.

THE MEASUREMENT OF RATE OF PHOTOSYNTHESIS AS A FUNCTION OF PHOTON FLUX DENSITY AND THE SIGNIFICANCE AND IMPLICATION OF THESE MEASUREMENTS

David Walker

Research Institute for Photosynthesis, University of Sheffield, Sheffield U.K. S10 2TN

Photosynthesis is driven by light. Particularly in low or moderate light, its rate is therefore largely determined by the rate at which photons reach the photosynthetic apparatus. When the rate of photosynthesis is plotted against the rate of arrival of photons (photon flux density or PFD) the relationship (Fig. 1a) is full of information. The intercept on the vertical axis is a measure of dark respiration. The intercept on the horizontal axis is the "light compensation point" (LCP). This, again, is largely determined by the rate of dark respiration because it is the PFD at which respiratory O_2 uptake and photosynthetic oxygen evolution come into balance. In the region between the LCP and 100 μmole quanta.$m^{-2}.s^{-1}$ the relationship is nearly linear and this initial slope gives the maximum efficiency of energy transduction; usually expressed as quantum yield (O_2 evolved per photon absorbed) or its reciprocal, quantum requirement (number of photons required to bring about the evolution of one molecule of oxygen). As will be seen, other information may be derived from the overall rate v PFD relationship by appropriate analysis. Although the importance of this relationship has been long recognised it is only recently that its measurement has become a practical possibility as a routine procedure in non–specialist laboratories.

Using a leaf–disc electrode designed by Delieu and Walker[5], Björkman and Demmig[1] recently determined the quantum requirement for 37 C_3 species. This was an important contribution for many reasons including the fact that, in the absence of stress, the quantum requirement was remarkably constant at about 9.3. This has obvious implications for plant breeders and eco–physiologists because it suggests that a majority of plants may be working at a photosynthetic efficiency which is close to the theoretical maximum (0.125 when expressed as quantum yield and 8 when expressed as quantum requirement). The fact that leaves were frequently unable to maintain these high efficiencies when subject to certain forms of stress, such as photoinhibition[13], also indicated the importance of measurements of this nature in evaluating environmental (and experimentally imposed) stress and the fact that consistent and reproducible measurements could be made with a very simple and readily available piece of apparatus also opened the way to wider applications. Although the procedure employed by Björkman and Demmig was clearly sound it did not recommend itself for routine measurement, involving, as it did, repeated and time–consuming manipulations with neutral–density filters. For this reason, a new system has been developed[19]. This is also based on the leaf–disc electrode but changes in PFD are accomplished by changing the electrical supply to a light–source composed of a bank of light–emitting diodes (Fig. 2). For LEDs the relationship between electrical supply and light emission is nearly linear and therefore lends itself to control in a way which would not be applicable to conventional tungsten sources. Moreover there is no sensible change in light quality as the electrical supply is varied. The LEDs which are used have a peak emission at about 660 nm, which is close to the absorption maxima of chlorophylls in the red. This facilitates light

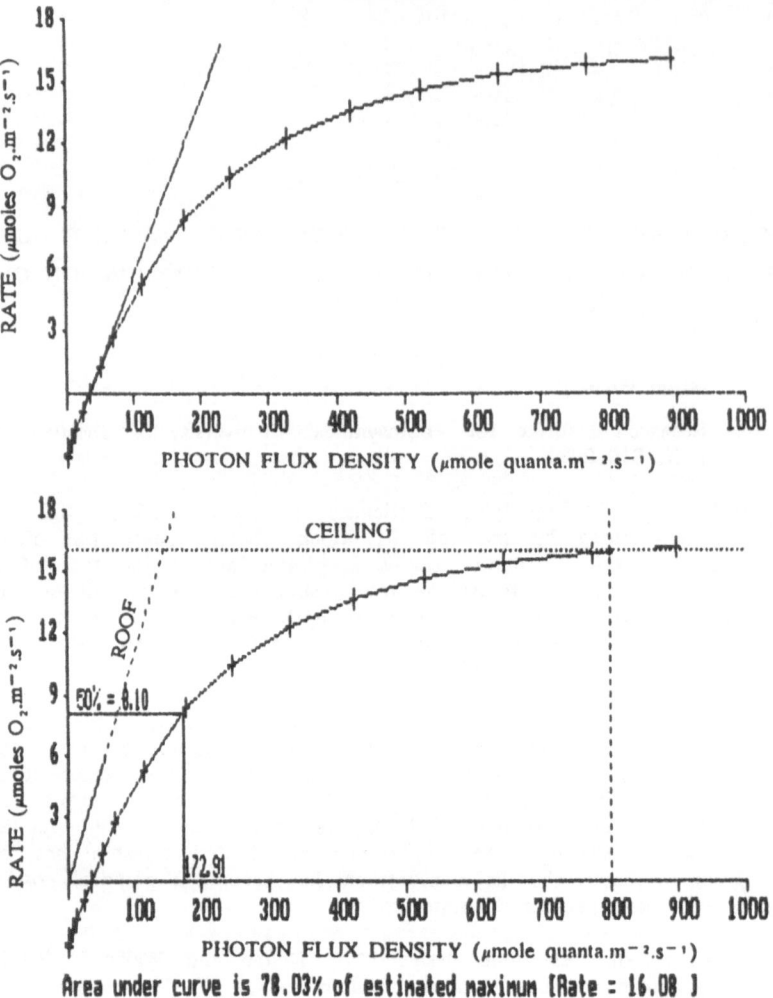

Y–Intercept = –2.80 X–Intercept = 33.42 Slope = 0.0838 Requirement = 11.94

Area under curve is 78.03% of estimated maximum [Rate = 16.08]

Figure 1. (a) Rate of photosynthetic oxygen evolution as a function of photon flux density; a typical plot, to illustrate the salient features. The intercept on the vertical axis is a measure of dark respiration. The intercept on the horizontal axis is the light compensation point. The relationship is linear is the zero to 100 μmole quanta.m^{-2}.s^{-1} range and slope the quantum yield, and its reciprocal, quantum requirement are derived from the initial slope. (b) Same data as in Fig. 1(a) but to illustrate the concept of "nominal light utilisation capacity" (see text). A computation of the PFD required to give half of the rate at 800 μmole quanta.m^{-2}.s^{-1} is also illustrated.

measurement, (made with a quantum sensor designed by Skye Instruments for this purpose) because all of the light which reaches the leaf is photosynthetically active. It now becomes a relatively simple matter to devise circuitry and software which enable a computer to undertake changes in light intensity and record oxygen evolution over a range of photon flux densities[14,20,21]. This not only removes unnecessary tedium from routine determinations but also makes such determinations highly reproducible. This is of considerable importance because a leaf will only behave in a consistent fashion throughout repeated changes in PFD if these are imposed in an entirely consistent way. Once the data has been derived the computer will, of course, also facilitate data handling and analysis as can be seen from the figures presented here.

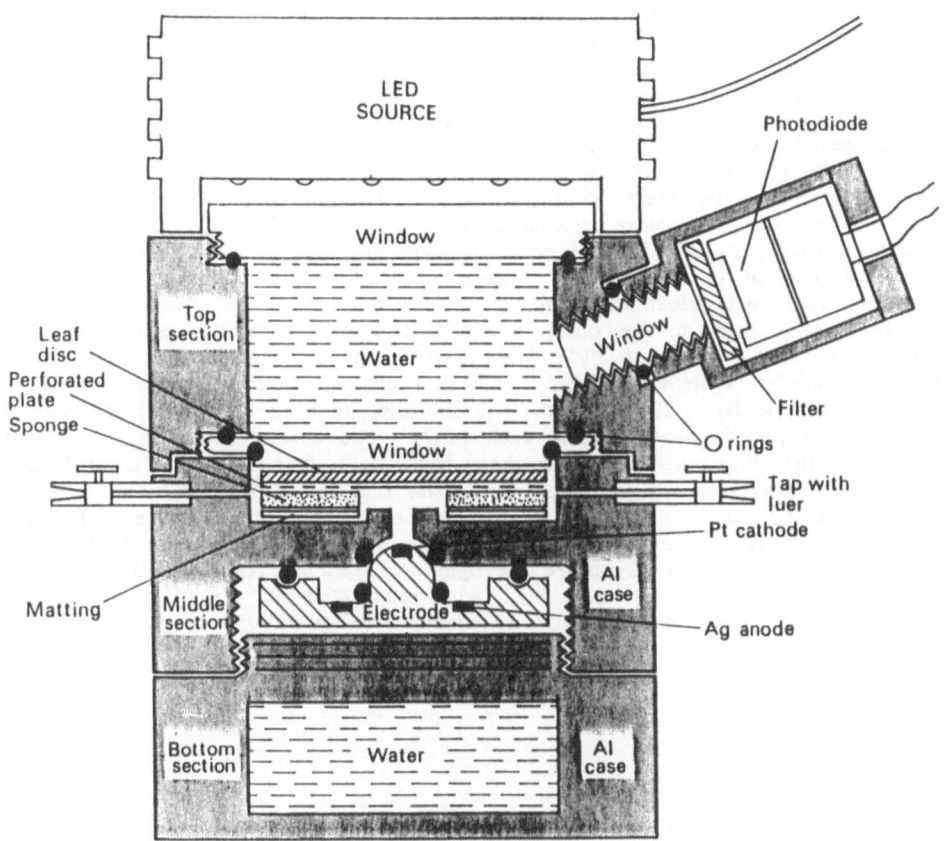

Figure 2. Schematic diagram of a gas–phase oxygen electrode and fluorescence probe[5,19].
The leaf–disc, or leaf pieces, are supported on a stainless steel mesh in a chamber which is located in the middle section of the apparatus. The O_2 sensor (Clark–type electrode) lies beneath the leaf chamber with its Pt cathode exposed to the atmosphere within it. The leaf tissue is pressed lightly against the temperature–controlled roof of the chamber by a foam disc which also separates it from carbonate–bicarbonate buffer carried on capillary matting. The leaf is illuminated through this window which also allows fluorescence to reach a probe (inserted at an angle of 40 degrees) where it is monitored by a photodiode. Actinic light is delivered to the top of the apparatus, by an array of light–emitting diodes, as shown, or from any other appropriate light source. The photodiode is protected from the actinic light by optical filter or filters. The clips which draw the top section on to the middle section (so that the roof of the leaf chamber is sealed against an O–ring) and the tubes which carry temperature–controlled water to the top and bottom sections are not shown. The taps (with luers) are for calibration and adjustment of the gas phase. The commercial version of this apparatus is manufactured by Hansatech[19].

In seeking to investigate the relationship between the rate of photosynthesis and photon flux density the investigator is immediately faced with a problem which is more or less inevitable in any similar enquiry. This is simply that measurement is rarely, if ever, completely non–intrusive. In this particular instance major changes in leaf physiology and biochemistry occur in a leaf as it accommodates to changes in PFD. If a dark–adapted leaf is suddenly and strongly illuminated, photosynthetic carbon assimilation does not commence at maximum rate and only approaches a maximum after an initial induction period which often lasts about 1–3 min but may take up to 30 or 60 min. Moreover, this "short" induction is often followed by a much slower upward adjustment (long "induction" which can take many minutes or even hours). Similarly, slow adjustments in rate may follow changes from low to high light or from high to low light intensities. For these and other reasons, emphasis has often been placed on the need to measure so-called "steady–state" photosynthesis. If there really is such a thing as steady–state photosynthesis this would obviously be desirable but much experience suggests that the steady–state is largely illusory and, if it occurs at all, may only represent an uncertain interval between the termination of short and long induction and the imposition of feedback limitations. An alternative approach, which has been used here, is to avoid, as far as possible, the major consequences of induction and, thereafter, to make a relatively large number of small changes in PFD over as short a time as the acquisition of sufficient data permits. This approach appears to be particularly useful in regard to the measurement of quantum yield. Figure 2 shows results obtained with spinach in which 20 rate determinations were made over the range from 0–125 μmole quanta.m^{-2}.s^{-1}. Prior to the onset of the first series of measurements the leaf was pre–illuminated at the highest PFD (i.e. at 125 μmole quanta.m^{-2}.s^{-1}) value for a period of 5 min. In spinach, which is not dark–adapted, this is sufficient to overcome the initial induction lag. The leaf was then subjected to each PFD, from the highest to the lowest, for periods of 45 secs at each PFD. The computer was also instructed to disregard the first 5 seconds of photosynthesis following each change to a lower PFD. In this particular experiment the entire procedure was repeated 3 times and Fig. 3 illustrates the remarkable reproducibility which can be obtained in these circumstances. It will also be seen that, when such measurements are made in this way with spinach, there is no change in slope or discontinuity below the light compensation point. Such discontinuities were first reported by Bessel Kok[10,11] who interpreted an abrupt change in slope beneath the light compensation point as an inhibitory effect of light on dark respiration.

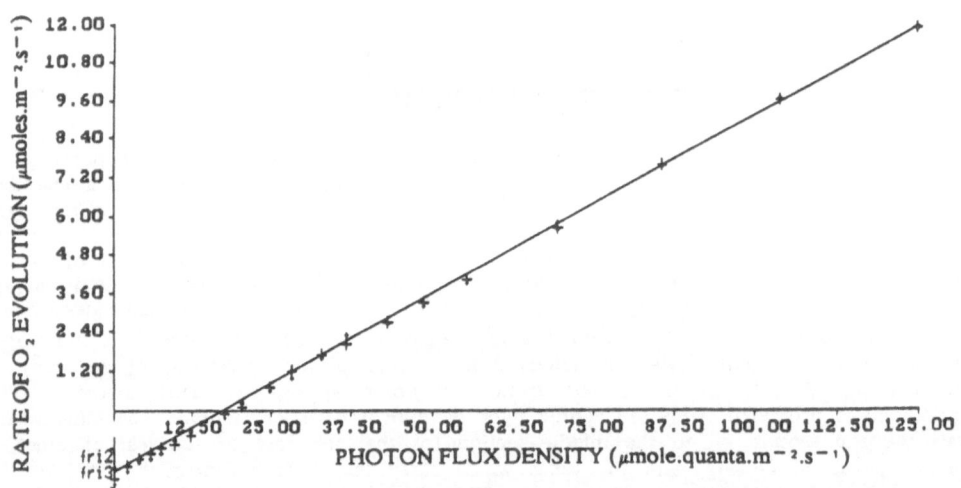

Figure 3. Determination of quantum requirement of spinach following pre–illumination in lower light. Twenty rate determinations were made over periods of 45 seconds and the first 5 seconds of each measurement were ignored in computation. The figure shows reproducibility which can be obtained (from Walker[20]).

The "Kok effect" has been repeatedly observed since[17,4,15] although both the ubiquity and the significance of this phenomenon has been questioned by Emerson[7]. Less abrupt but otherwise similar changes below the light compensation point can be observed with the leaf–disc electrode (Fig. 4) if leaves are illuminated in bright light immediately prior to a series of measurements in which the photon flux density is then increased in steps from darkness upwards. Whether or not these effects correspond to the Kok effect as such is still uncertain but in this situation they are clearly attributable to two causes. The first is simple artefact and has to do with the inevitable change in temperature within the chamber on going into darkness after a period of bright illumination. If a piece of black felt is substituted for a leaf, bright illumination produces a signal which corresponds to a small oxygen evolution and, conversely, darkening gives rise to a signal which corresponds to a small oxygen uptake. The latter will undoubtedly contribute to the apparent increase in the rate of dark respiration which is normally observed following a period of bright illumination.

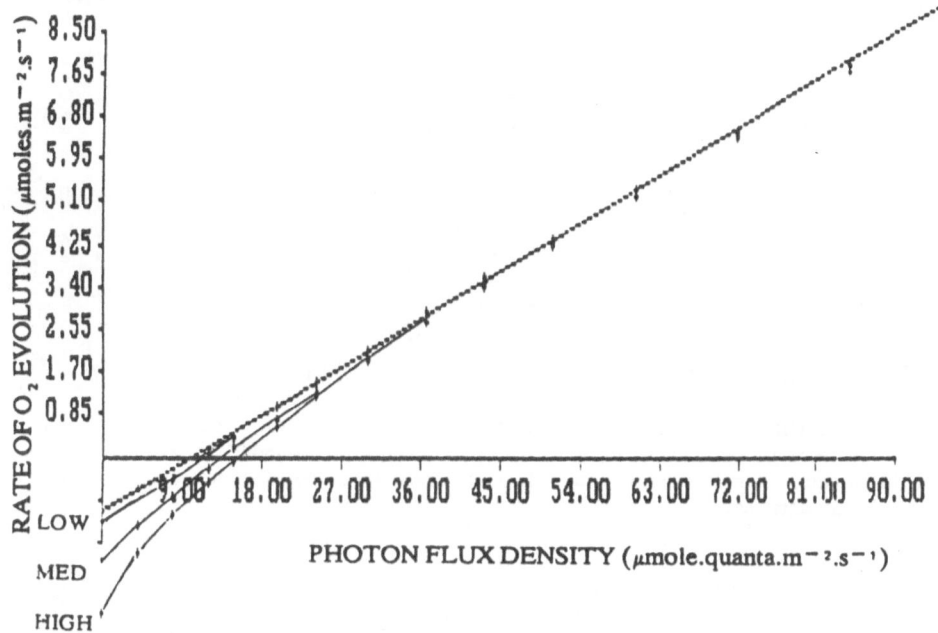

Figure 4. Changes in the slope of the rate v PFD relationship at low PFD's following pre-illumination at high light intensities. This departure from linearity, which is similar to the "Kok effect" is caused by light-enhanced respiration (see text). From Walker[20].

Figure 5 for example shows that, in spinach, respiration after 30 minutes or so of darkness was considerably slower than it was immediately following illumination at 500 μmole quanta.m^{-2}.s^{-1} but that, within about 5 min, much of this increase in rate was dissipated. When due allowance is made for the artefactual, temperature-related, component a light enhancement of dark respiration still remains. It seems clear, therefore, that the combined effects of light-enhanced dark respiration and artefact can account for changes in the slope of the rate v PFD relationship below the light compensation point and that these bear at least a superficial resemblance to the Kok effect. In addition, there is little doubt that the Kok effect itself is not invariably observed but relates very much to the manner in which measurements are made[20,21,7]. For practical purposes therefore, there is much to be said for using a procedure which, for whatever reason, does not give rise either to artefacts or genuine metabolic changes and can give a completely linear relationship over the whole of the 0-100 μmole quanta.m^{-2}.s^{-1} range[14,20]. To date, it has not been possible to apply this method to a large number of species but several diverse species have behaved in much the same way as spinach and, provided that the investigator is confident that his material does not display unusual induction behaviour, it would appear that this procedure might have broad applicability. It should be noted, however, that anomalous behaviour may result (for example) from the presence in a leaf of the dark inhibitor of ribulose bisphosphate carboxylase[14,18] and that, if it does, it may be necessary to pre-illuminate at higher light intensities (than 100 μmol quanta.m^{-2}.s^{-1}) for sufficiently long to remove the inhibitor prior to following the regime outlined above.

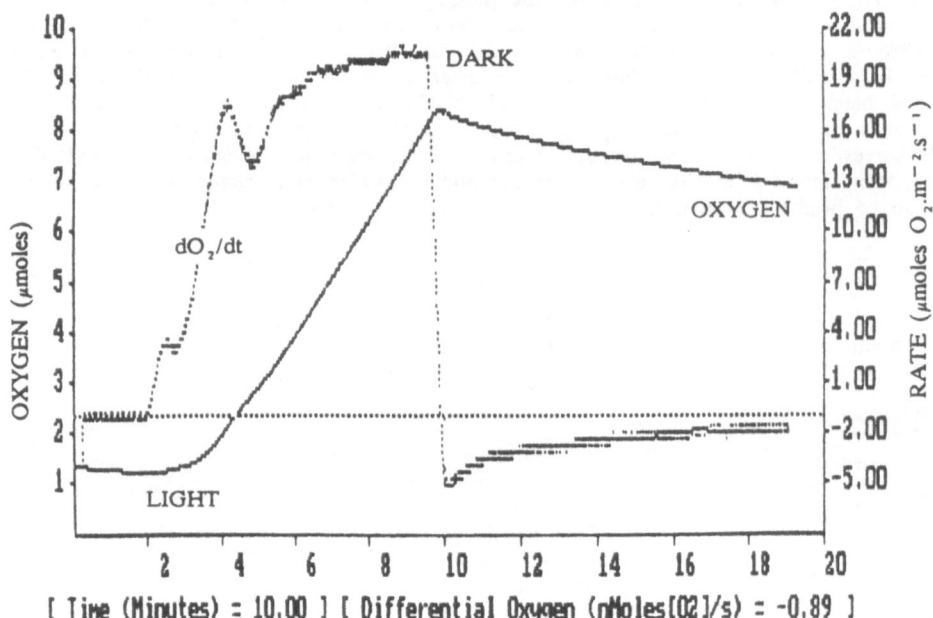

Figure 5. Light enhanced respiration in spinach. Oxygen uptake after prolonged darkness is seen to be much lower than that observed immediately after a period of bright illumination but the high rate which is then recorded declines rapidly during the first 5 min of darkness.

With the aid of a computer it is also possible to introduce an additional range of analytical procedures which would otherwise have been extremely laborious. One[14,20,21] is the concept of nominal light utilisation capacity (NLUC). For purposes of comparison, this accept the Blackman notion[2,3] of limiting factors. A thermodynamic "roof" is drawn through the origin with a pitch equivalent to a nominal quantum requirement of 9 (Figs. 1b, 6b,c). This thermodynamic constraint is imposed by the Z-scheme which demands an input of 8 photons to move 4 electrons through. to photosystems from water to NADP. (A quantum requirement of 9 was arbitarily chosen, rather than 8 in order to account for secondary, energy-consuming, reactions and to bring the value nearer to that of 9.3 determined by Björkman and Demmig[1]). A horizontal ceiling intersects this roof at a rate equal to that observed at a PFD of 800 μmole quanta.m^{-2}.s^{-1}. In reality, a leaf will approach but never reach these constraints (except for the intercept with the ceiling at the 800 μmole quanta.m^{-2}.s^{-1} point). This is partly a matter of structure and partly a matter of regulation. For example Laisk[12], Terashima[16] and others have pointed out that chloroplasts near to the upper part of a leaf will behave like those in a sun leaf and that those in the lower part of a leaf will behave like those in a shade leaf (Fig. 6a). For this and other reasons the rate v PFD relationship will be curvi-linear rather than two straight lines (the roof and ceiling of Blackman). Nevertheless the extent to which the curve approaches these constraints may be taken as a measure of the effectivenes with which a leaf utilises light – in this instance in the 0-800 μmole quanta.m^{-2}.s^{-1} range. Comparisons may of course be made within any range which an investigator cares to select. In the 0-100 μmole quanta.m^{-2}.s^{-1} range, for example, the NLUC for shade leaves will normally be much larger than the corresponding value for sun leaves (Fig. 6b,c). This is not because shade leaves are able to use light more effectively *in photosynthesis* and indeed it unlikely that there will be any difference in efficiency as indicated by the initial slope.

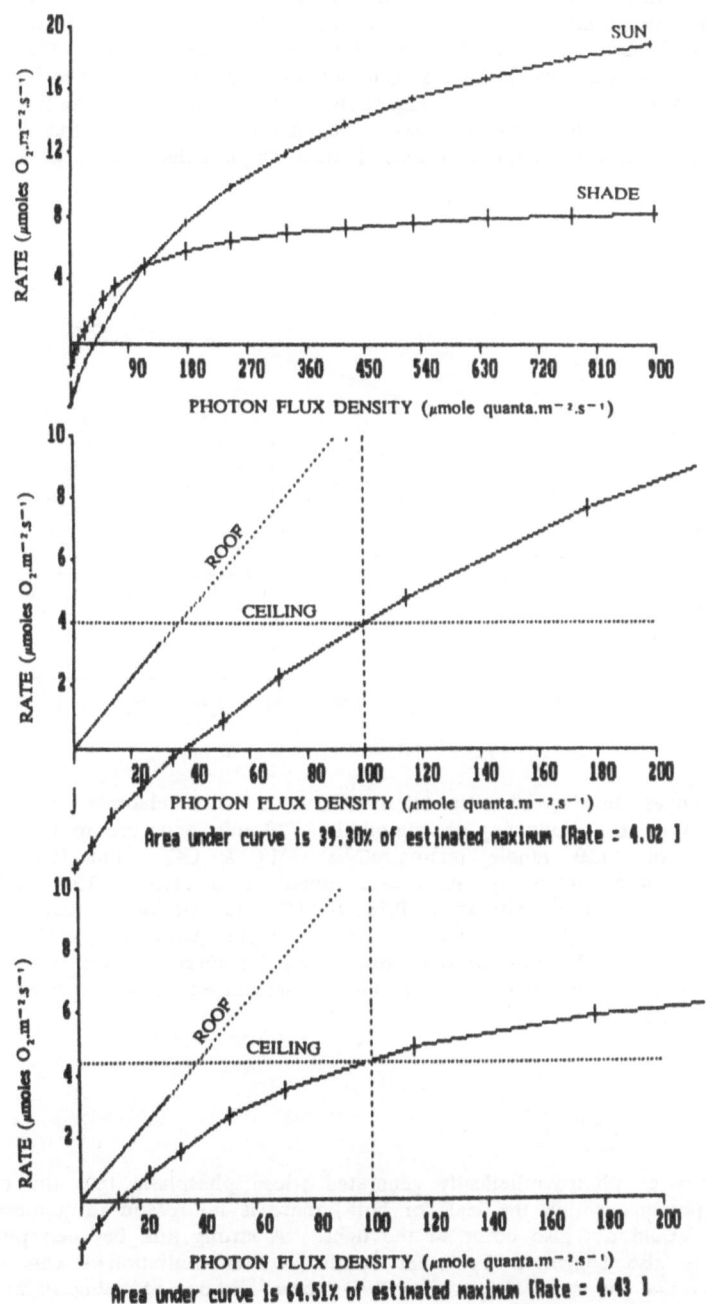

Figure 6. Rates of photosynthetic oxygen evolution of sun and shade leaves of avocado compared over the range 0 to 900 μmole quanta.m^{-2}.s^{-1}. The shade leaf (a) has a much higher (64%) NLUC than its sun (b) counterpart (39%) in the range of 0 to 100 μmole quanta.m^{-2}.s^{-1} but is should be noted that this is not as a result of more efficient photosynthesis at low light but because the light compensation point is lower as a consequence of diminished respiration.

What increases the NLUC in these circumstances is the fact that the light compensation (i.e. the intersect on the horizontal axis) is at a lower PFD in shade leaves than in sun leaves. Moreover, this decrease in the light compensation point is not an attribute of more effective photosynthesis, *per se*, but simply relates to the fact that the dark respiration value is lower in such shade leaves than it is in sun leaves. The relationship between photosynthesis and respiration is complex[8,9] and still very poorly understood. As we have already noted, there is the suggestion[10,11] that low light inhibits respiration. On the other hand, it seems clear that oxygen uptake is enhanced in the dark (Fig. 5) following a period of bright illumination and if such "light–enhanced dark respiration" is

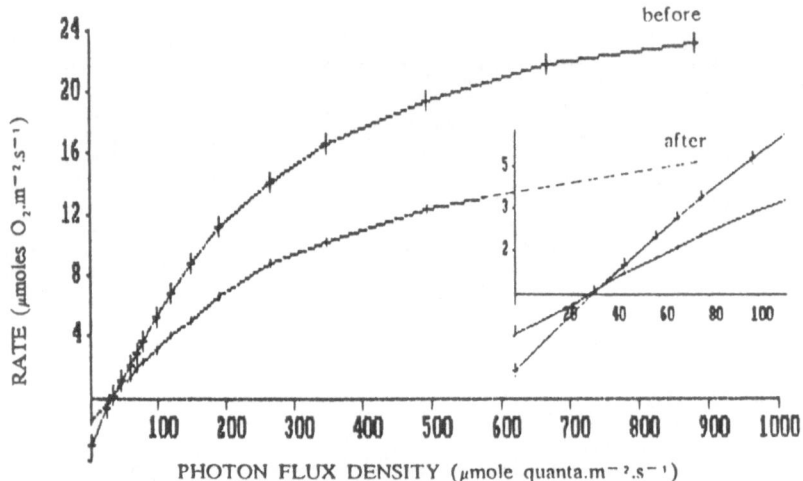

Figure 7. Changes in "dark" respiration associated with changes in the rate of photosynthesis in spinach. Following photoinhibition (70 min exposure to red light at 660 nm and a PFD of 1150 μmole quanta.m^{-2}.s^{-1} in a CO_2 and largely O_2 free atmosphere) which increased the quantum requirement from 11.3 to 19.8 and decreased the nominal light utilisation capacity from 70% to 51% the rate of photosynthesis (at 800 μmole quanta.m^{-2}.s^{-1} was lowered from 23 to 15 μmole quanta.m^{-2}.s^{-1}). There was an associated decrease in the rate of dark respiration (cf intercepts on the vertical axis) from 3 to 1.5 which maintained the LCP at an unchanged value of about 30 μmole quanta.m^{-2}.s^{-1}. Inset shows detail.

related to the flux of photosynthetically generated triose phosphate into the cytosol[6] (or to increased temperature within the leaf, or both) there is no reason to suppose that such faster respiration would not also occur in the light. A strong link between photosynthesis and respiration is also suggested by other results. Photoinhibition[22] can result in a decrease in the initial slope of the rate v PFD curve without affecting dark respiration, so that the LCP also shifts to a higher value. In some circumstances, however, the LCP remains unchanged (Fig. 7) because respiration is diminished, implying that the rate of oxygen uptake *is* then related to the rate of photosynthesis. Converse changes have also been observed in measurements on shade–grown *Phaseolus vulgaris* leaves (Fig. 8). This species contains the dark inhibitor of Rubisco[14,18] and will not photosynthesise at full capacity following a short period of illumination at similar PFD's to those that it has experienced during growth but only after an hour or so of stronger illumination. In some circumstances such activation (or acclimation, or presumed removal of the inhibitor) is associated with an increase in the initial slope of the rate v FPD relationship which does not affect the LCP. Accordingly there is an associated *increase* in dark respiration, again suggesting that the rate of respiration may have accommodated itself to increase substrate flux.

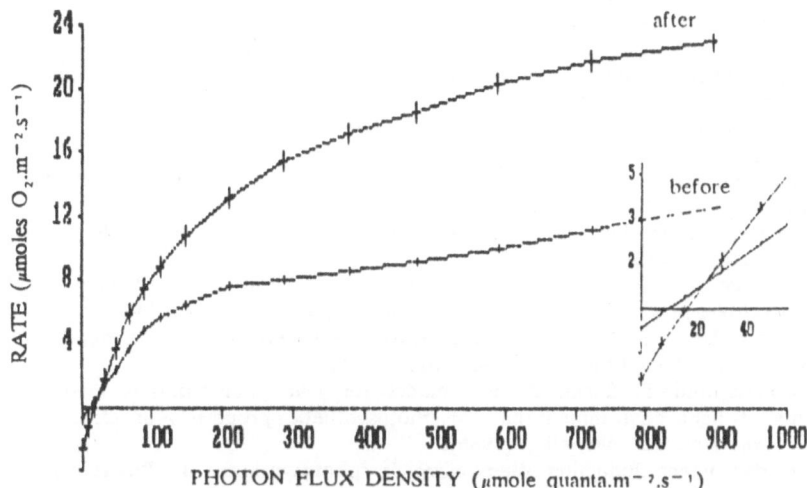

Figure 8. Changes in dark respiration associated with changes in the rate of photosynthesis in *Phaseolus vulgaris*. A leaf taken from a shaded environment was pre–illuminated for 5 minutes at about 200 μmole quanta.m^{-2}.s^{-1}. After initial measurement and illumination at 250 μmole quanta.m^{-2}.s^{-1} for 1 hour the measurement was repeated. Photosynthesis (at 800 μmole quanta.m^{-2}.s^{-1}) increased from 12.5 to 21 μmole quanta.m^{-2}.s^{-1} the NLUC from 44% to 74% (in the range 0–800 μmole quanta.m^{-2}.s^{-1}), the LCP from 12 to 18 μmole quanta.m^{-2}.s^{-1} and the quantum requirement *decreased* from 16 to 9.25. Associated with the effect (essentially the converse of that illustrated in Fig. 6) there was an increase in the rate of dark respiraiton from 0.74 to 2.6. Taken together Figs. 7 and 8 imply that the rate of dark respiration is linked to the rate of photosynthesis, rising and falling as the rate of photosynthesis rises and falls.

This article has attempted to illustrate the potential importance of rate v PFD measurements in the new context of computerisation. The combination[14] of hardware and software which is described allows the possibility of rate v PFD evaluation as a routine measurement even in non–specialist laboratories. Computer assisted analysis of results then greatly extends the usefulness of the results so obtained.

REFERENCES

1. O. Björkman and B. Demmig, Photon yield of O$_2$ evolution and chlorophyll fluorescence characteristics at 77K among vascular plant of diverse origins. *Planta* 170: 489–504 (1987).
2. F.F. Blackman, Experimental researches on vegetable assimilation and respiration. I. On a new method for investigating the carbon acid exchanges of plants. *Phil. Trans. Roy. Soc. B.* 186: 485–502, (1895a).
3. F.F. Blackman, Experimental researches on vegetable assimilation and respiration. II. On the paths of gaseous exchange between aerial leaves and the atmosphere. *Phil. Trans. Roy. Soc. B.* 186: 503–562, (1895b).
4. J.P. Decker, Further evidence of increased carbon dioxide production accompanying photosynthesis. *J. Solar Energy Sci. and Engng.* 1: 30–33, (1957).
5. T. Delieu and D.A. Walker, Polarographic measurement of photosynthetic oxygen evolution by leaf discs. *New Phytol.* 89: 165–178, (1981).

6. G.E. Edwards and D.A. Walker, "C$_3$, C$_4$, Mechanisms, and Cellular and Environmental Regulation of Photosynthesis". Blackwell Scientific Pubs., Oxford, (1983).

7. R. Emerson, The quantum yield of photosynthesis. *Ann. Rev. Plant Physiol.* 9: 1–24, (1958).

8. D. Graham, Effects of light on dark respiration. *in*: Biochemistry of Plants. Vol 2 (D.D. Davies, ed) Academic Press, New York, (1980).

9. D. Graham and D.A. Walker, Some effects of light on the interconversion of metabolites in green leaves. *Biochem. J.* 82: 554–650, (1962).

10. B. Kok, A critical consideration of the quantum yield of *Chlorella* – photosynthesis. *Enzymologia* 13: 1–56, (1948).

11. B. Kok, On the interrelation of photosynthesis and respiration in green plants. *Biochim. Biophys. Acta* 3: 625–631, (1949).

12. A. Laisk, "Kinetics of Photosynthesis and Photorespiration in C$_3$ Plants". Nauka, Moscow (In Russian), (1977).

13. S.B. Powles, Photoinhibition of photosynthesis induced by visible light. *Ann. Rev. Plant Physiol.* 35: 15–44, (1984).

14. J.C. Servaites, Binding of a phosphorylated inhibitor to ribulose bisphosphate carboxylase/oxygenase during the night. *Plant Physiol.* 78: 839–843, (1985).

15. R.E. Sharp, M.A. Matthews and J.S. Boyer, Kok effect and the quantum yield of photosynthesis. *Plant Physiol.* 75: 95–101, (1984).

16. I. Terashima and T. Saeki, A new model for leaf photosynthesis incorporating the gradients of light environment and of photosynthetic properties of chloroplasts within a leaf. *Ann. Bot.* 56: 489–499, (1985).

17. R. Van der Veen, Induction phenomena in photosynthesis, I. *Physiologia Plant.* 2: 217–234, (1949).

18. C.V. Vu, L.H. Allen Jr. and G. Bowes, Dark/light modulation of ribulose bisphosphate carboxylase activity in plants from different photosynthetic categories. *Plant Physiol.* 76: 843–845, (1984).

19. D.A. Walker, "The Use of the Oxygen Electrode and Fluorescence Probes in Simple Measurements of Photosynthesis (2nd Edition)". Oxygraphics Ltd, Sheffield (1988).

20. D.A. Walker, A new procedure for the measurement of the rate of photosynthesis as a function of photon flux density: methodology and implications, *in*: Proc. Phil. Trans. Roy. Soc. B. Meeting, London May 1988. In press, (1988).

21. D.A. Walker, Some aspects of the relationship between chlorophyll *a* fluorescence and photosynthetic carbon assimilation. *in*: Applications of Chlorophyll Fluorescence (In Photosynthesis Research, Stress Physiology, Limnology and Remote Sensing)" Kluwer Academic Pubs, Dordrecht. In press, (1988).

22. D.A. Walker and C.B. Osmond, Measurement of photosynthesis *in vivo* with a leaf disc electrode: correlations between light dependence of steady–state photosynthetic O$_2$ evolution and a chlorophyll *a* fluorescence transients. *Proc. R. Soc. Lond.* B 227: 267–280.

ASSIMILATORY FORCE AND REGULATION OF PHOTOSYNTHETIC CARBON REDUCTION IN LEAVES

K.-J. Dietz and U. Heber

Institute of Botany and Pharmaceutical Biology
University of Würzburg
D-8700 Würzburg
Germany

SUMMARY

The assimilatory force F_A is the product of the phosphorylation potential ATP/(ADP P_i) and the redox ratio NADPH/NADP. The light-dependent change in the free energy of F_A is the driving force of carbon assimilation. It is small compared to energy turnover in assimilation and not directly related to carbon flux. Deviations from proportionality between force and flux is explained by changes in the resistance offered to carbon flux. Enzyme modulation by the chloroplast thioredoxin system and by effectors and changes in substrate or metabolite concentrations and in activation energy modify resistance to carbon flux. As a result, low phosphorylation potentials and low ratios of NADPH to NADP simultaneously satisfy substrate requirements of thylakoid reactions and energy requirements of carbon assimilation over a wide range of fluxes. In illuminated leaves, high F_A values are indicative of flux limitations. However, even when stomata are closed under water stress so that access of CO_2 is retricted, F_A values are not excessive in the presence of air levels of oxygen. Under these conditions, control of photosystem II activity and NADPH and ATP consumption via photorespiratory CO_2 turnover prevent over-reduction of the electron transport chain, a condition leading to photoinactivation. Burdens of flux control are shared by different photosynthetic reactions. The extent of control exerted by a particular reaction changes as flux conditions change. The complex relationship between control coefficient, ΔG and activation energy of a reaction is discussed.

INTRODUCTION

In primary reactions of photosynthetic membranes, the electromagnetic energy of light is captured and used to oxidize water and reduce the soluble electron acceptor NADP. Simultaneously, ATP is synthesized from ADP and phosphate. As long as light strictly limits photosynthesis, as much as 50 % of the energy of absorbed red quanta is conserved in the form of the free energy of NADPH and ATP. These solutes can drive energy-requiring reactions such as carbon reduction. In the Calvin cycle, CO_2 is converted to sugars. When photorespiration is absent, one mole of CO_2 is reduced to the carbohydrate level at the expense of two moles of NADPH and three moles of ATP. At cellular sucrose and CO_2 concentrations, the free energy stored in sucrose is 504 kJ per mole of carbon. This compares to a package of about 584 kJ free energy available in two moles of NADPH and the slightly more than three moles of ATP per mole of carbon necessary to convert CO_2 to sucrose, if cellular concentrations of pyridine nucleotides, adenylates and phosphate are taken into account. Thus, 85 % of the free energy of NADPH and ATP is retained and stored in carbohydrate during light-limited and CO_2-saturated photosynthesis in C_3 plants such as spinach or beet, and only 15 % is expended to drive sugar formation.

In general, chemical reactions are driven by gradients in free energy, and their rate is determined by the magnitude of those gradients and by kinetic restrictions. In biological catalysis, enzyme properties define a large part of kinetic limitations. ATP and NADPH are produced in the chloroplasts of leaf cells not only by light-dependent reactions, but also in the dark by catabolic reactions such as glycolysis. Glucose-6-phosphate oxidation is also a source of chloroplast NADPH in the dark. In consequence, darkened chloroplasts still contain ATP and NADPH, although at levels which are lower than light levels. If the chloroplast NADP system is 10 % reduced in the dark, and the dark ratio of ATP to ADP is 1 at a phosphate concentration of 30 mmol l^{-1}, the free energy stored in two moles of NADPH and three moles of ATP is still about 559 kJoules. There is still a large gradient in free energy between the energy status of ATP and NADPH and the free energy level of carbohydrate, although carbon reduction is absent in the dark. Apparently, darkening produces an insurmountable kinetic barrier for carbon reduction, which is removed in the light. It should be noted that the well-known oxidative deactivation of photosynthetic enzymes such as fructose bisphosphatase (Buchanan, 1980) cannot solely be responsible for the absence of carbon reduction in the dark, because enzyme deactivation is slow after darkening (Laisk et al. 1988), whereas carbon reduction ceases rapidly as soon as NADPH and ATP have declined to new steady state levels in the dark. In this situation, it is of interest to know within which limits the levels of the light products ATP and NADPH are varied in chloroplasts *in vivo* to support photosynthesis at rates ranging from zero to more than 500 µmol CO_2 reduction per mg chlorophyll in one hour (Dietz et al., 1984).

In the following, we will first define the product of the phosphorylation potential (ATP)/(ADP)(P$_i$) and the redox ratio of the NADP system as assimilatory force F_A and consider the relationship of F_A to carbon reduction. We will then proceed to show how ATP and NADPH support widely varying carbon fluxes at small changes in free energy gradients.

1. ASSIMILATORY FORCE IN THE CALVIN CYCLE

In the Calvin cycle, 13 different biochemical reactions are integrated. CO_2 enters the cycle by carboxylating a pentosephosphate, ribulose bisphosphate ($Ru-1,5-P_2$). The product of the reaction, 3-phosphoglyceric acid (3-PGA), is reduced to the sugar level in an endergonic reaction at the expense of the light products ATP and NADPH. The reaction involves two enzymic steps catalyzed by phosphoglycerate kinase and glyceraldehyde-3-phosphate dehydrogenase, and leads to the formation of glyceraldehyde-3-phosphate which is isomerized by triosephosphate isomerase to dihydroxyacetone phosphate (DHAP). From 6 molecules of 3-PGA formed during the carboxylation of 3 molecules of $Ru-1,5-P_2$, at least 5 must be retained in the chloroplasts to ensure regeneration of the carbon acceptor molecule $Ru-1,5-P_2$, and no more than one can be exported to the cytosol for further processing.

3-PGA reduction may be written as

$$3-PGA^{3-} + ATP^{4-} + NADPH + H^+ \longrightarrow DHAP^{2-} + ADP^{3-} + P_i^{2-} + NADP^+ \quad (1)$$

and the mass action ratio M of this complex reaction is

$$\frac{(3-PGA)}{(DHAP)} \frac{(ATP)}{(ADP)(P_i)} \frac{(NADPH)}{(NADP^+)} (H^+) = M \quad (2)$$

As the reaction has been shown to be close to equilibrium even when photosynthetic carbon reduction is fast (Dietz and Heber 1984), it is usually permissible, at first approximation, to replace M by the equilibrium constant K which is 9.8×10^{-6} (M). Transformation of equation (2) then yields

$$\frac{(DHAP) K}{(3-PGA)(H^+)} = \frac{(ATP)}{(ADP)(P_i)} \frac{(NADPH)}{(NADP^+)} \quad (3)$$

Equation (3) shows that determinations of 3-PGA and DHAP in the chloroplast stroma permit the calculation of the product of the phosphorylation potential

$$P = \frac{(ATP)}{(ADP)(P_i)} \quad (4)$$

and the redox ratio of NADPH to $NADP^+$, if the H^+ concentration of the chloroplast stroma is known. This product is termed assimilatory force

$$F_A = \frac{(ATP)}{(ADP)(P_i)} \frac{(NADPH)}{(NADP^+)} \quad (5)$$

It is instrumental in driving carbon reduction.

In illuminated chloroplasts, H^+ is usually not far from 10^{-8} (mol l^{-1}), and in darkened chloroplasts it is close to 1.58×10^{-8} (mol l^{-1}) (Oja et al. 1986, Laisk et al. 1988). Under conditions of photosynthetic flux, $F_{A\ real}$ is somewhat larger than F_A calculated from DHAP and 3-PGA values because the mass action ratio M is somewhat larger than the equilibrium constant K of reaction (1). For practical purposes, and under normal conditions for photosynthesis of leaves, the difference is unimportant and quasi-equilibrium conditions may be assumed (Dietz and Heber 1984).

As it is very difficult and time-consuming to measure adenylates, phosphate and pyridine nucleotides in the chloroplast stroma, and as concentrations measured after extraction encompass both free and thermodynamically inactive bound solutes, whereas only the concentrations of free solutes are of interest, equation (3) offers an easy possibility to gain information on the status of the light products ATP and NADPH in their relationship to carbon reduction. In the light, Ru-1,5-P_2 carboxylase is usually saturated by Ru-1,5-P_2 and binding of 3-PGA and DHAP is insignificant compared to binding of pyridine nucleotides so that measured concentrations are similar to free concentrations (Ashton, 1982).

Table 1. Effect of freezing conditions on ATP to ADP ratios and ADP concentrations in spinach chloroplasts isolated non-aqueously from leaves photosynthesizing at high light and saturating CO_2 concentrations. Care was taken to freeze the leaves in the light.

Freezing medium	$[ATP][ADP]^{-1}$	[ADP]
Precooled petrolether		
−10°C	0.40	1.52 mmol l^{-1}
−40°C	0.92	1.22
−100°C	1.09	1.12
Liquid nitrogen	1.04	1.14

When information on the status of ATP and NADPH is required for leaves, it is necessary to isolate chloroplasts in a state which corresponds to the *in vivo* state under the particular set of external conditions which is of interest. Usually, the *in vivo* status can be literally frozen in liquid nitrogen or in heptane precooled to − 100°C to avoid the *Leidenfrost* phenomenon which slows freezing of leaves down. Table 1 illustrates the importance of a proper maintenance of temperature conditions during leaf freezing. If it is not rapid, and if temperatures are not low enough, artifacts may be encountered. The frozen leaves must be freeze-dried at carefully controlled low temperatures to make secondary translocation of metabolites impossible. Non-aqueous chloroplast isolation then yields chloroplasts which can be assayed for 3-PGA and DHAP (Heber 1957, Stocking 1959, Dietz and Heber 1984, Heber et al. 1986).

Often, it may actually be sufficient to measure 3-PGA and DHAP in leaf tissue instead of measuring these metabolites in non-aqueously isolated chloroplasts. Both 3-PGA and DHAP are transport metabolites. Both are exchanged between chloroplasts and cytosol of green leaf cells by the phosphate translocator of the chloroplast envelope (Heber and Heldt 1981), and both are absent from the large central vacuole of these cells (Kaiser et al., 1982). Since the ratio of DHAP to 3-PGA is lower in the chloroplast stroma during photosynthesis than in the cytosol (Heber and Heldt 1981), cellular ratios of DHAP to 3-PGA may in the green mesophyll be higher by a factor of about 1.4 than chloroplast values, but since leaves contain a significant percentage of non-green cells which contain PGA, but very little triosephosphate, the error is not as large as indicated by mesophyll data and may be unimportant particularly in measurements where different conditions for photosynthesis are compared.

In earlier work, it has been shown that illumination increases the assimilatory force F_A (Heber et al. 1986). This is not surprising. It is exactly what should be expected if photophosphorylation increases ATP at the expense of ADP, and if electron donation raises the level of NADPH at the expense of $NADP^+$. However, there was no proportionality between F_A or, as the logarithm of F_A represents a thermodynamic driving force, the expression

$$\Delta G_{FA} = RT \ln F_{A\ dark}/F_{A\ light} \qquad (6)$$

and rates of carbon assimilation, when photosynthesis was increased by either increasing the light intensity at constant CO_2, or by increasing CO_2 levels in air at constant light intensity (Heber et al. 1987). Proportionality between flux and driving force should be expected on the basis of a chemical flux equation which may be written as

$$\phi = \Delta G/R \qquad (7)$$

where ϕ represents a reaction rate, ΔG a thermodynamic driving force and R the sum of kinetic barriers opposing flux. Failure of the photosynthetic system to exhibit proportionality between flux of carbon and driving force immediately reveals that the kinetic barriers R are not constant but adjustable, and that flux of carbon in photosynthesis is controlled by controlling R.

As F_A is the product of the phosphorylation potential P and a pyridine nucleotide redox ratio, a light-dependent change in ΔG_{FA} is the sum of a light-dependent change in the free energies of ATP hydrolysis and NADPH oxidation

$$\Delta G_{FA} = RT\ln P_{dark}/P_{light} + RT\ln \frac{(NADPH)_{dark}\ (NADP^+)_{light}}{(NADPH)_{light}\ (NADP^+)_{dark}} \qquad (8)$$

Maximum rates of photosynthesis were observed at values of F_A often below 100 ($l\ mol^{-1}$) and rarely exceeding 200 ($l\ mol^{-1}$). After significantly increasing in the light at low rates of carbon assimilation, F_A has been shown to actually decrease in the light when carbon assimilation increased (Heber et al. 1986). When the difference of F_A between chloroplasts in darkened leaves of C_3 plants such as spinach ($F_A \sim 3$ ($l\ mol^{-1}$)) and chloroplasts photosynthesizing at maximum rates in illuminated plants ($F_A \sim 100$ ($l\ mol^{-1}$)) is expressed as a gradient in free energy according to equations (6) and (8), calculation reveals a driving force of only about 8.6 kJ mol^{-1} ATP or NADPH. This compares to an energy consumption during reduction of one mole carbon of about 580 kJ.

As both the adenylate system and the NADP system contribute to F_A, it is of interest to consider the states of the phosphorylation potential and the NADP system in chloroplasts of illuminated leaves separately.

2. THE CHLOROPLAST PHOSPHORYLATION POTENTIAL

At pH 8, which corresponds to the stroma pH of chloroplasts *in situ*, isolated thylakoids phosphorylate added ADP in the presence of a suitable electron acceptor such as ferricyanide or a mediator of (artificial) cyclic photophosphorylation such as phenazine methosulfate until the phosporylation potential (equation 4) is 100,000 (1 mol^{-1}) or more. At this level of phosphorylation, the free energy of ATP hydrolysis

$$\triangle G_{ATP} = \triangle G_{0\ ATP} - RT \ln P \tag{9}$$

is as high as 66.8 kJ mol^{-1} at pH 8 ($\triangle G_{0\ ATP} = -32$ kJ mol^{-1} at pH 7).

Chloroplast ATP is synthesized by the thylakoid ATP synthetase. The driving force for ATP synthesis is the proton motive force *pmf*, with the transthylakoid proton gradient and a transthylakoid membrane potential $\triangle E$ as its two components. At a membrane potential $\triangle E$ of 20 mV and a temperature of 20°C, the transthylakoid pH gradient is as high as 3.5 pH units, because 3 H$^+$ are consumed per ATP synthesized:

$$\triangle G_{ATP} = 3\ pmf = 3(5.6\ \triangle pH + F\ \triangle E) \tag{10}$$

However, phosphorylation potentials as high as 100.000 (1 mol^{-1}) have neither been observed in intact chloroplasts nor in leaves. Apparently, electron transport is tightly controlled *in vivo* so that even when ATP consumption is decreased, chloroplast phosphorylation potentials cannot rise to high levels. Even in the absence of CO_2, chloroplast ATP to ADP ratios do not exceed 3.5 in illuminated leaves. Apparently, neither the Mehler reaction nor cyclic electron flow are effective in increasing the phosphorylation potential of chloroplasts *in situ* to values even remotely similar to those observed with isolated thylakoids.

As the sum of esterified [P$_0$] and inorganic phosphate [P$_i$] is constant in chloroplasts, and phosphate esters accumulate in the light, inorganic phosphate decreases in the chloroplast stroma under illumination. Unfortunately, reliable data on chloroplast phosphate concentrations are scarce. Because non-aqueously isolated chloroplast fractions are frequently contaminated by vacuolar phosphate, phosphate determinations in these fractions may overestimate chloroplast phosphate. On the other hand, phosphate levels in aqueously isolated chloroplasts may be significantly below chloroplast phosphate levels *in situ* because of possible phosphate leakage during chloroplast isolation. To estimate the concentration of chloroplast phosphate *in situ*, the following approach was taken: The concentrations of the dominant phosphate esters ΣP_0 were measured in chloroplasts which had been isolated non-aqueously from illuminated spinach leaves. Because the large central vacuole of leaf cells does not contain organic phosphate esters, the contamination of non-aqueously isolated chloroplasts by extra-chloroplast phosphate esters is small and can be corrected for by measuring cytoplasmic cross contamination. Total phosphate is then approximated by

$$P_{total} = \Sigma P_o + P_i = 2[ADP] + 3[ATP] + [DHAP] + [Fru-6-P] +$$
$$2[Fru-1,6-P_2] + [Glc-6-P] + [3-PGA] + 2[Rul,5-P_2] + P_i \qquad (11)$$

$Fru-1,6-P_2$ stands for fructose bisphosphate, Fru-6-P and Glc-6-P for fructose and glucose monophosphate. Under different conditions for photosynthesis, the organic phosphate pool had been found to be maximum at low temperatures and saturating CO_2 concentrations. It was 30 mmol l^{-1} at 12°C in air containing 2000 µl CO_2 l^{-1}, when photosynthetically active light was 250 Wm^{-2}. Under these conditions, P_i is thought to limit photosynthesis (Leegood and Furbank, 1986; Sage and Sharkey, 1987; Sivak and Walker, 1986; Stitt, 1986). As the K_m of the ATP synthetase is 0.17 mmol l^{-1} (Huchzermeyer, B., 1988), P_i could scarcely have been higher and was likely to be lower than 1 mol l^{-1} at 12°C, if phosphate was really rate-limiting for photosynthesis. Assuming P_i to be 1 mmol l^{-1}, the sum of ΣP_o and P_i was then 31 mmol l^{-1}, and the phosphorylation potential calculated from an ATP/ADP ratio of 1.13 was 1130 l mol^{-1}.

In the same experiment, comparable leaves were illuminated at higher temperatures and at two different CO_2 concentrations (Table 2). As the temperature increased beyond 12°C, photosynthesis increased, and this increase was particularly pronounced when the CO_2 concentration in air was saturating. Simultaneously, ATP/ADP ratios increased and the concentration of organic phosphate esters decreased. The decrease in ΣP_o can be used to calculate the increase in

Table 2. Adenylate ratios, organic phosphate esters ΣP_o and calculated inorganic phosphate concentrations P_i in chloroplasts at two different CO_2 concentrations and various temperature. The leaves were illuminated (incident energy 250 Wm^{-2}) until steady state photosynthesis was established. Chloroplasts were isolated non-aqueously as described in Dietz and Heber (1984). For details see text. Rates of photosynthetic CO_2 fixation (PS) were determined by CO_2 gas exchange analysis (2000 µl l^{-1}). They are given in µmol CO_2 per mg chlorophyll and hour. Concentrations were calculated from measured metabolite contents by assuming a ratio of chlorophyll to chloroplast volume of 30 µl mg^{-1} chlorophyll (Heldt and Sauer, 1971).

CO_2 concentration µl l^{-1}	Temperature °C	$[ATP][ADP]^{-1}$	$[P_o]'$ mmol l^{-1}	$[P_i]'$ mmol l^{-1}	P l mol^{-1}	PS
330	12	2.35	26	5	470	−
330	16	2.50	21	10	250	−
330	20	2.47	21	10	247	−
330	24	2.55	18	13	200	−
330	28	2.43	19	12	203	−
2000	12	1.13	30	1	1130	80
2000	16	1.35	26	5	270	120
2000	20	1.58	24	7	226	155
2000	24	1.84	22	9	204	190
2000	28	2.25	20	11	205	215

chloroplast P_i, because the total content of phosphate plus
phosphate esters is constant in chloroplasts (Heber and Heldt
1981). For instance, if P_i is 1 mmol l^{-1} and ΣP_0 30 mmol l^{-1},
P_i must be 11 mmol l^{-1} after ΣP_0 has declined to 20 mmol l^{-1}.
Alternatively, if P_i is not 1 mmol l^{-1} but 0.5 or 2 mmol l^{-1}
at $\Sigma P_0 = 30$ mmol l^{-1}, then P_i is not 11, but 10.5 or 12 mmol
l^{-1}, when ΣP_0 is 20 mmol l^{-1}. It should be noted that even a
considerable error in the assumption of the lowest phosphate
level does not much influence calculations of phosphorylation
potentials at high phosphate concentrations. The large
increase in P_i concentration with increasing temperature must
reflect changes in the concentration of free phosphate as the
binding capacity of chloroplast constituents is largely
saturated by other phosphorylated metabolites in the light
when CO_2 is present.

Table 3. Stromal phosphate concentrations and phosphorylation
potentials P calculated under the assumption that P_i is
either 0.5, or 1, or 2 mmol l^{-1} at 12°C, 2000 μl l^{-1} CO_2,
in illuminated leaves. See also data of Table 2. The K_m
of NADP-dependent glyceraldehydephosphate dehydrogenase
for 1,3-DPGA is 1 μmol l^{-1} (Leegood et al., 1985).

T	P_i mmol l^{-1}	P l mol^{-1}	P_i mmol l^{-1}	P l mol^{-1}	P_i mmol l^{-1}	P l mol^{-1}
12°C	0.5	2230	1	1130	2	565
16°C	4.5	300	5	270	6	225
28°C	10.5	214	11	205	12	187

In the experiment of Table 2, phosphate concentrations
were calculated on the assumption of $P_i = 1$ mmol l^{-1} at $(CO_2) =$
2000 μl l^{-1}, T = 12°C and 250 Wm^{-2} light. From obtained values
and known ATP/ADP ratios phosphorylation potentials were
calculated. They decreased considerably as the temperature was
increased from 12 to 16°C (Table 2). The further decrease
observed with a further increase in temperature was small,
both when photosynthesis was limited by CO_2 and when it was
fast in the presence of saturating CO_2. Interestingly, under
both conditions, phosphorylation potentials were comparable,
although rates of photosynthesis, ATP/ADP ratios and phosphate
concentrations were different. Apparently, there was a tight
correlation between ATP production by the thylakoid system and
ATP consumption by reactions of the carbon cycle. Only when
consumption of ATP was considerably slowed down by low
temperatures did the phosphorylation potential increase
significantly.

2.1 Does phosphate limit ATP synthesis at low temperatures?

The observation that calculated phosphorylation
potentials were higher at low than at high temperatures is
puzzling. It contradicts the notion that phosphate limits

Table 4. Equilibrium concentrations of 1,3-diphosphoglycerate in the chloroplast stroma during photosynthesis of leaves at different temperatures in the presence of 2000 or 330 ppm CO_2. Calculations from the adenylate data shown in Table 2 and from measured 3-PGA concentrations.

T, °C	12	16	20	24	28
		A. 2000 μl l^{-1} CO_2			
1,3-DPGA, μmol l^{-1}	3.5	2.1	2.3	2.3	2.0
		B. 330 μl l^{-1} CO_2			
1,3-DPGA, μmol l^{-1}	2.3	1.6	1.4	1.1	0.7

photosynthesis at low temperatures by restricting ATP synthesis. If ATP synthesis were limited by low phosphate levels, phosphorylation potentials should decrease, not increase. Table 3 compares phosphorylation potentials calculated for three different temperatures (12, 16 and 28°C, 2000 μl l^{-1} CO_2, see Table 2) under the assumption, that the P_i concentration at the highest ΣP_0 at 12°C was either 0.5 or 1 or 2 mmol l^{-1}. It can be seen that in all cases the phosphorylation potentials increased with decreasing temperature. They were lowest under the assumption that P_i is 2 mmol l^{-1}. In such a situation, the low K_m of the ATP synthetase for phosphate precludes a limitation of ATP synthesis by phosphate. A similar conclusion can be drawn from the large increase in the phosphorylation potentials caused by lowering the temperature at the lower phosphate concentrations.

However, whereas the data do not support the idea of a phosphate limitation of ATP synthesis at low temperatures, it is still possible that low ATP/ADP ratios restrict 3-PGA reduction. They are a consequence of low phosphate concentrations and are observed particularly at low temperatures in the presence of high CO_2. Table 4 shows concentrations of 1,3-diphosphoglycerate in equilibrium with 3-phosphoglycerate, ATP and ADP. The concentrations are indeed low and not very far from the K_m of glyceraldehyde-3-phosphate dehydrogenase for 1,3-diphosphoglycerate which is 1 μmol l^{-1} (Leegood et al. 1985). However, as they decrease with increasing rates of photosynthesis at higher temperatures, even rate-limitations by low 1,3-diphosphoglycerate concentrations are unlikely.

The fact remains that feeding P_i to leaves can increase the rate of photosynthesis particularly at low temperatures. It appears that the action of P_i is complex in these circumstances, bringing about adjustments in the concentrations of several intermediates after increasing ATP/ADP ratios. As substrate levels can be important in determing flux resistance, this can decrease resistance to carbon flux and increase assimilation.

2.2 Is photosynthetic electron transport under photosynthetic control in leaves?

At a phosphorylation potential of 200 l mol^{-1} (Table 2), the ΔG_{ATP} is 51 kJ mol^{-1} at pH 8, i.e. about 16 kJ mol^{-1} lower than the maximum ΔG_{ATP} generated by isolated thylakoids. When the membrane potential is 20 mV, a Δ pH of not much more than 2.6 is sufficient to generate a phosphorylation potential of 200 l mol^{-1} which is capable of supporting fast photosynthesis. It is important to note that photosynthetic control, i.e. control of electron transport by the transthylakoid proton gradient, is sufficiently strong to inhibit electron flow only at large proton gradients. By permitting rapid photosynthesis at low phosphorylation potentials, the chloroplast avoids inhibition of linear electron transport by a large ΔpH. Simultaneously, high concentrations of ADP ensure saturation of the ATP synthetase with its substrate (Aflalo and Shavit, 1984; Tran-Anh and Rumberg, 1987).

3. THE NADP SYSTEM

The second component of the assimilatory force F_A is the NADP system (see equation 5). Its redox state is determined by the balance between electron donation to NADP$^+$ by the electron transport chain and oxidation of the resulting NADPH during the reduction of 1,3-diphosphoglycerate which is mediated by NADP-dependent glyceraldehyde-3-phosphate dehydrogenase. The chloroplast electron transport chain can reduce several electron acceptors. Oxygen reduction occurs in the Mehler reaction and during autoxidation of reduced ferredoxin. However, ferredoxin-NADP-reductase has a high affinity for electrons of the electron transport chain and for NADP$^+$. Isolated thylakoids reduce added NADP$^+$ efficiently even in the presence of high oxygen levels until the ratio of NADP$^+$ to NADPH is 50 or even higher.

However, neither in isolated chloroplasts nor in chloro-

Table 5. Chloroplast concentrations of 3-PGA and DHAP, assimilatory force F_A, phosphorylation potentials P and percent reduction of the chloroplast NADP system in illuminated spinach leaves. Photosynthesis in air, with 330 or 2000 µl l^{-1} CO_2 present. Light intensity 250 Wm^{-2}. The pH of the chloroplast stroma was assumed to be 8.

CO_2 µl l^{-1}	T,$^\circ$C	3-PGA mmol l^{-1}	DHAP mmol l^{-1}	F_A l mol^{-1}	P l mol^{-1}	NADP, % reduced
330	12	3.15	0.77	148	470	12 %
330	16	2.00	0.60	185	250	36 %
330	20	1.79	0.56	194	247	46 %
330	24	1.33	0.44	205	200	50 %
330	28	0.98	0.52	328	203	62 %
2000	12	9.88	0.91	57	1130	11 %
2000	16	4.98	0.82	102	270	29 %
2000	20	4.74	0.74	96	226	28 %
2000	24	3.95	0.74	116	204	37 %
2000	28	2.82	0.70	153	205	43 %

plasts *in situ* could such high ratios ever be observed. When intact chloroplasts are illuminated in the absence of substrates which can accept electrons from NADPH, NADP⁺ levels decrease rapidly. Maximum ratios of NADPH to NADP⁺ are usually not much higher than 3 (Heber et al. 1986). Addition of reducible substrates rapidly shifts the ratio towards oxidation (Takahama et al. 1981).

When dark-adapted leaves are illuminated, NADPH increases only transiently in the chloroplasts. It decreases again as the photosynthetic apparatus becomes light-activated (Heber and Santarius 1965). It is difficult to estimate thermodynamically active concentrations of NADPH and NADP⁺ in chloroplasts from measured concentrations because the pools of the pyridine nucleotides are small and a considerable proportion of them may be enzyme-bound *in vivo* (Borst, 1963). However, if values of the assimilatory force F_A (equation 5) and of phosphorylation potentials P are known, the percent reduction of the NADP system can be easily calculated. Table 5 lists chloroplast concentrations of 3-PGA and DHAP for the experiment shown already in Table 2. They are used to calculate F_A. Phosphorylation potentials P are taken from Table 2. Equation 5 then permits the calculation of the percent reduction of the NADP system. The data show that reduction of the NADP system increased with the increased rates of photosynthesis brought about by increased temperatures.

Under the same conditions, the phosphorylation potential actually decreased. Apparently, some degree of photosynthetic control was still operative, when low temperatures limited photosynthesis. Such control shifted the redox state of the NADP system towards oxidation.

3.1 Oxidation of the NADP system at low temperatures

The data of Table 5 show that even when photosynthesis is restricted by air levels of CO_2 or by temperatures as low as 12°C, significant NADP is always available as an electron acceptor. This permits charge separation at the reaction center of photosystem I even when photosynthesis is slow. Under high intensity illumination, cytochrome f is oxidized, and a large redox gradient is maintained between photosystem II and photosystem I (Krause and Behrend, 1986). This is a prerequisite for light-dependent changes in the photochemical system which permit efficient dissipation of excess excitation energy as heat and protect the thylakoid system against photoinactivation (Heber et al. 1988a). These changes have been suggested to involve the xanthophyll cycle (Demmig et al., 1987).

4. ROLE OF F_A AS A MEDIATOR OF ENERGY TRANSFER AND CHANGING FLUX CONTROL IN PHOTOSYNTHESIS

The difference in the temperature dependence of the phosphorylation potential (Table 2) and of the redox state of the NADP system (Table 5) reveals that control of electron flux in the thylakoid system changes with temperature. This may be illustrated by a simple example. Electron flow may be compared with water flow in a vertical pipe system which encloses a turning wheel with variable resistance to turning. The water flow drives the wheel and the hydrostatic pressure in the pipe determines how fast the wheel is spinning. It also responds to the resistance the wheel offers to the passage of water. When the wheel is stopped, hydrostatic pressure will increase as the outflow of water is decreased while water is still pouring into the pipe. The latter may actually burst when it can no longer withstand pressure. Pressure will decrease when the wheel is permitted to spin faster. Obviously, the resistance the wheel offers to spinning feeds back on hydrostatic pressure in the system. To prevent the pipe from bursting, it may actually be necessary to control water inflow into the pipe so that water is diverted from the pipe and excessive hydrostatic pressure cannot develop.

In terms of photosynthesis, it may be necessary to prevent light energy from reaching the reaction center of photosystem II or to control the activity of photosystem II, which feeds electrons into the electron transport chain, in order to avert damage to the electron transport chain which is sensitive to light when over-reduced. Such control has actually been demonstrated to occur (Demmig and Winter, 1988 a, b; Weis and Berry 1987) . As the phosphorylation potential increases with decreasing temperature (Table 2), reduction of the NADP system decreases. An increase in the phosphorylation potential indicates an increase in the proton motive force across the thylakoids. This restricts photosystem II activity (Weis and Berry 1987). It also indicates increased control of electron flow by the cytochrome b/f complex (Heber et al. 1988b). The decrease in the phosphorylation potential with increasing temperature (Table 5) indicates relaxation of this control. At the same time F_A and NADP reduction increase (Table 5) suggesting, at first sight, that now resistance to electron flux has shifted to carbon reduction and associated reactions. However, such a view may be misleading. The increase in F_A must be seen in relation to the increase in carbon flux. Increased hydrostatic pressure in the pipe will be in direct relation to a faster spinning of the wheel only as long as resistance to spinning remains unchanged. If the center axis of the wheel is lubricated so as to decrease resistance to spinning, faster spinning may be observed at an unchanged or even a decreased hydrostatic pressure in the pipe.

In an enzymic system, lubrication may be equated with the release of control, with enzyme activation or the removal of substrate or effector limitations. This decreases resistance to metabolic flux. F_A is higher when photosynthesis is restricted by CO_2 than when it is fast under CO_2 saturation (Table 5) because, at high CO_2, carboxylation resistance is decreased. This should be expected to increase flux (see rates of photosynthesis in Table 2) and decrease F_A wich mediates energy transfer from the thylakoids to the stroma enzymes. Although F_A drives carbon reduction, it is simultaneously consumed in this process. This explains why F_A may be lower when photosynthesis is fast than when it is slow (Heber et al. 1987). Often, enzyme activation may account for such behaviour.

Equation 7 shows that chemical flux is proportional to a gradient in free energy and inversely proportional to existing kinetic limitations. Flux of carbon is driven by light-dependent changes in the free energy of F_A which represents the sum of light-dependent changes in two different interacting thermodynamic driving forces, ΔG_P and ΔG_{NADPH}, as

352

shown in equation (8). In Table 6, flux data of Table 2 and F_A data of Table 5 are used to calculate resistance to carbon flux in photosynthesis from ΔG_{FA} as the temperature is increased from 12 to 28°C. The calculations reveal that resistance to carbon flux actually decreases with increasing temperature. This shows that the tentative conclusion drawn above from the increase of F_A is incorrect. Control does not shift with increasing temperature to carbon assimilation. Rather, at the light intensity at which the experiments were performed (see Table 2), considerable control still resides in the electron transport chain (Heber et al. 1988b) although the decreased phosphorylation potential indicates decreased

Table 6. Calculation of nominal values of flux resistance R in carbon assimilation as a function of temperature according to equation (7) from measured rates of photosynthesis (carbon flux Φ (in mol $(cm^2$ leaf area $s)^{-1}$); the calculation is based on the assumption that 1 mg chlorophyll corresponds to 30 cm^2) in the presence of 2000 $\mu l\ l^{-1}$ CO_2 and ΔG_{FA} (in kJ mol^{-1}, see equations 6 and 8; $F_{A\ dark} = 3\ l\ mol^{-1}$). Values of F_A are from Table 5. The light intensity was 250 Wm^{-2}

Temperature, °C	Φ x 10^{-10}	ΔG_{FA}	R x 10^9
12	7.4	7.1	9.6
16	11.1	8.6	7.7
20	14.3	8.6	6.0
24	17.6	9.2	5.2
28	19.9	10.0	5.0

control of electron flow by the transthylakoid proton gradient. The decrease in resistance to carbon flux observed with increasing temperature (Table 6) may be a simple temperature effect being due to a decrease in activation energy. Still, as the overall flux resistance in a complex system is always the sum of various resistances, closer inspection is required for additional information.

4.1 Free energy changes in Calvin cycle reactions

The Calvin cycle may be subdivided into the carboxylation of ribulose bisphosphate ($Ru-1,5-P_2$), the reduction of the resulting 3-PGA to triosephosphate, the condensation of two triosephosphates to fructose bisphosphate ($Fru-1,6-P_2$), the hydrolysis of $Fru-1,6-P_2$ to fructose 6-phosphate (Fru-6-P) and the regeneration of $Ru-1,5-P_2$ from Fru-6-P and triosephosphates. In pioneering investigations with the alga *Chlorella vulgaris*, Bassham and Krause (1969) have analyzed the thermodynamics of these reactions. Long before thioredoxin regulation of key chloroplast enzymes (Buchanan, 1980) was known, they have identified irreversible partial reactions and recognized the need for their metabolic control.

Table 7 shows an adaptation of their basic data to the situation existing in the chloroplast stroma of illuminated spinach leaves. The carboxylation of $Ru-1,5-P_2$ is irreversible, and considerable free energy is lost. In contrast, the reactions reducing 3-PGA are freely reversible.

Even under flux conditions, the thermodynamic gradient is small. Calculation reveals a positive free energy change for the formation of Fru-1,6-P$_2$. It is evident that this cannot reflect the real flux situation. It has been recognized earlier that part of the measured Fru-1,6-P$_2$ is protein-bound (Dietz and Heber 1984). This explains why the calculated gradient appears to oppose carbon flux. It is justified to assume quasi-equilibrium for the formation of Fru-1,6-P$_2$ from triosephosphates. Hydrolysis of Fru-1,6-P$_2$ is irreversible and

Table 7. Chloroplast concentrations of Calvin cycle intermediates, free energy changes of formation from the elements $\Delta G'$, free energy corrections for the *in situ*-concentrations $\Delta\Delta G$ and physiological free energy change of formation from the elements ΔG^a (cf. Bassham and Krause, 1969). Metabolite concentrations were measured in chloroplasts isolated non-aqueously from spinach leaves photosynthesizing in air containing 330 µl l^{-1} CO$_2$ at 200 Wm^{-2}. The CO$_2$ concentration at the site of carboxylation was estimated to 6 µmol l^{-1}. The stroma pH is considered to be 8.0, the NADPH to NADP$^+$ ratio is 1 (see Table 5) and the inorganic phosphate concentration 10 mmol l^{-1} (see Table 2). The free energy change of formation of H$^+$ as product of the Ru-1,5-P$_2$ regeneration reaction is taken into account in the free energy change of hydrolysis of ATP^{4-}+ H$_2$O -> ADP^{3-}+ P$_i^{2-}$+ H$^+$. The letter P represents the free energy of formation from the elements of phosphate, HPO$_4^{2-}$ minus the free energy of formation from the elements of water (cf. Bassham and Krause, 1969).

Metabolite	conc. (mmol l^{-1})	$\Delta G'$, kJ	$\Delta\Delta G$, kJ	ΔG^a, kJ	balance, kJ

Ru-1,5-P$_2$ carboxylation :

$$Ru-1,5-P_2{}^{4-}+ CO_2 + H_2O \rightarrow 2\ 3\text{-}PGA^{3-}+ 2\ H^+$$

Metabolite	conc.	$\Delta G'$	$\Delta\Delta G$	ΔG^a	balance
Ru-1,5-P$_2$	6.1	-733.3+2P	-12.6	-745.9+2P	+745.9
CO$_2$	6*10^{-3}	-396.1	-29.8	-425.9	+425.9
H$_2$O	/	-238.1	/	-238.1	+238.1
H$^+$	10^{-8}	- 39.9 (pH7)	-5.9	- 45.8	- 91.6
3-PGA	2.4	-661.5+P	-6.7	-676.6+P	-1353.2

$$\Delta G = - 34.9$$

continued

known to be under flux control (Buchanan 1980). The regeneration of Ru-1,5-P$_2$ from Fru-6-P and triosephosphates requires ATP. The overall free energy change for the formation of 3 mol Ru-1,5-P$_2$ is -62.2 kJ, i.e. about -21 kJ/mol.

3-PGA reduction:

$$3\text{-PGA}^{3-} + H^+ + ATP^{4-} + NADPH \rightarrow DHAP^{2-} + ADP^{3-} + NADP^+ + P_i^{2-}$$

3-PGA	2.4	−661.5+P	−6.7	−676.6+P	+676.6
H+	10^−8	− 39.9 (pH7)	−5.9	− 45.8	+ 45.8
NADPH/NADP+	1.0	−220.9	0	−220.9	−220.9
ATP/ADP	2.3	− 32.1	−2.1	− 34.1	− 34.1
P_i	10.0	P	−11.4		− 11.4
DHAP	0.7	−438.1+P	−18.1	−456.1+P	−456.1
GA1-3-P	0.032	−430.5+P	−25.6	−456.1+P	

$$\Delta G = -0.1$$

Fru-1,6-P₂ aldolase:

$$DHAP^{2-} + GA1\text{-}3\text{-}P^{2-} \rightarrow Fru\text{-}1,6\text{-}P_2^{4-}$$

DHAP	0.7	−438.1+P	−18.1	−456.1+P	+456.1
GA1-3-P	0.032	−430.5+P	−25.6	−456.1+P	+456.1
Fru-1,6-P₂	0.67	−890.4+2P	−18.2	−908.5+2P	−908.5

$$\Delta G = + 3.7$$

Fru-1,6-P₂ ase:

$$Fru\text{-}1,6\text{-}P_2^{4-} + H_2O \rightarrow Fru\text{-}6\text{-}P^{2-} + P_i^{2-}$$

Fru-1,6-P₂	0.67	−890.4+2P	−18.2	−908.5+2P	+908.5
Fru-6-P	1.5	−904.8+P	−16.2	−921.0+P	−921.0
P_i	10.0	P	−11.4		− 11.4

$$\Delta G = -23.9$$

Ru-1,5-P₂-regeneration:

$$DHAP^{2-} + 2\ GA1\text{-}3\text{-}P^{2-} + Fru\text{-}6\text{-}P^{2-} + 3\ ATP^{4-} \rightarrow$$
$$3\ Ru\text{-}1,5\text{-}P_2^{4-} + P_i^{2-} + 3\ ADP^{3-} + 3\ H^+$$

DHAP	0.7	−438.1+P	−18.1	−456.1+P	+456.1
GA1-3-P	0.032	−430.5+P	−25.6	−456.1+P	+912.2
Fru-6-P	1.5	−904.8+P	−16.2	−921.0+P	+921.0
Ru-1,5-P₂	6.1	−733.3+2P	−12.6	−745.9+2P	−2237.8
P_i	10.0	P	−11.4		− 11.4
ATP/ADP	2.3	− 32.1	− 2.1	− 34.1	−102.4

$$\Delta G = - 62.3$$

The total loss of free energy was calculated from these data for the fixation and reduction of 1 mol CO_2. The result is shown in Table 8. ΔG values for F-1,6-P₂ hydrolysis and Ru-1,5-P₂ regeneration had to be divided by 3, because during reduction and processing of one mole carbon only 1/3 of the energy loss recorded in Table 7 is encountered. The data show that the largest loss in free energy occurs during carboxylation of Ru-1,5-P₂, but regeneration of this molecule

Table 8. Calculation of the chemical free energy lost in the Calvin cycle (A) and expended (B) in the reactions of the Calvin cycle and of sucrose synthesis (1/12 th ATP per carbon fixed), as compared to the energy conserved in carbohydrates (C). The latter amounts to 503.6 kJ per mole carbon. The cellular concentration of sucrose was estimated to 30 mmol l^{-1}. Thus, the investment of 583 kJ minus the energy conserved in carbohydrates of 503.6 kJ gives the total amount of free chemical energy lost, i.e. 79.4 kJ. The energy balance of the Calvin cycle is based on one turnover, i.e. one carboxylation reaction, two 3-PGA reduction reactions, one third Fru-1,6-P_2ase reaction and one third of the Ru-1,5-P_2 regeneration reaction as described in Table 7.

A. Energy lost:
Ru-1,5-P_2 carboxylation	- 34.9
3-PGA reduction	- 0.2
Fru-1,6-P_2 hydrolysis	- 8.0
Ru-1,5-P_2 regeneration	- 20.8
	- 63.9 kJ

B. Energy expended:
2 NADPH, 3 ATP	578.8
1/12 ATP (for sucrose synthesis)	4.2
	583.0 kJ

C. Energy conserved: 503.6 kJ

D. Unaccounted for and probably lost in part during cytosolic processing of carbon to sucrose: -15.5 kJ

---

is also costly. 20 % of the energy lost could not be accounted for by reactions of the Calvin cycle. This may be due in part to losses encountered during the processing of reduced carbon to sucrose in the cytosol, but it is also possible that the free energy data used for calculating ΔG_{ATP} which is influenced by Mg^{++} are not fully correct.

4.2 Shifting control within the Calvin cycle

As all individual enzymes of a biochemical pathway are indispensible for its functioning, all must participate in the control of flux through the pathway. The analysis of flux control has been pioneered by Kacser and Burns (1973, 1979) and by Heinrich and Rapaport (1974). These authors have introduced the use of control coefficients which define the extent of control by individual catalysts. The sum of all control coefficients must be 1, as any one of the catalysts exercizes only partial control. If an enzyme is so abundant as to catalyze metabolic flux very close to the equilibrium between its substrates and its products, a small change in its concentration or its activity is unlikely to affect flux significantly. The control coefficient assigned to this enzyme must be small. On the other hand, if the enzyme catalyzes flux

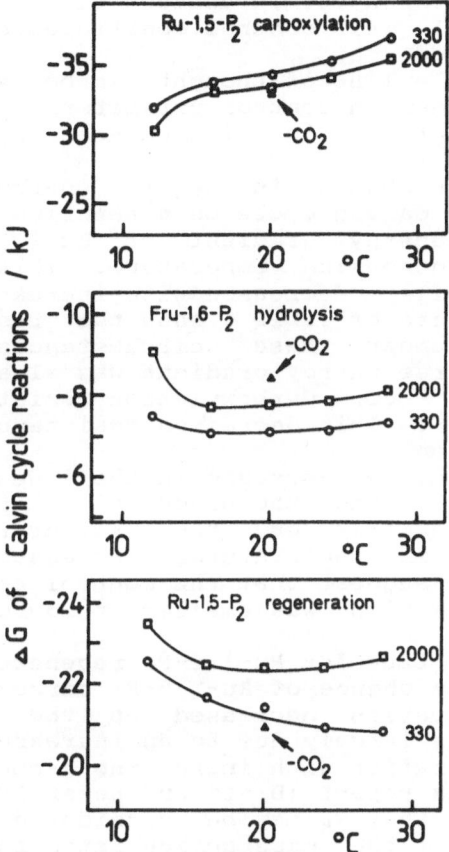

Figure 1. Changes in the physiological free energy of partial reactions of the Calvin cycle in dependence of the temperature at two different CO_2 concentrations. Data are based on one turnover, i.e. one carboxylation reaction, two 3-PGA reduction reactions, one third Fru-1,6-P_2ase reaction and one third of the Ru-1,5-P_2 regeneration reaction as described in Table 7. Asterisks indicate free energy changes determined when metabolic flux was slow at 20° C and 125 Wm^{-2} light in the absence of external CO_2. Chloroplast metabolite concentrations were in this case : 0.57 mmol l^{-1} DHAP, 0.42 mmol l^{-1} Fru-1,6-P_2, 0.81 mmol l^{-1} Fru-6-P, 1.07 mmol l^{-1} 3-PGA, 6.08 mmol l^{-1} Ru-1,5-P_2. The ratio of ATP to ADP was 2.92, the calculated [P_i] 7 mmol l^{-1}.

in a situation in which the reactants are far from equilibrium, its control coefficient is likely to be significant. If it is, disequilibrium will increase, when flux through the system is increased. In this situation, the control coefficient must adjust to the new situation. It will also increase. This causes changes in all other control coefficients of the system, because the sum of the control coefficients is 1. All control coefficients must change if only one is changed.

This simple line of thought can be exploited to gain insight into changes in control parameters of the Calvin cycle when photosynthetic flux is changed by changing the temperature.

Fig. 1 shows changes in the free energy of some partial reactions of the Calvin cycle as a function of temperature. Only the free energy gradient of Ru-1,5-P_2 carboxylase increased with increasing temperature. This was unexpected because CO_2 solubility decreases with increasing temperature, and because the data of Table 6 show that resistance to carbon flux decreases under these circumstances. However, as expected, the free energy gradient was always larger at 330 than at 2000 µl l^{-1} CO_2. Carboxylation resistance is decreased at 2000 µl l^{-1} CO_2. This decreases resistance to carbon flux in CO_2 assimilation.

After an initial decrease in the free energy gradient between Fru-1,6-P_2 and the products of the Fru-1,6-P_2ase reaction at low temperatures, the gradient remained constant as the increase in temperatures increased photosynthesis. Differences in ΔG suggest that the control coefficient of the Fru-1,6-P_2ase reaction was smaller at low CO_2 than at high CO_2.

The same was true for Ru-1,5-P_2 regeneration. In contrast to the free energy change of Ru-1,5-P_2 carboxylation, that of Ru-1,5-P_2 regeneration decreased as the temperature was increased. This is largely due to an increase in the inorganic phosphate concentration with increasing temperature (cf. Table 2). In a previous report (Dietz and Heber 1986), it has been wrongly assumed that P_i may be considered constant. Under these conditions, the mass action ratio for the reactions regenerating Ru-1,5-P_2 from triosephosphates and Fru-6-P has been observed to increase with increasing temperature.

The experiment shown in Fig. 1 was performed at a high light intensity. In this situation, the control coefficient of plastohydroquinone oxidation by the cytochrome b/f complex of the electron transport chain is large, when CO_2 is saturating (Heber et al. 1988b). It is smaller when CO_2 is rate-limiting. However, the data of Fig. 1 and Table 8 show that even under CO_2 saturation significant flux control resides in Ru-1,5-P_2 carboxylase. With increasing temperature, it appears to increase, whereas that of other Calvin cycle reactions seemed to decrease. Even in the absence of CO_2, flux control was recognizable (Fig.1), although carbon flux was very slow. It was supported only by photorespiratory CO_2 turnover.

It must be emphasized that changes in ΔG give only limited insight into the flexibility of control. As enzyme activity is subject to modulation, thermodynamic gradients are adjustable. The activation state of Fru-1,6-P_2ase is determined by several factors. Upon illumination of predarkened leaves, thiol modulation by the ferredoxin/thioredoxin-system leads to the reductive activation of the enzyme. The activation process is accelerated by stroma alkalization. Still, reductive activation of the enzyme is only one aspect of enzyme regulation during steady state photosynthesis (Leegood et al., 1985). Once the enzyme is activated, metabolite modulation becomes important (Leegood et al., 1982).

It should be noted that the data of Fig. 1 do not contradict the observation that resistance to carbon flux decreases by one half as the temperature is increased from 16 to 28°C (Table 6). Probably much of the decrease in flux resistance can be attributed to changes in the activation energy of enzyme reactions which do not find proper expression in the ΔG data of Fig. 1.

5. CARBON CYCLE TURNOVER UNDER WATER STRESS

When leaves are water-stressed, they close their stomata in the light. Net photosynthesis may approach zero, but Calvin cycle activity continues to a limited extent owing to photorespiratory CO_2 release inside the leaf and refixation of the evolved CO_2. Under these conditions, F_A increases to values not much higher than 600 or 900 (1 mol^{-1}) (Heber et al. 1988a), because charge separation at the reaction centers of photosystem II is regulated down by a large proton gradient (Weis and Berry 1987). Extensive photooxidation of P_{700} is still possible showing that the NADP system is partially oxidized in the light even though closed stomata restrict access of CO_2. Only when photorespiration is decreased by reducing the oxygen concentration from 21 to 1 % is photooxidation of P_{700} significantly decreased indicating reduction of the NADP system. In the absence of net photosynthesis, and with only 1% oxygen present, F_A is increased to about 3000 (1 mol^{-1}). The light-dependent change in the free energy of ATP or NADPH represented by this increase is less than 21 kJ mol^{-1}. Under these conditions, the electron transport chain is over-reduced and becomes sensitive to photooxidation. Partial oxidation of the NADP system by photorespiration protects the photosynthetic apparatus against photoinactivation (Powles, 1984; Heber et al. 1988a).

6. MOLECULAR BALANCE AND MOLECULAR COMPROMISE

The photosynthetic apparatus catalyzes flux of electrons to carbon under widely differing conditions and at widely differing rates. To make flux irreversible, free energy is sacrificed within the electron transport chain, the carbon cycle and in cytosolic reactions. When significant loss of free energy occurs in enzyme-catalyzed reactions, regulation is necessary to avoid substrate depletion. If, for instance, the chloroplast fructose bisphosphatase were not regulated, the excessive activities necessary for making high rates of photosynthesis possible would at lower rates deplete not only Fru-1,6-P_2, but also the triosephosphates which are very close to equilibrium with Fru-1,6-P_2. This would make operation of the Calvin cycle impossible. As a matter of fact, ratios of Fru-1,6-P_2 to Fru-6-P change within remarkably narrow limits while photosynthetic flux changes drastically (Dietz and Heber 1984). This proves tight control of fructose bisphosphatase.

Obviously, regulation of different irreversible enzyme reactions within wide limits of flux must be strictly coordinated. This is not a simple task as different proteins differ in their kinetic and regulatory properties, while conditions for photosynthesis change as the day progresses and wheather and seasons change. A major, but not the only, coordinator in the Calvin cycle is the thioredoxin system which activates different light-regulated enzymes in accordance with flux requirements. Because enzyme proteins differ in their properties which dictate their response to signals, burdens of flux control are bound to shift with changing conditions for photosynthesis (Heber et al. 1988b). Evolutionary forces have shaped the properties of catalytic proteins performing the same function in different plants. They differ subtly in their kinetic and regulatory properties. It appears that the differences enable them to balance changing burdens of control successfully only within the conditions of their particular environments thereby restricting their ecological distribution.

The photosynthetic apparatus is a heterogenous system which requires cooperation between different compartments. Whereas the thioredoxin system informs stroma enzymes on electron pressure in the thylakoids, the assimilatory force F_A transfers energy. Once again, signal transmission is required to maintain proper balance. As has been outlined above, a high assimilatory force may aid carbon reduction, but will be incompatible with high rates of electron flow, because the over-reduction accompanying high NADPH/NADP ratios, or the photosynthetic control accompanying high phosphorylation potentials, would curtail electron transport. Actually, over-reduction would expose the photosynthetic apparatus to photodestruction. It must be avoided particularly when photosynthetic flux is restricted as it is under water stress. Decreasing the electron pressure in the thylakoid system by regulation of photosystem II requires the presence of a large transthylakoid pH gradient (Krause and Behrend 1986) which can only be produced and maintained by electron flow. As the Mehler reaction is inadequate to support electron transport (Heber et al. 1988a) and external CO_2 is inaccessible under water stress, consumption of NADPH and ATP during intercellular photorespiratory CO_2 turnover provides the basis for charge separation at photosystem I which protects this photosystem against photodestruction, and for the maintenance of a transthylakoid proton gradient, which protects photosystem II. Photorespiratory carbon loss and reduced productivity is the price C_3 plants have to pay for protection. C_4 plants pay by increased energy requirements.

Thus, a precarious balance between inter-dependent reactions which is based on compromise ensures functionality of the photosynthetic appartus under widely varying environmental conditions as long as the system is not burdened beyond the limits of design and tolerance.

ACKNOWLEDGMENTS

This work has been performed within the Forschergruppe Ökophysiologie der Deutschen Forschungsgemeinschaft and the Sonderforschungsbereich 176 of the University of Würzburg.

LITERATURE

Aflalo, C., and Shavit, N., 1984, Limited access of

nucleotides to the active site of ATP synthetase during photophosphorylation. In: Advances in photosynthesis research, vol. 2: 559, Sybesma, C., ed., Martinus Nijhoff/Dr. Junk Publishers, The Hague.

Ashton, A.R., 1982, A role for ribulose-1,5-bisphosphate carboxylase as a metabolite buffer. FEBS letters 145, 1-7.

Bassham, J.A., and Krause, G.H., 1969, Free energy changes and metabolic regulation in steady-state photosynthetic carbon reduction. Biochim. Biophys. Acta, 189: 207.

Borst, P., 1963: Hydrogen transport and transport metabolites. In: Funktionelle and morphologische Organisation der Zelle, pp 137, P. Karlson, ed., Springer, Berlin.

Buchanan, B., 1980, Role of light in the regulation of chloroplast enzymes. Ann. Rev. Plant Physiol. 31: 341.

Demmig, B., Winter, K., Krüger, A., and Czygan, F.-C., 1987, Photoinhibition and zeaxanthin formation in intact leaves. A possible role of the xanthophyll cycle in the dissipation of excess light energy. Plant Physiol. 84, 218-224.

Demmig, B., and Winter, K., 1988a, Characterization of three components of non-photochemical fluorescence quenching and their response to photoinhibition. Austr.J.Plant Physiol. 15: 163.

Demmig, B., and Winter, K., 1988b, Light response of CO_2 assimilation, reduction state of Q and radiationless energy dissipation energy dissipation in intact leaves. Austr.J.Plant Physiol. 15: 151.

Dietz, K.-J., and Heber, U., 1984, Rate-limiting factors in leaf photosynthesis. 1. Carbon fluxes in the Calvin cycle, Biochim. Biophys. Acta 767: 432.

Dietz, K.-J., Neimanis, S., and Heber, U., 1984, Rate-limiting factors in leaf photosynthesis. 2. Electron transport. Biochim. Biophys. Acta 767: 444.

Dietz, K.-J., and Heber, U., 1986, Light and CO_2 limitation of photosynthesis and states of the reactions regenerating ribulose-1,5-bisphosphate or reducing 3-phosphoglycerate, Biochim. Biophys. Acta 848: 392.

Heber, U., 1957, Zur Frage der Lokalisation von löslichen Zuckern in der Pflanzenzelle. Ber. Deutsche Bot. Ges. 70: 371.

Heber, U., and Santarius, K. A., 1965, Compartmentation and reduction of pyridine nucleotides in relation to photosynthesis, Biochim. Biophys. Acta 109: 390.

Heber, U., and Heldt, H.W., 1981, The chloroplast envelope: structure, function and role in leaf metabolism. Ann. Rev. Plant Physiol 32: 139.

Heber, U., Neimanis, S., Dietz, K.-J., and Viil, J., 1986, Assimilatory power as driving force in photosynthesis, Biochim. Biophys. Acta 852:144.

Heber, U., Neimanis, S., Dietz, K.-J., and Viil, J., 1987, Assimilatory force in relation to photosynthetic flux. In: Progress in Photosynthesis Research., vol. III: 293, Biggins, J., ed., Martinus Nijhoff, Dordrecht.

Heber, U., Neimanis, S., Setliková, E., and Schreiber, U., 1988a, Why is photorespiration a necessity for leaf survival under water stress? In: Proceedings of the International Congress of Plant Physiol., New Delhi, in the press.

Heber, U., Neimanis, S., and Dietz, K.-J., 1988b, Fractional control of photosynthesis by the Q_B protein, the cytochrome f/b₆ complex and other components of the photosynthetic apparatus. <u>Planta</u> 173: 267.

Heinrich, R., and Rapaport, T.A., 1974, A linear steady-state treatment of enzymatic chains. General properties, control and effector strength. <u>Eur. J. Biochem.</u> 42: 107.

Heldt, H.W., and Sauer, 1971, The inner membrane of the chloroplast envelope as the site of specific transport. <u>Biochim. Biophys. Acta</u> 234: 83.

Huchzermeyer, B., 1988, Phosphate binding to isolated chloroplast coupling factor (CF₁). <u>Z. Naturforschg.</u> 43c: 213.

Kacser, H., and Burns, J.A., 1973, <u>Symp. Soc. Exp. Biol.</u> 27: 65.

Kacser, H., and Burns, J. A., 1979, Molecular democracy: who shares the control. <u>Biochem. Soc. Transact.</u> 7: 1149.

Kaiser, G., Martinoia, E. and Wiemken, A., 1982, Rapid appearance of photosynthetic products in the vacuoles isolated from barley mesophyll protoplasts by a new fast method. <u>Z. Pflanzenphysiologie</u> 107: 103.

Krause, G.H., and Behrend, U., 1986, pH-dependent chlorophyll fluorescence quenching indicates a mechanism of protection against photoinhibition of chloroplasts. <u>FEBS Letters</u> 200: 298.

Laisk, A., Oja, V., Kiirats, O., Raschke, K., and Heber, U., 1988, The state of the photosynthetic apparatus in leaves as analyzed by rapid gas exchange and optical methods: the pH of the chloroplast stroma, activation and deactivation of enzymes and force resistance relationships in the Calvin cycle. <u>Planta</u>, submitted.

Leegood, R. C., and Furbank, R. T., 1986, Stimulation of photosynthesis by 2% oxygen at low temperatures is restored by phosphate, <u>Planta</u> 168: 84.

Leegood, R. C., Kobayashi, Y., Neimanis, S., Walker, D. A., and Heber, U., 1982, Co-operative activation of chloroplast fructose-1,6-bisphosphatase by reductant, pH and substrate. <u>Biochim. Biophys. Acta</u> 682: 168.

Leegood, R. C., Walker, D. A., and Foyer, C. H., 1985, Regulation of the Benson-Calvin cycle. <u>In</u>: "Photosynthetic Mechanisms and the Environment", Barber, J., and Baker, N.R., eds., Elsevier Biomedical Press, Amsterdam.

Oja, V., Laisk, A., and Heber, U., 1986, Light induced alkalization of the chloroplast stroma <i>in vivo</i> as estimated from the CO₂ capacity of intact sunflower leaves. <u>Biochim. Biophys. Acta</u> 849: 355.

Powles, S.B., 1984, Photoinhibition of photosynthesis induced by visible light. <u>Ann. Rev. Plant Physiol.</u> 35

Sage, R., and Sharkey, T. D., 1987, The effect of temperature on the occurrance of O₂ and CO₂ insensitive photosynthesis in field grown plants. <u>Plant Physiol.</u> 84: 658.

Sivak, M.N., and Walker, D.A., 1986, Photoynthesis <i>in vivo</i> can be limited by phosphate supply. <u>New Phytologist</u> 102: 499.

Stitt, M., 1986, Limitation of photosynthesis by carbon assimilation. 1. Evidence for excess electron transport capacity in leaves carrying out photosynthesis in saturating light and CO₂. <u>Plant Physiol.</u> 81: 1115.

Stocking, C.R., 1959, Chloroplast isolation in nonaqueous media. <u>Plant Physiol.</u> 34: 56.

Takahama, U., Shimizu-Takahama, M., and Heber, U., 1981, The redox state of the NADP system in illuminated chloroplasts. <u>Biochim. Biophys. Acta</u> 637: 530.

Tran-Anh, T., and Rumberg, B., 1987, Coupling mechanism between proton transport and ATP synthesis in chloroplasts. <u>In</u>: Progress in photosynthesis research, Vol 3: pp.185, Biggins, J., ed., Martinus Nijhoff Publishers, Dortrecht, Boston.

Weis, E., and Berry, J., 1987, Quantum efficiency of photosystem II in relation to "energy"-dependent quenching of chlorophyll fluorescence. Biochim. Biophys. Acta 849:198-208

CONTROL OF SUCROSE SYNTHESIS: ESTIMATION OF FREE ENERGY CHANGES, INVESTIGATION OF THE CONTRIBUTION OF EQUILIBRIUM AND NON-EQUILIBRIUM REACTIONS, AND ESTIMATION OF ELASTICITIES AND FLUX CONTROL COEFFICIENTS

Mark Stitt

Lehrstuhl für Pflanzenphysiologie
Universität Bayreuth
8580 Bayreuth, FRG

APPROACHES TO THE STUDY OF CONTROL

In this chapter I will review how ideas about the control of metabolism have developed in other fields, and will then apply these ideas to the regulation of photosynthetic sucrose synthesis. My main aim is to explore how control is organised and distributed between various enzymes, and how we can make quantitative statements about their response to effectors and their contribution to control in vivo. I will argue that a proper understanding of the nature of control is an essential prerequisite for the evalution of changes in flux and metabolite levels. In particular, it is essential to escape from an oversimplified use of the notion of 'limitation' in photosynthetic research.

The earliest approach to control was the concept of a "pacemaker" reaction, in which one enzyme was seen as controlling or "limiting" flux through the pathway. This is reminiscent of the concept of "limiting factors" developed by Blackman at the beginning of this century. A serius of criteria were proposed for identifying such "regulatory" reactions (Newsholme and Start, 1973; Rolleston, 1972)

(a) The enzyme should possess the requisite 'regulatory' properties eg. allosteric regulation, substrate cooperativity.
(b) The enzyme should be present at activites which are not greatly above ("not in excess of") the required flux through the pathway ("bottleneck")..
(c) The enzyme should catalyse a "non-equilibrium" or "irreversible" reaction.
(d) There should be characteristic changes of the substrate concentration which are reciprocal to the change of flux.

Before discussing the usefulness of these criteria, I shall briefly enlarge on the last two points. An enzyme, as a catalyst, will accalerate the forward and reverse reactions in a non-discriminatory manner. This means that when a reaction lies close to its thermodynamic equilibrium, the foreward and reverse reactions will be occuring at similar rates (v^{+1} and v^{-1}, respectively) and the net flux (v, $= v^{+1} - v^{-1}$) will only represent a small fraction of the unidirectional flux which the enzyme could catalyse if the product were to be absent ($= v^{+1} + v^{-1}$). Provided the enzyme is not saturated, v^{+1} will depend on the substrate concentration and v^{-1} will depend on the product concentration. This means that the net flux ($v^{\rightarrow} - v^{\leftarrow}$) can be increased by a relative small shift of the product substrate ratio (termed the mass action ratio, Γ) away from its equilibrium position (keq). Fig. 1 shows the relation between the Γ/keq ratio (the disequilibrium ratio), the free energy change of the reaction, and the relative rates of the forward and reverse reactions. It can be seen, for example, that a change of the disequilibrium ratio from 0.8 to 0.6 (equivalent to a free energy change of -0.14 and -0.31, respectively) will more than double the net flux; expressed another way, it would effectively compensate for a halving of the enzyme activity due some other factor. For this reason, reactions which are near to equilibrium were considered to be unsuitable sites for control. It was tacitly assumed that an enzyme catalysing a non-equilibrium reaction would be a suitable site for regulation, because the reverse reaction is negligible (see Fig. 1), and decreased enzyme activity would impact directly on the net flux. The validity of this assumption will be discussed later.

It was also suggested that 'regulatory' and 'non-regulatory' reactions could be distinguished by comparing the way in which their substrates and products change when the flux is altered in vivo. When an enzyme responds in a passive manner to an increased flux at some other "pacemaker", the increased flux will be accompanied by a higher substrate and a lower product concentration i.e. disequilibrium will increase. In contrast, if an enzyme is activated via regulation, its substrate concentration will fall as the flux increases. This resembles the logic of the "crossover" theorem (Chance, 1956) although it should be noted that the product concentration will not nessecarily change when we are dealing with non-equilibrium reactions (Rolleston, 1972).

As these criteria were applied, a serius of problems emerged. Firstly, pathways often contain more than one enzyme with the requisite "regulatory" properties. For example, the above criteria were often fulfilled by hexokinase, phosphofructokinase and pyruvate kinase in glycolysis (see Rolleston, 1972 for references) and by Rubisco, the stromal Fru1,6Pase, Sedu1,7Pase and Ru5P-kinase in the calvin cycle (Bassham and Krause, 1969). Secondly, problems arose with the simplifying notion that enzymes catalysing equilibrium reactions could be ignored. This distinction is somewhat arbitrary, anyway, because it is not obvious when a reaction becomes 'irreversible'. It was also realised that marked changes of the individual reactants could be generated in equilibrium reactions involving pairs of substrates and products. During photosynthetic PGA reduction, for example, a decrease in the supply of ATP and /or NADPH will lead to an increased PGA/ trioseP ratio because $[PGA]/[triose\ P] = keq \cdot \dfrac{[ADP] \cdot [Pi] \cdot [NADP]}{[ATP] \cdot [NADPH] \cdot [H^+]}$. This in turn, will influence other reactions because high PGA inhibits Rubisco (Foyer et al., 1987) and phosphoribulokinase (Gardeman et al. 1983), while low triose P will restrict the activity of the stromal Fru1,6Pase and other enzymes involved in the regeneration of Ru1,5bisP (Scheibe, 1987).

Thus, control actually involves an interplay between several enzymes. Unfortunately, this conclusion leads to obvious complications it we want to interpret the significance of changes in metabolite levels, or assess what contributions the various enzymes are making to control. Kacser and Burns (1973) and Heinrich and Rapoport (1974) made the decisive step of explicitly recognising (a) that control is a property of an entire pathway rather than the individual enzymes, and (b) that control is shared between the enzymes in a pathway. Even more important, they developed new concepts which allowed precise statements about control in these complex systems. Central to their approach is a clear separation between the properties of the parts of the system (eg.

the properties of a particular enzyme) and the properties of the system, which emerge from the interaction of all the enzymes and metabolites.

The properties of the enzymes are expressed as elasticity coefficients. This is a quantitative measure for the fractional change in activity (dv/v) which results from a fractional change in the concentration of a particular substrate (ds/s), when all other potential substrates and effectors are held constant. The activity of the enzyme will ultimately depend upon the sum of the interactions between the enzyme and all of its other substrates and effectors, each of which has an elasticity coefficient (dv/v = $\varepsilon_1 \cdot ds_1 / s_1$ + $\varepsilon_2 \cdot ds_2 / s_2$ ... $\varepsilon_n \cdot ds_n / s_n$).

The properties of the system are expressed as the flux control coefficient, $C = dJ/J \div dE/E$. This is defined with respect to one particular flux, and one particular enzyme. It

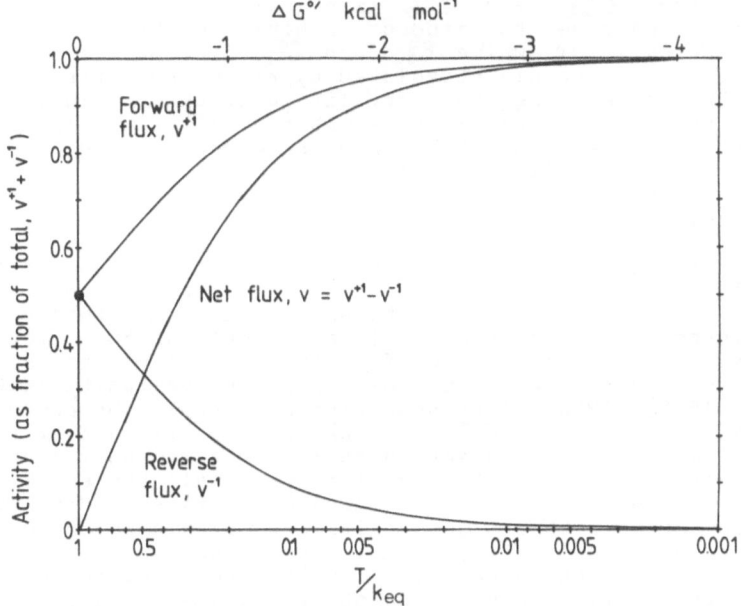

Fig. 1. Relation between the Disequilibrium ratio, the free energy change, the rate of the forward and reverse reaction, and the net flux. The figure is reformulated from Rolleston, 1972 and is based on the relation v /v = /keq. This expression is derived from a rate equation based on Michaelis-Menten kinetics for hyperbolic reactions in the steady state. The net flux (v) = forward flux (v^{+1}) - reverse flux (v^{-1}). The free energy change (Δ G) accompanying a reaction at a given disequilibrium was estimated as ΔG= 1.36 $\log_{10}$ (Γ/keq).

describes the fractional change of flux through the pathway (dJ/J) which results from a fractional change of enzyme activity (dE/E). It provides a quantitative expression for the contribution of the enzyme to control of flux through this pathway.

The flux control coefficient and the elasticity coefficient are related because, in a pathway, any one enzyme is flanked by metabolites which interact with this enzyme in a concentration dependent manner, and which also interact with further enzymes in the pathway. This means that a peturbation at one point in the system will be transmitted throughout the system, as can be illustrated by taking a simple pathway,

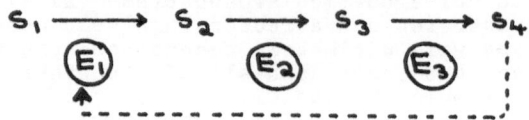

and peturbing the activity (dE_1/E_1) of the first enzyme. This will generate changes in the concentrations of the metabolites (s_1, s_2, s_3, s_4) which in turn, will affect activity of E_1, and will also affect the activity of other enzymes (E_2, E_3). These interactions depend on the elasticity coefficient for each particular enzyme and metabolite and they will determine how much of the initial change of E_1 activity is transmitted as a change of flux (dv/v) through the whole pathway. For example, suppose that $\mathcal{E}^{E_1}_{s_2}$ and $\mathcal{E}^{E_1}_{s_4}$ are large and negative (i.e. E_1 is very sensitive to product inhibition by s_2 and feedback inhibition by s_4), while $\mathcal{E}^{E_2}_{s_2}$ and $\mathcal{E}^{E_3}_{s_3}$ are small (i.e. E_2 and E_3 are only weakly stimulated by rising substrate concentrations). We can envisage how very little of the original increment of E1 activity remains as a change of flux. The resulting increase of the concentration of s_2, s_3 and s_4 just reinhibits E_1, without significantly affecting the flux through the subsequent enzymes. The flux control coefficient of E_1 would be very low in this case.

It should now be apparent why the magnitude of the flux control coefficient is not a property of the enzyme per se. Rather, it depends on the response of all the other enzymes in the system. In this case, it is not suprising that, for a given pathway, the flux control coefficient for all of its enzymes sum to a finite value, which has been shown to be unity (Kacser and Burns, 1973; 1979; flux summation property). The above example also stresses how the flux control coefficient of an enzyme emerges out of an interaction between the metabolite concentrations and the elasticity coefficients of all the enzymes in the pathway. In particular, when a metabolite is shared by two enzymes, an enzyme with a high sensitivity to this shared metabolite (i.e. a high elasticity) will have a low flux control coefficient. The relation between the flux control coefficients and the elasticity coefficients is quantitatively expressed in the connectivity theorem (Kacser and Burns, 1973), which states that the flux control coefficient of two adjacent steps is inversly related to the ratio of their elasticity coefficients for the shared substrate i.e. $C_{E_1}/C_{E_2} = -\mathcal{E}^{E_2}_{s_2}/\mathcal{E}^{E_1}_{s_2}$

These proposals have been subject to considerable criticism (see Crabtree and Newsholme, 1987) but, probably, the only valid reservation is that they are difficult to apply. Nevertheless, direct determination of flux control coefficients is possible, provided dosage mutants are available for a particular enzyme (see Kacser and Porteus, 1987 for examples). Alternatively, Groen et al. (1982) have developed a methodology for using inhibitors to estimate flux control coefficients. It should also be noted that, provided one flux control coefficient is known, other can be estimated by measuring the shared metabolite pools and then using the connectivity theorem (Kacser and Burns 1979, Groen et al. 1982).

Table 1. Cytosolic metabolite levels during photosynthetic sucrose synthesis in spinach leaves. The data are from Gerhardt et al (1987) except for PPi (Weiner et al., 1987), UTP and UDP (Neuhaus E., Quick, P., and Stitt, M., unpublished) sucrose 6P (Isherwood and Selvandran, 1972), and Pi which is estimated from Stitt et al $_1$ (1985) as in text. A cytosolic volume of 20µl mg Chl^{-1} is assumed.

Metabolite	Level in cytosol		% of total found in in the cytosol
	nmol mgChl^{-1}	mM	
PGA	120	0.8	30
GAP	2	0.05	-
DHAP	16	0.8	53
Fru1,6bisP	4	0.2	10
Fru6P	55	2.7	60
Glc6P	167	8.4	81
Glc1P	19	0.9	75
UTP	15	0.7	
UDPGlc	40	2.0	94
PPi	5	0.25	90
UDP	6	0.3	-
Sucr6P	4	0.2	-
Sucrose	800	40	35
Fru2,6bisP	0.2	0.01	98
Pi		5	5

In applying these ideas to the regulation of sucrose synthesis I shall ask the following questions
(a) can we identify clear cases of equilibrium and non equilibrium reactions, using measured levels of metabolites in the cytosol.
(b) To what extent can we use qualitative comparisons of fluxes and metabolites to identify sites for regulation or control? By 'control', I mean the ability of an enzyme to alter the flux through a pathway. By 'regulation', I will be refering to enzymes whose activity can be altered by mechanisms other than a change of the substrate or product concentration. "Regulation", however, does not necessarily mean that the enzyme is controlling flux through the pathway.
(c) Is it really true that equilibrium reactions make no contribution to control?
(d) Is it possible to obtain values for the elasticity coefficients and flux control coefficients for at least some of the enyzmes in the cytosol?

2. DISEQUILIBRIUM RATIOS ESTIMATED FROM CYTOSOLIC METABOLITE LEVELS

Table 1 summarises estimates for the cytosolic concentrations of metabolites involved in photosynthetic sucrose synthesis. They were obtained by non-aqueous fractionation of spinach leaves (Gerhardt et al., 1987; Weiner et al., 1987) except for UTP, UDP and UMP, which still represent overall levels in spinach (Quick, P., Neuhaus, H.E. and Stitt, M. unpublished) and sucrose-6-P, which represents overall levels in strawberry leaves (Isherwood and Selvandran, 1972). Cytosolic Pi was estimated as 5mM, because the value in spinach protoplasts is about 20 mM in the dark (Stitt et al., 1985), of which 10-15 mM is likely be sequestered into rising pools of phosphorylated intermediates in the

Table 2. Estimate of the mass action ratio, the disequilibrium ratio and free energy for each reaction involved in photosynthetic sucrose synthesis. The molar mass action ratios are estimated from Table 1, and the free energy change, G, is estimated as $\Delta G = \Delta G^\circ + 1.36$ log $\frac{\Gamma}{\text{keq}}$. The equilibrium constants, keq, and standard free energy changes are from Lehninger (1970), Barber (1985), and Dyson and Noltman (1963).

	Measured Molar Mass action ratio = Γ	Theoretical Equilibrium constant	$\frac{\Gamma}{\text{keq}}$	ΔG° (kcal·mol⁻¹)	ΔG in vivo (kcal·mol⁻¹)
Triose P isomerase	0.060	0.045	1.1	+1.8	+ 0.1
Aldolase	5.000	10100	0.5	-5.5	-0.5
Fru1,6Pase	0.07	875	0.0008	-4.0	-5.6
Phosphoglucose isomerase	3.0	3.3	0.8	-0.7	-0.1
Phosphoglucose mutase	0.11	0.08	1.2	+1.5	+0.2
UDPGlc pyrophosphorylase	0.7	0.4	1.4	+0.6	+0.4
Sucrose P synthase	0.01	1.7	0.006	-0.3	-3.0
Sucrose phosphatase	1.0	266	0.004	-3.3	-3.3
PPi hydrolysis to Pi	0.1	2410	0.0004	-4.6	-6.0

light (Stitt et al. 1985; Gerhardt et al., 1987). Table 1 also
shows the estimated percentage of each metabolite which is found
in the cytosol. It should be noted that our estimated values will
become increasingly inaccurate when only a small proportion is
present in the cytosol (e.g. Fru1,6bisP) because the correction
for cross-contamination from other compartments becomes progressively larger.

The mass action ratio (Υ) for each reaction is shown in
Table 2. Division of by the theoretical equilibrium constant
(keq) yields the disequilibrium ration (Υ/keq). Most of the
values are very close to unity, revealing these reactions are
close to equilibrium. Only three reactions on the direct route to
sucrose are removed from equilibrium: the Fru1,6Pase, SPS and
sucrose-6-phosphatase. The hydrolysis of PPi to Pi is also removed from equilibrium but (see Weiner et al., 1987) it remains an
open question whether PPi is actually removed via pyrophosphatase, or whether this energy is conserved by using another route
to remove PPi. The values for ΔG , the free energy change in
vivo, are also computed in Table 2 from the value for Υ and the
standard free energy for each reaction. By definition, this
yields the same answer as the disequilibrium ratio. Visually, the
differences between the ΔG values are less striking, but it
must be remembered that the free energy is a logarithmic function
(compare the upper and lower axis of Fig. 1). Thus, the reactions
in the cytosol are clearly divided into two classes; those which
are very far from equilibrium, and those which are near to
equilibrium.

3 RELATION BETWEEN FLUXES AND METABOLITE LEVELS

I shall now discuss how measurements of metabolite levels
and fluxes can be used to show that two of the non-equilibrium
reactions, the cytosolic Fru1,6Pase and SPS, are being regulated.
To do this, we need to find ways of increasing or decreasing the
rate of sucrose synthesis. We can then measure the resulting
changes of fluxes and metabolite levels. According to the qualitative criteria outlined in the introduction, the substrates of
'regulatory' enzymes should change in the opposite direction to
the flux.

The flux to sucrose can be decreased by allowing sucrose to
accumulate in the leaf (Fig. 2). The redirection of photosynthate
away from sucrose synthesis towards starch accumulation in the
chloroplast is accompanied by an increase of the substrates of
SPS (Fru6P and Glc6P) and the cytosolic Fru1,6Pase (Fru1,6bisP).
This provides evidence that both enzymes are down-regulated in
response to a falling demand for sucrose (Gerhardt et al. 1987).

The flux to sucrose can be stimulated by increasing the
light intensity or the crabon dioxide concentration. In this
case, the picture is less clear. Fig. 3 summarises changes of
cytosolic metabolites in spinach protoplasts after illumination
(Stitt et al , 1980); similar results have also been obtained
in spinach leaves (see Fig. 2). The substrates of SPS decline as
sucrose synthesis increases, showing SPS has been activated (for
further details, see below). However, the substrates for the
Fru1,6Pase rise as sucrose synthesis increases.

Does this mean that the cytosolic Fru1,6Pase is not involved
in the regulation of sucrose synthesis in response to rising
rates of photosynthesis? This experiment illustrates the problems
inherent in analysing the relation between metabolites and fluxes
in complex branching pathways. Triose P increase after illumination because (a) the rising supply of ATP and NAD(P)H leads to a
marked alteration of the cytosolic triose P/PGA ratio from under

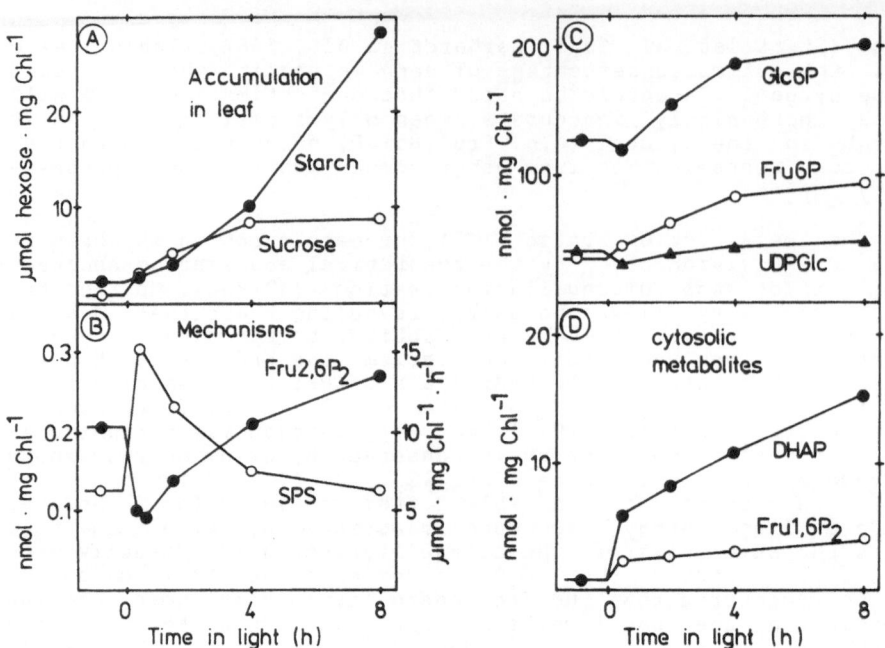

Fig. 2. Changes of cytosolic metabolite levels during the photoperiod in spinach leaves. The data are from Gerhardt et
al. (1987) and Stitt et al. (1988). Accumulation of sucrose (○) and starch (●). B, Fru2,6bisP (◕)
and SPS activity (○). C, Glc6P (●), Fru6P (○)
and UDPGlc (▲). D, Fru1,6bisP (○) and triose P
(●).

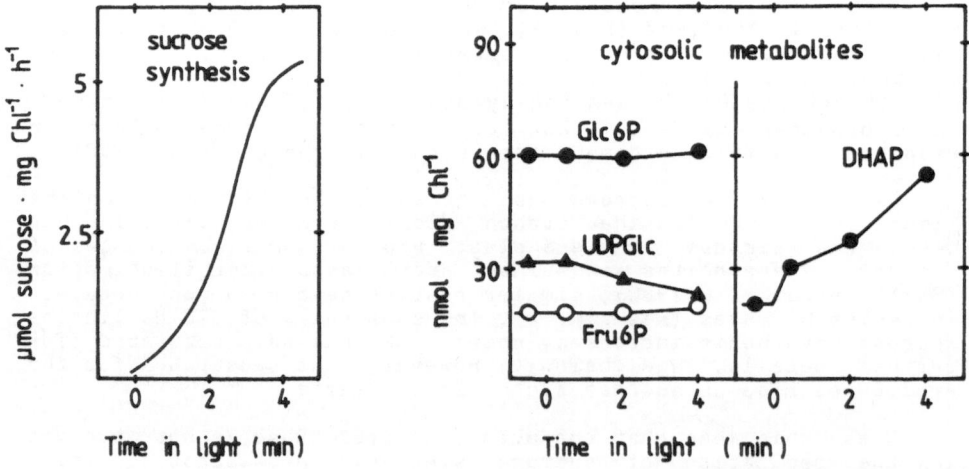

Fig. 3. Changes of cytosolic metabolites after illuminating
spinach protoplasts. The data are from Stitt et al.
(1980), A, sucrose synthesis. B, Glc6P(●), Fru6P(○) and
UDPGlc (▲), C,   triose P (●).

372

0.05 in the dark to over 0.25 in the light (Gerhardt et al., 1987) and (b) the crucial mechanisms which regulate the Fru1,6Pase are actually closely related to rising levels of triose P and PGA. It can be shown that the cytosolic Fru1,6Pase is deactivated by marginal changes of the triose P concentration within seconds when leaves are darkened, and long before we could expect on the basis of its substrate affinity alone (Stitt et al, 1983). However, this regulation does not make itself visible in the experiment of Fig. 3. In other words, these kinds of experiment provide a useful way of finding possible sites for regulation, but they may not reveal all of them.

Nevertheless, this kind of experiment provides a good way of identifying a need, and stimulating a search, for particular regulatory mechanisms. For example, the reciprocal behaviour of hexose P or UDPGlc and the rate of sucrose synthesis predicted that there were additional control mechanisms at SPS. They can now be explained because SPS is subject to 'coarse' control, involving rapid alterations of the kinetic properties of the enzyme which are probably due to covalent modification of this protein (Stitt et al., 1988). When photosynthesis increases, activation of SPS allows fluxes to be increased without this requiring an increased level of its substrates; conversely as sucrose accumulates, deactivation of SPS allows the flux to be restricted and hexose P accumulate. The key to understanding regulation of the cytosolic Fru1,6Pase was provided the discovery of the regulatory metabolite Fru2,6bisP. As discussed in detail elsewhere (Stitt et al, 1987) Fru2,6bisP is a very potent inhibitor of the cytosolic Fru1,6Pase. It is synthesized and degraded by Fru6P,2-kinase and Fru2,6PbisPase, both of which are subject to modulation by metabolites. When photosynthesis increases, rising levels of triose P and 3PGA and falling Pi inhibit Fru2,6bisP synthesis. This leads to a decrease of Fru2,6bisP and the Fru1,6Pase is activated. Conversely, rising Fru6P stimulates the synthesis and inhibits the degradation of Fru2,6bisP. The resulting increase of Fru2,6bisP will inhibit the cytosolic Fru1,6Pase and prevents overproduction of hexose P.

However, although we can now describe the mechanisms which are involved, we still cannot say what quantitative contribution each is making to the control of flux. This problem can be illustrated by considering how several studies have seen that SPS activity and Fru2,6bisP change reciprocally (Sicher et al., 1986; Kerr and Huber, 1987; Stitt et al., 1987). This has given rise to the notion that there is "coordinate control" at SPS and the Fru1,6Pase (remember rising Fru2,6bisP inhibits the Fru1,6Pase). This conclusion is actually trivial. Since the Fru1,6Pase and SPS catalyse flux into, and out of a linear sequence from which there is no other significant outlet, flux at both enzymes will, by definition, have to be the same in steady state conditions. Indeed, the flux will have to be the same at all the enzymes in the pathway. This obvious conclusion does not tell us anything about how control is distributed between the enzymes: which is the 'master' and which is the 'slave', or are the enzymes all a bit of both, or do they indulge in role-swapping.

4 CONTRIBUTION OF NEAR-EQUILIBRIUM REACTIONS TO CONTROL

Phosphoglucoisomerase I shall first consider whether the equilibrium reactions really make no contribution to control, taking phosphoglucoisomerase and UDPGlc pyrophosphorylase as examples.

Using dosage mutants for the cytosolic phosphoglucoisomerase (PGI) in Clarkia xantiana (Jones et al., 1986), we have recently been able to directly determine the flux control coefficient of

this enzyme for sucrose synthesis. Fig 4 shows the relation between the rate of sucrose synthesis and the dosage of PGI (Kruckeberg R.L., Neuhaus, H.E., Gottlieb, L.A., Feil, R. and Stitt, M.,
unpublished data); C_{PGI} is given by the initial slope (Kacser and
Burns, 1973) and is almost zero. We therefore concluded that PGI
does not make a major contribution to control in the wildtype.

Nevertheless, it is intriguing to find that a 3-5 fold lower
PGI dosage does lead to a lower rate of sucrose synthesis.

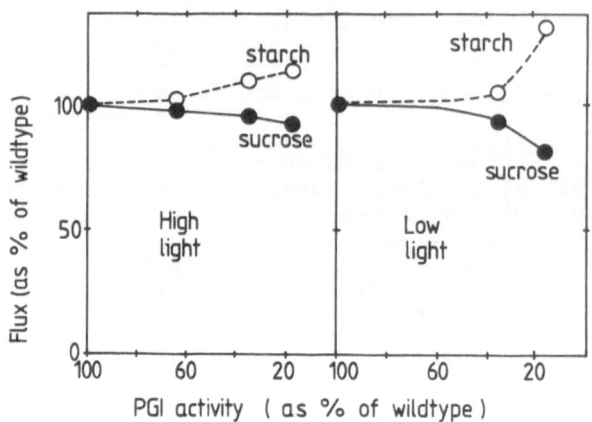

Fig. 4. Flux control coefficients for the cytosolic phosphoglucose isomerase in Clarkia xantiana. The data are from
Kruckeberg A.L., Neuhaus, E., Feil, R., Gottlieb, L.D.
and Stitt, M. (unpublished). A, Limiting light (125
μmol m^{-2} s^{-1}). B, Saturating light (1000 μmol m^{-2} s^{-1}).

Clearly, PGI is not present in a large "excess" for the pathway.
Independent evidence for this conclusion is given in Table 3
which compares the theoretical and in vivo mass action ratios
during rapid sucrose synthesis in each of these genotypes. The
theoretical values provide an estimate of the Glc6P/Fru6P ratio
which would be needed to generate the observed in vivo flux,
using the PGI activity present in each genotype. The calculation
uses the relation between total activity, disequilibrium ratio
and net flux which is shown in Fig. 1. It predicts that the
amount of PGI in the cytosol of the wildtype is barely enough to
keep the reactants at near-equilibrium, and that a relatively
small drop in the dosage of PGI should lead to a marked disequilibrium at this reaction. The predicted behaviour is quite similar to the shift of the mass action ratio which we actually
measured in these leaves (Table 3). The ratio in leaves does not
change quite as far as predicted. However, these represent overall measurements, and it is almost certain that the change in the
cytosol will have been larger, but is masked by the metabolites P
in the chloroplast (see Table 1).

Two further important points emerge from Fig. 4. First, the
impact of decreased PGI varies, being greater in low light than
in high light. Second, the mutants with low PGI activity have an
increased rate of starch synthesis. These results are completely
consistent with the analysis of Kacser and Burns, which predicts
(a) that the value of a flux control coefficient will vary depen-

ding on the precise metabolic condition (Kacser and Burns 1973, 1979) and (b) that when there is a branch point, an enzyme in one pathway can have a negative flux control coefficient (ie decreased enzyme gives increased flux) for the flux through the other pathway (Kacser and Burns, 1979).

Finally, Table 4 illustrates the mechanism by which an increase of Fru6P and decrease of Glc6P acts to lower the activity of other enzymes in the cytosol and, hence, to lower the rate of sucrose synthesis. SPS activity will be decreased because Glc6P is an allosteric activator (Doehlert and Huber, 1983); lower levels of the activator presumably outweigh the higher concentrations of one of the substrates (Fru6P). The cytosolic Fru1,6Pase is restricted via an increased level of Fru2,6bisP. This results from activation of Fru6P,2-kinase and inhibition of Fru2,6bisP by Fru6P (Stitt et al., 1984; 1987). These interactions will be analysed more quantitatively in the later sections.

UDPGlc pyrophosphorylase Recently we have observed that moderate concentrations of fluoride inhibit photosynthesis, acting selectively on sucrose synthesis (Quick, P., Neuhaus, E., Feil, R. and Stitt, M., in preparation). This inhibition is due to an accumulation of PPi and its effect on the mass action ratio of UDPGlc pyrophosphorylase. (Table 5). A 3-fold increase of PP_i leads to a halving of the other product (UDPGlc) and an increase of the two substrates (UTP and Glc1P) of UDPGlc pyrophosphorylase. These peturbations of UDPGlc and Glc1P transmit the restriction to the non-equilibrium reactions in the cytosol. Falling UDPGlc restricts SPS because this enzyme has a low affinity for UDPGlc, which cannot be increased by accumulating Glc6P or Fru6P (Doehlert and Huber, 1983, see also below). The increase of Glc1P is transmitted through the hexose P pool, and the resulting increase of Fru6P leads to higher Fru2,6bisP (see above) and restricts the cytosolic Fru1,6Pase.

This example shows how accumulation of one of the products of a 2nd order equilibrium reaction can lead, via "mass action effects", to a shift of the other products and substrates which, in turn, act to restrict flux at the 'non-equilibrium' reactions in the pathway. These results also underline the need to understand more about the control of PP_i metabolism in the cytosol, and raise the possibility that removal of PP_i could contribute to the regulation of sucrose synthesis.

Elasticities of near-equilibrium reactions. These two examples suggest the following qualitative conclusions about the contribution of near-equilibrium reactions to control. (a) The enzymes are not present in "excess" over what we need to keep their reactants reasonably close to equilibrium. Indeed, it is not easy to see how such an "excess" could be generated by natural selection. Random mutations would tend to decrease the catalytic effeciency or the expression down to a point where an enzyme starts to be 'needed'. (b) Near-equilibrium reactions provide an excellent way of transmitting flux changes through a serius of reactions i.e. of linking the reactions which are further removed from equilibrium. (c) A near-equilibrium reaction can influence the flux through a pathway, provided other enzymes are sensitive to the shift of metabolites which results when it is slightly displaced from equilibrium. (d) A similar influence can be exerted by "mass action" effects at a reaction with pairs of substrates and products, even when the reaction itself stays at equilibrium.

These results also emphasise that, for control, the crucial question is not just whether a reaction is near to equilibrium or not. When PGI dosage was reduced to 18% of the wild-type dosage, it had a significant flux control coefficient (about 0.2, Fig.

Table 3. Comparison of PGI activity, fluxes, and the estimated and measured mass action ratio in cytosolic PGI dosage mutants of _Clarkia xantiana_. Cytosolic PGI activity in leaves was measured under substrate saturation in the direction of Glc6P formation immediately after preparing an extract with albumin, thiol reagents and protease activity to minimise loss of activity. The net activity of PGI required in vivo is estimated from the fluxes in Fig. 4, remembering one molecule of Glc6P is needed per molecule of sucrose. The net activity ($v = v^{+1} - v^{-1}$) was then divided by the total activity ($v^{+1} + v^{-1}$), and the relation in Fig. 1 was then used to estimate the disequilibrium ratio (Γ/keq) which would be necessary to obtain the observed flux at each dosage level of PGI. The mass action ratio was then estimated assuming $keq = 3.3$. These estimated values are compared with the Glc6P/Fru6P ratios actually measured in the leaves. Leaves were illuminated in saturating light and CO_2 for 10 min, quenched in liquid N_2, and then extracted for assay of Glc6P and Fru6P. There values are not corrected for metabolites deriving from the stroma. The results are from Knuckeberg R.L., Neuhaus, E. Feil, R., Gottlieb, L.A. and Stitt, M. (submitted).

PGI activity as % of wildtype	Cytosolic PGI activity (mol mgChl^{-1} h^{-1})	Net activity required in vivo	Net activity as fraction of total activity	Γ/keq	Glc6P/Fru6P ratio Estimated	Glc6P/Fru6P ratio Measured
100	143	12.5	0.09	0.82	2.71	2.63
64	91	12.3	0.14	0.76	2.50	2.43
36	51	11.9	0.25	0.60	1.98	2.22
18	26	11.6	0.45	0.36	1.2	1.72

Table 4. Changes of fluxes and metabolite levels in response to lower dosage of the cytosolic PGI in Clarkia xantiana (H.E. Neuhaus, R.L. Kruckeberg, R. Feil, M. Stitt, unpublished results) The fluxes and metabolites were measured in saturating CO_2 at 15°C at a light intensity of 125 μmol m^{-2} s^{-1}.

Cytosolic PGI (as % of wild-type)	Fluxes (μmol CO_2 mgChl^{-1}h^{-1})			Metabolites (nmol mgChl^{-1})				
	CO_2 fixation	Sucrose	starch	Glc6P	Fru6P	Fru2,6bisP	triose P	PGA
100	111	39	40	393	137	0.16	35	282
18	121	33	56	363	176	0.26	34	386

Table 5: Inhibition of sucrose synthesis via an increase of PPi and the following mass action effect on UDPGlc pyrophosphorylase. The data are from Quick, P., Neuhaus, E., Feil, R. and Stitt, M. (unpublished) and are for leaf discs in saturating light and CO_2. NaF or (control) NaCl were supplied at 5 mM.

	Fluxes (mol mgChl^{-1} h^{-1})			Metabolites (nmol mgChl^{-1})								SPS activity (mol mgChl^{-1} h^{-1})
	O_2 evolution	Sucrose synthesis	Starch synthesis	PPi	UDPGlc	UTP	Glc1P	Glc6P	Fru6P	Fru2,6P	triose P	
Chloride	180	81	32	4.8	62	13	17	249	80	0.07	35	12.1
Fluoride	122	50	28	12.6	30	20	25	498	160	0.10	56	8.3

4). However, the measured Glc6P:Fru6P ratio (2.2) suggests a very moderate disequilibrium ratio of about 0,66. This is negligible compared to the truly non-equilibrium reactions in the cytosol (see Table 2). In high light, the Glc6P:Fru6P ratio was further displaced from equilibrium (1.7) but the flux control coefficient actually lower (Fig. 4). Obviously, control does not just depend on the extent of disequilibrium. Rather, it depends on the balance between (a) the effectiveness of shifts of Fru6P and Glc6P in maintaining flux at PGI and (b) the sensitivity of other reactions in the pathway to the (inhibiting) effect of higher Fru6P and lower Glc6P. At least for PGI, the pathway has evolved to operate with PGI at near-equilibrium and a relatively minor shifts away from equilibrium has adverse effects on the operation of the remainder of the pathway. The sensitivity of the other enzymes to these changes of Fru6P and Glc6P also seems to vary, depending on the conditions.

This can be stated more precisely as follows: Near-equilibrium reactions have a high elasticity coefficient for their substrate and product. Since the flux control coefficient is inversly related to the elasticity coefficient for a shared metabolite (the connectivity theorem, see above), near-equilibrium reactions usually exert little control. However, they can start to exert control if their elasticity coefficients for their substrate or product fall to values which resemble those of other enzymes for these metabolites.

Why do enzymes like PGI have high elasticities? Enzymes catalysing near-equilibrium reactions are usually relatively simple, have few effectors, and have hyperbolic substrate saturation kinetics. Provided the enzyme is partially saturated, the slope of the michaelis-menten relation might lead us to expect a relatively low elasticity (about 0.5) for the substrate. However, we must remember that the net activity, v, is the difference between the rate of the forward (v^{+1}) and the reverse (v^{-1}) reactions (see Fig. 1). When a reaction is near equilibrium, v^{+1} and v^{-1} will be of a comparable magnitude, and both will be far larger than v. Further, v^{+1} and v^{-1} will depend on the substrate and product concentration, respectively. This means that relatively small changes of the substrate or product concentration will lead to large changes of the net _in vivo_ activity, i.e. the elasticity coefficients for the substrate and product will be very high.

This has been rigourously treated by Groen et al. (1982a). They have described a serius of expressions for calculating the elasticities for the substrate and product from the measured mass action ratio, the theoretical equilibrium constant, the measured flux, and the Vmax of the enzyme. In Table 6, I have used these expressions to estimate the elasticity coefficients of PGI for Fru6P and Glc6P, using measurements made with the Clarkia dosage mutants. Since they are estimated for sucrose synthesis, the elasticities for Glc6P are negative (i.e. rising Glc6P would tend to decrease Glc6P formation by PGI). The estimated elasticities are high in the wildtype, lying between 5 and 8. As we will see, the cytosolic Fru1,6Pase and SPS have much lower elasticities for Fru6P and Glc6P. This explains why PGI exerts little control in the wild type (Fig. 4). When PGI dosage is decreased, there is a progressive decrease of the elasticity coefficients for Fru6P and Glc6P. This implies that PGI could begin to exert more control in these mutants. Whether this actually occurs depends upon the response of other enzymes. (see below). For reference, it might be noted that PGI does not have a measurable impact on fluxes (i.e. does not have a significant flux control coefficient) until the elasticities for Fru6P and Glc6P fall below about 2.5. It is also interesting that the elasticity coefficients are lower in saturating light than limiting light. This can be explained because in saturating light there are higher fluxes, so the reaction has to be further from equilibrium, and because metabolite concentrations are generally higher (Stitt et al 1983; 1987), so PGI is more saturated with (and less responsive to) its substrate and product.

In vivo elasticity coefficients can also be estimated from measurements of fractional changes of fluxes and metabolites, without any asumptions about the keq or the Vmax of an enzyme (Kacser and Burns, 1979). We have already seen in the introduction how the activity of an enzyme can be described as the sum of the interactions with all of its effectors, each effector having its own elasticity coefficient. If we decrease the amount of enzyme available in the leaf another term, dE/E, has to be introduced to describe the fractional decrease in the amount of enzyme. For PGI we can write dv/v = ε_{F6P}· dF6P/F6P + ε_{G6P}· dG6P/G6P + dPGI/PGI where dv/v is the fractional change in the net activity (equivalent here to the fractional change in the flux to sucrose), dPGI/PGI is the fractional change in the amount of PGI, dF6P/F6P and dG6P/G6P are the fractional changes of substrate and product and ε_{F6P} and ε_{G6P} are the elasticities for Fru6P and Glc6P. This equation contains two unknowns - the elasticities for Fru6P and Glc6P. The other terms can all be

Table 6. Estimated in vivo elasticities of PGI for Fru6P and Glc6P. The calculation is based on the equations derived by Groen et al (1982) ε_{F6P} = $(1 - \tau/keq)^{-1}$· $(1 - J/V_{max})$, and ε_{G6P} = (τ/keq) $(1 - \tau/keq)^{-1}$· $(1 + J/V_{max})$, where τ is the measured Glc6P/Fru6P ratio, J is the flux to sucrose, and V_{max} is the maximum activity of PGI measured in extracts of that particular genotype. The fluxes and mass action ratios were measured in limiting light (125 μmol m^{-2} s^{-1}) and saturating light (1000 μmol m^{-2} s^{-1}).

PGI activity (as a % of wild-type)	ε_{F6P}		ε_{G6P}	
	High light	Low light	High light	Low light
100	4.6	7.6	−4.6	−6.9
64	3.2	8.4	−3.3	−8.4
36	2.3	5.2	−2.5	−4.9
18	1.2	2.6	−1.5	−2.1

measured. If we have two independent sets of experimental data, we can write two simultaneous equasions, and then solve them to provide estimates for the in vivo values of the two elasticity coefficients.

Using Clarkia xantiana PGI mutants we investigated the effect of lowering PGI dosage to 36% of the wildtype level in high and in low light. These experiments, which will be presented elsewhere. (Kruckeberg A.L., Neuhaus, H.E., Feil, R., Gottlieb, L. and Stitt, M., in preparation), allowed us to estimate elasticities of + 5.5 and -5.8 for Fru6P and Glc6P, respectively. When the PGI dosage was reduced to 18%, much lower elasticities were obtained, 1.2 and -2.0, respectively. These estimates are similar to those obtained by the approach used in Table 6. For practical reasons, fairly large changes have to be measured to allow us to measure the fractional changes of fluxes and metabolites with reasonable accuracy. This means that the values estimated for the elasticities in both these approaches are approximations. Errors also arise because these estimates are based on overall metabolite levels. Nevertheless, they provide insights into the way in which a near equilibrium enzyme will respond to changes of its substrate and product in vivo. They also emphasise how the elasticity coefficients can change quite markedly of the flux through the pathway is altered, or if the amount of an enzyme is altered.

5. QUANTITATIVE ANALYSIS OF FLUX CONTROL AT THE NON-EQUILIBRIUM REACTIONS

I will now turn to the non-equilibrium reactions, and consider how control is distributed between them during sucrose synthesis. Empirical models are already available which describe the regulation of the cytosolic Fru1,6Pase (Herzog et al., 1984; Stitt and Heldt, 1985; Stitt et al., 1987) and SPS (Stitt et al., 1987, 1988). These describe (a) how each of these enzymes is stimulated by a rising supply of photosynthate once a 'threshold' concentration is exceeded and (b) how this threshold can be modified to allow e.g. alterations of photosynthate partitioning (Stitt et al., 1987, 1988) or adaptation to temperature (Stitt and Große, 1988). These studies also indicate how flux at the Fru1,6Pase and SPS could be integrated (see also Kerr and Huber, 1987). Imagine that a peturbation (eg increased flux at the Fru1,6Pase or inhibition of SPS) leads to an increase of the hexose P pool, which links these two enzymes. SPS would be stimulated because one of its substrates (Fru6P) and an activator (Glc6P) rise; simultaneously the increased Fru6P would lead to an increase of Fru2,6P and, hence, reinhibit cytosolic Fru1,6Pase. In this way, the flux at the two enzymes would be brought back into step. This account, however, is still strictly qualitative and does not reveal how control is shared.

As discussed in the introduction, the distribution of control is inversely related to the elasticity of two enzymes for a shared metabolite pool (the connectivity theorem). It should therefore be possible to develop a quantitative account of control in the cytosol if we know (a) how SPS activity responds to a change of the hexose P pool and can compare this with (b) the response of Fru2,6bisP to a change of Fru6P and (c) the response of the cytosolic Fru1,6Pase to a change of Fru2,6bisP. In the following section I will set up a simplified expression to describe each of these interactions, and will then provide estimates for the relevant elasticity coefficients.

Fig. 5 summarises the interactions for which I will try to estimate elasticity coefficients. I will make two simplifiying assumptions in the following treatment. Firstly, I will assume that other near-equilibrium reactions are like PGI, and do not contribute significantly to control in the wildtype. Secondly, I will group SPS, sucrose-phosphatase, and the hydrolysi of PPi in one block and will, for brevity, refer to this block as SPS. This is necessary because we do not have enough information about sucrose phosphate and uridine nucleotide levels to allow these reactions to be seperately analysed.

The approach I will use is similar to that used to estimate in vivo elasticity coefficients for PGI. For each enzyme we can write an expression of the form

$$dv/v = \varepsilon_{s1} \cdot ds_1 / s_1 \; + \; \varepsilon_{s2} \cdot ds_2 / s_2 \; \ldots \ldots \varepsilon_{sn} \cdot ds_n / s_n \; .$$

We can carry out a serius of experiments to obtain empirical values for the fractional changes of flux (dJ/J, which at steady state is equivalent to the change in activity of the individual enzyme, dv/v) and the metabolites (ds_1 / s_1, ds_2 / s_2 .....). Provided we have a finite number of effectors, it is possible to

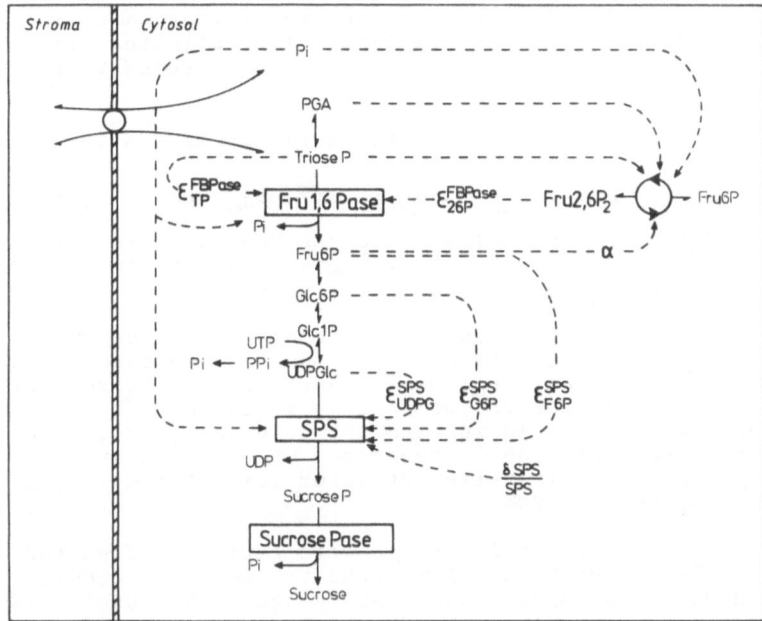

Fig. 5. The control of photosynthetic sucrose synthesis. The scheme shows the metabolic fluxes to carbon (———▶) and the regulatory interactions (---▶). Typical values for the elasticity coefficents are given in the text.

set up a serius of simultaneous equations and solve them to obtain values for the elasticities. A more detailed account of this analysis has already been published (Stitt, In press). In the following treatment, I will summarise the expressions, briefly note the major assumptions and simplifications involved in them, and will then discuss the significance of the elasticities which we obtain using them.

The cytosolic Fru1,6Pase. I represent the fractional response of Fru1,6Pase activity (dv/v) to a fractional change of triose P and Fru2,6bisP as

$$dv/v = \mathcal{E}^{FBPase}_{TP} \cdot dTP/TP + \mathcal{E}^{FBPase}_{26P} \cdot d26P/26P$$

where $\mathcal{E}^{FBPase}_{TP}$ and $\mathcal{E}^{FBPase}_{26P}$ represent the elasticities for triose P and Fru2,6bisP. The expression treats Fru1,6bisP and triose P as a composite pool, and replaces the true substrate by triose P because the latter alters in equilibrium with Fru1,6bisP (Table 2) and can be more readily estimated for the cytosol (see above). Since Fru6P has negligible effect on activity in the presence of Fru2,6bisP (Stitt et al., 1985), i.e. $\mathcal{E}^{FBPase}_{F6P} \longrightarrow 0$, we can omit the corresponding term for the reaction product. This simplified expression, of course assumes that other potential effectors of the Fru1,6Pase (eg AMP, magnesium, pH) are not changing.

Table 7 summarises the fractional changes in the rate of sucrose synthesis and the level of various metabolites when sucrose synthesis is restricted by a falling rate of CO_2 fixation, or by accumulation of sucrose in the leaf. The values are derived from Stitt and Heldt (1985) and allow elasticities of 1.0 for triose P and -0.5 for Fru2,6bisP. to be estimated (Table 8). Table 7 also shows the fractional changes of sucrose synthesis and metabolites in Clarkia xantiana when the cytosolic PGI is decreased to 18% of the wildtype level. They were measured in saturating light, and in low light. The estimated elasticities for triose P and Fru2,6bisP are 1.6 and -0.4, respectively (Table 8).

Sucrose P synthase A change in the activity of SPS (dv/v) can be described as

$$dSPS/SPS + \varepsilon_{F6P}^{SPS} \cdot dFru6P/ Fru6P + \varepsilon_{G6P}^{SPS} \cdot dGlc6P/Glc6P + \varepsilon_{G}^{SPS} \cdot dUDPGlc/UDPGlc + dPi/Pi \cdot \varepsilon_{Pi}^{SPS}$$

The term dSPS/SPS is included because SPS is subject to 'coarse' control. This is treated as equivalent to changes in the amount of the enzyme, with a enzyme concentration elasticity coefficient of unity (Kacser and Burns, 1987). This approach is valid provided the 'coarse' control involves a change in the amount of the enzyme, or an interconversion between an active form and an inactive form which has negligible activity in vivo. Available evidence suggests the former is true in soybean (Kerr and Huber, 1987) and the latter in spinach (Stitt et al., 1988). The remaining terms reflect the dependence on the substrates (Fru6P, UDPGlc), and the allosteric activator (Glc6P) and inhibitor (Pi) (Doehlert and Huber, 1984).

In vivo elasticities of SPS were estimated from the data in Table 7. For spinach, we eliminated the term for UDPGlc. This is justified because UDPGlc does not change significantly ($dUDPG/UDPG \rightarrow 0$) The terms for Glc6P, Fru6P and Pi were replaced by a composite term ε_{HP}^{SPS} . dHP/HP, which describes the response to rising levels of hexose P. This is valid, provided the reaction catalysed by PGI is reasonably close to equilibrium. This expression tacitly includes the effect of falling Pi as metabolic intermediates like Fru6P and Glc6P rise. We also included a notional term, $\varepsilon_{sucrose}$, for restriction of the enzyme by sucrose or sucrose phosphate when sucrose accumulates in the leaf. To obtain values for the changes of SPS activity which result from 'coarse' control SPS was assayed in extracts using conditions which are selective for the active form of SPS (Stitt et al., 1988). Using the resulting equasion

$$dv/v = dSPS/SPS + \varepsilon_{HP}^{SPS} \cdot dHP/HP + \varepsilon_{sucr.}^{SPS} \cdot dsucrose/sucrose$$

we estimated an elasticity for hexose P of + 1.0. The notional elasticity for the inhibition by sucrose was negligible.

The Clarkia xantiana PGI dosage mutants also allowed estimation of the seperate elasticities for Fru6P and Glc6P, yielding values of + 0.4 and + 1.6 respectively. The changes of metabolites after supplying fluoride to spinach leaves (Fig. 5) allowed a elasticity for UDPGlc of about 0.8 to be estimated, assuming that ε_{HP}^{SPS} was about unity.

Interaction between Fru6P and Fru2,6bisP The response of Fru2,6bisP to rising Fru6P will depend on (a) the stimulation of Fru6P, 2-kinase and (b) the inhibition of Fru2,6bisPase. The change of Fru2,6bisP will be even greater than the changes in the individual enzyme activities, because they operate as a cycle. Taking the simplest case, where other effectors do not change, it can be shown (Stitt, in press) that

$$\alpha = \frac{dFru2,6bisP}{Fru2,6bisP} \bigg/ \frac{dFru6P}{Fru6P} = \frac{\varepsilon_{F6P}^{2-k} - \varepsilon_{F6P}^{2-P}}{\varepsilon_{26P}^{2-P} - \varepsilon_{26P}^{2-k}}$$

where ε_{F6P}^{2-k} and ε_{26P}^{2-k} are the elasticities of Fru6P, 2-kinase for Fru6P and Fru26bisP, and ε_{F6P}^{2-P} and ε_{26P}^{2-P} are the elasticities of Fru2,6bisPase for Fru6P and Fru2,6bisPase. This relation allows is to relate an amplification factor, α , to the individual elasticites of the synthesing and degrading enzymes for Fru6P and Fru2,6bisP. As discussed in Stitt (1989), Fru6P,2-kinase and Fru2,6bisPase are regulated in a complex manner, and α probably adopts a wide range of values. A value of about 4 might be reasonable if Pi is low; at 5 - 10 mM Pi, a lower value might be expected. Table 9 lists several empirically measured values for α in leaves, in conditions where rising Fru6P is leading to an increase of Fru2,6bisP. The measured values lie between 2 - 3. These suggest our theoretical estimate is not unrealistic; indeed somewhat lower values would be expected because Pi will be present, and because rising Fru2,6bisP will inhibit the cytsolic Fru1,6Pase and this, in turn, would lead to higher triose P and PGA which themselves act to inhibit Fru6P,2-kinase and dampen the response to rising Fru6P.

Estimation of flux control coefficients The connectivity theorem can now be applied to estimate how control is shared between the Fru1,6Pase and SPS. In doing this, a modification is needed, because the shared pool (hexoseP) acts indirectly via Fru2,6bisP on the cytosolic Fru1,6Pase. The distribution of flux control for SPS (C_{SPS}) and the Fru1,6Pase (C_F) during sucrose synthesis will be given as

$$ C_F \Big/ C_{SPS} \quad = \quad - \quad \varepsilon^{SPS}_{HP} \Big/ \varepsilon^{FBPase}_{26P} \cdot \alpha $$

reflecting the fact that a fractional change dHP/HP of hexose P will be accompanied by a fractional change, $\alpha \cdot$ dHP/HP of Fru2,6bisP. Fig. 6 shows how the relative control strengths of SPS and Fru1,6Pase vary as is increased at three different values (-1, -2.5, -5) of the $\varepsilon^{SPS}_{HP}/\varepsilon^{FBPase}_{26P}$ ratio. At the in vivo values during rapid photosyntnesis (ε^{SPS}_{HP} = 1.0; $\varepsilon^{FBPase}_{26P}$ = 0.4; α = 2-3), control is shared almost equally between SPS and the cytosolic Fru1,6Pase (see Stitt, 1989, for more details).

Interactions between enzymes controlling the flux to sucrose
Elasticity coefficients can adopt a wide range of values and we must therefore expect control will shift between the Fru1,6Pase and SPS depending on the conditions. Our model predicts that control will shift towards the cytosolic Fru1,5Pase if (a) SPS becomes more sensitive to activation by hexose P i.e. ε^{SPS}_{HP} increases, (b) Fru1,6Pase becomes more sensitive to inhibition by Fru2,6bisP i.e. $\varepsilon^{FBPase}_{26P}$ decreases, or (c) Fru2,6bisP responds less sensitively to a change of Fru6P i.e. α decreases. For example, high Pi will decrease the response of Fru2,6bisP to Fru6P (see Stitt, 1989). This means than the flux control coefficient of the Fru1,6Pase may increase in conditions when metabolites are low (e.g. low light or carbon dioxide). This would ensure a first priority for maintaining high metabolite pools in the stroma. Conversly, during very rapid photosynthesis and starch accumulation, metabolite levels are high and Pi is probably low (Gerhardt et al., 1987). In these conditions, Fru2,6bisP may respond more sensitively to changes of Fru6P, and control would shift towards SPS.

This analysis has concentrated on hexose P and Fru2,6bisP, because these represent the direct link between the cytosolic Fru1,6Pase and SPS. However, both these enzymes will also be influenced by further factors. For SPS, we have seen that the elasticity for UDPGlc is relatively high. This can be understood because, although the saturation curve is hyperbolic, high con-

Table 7. Fractional changes of metabolites, SPS activation and fluxes to sucrose The results in spinach leaf discs are derived from the slope of the model shown in Heldt and Stitt (1985) in the region where carbon dioxide fixation decreases due to lower light or carbon dioxide, and the region where sucrose synthesis decreases in response to accumulation of sucrose. The changes of SPS activity (ΔSPS/SPS) are estimated from Stitt et al. (1988) and unpublished results which compare the change of SPS activity in selective assay conditions (2 mM Fru6P, 10 mM Glc6P, 5 mM Pi) as sucrose synthesis decreases in response to falling light, or accumulation of sucrose. The data for Clarkia xantiana are from Neuhaus E., Kruckeberg, A.L., Feil, R. and Stitt, M. (unpublished data) and show the response of metabolite levels and fluxes to a decrease of cytosolic phosphoglucose isomerase (PGI) activity to 18% of the wildtype level (Jones et al. 1986) in low light (125 μmol m^{-2} s^{-1}) or high light (1000 μmol m^{-2} s^{-1}). The pools of Fru6P (*) and Glc6P(**) are shown seperately.

Material	Treatment	Measured fractional change					
		ΔJ	ΔTP	Δ26P	ΔHP	ΔSucrose	ΔSPS
Spinach	Decreased CO$_2$ fixation	−0.33	−0.2	+0.3	−0.2	0	−0.15
	Accumulation of sucrose	−0.33.	+0.5	+1.7	+0.5	+8	−0.55
Clarkia	Decreased PGI (high light)	−0.08	+0.22	+1.1	+0.5** −0.1*	–	–
	Decreased PGI (low light)	−0.23	−0.01	+0.6	+0.3** −0.13*	–	–

Table 8. Experimental estimates of elasticities for SPS and Fru1,6Pase. They are estimated from the fractional changes of fluxes, and metabolites shown in Table 3, using two simultaneous equations of the form dJ/J = ε_{61} · dS1/S1 + ε_{62} · dS2/S2 as described in the text. For Clarkia, seperate elasticities of SPS to Glc6P (**) and Fru6P (*) are shown.

	$\varepsilon_{TP}^{FBPase}$	$\varepsilon_{26P}^{FBPase}$	ε_{HP}^{SPS}	$\varepsilon_{sucrose}^{SPS}$
Spinach	+1.0	−0.5	+1.0	0.03
Clarkia	+1.5	−0.4	+2.0** +0.4*	–

Table 9. Experimental values for α, the amplification factor for the relation between changes of Fru6P and of Fru6bisP. The results are taken from (a) the rise of Fru2,6bisP and Fru6P as sucrose accumulates in leaf discs during photosynthesis in saturating light and carbon dioxide (Stitt and Heldt, 1982) (b) the relation between Fru2,6bisP (Stitt et al., 1983) and cytosolic Fru6P (Gerhardt et al., 1987) level in spinach leaves in ambient conditions. (c) the relation between Fru6P and Fru2,6bisP in Clarkia mutants with decreased PGI activity (Neuhaus, E., Kruckeberg, A., Feil, R. and Stitt, M. unpublished), in high (1000 μmol m^{-1} s^{-1}) light and (d) low (125 μmol m^{-2} s^{-1}) light.

Treatment	$\dfrac{d\ \text{Fru2,6bisP}}{\text{Fru2,6bisP}} \bigg/ \dfrac{d\ \text{Fru6P}}{\text{Fru6P}}$
a) Sucrose accumulation in saturating light and CO_2	3.1
b) Sucrose accumulation in ambient conditions	1.8
c) Decreased PGI activity in high light	2.2
d) Decreased PGI activity in low light	2.0

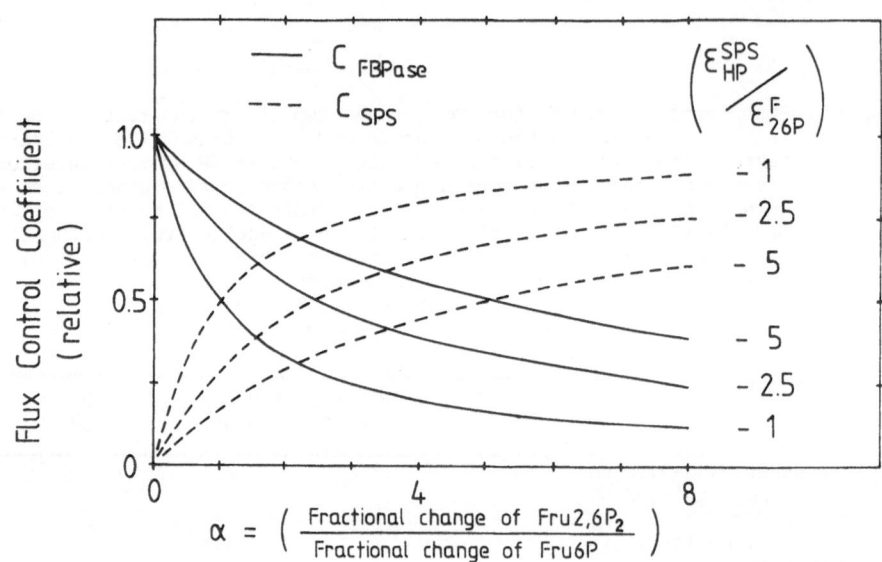

Fig. 6. Model for the relative control strengths of SPS and the cytosolic Fru1,6Pase during photosynthetic sucrose synthesis. C_{FBPase} (———) and C_{SPS} (----) are shown for three different values of the ε^{SPS}_{HP} / $\varepsilon^{FBPase}_{26P}$ ratio (see Figure) as the value of is increased (lower axis). The values are computed as C_{FBPase}= x/(1+ x) where x = ε^{SPS}_{HP}/α. $\varepsilon^{FBPase}_{26P}$, the expression being derived from the modified connectivity theorem C_F / C_{SPS} = ε^{SPS}_{HP} / $\varepsilon^{FBPase}_{26P} \cdot \alpha$.

centrations are needed to saturate (Doehlert and Huber, 1984; Stitt et al., 1988). As a result, SPS will be operating at UDPGlc concentrations which are well below saturating, explaining why a relatively small increase of PPi and decrease of UDPGlc has quite a marked effect on the flux to sucrose. The 'coarse' changes of activity will be another important factor effecting SPS, of course, and this requires further study.

For the Fru1,6Pase, this simplified analysis has already shown that the enzyme will be very dependent on trioseP, as well as Fru2,6bisP (see also Stitt and Heldt 1985; Stitt et al., 1987). The elasticity for trioseP is far larger than for Fru2,6bisP (Stitt, 1989). Even after allowing for the amplification produced by the regulator cycle, it appears likely that the elasticity for trioseP will be larger than the notional response to accumulating Fru6P (= $\alpha \cdot \varepsilon^{FBPase}_{FCP}$). The high elasticity to trioseP can be explained by (a) the sigmoidal substrate response of the cytoslic Fru1,6Pase (Herzog et al., 1984) and (b) the second order relation between trioseP and Fru1,6bisP, which means that the notional elasticity for trioseP will be twice as large as the elasticity for Fru1,6bisP (Stitt, 1989). This high elasticity for trioseP has important theoretical and experimental concequences. Theoretically, it means that a restriction of the cytosolic Fru1,6Pase by high Fru2,6bisP will be overcome if trioseP rise, for example because the rate of photosynthesis rises or because the triose P cannot be diverted to starch. This means that rising Fru2,6P can be used to control partitioning, but can be overridden, if neccesary, to allow rapid photosynthesis to continue, and has obvious functional advantages. Experimentally, the high elasticity for trioseP means that extreme caution is needed in interpreting numerus studies which have tried to relate Fru2,6bisP and photosynthate partitioning, but to omitted measure trioseP.

We can also integrate this treatment of SPS and the cytosolic Fru1,6Pase with the earlier treatment of PGI, by comparing the elasticities of all these enzymes for their shared metabolites. In the wildtype, PGI had elasticities of 5 to 8 for Fru6P and Glc6P. The notional elasticity of the cytosolic Fru1,6Pase for Fru6P (= $\alpha \cdot \varepsilon^{FBPase}_{2CP}$) is about 1. The elasticity of SPS for Glc6P (1.5) is larger than that for Fru6P (0.4) by a similar factor. Applying the connectively theorem, we would expect little control at PGI in the wildtype. On the other hand, PGI does start to adopt a measurable control coefficient in the Clarkia mutants, when the elasticity coefficient falls to around 2. (see Fig. 4 Table 6 and accompanying text). At these values, we can estimate that PGI might adapt a control coefficient about half that of the Fru1,6Pase or SPS (about, because their elasticities probably also change). This would indicate control strengths of about 0.4, for Fru1,6Pase and SPS, and about 0.2 for PGI. Direct determination of the flux control coefficient from the slope of Fig. 4 provides a value in the same range when PGI dosage is reduced to 18% in limiting light.

In saturating light, however, the situation becomes more complex. The flux control coefficient of PGI is less (Fig. 4B) than in limiting light, and there is a large accumulation of PGA and triosP (E. Neuhaus, A.L. Kruckeberg and M. Stitt, unpublished). The high levels of PGA may be acting to dampen the response of Fru2,6bisP to rising Fru6P, and the increased triose P will effectively counteract the effect of rising Fru2,6bisP on Fru1,6-Pase. Thus, in these conditions, control in the cytosol is being influenced by events in the chloroplast.

Interaction with the chloroplast Clearly, the model will need to be extended to cover the interaction between sucrose synthesis and the reactions in the chloroplast. Metabolism in these compartments is mutually dependent, because the chloroplast provides the triose P for sucrose synthesis, while sucrose synthesis recycles Pi to allow further photosynthesis.

An analysis by Woodrow (1986) provides a starting point by describing the interaction between the six major irreversible reactions in the stroma, and introducing the idea of phosphate as a conserved moiety which is distributed between free Pi and the pools of phosphorylated intermediates. However, it will probably be necessary to extend this model to include (a) the detailed action of three-carbon metabolites and Pi on Fru2,6bisP levels and (b) factors which affect the supply of ATP and NADPH, and their effect on PGA reduction, and the PGA/trioseP ratio.

The PGA/triose P ratio varies greatly depending not on the availability of light and CO2, and also on the availability of Pi. Changes of PGA or triose P will have differing effects on fluxes. Triose P is the starting substrate for sucrose synthesis, and will also increase the activity of several enzymes in the calvin cycle via increased substrate concentration, and their action on the thioredoxin activation of the stromal bisphosphatases (Scheibe, 1987). PGA will act as an activator of sucrose synthesis because it inhibits Fru2,6bisP synthesis, but it will also activate starch synthesis (Preiss, 1980) and will inhibit the calvin cycle via its action on Rubisco (Foyer et al., 1987) and Ru5P kinase (Gardeman et al., 1983). This means that the PGA/triose P ratio will exert complex effects on flux and its control.

CONCLUSIONS

(a) Sucrose synthesis can be divided into reactions which are near-equilibrium and very far from equilibrium. (b) The first order equilibrum ractions may not make a significant contribution to control, at least in the case of phosphoglucoisomerase. However, it is misleading to think that these enzymes are present in a massive excess over that required to maintain their reactants close to equilibrium. Further, a relatively small displacement from equilibrium can have a marked influence on flux, if other reactions have a comparable elasticity to these shared metabolites. Indeed, proper functioning of the pathway may depend upon these reactions being very close to equilibrium. (c) Second order equilibrium reactions can influence flux by "mass action", one example being the displacement of UDPGlc and hexose P which results if the removal of PPi is inhibited (d). There are several non-equilibrium reactions, and control is shared between these reactions. The distribution of control probably shifts continually, as these enzymes have a large range of elasticities for any given effector, depending on the conditions. (e) This complex interaction means it is essential to study control within a framework which dictates a rigourous comparison of flux, metabolite concentrations and response of the enzymes, as is provided by the determination of elasticities and control coefficients.

Systems at this level of complexity cannot be adequately understood by ad hoc comparisons of a few selected metabolites or enzymes and prima facie not by studies attempting to correlate control with a single factor or 'limitation.

ACKNOWLEDGEMENTS

the unpublished work of the author discussed in this chapter was carried out in cooperation with L. Gottlieb, A.L. Kruckeberg, E. Neuhaus and R. Feil, and was supported by the Deutsche Forschungsgemeinschaft (SFB 137). I am indebted to Fr. U Küchler for her help in preparing the manuscript.

REFERENCES

Barber, T., G. 1985. The equlibrium of the reaction catalysed by sucrose P synthase, Plant Physiol 79, 1127-1128.

Bassham J.A. and Krause, G.H. (1969). Free energy changes and metabolic regulation in steady state photosynthetic carbon metabolism. Biochim. Biophys. Acta 189, 207-221.

Crabtree, B. and Newsholme, E.A. 1987. A systematic approach to describing and analysing metabolic control systems. Trends in Biochem. Sciences 12, 4-12.

Doehlert D.C. and Huber, S.C. 1988. Regulation of spinach leaf sucrose phosphate synthase by gluocose-6-phosphate, inorganic phosphate and pH.

Doehlert, D.C. and Huber S.C. 1984. Phosphate inhibition of spinach leaf sucrose phosphate synthase as affected by glucose-6-phosphate and phosphoglucoisomerase. Plant Physiol. 76, 250-253.

Dyson, J.E.D. and Noltman, E.A. 1963. The effect of pH and temperature on the kinetic parameters of phosphoglucose isomerase. J. Biol. Chem. 243, 1401-1414.

Foyer, C.H., Furbank, R.T. and Walker, D.A., 1987. Interactions between ribulose-1,5-bisphosphate carboxylase and stromal metabolites I. Modulation of enzyme activity by Benson-Calvin cycle intermediates Biochim. Biophys. Acta 894, 157-164.

Gardemann, A., Stitt, M. and Heldt, H.W., 1983. Control of carbon dioxide fixation. Regulation of spinach ribulose-5-phosphate kinase by stromal metabolite levels. Biochim Biophys. Acta 722, 51-60.

Gerhardt, R., Stitt, M. and Heldt, H.W. 1987. Subcellular metabolite levels in spinach leaves. Plant Physiol. 83, 399-407.

Groen, A.K., Van der Meer, R., Westerhoff, H.V., Wanders, R.T.A., Akerboom, T.P.M., Tager, J.M. 1982a. Control of Metabolic Fluxes. In Metabolic Compartmentation, (ed. H. Sies) p 937. London Academic Press.

Groen, A.K., Wanders, R.T.A., Westerhoff, H.V., van der Meer, R. and Tager, T.M., 1982b. Quantification of the contribution of various steps to the control of mitochondrial respiration. J. Biol. Chem. 257, 2754-2757.

Heinrich, R. and Rapaport, T.A., 1974. A linear steady state treatment of enzymatic chanis. Eur. J. Biochem. 42, 89-120.

Herzog, B., Stitt, M., and Heldt, H.W., 1984. Control of photo-
    synthetic sucrose synthesis by fructose-2,6-bisphosphate III
    Properties of the cytosolic Fructose-1,6-bisphosphatase.
    Plant Physiol. 75, 561-565.
Isherwood, F.A. and Selvandran, R.C. 1970. A note of the occuren-
    ce of nucleotides in strawberry leaves. Phytochemistry,
    2265-2267.
Jones, T.W.A., Pichersky, E. and Gottlieb, L.D., 1986. Enzyme
    activity in EMS-induced null mutations of duplicated genes
    encoding phosphoglucose isomerase in Clarkia. Genetics 113,
    101-114.
Kacser, H. and Burns, J.A. 1973. The control of flux. Symp. Soc.
    Exp. Biol. 27, 65-107.
Kacser, H. and Burns, J.A. 1979. Molecular Democracy : who shares
    the controls? Biochem. Soc. Trans 7, 1149-1161.
Kacser, H. and Porteus, J.W., 1987. Control of metabolism : what
    do we have to measure? Trends in Biochem. Sciences 12, 5-14.
Kerr,P.S. and Huber, S.C., 1987. Coordinate control of sucrose
    formation in soybean leaves by sucrose phosphate synthase
    and fructose-2,6-bisphosphate. Planta 170, 197-204.
Lehringer, A.L., 1970. Biochemistry, Worth Publ., New York.
Newsholme, E.A. and Start, C., 1973. Regulation in Metabolism.
    Wiley and Sons.
Rolleston, F.S., 1972. A theoretical background to the use of
    measured intermediates in the study of the control of inter-
    mediary metabolism. Curr. Top Cell. Reg. 5, 47-75.
Scheibe, R., 1987 NADP malate dehydrogenase in C-3 plants: regu-
    altion and role of a light-activated enzyme. Physiol. Plan-
    tarum 71, 393-400.
Sicher, R.C., Kremer, D.F. and Harris, W.G., 1986. Control of
    photosynthetic sucrose synthesis in barley primary leaves.
    Plant Physiol. 82, 15-18.
Stitt, M., Wirtz, W. and Heldt, H.W., 1980. Metabolite levels in
    the chloroplast and extrachloroplast compartments of spinach
    protoplasts Biochim. Biophys. Acta 593, 85-102.
Stitt, M., Wirtz, W. and Heldt, H.W., 1983. Regulation of sucrose
    synthesis by cytoplasmic fructose bisphosphatase and sucrose
    phosphate synthase during photosynthesis in response to
    varying light and carbon dioxide. Plant Physiol. 72,
    767-774.
Stitt, M., Cseke, C. and Buchanan, B.B., 1984. Regulation of
    fructose-2,6-bisphosphate concentration in spinach leaves.
    Eur. T. Biochem. 143, 89-93.
Stitt, M., Herzog, H. and Heldt, H.W., 1985a. Control of photo-
    synthetic sucrose synthesis by fructose-2,6-bisphosphate V.
    Modulation of the cytosolic fructose-1,6-bisphosphatase in
    vitro Plant Physiol. 79, 590-598.
Stitt, M. and Heldt, H.W., 1985. Control of photosynthetic suc-
    rose synthesis by fructose-2,6-bisphosphate VI. Regulation
    of the cytosolic fructose-1,6-bisphosphatase in vivo. Plant
    Physiol. 79, 599-608.
Stitt, M., Wirtz, W., Gerhardt, R., Heldt, H.W., Spencer, C.,
    Walker, D.A., Foyer, C.H., 1985b. A comparative study of
    metabolite levels in plant leaf material in the dark. Planta
    166, 354-364.
Stitt, M., Gerhardt, R., Wilke, I. and Heldt, H.W., 1987. The
    contribution of fructose-2,6-bisphosphate to the regulation
    of sucrose synthesis during photosynthesis. Physiol. Planta-
    rum 69, 377-386.
Stitt, M., Wilke, I., Feil, R. and Heldt, H.W. (1988). Coarse
    control of sucrose phosphate synthase in leaves. Planta 174,
    217-230.
Stitt, M. and Große, H., 1988. Interaction between sucrose syn-
    thesis and photosynthesis IV. Temperature dependent adjust-
    ment of the relation between sucrose synthesis and CO_2 fixa-
    tion. J. Plant Physiol. (in press).
Stitt, M., 1989. Control analysis of photosynthesis sucrose
    synthesis: assignment of elasticity coefficients and flux

control coefficients to the cytosolic fructose-1,6-bisphosphatase and sucrose phosphate synthase. Phil. Trans. Roy. Soc. Lond. (In Press).

Woodrow, I.E., 1986. Control of the rate of photosynthetic carbon dioxide fixation. Biochim. biophys. Acta 851, 181-192.

Weiner, H., Stitt, M. and Heldt, H.W., 1987. Subcellular compartmentation of pyrophosphate and alkaline pyrophosphatase in leaves. Biochim. Biophys. Acta 893, 13-21.

THE PHOTOSYNTHETIC ASSIMILATION OF NITRATE AND ITS INTERACTIONS WITH CO_2 FIXATION

Catalina Lara and Miguel G. Guerrero

Instituto de Bioquímica Vegetal y Fotosíntesis
Universidad de Sevilla-CSIC
Apartado 1113, 41080 Sevilla, Spain

INTRODUCTION

Photosynthesis is usually identified with the light-driven formation of carbohydrates and oxygen from CO_2 and water. This formulation ignores, however, the basic fact that in the photosynthetic process not only CO_2, but also the oxidized forms of other primordial bioelements are reduced and incorporated into cell material. Actually, photosynthesis drives a number of biosynthetic pathways involved in the assimilation of inorganic carbon, nitrogen and sulfur. At the expense of sunlight energy, unstable energy-rich products -cell material and oxygen- are synthesized from fully oxidized substrates with no useful chemical potential, namely water, carbon dioxide, nitrate, sulfate and phosphate (Losada and Guerrero, 1979; Losada et al., 1987).

The primary nitrogen source for plants is nitrate. The assimilation of nitrate by higher plants and algae is a fundamental photosynthetic process in which a highly oxidized form of inorganic nitrogen is reduced to ammonium, which is in turn combined with carbon skeletons to form the different organic nitrogenous compounds. It has been estimated that the process accounts for the assimilation of about 10^{10} tons nitrogen on a yearly basis. In this chapter we will attempt to summarize the present status of knowledge on nitrate assimilation in green cells and its integration with carbon metabolism. Although the relevance of a balanced assimilation of carbon and nitrogen is self-evident, little is known about the underlying control mechanisms. Interactions between photosynthetic nitrate assimilation and CO_2 fixation will be considered with respect to two different aspects: a) incidence of nitrate (and ammonium) assimilation on CO_2 fixation; and b) effects of CO_2 fixation on nitrate assimilation.

Abbreviations: DLG, D,L-glyceraldehyde; GOGAT, glutamate synthase; GS, glutamine synthetase; MSX, L-methionine, D,L, sulfoximine; NiR, nitrite reductase; NR, nitrate reductase.

PATHWAY AND ENZYMES OF NITRATE ASSIMILATION

Three main steps will be considered in the nitrate assimilation pathway, namely, nitrate transport into the cell, reduction to ammonium, and subsequent ammonium incorporation to carbon skeletons yielding amino acids. Intermediate stages can, however, exist, depending on the organism, between nitrate uptake and reduction. These include storage of the ion in vacuoles and/or translocation from the roots to the leaves where primary nitrate assimilation occurs in most higher plants. The proportion of nitrate that is assimilated in the root or transported to the leaf varies, depending on the species, age and nitrogen status of the plant.

Nitrate transport

Thermodynamic considerations have brought to the general belief that nitrate transport through a polarized plasma membrane (negative inside) must be an active process. However, little is known of the biochemistry of nitrate transport in photosynthetic organisms. Studies carried out with Lemna and some higher plant roots on depolarization of plasmalemma upon addition of nitrate suggest the possible operation of a NO_3^-/H^+ cotransport (Ullrich, 1987). Several polypeptides of the plasma membrane of corn roots have been shown to be synthesized upon exposure to nitrate and their possible implication in nitrate transport has been suggested (Dhugga et al., 1988). Intracellular accumulation of nitrate against the electrochemical gradient of the ion has been measured in the blue-green alga (cyanobacterium) Anacystis nidulans (Lara et al., 1987a). A parallelism exists between the amount of a 47-kDa polypeptide of the plasma membrane and the nitrate transport activity of A. nidulans cells, suggesting the involvement of this adaptive protein in nitrate transport (M.N. Sivak, C. Lara, J.M. Romero, R. Rodríguez and M.G. Guerrero, submitted for publication).

Nitrate reduction

The assimilatory reduction of nitrate to ammonium occurs in two sequential exergonic reactions, the 2-electron reduction of nitrate to nitrite, catalyzed by nitrate reductase (NR), and the 6-electron reduction of nitrite to ammonium, catalyzed by nitrite reductase (NiR).

$$NO_3^- \xrightarrow[\text{NR}]{2e} NO_2^- \xrightarrow[\text{NiR}]{6e} NH_4^+$$

In photosynthetic eukaryotes NR is a multicomponent enzyme complex containing FAD, a b-type heme (cytochrome b_{557}) and molybdenum, the latter present as a pterin-Mo complex similar to the Mo-cofactor isolated from other molybdoenzymes (Solomonson and Barber, 1987). The enzyme from most origins is either a dimer (of about 200 kDa) or a tetramer (of about 400 kDa) composed by identical ca. 100-kDa polypeptide chains containing one each of the cofactors (Howard and Solomonson, 1982; Nakagawa et al., 1985; Solomonson and Barber, 1987). Some significant deviations from this pattern have been reported, however. Most notable is the case of the enzyme from the green alga Ankistrodesmus braunii, which has been found to be composed by eight subunits of 59 kDa containing four FAD, four heme groups and two molybdenum atoms per enzyme molecule (de la Rosa et al., 1981). Eukaryotic NR uses NADH as electron donor, although in some cases NADPH is also utilized (Guerrero et al., 1981). During catalysis electrons from NAD(P)H are sequentially transferred to the flavin, the heme and the molybdenum cofactors, and eventually to nitrate.

There is a long standing controversy about the intracellular location of NAD(P)H-NR (see Losada et al., 1981 for a review). Although many authors assume a cytoplasmic location of the enzyme, recent immunocytochemical studies have shown, however, that the problem is far from solved. Thus, studies with a number of green algal species indicate the location of NR in the pyrenoid region of the chloroplast (López-Ruíz et al., 1985a,b). In spinach leaves, NR has been reported to be located in both the chloroplast and the cytoplasm (Roldán et al., 1987) or solely in the chloroplast (Kamachi et al., 1987).

Nitrate reductase from photosynthetic prokaryotes differs substantially from its eukaryotic counterpart. In cyanobacteria, NR is a smaller protein (75-85 kDa) consisting of a single polypeptide chain containing one molybdenum, four non-heme iron and four acid-labile sulphur atoms per enzyme molecule (Guerrero et al., 1981; Mikami and Ida, 1984). Circular dichroism spectra suggest an arrangement of the iron and sulfur atoms as two $[2Fe-2S]$ clusters (Mikami and Ida, 1984). The physiological electron donor of cyanobacterial NR is reduced ferredoxin. The enzyme seems to be tightly attached to thylakoid membranes, prolonged sonication or detergent treatment being required to release the enzyme in a soluble form (see Guerrero and Lara, 1987 for a review).

Nitrite reductase in all photosynthetic organisms tested is a ferredoxin-dependent enzyme (Vega et al., 1980). The structure of NiR from different origins has been traditionally considered as composed by a single polypeptide chain of about 60 kDa containing one $[4Fe-4S]$ iron-sulfur cluster and one siroheme group per enzyme molecule (Vega and Kamin, 1977; Cammack et al., 1978; Ida and Mikami, 1986; Kamin and Stein Privalle, 1987). However, it has also been reported that the native enzyme from spinach (Hirasawa et al., 1984; Hirasawa and Knaff, 1985) and Chlamydomonas (Romero et al., 1987) contains an additional 25-kDa polypeptide, exhibiting a M_r of about 85000. The small subunit has been proposed to function in the coupling of the iron protein to reduced ferredoxin, being thus responsible for the ferredoxin-dependent activity (Hirasawa et al., 1984; Hirasawa and Knaff, 1985). Actually, the 85-kDa NiR forms a complex with ferredoxin exhibiting an apparent M_r of 110000 (Hirasawa et al., 1986). Present evidence supports the idea that the iron-sulfur cluster would receive electrons from reduced ferredoxin and donate them to the siroheme iron, which forms a complex with nitrite (Aparicio et al., 1975; Cammack et al., 1978). In this complex, the nitrite nitrogen would be progresively reduced and protonated, ammonium being eventually released at the end of the catalytic cycle (Cammack et al., 1987). Nitrite reductase is located in the chloroplast (Wallsgrove et al., 1979).

Incorporation of ammonium to carbon skeletons

Ammonium taken up from the outer medium or internally generated by nitrate reduction is assimilated through its combination with carbon skeletons yielding amino acids. In parallel to this primary assimilation, the ammonium resulting from the photorespiratory conversion of glycine to serine and from protein degradation is also reassimilated (Wallsgrove et al., 1983; Wallsgrove and Lea, 1985).

It is now firmly established that ammonium assimilation occurs via the glutamine synthetase (GS)-glutamate synthase (GOGAT) cycle (Miflin and Lea, 1982). This pathway consists in the incorporation of ammonium to glutamate forming glutamine and the subsequent transfer of the amide group to α-ketoglutarate yielding two glutamate units, one of which recycles, the other being the final product. One ATP and two electrons are required for the net formation of one glutamate unit from ammonium and α-ketoglutarate.

$$NH_4^+ \; + \; glutamate \; \xrightarrow[\text{GS}]{\sim P} \; glutamine$$

$$glutamine \; + \; \alpha\text{-ketoglutarate} \; \xrightarrow[\text{GOGAT}]{2e} \; 2 \; glutamate$$

Glutamine synthetase is a multimeric enzyme. In photosynthetic eukaryotes native GS has a M_r of 340000-380000 and is composed by eight 43-48 kDa subunits (Hirel et al., 1984a; Ericson, 1985). Two isoforms of GS, one located in the cytosol (GS_1) and another in the chloroplast (GS_2) exist in most higher plants and green algae (Kretovich et al., 1981; McNally et al., 1983; Hirel et al., 1984b; Beudeker and Tabita, 1985). They can be separated by ion-exchange chromatography and have different kinetic and immunological properties. The relative amount of each isoform varies widely between plant species. Apparently both GS_1 and GS_2 are synthetized in the cytoplasm, GS_2 being further transported to the chloroplast in a light-dependent process (Hirel et al., 1982). Prokaryotic type GS has been isolated from blue-green algae as a dodecamer of 600 kDa composed by apparently identical 50-kDa subunits (Sampaio et al., 1979; Orr et al., 1981; Florencio and Ramos, 1984).

Glutamate synthase of green cells was originally found as a ferredoxin-dependent enzyme located in the chloroplast (Lea and Miflin, 1974; Wallsgrove et al., 1979). More recently, NAD(P)H-dependent GOGAT has also been found to be present in green algae (Cullimore and Sims, 1981) and higher plants (Matoh et al., 1980; Wallsgrove et al., 1982; Suzuki et al., 1982, 1984). Nevertheless, the predominant role in photosynthetic ammonium assimilation is played by the ferredoxin-dependent enzyme. Etiolated shoots exhibit similar levels of both GOGAT activities but upon greening, only the

ferredoxin-dependent activity develops, being the
NAD(P)H-GOGAT in mature green leaves less than 3% of total
activity (Matoh and Takahashi, 1981, 1982; Wallsgrove et al.,
1982). In blue-green algae only the Fd-dependent activity has
been detected (Lea et al., 1982).

Ferredoxin-dependent GOGAT has been characterized as a
single polypeptide chain of 140-180 kDa containing iron and
flavins (Hirasawa and Tamura, 1984; Márquez et al., 1986).
The spinach enzyme appears to contain four non-heme iron and
four acid-labile sulfur atoms, one FAD and one FMN groups per
enzyme molecule (Hirasawa and Tamura, 1984), and to form an
electrostatic complex with ferredoxin (Hirasawa et al.,
1986).

Amino acid biosynthesis

After ammonium-N has been assimilated to the level
of the α-amino group of glutamate, it may be transferred to
different α-ketoacids in a series of reactions catalyzed by
multiple amino transferase enzymes yielding amino acids.
Glutamate itself is also a direct presursor of the
biosynthesis of proline, arginine and ornitine. Besides this,
glutamine can donate its amido group for the biosynthesis of
asparagine, arginine, tryptophan or histidine in reactions
catalyzed by the corresponding amido transferase enzymes (see
Miflin and Lea, 1982; Wallsgrove et al., 1983 for reviews).

Green cells have the capacity of synthesize all of the
nutritionally essential amino acids from inorganic carbon,
nitrogen and sulfur. In eukaryotic cells, the chloroplast is
the major site for amino acid biosynthesis and, according to
enzyme localization studies, for many amino acids the
chloroplast is the only site of their synthesis (Wallsgrove
et al., 1983). The limitation of the chloroplast is the
ability to synthesize α-keto acids and other carbon
skeletons, which must be imported from the cytoplasm. This is
the case of α-ketoglutarate and oxaloacetate which enter the
chloroplast and exchange for amino acids, transport being
mediated by specific dicarboxylate carriers of the inner
envelope membrane (Lehner and Heldt, 1978; Woo et al.,
1987a,b). Thus, the chloroplast behaves as an organic acid
importer-amino acid exporter organelle.

PHOTOSYNTHETIC NATURE OF NITRATE ASSIMILATION

In green cells, a direct utilization of
photosynthetically generated reducing power appears evident
for both nitrite reduction and ammonium assimilation. On the
one hand, there is the ferredoxin-dependent character of NiR
and GOGAT and the clear evidence of their localization in the
chloroplast. On the other hand, many studies have
shown the ability of isolated chloroplasts to reduce nitrite
to ammonium in a light-dependent process coupled to water
photolysis (Neyra and Hageman, 1974; Anderson and Done, 1978;
Anderson, 1981). The same applies to ammonium assimilation,
which can be followed in intact illuminated chloroplasts as
(ammonium plus α-ketoglutarate)-dependent O_2 evolution
(Anderson and Done, 1977; Woo and Osmond, 1982; Anderson and
Walker, 1983).

The situation is not that clear for nitrate reduction. The piridin nucleotide dependence of eukaryotic NR, its uncertain localization and the apparent inability of purified intact chloroplast to reduce nitrate (Swader and Stocking, 1971; House and Anderson, 1980) have led to consider that the reduction of nitrate may not be a direct photosynthetic process, but dependent on organic substrates as the immediate source of reducing power (see Losada et al., 1981; Syrett, 1981 for reviews). Nevertheless, a pioneering study of Van Niel et al., (1953) with intact <u>Chlorella</u> cells showed that, at saturating light intensity , the rate of O_2 evolution by cells supplied with non-limiting concentrations of CO_2 was increased by addition of nitrate while that of CO_2 fixation was unchanged. This was a clear indication that the extra O_2 evolution induced by nitrate was due to a direct photochemical reduction of nitrate, regardless the intracellular location of NR or the nature of its electron donor. Although some attemps to reproduce these results with other green algae were not successful (Bongers, 1958; Grant and Turner, 1969), these observations have been confirmed for <u>Scenedesmus</u>, where nitrate induces a decrease in the CO_2/O_2 quotient at saturating light intensity (Larsson et al., 1982).

A recent study with the blue-green alga <u>Anacystis</u> (Romero and Lara, 1987) has also shown that the rate of light-dependent O_2 evolution by intact cells supplied with $NaHCO_3$ at saturating concentrations was enhanced for a wide range of light intensities following the addition of nitrate (Fig. 1). This indicates that when CO_2 is the only electron acceptor available, the rate of noncyclic electron flow is not limited by light but rather by the rate of CO_2 fixation. Addition of nitrate as a second electron acceptor would release noncyclic electron flow from this limitation, stimulating the rate of O_2 evolution and, hence, of electron transport from water to ferredoxin. As Fig. 1 shows, the extent of the stimulation of O_2 evolution induced by nitrate was dependent on light intensity, being maximal under light saturating conditions. Moreover, the ratio between extra O_2 evolved and nitrate taken up was very low at low light intensity, but it increased with photon flux, becoming higher than 2 at saturating light intensity (Table I). This indicates that the stimulation of noncyclic electron flow induced by nitrate at saturating light may suffice to support the reductant requirements of nitrate assimilation, while at low light intensities it will be not. Accordingly, nitrate depressed the rate of CO_2 fixation at limiting but not at saturating light, this depression reflecting the competition between both processes for photosynthetically generated assimilatory power (Romero and Lara, 1987, see below).

Strong evidence on the association of non-cyclic electron flow with nitrate reduction has also been obtained in algae under less physiological conditions. Preventing ammonium assimilation and CO_2 fixation, quantitative photochemical conversion of nitrate to ammonium has been demonstrated in whole cells of several green algal species (Syrett and Morris, 1963; Thacker and Syrett, 1972; Larsson et al., 1982) and in the blue-green alga <u>Anacystis</u> (Flores et al., 1983a), with stoichiometry values of 2 mol O_2 per mol

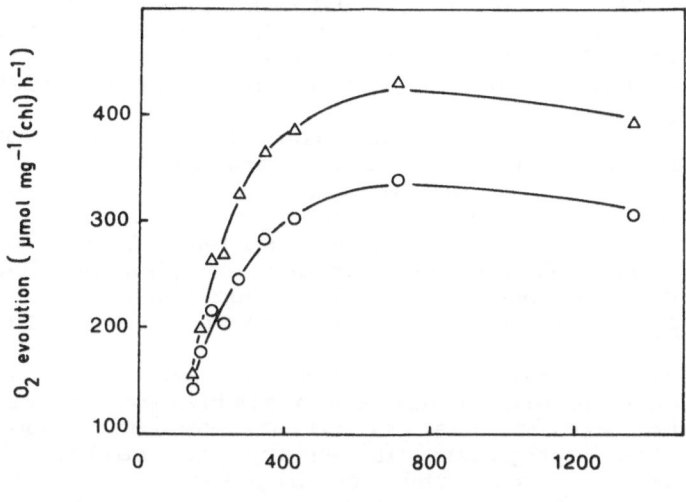

Figure 1. Light-dependent oxygen evolution by intact <u>Anacystis nidulans</u> cells upon addition of bicarbonate and nitrate as electron acceptors. At every light intensity, linear rates of O_2-evolution were measured in reaction mixtures containing 10 mM $NaHCO_3$ before (◯) and after the addition of 0.25 mM KNO_3 (△). From Romero and Lara, 1987, with permission.

TABLE I. Dependence of the in vivo O_2/NO_3^- ratio upon light intensity in intact <u>Anacystis nidulans</u> cells

Incident photon flux (μE m^{-2} s^{-1})	Nitrate taken up (μmol mg^{-1} Chl)	Extra oxygen evolved[a] (μmol mg^{-1} Chl)	O_2/NO_3^- ratio
57	9.45	8.36	0.88
92	14.18	12.00	0.85
204	17.45	36.73	2.10
260	21.45	51.27	2.39

[a]The extra oxygen evolved was estimated manometrically as the difference between the O_2 produced during 20 min in vessels containing saturating amounts of CO_2 plus nitrate as electron acceptors and in those containing CO_2 alone. From Romero and Lara, 1987, with permission.

nitrate reduced. Thylakoid membrane preparations from the blue-green algae Anabaena and Anacystis have been used to photoreduce nitrate with water as electron donor (Ortega et al., 1976; Candau et al., 1976) taking advantage of the ferredoxin-dependent character of NR and of its association to thylakoids. This can be considered as one of the simplest examples of photosynthesis (Losada and Guerrero, 1979).

Additional evidence of the direct relationship of nitrate assimilation with the photochemical reactions of photosynthesis is provided by observations of nitrate acting as a quencher of chlorophyll fluorescence in blue-green algal cells (Serrano et al., 1981) and spinach leaves (Sivak and Walker, 1985). Increased non-cyclic photophosphorylation coupled to the photochemical reduction of nitrate in Hydrodictyon cells has also been reported (Raven, 1977).

In summary, experimental evidence indicates clearly the photosynthetic nature of nitrate assimilation in green cells. This does not exclude that if nitrate reduction does in fact occur in the cytoplasm of eukaryotic cells, export of reducing power from the chloroplast via the malate--oxaloacetate or the triose phosphate shuttle (Rathnam, 1978; House and Anderson, 1980) should be involved between the photosynthetic generation of reductant and nitrate reduction.

INCIDENCE OF NITROGEN ASSIMILATION ON CO_2 FIXATION

The assimilation of inorganic nitrogen consumes assimilatory power and generates organic nitrogenous compounds, utilizing CO_2 fixation products as nitrogen acceptors. CO_2 fixation is, therefore, affected by nitrogen assimilation at two different levels. A competition for reducing power and/or ATP results, under certain conditions, in decreased rates of CO_2 fixation as a result of simultaneous nitrogen assimilation. On the other hand, the utilization of -ketoacids for ammonium assimilation affects the extent of CO_2 fixation and the pattern of carbon assimilation products (Lara et al., 1987b).

A useful experimental approach to the study of the influence of nitrogen assimilation on CO_2 fixation is to analyze these effects as a function of light intensity under conditions of saturating CO_2 supply, using N-sufficient cells. In spinach protoplasts (Rathnam, 1978), the green alga Scenedesmus (Larsson et al., 1985) and the blue-green alga Anacystis (Romero et Lara, 1987) nitrate depresses CO_2 fixation at low light intensities but has no negative effects at saturating light (Fig. 2). This can be explained in terms of competition for assimilatory power. As discussed above for Anacystis (see Table I), no competition for assimilatory power exists at saturating light, the photosynthetic apparatus being able to generate enough assimilatory power for the simultaneous assimilation of carbon and nitrogen. At limiting light intensities, a moderate competition between nitrate utilization and CO_2 fixation is apparent in consistency with the moderate extent of the nitrate-induced stimulation of O_2 evolution observed under these conditions (Table I).

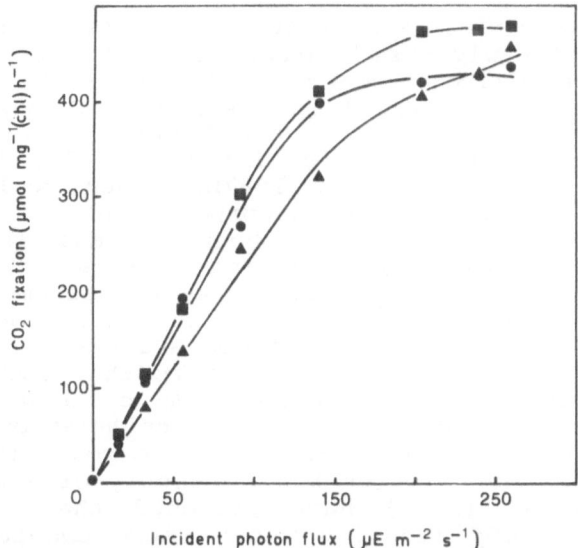

Figure 2. Changes in the light-saturation profile of ^{14}C $NaHCO_3$ assimilation by intact <u>Anacystis nidulans</u> cells in response to concomitant nitrate or ammonium utilization. The rates of ^{14}C incorporation to acid-stable cell material were measured at every indicated light intensity in the absence of a nitrogen source (●); in the presence of 0.25 mM KNO_3 (▲); and in the presence of 0.25 mM NH_4Cl (■). From Romero and Lara, 1987 with permission.

In contrast, no competition is evident when ammonium is the nitrogen source available to <u>Anacystis</u> cells. In this case, even a consistent stimulation of CO_2 fixation is observed under a wide range of light intensities, being particularly noticeable under light saturating conditions (Fig. 2). Stimulation of CO_2 fixation by ammonium has also been reported for the blue-green alga <u>Anabaena</u> (Lawrie et al., 1976) and for a variety of eukaryotic green algae (Kanazawa et al., 1972) and isolated higher plant cells (Rehfield and Jensen, 1973; Paul et al., 1978; Woo and Canvin, 1980). Several mechanisms have been proposed to explain this phenomenon. Most notable are the proposals of phosphoenol pyruvate carboxylase activation by ammonium ions (Bassham et al., 1981) and that of ammonium-induced alkalinization of the chloroplast stroma (Tillberg et al., 1977). In <u>Anacystis</u>, however, the positive effect of ammonium on CO_2 fixation results abolished when ammonium assimilation is specifically inhibited, indicating that the stimulation is not caused by the cation itself but it is a consequence of its incorporation to carbon skeletons (Romero and Lara, 1987). It would, therefore, appear that when nitrate is the nitrogen source utilized by the cells, the actual rate of CO_2 fixation is in fact the balanced result of the negative effect of nitrate reduction and the positive effect of the resulting ammonium, the predominance of one effect over the other being dependent on the light intensity.

Active ammonium assimilation induces changes in the distribution of newly fixed carbon into metabolite fractions, regardless the nitrogen source utilized by the cells. As a general phenomenon observed in eukaryotic (Kanazawa et al., 1972; Paul et al., 1978; Woo and Canvin, 1980; Elrifi and Turpin, 1987) and prokaryotic (Lawrie et al., 1976; Romero and Lara 1987; Lara et al., 1987b) green cells, ammonium assimilation decreases the incorporation of newly fixed carbon into sugar and sugar phosphate and increases the incorporation into organic acids and amino acids. In isolated spinach leaf cells, the stimulatory effect of ammonium on CO_2 fixation is dependent on the CO_2 concentration available, being maximal at saturating CO_2 (Woo and Canvin, 1980). This behavior, which appears very similar to that shown in Fig. 2 for A. nidulans cells as a function of light intensity, suggesst that the ammonium-induced stimulation of carbon assimilation is specially relevant under conditions in which CO_2 fixation is not limited by the supply of light and CO_2. Walker and Sivak (1986) have proposed that an intrinsic factor limiting CO_2 fixation under these conditions is the rate of orthophosphate (P_i) recycling. During active CO_2 fixation, trioses phosphate are utilized for the synthesis of non-phosphorylated sugars releasing P_i which is needed to support photophosphorylation. Under conditions of active ammonium (or nitrate) assimilation, synthesis of organic and amino acids from triose phosphate is enhanced, thus allowing a second pathway of P_i regeneration in addition to that of sucrose or non-phosphorylated sugar synthesis. According to this view, active ammonium assimilation would stimulate CO_2 fixation by increasing the rate of P_i recycling. In this regard, ammonium or nitrate feeding to barley leaves prevented oscillations in chlorophyll fluorescence which otherwise rise upon transition from low to high CO_2, mimicking the effect of P_i feeding. Conversely, in nitrate or ammonium fed leaves, oscillations were restored by simultaneous supply of -ketoglutarate or by feeding P_i-sequestering sugars (C. Lara, M.N. Sivak and D.A. Walker, unpublished observations).

DEPENDENCE OF NITRATE ASSIMILATION ON CO_2 FIXATION

In the assimilation of any form of inorganic nitrogen an obligate requirement for CO_2 fixation products arises at the level of ammonium assimilation as a consequence of the substrate demand for carbon skeletons. Clear differences exist, however, between nitrate and ammonium assimilation with regard to their dependence upon the provision of fixed carbon (Lara et al., 1987b). In the blue-green alga Anacystis ammonium utilization reaches its maximum rate at photon fluxes allowing about one third of the maximum rate of CO_2 fixation. In contrast, the rates of nitrate uptake and CO_2 fixation at every light intensity correlate very tighly (correlation coefficient, 0.996; ß 0.001). The strict dependence upon CO_2 fixation exhibited by nitrate utilization cannot be explained in terms of -ketoacid demand for ammonium assimilation and indicates an additional carbon requirement for this process (Lara and Romero, 1986).

Other observations illustrate the tight correlation between nitrate utilization and CO_2 fixation in __Anacystis__. Nitrate uptake determined as a function of CO_2 availability exhibits saturation kinetics with the CO_2 concentration allowing half-maximum rate of nitrate uptake (3.7 µM) being very close to the Km for CO_2 of CO_2-dependent O_2 evolution (4 µM) (Flores et al., 1983b). Also, the progressive inhibition of CO_2 fixation by increasing concentrations of DLG results in a gradual decrease in the rate of nitrate utilization (Romero et al., 1985). A strict requirement for CO_2 of nitrate utilization has also been reported for green algae (Grant, 1968; Thacker and Syrett, 1972; Syrett, 1981; Larsson et al., 1985).

As a feature common to both prokaryotic and eukaryotic microalgae, the CO_2-requirement of nitrate utilization can be alleviated or supressed by altering the C/N ratio of the cells. In this respect, N-starved microalgae exhibiting a high C/N ratio are able to take up nitrate in the absence of CO_2, indicating that carbohydrate accumulated during the N-starvation period can substitute for fresh products of CO_2 fixation in their positive effect on nitrate uptake (see Syrett, 1981; Guerrero and Lara, 1987, for reviews).

The strict carbon dependence of nitrate utilization seems to be related to the well-known phenomenon of the ammonium-promoted inhibition of nitrate uptake (see Guerrero et al., 1981; Syrett, 1981; Guerrero and Lara, 1987, for reviews), being regulatory in nature. Studies with different inhibitors of ammonium assimilation have shown that prevention of ammonium assimilation protects nitrate uptake from the negative effects of ammonium and allows the process to proceed in the complete absence of CO_2. Under these conditions, nitrate is taken up by the cells in an uncontrolled manner and reduced to ammonium, which is excreted to the medium (Syrett, 1981; Guerrero and Lara, 1987). These observations suggest a common link for the mechanism underlying the positive effect of CO_2 and the negative effect of ammonium on nitrate uptake in microalgae. It has been proposed (Flores et al., 1983b) that nitrate utilization is regulated by a negative feed-back control exerted by certain organic nitrogenous compounds produced during nitrate assimilation via ammonium. The intracellular concentration of these inhibitory metabolites, whose identity is still unknown, is determined partly by the supply of ammonium from which they are generated and partly by the supply of certain CO_2-fixation products, which in the course of normal metabolism would combine with the inhibitory ammonium derivatives, thus removing them while generating other non-inhibitory C,N-metabolites. By either withdrawing CO_2 or impeding its assimilation, accumulation of the inhibitory metabolites is favored and inhibition of nitrate uptake occurs. On the other hand, by inhibiting ammonium assimilation and the generation of negative effectors, nitrate uptake is relieved from the requirement for CO_2 fixation products. Such a regulatory system would allow a fine modulation of nitrate utilization exerted not only by accumulation of some of its own products, but also through the availability of carbon metabolites, exerting a close control of nitrate utilization by carbon assimilation (Flores

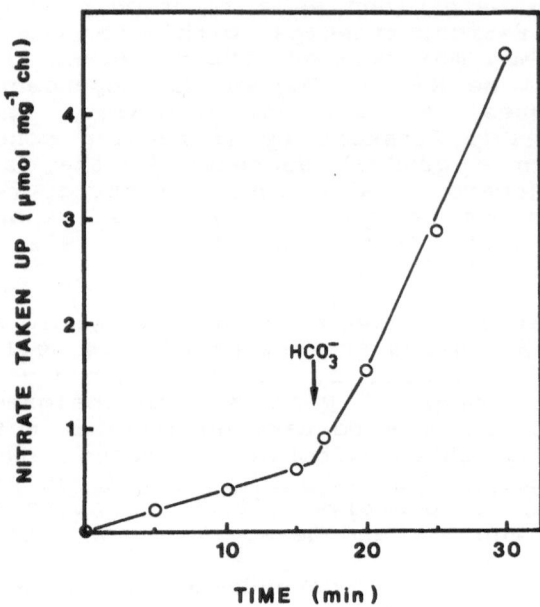

Figure 3. Reversion by bicarbonate of the inhibitory effect of ammonium on nitrate uptake in 5-hydroxylysine-treated <u>Anacystis nidulans</u> cells. Nitrate disappearance from the medium was measured in suspensions of cells pretreated with 1 mM 5-hydroxylysine containing 0.25 mM KNO_3 and 0.5 mM NH_4Cl before and after the addition of 10 mM $NaHCO_3$.

Table II. Inhibition of nitrate transport in <u>Anacystis nidulans</u> by ammonium and DLG and its prevention by MSX[a].

Supplement[b]	Intracellular nitrate (μM)
None	208
MSX	225
NH_4^+	0
NH_4^+ plus MSX	211
DLG	0
MSX plus DLG	196

[a]Assays were performed on cells exposed for 1 min to 50 μM KNO_3.
[b]$(NH_4)_2SO_4$, (0.5 mM) was added 5 min before nitrate. For inhibiting ammonium assimilation and/or CO_2 fixation, cells were preincubated for 15 min with MSX (1 mM) and/or DLG (30 mM), respectively. After Lara et al., 1987a.

et al., 1983b; Larsson and Larsson, 1987). In this respect, the existence of an antagonism between ammonium and CO_2 in the control of nitrate utilization has become evident in <u>Anacystis</u> cells with moderate ammonium assimilation capacity after treatment with 5-hydroxylysine, an inhibitor of GS. In such cells (Fig. 3), acceleration of CO_2 fixation induced by transition from CO_2-limiting to CO_2-saturating conditions results in abolishment of the negative effect of ammonium (Romero et al., 1987).

This concerted control of nitrate assimilation by products of both ammonium assimilation and CO_2 fixation is exerted primarily on nitrate transport. Thus, ammonium prevents intracellular nitrate accumulation in the vacuolated diatom <u>Phaeodactylum</u> (Cresswell and Syrett, 1979). In the blue-green alga <u>Anacystis</u> nitrate transport is inhibited by ammonium and is dependent upon CO_2 fixation (Lara et al., 1987a). As illustrated in Table II, intracellular nitrate accumulation in <u>Anacystis</u> is effectively inhibited by either ammonium addition of by inhibiting CO_2 fixation with DLG. The inhibition of nitrate transport in either case is prevented by blocking ammonium assimilation with MSX, indicating also the common basis of the mechanism underlying the effects of ammonium and CO_2 on nitrate transport (Lara et al., 1987a).

In summary, present knowledge indicate that, in microalgae, the balance between ammonium assimilation and CO_2 fixation controls nitrate assimilation at the level of substrate supply to the cell. This can account for the tight dependence of nitrate assimilation on CO_2 fixation observed in these organisms and may perhaps represent a general mechanism of regulation of nitrate assimilation in green cells.

Acknowledgements

Research from the author's laboratory has been supported by grants from Comisión Asesora de Investigación (Spain). We are grateful to A. Friend for skilful secretarial assistance.

REFERENCES

ANDERSON, J.W. (1981). Light-energy-dependent processes other than CO_2 assimilation. In: "The Biochemistry of Plants", vol. 8 (M.D. Hatch and N.K. Boardman, eds.), pp. 473-500. Academic Press, New York.

ANDERSON, J.W. and DONE, J. (1977). Polarographic study of ammonia assimilation by isolated chloroplasts. Plant Physiol. 60:504-508.

ANDERSON, J.W. and DONE, J. (1978). Light-dependent assimilation of nitrite by isolated pea chloroplasts. Plant Physiol. 61:692-697.

ANDERSON, J.W. and WALKER, D.A. (1983). Ammonia assimilation and oxygen evolution by a reconstituted chloroplast system in the presence of 2-oxoglutarate and glutamate. Planta 159:247-253.

APARICIO, P.J., KNAFF, D.B. and MALKIN, R. (1975). The role of an iron-sulfur center and siroheme in spinach nitrite reductase. Arch. Biochem. Biophys. 169:102-107.

BASSHAM, J.A., LARSEN, P.O., LAWYER, A.L. and CORNWELL, K.L. (1981). Relationship between nitrogen metabolism and photosynthesis. In: Nitrogen and Carbon Metabolism (J.D. Bewley, ed.). Nijhoff/Junk, The Hague, pp. 135-163.

BEUDEKER, R. and TABITA, F.R. (1985). Characterization of glutamine synthetase isoforms from Chlorella. Plant Physiol. 77:791-794.

BONGERS, L.H.J. (1958). Kinetic aspects of nitrate reductase. Neth. J. Agric. Sci. 6:70-88.

CAMMACK, R., HUCKLESBY, D.P. and HEWITT, E.J. (1978). Electron paramagnetic-resonance of the mechanism of leaf nitrite reductase: signals from the iron-sulphur centre and haem under turnover conditions. Biochem. J. 171:519-526.

CAMMACK, R., FRY, I.V. and PAYNE, M.J. (1987). The significance of iron-nitrosyl complexes in biology and in the reaction of assimilatory nitrite reductase. In: Inorganic Nitrogen Metabolism (W.R. Ullrich, P.J. Aparicio, P.J. Syrett and F. Castillo, eds.). Springer-Verlag, Berlin, pp. 192-194.

CANDAU, P., MANZANO, C and LOSADA, M. (1976). Bioconversion of light energy into chemical energy through reduction with water of nitrate to ammonia. Nature 262:715-717.

CRESSWELL, R.C. and SYRETT, P.J. (1979). Ammonium inhibition of nitrate uptake by the diatom Phaeodactylum tricornutum. Plant Sci. Lett. 14:321-325.

CULLIMORE, J.V. and SIMS, A.P. (1981). Occurrence of two forms of glutamate synthase in Chlamydomonas reinhardii. Phytochemistry 20:597-600.

DE LA ROSA, M.A., VEGA, J.M. and ZUMFT, W.G. (1981). Composition and structure of assimilatory nitrate reductase from Ankistrodesmus braunii. J. Biol. Chem. 256:5814-5819.

DHUGGA, K.S. WAINES. J.G. and LEONARD, R.T. (1988). Correlated induction of nitrate uptake and membrane polypeptides in corn roots. Plant Physiol. 87:120-125.

ELRIFI, I.R. and TURPIN, D.H. (1987). The path of carbon flow during NO_3^--induced photosynthetic suppression in N-limited Selenastrum minutum. Plant Physiol. 83:97-104.

ERICSON, M.C. (1985). Purification and properties of glutamine synthetase from spinach leaves. Plant Physiol. 79:923-927.

FLORENCIO, F.J. and RAMOS, J.L. (1984). Purification and characterization of glutamine synthetase from the unicellular cyanobacterium Anacystis nidulans. Biochim. Biophys. Acta 838:39-48.

FLORES, E., GUERRERO, M.G. and LOSADA, M. (1983a). Photosynthetic nature of nitrate uptake and reduction in the cyanobacterium Anacystis nidulans. Biochim. Biophys. Acta 722:408-416.

FLORES, E., ROMERO, J.M., GUERRERO, M.G. and LOSADA, M. (1983b). Regulatory interaction of photosynthetic nitrate utilization and carbon dioxide fixation in the cyanobacterium Anacystis nidulans. Biochim. Biophys. Acta 725:529-532.

GRANT, B.R. (1968). Effect of carbon dioxide concentration and buffer system on nitrate and nitrite assimilation in Dunaliella tertiolecta. J. Gen Microbiol. 54:327-336.

GRANT, B.R. and TURNER, I.M. (1969). Light-stimulated nitrate and nitrite assimilation in several species of algae. Comp. Biochem. Physiol. 29:995-1004.

GUERRERO, M.G. and LARA, C. (1987). Assimilation of inorganic nitrogen. In: The Cyanobacteria (P. Fay and C. Van Baalen, eds.) Elsevier Science Publishers BV (Biomedical Division), Amsterdam, pp. 163-186.

GUERRERO, M.G., VEGA, J.M. and LOSADA, M. (1981). The assimilatory nitrate-reducing system and its regulation. Annu. Rev. Plant Physiol. 32:169-204.

HIRASAWA, M. and KNAFF, D. (1985). Interaction of ferredoxin--linked nitrite reductase with ferredoxin. Biochim. Biophys.Acta 830:173-180.

HIRASAWA, M. and TAMURA, G. (1984). Flavin and iron-sulfur containing ferredoxin-linked glutamate synthase from spinach leaves. J. Biochem. 95:983-994.

HIRASAWA, H., FUKUSHIMA, K., TAMURA, G. and KNAFF, D.B. (1984). Biochim. Biophys. Acta 719:145-154.

HIRASAWA, M., BOYER, J.M., GRAY, K.A., DAVIS, D.J. and KNAFF, D.B. (1986). The interaction of ferredoxin with chloroplast ferredoxin-linked enzymes. Biochim. Biophys. Acta 851:23-28.

HIREL, B., VIDAL, J. and GADAL, P. (1982). Evidence for a cytosolic-dependent light induction of chloroplastic glutamine synthetase during greening of etiolated rice leaves. Planta 155:17-23.

HIREL, B. WEATHERLEY, C., CRETIN, C. BERGOUNIOUX, C. and GADAL, P. (1984a). Multiple subunit composition of chloroplastic glutamine synthetase of Nicotiana tabacum L. Plant Physiol. 74:448-450.

HIREL, B., McNALLY, S., GADAL, P., SUMAR, N. and STEWART, G.R. (1984b). Cytosolic glutamine synthetase in higher plants. A comparative inmmunological study. Eur. J. Biochem. 138:63--66.

HOUSE, C.H. and ANDERSON, J.M. (1980). Light-dependent reduction of nitrate by pea chloroplasts in the presence of nitrate reductase and C_4-dicarboxylic acids. Phytochemistry 19:1925-1930.

HOWARD, W.D. and SOLOMONSON, L.P. (1982). Quaternary structure of assimilatory NADH:nitrate reductase from Chlorella. J. Biol. Chem. 257:10243-10250.

IDA, S. and MIKAMI, B. (1986). Spinach ferredoxin-nitrite reductase: A purification procedure and characterization of chemical properties. Biochim. Biophys. Acta 871:167-176.

KAMACHI, K., AMEMIYA, Y., OGURA, N. and NAKAGAWA, H. (1987). Immuno-gold localization of nitrate reductase in spinach (Spinacea oleracea) leaves. Plant Cell Physiol. 28:333-338.

KAMIN, H. and STEIN PRIVALLE, L. (1987). Nitrite reductase. In: Inorganic Nitrogen Metabolism (W.R. Ullrich, P.J. Aparicio, P.J. Syrett and F. Castillo, eds.). Springer-Verlag, Berlin, pp. 112-117.

KANAZAWA, T.K., KANAZAWA, M.R., KIRK, M.R. and BASSHAM, J.A. (1972). Regulatory effects of ammonia on carbon metabolism in Chlorella pyrenoidosa during photosynthesis and respiration. Biochim. Biophys. Acta 265:656-669.

KRETOVICH, W.L., EVSTIGNEEVA, Z.G., PUSHKIN, A.V. and DZHOKHARIDZE, T.Z. (1981). Two forms of glutamine synthetase in leaves of Cucurbita pepo. Phytochemistry 20:625-629.

LARA, C. and ROMERO, J.M. (1986). Distinctive light and CO_2--fixation requirements of nitrate and ammonium utilization by the cyanobacterium Anacystis nidulans. Plant Physiol. 81: 686-688.

LARA, C. ROMERO, J.M. and GUERRERO, M.G. (1987a). Regulated nitrate transport in the cyanobacterium Anacystis nidulans. J. Bacteriol. 169:4376-4378.

LARA, C., ROMERO, J.M., CORONIL, T. and GUERRERO, M.G. (1987b). Interactions between photosynthetic nitrate assimilation and CO_2 fixation in cyanobacteria. In: Inorganic Nitrogen Metabolism (W.R. Ullrich, P.J. Aparicio, P.J. Syrett and F. Castillo, eds.) Springer-Verlag, Berlin, pp. 45-52.

LARSSON, C.-M. and LARSSON, M. (1987). Regulation of nitrate utilization in green algae. In: Inorganic Nitrogen Metabolism (W.R. Ullrich, P.J. Aparicio, P.J. Syrett and F. Castillo, eds.). Springer-Verlag, Berlin, pp. 203-207.

LARSSON, M., INGERMARSSON, B. and LARSSON, C.-M. (1982). Photosynthetic energy supply for NO_3^-assimilation in Scenedesmus. Physiol. Plant. 55:301-308.

LARSSON, M., OLSSON, T. and LARSSON, C.-M. (1985). Distribution of reducing power between photosynthetic carbon and nitrogen assimilation in Scenedesmus. Planta 164:246--253.

LAWRIE, A.C., CODD, G.A. and STEWART, W.D.P. (1976). The incorporation of nitrogen into products of recent photosynthesis in Anabaena cylindrica Lemm. Arch. Microbiol. 107:15-24.

LEA, P.J. and MIFLIN, B.J. (1974). An alternative route for nitrogen assimilation in higher plants. Nature 251:614-616.

LEA, P.J., MILLS, R., WALLSGROVE, R.M. and MIFLIN, B.J. (1982). Assimilation of nitrogen and synthesis of amino acids in chloroplast and cyanobacteria (blue-green algae). In: Origin of Chloroplasts (J.A. Schiff, ed.). Elsevier North Holland, New York, pp. 149-178.

LEHNER, K. and HELDT, H.W. (1978). Dicarboxylate transport across the inner membrane of the chloroplast envelope. Biochim. Biophys. Acta 368:269-278.

LOPEZ-RUIZ, A., ROLDAN, J.M., VERLEBEN, J.P. and DIEZ, J. (1985a). Nitrate reductase from Monoraphidium braunii. Immunocytochemical localization and immunological characterization. Plant Physiol. 78:614-618.

LOPEZ-RUIZ, A., VERLEBEN, J.P., ROLDAN, J.M. and DIEZ, J. (1985b). Nitrate reductase of green algae is located in the pyrenoid. Plant Physiol. 79:1006-1010.

LOSADA, M. and GUERRERO, M.G. (1979). The photosynthetic reduction of nitrate and its regulation. In: Photosynthesis in Relation to Model Systems (J. Barber, ed.) Elsevier/North Holland Biomedical Press, Amsterdam, pp. 365-408.

LOSADA, M., GUERRERO, M.G. and VEGA, J.M. (1981). The assimilatory reduction of nitrate. In: Biology of Inorganic Nitrogen and Sufur (H. Bothe and A. Trebst, eds.) Springer, Berlin, pp. 30-63.

LOSADA, M., HERVAS, M. and ORTEGA, J.M. (1987). Photosynthetic assimilation of the primordial bioelements. In: Inorganic Nitrogen Metabolism (W.R. Ullrich, P.J. Aparicio, P.J. Syrett and F. Castillo, eds.). Springer--Verlag, Berlin, pp. 3-15.

MARQUEZ, A.J., GOTOR, C., ROMERO, L.C., GALVAN, F. and VEGA, J.M. (1986). Ferredoxin-glutamate synthase from Chlamydomonas reinhardii. Prosthetic groups and preliminary studies of mechanism. Int. J. Biochem. 18:531-535.

MATOH, T. and TAKAHASHI, E. (1981). Glutamate synthase in greening pea shoots. Plant Cell Physiol. 22:727-731.

MATOH, T. and TAKAHASHI, E. (1982). Changes in the activies of ferredoxin and NADH-glutamate synthase during seedling development of peas. Planta 154:289-294.

MATOH, T., IDA, S., and TAKAHASHI, E. (1980). Isolation and characterization of NADH-glutamate synthase from pea (Pisum sativum L.). Plant Cell Physiol. 21:1461-1474.

McNALLY, S.F., HIREL, B., GADAL, P., MANN, F. and STEWART, G.R. (1983). Glutamine synthetase of higher plants. Evidence for a specific isoform content related to their possible physiological role and their compartmentation within the leaf. Plant Physiol. 72:22-25.

MIFLIN, B.J. and LEA, P.J. (1982). Ammonia assimilation and amino acid metabolism. In: Encyclopedia of Plant Physiology, New Series, Vol. 14A (D. Boulder and B. Parthier, eds.) Springer, Berlin. pp 3-64.

MIKAMI, B. and IDA, S. (1984). Purification and properties of ferredoxin-nitrate reductase from the cyanobacterium Plectonema boryanum. Biochim. Biophys. Acta 791:294-304.

NAKAGAWA, H., YONEMURA, Y., YAMAMOTO, H., SATO, T., OGURA, N. and SATO, R. (1985). Spinach nitrate reductase: Purification, molecular weight and subunit composition. Plant Physiol. 77:124-128.

NEYRA, C.A. and HAGEMAN, R.H. (1974). Dependence of nitrite reduction on electron transport in chloroplasts. Plant Physiol. 54:480-483.

ORR, J., KEEFER, L.M., KEIM, P., DINH NGUYEN, T., WELLEMS, Th., HEINRIKSON, R.L. and HASELKORN, R. (1981). Purification, physical characterization and NH_2-terminal sequence of glutamine synthetase from the cyanobacterium Anabaena 7120. J. Biol. Chem. 256:13091-13098.

ORTEGA, T., CASTILLO, F. and CARDENAS, J. (1976). Photolysis of water coupled to nitrate reduction by Nostoc muscorum subcellular particles. Biochem. Biophys. Res. Commun. 71:885-891.

PAUL, J.S., CORNWELL, K.L. and BASSHAM, J.A. (1978). Effects of ammonia on carbon metabolism in photosynthesizing isolated cells from Papaver somniferum L. Planta 142:49-54.

PLATT, S.T., PLAUT, Z. and BASSHAM, J.A. (1977). Ammonia regulation of carbon metabolism in phtosynthesizing leaf discs. Plant Phsyiol. 60:739-742.

RATHNAM, C.K.M. (1978). Malate and dihydroxyacetone phosphate-dependent nitrate reduction in spinach leaf protoplasts. Plant Physiol. 62:220-223.

RAVEN, J.A. (1977). ATP synthesis coupled to nitrate photoreduction in the alga Hydrodictyon africanum. J. Exp. Bot. 28:314-319.

REHFIELD, D.W. and JENSEN, R.G. (1973). Metabolism of separated leaf cells. III. Effect of calcium and ammoniun on product distribution during photosynthesis with cotton cells. Plant Physiol. 52:17-22.

ROLDAN, J.M., ROMERO, F., LOPEZ-RUIZ, A., DIEZ, J. and VERLEBEN, J.P. (1987). Immunological approaches to inorganic nitrogen metabolism. In: Inorganic Nitrogen Metabolism (W.R. Ullrich, P.J. Aparicio, P.J. Syrett and F. Castillo, eds.). Springer-Verlag, Berlin, pp. 94-98.

ROMERO, J.M. and LARA, C. (1987) Photosynthetic assimilation of NO_3^- by intact cells of the cyanobacterium Anacystis nidulans. Influence of NO_3^- and NH_4^+ assimilation on CO_2 fixation. Plant Physiol. 83:208-212.

ROMERO, J.M., LARA, C. and GUERRERO, M.G. (1985). Dependence of nitrate utilization upon active CO_2 fixation in Anacystis nidulans: A regulatory aspect of the interaction between photosynthetic carbon and nitrogen metabolism. Arch. Biochem. Biophys. 237:396-401.

ROMERO, J.M., CORONIL, T., LARA, C. and GUERRERO, M.G. (1987). Modulation of nitrate uptake in Anacystis nidulans by the balance between ammonium assimilation and CO_2 fixation. Arch. Biochem. Biophys. 256:578-584.

ROMERO, L.C., GALVAN, F. and VEGA, J.M. (1987). Purification and properties of the siroheme-containing ferredoxin-nitrite reductase from Chlamydomonas reinhardtii. Biochim. Biophys. Acta 914:55-63.

SAMPAIO, M.J.A.M., ROWELL, P. and STEWART, W.D.P. (1979). Purification and some properties of glutamine synthetase from the nitrogen-fixing cyanobacteria Anabaena cylindrica and a Nostoc sp. J. Gen. Microbiol. 111:181-191.

SERRANO, A., RIVAS, J. and LOSADA, M. (1981). Nitrate and nitrite as "in vivo" quenchers of chlorophyll fluorescence in blue-green algae. Photosyn. Res. 2:175-184.

SIVAK, M.N. and WALKER, D.A. (1985). Can in vivo photosynthesis be modified?. Ann. Proc. Phytochem. Soc. Eur. 26:29-44.

SOLOMONSON, L.P. and BARBER, M.J. (1987). Structure-function relationships of assimilatory nitrate reductase. In: Inorganic Nitrogen Metabolism (W.R. Ullrich, P.J. Aparicio, P.J. Syrett and F. Castillo, eds.). Springer-Verlag, Berlin, pp. 71-75.

SUZUKI, A., VIDAL, J. and GADAL, P. (1982). Glutamate synthase isoforms in rice. Inmunological studies of enzymes in green leaf, etiolated leaf and root tissue. Plant Physiol. 70:827-832.

SUZUKI, A., NATO, A. and GADAL, P. (1984). Glutamate synthase isoforms in tobacco cultured cells. Plant Sci. Lett. 33:93--101.

SWADER, J.A. and STOCKING, C.R. (1971). Nitrate and nitrite reduction by Wolffia arrhiza. Plant Physiol. 47:189-191.

SYRETT, P.J. (1981). Nitrogen metabolism in microalgae. Can. Bull. Fish. Aquat. Sci. 210:182-210.

SYRETT, P.J. and MORRIS, I. (1963). The inhibition of nitrate assimilation by ammonium in Chlorella. Biochim. Biophys. Acta 67:566-575.

THACKER, A. and SYRETT, P.J. (1972). The assimilation of nitrate and ammonium by Chlamydomonas reinhardii. New Phytol. 71:423-433.

TILLBERG, J.E., GIERSCH, C. and HEBER, U. (1977). CO_2 reduction by intact chloroplasts under a diminished proton gradient. Biochim. Biophys. Acta 461:31-47.

ULLRICH, W.R. (1987). Nitrate and ammonium uptake in green algae and higher plants: Mechanism and relationship with nitrate metabolism. In: Inorganic Nitrogen Metabolism (W.R. Ullrich, P.J. Aparicio, P.J. Syrett and F. Castillo, eds.). Springer-Verlag, Berlin, pp. 32-38.

VAN NIEL, C.B., ALLEN, M.B. and WRIGHT, B.E. (1953). On the photochemical reduction of nitrate by algae. Biochim. Biophys. Acta 12:67-74.

VEGA, J.M. and KAMIN, H. (1977) Spinach nitrite reductase: Purification and properties of siroheme-containing iron--sulfur enzyme. J. Biol. Chem. 252:896-909.

VEGA, J.M., CARDENAS, J. and LOSADA, M. (1980). Ferredoxin nitrite reductase. Meth. Enzymol. 69:255-270.

WALKER, A. and SIVAK, M.N. (1986). Photosynthesis and phosphate: A cellular affair?. Trends Biochem. Sci. 11:176--179.

WALLSGROVE, R.M. and LEA, P.J. (1985). Photosynthetic nitrogen metabolism. In: Photosynthetic Mechanisms and the Environment (J. Barber and N.R. Baker, eds). Elsevier Science Pub. B.V. (Biomedical Division), Amsterdam, pp. 389--418.

WALLSGROVE, R.M., LEA, P.J. and MIFLIN, B.J. (1979). Distribution of the enzymes of nitrogen assimilation within the pea leaf cell. Plant Physiol. 63:232-236.

WALLSGROVE, R.M., LEA, P.J. and MIFLIN, B.J. (1982). The development of NAD(P)H-dependent and ferredoxin-dependent glutamate synthase in greening barley and pea leaves. Planta 154:473-476.

WALLSGROVE, R.M., KEYS, A.J., LEA, P.J. and MIFLIN, B.J. (1983) Photosynthesis, photorespiration and nitrogen metabolism. Plant Cell Environ. 6:301-309.

WOO, K.C. and CANVIN, D.I. (1980). Effect of ammonia on photosynthetic carbon fixation in isolated spinach leaves cells. Can. J. Bot. 58:505-510.

WOO, K.C. and OSMOND, C.B. (1982). Stimulation of ammonia and 2-oxoglutarate-dependent O_2 evolution in isolated chloroplasts by dicarboxylates and the role of the chloroplast in photorespiratory nitrogen recycling. Plant Physiol. 69:591-596.

WOO, K.C., BOYLE, F.A., FLUGGE, I.U. and HELDT, H.W. (1987a) ^{15}N-ammonia assimilation, 2-oxoglutarate transport, and glutamate export in spinach chloroplasts in the presence of dicarboxylates in the light. Plant Physiol. 85:621-625.

WOO, K.C., FLUGGE, I.U. and HELDT, H.W. (1987b). A two translocator model for the transport of 2-oxoglutarate and glutamate in chloroplasts during ammonia assimilation in the light. Plant Phsyiol. 84:624-632.

STUDIES OF THE Co^{2+}-ACTIVATED RIBULOSE-1,5-BISPHOSPHATE CARBOXYLASE/OXYGENASE BY THE USE OF SPECTROPHOTOMETRY

Rolf Brändén, Kristina Janson and Peter Nilsson

Department of Biochemistry and Biophysics

University of Göteborg and Chalmers Institute
of Technology, S-412 96 Göteborg, Sweden

SUMMARY

Transient absorption bands are formed upon addition of ribulose-1,5-bisphosphate (RuBP) to the Co^{2+}-activated ribulose-1,5-bisphosphate carboxylase/oxygenase, RubisCO. In the visible region the prominent absorption band during steady state has a maximum at 610 nm. Stopped flow technique was used to study the increase in absorbance at this wavelength and two distinct phases in the progress curve for the approach to steady state absorbance were observed. The rates for these two phases, respectively, were similar to those found earlier for the two enzyme-Co^{2+}-bound intermediates using EPR technique (1). It is therefore proposed that most of the transient optical absorption originates from an enzyme-Co^{2+}-coordinated RuBP molecule and an enzyme-Co^{2+}-coordinated enediolate anion of it, where bound RuBP appears first. Furthermore, the most rapid phase in the progress curve is a first order reaction, independent of the concentration of RuBP. This indicates that the formation of enzyme-Co^{2+}-coordinated RuBP is preceeded by another reaction in which RuBP binds to the enzyme, probably without metal coordination.

INTRODUCTION

RubisCO initiates two different metabolic pathways in all photosynthetic organisms (2): assimilation of CO_2 and photorespiration.

$RuBP + CO_2 \longrightarrow 2$ 3-phospho-D-glycerate $\longrightarrow$ BIOSYNTHESIS

$RuBP + O_2 \longrightarrow$ 3-phospho-D-glycerate + 2-phosphoglycolate

PHOTORESPIRATION

Both reactions occur at the same site and are kinetically treated as competitive (3). The active form of the enzyme consists of a ternary enzyme-CO_2-Me(11)-complex, where CO_2 (other than the substrate CO_2) and a lysine residue form a carbamate which is stabilized by the metal ion (4). Mg(11) is the activator found in vivo but several other divalent metal ions support activity with varying efficiency (5,6).

By the use of EPR spectroscopy inner sphere substrate- or intermediate metal complexes have been found to occur during catalysis for the Co(11)-activated enzyme (1,7). Proofs have also been presented for a direct coordination between the metal ion and the following substances: RuBP, the product 3-phospho-D-glycerate (3-PGA), 2-carboxy-arabinitol-1,5-bisphosphate (CABP) which is an intermediate analogue, and various substrate analogues (8-11). All these results strongly indicate that the metal ion participates and plays a central part in catalysis.

RESULTS

The optical properties of Co(11)- activated RubisCO in absence and presence of RuBP are shown in figure 1. Figure 1A is the spectrum of Co(11) and $NaHCO_3$; when 10 mM RuBP was added, no changes was observed. The spectrum of RubisCO, activated with $CoCl_3$ and $NaHCO_3$ is shown in figure 1B. When RuBP was added, transient absorption bands are obtained, seen in figure 1C. Two distinct absorption peaks are seen at 530 nm and 610 nm. When all added RuBP had been consumed, the spectrum of figure 1D was obtained. Most of the two prominent bands had now vanished, and no further changes were obtained at longer reaction times.

It is known that CABP forms a very strong enzyme-CO_2-Me(11)-CABP-complex (2) where CABP is directly coordinated to the metal at the activator site. For the Co(11)-activated enzyme an homogenous enzyme-Co(11)-CABP-complex is formed as seen by EPR. Figure 2B shows the optical absorption spectrum for that complex.

From EPR experiments it is also known that PGA coordinates to the metal ion in Cu(11)-activated RubisCO. Figure 2C shows the spectrum of Co(11)-activated RubisCO in presence of PGA.

In order to obtain a better resolution of individual absorption bands the spectra in figure 2A-C are recorded with the spectrum of Co(11)-activated RubisCO in absence of effector stored in the baseline memory. The spectra in figure 1 can therefore not be directly compared to those in figure 2.

We also tested if a minor amount of unspecifically bound Co(11) could coordinate RuBP and produce some of the optical absorptions observed during turnover. Therefore the spectrum of figure 2B was stored in the baseline memory and RuBP was added to the enzyme solution. As seen from figure 2D, no absorption changes were obtained.

Furthermore, when RuBP was added to RubisCO activated with Co(11) in absence of $NaHCO_3$ no optical absorption changes were observed (not shown). This means that the carbamate formation is essential for the formation of turnover absorptions.

Stopped flow experiments were made with Co(11)-activated RubisCO in presence of varying amounts of RuBP. The spectra obtained are shown in figure 3. As can be seen in figure 3b, there are two distinct phases in the approach to stationary state absorbance, and they are both independent of the RuBP-concentration. This means that none of them can reflect the very first catalytic step - RuBP must of course bind to the enzyme in a RuBP dependent fashion. Therefore we propose that RuBP initially binds to the enzyme without metal coordination and that the first phase

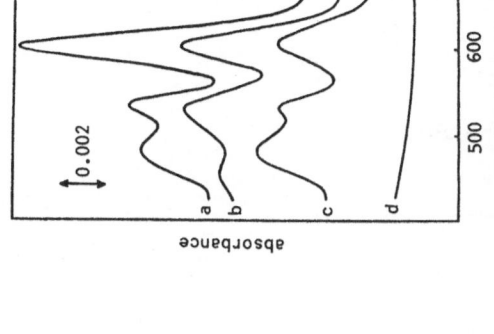

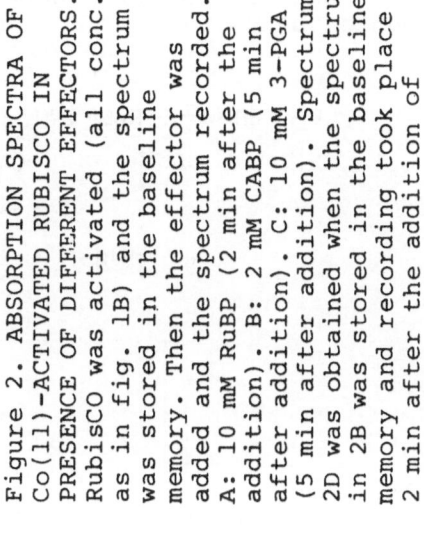

Figure 1. OPTICAL ABSORPTION
SPECTRA FOR Co(ll)-ACTIVATED
RUBISCO DURING TURNOVER.
A: 1 mM $CoCl_2$ and 10 mM $NaHCO_3$
in 50 mM HEPPS buffer, pH 8.2.
B: RubisCO (0.9 mM protomer)
activated for 5 min in 50 mM
$NaHCO_3$ and 0.6 mM $CoCl_2$ before
the spectrum was recorded.
C: The spectrum obtained after
addition of 20 ul 0.5 M RuBP
to 1.0 ml of the enzyme solution in B.
D: The spectrum recorded when
added RuBP had been consumed,
10 min after addition of RuBP.

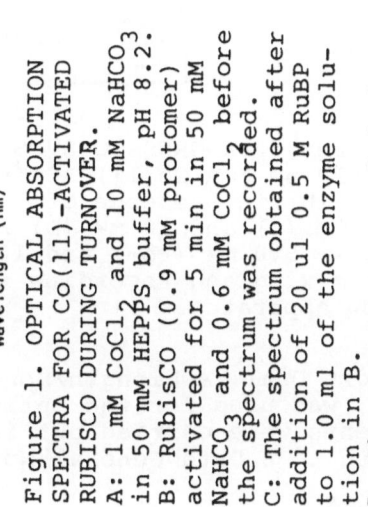

Figure 2. ABSORPTION SPECTRA OF
Co(ll)-ACTIVATED RUBISCO IN
PRESENCE OF DIFFERENT EFFECTORS.
RubisCO was activated (all conc.
as in fig. 1B) and the spectrum
was stored in the baseline
memory. Then the effector was
added and the spectrum recorded.
A: 10 mM RuBP (2 min after the
addition). B: 2 mM CABP (5 min
after addition). C: 10 mM 3-PGA
(5 min after addition). Spectrum
2D was obtained when the spectrum
in 2B was stored in the baseline
memory and recording took place
2 min after the addition of
10 mM RuBP.

in the progress curve results from the metal coordination of enzyme-bound RuBP. The second phase may then result from the formation of an enzyme-Co(11)-coordinated enediolate anion of RuBP.

From figure 3a it is seen that the stationary state absorbance decreases with increasing concentrations of RuBP. We propose that this is due to a competition between RuBP in solution and an active site-bound enediolate anion of RuBP.

A possible reaction scheme, including this inhibition, is presented below.

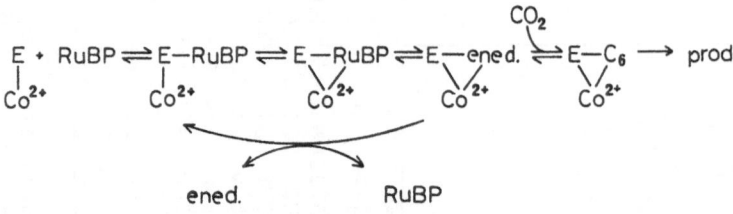

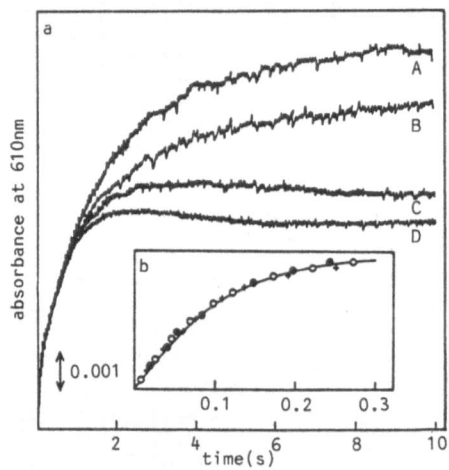

Figure 3. PROGRESS CURVES FOR THE ABSORPTION INCREASE AT 610 nm FOR Co(11)-ACTIVATED RUBISCO (conc. as in fig. 1B) in presence of 1.25 mM (A), 2.5 mM (B), 5.0 mM (C) and 10 mM (D) RuBP in figure 3a; 2.5 mM (-●-), 5.0 mM (-o-) and 10 mM (-+-) RuBP in fig. 3b. The mixing and recording took place in a stopped flow apparatus.

A Shimadzu Dual Vawelength/Double Beam Spectrophotometer, model UV-3000, was used for the optical measurements. Stopped flow experiments were carried out in an apparatus described in (12), connected to a Data General Nova minicomputer.

References

1. Brändén R., Janson K., Nilsson P. & Aasa R., Biochem. Biophys. Acta 916, 298-303,1987

2. Miziorko H.M. & Lorimer G.H., Ann. Rev. Biochem. 52, 507-535, 1983

3. Andrews T.J., Lorimer G.H. & Tolbert N.E., Biochemistry 12, 11-18, 1973

4. Lorimer G.H. & Miziorko H.M., Biochemistry 19, 5321-5328, 1980

5. Horecker J.M., Hurwitz J. & Weissbach A., J. Biol. Chem. 218, 785-810, 1956

6. Christeller J.T., Biochem. J. 193, 839-844, 1981

7. Nilsson T., Brändén, R. & Styring, S., Biochem. Biophys. Acta 788, 274-280, 1984

8. Brändén, R., Nilsson, T. & Styring S., Biochemistry 23, 4378-4382, 1984

9. Miziorko H.M. & Sealy R.C., Biochemistry 23, 479-485, 1984

10. Styring S. & Brändén R., Biochemistry 24, 6011-6019, 1985

11. Styring S. & Brändén R., Biochem. Biophys. Acta 832, 113-118, 1985

12. Andréasson L-E, Brändén R., Malmström B.G., Strömberg C. & Vänngård T., "Oxidases and Related Redox Systems", pp.87-95, 1973, University Park Press, Baltimore

COARSE CONTROL OF SUCROSE-PHOSPHATE SYNTHASE

Gabriele Siegl and Mark Stitt

Universität Bayreuth, Pflanzenphysiologie
Postfach 101251, D-8580 Bayreuth

In green plant tissue the photosynthate is transported to the sink regions of the plant as sucrose. In order to investigate plant physiology it is of great interest to look at the regulation of sucrose synthesis. In the reaction sequence leading to sucrose one of the steps that are thought to be regulated is the reaction catalyzed by sucrose-phosphate synthase in which fructose-6-phosphate and UDP-glucose are converted to sucrose-phosphate and UDP. When people measure sucrose-phosphate synthase activity in spinach leaves at different times of a day they find a diurnal rhythm when the reaction is assayed using conditions that correspond to in vivo concentrations. A diurnal rhythm measured under Vmax conditions, however, results in constant activities over 24 hours. Thus the properties of sucrose-phosphate synthase change during the day: In the morning the enzyme shows high activity that is hardly influenced by physiological phosphate concentrations and in the evening it has low activity that is inhibited by physiological phosphate concentrations, only leaving a residual activity of about 10%. We now wish to characterize these two forms of sucrose-phosphate synthase, whose relative amount change in leaves and produce the diurnal changes. We aim to purify the enzyme so far that kinetic studies can be done. The extreme properties are obtained when spinach leaf discs are either incubated in 0.2M mannose to induce the high activity form or in 0.2M sucrose to obtain the low activity form. By means of Fast Protein Liquid Chromatography (anion exchange followed by gelfiltration) the specific properties of the two enzyme forms stay the same. The selective assay thereby works as a tool to check on the stability of the sucrose-phosphate synthase properties during purification. The kinetic studies with the tow purified forms showed that:

	high activity form	low activity form
F6P dependence	partial	complete
Pi sensitivity	low	high
F6P affinity	high	low

Thus the high activity form will work in the morning when F6P and G6P concentrations are low and fluxes into sucrose are high. The low activity form however might contribute to the sucrose-phosphate synthase activity in the evening at high F6P and G6P pools and low fluxes into sucrose.

In our current work we are trying to convert one of the two forms into the other in order to investigate the mechanism of activation of the sucrose-phosphate synthase.

LITERATURE

Stitt, M., Wilke, I., and Heldt, H.W., (1988), Coarse Control of Sucrose-Phosphate Synthase in Leaves: Alterations of the kinetic properties in response to the rate of photosynthesis and the accumlation of sucrose. Planta 174:217-230

Stitt, M., Huber, S., and Kerr, P., (1987), Control of photosynthetic sucrose formation In: The Biochemistry of Plants, Vol. 10, pp. 327-409, Hatch, M.D. and Boardman, N.K., eds.

SECTION IV

MOLECULAR BIOLOGY, GENETIC ENGINEERING AND BIOTECHNOLOGY

GENES AND POLYPEPTIDES OF PHOTOSYSTEM II

John C Gray, Sean M Hird, Richard Wales,
Andrew N Webber and David L Willey

Botany School
University of Cambridge
Downing Street, Cambridge, CB2 3EA, UK

INTRODUCTION

Photosystem II catalyses the light-driven transfer of electrons from water to plastoquinone, producing oxygen and generating a proton gradient across the thylakoid membrane. The complex may be regarded as made up of three assemblies of polypeptides: a light-harvesting complex (LHCII), a core complex containing the reaction centre and two antenna chlorophyll proteins, and an extrinsic complex concerned with oxygen evolution. Photosystem II is structually the most complex of the supramolecular assemblies of the thylakoid membrane and is currently recognised to be composed of at least 20 different polypeptides.[1]

The genes for most of these polypeptides have been isolated and characterised in the last few years, and this has generated a large amount of information on the primary structures of the polypeptides. The aim of this article is to review the information available on the genes and polypeptides of photosystem II in higher plants and to discuss the organisation of the polypeptides in the photosystem II complex.

PHOTOSYSTEM II CORE COMPLEX

The core complex of photosystem II comprises two antenna chlorophyll a proteins of 47kDa and 44kDa, two reaction centre polypeptides D1 and D2 which bind P680, phaeophytin a, chlorophyll a and the plastoquinone acceptors Q_A and Q_B, two polypeptides of cytochrome b-559 and a number of small polypeptides of unknown function. All these polypeptides appear to be encoded in the chloroplast genome of higher plants. The genes are listed in Table 1 and their locations in the chloroplast genomes of tobacco, wheat and pea are shown in Figure 1.

The 47kDa and 44kDa chlorophyll a-proteins are encoded by the chloroplast genes psbB and psbC, and nucleotide sequence analysis indicates that they are related proteins[4-6]. Each polypeptide may be predicted to fold with six hydrophobic membrane-spanning regions (Fig 2) containing conserved histidine residues which may be involved in binding approximately 15 chlorophyll a molecules per polypeptide. Previous models of the topology of the polypeptides suggested seven membrane-spanning regions[4-7] but closer inspection of the sequence proposed to form membrane-span II suggests that it is unlikely to form a hydrophobic alpha-helical structure. The region contains a large number of polar side-chains and helix-destabilising residues. The conserved histidine residues are located within the hydrophobic region of five of the proposed membrane spans, approximately one turn of an alpha-helix into the lipid bilayer. The phosphorylation of the N-terminal threonine residue of 44kDa polypeptide[8] suggests that

Table 1. Chloroplast genes for photosystem II polypeptides

Gene Designation	Gene Product	Codons
psb A	32kDa Q_B protein, D1	353
psb B	47kDa chlorophyll *a* protein	508
psb C	44kDa chlorophyll *a* protein	461
psb D	34kDa protein, D2	353
psb E	9kDa cytochrome *b*-559	83
psb F	4kDa cytochrome *b*-559	39
psb G	24kDa polypeptide	248
psb H	10kDa phosphoprotein	73
psb I	4.5kDa reaction centre polypeptide	36
psb J	hypothetical protein	53
psb K (*lhc*A)	2kDa polypeptide	37
psb L	3.2kDa polypeptide	38

the *N*-termini of both polypeptides are located on the stromal side of the membrane, and this would place the large hydrophilic sequences between membrane spans V and VI of both polypeptides in the thylakoid lumen where they may interact with the extrinsic oxygen-evolving complex.

The coding region of the *psb*C gene overlaps the coding region of the *psb*D gene for the D2 polypeptide (Fig 3) but probably not by 50bp as reported previously[5,6]. Translation initiation at GTG is suggested for transcripts of the cyanobacterial *psb*C gene[9] and the conservation of this sequence and a putative ribosome-binding site GAGGAGG 10-16bp upstream from the GTG codon suggests a similar initiation site for chloroplasts. Initiation at GTG would produce a 14bp overlap with the *psb*D gene, and would require the removal of two residues, Met-Glu, to generate the *N*-terminal threonine residue that is modified by *N*-acetylation and *O*-phosphorylation[8].

The polypeptides D1 and D2 which form the reaction centre of photosystem II are encoded by the chloroplast genes *psb*A and *psb*D. Both genes encode polypeptides of 353 amino acid residues which are predicted to form five membrane-spanning regions (Fig 2) similar to the L and M subunits of the reaction centre of *Rhodopseudomonas viridis* [10]. However it is probably that the mature D1 polypeptide is smaller than the mature D2 polypeptide due to the removal of 12-16 residues at the *C*-terminus of D1[11]. Both polypeptides are

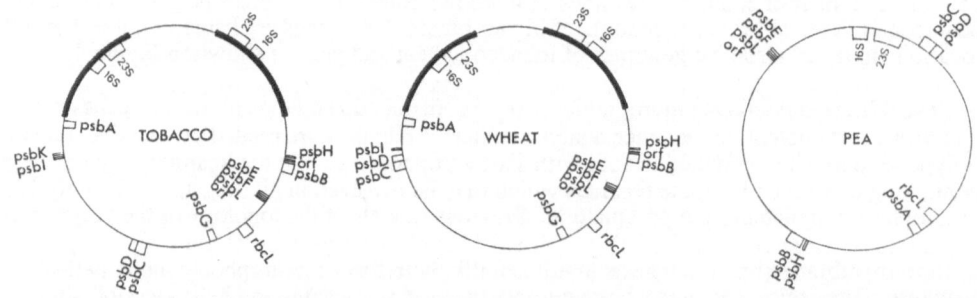

Fig 1   Location of genes for photosystem II polypeptides in the chloroplast genomes of tobacco, wheat and pea. The tobacco map is redrawn from Sugiura[2]; the wheat and pea maps are updated from those presented previously[3].

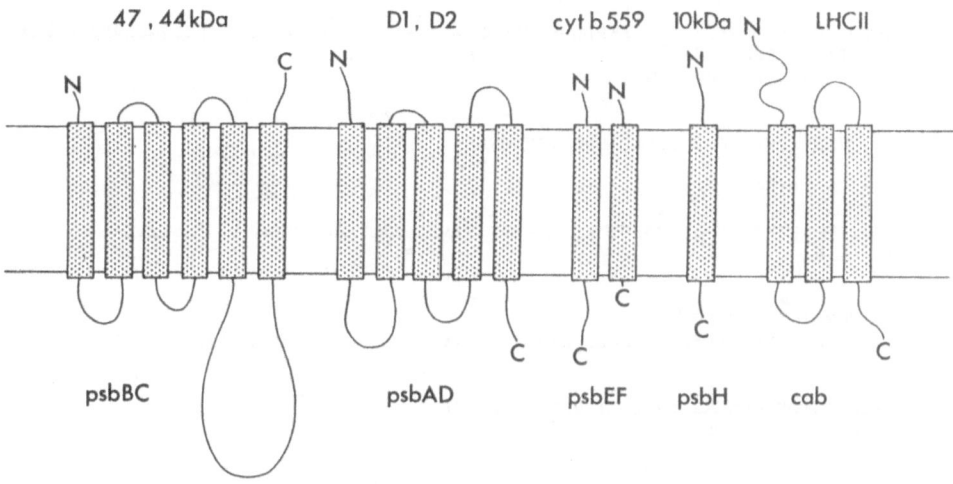

Fig 2. Proposed models for the topology of photosystem II polypeptides in the thylakoid membrane. Membrane-spanning alpha-helical regions are drawn as cylinders.

modified at the *N*-terminus by the removal of *N*-formyl methionine and *N*-acetylation of the threonine reidues. Both the D1 and D2 polypeptides may be phosphorylated on the *N*-terminal threonine residue[8].

Residues in the D1 and D2 polypeptides involved in binding P680, Q_A, Q_B and the non-haem Fe ion have been predicted.[12,13] The *psb*A gene for the D1 polypeptide has been shown to be the locus of mutations producing resistance to the triazine group of herbicides in a number of higher plants[14]. Atrazine has been shown to bind at the Q_B site and prevent electron transfer out of photosystem II[15]. Characterisation of the mutant genes has indicated that a single point mutation is responsible for the change in herbicide susceptibility. In all atrazine-resistant higher plants examined, including *Amaranthus hydridus*,[16] *Solanum nigrum*,[17,18] *Chenopodium album*[19] and *Senecio vulgaris*,[20] an A-G transition results in the replacement of a serine residue with a glycine residue at position 264 in the loop joining membrane spans IV and V. This region of the polypeptide is predicted to be involved in forming the Q_B site.The secondary electron donor D has recently been shown to be a tyrosine residue (Tyr 160) on the D2 polypeptide[21,22] and this suggests that Z is also a tyrosine residue, but located in the D1 polypeptide. A conserved tyrosine (Tyr 161) is predicted from all *psb*A sequences determined.

The genes (*psb*E and *psb*F) for the polypeptides of cytochrome *b*-559 are located in chloroplast DNA and co-transcribed in all plants examined[23-25]. The amino acid sequences deduced for the 9kDa and 4kDa polypeptides of cytochrome *b*-559 indicate that both polypeptides contain a single histidine residue located in a putative membrane-spanning region. The requirement for two histidine residues as ligands to the haem[26] indicates that cytochrome *b*-559 must consist of, at least, a dimer of polypeptides. The simplest model is a heterodimer of

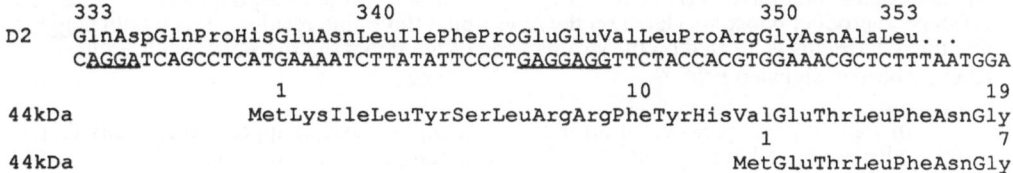

Fig 3. Overlap of the coding regions of the *psb*D and *psb*C genes from wheat. The two alternative translations of the *psb*C gene are shown. Putative ribsome-binding sites are underlined.

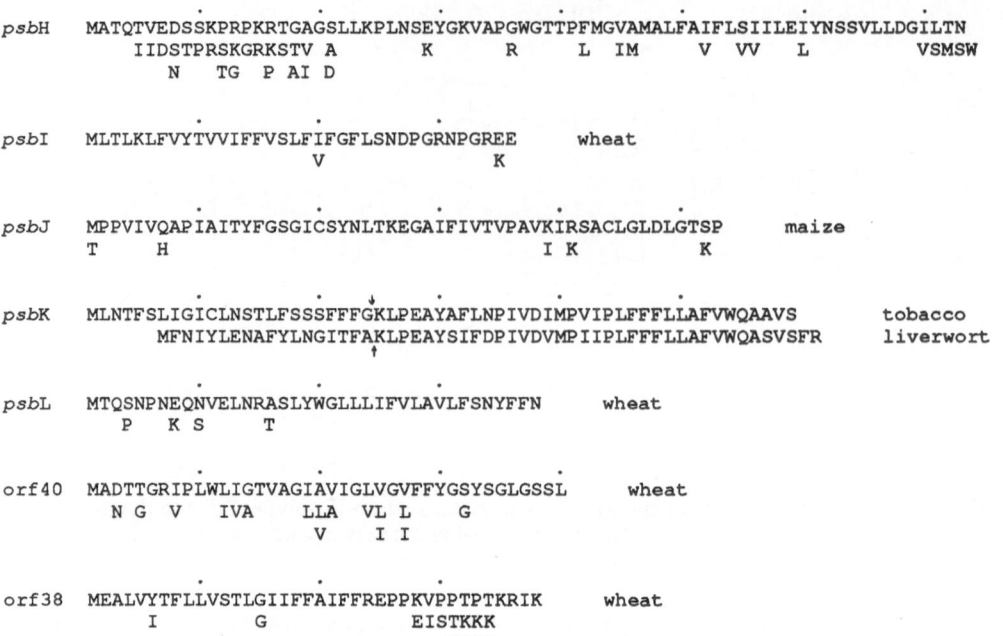

```
psbH    MATQTVEDSSKPRPKRTGAGSLLKPLNSEYGKVAPGWGTTPFMGVAMALFAIFLSIILEIYNSSVLLDGILTN
        IIDSTPRSKGRKSTV A           K       R       L   IM        V  VV     L                    VSMSW
          N      TG   P AI  D


psbI    MLTLKLFVYTVVIFFVSLFIFGFLSNDPGRNPGREE    wheat
                    V              K


psbJ    MPPVIVQAPIAITYFGSGICSYNLTKEGAIFIVTVPAVKIRSACLGLDLGTSP      maize
          T      H                            I K          K


psbK    MLNTFSLIGICLNSTLFSSSFFFGKLPEAYAFLNPIVDIMPVIPLFFFLLAFVWQAAVS      tobacco
        MFNIYLENAFYLNGITFAKLPEAYSIFDPIVDVMPIIPLFFFLLAFVWQASVSFR      liverwort


psbL    MTQSNPNEQNVELNRASLYWGLLLIFVLAVLFSNYFFN    wheat
           P   K S     T


orf40   MADTTGRIPLWLIGTVAGIAVIGLVGVFFYGSYSGLGSSL     wheat
          N  G  V   IVA      LLA   VL L     G
                        V       I I


orf38   MEALVYTFLLVSTLGIIFFAIFFREPPKVPPTPTKRIK    wheat
            I       G              EISTKKK
                                    SEGN
```

Fig 4   Amino acid sequences of polypeptides deduced from small open reading frames in chloroplast DNA. For each open reading frame the upper sequence represents a complete polypeptide from one species, with all known variations from other species in the lower lines. Orf38 is the open reading frame located between *psb*B and *psb*H.

9kDa and 4kDa polypeptides with the *N*-termini located on the same side of the membrane (Fig 2). If the N-termini are located on the stromal side of the membrane, as predicted for all other intrinsic proteins of photosystem II, this would position the haem group on the stromal side where it could act as an electron acceptor from Q_B. The stoichiometry of the cytochrome b-559 polypeptides with respect to other photosystem II core components[27] does not support more complex models for cytochrome b-559 involving two haem groups on opposite sides of the membrane.

Two small open reading frames, of 38 and 40 codons, are present in the transcriptional unit for the cytochrome b-559 polypeptides, and are predicted to encode hydrophobic proteins with single membrane-spanning regions[24,25]. The sequences derived from these and other small open reading frames are shown in Fig 4. The product of orf 38 has recently been located in photosystem II preparations and characterised by *N*-terminal sequencing[25]. The initiating *N*-formyl-methionine residue of the orf38 gene product is removed from the protein to produce a mature polypeptide of 37 amino acid residues with an *N*-terminal threonine residue. The orf38 product has the same electrophoretic mobility as a small phosphoprotein recently identified in wheat etioplast membranes (A N Webber, unpublished). If the phosphoprotein represents the orf38 gene product phosphorylated on the *N*-terminal threonine residues, this would suggest that the *N*-terminus is located on the stromal side of the thylakoid membrane. Orf38 has recently been designated *psb*L.[25]

In view of the expression of orf38 in wheat chloroplasts, it appears very likely that orf40 is also expressed at the protein level and is a component of photosystem II. The orf40 product is a hydrophobic polypeptide (Fig 4) which is predicted to fold with a single membrane-spanning region. Neither the orf40 or orf38 products contain any amino acid residues which might provide a clear indication to their function. However all four polypeptides encoded by the *psb*EFL transcription unit contain a tryptophan residue in a similar position, bordering the start of the predicted membrane-spanning alpha-helices. This is a feature also observed in the alpha and beta chains of the bacterial light-harvesting complex[28], although the significance of this occurrence is unknown.

The products of several other small open reading frames in chloroplast DNA have recently been identified as polypeptides associated with photosystem II preparations. The 9-10kDa phosphoprotein of photosystem II was identified as the product of the *psb*H gene by comparison of the determined *N*-terminal amino acid sequence[29] and the sequence deduced from the nucleotide sequence[30,31] (Fig 4). The mature polypeptide of 72 amino acid residues is predicted to form a single membrane-spanning region with the *N*-terminal threonine residue on the stromal side of the membrane where it may be phosphorylated by a protein kinase. The *psb*H gene is cotranscribed with *psb*B[31]. There is a conserved open reading frame of 38 codons located between *psb*B and *psb*H in wheat chloroplast DNA (S M Hird, unpublished). A similar conserved open reading frame of 33-35 codons is present in the chloroplast genomes of maize, spinach, tobacco and liverwort. There is as yet no evidence for the presence of a protein product derived from this open reading frame in chloroplasts. The product of the open reading frame is predicted to form a hydrophobic membrane-spanning region, with a highly charged *C*-terminal sequence (Fig 4).

The product of a gene designated *psb*I was first identified in photosystem II preparations by antibodies raised against a synthetic peptide deduced from the sequence of an open reading frame in maize chloroplast DNA[32]. A polypeptide with an *N*-terminal amino acid sequence similar to the *psb*I gene product has recently been identified in a photosystem II reaction centre complex from pea (A.N. Webber, unpublished). *psb*I is located upstream of the *psb*DC genes in maize, and a similar open reading frame is present in a similar position in wheat chloroplast DNA (S M Hird, unpublished). In the tobacco chloroplast genome *psb*I is not located near *psb*DC due to the inversion of approximately 20kbp with respect to wheat and maize[2]. In these species, and in the liverwort *Marchantia polymorpha*,[33] *psb*I encodes a highly conserved polypeptide of 36 amino acid residues (Fig 4). The product of this gene is predicted to form a single membrane-spanning region.

The product of another gene, *psb*K, has been identified by *N*-terminal sequencing of a small polypeptide associated with spinach photosystem II preparations[34]. The gene is located just upstream of *psb*I in tobacco and liverwort chloroplast DNA.[2,33] In tobacco an open reading frame of 98 or 61 codons encodes the sequenced polypeptide which corresponds to the *C*-terminal 37 amino acid residues (Fig 4). In liverwort an open reading frame of 55 codons encodes a very similar *C*-terminal 37 residue polypeptide although there is little similarity between the *N*-terminal sequences predicted from the open reading frame but not present in the mature polypeptide.[33] The similarity between the proteins deduced from the tobacco and liverwort genes starts at precisely the *N*-terminal residue of the mature polypeptide. The liverwort gene has been called *lhc*A[35] because it shows limited sequence homology with the small light-harvesting polypeptides of photosynthetic bacteria. The chloroplast polypeptide contains a tryptophan residue near the end of a hydrophobic region which has been predicted to form a single membrane-spanning segment.

Two other open reading frames in chloroplast DNA have been described as genes for photosystem II components on the basis of immunochemical studies with antibodies raised against synthetic peptides predicted from the sequences of the open reading frames in maize[36,32]. A 24kDa polypeptide in a spinach photosystem II preparation was identified as the product of an open reading frame of 248 or 255 codons.. This gene has been called *psb*G[36]. However it has not been possible to repeat this result with antibodies raised against the product of a gene-fusion including the analogous wheat open reading frame (P. Nixon & J. Barber, unpublished). As the *psb*G open reading frame overlaps with the putative *ndh*C gene for a component of NADH dehydrogenase it is possible that *psb* G does not constitute an authentic photosystem II gene. The gene designation *psb*J has been suggested (M Sugiura, personal communication) for an open reading frame of 53 codons located on the opposite strand to *psb*G in maize chloroplast DNA. A similar sequence is located in tobacco and liverwort chloroplast DNA, but there is as yet no published evidence that the product of this open reading frame is associated with photosystem II. Further experimentation is necessary to authenticate the products of the *psb*G and *psb*J genes as components of photosystem II.

EXTRINSIC PROTEINS

Several extrinsic proteins are associated with the oxidising side of photosystem II in the thylakoid lumen. These polypeptides all appear to be encoded by nuclear genes (see Table 2). The best characterised of these polypeptides are the 33kDa, 23kDa and 16kDa hydrophilic proteins which are responsible for maintaining a suitable environment for water oxidation. These proteins are present in equimolar amounts [37,38] although there is some controversy concerning the stoichiometry with respect to P680 and the polypeptides of the reaction centre complex. One or two copies of each of the 33kDa, 23kDa and 16kDa polypeptides per reaction centre have been reported.[37,38]

cDNA clones for the 33kDa, 23kDa and 16kDa polypeptides have been isolated from a number of higher plants and their characterisation has given information on the primary sequence of the mature proteins and the sequences involved in targeting the proteins to the thylakoid lumen. The 33kDa polypeptide is synthesised initially as a precursor of 331 amino acid residues in spinach[39] or 329 residues in pea[40]. The presequences of 84 residues (spinach) or 81 residues (pea) appear to be composed of a hydrophilic chloroplast import domain and a hydrophobic thylakoid transfer domain (see Fig 5), as has been demonstrated for the plastocyanin presequence[41,42]. The mature polypeptide of 247 (spinach)[39,43] or 248 (pea)[40] residues is highly conserved, with identical amino acid residues at 83% of the positions in pea and spinach. Some variation at the *N*- and *C*-termini of the polypeptide from a single species is indicated by variation in the sequences obtained from protein sequencing[43,45] and from the nucleotide sequences of cDNA clones. This variation may be due to the presence of a multi-gene family for the 33kDa polypeptide[40]. In pea, hybridisation to Southern blots of nuclear DNA gives 7-9 bands which may represent separate genes[40]. A small gene family has also been suggested for the 33kDa polypeptide in spinach.[47]

A sequence in the 33kDa protein showing some similarity to a region of Mn-superoxide dismutase has been suggested[43] to have a possible role in Mn binding. However, X-ray crystallographic studies on the Mn-superoxide dismutase from *Thermus thermophilus*[48] indicate that the sequence is not involved in Mn-coordination. A sequence of approximately 15 amino acid residues in the spinach and pea proteins shows strong homology to the calcium-binding site of mammalian intestinal calcium-binding proteins, calmodulin and related proteins.[40] This sequence in the 33kDa polypeptide is predicted to fold to form an EF hand structure typical of calmodulin.[49] Calcium binding by a 33kDa polypeptide has recently been demonstrated (A N Webber, unpublished) by [45]Ca binding to photosystem II polypeptides immobilised on nitrocellulose.

The 23kDa and 16kDa polypeptides are also synthesised initially as larger precursors, with the presequences composed of putative chloroplast import and thylakoid transfer domains[40] (R Wales, unpublished) (see Fig 5). The mature 23kDa polypeptides of spinach[59] and pea (R Wales, unpublished) both consist of 186 amino acid residues, with 83%

Table 2	Nuclear genes for photosystem II polypeptides		
Gene Designation	Gene product	Processor	Mature
psb 1	33kDa polypeptide	329-331	247-248
psb 2	23kDa polypeptide	266-267	186
psb 3	16kDa polypeptide	232	149
st-ls1	10kDa polypeptide	138	99
cab type I	26-28kDa LHCII	265-269	232-233
cab type II	24kDa LHCII	264-265	224-225

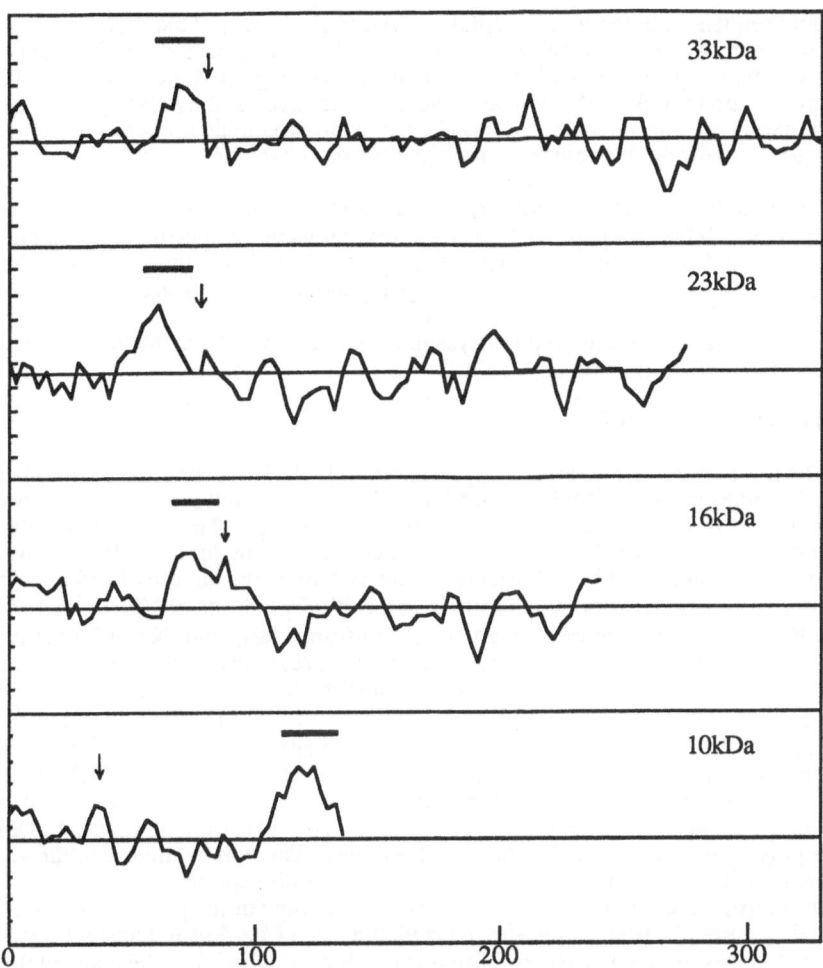

Fig 5   Hydropathy plots of the precursor proteins of the extrinsic polypeptides of
         photosystem II. The single hydrophobic region, marked with a black bar, in each
         protein is believed to constitute a thylakoid-transfer domain. The cleavage site between
         the presequence and the mature polypeptide is indicated with an arrow.

conservation of the amino acid sequence. The mature protein is clearly hydrophilic with no
extended hydrophobic sequences. Near the C-terminus there is a cluster of basic residues
which may play a part in chloride binding or in protein - protein interactions. The mature 16kDa
polypeptide is a hydrophilic protein of 149 residues[50], but it is not possible to identify any
amino acid residues involved in photosystem II functions.

Several other polypeptides are associated with the oxidising side of photosystem II.
Polypeptides of 10kDa, 22kDa and 24kDa were shown to be immunoprecipitated from
detergent-solubilised photosystem II preparations with antibodies to the 33kDa and 23kDa
polypeptides[51]. The 10kDa and 22kDa polypeptides appear to be extrinsic polypeptides located
in the thylakoid lumen because they are removed from everted photosystem II preparations by
washing with chaotropic agents, although they are both clearly hydrophobic proteins.[52] The
single copy gene for the 10kDa polypeptide has been isolated from potato and shown to contain
four introns[53]. The protein is initially synthesised as a precursor of 138 amino acid residues,
and is cleaved to a mature protein of 99 amino acid residues. The cleavage site has been
predicted from the N-terminal sequence of the wheat 10kDa polypeptide[54] and this suggests that
the transit peptide consists only of sequences for transfer across the chloroplast envelope.

There are no sequences in the transit peptide similar to the hydrophobic thylakoid transfer domains of the presequences of plastocyanin or the 33kDa, 23kDa and 16kDa polypeptides. However the mature protein contains a large hydrophobic region at the *C*-terminus (Fig 5) which may act as a non-cleavable signal sequence for transfer of the polypeptide across the thylakoid membrane into the lumen. This *C*-terminal hydrophobic sequence shows some similarity to the hydrophobic thylakoid transfer domains of the lumen proteins.

The identity of the 24kDa polypeptide which was immunoprecipitated in a complex with antiserum to the 33kDa polypeptide[51] is not known. However it appears to comigrate in SDS-polyacrylamide gels with the 24kDa LHC II polypeptide which is the product of the type II *cab* genes. Further information on the 22kDa polypeptide will be available when the nucleotide sequence of a cDNA clone is published. There is little available information on the hydrophilic 5kDa polypeptide associated with photosystem II. It seems probable that the 5kDa polypeptide is located in the thylakoid lumen.

LIGHT-HARVESTING COMPLEX

The light-harvesting complex of photosystem II (LHCII) comprises a number of polypeptides in the range 21-29kDa which bind 13-14 chlorophyll molecules per polypeptide. The polypeptides differ in the ratio of chlorophyll *a* to chlorophyll *b* and the amount and identity of carotenoids bound. A nuclear location of the genes for the major LHCII polypeptides (LHCIIb) was first indicated by the Mendelian inheritance of the pattern of tryptic peptides of the chlorophyll protein CPII obtained from interspecific F_1 hybrids of *Nicotiana* [56]. Genes (*cab*) for the LHCII polypeptides have been characterised from a large number of higher plants, and they may be divided into two classes depending on the presence or absence of an intron which interrupts the coding sequence in the region encoding the *N*-terminus of the mature protein[57]. The genes make up a small gene family in all plants examined; in petunia,[58] tomato[59] and lemna[60] there are approximately 12-16 genes and a similar number is suggested by Southern hybridisation in pea nuclear DNA. The smallest number of genes is found in *Arabidopsis thaliana*; three *cab* genes have been identified and characterised, although it is probable that an additional gene remains to be characterised[62]. Type I *cab* genes are uninterrupted and encode precursor polypeptides of 265-269 amino acid residues. The polypeptides contain an *N*-terminal transit sequence necessary for transfer of the polypeptide across the chloroplast envelope, but the exact site of cleavage of the presequence has not been determined (see Fig 6). A cleavage site in the vicinity of a Met-Arg sequence has frequently been assumed based on the sequences of tryptic peptides of LHCII[63]. Cleavage at this site would remove a transit peptide of 34-36 residues to produce a mature protein of 232-233 residues. *N*-acetylation of the *N*-terminal residue has so far prevented the determination of the *N*-terminal amino acid sequence of the mature protein[64].

cab type I

```
                     .         .           .    ↓↓      .           .          .
Petunia   MAAATMALSSPSFAGKAVKFSPSSSEITGNGKATMRKTVT-KAKPVSSGSPWYGPDRVKYLGPFS
Tomato    MAAATMALSSPSFAGQAVKLSPSASEISGNGRITMRKAVA-KSAP--SSSPWYGPDRVKYLGPFS
Lemna     MAA-SMALSSPSLVGKAVKLAPAASEVFGERIVSMRKATAGKPKPVSSGSPWYGPDRVKYLGPFS
Wheat     MAATTMSLSSSSFAGKAVKNLPSSA-LIGDARVNMRKTAAKAKQ-VSSSSPWYGSDRVLYLGPLS
```

cab type II

```
                     .         .           .       .    ↓     .          .
Petunia   MATSAIQQSAFAGQTALKSQNELVRKIGSFGGGRATMRRTVKSAPQSIWYGEDRPKYLGPFSEQT
Tomato    MATCAIQQSAFVGQAVGKSQNEFIRKVGNFGEGRITMRRTVKSAPQSIWYGEDRPKYLGPFSEQT
Lemna     MAASAIQSSAFAGQTALKQREELVRKVGVS-DGRFSMRRTVKAVPQSIWYGADRPKFLGPFSEQT
Wheat                                             GNDLWYGPDXVKYLGPFSAQT
```

Fig 6   *N*-terminal sequences of precursor proteins for LHCII polypeptides derived from type I and type II *cab* genes. The determined *N*-terminal sequence of the wheat 24kDa polypeptide is also shown. Proposed cleavage sites between presequences and the mature polypeptides are indicated with arrows.

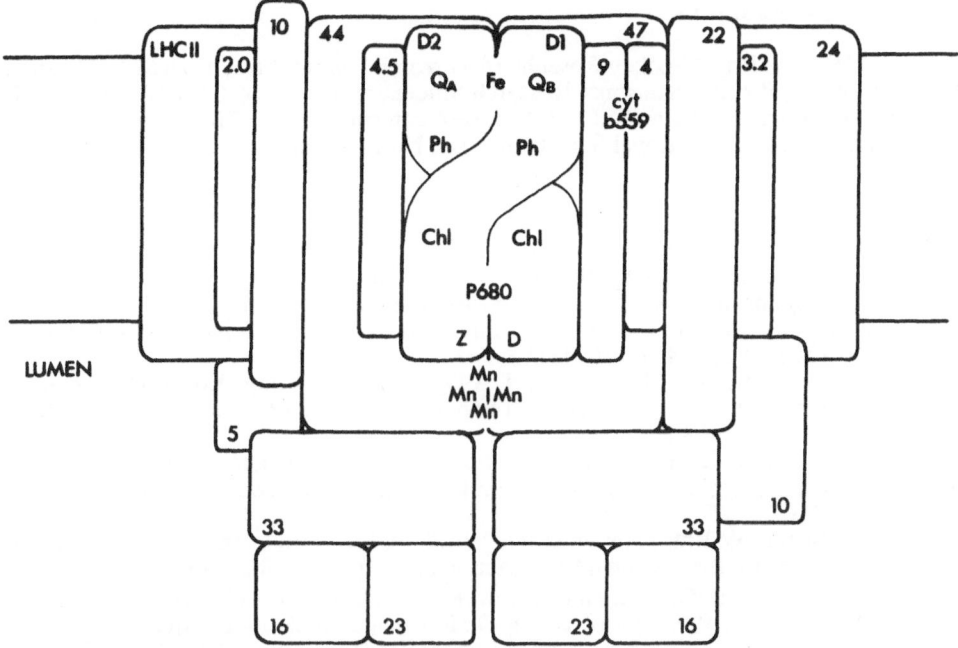

Fig 7   Schematic view of the structure of photosystem II in the chloroplast thylakoid membrane. The location of polypeptides in the thylakoid membrane or in the lumenal space is shown diagramatically. Polypeptides are identified by their approximate molecular mass in kDa.

Type II *cab* genes are interrupted by an intron and encode polypeptides of 264-265 amino acid residues[57,59,60]. The processed gene product has recently been identified as a 24kDa polypeptide by *N*-terminal sequencing of the isolated polypeptide (A N Webber, unpublished). This suggests a cleavage site in a different position to the type I gene products (Fig 6), generating a transit peptide of 40 residues and a mature protein of 224-225 amino acid residues. The product of the type II *cab* gene is therefore predicted to be approximately 1kDa smaller than type I *cab* gene products. These 24kDa LHCII polypeptides do not contain threonine residues near the *N*-terminus of the mature protein and are not phosphorylated.

All LHCII *cab* genes characterised to date encode very similar mature proteins which have been predicted to fold with three hydrophobic membrane-spanning regions[65] (see Fig 2) The *N*-terminus is located on the stromal side of the membrane, and the *C*-terminal tyrosine residues have been located in the thylakoid lumen[64] in accord with the model of three membrane spans.

Little detailed information is available for the other LHCII polypeptides. Considerable differences in electrophoretic mobility of the polypeptides CP29 (LHCII*a*), CP24 (LHCII*c*) and 21kDa LHCII*d* have been observed. The basis for the differences between these polypeptides should be apparent when nucleotide sequences of cDNA or genomic clones become available.

CONCLUSION

A considerable amount of structural information on the polypeptides of photosystem II is now available and it is possible to present a rational model for the organisation of photosystem II (Fig 7). However this is essentially a two-dimentional model and we have no clear indications of the three-dimensional arrangement of the polypeptides. The future must lie with a high-resolution crystal structure, such as has been obtained for the bacterial reaction centre.[10]

ACKNOWLEDGEMENTS

DLW is the Royal Society Rosenheim Research Fellow, ANW is an SERC Postdoctoral Research Fellow, RW is a Sainsbury Research Student of the Gatsby Charitable Foundation, and SMH was supported by an SERC-CASE research studentship. This work was supported by grants from SERC, The Royal Society and The Nuffield Foundation.

REFERENCES

1. J C Gray, Genetics and synthesis of chloroplast membrane proteins, *in*: "Photosynthesis", J Amesz, ed., Elsevier, Amsterdam, pp319-342 (1987).
2. M Sugiura, Structure and function of the tobacco chloroplast genome, *Bot Mag (Tokyo)* 100: 407-436 (1987).
3. G R M Courtice, C M Bowman, T A Dyer, and J C Gray, Location of genes for components of photosystem II in pea and wheat chloroplast DNA, *Curr Genet* 10: 329-333 (1985).
4. J Morris and R G Herrmann, Nucleotide sequence of the gene for the P_{680} chlorophyll *a* apoprotein of the photosystem II reaction center from spinach, *Nucleic Acids Res* 12: 2837-2850 (1984).
5. K Holschuh, W Bottomley, and P R Whitfeld, Structure of the spinach chloroplast genes for the D2 and 44-kd reaction centre polypeptides of photosystem II and for tRNA^{ser} (UGA), *Nucleic Acids Res* 12: 8819-8834 (1984).
6. J Alt, J Morris, P Westhoff, and R G Herrmann, Nucleotide sequence of the clustered genes for the 44kd chlorophyll *a* apoprotein and the "32kd"-like protein of the photosystem II reaction center in the spinach plastid chromosome, *Curr Genet* 8: 597-606 (1984).
7. J C Gray, P P J Dunn, C J Eccles, S M Hird, A S Hoglund, D I Last, B J Newman, and D L Willey, Biogenesis of thylakoid membrane proteins, *in*: "Plant Membranes: Structure, Function, Biogenesis", C J Leaver and H Sze, eds, Alan R Liss, New York, pp 163-179 (1987).
8. H Michel, D F Hunt, J Shabanowitz, and J Bennett, Tandem mass spectrometry reveals three photosystem II proteins of spinach chloroplasts contain N-acetyl O-phosphothreonine at their NH_2 termini, *J Biol Chem* 263: 1123-1130 (1988).
9. S S Golden and G W Stearns, Nucleotide sequence and transcript analysis of three photosystem II genes from the cyanobacterium *Synechococcus* sp PCC 7942, *Gene* 67: 85-96 (1988).
10. J Deisenhofer, O Epp, K Miki, R Huber, and H Michel, Structure of the protein subunits in the photosynthetic reaction centre of *Rhodopseudononas viridis* at 3Å resolution, *Nature* 318: 618-624 (1985).
11. B Marder, P Goloubinoff, and M Edelman, Molecular architecture of the rapidly metabolised 32-kilodalton protein of photosystem II. Indications for COOH-terminal processing of a chloroplast membrane polypeptide, *J Biol Chem* 259: 3900-3908 (1984).
12. J Barber, Photosynthetic reaction centres: a common link, *Trends Biochem Sci* 12: 321-326 (1987).
13. A Trebst, The three-dimentional structure of the herbicide binding niche on the reaction center polypeptides of photosystem II, *Z Naturforsch* 42c: 742-750 (1987).
14. K E Steinback, L McIntosh, L Bogorad, and C J Arntzen, Identification of the triazine receptor protein as a chloroplast gene product, *Proc Natl Acad Sci USA* 78: 7463-7467 (1981).
15. K Pfister, K E Steinback, G Gardner, and C J Arntzen, Photoaffinity labelling of a herbicide receptor protein in chloroplast membranes, *Proc Natl Acad Sci USA* 78:981-985 (1981).
16. J Hirchberg and L McIntosh, Molecular basis of herbicide resistance in *Amaranthus hybridus*, *Science* 222: 1346-1349 (1983).
17. P Goloubinoff, M Edelman, and R B Hallick, Chloroplast-coded atrazine-resistance in *Solanum nigrum*:: *psb* A loci from susceptible and resistant biotypes are isogenic except for a single codon change, *Nucleic Acids Res* 12: 9489-9496 (1984).

18. J Hirchberg, A Bleecker, D L Kyle, L McIntosh, and C J Arntzen, The molecular basis of triazine-herbicide resistance in higher plant chloroplasts, *Z Naturforsch* 39C: 412-4420 (1984).

19. P Bettini, S McNally, M Sevignac, H Darmency, J Gasquez, and M Dron, Atrazine resistance in *Chenopodium album*. Low and high levels of resistance to the herbicide are related to the same chloroplast *psb*A gene mutation, *Plant Physiol* 84: 1442-1446 (1987).

20. E R Blyden and J C Gray, The molecular basis of triazine herbicide resistance in *Senecio vulgaris L., Biochem Soc Trans* 14: 62 (1986).

21. B A Barry and G T Babcock, Tyrosine radicals are involved in the photosynthetic oxygen-evolving system, *Proc Natl Acad Sci USA* 84: 7099-7103 (1987).

22. R J Debus, B A Barry, G T Babcock, and L McIntosh, Site-directed mutgenesis identifies a tyrosine radical involved in the photosynthetic oxygen-evolving system, *Proc Natl Acad Sci USA* 85:427-430 (1988).

23. R G Herrmann, J Alt, B Schiller, W R Widger, and W A Cramer, Nucleotide sequence of the gene for apocytochrome *b*-559 on the spinach plastid chromosome: implications for the structure of the membrane protein, *FEBS Lett* 176: 239-244 (1984).

24. D L Willey and J C Gray, Two small open readings are co-transcribed with the pea chloroplast genes for the polypeptides of cytochrome *b*-559, *Mol Gen Genet,* submitted.

25. A N Webber, S M Hird, L Packman, T A Dyer, and J C Gray, A photosystem II polypeptide is encoded by an open reading frame co-transcribed with genes for cytochrome *b*-559 in wheat chloroplast DNA, *Plant Mol Biol,* in press.

26 G T Babcock, W R Widger, W A Cramer, W A Oertling, and J G Metz, Axial ligands of chloroplast cytochrome *b*-559: identification and requirement for a heme-cross-linked polypeptide structure, *Biochemistry* 24: 3638-3645 (1985).

27. K Gounaris, U Pick, and J Barber, Stoichiometry and turnover of photosystem II polypeptides, *FEBS Lett* 211: 94-98 (1987).

28. H Zuber, R Brunisholz, and W Sidler : Structure and function of light-harvesting pigment-protein complexes,*in* : "Photosynthesis", J Amesz, ed, Elsevier, Amsterdam, pp 232-271 (1987).

29. J Farchaus and R L Dilley, Purification and partial sequence of the M_r 10,000 phosphoprotein from spinach thylakoids, *Arch Biochem Biophys* 244: 94-101 (1986).

30. S M Hird, T A Dyer, and J C Gray, The gene for the 10kDa phosphoprotein of photosystem II is located in chloroplast DNA, *FEBS Lett* 209, 181-186 (1986).

31. P Westhoff, J Farchaus, and R G Herrmann, The gene for the M_r 10,000 phosphoprotein associated with photosystem II is part of the *psb* B operon of the spinach plastid chromosome, *Curr Genet* 11: 165-169 (1986).

32. K Kato, R T Sayre, and L Bogorad, Expression of the PSB-I gene in maize chloroplasts, *Proc Ann Meeting Jap Soc Plant Physiol,* Urawa p208 (1987).

33. K Ohyama, H Fukuzawa, T Kohchi, H Shirai, T Sano, S Sano, K Umesono, Y Shiki, M Takeuchi, Z Chang, S Aota, H Inokuchi, and H Ozeki, Chloroplast gene organisation deduced from complete sequence of liverwort *Marchantia polymorpha* chloroplast DNA, *Nature* 322: 572-574 (1986).

34. N Murata, M Miyao, N Hayashida, T Hidaka, and M Sugiura, Identification of a new gene in the chloroplast genome encoding a low-molecular-mass polypeptide of photosystem II complex, *FEBS Lett* 235: 283-288 (1988).

35. K Ohyama, T Kohchi, T Sano, and Y Yamada, Newly identified groups of genes in chloroplasts, *Trends Biochem Sci.* 13:19-22 (1988).

36. A A Steinmetz, M Castroviejo, R T Sayre, and L Bogorad, Protein PSII-G. An additional component of photosystem II identified through its plastid gene in maize, *J Biol Chem* 261: 2485-2488 (1986).

37. N Murata and M Miyao, Extrinsic membrane proteins in the photosynthetic oxygen-evolving complex, *Trends Biochem Sci* 10: 122-124 (1985).

38. B Anderson, C Larsson, C Jansson, U Ljungberg, and H E Akerlund, Immunological studies on the organisation of proteins in photosynthetic oxygen evolution, *Biochem Biophys Acta* 766: 21-28 (1984).

39 A Tyagi, J Hermans, H Steppuhn, C Jansson, F Vater, and R G Herrmann, Nucleotide sequence of cDNA clones encoding the complex "33kDa" precursor

protein associated with the photosynthetic oxygen-evolving complex from spinach, *Mol Gen Genet* 207:288-293 (1987).

40. R Wales, B J Newman, D Pappin, and J C Gray, The extrinsic 33kDa polypeptide of the oxygen-evolving complex of photosystem II is encoded by a multi-gene family in pea, *Plant Mol Biol,* submitted.

41. S Smeekens, S de Groot, J van Binsbergen, and P Weisbeek, Sequence of the precursor of the chloroplast thylakoid lumen protein plastocyanin, *Nature* 317:456-458 (1985).

42. S Smeekens, C Bauerle, J Hageman, K Keegstra, and P Weisbeek, The role of the transit peptide in the routing of precursors towards different chloroplast compartments, *Cell* 46: 365-375 (1986).

43. H Oh-Oka, S Tanaka, K Wada, T Kuwabara, and N Murata, Complete amino acid sequence of the 33kDa protein isolated from spinach photosystem II particles, *FEBS Lett* 197: 63-66 (1986).

44. F P Wolter, J M Schmitt, H J Bohnert, and A Tsugita, Simultaneous isolation of three peripheral proteins - a 32kDa protein, ferredoxin-NADP$^+$ reductase and coupling factor - from spinach thylakoids and partial characterisation of a 32kDa protein, *Plant Sci Lett* 34: 322-334 (1984).

45. Y Yamamoto, M A Hermodson, and D W Krogmann, Improved purification and *N*-terminal sequence of the 33-kDa protein in spinach PSII, *FEBS Lett* 195:155-158 (1986).

46. A Watanabe, E Minami, M Murase, K Shinohara, T Kuwabara, and N Murata, Biogenesis of photosystem II complex in spinach chloroplasts, *in*: "Progress in Photosynthesis Research", J Biggins, ed., Martinus-Nijhoff, The Hague, Vol 4, pp 629-636 (1987).

47. J Tittgen, J Hermans, J Steppuhn, T Jansen, C Jansson, B Andersson, R Nechushtai, N Nelson, and R G Herrmann, Isolation of cDNA clones for fourteen nuclear-encoded thylakoid membrane proteins, *Mol Gen Genet* 204: 259-265 (1986).

48. W C Stallings, K A Pattridge, R K Strong and M L Ludwig, Manganese and iron superoxide dismutases are structural homologs, *J Biol Chem* 259: 10695-10699 (1984).

49. R M Tufty and R H Kretsinger, Troponin and parvalbumin calcium-binding regions predicted in myosin light-chain and T4 lysozyme, *Science* 187: 167-169 (1975).

50. T Jansen, C Rother, J Steppuhn, H Reinke, K Beyreuther, C Jansson, B Andersson, and R G Herrmann, Nucleotide sequence of cDNA clones encoding the complete '23kDa' and '16kDa' precursor proteins associated with the photosynthetic oxygen-evolving complex from spinach, *FEBS Lett* 216: 234-240 (1987).

51. U Ljungberg, H E Akerlund, C Larsson, and B Andersson, Identification of polypeptides associated with the 23 and 33kDa proteins of photosynthetic oxygen evolution, *Biochem Biophys Acta* 767: 145-152 (1984).

52. U Ljungberg, H E Akerland, and B Andersson, Isolation and characterisation of the 10kDa and 22kDa polypeptides of higher plant photosystem 2. *Eur J Biochem* 158: 477-482 (1986).

53. P Eckes, S Rosahl, J Schell, and L Willmitzer, Isolation and characterisation of a light-inducible, organ-specific gene from potato and analysis of its expression after tagging and transfer into tobacco and potato shoots, *Mol Gen Genet* 205: 14-22 (1986).

54. A N Webber, L C Packman, and J C Gray, A 10kDa polypeptide associated with the oxygen-evolving complex of photosystem II has a putative C-terminal non-cleavable thylakoid transfer domain, *FEBS Lett,* submitted.

55. U Ljungberg, T Henrysson, C P Rochester, H E Akerlund, and B Andersson, The presence of low molecular-weight polypeptides in spinach photosystem II core preparations. Isolation of a 5kDa hydrophilic polypeptide, *Biochem Biophys Acta* 849: 112-120 (1986).

56. S D Kung, J P Thornber and S G Wildman, Nuclear DNA codes for the photosystem II chlorophyll-protein of chloroplast membranes, *FEBS Lett* 24: 185-188 (1972).

57. M M Stayton, M Black, J Bedbrook, and P Dunsmuir, A novel chlorophyll *a/b*-binding (*cab*) protein gene from petunia which encodes the lower molecular

weight *cab* precursor protein, *Nucleic Acids Res* 14: 9781-9796 (1987).

58. P Dunsmuir, S M Smith, and J Bedbrook, The major chlorophyll *a/b* binding protein of petunia is composed of several polypeptides encoded by a number of disctint nuclear genes, *J Mol Appl Genet* 2: 285-300 (1983).

59. E Pichersky, R Bernatzky, S D Tanksley, R B Briedenback, A P Kausch, and A R Cashmore, Molecular characterization and genetic mapping of two clusters of genes encoding chlorophyll *a/b*-binding proteins in *Lycopersicon esculentum* (tomato), *Gene* 40: 247-258 (1985).

60. G A Karlin-Neumann, B D Kohorn, J P Thornber, and E M Tobin, A chlorophyll *a/b*-binding protein encoded by a gene containing an intron with characteristics of a transposable element, *J Mol Appl Genet* 3: 45-61 (1985).

61 G Coruzzi, R Broglie, A Cashmore, and N H Chua, Nucleotide sequences of two pea cDNA clones encoding the small subunit of ribulose 1,5-bisphosphate carboxylase and the major chlorophyll *a/b*-binding thylakoid polypeptide, *J Biol Chem* 258: 1399-1402 (1983).

62. L S Leutwiler, E M Meyerowitz and E M Tobin, Structure and expression of three light-harvesting chlorophyll *a/b*-binding protein genes in *Arabidopsis thaliana*, *Nucleic Acids Res* 14: 4051-4076 (1986).

63. J E Mullett, The amino acid sequence of the polypeptide segment which regulates membrane adhesion (grana stacking) in chloroplasts, *J Biol Chem* 258: 9941-9948 (1983).

64. R Burgi, F Suter and H Zuber, Arrangement of the light-harvesting chlorophyll *a/b* protein complex in the thylakoid membrane, *Biochim Biophys Acta* 890:346-351 (1987).

THE USE OF TRANSGENIC PLANTS TO MANIPULATE PHOTOSYNTHETIC PROCESSES

Tristan A. Dyer

Institute of Plant Science Research (Cambridge Laboratory)
Maris Lane, Trumpington
Cambridge CB2 2JB, U.K.

The techniques of molecular biology have undoubtedly given us great insight into the structure of the proteins involved in photosynthetic processes. The main reason is that it is far easier to first isolate and sequence the gene for a particular protein (and from this derive its amino acid sequence) than it is to directly sequence the protein itself. Nearly all the sequences for proteins known to be involved in photosynthesis have been derived in this way. Now that we are armed with all this information about the structure of these proteins and have their coding sequences in our possession, the question is whether we can use this to gain yet greater insight into how photosynthesis takes place and perhaps also perturb it in a beneficial way. What will be considered here is how transgenic plants could perhaps be used to accomplish these objectives. Some of the techniques involved will be described first, followed by a discussion of the type of change that we might wish to make.

Plant transformation

A fundamental requirement of being able to do any genetic manipulation of plants using molecular techniques is to be able to introduce isolated genes into plant cells and subsequently get the altered cells to grow into normal plants. Although this may be readily achieved with a relatively few types of plant, with most it cannot. This constitutes a substantial barrier to progress. Consequently a great deal of effort is being devoted to developing new techniques for plant transformation and regeneration.

Numerous methods have been devised for physically inserting new genes into plants. These include microinjection, using DNA-coated microprojectiles, uptake induced electrically or by chemicals or facilitated by viral or Agrobacterium vectors.

There are some factors which must be taken into account in all such transformation systems. The principal one is that it should be possible to readily identify the plant cells which have been transformed. However, the genes that one might wish to insert usually do not have characteristics which would make this possible. Therefore it is necessary to insert with them a selectable marker gene to permit such an identification.

Selectable marker genes. To be generally useful a selectable marker gene must meet certain criteria. Unless the transformation efficiency is very high, selection must be stringent with a minimum of non-transformed plant tissue escaping selection. The selection should result in a large number of independent transformation events and not significantly interfere with regeneration. The marker should also work well within a larger number of plant species. Ideally there should also be an easy assay for the expression of the marker gene so that its presence can be confirmed.

Several marker genes have been used in plant transformation studies. Most function by coding for enzymes which inactivate phytotoxic compounds that are included in the growth medium and which kill untransformed cells. Hitherto, the most useful marker genes have been the aminoglycoside acetyltransferases and of these, NPT II (neomycin phosphotransferase) has been used most often. This permits selection of transformed plants on the antibiotic kanamycin or its analogue G418. However, it does have some limitations such as not being suitable for all plant species and this has necessitated the development of other selectable marker systems. Also, there are many situations where the availability of second complementary markers is needed for selection when introducing additional DNA into plants that have already been transformed or for providing double selection markers. Other aminoglycoside acetyltransferases such as bacterial AAC (3)-III and -IV (which confer resistance to gentamycin, Hayford et al., 1988) have been employed and the bacterial hygromycin phosphotransferase gene (aph II) is also proving useful. In addition bacterial and mouse genes coding for methotrexate-resistant dihydrofolate reductase have been used (Rogers et al., 1987), as have some genes which confer herbicide resistance (Bevan and Goldsbrough, 1988). The bacterial gene encoding chloramphenicol acetyltransferase (CAT) which was used at one time quite extensively to provide a selectable marker, is now used mainly just as a reporter gene.

In each instance where a bacterial gene is made into a marker, the naturally occurring regulatory sequences must be replaced by promoters and terminators active in plants.

Transformation systems

Agrobacterium-mediated gene transfer. Techniques involving the use of Agrobacterium-based vectors have been by far the most successful for the stable introduction of DNA into plants. Wild-type Agrobacterium spp. infect most dicotyledonous and some monocotyledonous plants by entry through wounds (see Melchers and Hooykaas, 1987). The infection process involves a complicated interaction between the bacterium resulting in the integration of a portion of the bacterial DNA (the T-DNA) into the nuclear genome of the plant.

In wild-type Agrobacterium most of the genetic information relating to the transfer of DNA to the plant is sited in a large circular plasmid 190-240 kbp in size [the tumor-inducing (Ti) plasmid]. Two regions of the Ti plasmid are essential for the mobilization and integration of the transferred DNA. The T-DNA itself is located in the plasmid as is the vir region which encodes a group of genes whose products are required for, amongst other things, the excision of the T-DNA from the Ti plasmid and its transfer into and integration into the plant genome (Schell, 1988). The expression of most of the vir genes is induced by compounds released from plant cells damaged by wounding. The region of the Ti plasmid to be transferred is delineated by border sequences which are two almost perfect 25 bp direct repeats. The border sequences apparently function as sites both for excision of the T-DNA and integration into the host.

Two observations have been critical in the development of suitable vectors for plant transformation. Firstly, no portion of the T-DNA other than the borders is essential for the transfer of DNA. This has meant that

438

the endogenous T-DNA of the wild-type plasmids could be replaced by other more appropriate DNA. A second critical observation was that the vir gene products could function in trans. This means that the T-DNA can be on a different plasmid to the vir genes and still be transferred into plants. The advantage of this is that the cloning vector itself can be made relatively small. The large size of the Ti plasmids makes them difficult to handle in vitro because DNA itself is very fragile and easily sheared and also the larger the DNA is the less likelihood there is of finding unique restriction sites in the T-DNA into which exogenous DNA can be cloned.

To take advantage of both these factors a system of binary vectors have been devised which are now in widespread use. Initial cloning is into a relatively small plasmid which can replicate both in E. coli and in Agrobacterium (see Deblaere et al., 1987). This plasmid usually carries the border regions between which there is the cloning site which is a segment of DNA in which there are several recognition sites for restriction enzymes. This may be flanked on the 5' side by a promoter sequence and on the 3' side by a transcription terminator sequence both of which can function in plants. These sequences make it possible for a coding sequence inserted into the cloning site to be expressed. Alongside the cloning region and between the borders there is also usually a selectable marker gene of the type described above. Furthermore, because the vector is cloned first into E. coli and later transferred into Agrobacterium, the vector must also carry origins of replication which permit it to exist in both types of bacteria and an additional selectable marker which allows only those bacterial cells which contain the vector to survive when in the presence of an appropriate antibiotic. The Ti plasmid of the Agrobacterium cells into which the small vector is transferred is usually modified so that it no longer has a T-region itself and so only mediates in the transfer of the T-DNA from the small vector.

There are many variants of the binary type vector system described, each of which has particular advantages and drawbacks as described in some detail by Bevan and Goldsbrough (1988) and by Rogers et al. (1987).

Most of the plants obtained by Agrobacterium-induced transformation have a small number of insertion loci (usually only one). One or more closely linked copies can be present at each locus and the component genes are inherited in a Mendelian fashion. The level of expression in the T-DNA is not necessarily proportional to the number of copies of this DNA present and expression of the genes is somewhat variable. Consequently, transformed plants have to be assayed individually to determine how active the inserted genes are.

Direct gene transfer. Protoplasts made from plant cells can take up DNA through their plasma membranes and integrate it into their genomes. No special vectors are needed for this form of transformation. Expression of genes taken up in this way requires no more than that they should be under the control of plant expression signals (Paszkowski et al., 1984; Potrykus et al., 1985). Passage of the DNA into protoplasts is facilitated by polyethylene glycol in their suspension medium or by subjecting the protoplasts to a brief high voltage shock (Fromm et al., 1985). The main limitation of this form of transformation is that it must be possible to regenerate plants from the protoplasts and this is only possible as yet with a relatively few species. However, the method has been successfully used recently for the transformation of maize cells (Rhodes et al., 1988) using the NPT II gene as a selectable marker (see above).

DNA transfer using micro-projectiles. A surprisingly effective method of getting DNA into plant cells, tissues and even into chloroplasts and mitochondria is with DNA-coated micro-projectiles. These are fired at the plant tissue using a gun-like apparatus or using an electrical discharge (McCabe et al., 1988). With such a system DNA to replace that deleted by mutation has been transported into chloroplasts of <u>Chlamydomonas</u> (Boynton et al., 1988). Subsequently recombination in some transformants resulted in the wild type chloroplast genome being reconstituted. Although there have been previous indications that it was possible to get chloroplast transformation with DNA introduced by <u>Agrobacterium</u> (De Block et al., 1985) or by direct gene transfer (Potrykus et al., 1985), this represents the first unequivocal demonstration of chloroplast transformation.

<u>Microinjection</u>. Microinjection has been a very successful method of getting macromolecules and even organelles and chromosomes into animal cells. Considerable effort has also been devoted to developing it too as a method for plant transformation and it has successfully been used with alfalfa (Reich et al., 1986) and rape (Neuhaus et al., 1987). However, it will probably be used mainly for the introduction of large structures into cells rather than for getting in isolated genes for which the other techniques available are rapidly improving (Mathias, 1987).

<u>Pollen-tube pathway</u>. A method of plant transformation which has been successfully used with several different types of cereal is known as the "pollen-tube pathway" and was first reported by Duan and Chen (1985). The stigmas of florets that have been pollinated a short while before are cut off and a drop of solution containing donor DNA applied to the cut end of the styles. A comparatively large proportion of the seeds formed have the DNA integrated in their nuclear genome, presumably after having reached the ovule initially by flowing down the pollen tube. The method has been described in detail recently by Luo and Wu (1988).

Altering the levels of specific gene products

The use of mutants has for long been a powerful tool of geneticists for altering gene expression. However, there are many difficulties involved in the production of mutants using non-targeted methods such as chemical mutagens. The techniques of molecular genetics offer the possibility of making such mutants in a controlled fashion to increase or decrease the amounts of a particular gene product or to alter the composition of the product. In many instances one would wish to achieve this by replacing an existing gene by a modified one by transformation. Although this can be readily achieved in yeast and bacteria the same is not so for plants. Transforming DNA integrates randomly so that the products of both the incoming and endogenous genes are present. To overcome these problems ways are being devised of reducing the levels of endogenous mRNAs. It may also be desirable to reduce the level of a mRNA to determine the function of its product (reverse genetics).

A stratagem for trying to reduce the levels of a particular gene product is either to destroy or inactivate its mRNA. This can be achieved by making transgenic plants in which the coding sequence is reversed with respect to the transcriptional control sequences so that antisense RNA is produced (Ecker and Davis, 1986; van der Krol et al., 1988). This hybridizes with the functional mRNA and this either blocks its translation or to facilitates its destruction. A possible alternative way of destroying specific mRNAs which has been described recently involves the use of ribozymes (hydrolytically active RNAs) which can be designed to cleave specific RNA sequences (Haseloff and Gerlach, 1988). The use of these may be more efficient than antisense methods for mRNA destruction because the cleavage reaction is catalytic.

An increase in the level of a particular gene product can be accomplished by transforming plants with the coding sequence of the gene under the influence of a strong promoter. One of the most useful promoters of this type is the "35S" constitutive promoter of the cauliflower mozaic virus and this is a component of many transformation vectors.

Post-translational targeting of chloroplast proteins

Proteins that are located at sub-cellular sites other than where they were synthesized must contain the information which causes them to move from their site of synthesis to their site of function. In chloroplasts alone there are at least six different locations (the inner and outer envelope membranes and the space between them, the stroma, the thylakoids and the thylakoid lumen) where proteins are found. Mechanisms are also present which respond specifically to the targeting information in the proteins so that they are guided to their correct destination after they have been synthesized in either the cytosol or stroma. A knowledge of the principles which determine how this translocation occurs should make it possible to decide whether chloroplasts can be modified appropriately by the introduction of new genes. Several reviews devoted to the topic of protein transport in chloroplasts have recently been published (Schmidt and Mishkind, 1986; Lubben et al., 1988; Keegstra and Bauerle, 1988).

So far very little is known about the factors which guide proteins into the chloroplast envelope but the principles governing the location of proteins in the stroma and thylakoids are gradually emerging. Nuclear-encoded chloroplast proteins destined for these sites are synthesized in the cytosol as precursors. These precursors have an amino-terminal extension (the import or transit peptide) which facilitates the uptake of the precursors by chloroplasts. Proteins only travelling as far as the stroma have a transit peptide with a single functional domain but those proceeding further to the thylakoids or to the lumen have additional targeting information (Smeekens and Weisbeek, 1988).

The chloroplast import peptides of various precursors that have been studied are very diverse both with respect to their length (32-80 amino acids) and sequence (Back et al., 1988). Importantly in this context, is that a variety of foreign passenger proteins can be imported into the stroma when a chloroplast import peptide is attached to their amino termini. However, 'foreign' proteins are not imported as efficiently as natural ones.

Import of precursor proteins into the stroma occurs after their synthesis is complete. Post-translational import also occurs in mitochondria and many similarities exist between the import of proteins into mitochondria and chloroplasts (Keegstra and Bauerle, 1988). The importation of proteins into both these organelles is different from that into the endoplasmic reticulum which generally occurs as the polypeptides are being synthesized by membrane-bound ribosomes.

Several distinct processes can be distinguished in the importation of proteins into chloroplasts. The first step is the binding of the precursor protein to the chloroplast surface. It has been suggested that this binding is mediated by a receptor protein located at contact points between the inner and outer envelope membranes (Pain et al., 1988). The second step in the importation process is the translocation of the precursor protein through the envelope. This step requires energy in the form of ATP. The third step which occurs as soon as the precursor proteins get into the cytosol is the proteolytic cleavage of the precursor to remove the chloroplast import peptide. This cleavage occurs in the stroma. What happens from then on differs from protein to protein. Some have to assemble with similar or heterologous peptides to form active enzymes or they may then need to be inserted into or across the thylakoid membrane as well. As mentioned above, additional targeting information is required for this.

Efforts to define critical regions in or characteristics of the chloroplast import peptides have been somewhat inconclusive. Proposals that these peptides have a "common framework" were based on the comparison of a limited number of precursors. Three regions with some similarities were identified, one at the amino terminus, a second in the middle and a third at the carboxy terminus (Schmidt and Mishkind, 1986; Karlin-Neumann and Tobin, 1986). The first two were thought to be involved in the binding and/or translocation steps whereas the region near the carboxy terminus was thought to be involved in proteolytic cleavage of the import peptide from the mature protein. However, further comparisons and the results of deletion experiments indicate that this view is too simplistic. The differences which exist in size and primary structure of the import peptides suggests that features of secondary or tertiary structure rather than amino acid sequence as such is important. However, they do have a very characteristic composition - a large number of hydroxylated amino acids (serine and threonine), several basic amino acids such as arginine and lysine giving an overall basic charge and infrequently any negatively charged amino acids. Also, in general, they are rather hydrophobic. In contrast to the mitochondrial import peptides (von Heijne, 1986) they have no apparent tendency to form amphipathic alpha-helixes.

A specific protease in the stroma removes the chloroplast import peptide. This peptidase has been shown to be highly specific for chloroplast precursors. It can be assumed therefore that all the precursors have a characteristic structural element which the peptidase recognises.

Nuclear-encoded chloroplast proteins such as plastocyanin and the 33, 23 and 16 kDa components of the water-splitting complex which are transferred into the thylakoid lumen subsequent to their import into the stroma, require a second amino terminal import sequence in tandem to facilitate the additional transfer (Smeekens and Weisbeek, 1988). The main characteristics of this sequence is that it is very hydrophobic and, like the chloroplast import peptide, has a net positive charge. Interestingly, the apoprotein of the chloroplast-encoded thylakoid component cytochrome f is synthesized with a similar transfer sequence. This facilitates the transfer of the body of the protein into the lumen but the complete transfer of the protein into this compartment is arrested by a 'stop-transfer' sequence near its carboxy terminus. In cyanobacteria a similar thylakoid transfer sequence is found on the precursor of the plastocyanin apoprotein but, as might have been suspected, no sequence equivalent to the chloroplast import peptide is present.

Thylakoids contain a second peptidase which removes the thylakoid transfer sequence (Kirwin et al., 1987). As with the stromal peptidase it is highly specific for the protein precursors suggesting that there is a characteristic processing site for the thylakoid transfer sequence too.

Whereas it seems possible to get nearly all proteins into the stroma by fusing a chloroplast import peptide onto their amino termini, the transit peptide of a lumenal protein precursor does not target the alien passenger proteins such as ferredoxin or mitochondrial superoxide dismutase into the lumen. However, partially processed molecules of each can be found in the stroma. It would thus appear that there are more stringent requirements for the importation of proteins into the stroma than into the lumen.

Manipulating photosynthesis

It seems a reasonable assumption that if we could increase the efficiency of photosynthesis of plants, then they would grow better and be higher yielding. If this is indeed true [and surprisingly it remains to be proven - see Nelson (1988)] then if we could recognise the limiting factors in photosynthesis and raise the threshold levels at which they became limiting, then we could expect to exert beneficial effects. What then limits the rate of photosynthesis and can we isolate and manipulate the genes which code for the components involved?

A way in which to provide experimental data as to what the limiting factors are is to make a series of plants which are isogenic in all but the component being evaluated which is varied in amount or activity. Modern control theory provides the theoretical framework in which to assess the consequences of the differences and upon which to build a profile of the changes exert their effect in different conditions (Kacser and Burns, 1973; Heinrich and Rapoport, 1974; Kacser and Porteous, 1987).

Although it might be possible to make some of the desired changes by simple mutagenesis using chemical mutagens (Jones et al., 1986a,b), there are many problems involved in this approach such as the identification of the appropriate mutant plants and because many mutants of the photosynthetic components are lethal.

What processes in plants limit photosynthesis then and could possibly be manipulated? Several suggestions have been made as to what they might be and the answer cannot be a clearcut one as different components are likely to be limiting under different conditions and also control may be shared by several components. Also, there is very little prospect of being able to improve the efficiency of some processes as they are probably already operating optimally. This may include the light driven processes which occur in thylakoids which result in the production of ATP and NADPH. Further reasons for considering the thylakoid proteins as unsuitable targets for manipulation is that many of them are encoded by the chloroplast genome and are consequently not readily accessible as yet for manipulation by transformation, and most are highly conserved in sequence so there is little genetic variation to suggest how qualitative changes could be made. The amount of specific thylakoid complexes are probably controlled by nuclear genes which could be manipulated to change the relative amounts of each, as could the genes for the chlorophyll a/b–binding antenna proteins of photosystems I and II.

Most attention has been devoted to the properties of ribulose–1,5–bisphosphate carboxylase/oxygenase (Rubisco) as there is clearcut evidence derived by growing plants in CO_2–enriched environments (Kimball, 1983) to suggest that if the initial stage of CO_2 fixation which it catalyses could be made more efficient this would have a beneficial effect on yield. Rubisco is both inefficient so that large quantities of the enzyme is packed into chloroplasts (Walker et al., 1986) and it catalyses an oxygenation as well as a carboxylation reaction. This oxygenation reaction leads to the loss in photorespiration of a substantial proportion of the CO_2 fixed. Therefore it is apparently a very wasteful reaction (Keys, 1986) and whether or not it has a useful role in dissapating excess energy is still a matter for debate. Rubisco therefore seems to be a prime target for manipulation. However, the catalytic activity of the enzyme resides in the large subunit which is chloroplast encoded. One way of trying to overcome this problem is to modify the gene so that its product has a chloroplast transport peptide as well as the coding sequence of the large subunit. The gene can then be inserted into the nuclear genome and its product synthesized in the cytosol. It seems likely that such a strategy would work as it has already been shown that it is possible to introduce a large subunit precursor of this type into isolated chloroplasts (Gatenby et al., 1988). Although technically it may be possible to manipulate the composition of the large subunit protein, in higher plants there is little naturally occurring variation in its activity to indicate how this should be done. All the Rubisco's of higher plants are quite similar both with respect to their specific activities and relative carboxylation/oxygenation activities (measured as their 'specifity factor') (Parry et al., 1987), while the Rubisco's of the cyanobacteria and algae have worse specificity factors (Jordan and Ogren, 1981). This suggests that within the limits imposed on the structure of the enzyme by its evolutionary history, and by the nature of the reaction which it catalyses, its structure may already have been optimized in higher plants.

The results of several studies have suggested that enzymes other than Rubisco might have important roles in regulating aspects of CO_2 fixation (Woodrow and Berry, 1988; Pettersson and Ryde-Pettersson, 1988). These include fructose-1,6-bisphosphatase, phosphoribulokinase and sedoheptulose bisphosphatase in the photosynthetic carbon reduction (PCR) cycle. They bring about the regeneration of ribulose-1,5-bisphosphate while allowing some of the photosynthate to be diverted into sugars and starch. Each of these enzymes catalyses an essentially irreversible reaction and all are highly regulated. We are now isolating the genes for these enzymes (Raines et al., 1988) to make transgenic plants with altered levels of each. Photosynthetic carbon flux will be measured in plants with different amounts of the enzymes. By doing this in a variety of environments, it should be possible to obtain quantitative estimates of the relative contribution of each enzyme to the control of photosynthetic carbon assimilation.

Another important regulatory component of chloroplasts which it should also be possible to investigate in this way is the phosphate translocator (see Pettersson and Ryde-Pettersson, 1988) but its gene is proving to be very difficult to isolate. This component of the chloroplast envelope controls, in a strictly stoichiometric fashion, the passage of photosynthate (in the form of triose phosphates) out of and the passage of inorganic phosphate into chloroplasts.

There are many examples already of advances achieved by the application of control analyses of this type to the understanding of metabolic pathways (see Kacser and Porteous, 1987). This approach has already been used successfully in plants in which the amounts of the cytosolic and chloroplastic versions of phosphoglucomutase had been altered (Kruckeberg et al., 1988). Furthermore, the advent of transgenic plants should make it generally applicable to the study of various metabolic pathways in addition to photosynthetic ones.

Another strategy which some plants have evolved to overcome the deficiencies of Rubisco is to raise the concentration of CO_2 available to the enzyme either using a pumping mechanism or by taking advantage of modified leaf morphology and metabolic pathways. It remains to be seen whether the same sort of modifications could be introduced into plants which currently do not have them. Alternatively, it may be possible to limit the loss of CO_2 which occurs in photorespiration by introducing new enzymes or by increasing the levels of existing ones so that the production of metabolically useful substances from the products of the oxygenase reaction would be catalysed before CO_2 was formed from them.

One of the most promising ways in which to bring about enhanced net photosynthesis in plants is simply to delay the senescence of their leaves (Nelson, 1988). Plants which stay green longer have in several instances been found to be higher yielding. It has been known for some time that the external application of kinetin can delay senescence. In some preliminary experiments we have shown that it is possible to do the same by introducing into transgenic plants a cytokinin-producing gene which can be activated at an appropriate stage in development (T.A. Dyer, S.R. Scofield, M.W. Bevan, unpublished results). The leaves in such plants stay green far longer than they would do normally.

CONCLUSIONS

There can now be little doubt that the use of transgenic plants will provide a very powerful tool for the assessment of the regulatory processes involved in photosynthesis. Although it may be easier to evaluate the functional role of individual components by the use of micro-organisms (especially the cyanobacteria) due to the relative ease and speed with which they can be transformed by homologous recombination (see Shestakov and Reaston, 1987), when it comes to the assessment of how the integrated photosystems operate in eukaryotes, we must study these in higher plants. Anyway, if we are to put to practical use what we find out about photosynthesis, it is ultimately crop plants which must be modified.

According to Nasyrov (1978) it is theoretically possible to increase leaf photosynthesis by genetic means, and that this would increase canopy photosynthesis. However, whether or not this would be translated into increased crop yield may be masked by many other factors such as leaf area index, leaf orientation, leaf aging and sink size and number. The conclusion reached by Nelson (1988) is that progress in improving crop yield by improving leaf photosynthesis has not been rapid or encouraging and that improving leaf photosynthesis rate alone would probably not be very effective. He concludes also that a decrease or elimination of photorespiration in C_3 plants would give a potential for improvement that would be similar to or exceed the expected genetic gain which could be expected for better leaf photosynthesis. Increasing leaf duration, especially in determinate crops, could also be an important factor in increasing yield.

From the foregoing discussion it should be clear that many of these issues are now being addressed using transgenic plants. These will provide not only a way of testing some of the proposals of agronomists and physiologists as to how to increase crop yields could be increased, but also a means of doing so.

REFERENCES

Back, E., Burkhart, W., Mayer, M., Privalle, L., and Rothstein, S., 1988, Isolation of cDNA clones coding for spinach nitrite reductase: complete sequence and nitrate induction, Mol. Gen. Genet., 212:20-26.

Bevan, M., and Goldsbrough, A., 1988, Design and use of Agrobacterium transformation vectors, in: "Genetic Engineering. Principles and Methods" J. K. Setlow and A. Hollaender, eds., Plenum Publishing Co., New York, pp. 123-140.

Boynton, J. E., Gillham, N. W., Harris, E. H., Hosler, J. P., Johnson, A. M., Jones, A. R., Randolf-Anderson, B. L., Robertson, D., Klein, T. M., Shark, K. B., and Sanford, J. C., 1988, Chloroplast transformation in Chlamydomonas with high velocity microprojectiles, Science, 240:1534-1538.

Deblaere, R., Reynaerts, A., Höfte, H., Hernalsteens, J.-P., Leemans, J., and van Montagu, M., 1987, Vectors for cloning in plant cells, Methods in Enzymol., 153:277-292.

De Block, M., Schell, J., and van Montagu, M., 1985, Chloroplast transformation by Agrobacterium tumefaciens, EMBO J., 4:1367-1372.

Duan, X., and Chen, S., 1985, Variation of the characters in rice (Oryza sativa) induced by foreign DNA uptake, China Agric. Sci., 3:6-9.

Ecker, J. R., and Davis, R. W., 1986, Inhibition of gene expression in plant cells by expression of antisense RNA, Proc. Natl. Acad. Sci. USA, 83:5372-5376.

Fromm, M. E., Taylor, L. P., and Walbot, V., 1985, Expression of genes transferred into monocot and dicot plant cells by electroporation, Proc. Natl. Acad. Sci. USA, 82:5824-5828.

Gatenby, A. A., Lubben, T. H., Ahlquist, P., and Keegstra, K., 1988, Imported large subunits of ribulose bisphosphate carboxylase/oxygenase, but not imported ß-ATP synthase subunits, are assembled into holoenzyme in isolated chloroplasts, EMBO J., 7:1307-1314.

Haseloff, J. and Gerlach, W. L., 1988, Simple RNA enzymes with new and highly specific endoribonuclease activities, Nature 344:585-591.

Hayford, M. B., Medford, J. I., Hoffman, N. L., Rogers, S. G., and Klee, H. J., 1988, Development of a plant transformation selection system based on expression of genes encoding gentamycin acetyltransferases, Plant Physiol., 86:1216-1222.

Heinrich, R. and Rapoport, T. A., 1974, A linear steady state treatment of enzymatic chains, Eur. J. Biochem, 42:107-120.

Jones, T. W. A., Pichersky, E., and Gottlieb, L. D., 1986a, Enzyme activity in EMS-induced null mutations of duplicated genes encoding phosphoglucose isomerase in Clarkia, Genetics 113:101-114.

Jones, T. W. A., Gottlieb, L. D., and Pichersky, E., 1986b, Reduced enzyme activity and starch level in an induced mutant of chloroplast phosphoglucose isomerase, Plant Physiol., 81:367-371.

Jordan, D. B., and Ogren, W. L., 1981, Species variation in the specificity of ribulose bisphosphate carboxylase/oxygenase, Nature, 291:513-515.

Kacser, H., and Burns, J. A., 1973, Control of flux, Symp. Soc. Exp. Biol., 27:65-107.

Kacser, H., and Porteous, J. W., 1987, Control of metabolism: what do we have to measure?, TIBS, 12:5-14.

Karlin-Neumann, G. A., and Tobin, E. M., 1986, Transit peptides of nuclear-encoded chloroplast proteins share a common amino acid framework, EMBO J., 5:9-13.

Keegstra, K., and Bauerle, C., 1988, Targeting of proteins into chloroplasts, BioEssays, 9:15-19.

Keys, A. J., 1986, Rubisco: its role in photorespiration, Phil. Trans. Roy. Soc. Lond. B, 313:325-336.

Kimball, B. A., 1983, Carbon dioxide and agricultural yield: an assemblage and analysis of 430 prior observations, Agronomy J., 75:779-788.

Kirwin, P. M., Elderfield, P. D., Robinson, C., 1987, Transport of proteins into chloroplasts. Partial purification of a thylakoidal processing peptidase involved in plastocyanin biogenesis, J. Biol. Chem., 262:16386-16390.

Kruckeberg, A. L., Neuhaus, H. E., Feil, R., Gottlieb, L., and Stitt, M., 1988, Reduced activity of photophoglucose isomerase in the cytosol and chloroplast of Clarkia xantiana. I. Impact on mass action ratios and fluxes to sucrose and starch, In press.

Lubben, T. H., Theg, S. M., and Keegstra, K., 1988, Transport of proteins into chloroplasts, Photosyn. Res., 17:173-194.

Luo, Z.-X., and Wu, R., 1988, A simple method for the transformation of rice via the pollen-tube pathway, Plant Mol. Biol. Rep. 6:165-174.

McCabe, D. E., Swain, W. F., Martinell, B. J., and Christou, P., 1988, Stable transformation of soybean (Glycine max) by particle acceleration, BioTechnology, 6:923-926.

Mathias, R. J., 1987, Plant microinjection techniques, in: "Genetic Engineering" Vol. 9, J. K. Setlow, ed., Plenum Publishing Co., New York, pp. 199-227.

Melchers, L. S., and Hooykaas, P. J. J., 1987, Virulence in Agrobacterium, Oxford Surveys Plant Mol. Cell Biol., 4:167-220.

Nasyrov, Y. S., 1978, Genetic control of photosynthesis and inproving of crop productivity, Ann. Rev. Plant Physiol., 29:215-237.

Nelson, C. J., 1988, Genetic associations between photosynthetic characteristics and yield: Review of evidence, Plant Physiol. Biochem., 26:543-554.

Neuhaus, G., Spangenberg, G., Mittelsten Scheid O., and Schweiger, H.-G., 1987, Transgenic rapeseed plants obtained by the microinjection of DNA into microspore-derived embryos, Theor. Appl. Genet., 75:30-36.

Pain, D., Kanwar, Y.S., and Blobel, G., 1988, Identification of a receptor for protein import into chloroplasts and its localization to envelope contact zones, Nature, 331:232-237.

Parry, M. A. J., Schmidt, C. N. G., Cornelius, M. J., Millard, B. N., Burton, S., Gutteridge, S., Dyer, T. A., and Keys, A. J., 1987, Variation in properties of ribulose-1,5-bisphosphate carboxylase from various species related to differences in amino acid sequences, J. Exp. Bot., 38:1260-1271.

Paszkowski, J., Shillito, R. D., Saul, M., Mandak, V., Hohn, T., Hohn, B., and Potrykus, I., 1984, Direct gene transfer to plants, EMBO J., 3:2717-2722.

Pettersson, G., and Ryde-Pettersson, U., 1988, A mathematical model of the Calvin photosynthesis cycle, Eur. J. Biochem., 175:661-672.

Potrykus, I., Shillito, R. D., Saul, M. W., Paszkowski, J., 1985, Direct gene transfer. State of the art and future potential, Plant Mol. Biol. Rep., 3:117-128.

Raines, C. A., Lloyd, J. C., Longstaff, M., Bradley, D., and Dyer, T. A., 1988, Chloroplast fructose-1,6-bisphosphatase: the product of a mozaic gene, Nucleic Acids Res. 16:7931-7942.

Reich, T. J., Iyer, V. N., and Miki, B. L., 1986, Efficient transformation of alfalfa protoplasts by intranuclear microinjection of Ti plasmids, BioTechnology 4:1001-1004.

Rhodes, C. A., Pierce, D. A., Mettler, I. J., Mascarenhas, D., and Detmer, J. J., 1988, Genetically transformed maize plants from protoplasts, Science, 240:204-207.

Rogers, S. G., Klee, H. J., Horsch, R. B., and Fraley, R. T., 1987, Improved vectors for plant transformation: Expression cassette vectors and new selectable markers, Methods in Enzymol., 153:253-277.

Schell, J., 1988, Transfer of T-DNA from Agrobacterium into plants, Sym. Soc. Gen. Microbiol., 43:355-365.

Schmidt, G. W., and Mishkind, 1986, The transport of proteins into chloroplasts, Ann. Rev. Biochem., 55:879-912.

Shestakov, S. V., and Reaston, J., 1987, Gene-transfer and host-vector systems of cyanobacteria, Oxford Surveys Plant Mol. Cell Biol. 4:137-166.

Smeekens, S., and Weisbeek, P., 1988, Protein transport towards the thylakoid lumen: post-translational translocation in tandem, Photosyn. Res., 16:177-186.

van der Krol, A. R., Lenting, P. E., Veenstra, J., van der Meer, I. M., Koes, R. E., Gerats, A. G. M., Mol, J. N. M., and Stuitje, A. R., 1988, An anti-sense chalcone synthase gene in transgenic plants inhibits flower pigmentation, Nature, 333:866-869.

von Heijne, G., 1986, Mitochondrial targeting sequences may form amphiphilic helixes, EMBO J., 5:1335-1342.

Walker, D. A., Leegood, R. C., and Sivak, M. N., 1986, Ribulose bisphosphate carboxylase-oxygenase: its role in photosynthesis, Phil. Trans. Roy. Soc. Lond. B, 313:305-324.

Woodrow, I. E., and Berry, J. A., 1988, Enzymatic regulation of photosynthetic CO_2 fixation in plants, Ann. Rev. Plant Physiol. Plant Mol. Biol., 39:533-594.

PERMEABILIZED CYANOBACTERIA:

A MODEL SYSTEM FOR PHOTOSYNTHETIC AND BIOTECHNOLOGICAL STUDIES

George C. Papageorgiou

National Reseach Center Demokritos
Institute of Biology
Athens, Greece

INTRODUCTION

The cyanobacteria (blue-green algae) is the largest group of photo-trophs in the superkingdom Prokaryotae (Stanier and Cohen Bazire, 1977). Micropalaeontology places their origin in the middle of the Precambrian Eon, some 2.8-3.5 billion years ago, but it has not been able to decide yet whether filamentous or unicellular varieties appeared first (Schopf and Walter, 1982). Those archaeic cyanobacteria were the first photoauto-trophs on our planet that were endowed with an oxygenic photosynthetic machine, and so they are credited with the enrichment of Earth's atmo-sphere in oxygen.

According to Stanier's (1982) definition, cyanobacteria are "microorga-nisms that harbor, within a typically prokaryotic cell, a photosynthetic apparatus closely similar in structure and function to that located within chloroplasts of photosynthetic prokaryotes." Because of this similarity, the cyanobacteria, and particularly the unicellular species, have long been viewed as the equivalent of higher plant chloroplasts and have been used extensively for investigations relating to the mechanism of photosyn-thesis. Advantages are the greater stability of bacterial suspensions com-pared to suspensions of isolated chloroplasts, the fact that photosynthe-sis is studied in integral organisms rather than in membranes suspended in artificial media, and the fact that unicellular cyanobacteria are virtual-ly free of the autoprecipitation problem that complicates optical measure-ments in suspensions of the larger algal cells. The principal disadvantage is the impermeability of the bacterial cell envelope to ions. Thus, they cannot be used in experiments which include ionic reactants.

The cyanobacteria are themselves highly interesting photosynthetic mi-cro-organisms as they differ in several noteworthy aspects from eukaryotic cell chloroplasts. Some of the most striking differences are: their gene-tic autonomy; the Gram negative cell wall which together with the plasma membrane constitute an ion-impermeable cell envelope (Drews and Weckesser, 1982); thylakoids which do not form grana and lack Chl $\underline{b}$; light gathering organelles (phycobilisomes) attached on the stromal thylakoid face in regu-larly spaced arrays (Glazer, 1984); adaptation to the quantity and quality of actinic light by different mechanisms than higher plants (Williams and Allen, 1987); and the ability of some filamentous cyanobacteria to photore-duce hydrogen ions and dinitrogen in aerobic environment (Lambert and Smith, 1981).

Perhaps, the most popular cyanobacterium is the unicellular Anacystis

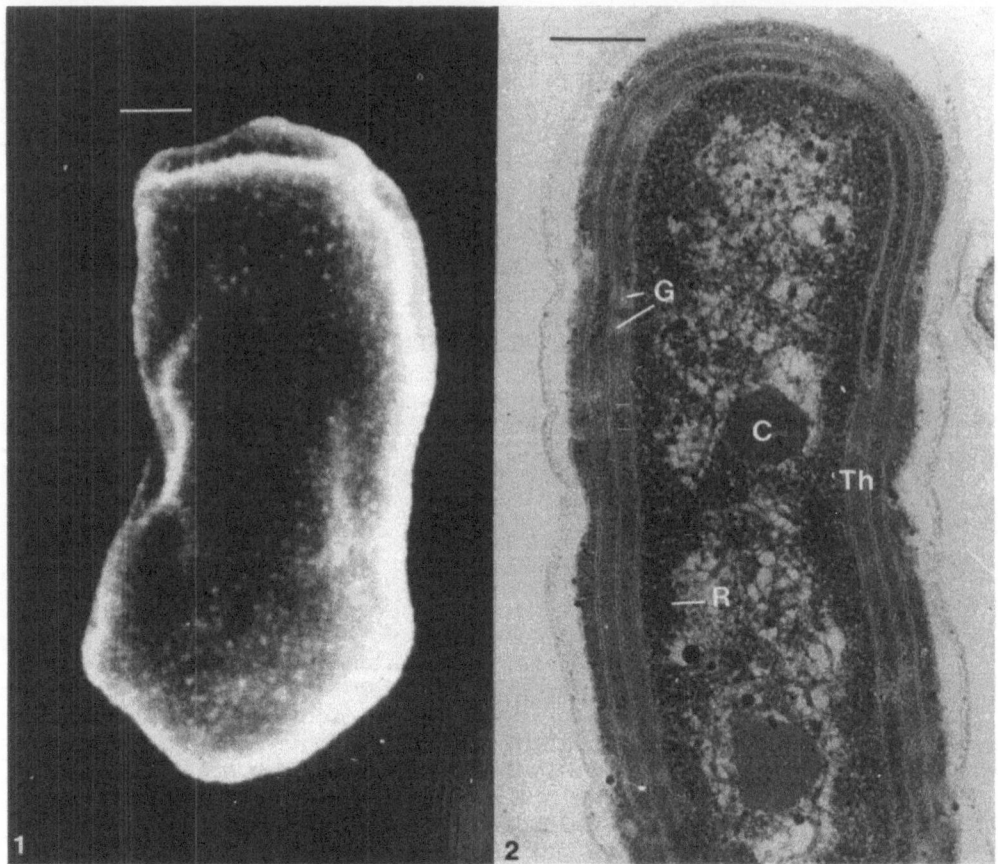

Fig. 1   Scanning electron micrograph of <u>Anacystis</u> <u>nidulans</u> cell. Bar,
          200 µm. A JEOL JSM 840 scanning electron microscope was
          used. Samples prepared as described by Barbotin and Tho-
          masset (1980).

Fig. 2   Transmission electron micrograph of <u>Anacystis</u> <u>nidulans</u> cell.
          Th, thylakoid; C, carboxysome; G, glycogen granules; R, ribo-
          somes. Bar, 200 µm. A JEOL 100 C transmission electron micro-
          scope was used. Details as in Barbotin and Thomasset (1980).

<u>nidulans</u> (Synechococcus). Fig. 1 shows a scanning electron micrograph of
this bacterium and Fig. 2 a transmission electron micrograph. It has a
smooth surface and a characteristic elongated shape which resists deformation in hypo-osmotic media. As a prokaryote, the cell lacks nucleus, mitochondria and chloroplasts.
      Morphological features include few large thylakoids running parallel
to the cell periphery, nearly perfectly equidistant (35-45 nm) from each
other, carboxysomes in the middle part and glycogen granules in between
thylakoids. The cell is about 2 µm long and 1 µm wide. By modelling it as
a cylinder, we may estimate its volume to be 1.6 µm^3, out of which more
than 60% is occupied by the thylakoid area. Assuming thylakoids to be con-
centric cylinders, we may estimate a total thylakoid surface equal to
about 35 µm^2.
      The main photosynthetic pigments of Anacystis, Chl <u>a</u>, C-phycocyanin
and allophycocyanin, show up in the absorption spectrum of Fig. 3. The a-
mount of phycobiliproteins depends on the quality of illumination at the

culturing stage, a phenomenon known as complementary adaptation. Fig. 3 displays also the inverted second derivative absorption spectrum, of Anacystis. Second derivatives of Gaussians have their major maximum coinciding with that of the parent curve, but pointing in the opposite direction. Inverted second derivatives, on the other hand, have maxima coinciding both in location and in direction with the parent curve maxima. The great advantage of the second derivative absorption spectrum lies in its power to resolve overlapping spectral bands. This has been employed to resolve phycobiliprotein spectra (Erokhina et al. 1982, Papageorgiou and Lagoyanni, 1983). Major absorption contributions resolved in Fig. 3 are Chl a (A442, A684), C-phycocyanin (A636), carotenoids (469, 502). The shoulders at 670 nm and 615 nm in the second derivative spectrum, must be assigned to allophycocyanin B, a minor Anacystis phycobiliprotein with absorption maxima at 618 and 671 nm (Glazer and Bryant, 1975). Allophycocyanin B is not resolved in the direct absorption spectrum.

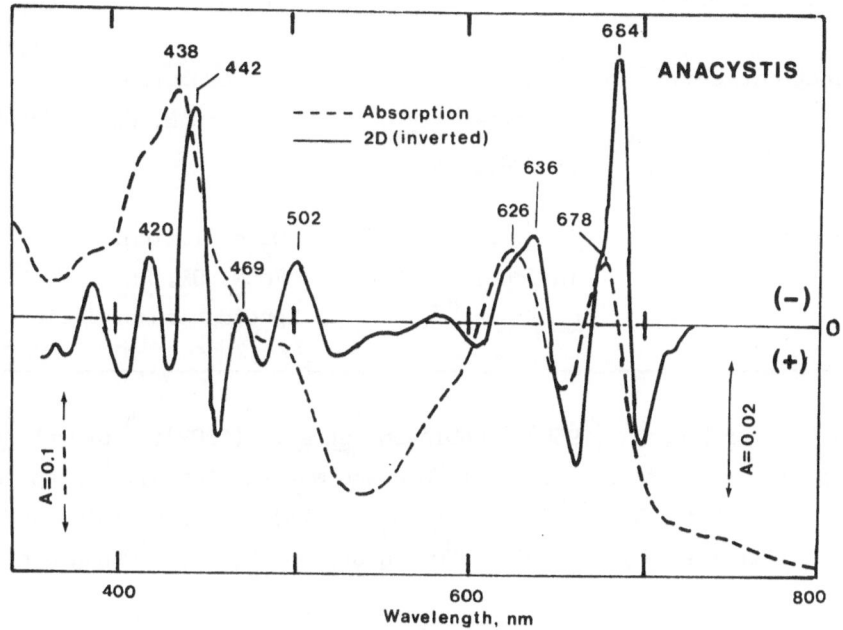

Fig. 3   Inverted second derivative absorption spectrum (———) and direct absorption spectrum (- - -) of Anacystis nidulans. Differentiation interval, 4 nm; scanning bandwidth, 2nm.

PERMEABILIZATION OF THE CYANOBACTERIAL CELL ENVELOPE

Cyanobacteria are virtually impermeable to ions, with the exception perhaps of H^+ and OH^-. Lipid bilayers exhibit anomalously high permeability to these ions (Nichols and Deamer, 1980). Ion impermeability restricts the usefulness of cyanobacteria as model photosynthetic systems to experiments in which permeant only reactants are included. In other cases, either the cells must be broken and thylakoid fragments must be isolated, or the cell envelope must be permeabilized.

Since cyanobacteria have no chloroplasts, it suffices to permeabilize

Table 1. Methods of permeabilizing cyanobacteria

Method	Species to which the method was applied	Surviving activities
Freeze drying	A. nidulans[1]	O_2 evolution, FeCN, BQ, $NADP^+$ reduction, photophosphorylation
Washing with hypertonic KCl plus EDTA	S. platensis[2]	O_2 evolution, FeCN, PD reduction
Transient chilling	A. nidulans,[3-7] A. variabilis[3] N. muscorum[3]	O_2 evolution, BQ, FeCN, CO_2 reduction
Lysozyme treatment	F. diplosiphon,[6] P. luridum,[6-9] A. nidulans,[8-11]	O_2 evolution, FeCN, BQ, CO_2 reduction, photophosphorylation

[1]Gerhardt and Trebst (1965); [2]Robinson et al. (1982); [3]Forrest et al. (1957); [4]Jansz and Maclean (1972); [5]Ono and Murata (1981a); [6]Crespi et al. (1962); [7]Biggins (1967); [8]Binder et al. (1975); [9]Papageorgiou and Tzani (1980); [10]Ward and Myers (1972); [9]Papageorgiou (1977); [11]Papageorgiou and Lagoyanni (1985)

the cell envelope to enable ionic solutes to access the thylakoids. In contrast, two permeability barriers, the cell envelope and the chloroplast envelope, need to be overcome for the same purpose in the case of higher plant and algal cells.

Permeabilization of cyanobacteria has been achieved by several methods (Table 1).

(a) Enzymic permeabilization

Enzymic permeabilization of cyanobacteria depends on the hydrolytic cleavage of cell wall peptidoglycan by lysozyme and results either in osmo-resistant cells (permeaplasts; partial peptidoglycan hydrolysis) or in os-mosensitive cells (spheroplasts; total peptidoglycan hydrolysis). Peptido-glycan is a branched polysaccharide consisting of alternating N-acetyl glu-cosamine and N-acetyl muramic acid residues, and of oligopeptides which on one hand crosslink the polysaccharide chains and on the other connect them to outer membrane lipoproteins. As an open network, it offers no resistance to the passage of small molecules or ions. It is the combination of the plasma membrane and its structural support, the peptidoglycan network, which serves the cell as an effective permeability barrier. Upon hydrolytic cleavage of the polysaccharide backbone, the plasma membrane becomes unable to control permeability.

Lysozyme, a sphere 3.2 nm in diameter and 8 nm^2 in cross-section (Phillips, 1967), encounters significant steric inhibition on the way to its substrate. Being able to cleave glucosidic bridges by attacking peptidoglycan only from the plasma membrane side (Braun et al., 1970) it must cross not only the outer membrane but also the peptidoglycan layer itself in order to perform its task. Abnormally high enzyme concentrations and auxiliary treatments, such as incubation with EDTA (Ward and Myers, 1973) and hypo-osmotic reaction media (Witholt, 1976), are often necessary in order to overcome the steric inhibition by the cell wall layers.

Lysozyme digestion of cyanobacterial cell walls has been extensively used as a preliminary step for the preparation of cell-free thylakoid fragments (for example, Biggins, 1967b, Binder et al., 1976; Spiller and Boeger, 1980). In other cases, the desired product is an ion-permeable cell which retains the largest possible fraction of the parent cell photosynthetic activity. Preparation of spheroplasts by treating cyanobacteria with lysozyme in the presence of 3% ficoll was first reported by Crespi et al. (1962). The osmosensitive spheroplasts tended to rupture, but they retained phycobilins. Activities were not tested. Subsequently, Biggins (1967a) prepared spheroplasts from Phormidium luridum, which were able to photoassimilate CO_2. Anacystis nidulans, a much tougher organism to permeabilize than the filamentous species, was first permeabilized by Ward and Myers (1973) who combined lysozyme and EDTA treatments. The products, osmoresistant permeaplasts, were more active in FeCN photoreduction and less active in p-benzoquinone and carbon dioxide photoreduction than intact cells.

By carrying out the lysozyme-EDTA treatment of Anacystis in hypo-osmotic low ionic strength reaction medium, it is possible to speed up the permeabilization of Anacystis. The resulting cells are able to photoreduce ferricyanide and methylviologen at high rates. Furthermore, the electron transport rate more than doubles in the presence of the photophosphorylation cofactors (ADP, inorganic phosphate and Mg^{2+}. This implies ability to generate and maintain electrochemical gradients, sufficient to drive photophosphorylation, across the thylakoid membrane, suggesting that the permeaplast thylakoids are structurally and functionally intact (Papageorgiou and Lagoyanni, 1985; Papageorgiou, 1988a).

Phycobilisomes are made up of phycobiliprotein heterodimers ($\alpha\beta$) and linker polypeptides held together noncovalently by forces originating from hydrophobic and ionic interactions. Noncovalent, also, is the attachment of phycobilisomes to the stromal surface of the thylakoid membrane. Because of the nature of the association forces, phycobilisomes are stable only in high ionic strength media of kosmotropic (antichaotropic) electrolytes. In low ionic strength media, or in the presence of chaotropic electrolytes, they tend to detach from the membrane and to dissociate into oligomeric phycobiliprotein subunits.

Anacystis phycobilisomes, consisting of C-phycocyanin, allophycocyanin and allophycocyanin B, transfer electronic excitation and sensitize Chl a. As Fig. 4 shows, intact Anacystis emits primarily Chl a fluorescence, either upon direct excitation of the pigment (left panel, shaded area), or via phycobiliprotein sensitization (right panel, shaded area). When Anacystis is permeabilized, the low ionic strength of the suspension medium is communicated to the cytoplasm, destabilizing the phycobilisomes and interrupting the excitation transfer train. As a result, excitation with 436 nm light, absorbed primarily by Chl a, and to a lesser extent by the blue tail of C-phycocyanin absorption (left panel), results in the emission of both of C-phycocyanin fluorescence (F658) and of Chl a fluorescence

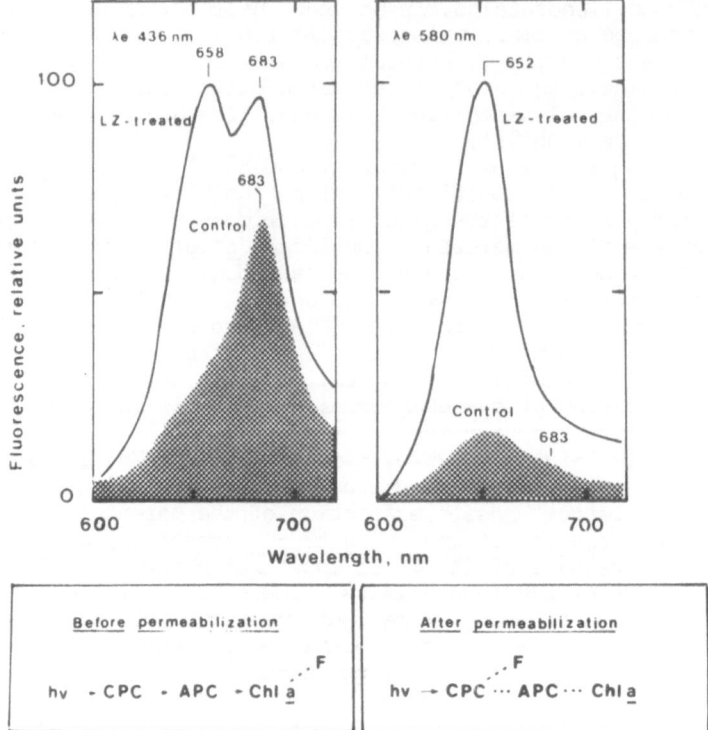

Fig. 4   Fluorescence spectra of intact Anacystis (shaded areas) and
of Anacystis permeaplasts. Left, excitation at 436 nm;
right excitation at 580 nm. The spectra have been adjusted
to equal heights. Excitation and detection half-band widths,
2 nm. Spectra were recorded with a Perkin-Elmer MPF-3L spec-
trofluorometer.

(F683). Excitation with 580 nm light, absorbed exclusively by C-phycocyanin (right panel), results in the emission primarily of phycobilin fluorescence (F652) plus a small contribution from phycobilin-sensitized Chl a
fluorescence (F683).

The relative proportions of phycobilin and Chl a fluorescence influence the location of the band maxima in Fig. 4. The main phycobilin emit-
ter of permeaplasts appears to be C-phycocyanin (Fmax = 645 nm). This suggests that phycobilisomes are not only detached but they are also dissociated. Detachment alone would result in a stronger contribution by allophycocyanin B (Fmax = 680 nm) to the fluorescence of permeaplasts.

The more intense fluorescence of permeaplasts can be used to follow the
kinetic of permeabilization (Fig. 5). In the experiment shown in Fig. 5, a-
liquots were withdrawn from the lysozyme reaction mixture and after resuspension in fresh, lysozyme-free medium (250 mM sorbitol, 250 mM KCl, 50 mM
Hepes.NaOH, pH 7.5) they were assayed for 685 nm fluorescence. Fluorescence increases with the duration of the enzymic reaction becoming maximal in 15 to 20 min.

Photosynthetic oxygen evolution in the presence of ferricyanide in the
suspension medium (ferricyanide Hill reaction) is another way to monitor
the progress of Anacystis permeabilization. As a trivalent anion, ferricyanide cannot enter the intact Anacystis cell and therefore it does not sup-

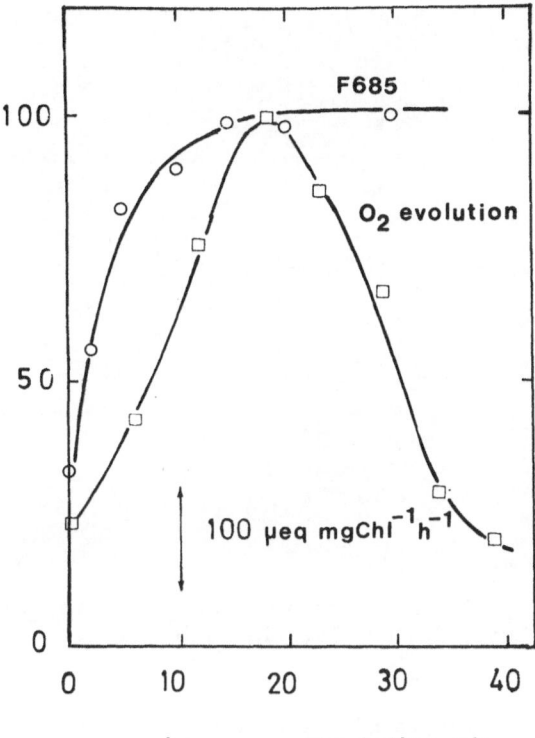

Fig. 5 Fluorescence intensity at 685 nm and rate of FeCN supported oxygen evolution as a function of the duration of lysozyme and EDTA treatment. Excitation, 580 nm. Other details, as in Fig. 4.

port photosynthetic oxygen evolution. Oxygen evolution accelerates as the fraction of permeabilized cells rises, but instead of reaching a plateau, it passes through a maximum and then declines (Fig. 5). The decline is due to inhibition of electron transport at a site between the water-splitting function and the pre-photosystem II donor known as Z and reflects the invasion of EDTA into the cytoplasm (Papageorgiou and Lagoyanni 1985). The non-zero rate of oxygen evolution at the beginning of the lysozyme reaction is due to carbon dioxide present in the reaction mixture. In spite of the formally different courses of 685 nm fluorescence and of the Hill reaction kinetics, the curves of Fig. 5 provide a clear and quantititave picture of the progress and the extent of Anacystis permeabilization.

After prolonged treatment with lysozyme and EDTA, electron transport in Anacystis becomes dominated by a DCMU and KCN insensitive photoinduced oxygen uptake, mediated by an endogenous autoxidizable post-photosystem I intermediate. The oxygen uptake is subject to control by electrostatic forces originating from fixed negative surface charges since screening and neutralization by cations suppress it. The chemical identity of the endogenous intermediate is unknown (Sotiropoulou et al., 1984).

Figure 6 presents a scanning electron view of <u>Anacystis</u> <u>nidulans</u> cells after 20 min incubation with lysozyme and EDTA. The permeabilized cells retain the typical elongated shape of Anacystis but their surface acquires a

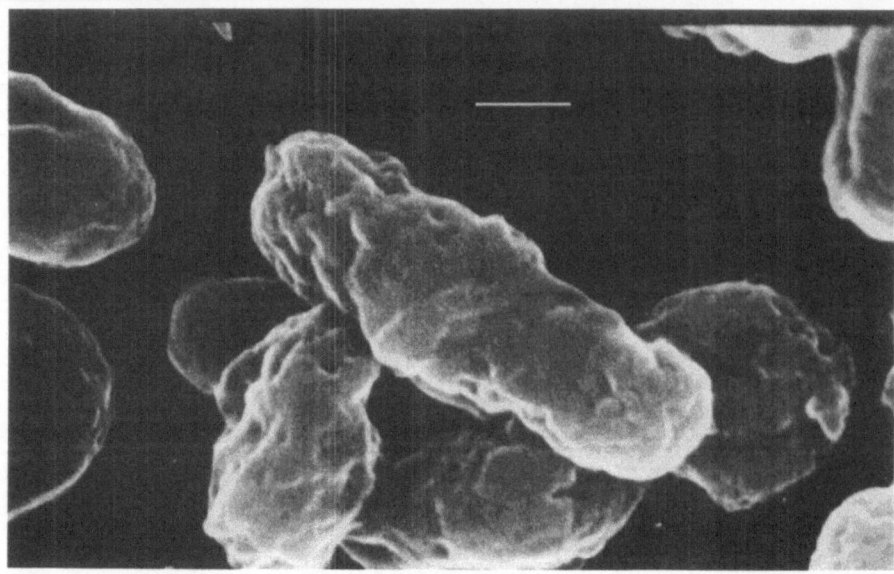

Fig. 6   Scanning electron micrograph of <u>Anacystis</u> <u>nidulans</u> permeaplast. Bar, 0.5 µm. Details as in Fig. 1.

rough appearance as a result EDTA activity on the outer membrane. Degranu-
lation of the outer membrane by EDTA treatment has been described in other
Gram negative bacteria (Gilleland <u>et al</u>., 1973). Transmission electron micrographs (not shown) of similarly treated cells reveal little change in
internal morphology, except for some distentionn of thylakoids.

(b) <u>Permeabilization</u> <u>by</u> <u>transient</u> <u>chilling</u>.

    <u>Anacystis</u> <u>nidulans</u> stands out, among cyanobacteria, for its unique re-
sponse to mild chilling. Chilling changes the physical state of membrane
lipids, and likely of other constituents, in the two kinds of membranes of
the cyanobacterium, the plasma membrane and the thylakoid membrane.
    Briefly reviewed, the chilling-induced phenomena are as follows: After
a relatively short incubation (typically 30 min) at near 0 ^{o}C, the viabili-
ty of Anacystis is impaired, photosynthetic electron transport is partially inactivated, and dialysable substances, such as pteridines (Forrest <u>et</u>
<u>al</u>. 1955), aminoacids (mainly glutamate; Jansz and Maclean, 1973; Rao <u>et</u>
<u>al</u>., 1977) and K^{+} ions (Ono and Murata, 1981a) are massively released.
Chilling causes bleaching at 458 nm (Brand, 1977) which was assigned to an
aggregation chánge of zeaxanthin, a major plasma membrane xanthophyll (Ono
and Murata, 1981a) and stimulates C-phycocyanin fluorescence, presumably
because of the functional detachment of phycobilisomes from the Chl <u>a</u> complexes of the thylakoid membrane (Schreiber <u>et</u> <u>al</u>., 1979; Ono and Murata,
1981a). Media protecting bacteria from cold shock had no effect on Anacystis (Jansz and Maclean, 1973). On the other hand, Ono and Murata (1981b)
found that monovalent or divalent metal cations (e. g. 20 mM KCl) selectively protected the water ⟶ p-benzoquinone electron transport activity,
but not the water ⟶ carbon dioxide activity. Protection was unrelated to
the osmotic pressure, since presence of 300 mM sucrose in the suspension
exerted no protective effect.
    Chilling is attended also by morphological changes. Freeze-fracture
microscopy gives visual evidence of particle displacements upon chilling
Anacystis below the temperature at which membrane lipids transit from the

Table 2. Transition temperatures for lipid state
change in Anacystis (liquid state to
phase separation state); in $^{\circ}$C.

Growth temperature	Transition temperature	
	Thylakoids[1]	Plasma membrane[2]
38	18-24	15
28	10-13	5

[1]Murata et al. (1975); [2]Ono and Murata (1981a)

liquid state to a mixed liquid-solid state (or phase separation state; Armond and Staehelin, 1979; Furtado et al., 1979; Ono and Murata, 1981c). When cells are frozen from above the transition temperature, particles appear uniformly distributed in the EF and the PF faces of the plasma and the thylakoid membrane. In contrast, particle-free domains appear on the fracture faces when the cells are equilibrated below the transition temperature before freezing. Both lateral and vertical movements have been invoked to account for the particle free areas, which are more extensive in the thylakoid than in the plasma membranes (Armond and Staehelin, 1979).

Chilling of higher plant chloroplasts of algae and of various filamentous cyanobacteria is not attended by similar phenomena. What sets Anacystis apart from the other photosynthetic organisms is the high content of its membranes in saturated fatty acids. Anacystis is devoid of the tri-unsaturated linolenic acid which is abundant in higher plants. The physical consequence is that changes in the physical state of membrane lipids of the type

Liquid State ⟶ Phase Separation State ⟶ Solid State)

occur at temperatures above 0 $^{\circ}$C, while in organisms with more unsaturated fatty acids they occur 30 to 50 degrees below zero (Murata et al, 1975). It was further established that the degree of fatty acid saturation and the manifestation of chilling phenomena are closely related to the culture temperature. Cells grown at elevated temperatures are characterized by a higher content in saturated fatty acids and are more cold-labile than those grown at lower temperatures.

The temperature at which thylakoid lipids in Anacystis transit from the liquid state to the phase separation state was determined by Murata et al. (1975) from breaks in the Arrhenius plots of such variables as the maximal hyperfine splitting of spin labeled fatty acids, Chl a fluorescence, O_2 evolution and State 1-State 2 transitions. For plasma membrane lipids, the temperature dependence of chilling inactivation of photosynthetic electron transport was used (Ono and Murata, 1981a). The results, displayed in Table 2, show that the thermotropic transition temperature for plasma membrane lipids lies approx. 10 $^{\circ}$C below that of thylakoid membrane lipids.

Thylakoid and plasma membranes become leaky to small molecules and ions presumably as a result of structural imperfections that are inherent in membranes in which solid and liquid lipid domains coexist. In addition, the different textures of the particle-free regions in the fracture faces of thylakoid and plasma membranes suggest that chilling may segregate also different amphipathic lipids in different membrane domains (Armond and Staehelin, 1979).

One important feature is that on lowering the temperature, thylakoids lose permeability control ahead of plasma membranes (i. e. at a higher temperature; see Table 2). Thus, the first consequence of chilling is the dissipation of the electrochemical potential across the thylakoid membrane (decoupling) and subsequently the inactivation of photosynthetic electron transport due to cell envelope permeabilization.

Permeabilization of the cytoplasmic and the thylakoid membranes of Anacystis by means of a chilling treatment is temporary. The membranes reseal when the cells are re-equilibrated above the transition temperature. Such reversible permeabilization could be a suitable way to load Anacystis cells with normally impermeant substances. No application, however, has been reported.

Schreiber (1979) attributes the stimulation of C-phycocyanin fluorescence upon chilling Anacystis cells to lateral displacements of Chl a complexes in the membrane. This increases the distance between the terminal phycobilisome chromophore (allophycocyanin B) and Chl a and decreases the probability of electronic excitation transfer (Foerster's R^{-6} law). A high negative thylakoid surface charge extends the liquid state of membrane lipids to lower temperatures and facilitates more extensive particle displacements. This is manifested by a further increase in C-phycocyanin fluorescence. It should be noted, however, that if the chilling resulted only in an increase of the Chl a - phycobilisome distance, then we would expect stimulation of allophycocyanin fluorescence (680 nm) and not as the experiment shows stimulation of C-phycocyanin fluorescence. (Schreiber et al., 1979; Ono and Murata, 1981a). We must infer, therefore, that chilling interrupts both the C-phycocyanin to allophycocyanin and the allophycocyanin to Chl a excitation transfers. This implies both detachement of phycobilisomes and dissociation into oligomeric phycobiliproteins.

The physico-chemical picture of the chilling-induced phenomena in Anacystis is still incomplete. One overlooked factor is the destabilization of hydrophobically-stabilized supramolecular structures at low temperature. A notable candidate is the phycobilisome, whose numerous subunits are held together by forces arising primarily from hydrophobic interactions.

APPLICATIONS OF ENZYMICALLY PERMEABILIZED CYANOBACTERIA

Chilling permeabilizes both the cell envelope and the thylakoid. As a result of the general uncontrolled permeability, only a small fraction of photosynthetic activity survives. In contrast, the rapid enzymic procedure permeabilizes the cell envelope alone, leaving the thylakoids structurally intact.

In enzymically permeabilized cyanobacteria, only the outer thylakoid surface is accessible to ionic solutes existing in the suspension medium, while the inner surface can be accessed with the help of specific ionophores. This characteristic makes permeabilized cyanobacteria a convenient model to investigate properties depending on the sidedness of the thylakoid membrane. In fact, enzymically permeabilized cyanobacteria are superior to preparations of oriented thylakoid vesicles (right-side-out and inside--out) since in the latter neither the sealing nor the orientation of the vesicles can be fully guarranteed. Furthermore, due to their ability to import, photoreduce and export ionic compounds, permeaplasts may find applications in biotechnological reaction schemes. In our laboratory, we have used Anacystis permeaplasts in order first to investigate the control

Table 3. Surface charge density of higher plant and
cyanobacterial thylakoid fragments.

Source of thylakoids	Coulomb.m^{-2}
Higher plants	
<u>Pisum sativum</u>[1]	- 0.035
<u>Spinacea oleracea</u>[2]	- 0.030
Cyanobacteria[3]	
<u>Anacystis nidulans</u>	- 0.100
<u>Phormidium luridum</u>	- 0.240
<u>Nostoc muscorum</u>, vegetative cells	- 0.105
heterocysts	- 0.137

[1]Chow and Barber (1980); [2]Kalosaka and Papageorgiou (1984);
[3]Kalosaka (1987).

of photosynthetic electron transport activity by electrolytes, and secondly to achieve sustained production and exportation of photoreducing power.

(a) <u>Ionic control of photosynthetic electron transport</u>

Thylakoid surfaces are negatively charged because the number of exposed anionic groups on them exceeds the number of exposed cationic groups.
Contributors are aminoacid carboxylates and phospholipid phosphates. Surface electricity magnitudes have been has estimated in thylakoid membrane
fragments, using methodology based on (a) the stimulation of 9-aminoacri-
dine fluorescence in thylakoid suspensions by metal cations (b) on reaction rates involving ionic electron acceptors and donors and (c) by particle electrophoresis. Results of these measurements were interpreted in
the frame of the Gouy-Chapman-Stern theory, which was developed for ideally planar and homogeneously charged surfaces and which views diffusible
ions as point charges. Biological membranes do not conform to the two-dimensional concept, their charges are discreet and localized, whereas ionic
solutes are always solvated. A great deal has been learned, nevertheless,
about the surface electric properties of biomembranes by applying the theory to artificial and to natural systems (McLaughlin, 1977; Barber, 1980,
1982).

The surface charge forces redistribution of ions in the liquid phase
in contact with the membrane. For an observer positioned a few nm away,
cations gathering near the surface appear to form a positive screen in
front of it. As these cations are visualized to make no permanent contact
with surface anions, formation of the screen amounts to the suppression
only of the surface potential, while the surface charge remains the same.
Permanent metal cation contacts with negatively charged groups of the surface are likely, however, in view of the appreciably large association constants of alkalis with carboxylates and sulfonates and of alkaline earths

with phosphates. Such contacts lower both the thylakoid surface charge and the thylakoid surface potential.

Screening and neutralization by monovalent cations may push the surface charge and potential asymptotically to zero, as their bulk phase concentration increases. In that limit, ion concentrations near the surface are no longer subject to electrostatic control and equal the bulk phase concentrations. In the case of divalent and of higher valence cations, however, sign inversion of membrane charge is conceivable since the fixed membrane surface anions are monovalent. When this happens, the concentrations of diffusible anions in the surface area may exceed the respective bulk phase concentrations.

Cyanobacteria thylakoids carry 3 times, or more, as much negative surface charge as higher plant thylakoids (Table 3). Accordingly, the requirement for counterion charge compensation in order to evoke photosynthetic oxygen evolution in the presence of anionic oxidants, such as ferricyanide, is more severe in of cyanobacteria than in isolated higher plant chloroplasts.

As reported by Sotiropoulou and Papageorgiou (1985) metal counterions stimulate the ferricyanide Hill reaction with concentration and valence dependencies conforming to the predictions of the Gouy-Chapman theory only up to a certain concentration. Above that, the Hill rate declines. Concentrations corresponding to peak rates are approx. 1 mM for $LaCl_3$, 30 mM for $CaCl_2$, and 300 mM for KCl. When the same data are plotted against the thylakoid surface potential, then the curves for the mono-, di- and trivalent metal cations cluster together in a narrow range of potentials. The form of these curves clearly indicates two competing processes: a rate stimulating process and a rate suppressing process. Both appear to depend primarily on membrane surface potential and to a lesser extent on the chemical nature of the cations(Papageorgiou, 1988b).

Although the rate stimulating process can be rationalized in terms of the counterion charge compensation, the rate suppressing one defies interpretation on simple electrostatic terms. As noted above, one of the oversights of the Gouy-Chapman-Stern theory is that it considers ions as point charges. In reality, however, ions occupy non-negligible volumes which are further enlarged by solvation. According to Hatefi and Hanstein (1969), the hydrogen atoms water molecules in the hydration sphere of an anion are directed in, while those in the hydration sphere of a cation are directed out. This makes the hydrated anion surface more hydrophobic and the hydrated cation surface more hydrophilic than the mass of pure water. Furthermore, an anion is the more hydrophobic the larger its volume and the lower its electric charge (i.e. the lower the electric field strength on its surface). According to Flewelling and Hubbel (1986) hydrophobic anions bind several orders of magnitude more strongly to lipid bilayers and translocate several orders of magnitude more rapidly through them than cations of similar structures

Solvated ions with low electric charge spread over a relatively large volume are called chaotropes (structure breakers) while those with high electric charge in a relatively small volume are called polar kosmotropes (structure makers; see Collins and Washabaugh, 1985). Chaotropic anions disrupt hydrophobically stabilized biological structures by interfering with and by rearranging hydrogen bonds. The ranking of anions according to chaotropic potency follows their order in the Hoffmeister series of neutral salt effectiveness in solubilizing proteins. In the example shown below, chaotropic potency increases from left to right.

$$SO_4^{2-} = HPO_3^{2-} < Cl^- < Br^- < NO_3^- < SCN^-$$

In the experiment shown in Table 4, we examined the effect of 0.6 M potassium ions on the photoreduction rate of DCIP by Anacystis permeaplasts with water and diphenylcarbazide (DPC) as electron donors. As the table illustrates, it depends on the associated anion whether K^+ ions will

Table 4. Effect of the chemical nature of the anion on the stimulation or inhibition of photoinduced electron transport in Anacystis permeaplasts by 0.6 M of K^+ ions.[1,2]

Added anion as K^+ salt	Relative electron transport rate			
	H_2O	DCIP	DPC	DCIP
None	100^3		100^4	
HPO_4^{2-}	195		171	
SO_4^{2-}	190		144	
Cl^-	149		110	
Br^-	139		91	
NO_3^-	63		42	
SCN^-	0		0	

[1]Sotiropoulou and Papageorgiou, 1986. [2]Sotiropoulou, 1987. [3]Actual rate, 50 uequiv.mgChl^{-1}.h^{-1}. [4]Actual rate, 100 uequiv.mgChl^{-1}.h^{-1}.

stimulate or will inhibit DCIP photoreduction. In terms of their effect, the examined anions rank in the same order as for chaotropic potency. Chaotropic effects are expressed at anion concentrations ranging from 0.01 to 1 M and they are additive over all species present in the solution. Surface area anion concentrations in that range are possible given the strong compensation of thylakoid surface charge in this experiment. Structural perturbation of thylakoid membranes by the anion chaotropes clearly emerges as the most likely physical cause for the observed inhibition of the photoinduced electron transport.

According to Collins and Washabaugh (1985) anions positioned left of chloride (in the sequence shown above) are polar kosmotropes, those positioned after it are chaotropes, while chloride itself has no effect the structure of water. Our results do not fully agree with this. Relative to sulfate and to phosphate, chloride clearly appears to have a chaotropic effect. Chaotropic inhibition is also evident in the absence of potassium salt. It must reflect destabilization of hydrophobically-stabilized thylakoid domains by the low ionic strength (Hepes.NaOH, 50 mM; pH 7.5) suspension medium.

A suitable marker for the inner thylakoid surface is the reaction center chromophore of photosystem I, known as P700. Its photo-oxidation is reported by a bleaching at 700 nm, and its dark reduction by intersystem intermediates by a recovery of the absorption loss. In cyanobacteria, the immediate electron donor to P700 is Cyt c-550, and to a lesser extent plastocyanin. Both these electron transporting proteins are negatively charged, mobile and dissociate easily from the membrane. In subthylakoid fragments,

Table 5. Effect of compensation of the negative electric charge on the inner thylakoid surface of Anacystis permeaplasts on the rate of electron donation to P700$^+$ in the dark by endogenous donors.[1]

| | Relative rate | |
Additions	Intact cells	Permeaplasts
None	100	100
KCl, 50 mM	100	123
KCl, 50 mM; valinomycin 3 µM	126	250

[1]Kalosaka et al., 1985

cations stimulate electron donation from plastocyanin to P700. Apparently the process is subject to electrostatic control. Permeaplasts are a suitable system to investigate whether this is true also in the case of intact thylakoids.

Employing Anacystis permeaplasts, we investigated whether potassium ions accelerate the dark reduction of photo-oxidized P700 (Table 5). Acceleration was observed only when KCl (50 mM) and the potassium ionophore valinomycin (2 µM) were present simultaneously. Acceleration depended on the concentration of KCl in a way suggesting a positive shift of the inner surface potential as the the cause. Added separately, valinomycin and KCl had no effect. The experiment clearly proves electrostatic control of electron donation to P700 in the thylakoid interior.

In summary, the principal mechanisms by which diffusible ions in the bathing liquid interact with thylakoid membranes are:

(i) Compensation of the negative surface charge by screening and by neutralization by cations.

(ii) Chaotropic and kosmotropic interactions of the associated anions with surface structures. Anion effects manifest only after drastic suppression of the negative surface potential.

Conceivably, an additional mechanism exists by which chaotropic and kosmotropic effects may be brought about by electrolytes present in the liquid media bathing thylakoid membranes. It pertains to the absorption of solvated anions to electroneutral domains of the membrane surface. No experiments in favor or against this mechanism have been reported.

(b) Biotechnological potential of permeabilized cyanobacteria

The photosynthetic apparatus of green plants, algae and cyanobacteria is a unique generator of low potential reducing power and of oxygen that essentially runs on sunlight and water. Because of its undisputable biotechnological potential, several laboratory models, capable of diverting photosynthetically produced reductants to the production of fuels, ammonia, special chemicals and electricity have been described and studied (see reviews by Rao and Hall, 1984, and Papageorgiou, 1987). None of them, however, evolved to a major application until now.

Table 6. Immobilization methods applied to enzymically
permeabilized cyanobacteria

Method	Species
Crosslinking with glutaraldehyde[1]	Phormidium luridum
Covalent attachment to glutaralde-hyde-crosslinked albumin[2]	Phormidium luridum, Anacystis nidulans
Entrapment in calcium alginate[3]	Anacystis nidulans

[1]Papageorgiou and Tzani (1979); [2]Papageorgiou and Lagoyanni (1986); [3]Papageorgiou et al. (1988)

Two salient problems complicate the prospect of the technological ex-ploitation of oxygenic photosynthesis: reactor compatibility and biocata-lyst stability. Reactor compatibility refers to the accessibility of cata-lytic sites by reactants present in the reactor and stability to the pre-servation of the catalytic function both in conditions of storage and in conditions of continuous activity. Since, as a rule, cells are not freely permeable to ions, the photosynthetic membranes which reside in the cyto-plasm are not readily accessible to reactants of that nature. To overcome this difficulty, it becomes necessary to use isolated chloroplast frag-ments. Indeed, such cell-free preparations were the first to be used in the early explorations of the technological potential of photosynthesis (Papageorgiou, 1979).

Thylakoid membranes are quite unstable in vitro but stability improves when their spatial degrees of freedom are partially or totally suppressed. This is known as "biocatalyst immobilization" and can be achieved either by chemical means, such as by the establishment of intramolecular and in-termolecular covalent crosslinks using bifunctional reagents (e.g. dialde-hydes, diimidoesters), or physically, such as by the adsorption of bioca-talysts on inert solid supports or by their entrapment in cavities of hy-drophilic gels.

The total reductant output (time integral of electron transport rate) of isolated chloroplasts improved significantly after glutaraldehyde cross-linking (Papageorgiou 1979, 1980). Even better results were achieved when chloroplasts were reacted with glutaraldehyde in the presence of bovine serum albumin at subzero temperature (Barbotin et al.m 1987). The increase in photoreductant output was roughly one order of magnitude, but it was still short of desired performance for a photosynthetic biocatalyst dedi-cated to the production of high volume-low value chemicals, such as hydro-gen and ammonia.

Inevitably, attention was turned to the inherently stabler intact pho-tosynthetic microorganisms, the algae, the cyanobacteria and the photosyn-thetic bacteria. The functional stability of intact photosynthetic micro--organisms, as well as their reactor handling, may be enhanced by immobilization. A wide variety of techniques is already available (see for example Fukui and Tanaka, 1982, and Papageorgiou, 1987, as well as references cited therein). Being incompatible with ionic reactants intact cells must be permeabilized before they are put into use.

Cyanobacteria have certain advantages, relative to other kinds of photosynthetic micro-organisms, as producers of exportable reducing power. First, being oxygenic, they are capable of using water in order to produce low potential reductants photosynthetically. Indeed, they are the only class of phototrophic prokaryotes thus endowed, since the green and the purple photosynthetic bacteria lack the water oxidizing complex. Secondly, they are easily permeabilized to ions by partial enzymic hydrolysis of the thin peptidoglycan layer, present at the base of their Gram negative cell wall, next to the cell membrane. Since they have no chloroplasts, as soon as the cell envelope is permeabilized, their thylakoids become accessible to the ionic solutes of the medium. In contrast, two permeability barriers need to be overcome in the case of the eukaryotic microalgae in order to achieve the same objective, the cell envelope and the chloroplast envelope.

In our laboratory, we applied a number of techniques to immobilize intact and permeable cyanobacteria (Table 6). The products were tested for survival of the photosynthetic activity, compatibility with ionic oxidants in batch and in continuous (open) reactors, functional longevity and temperature tolerance. Immobilized permeable cyanobacteria were able to import ionic oxidants, photoreduce them and export the products. Compared to free bacteria and to free or immobilized chloroplasts, the immobilized permeable cyanobacteria exhibit superior storage and working stability and tolerate higher ambient temperatures. Finally, since the immobilized biocatalysts are in the form of macroscopic solid particles (glutaraldehyde-albumin foams and calcium alginate beads) they are retained by sintered glass filters, and therefore they are suitable for continuously stirred open reactors.

Abbreviations BQ, p-benzoquinone; Chl, chlorophyll, EDTA, ethylenediamine tetra-acetate; FeCN, ferricyanide; Hepes, PD, N-2-hydroxyethylpiperazine-
-N'-2-ethanesulfonic acid; p-phenylenediamine.

Acknowledgements Partial financial support by the French-Greek Binational Scientific and Technological Co-operation Agreement and by the Community of Mediterranean Universities is gratefully acknowledged.

REFERENCES

Armond, P. A. and Staehelin, L. A., 1979, Lateral and vertical displacements of integral membrane proteins during lipid phase transition in Anacystis nidulans, Proceed. Natl. Acad. Sci. US., 76: 1901.
Barber, J., 1980, Membrane surface charges and potentials in relation to photosynthesis, Biochim. Biophys. Acta, 594: 523.
Barber, J., 1982, Influence of surface charges on thylakoid structure and function, Ann. Rev. Plant Physiol., 33:261.
Barbotin, J. N., Cocquempot, M. F., Larreta-Garde, V., Thomasset, B., Gellf, G., Clement-Metral, J. D., and Thomas, D., 1987, Meth. Enzymol., 135: 454.
Barbotin, J. N., and Thomasset, B., 1980, Immobilized organelles and whole cells into protein foam structures. Scanning and transmission electron microscopy observations, Biochimie, 62: 359.
Biggins, J., 1967a, Preparation of metabolically active protoplasts from the blue-green alga Phormidium luridum, Plant Physiol., 42: 1442.

Biggins, J., 1967b, Photosynthetic reactions by lysed protoplasts and par-
    ticle preparations from the blue-green alga Phormidium luridum,
    Plant Physiol. 42: 1447.
Binder, A., Tel-Or, E. and Avron, M., 1976, Photosynthetic activities of
    membrane preparations of the blue-green alga Phormidium luridum,
    Eur. J. Biochem. 67: 187.
Brand, J. J., 1977, Spectral changes in Anacystis nidulans induced by
    chilling, Plant Physiol., 59: 970.
Braun, V., Rehn, K., and Wolff, H., 1970, Supramolecular structure of the
    rigid wall of Salmonella, Serratia, Proteus, and Pseudomonas
    fluorescens: number of lipoprotein molecules in a membrane layer,
    Biochemistry, 9: 5041.
Chow, W. S., and Barber, J., 1980, Salt dependent changes of 9-aminoacri-
    dine fluorescence as a measure of charge densities of membrane
    surfaces. J. Biochem. Biophys. Methods, 3: 173.
Collins, K. D., and Washabaugh, M. W., 1985, The Hofmeister effect and the
    behavior of water ate interfaces, Quart. Rev. Biophysics, 18:
    323.
Crespi, H. L., Mandeville, S. E., and Katz, J. J., 1962, The action of ly-
    sozyme on several blue-green algae, Biochem. Biophys. Res. Comm-
    un., 9: 569.
Drews, G., and Weckesser, J., 1974, Function, structure and composition of
    cell walls and external layers, in: "The Biology of Cyanobacte-
    ria," N. G. Carr and B. A. Whitton, eds., Blackwell, Oxford.
Erokhina, L., Shubin, L., Klimov, V., and Proskuryakov, I., 1982, Rever-
    sible photoinduced changes of absorption and fluorescence yield
    of phycobilisomes related to the photoreduction of allophycocya-
    nin B, in: "Photosynthetic Prokaryotes," G. C. Papageorgiou and
    L. Packer, eds., Elsevier, New York.
Flewelling, R. F., and Hubbel, W. L., 1986, Hydrophobic ion interactions
    with membranes. Thermodynamic analysis of tetraphenylphosphonium
    binding to vesicles, Biophys. J., 49: 531.
Forrest, H. S., Van Baalen, C., and Myers, J., 1957, Occurence of pteri-
    dines in a blue-green alga, Science, 125: 699.
Fukui, S., and Tanaka, A., 1982, Immobilized microbial cells, Ann. Rev.
    Microbiol. 36: 145.
Furtado, D., Williams, W. P., Brain, A. P. R. and Quinn, P., 1979, Phase
    separation in membranes of Anacystis nidulans grown at different
    temperatures, Biochim. Biophys. Acta, 555: 352.
Gerhardt, B. and Trebst, A., 1965, Photosynthetische Reaktionen in lyo-
    philisierten Zellen des Blaualge Anacystis, Zeit. Naturforsch.,
    20b: 879.
Gilleland, H. E., Stinnet, J. D. Jr., Roth, I. L. and Eagon, R. G., 1973,
    Freeze-etch study of Pseudomonas aeruginosa: localization within
    the cell wall of an ethylene diamine tetracetate extractable com-
    pound. J. Bacteriol., 113: 417.
Glazer, A. N., 1984, Phycobilisome: a macromolecular complex optimized for
    light energy transfer. Biochim. Biophys. Acta, 768: 29.
Glazer, A. N., and Bryant, D. A., 1975, Allophycocyanin B (max 671, 618
    nm). A new cyanobacterial phycobiliprotein. Arch. Microbiol.,
    104: 11.
Hatefi, Y., and Hanstein, W. G., 1969, Solubilization of particulate pro-
    teins and nonelectrolytes by chaotropic agents, Proceed. Natl.
    Acad. Sci. U. S., 62: 1129.
Jansz, E. R., and Maclean, F. I., 1973, The effect of cold shock on the
    blue-green alga Anacystis nidulans, Can. J. Microbiol., 19: 381.
Kalosaka, K., 1987, Surface electric properties of photosynthetic mem-
    branes from cyanobacteria (cyanophytes), Ph. D. thesis, Univ. of
    Patras, Greece.
Kalosaka, K. and Papageorgiou, G. C., 1984, Surface electric properties of
    thylakoid fragments isolated from vegetative and heterocystous

cyanobacteria, in "Proceedings of the VIth International Congress on Photosynthesis," C. Sybesma, ed., Martinus Nijhoff/Dr W. Junk, The Hague.

Kalosaka, K., Sotiropoulou, G. and Papageorgiou, G. C., 1985, Retardation of electron donation to photosystem I in aged cyanobacteria and its reversal by metal cations. Biochim. Biophys. Acta, 808: 273.

Lambert, G. R, and Smith, G. D., 1981, The hydrogen metabolism of cyanobacteria (blue-green algae), Biol. Rev., 56: 589.

McLaughlin, S., 1977, Electrostatic potentials at membrane solution interfaces, in "Current Topics in Membranes and Transport, Bronner, F., and Kleinzeller, A., eds., Academic Press, New York.

Murata N., Troughton, J. H. and Fork, D. C., 1975, Relationships between the transition of the physical phase of membrane lipids and photosynthetic parameters in Anacystis nidulans, Plant Physiol., 56: 508.

Nichols, J. W., and Deamer, D. W., 1980, Net proton-hydroxyl permeability of large unilamellar liposomes measured by an acid-base titration technique, Proceed. Natl. Acad. Sci. US. 77: 2038.

Ono, T. A., and Murata, N., 1981a, Chilling susceptibility of the blue-green alga Anacystis nidulans: I. Effect of growth temperature. Plant Physiol., 67: 176.

Ono, T. A. and Murata, N., 1981b, Chilling susceptibility of the blue-green alga Anacystis nidulans. II. Stimulation of the passive permeability of the cytoplasmic membrane at chilling temperatures. Plant Physiol., 67: 182.

Ono, T. A. and Murata, N., 1981c, Chilling susceptibility of the blue-green alga Anacystis nidulans. III. Lipid phase of the cytoplasmic membrane, Plant Physiol., 69: 125.

Papageorgiou, G. C., 1977, Photosynthetic activity of diimidoester-modified cells, permeaplasts, and cell-free fragments of the blue-green alga Anacystis nidulans, Biochim. Biophys. Acta, 461: 379.

Papageorgiou, G. C., 1979, Molecular and functional aspects of immobilized chloroplast membranes, in "Photosynthesis in Relation to Model Systems", Barber, J., ed., Elsevier, Amsterdam.

Papageorgiou, G. C., 1980, Stabilization of chloroplast and subchloroplast particles, Meth. Enzymol., 69: 613.

Papageorgiou, G. C., 1987, Immobilized photosynthetic microorganisms. Photosynthetica, 21: 367.

Papageorgiou, G. C., 1988a, Rapid permeabilization of Anacystis nidulans to electrolytes, Meth Enzymol, Vol 167 (in press).

Papageorgiou, G. C., 1988b, Interactions of inorganic ions with oriented cyanobacterial thylakoids, in "Water and Ions in Biological Systems," P. Lauger, L. Packer, and V. Vasilescu, eds., Birkhauser Verlag, Basel.

Papageorgiou, G. C., and Lagoyanni, T., 1983, Effects of chaotropic electrolytes on the structure and electronic excitation of glutaradehyde and diimidoester-crosslinked phycobilisomes, Biochim. Biophys. Acta, 714: 323.

Papageorgiou, G. C. and Lagoyanni, T., 1985, Photosynthetic properties of rapidly permeabilized cells of the cyanobacterium Anacystis nidulans. Biochim Biophys Acta, 807: 230.

Papageorgiou, G. C. and Lagoyanni, T., 1986, Immobilization of photosynthetically active cyanobacteria in glutaraldehyde-crosslinked albumin matrix, Appl. Microbiol. Biotechnol., 23: 417.

Papageorgiou, G. C., Kalosaka, K, Sotiropoulou, G., Barbotin, J. N., Thomasset, B., and Thomas, D., 1988, Entrapment of ion-permeable cyanobacteria (Anacystis nidulans) in calcium alginate, Appl. Microbiol. Biotechnol., in press.

Papageorgiou, G. C. and Tzani, H., 1980, The action of lysozyme on glutaraldehyde-treated cells of the cyanobacterium Phormidium luridum, J. Appl. Biochem., 2: 230.

Philips, D. C., 1967, The hen egg-white lysozyme molecule. <u>Proceed. Natl. Acad. Sci.</u>, 57: 484.

Rao, K. K., and Hall, D. O., 1984, Photosynthetic production of fuels and chemicals in immobilized systems, <u>Trends in Biochem. Sci.</u> 2:1.

Rao, V. S. K., Brand, J., and Myers, J., 1977, Cold-shock syndrome of <u>Anacystis nidulans</u>, <u>Plant Physiol.</u>, 59: 965.

Robinson, S. J., DeRoo, C. S. and Yocum, C. F., 1982, Photosynthetic electron transfer in preparations of the cyanobacterium <u>Spirulina platensis</u>, <u>Plant Physiol.</u> 70, 154-161.

Schopf, J. W. and Walter, M. R. (1982) Origin and early evolution of cyanobacteria: the geological evidence, <u>in</u> "The Biology of Cyanobacteria," N. G. Carr and B. A. Whitton, eds., Blackwell, Oxford.

Schreiber, U., 1979, Cold-induced uncoupling of energy transfer between phycobilins and chlorophyll in <u>Anacystis nidulans</u>. <u>FEBS Letters</u>, 107: 4.

Schreiber, U., Rijgersberg, C. P., and Amesz, J., 1979, Temperature-dependent reversible changes in phycobilisome-thylakoid attachment, <u>FEBS Letters</u>, 104: 327

Sotiropoulou, G., 1987, Interactions of Ions with Oriented Cyanobacterial Thylakoids, Ph. D. thesis, University of Thessaloniki, Greece.

Sotiropoulou, G., Lagoyanni, T., and Papageorgiou, G. C. (1984) Effects of Ca^{2+} ions on light-induced electron transport activities of <u>Anacystis nidulans</u> permeaplasts and spheroplasts, <u>in</u> "Proceedings of the VIth International Congress on Photosynthesis," C. Sybesma, ed., Martinus Nijhoff/Dr W. Junk, The Hague.

Sotiropoulou, G., and Papageorgiou, G. C., 1985, Modulation of the Hill reaction rates by ions iteracting with the outer surface of cyanobacterial thylakoids, <u>in</u> "Ion Interactions in Energy Transfer Biomembranes", Papageorgiou, G. C., Barber, J., and Papa, S, eds., Plenum Press, New York.

Sotiropoulou, G. and Papageorgiou, G. C., 1986, Stimulation and inhibition of photosystem II electron transport in cyanobacteria by ions interacting with the cytoplasmic face of thylakoids. <u>Photosynt. Res.</u>, 10: 445.

Spiller, H., and Boeger, P., 1980, Photosynthetically active algal preparations, <u>Meth. Enzymol.</u> 69: 105.

Stanier, G. (1982) Foreword, <u>in</u> "The Biology of Cyanobacteria," Carr, N. G., and Whitton, B. A., eds., Blackwell, Oxford.

Stanier, G., and Cohen-Bazire, G., 1977, Phototrophic prokaryotes: the cyanobacteria, <u>Ann. Rev. Microbiol.</u>, 31: 225.

Ward, B. and Myers, J., 1972, Properties of permeaplasts of Anacystis, <u>Plant Physiol.</u>, 50: 547.

Williams, W. P., and Allen, J. F., 1987, State 1/State 2 changes in higher plants and algae, <u>Photosynth. Res.</u>, 13: 19.

Witholt, B., Heerikhuizen, van H., and De Leij, L., 1976, How lysozyme penetrates through the bacterial outer membrane, <u>Biochim. Biophys. Acta</u>, 443: 534.

POSSIBLE ROLE OF *psbJ* IN PHOTOSYSTEM II: SITE-DIRECTED MUTAGENESIS IN THE CYANOBACTERIUM, *SYNECHOCYSTIS* 6803

Karin J. Nyhus and Himadri B. Pakrasi
Plant Biology Program, Biology Department, Box 1137
Washington University, St. Louis, MO 63130, USA

INTRODUCTION

The unicellular transformable cyanobacterium Synechocystis 6803 is ideally suited for genetic analysis of the photosystem II complex in thylakoid membranes (1-3). Initially, we used a homologous probe from spinach chloroplasts to clone two neighboring genes psbE and psbF, encoding the α and the ß subunits of the cytochrome b_{559} protein of PS II. During our examination of the organization of these two genes and the neighboring DNA regions, we have discovered the presence of two open reading frames downstream of the psbF gene. Northern analysis indicate that a ~1kb mRNA, transcribed from the entire psbEFIJ region is present in spinach (4), Euglena (5), as well as Synechocystis 6803 (H. Pakrasi, unpublished observation). Since cotranscribed genes usually encode proteins of the same complex, these ORF's are thought to be related to PSII and are thus termed psbI and psbJ, respectively.

In this report we present genetic evidence for the possible involvement of the psbJ gene product in PS II function.

MATERIALS AND METHODS

Translational stop codons were created at the beginning of the psbJ ORF by site-directed mutagenesis of the cloned gene in the plasmid pSL136 (6). The gene region was tagged by a chloramphenicol (Cm) resistance gene cartridge inserted at the unique Nhe I site downstream of the psbEFIJ gene cluster (see fig. 1). It is known that a kanamycin resistance cartridge in the same position does not affect PSII function (1). Sequencing for the confirmation of the mutation was done by the Sanger dideoxy method the Sequenase (modified T7 Polymerase) enzyme (U.S. Biochemical). The mutant plasmids were transformed into the psbEFIJ-deficient Synechocystis strain T1297 (1), where the cloned gene integrated into the chromosome by homologous recombination. Transformants were selected by the addition of Cm to the growth medium.

Growth of the cultures (light intensity ~28 $\mu E/m^2/sec$) was monitored by absorbance at 730 nm (1). Oxygen evolution from whole cells was measured on a Clark-type electrode with ferricyanide and DMBQ as artificial electron acceptors (1). Addition of DCMU, when needed, was at 10 μM. Thylakoid proteins were analyzed on LDS-PAGE and used for immunoblotting with monospecific antibodies raised against the PSII reaction center proteins D1 and D2 (kind gifts of Dr. Y. Inoue, 7).

RESULTS

The organization of the psbE, psbF, psbI, and psbJ genes in Synechocystis 6803 is shown in Fig. 1. This arrangement is similar in all of the photosynthetic organisms examined so far (5). The sequence of the psbEFIJ region is shown in Fig. 2. The underlined sequences correspond to the consensus Shine-Dalgarno ribosome binding site in bacteria. The gene products of both psbI and psbJ are 39 amino acids long (~4kDa).

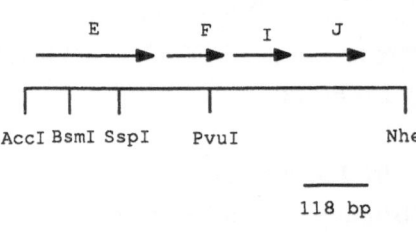

Figure 1. Arrangement of the psbEFIJ genes in Synechocystis 6803.

In Fig. 3, the predicted gene products are compared to those from other photosynthetic organisms. The underlined region corresponds to the hydrophobic domain revealed by hydropathy analysis (12). Evolutionarily, these proteins are highly conserved: Synechocystis and pea are 70% identical in the psbI product, and 64% in psbJ. The psbI protein contains a conserved possible glycosylation site (asn-arg-thr, 9). Interestingly, a number of proteins in cyanobacterial phycobilisomes are glycosylated (10). However, the role of such glycosylation in bacterial systems is currently unknown.

```
AGGAGTTTTTCATGGACAGAAATTCAAACCCAAACCGCCAACCGGTGGAATTGAACCGCACTTCT
        M  D  R  N  S  N  P  N  R  Q  P  V  E  L  N  R  T  S

TTATACCTGGGTCTATTGTTGGTGGCTGTGTTGGGGATTTTGTTCTCCAGCTATTTCTTT      psbI
 L  Y  L  G  L  L  L  V  A  V  L  G  I  L  F  S  S  Y  F  F

AACTAAACTTTTTTAATACGCAATTTAGGAGGCATGGTATGTTCGCAGAAGGCAGAATCC
 N  *                                  M  F  A  E  G  R  I  P

CTTTGTGGGTGGTGGGTGTAGTGGCCGGTATTGGCGCCATTGGTGTTCTAGGATTATTTT      psbJ
 L  W  V  V  G  V  V  A  G  I  G  A  I  G  V  L  G  L  F  F

TCTACGGAGCCTATGCTGGTTTAGGTTCTTCCATGTAATCGA
 Y  G  A  Y  A  G  L  G  S  S  M  *
```

Figure 2. Nucleotide and derived amino acid sequences of psbI and psbJ genes.

```
psbI
                     1          10          20          30          39
Synechocystis 6803   M - D R N S N P N R Q P V E L N R T S L Y L G L L L V A V L G I L F S S Y F F N
Euglena (ref. 5)     A Q T     K K - T -             W       I F     A V         I
Cyanophora (8)       V S Q   P - -     K             F W     I F     A           I
Pea (5)              T Q S     - - E   N             W       I F     A V     N F

psbJ
                     1          10          20          30          39
Synechocystis 6803   M - F A E - - G R I P L W V V G V V A G I G A I G V L G L F F Y G A Y A G L G S S M
Euglena (5)          S N S   N T     L   L T I I   L A   A L A       S   S         L
Cyanophora (8)       A N T - - G       A T         L A       I       G   S         I
Pea (5)              A N T - - T   I I   T         V V   L I         S   S         V
```

Figure 3. Homology among the protein products of the psbI and psbJ genes from various organisms.

The nucleotide sequence of the mutated region of the cloned psbJ gene is shown in Fig. 4. A translation stop codon was created at codon position four, normally coding for a glutamic acid. Thus only a very short peptide is expected to be produced, if at all, and its

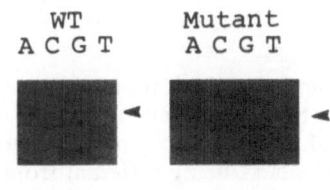

WT   Mutant
A C G T   A C G T

Figure 4. Site-directed mutation in the psbJ gene to introduce a stop codon (TAA) downstream of the start codon (ATG)

CAT GGT ATG TTC GCA GAA GGC AGA AT

CAT GGT ATG TTC GCA TAA GGC AGA AT

hydrophobic domain would be missing. This is expected to eliminate the ability of the protein to function.

The resultant mutant cyanobacterium, T203, was able to grow without glucose in the growth medium; hence, its PSII is functional . In Fig. 5, the growth of this mutant in liquid culture is compared with that of wild type cells. In the absence of glucose in the growth medium, the growth rate of T203 was ~55% of wild type, corresponding to doubling times of approximately 18 h for wild type and 32 h for T203. (This result was observed in three separate growth experiments). Addition of glucose returned the growth rate of the mutant to within 90% of that of wild-type cells. In comparison, the psbEFIJ deletion strain T1297 does not grow without glucose (1). Thus, as compared to the wild type cells, T203 has somewhat impaired PSII activity.

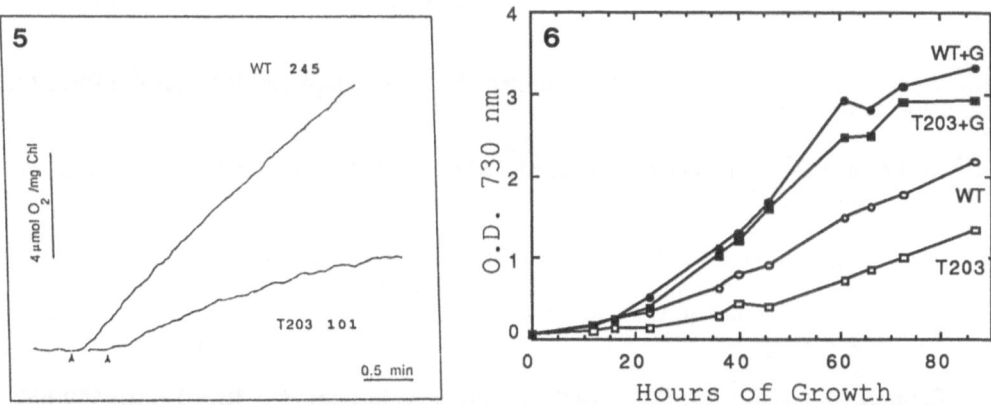

Figure 5 (left). Growth of wild type and T203 cells in liquid BG-11 media with and without glucose.
Figure 6 (right). PSII-mediated O_2 evolution (water to FeCN/ DMBQ) from wild type and T203 cells.

This difference between the wild type and the mutant strains prompted us to measure the light-induced PS II-mediated electron transfer rates from the normal donor water to the artificial electron acceptors ferricyanide and DMBQ. As shown in Fig. 6, the decrease in growth rate could be correlated with decreased O_2 evolution . T203 showed a 60% lower rate of O_2 evolution as compared to that of wild type. Both wild type and mutant PS II activities were completely inhibited by 10 µM DCMU.

DISCUSSION

Our results indicate that the psbI gene product is involved in, but not essential for, PS II activity. Site-directed mutagenesis has proven to be an excellent method for the study of the function of this protein, since traditional random mutagenesis methods would probably not detect such weakly photosynthetic mutants as being different from wild-type

Currently, we are also using similar methods to create mutations in the psbI gene. The protein encoded by this gene has recently been found in PS II preparations (11), but its functional role in PSII is still unknown.

Acknowledgments

The authors would like to thank Pamela De Ciechi, Richard Hodgson, Thomas Hartnett and Alan Eggenberger. This work was supported by the National Science Foundation Grant DMB-8704206 to H. B. P. K. J. N. is supported by a NSF predoctoral fellowship, an NSF travel grant, and an Olin fellowship from Washington University.

REFERENCES

1. Pakrasi, H. B., William, J.G.K. and Arntzen, C. J. (1988) EMBO J. 7, 325-32.

2. Vermaas, W.F.J., Williams, J.G.K., Rutherford, A.W., Mathis, P., and Arntzen, C.J. (1986) Proc. Natl. Acad. Sci. USA 83, 9474-77.

3. Janson, C., Debus, R.J., Osiewacz, H.D., Gurevitz, M. and McIntosh, L. (1987) Plant Physiol. 85, 1021-25.

4. Westhoff, P., Alt, J., Widger, W., Cramer, W.A., and Herrman, R.G. (1985) Plant Mol. Biol. 4, 103-110.

5. Cushman, J.C., Christopher, D.A., Little, M.C., Hallick, R.B. and Price, C.A. (1988) Curr. Genet. 13, 173-80.

6. Eggenberger, A. and Pakrasi, H.B. (1988) Anal. Bioch., submitted for publication.

7. Pakrasi, H.B., Riethman, H.C., and Sherman, L.A. (1985) Proc. Natl. Acad. Sci. USA 82, 6903-907.

8. Cantrell, A. and Bryant, D.A. (1987) In: Progress in Photosyn. Res. IV 10: 659-662.

9. Kornfeld, R. and Kornfeld, S. (1985) Ann. Rev. Bioch. 54, 631-664.

10. Riethman, H.C., Mawhinney, T.P., and Sherman, L.A. (1988) J. Bact. 170, 2433-2440.

11. Gray, J. (May, 1988) Proceedings of the Pennsylvannia State University Symposium, In press.

12. Kyte, J., and Doolittle, R. F. (1982) J. Mol. Biol. 157, 105-132.

CHARACTERIZATION OF cDNA CLONES ENCODING THE PEA CHLOROPLAST
RIESKE Fe-S PROTEIN

A. Hugh Salter, Barbara J. Newman and John C. Gray

Botany School, University of Cambridge, U.K CB2 3EA

INTRODUCTION

The cytochrome *b-f* complex of the chloroplast thylakoid membrane operates as a plastoquinol/plastocyanin oxido-reductase and is thus analogous to the mitochondrial cytochrome bc_1 complex with which it bears a large amount of structural and functional similarity[1] It consists of four components, which are cytochrome *f* ,cytochrome *b*-563, the Rieske Fe-S protein, and a 17kDa protein (subunit IV). In addition a number of small (~5kDa) have been suggested to be part of the complex[2],though this requires confirmation.The Rieske Fe-S protein may be removed from the isolated complex by hydroxyapatite chromatography in the presence of Triton X-100[3] and may be identified by a characteristic EPR signal[4] as well as by a size of 20kDa on SDS-PAGE.

The genes that encode the components of the complex are distributed between the nucleus and the chloroplast.The chloroplast genes *pet*A, *pet*B and *pet*D encode cytochrome *f*, cytochrome *b*-563 and subunit IV respectively.The Rieske Fe-S protein has been identified as the product of a nuclear gene. The protein is synthesised from poly-A RNA as a 26kDa precursor which is processed to the 20kDa mature protein,suggesting a cytoplasmic site of synthesis[5]. The mRNA for the Rieske Fe-S protein can be hybrid-selected using a cDNA clone and translated in rabbit reticulocyte lysate to give the 26kDa precursor, which can be imported and processed to the mature size in isolated spinach chloroplasts[6]. Although the nuclear gene itself has not yet been identified, cDNA clones from spinach[6,7] have been isolated and shown to encode the Rieske Fe-S protein by comparison of the derived amino acid sequence with partial protein sequence[8].

The aim of the work described in this paper was to isolate and characterise cDNA clones encoding the pea Rieske Fe-S protein.The cDNA clones are being used to study the organisation and expression of the nuclear gene that encodes the protein.

METHODS AND MATERIALS

Rieske Fe-S protein was purified from isolated pea cytochrome *b-f* complex using published methods[3,9] and antibodies were raised in rabbits. These were used to screen a cDNA library in λgt11 as described previously[10]. Inserts were subcloned into pUC and M13 vectors for restriction mapping and sequencing by the dideoxy method[11]. A clone encoding the Rieske Fe-S protein was used to rescreen the cDNA library to obtain further clones.

RESULTS

A cDNA library prepared from pea leaf poly-A RNA in λgt11 was screened with antibodies to the purified protein and ^{125}I protein A.Forty positive plaques from a total of 150,000 were identified from the cDNA library.The inserts from five of these were removed by digestion with *Eco* R1 and subcloned into pUC18. These divide into two classes containing 450bp and 850bp inserts.Clone R4 containing an 850bp insert was sequenced and identified as coding for the Rieske Fe-S protein on the basis of similarity to known amino acid[7] and nucleotide[8] sequence from spinach.This clone was not full-length, being short by approximately 50bp relative to spinach clones, and was used as a probe to rescreen the cDNA library. A total of 15 positive clones were identified from 75,000 plaques screened, and the inserts from 10 were subcloned. In addition to the two classes previously obtained, clones containing 900bp and 1000bp inserts were identified. A clone containing the 1000bp insert was sequenced and shown to be a longer, possibly full-length cDNA clone for the Rieske Fe-S protein encoding a precursor of 5.3kDa and a mature protein of 18.5 kDa and 178aa long. The deduced amino acid sequence is shown in figure 1 together with the comparison of

```
Ps                      MSSTIYLPQSPSQLCSGKSGISCPSIALLVKPTRTQMTG-RGNKGMKITCQ
So   MIISIFNQLHLTENSSLMASFTLSSATPSQLCSSKNGMFAPSLALAKAGRVNVLISKERIRGMKLTCQ

Ps   ATSIPADRVPDMSKRKTLNLLLLGALSLPTAGMLVPYGSFLVPPGLGSSTGGTVAKDAVGNDVVATEW
So   ATSIPADNVPDMQKRETLNLLLLGALSLPTGYMLLPYASFFVPPGGGAGTGGTIAKDALGNDVIAAEW

Ps   LKTHATGDVLYTGLKGDPAT-LVVGKDRTLATFAINAVCTHLGCVVPFNQAGNKFICPCHGSGTNDQG
So   LKTHAPGDRTLTQGLKGDPTYLVVESDKTLATFGINAVCTHLGCVVPFNAAENKFICPCHGSQYNNQG

Ps   RVVRGPAPLSLALAHCDVGVEDGKVVFVPWVGTDFRTGDAPWWS
So   RVVRGPAPLSLALAHCDVD--DGKVVFVPWTETDFRTGEAPWWSA
```

Figure 1. Comparison of pea(Ps) and spinach (So) Rieske Fe-S proteins. The proposed Fe-S binding domains are shown boxed and the *N*-terminal hydrophobic domains are underlined.The arrow indicates the processing site between the mature protein and the presequence.

the spinach and pea sequences. There was no significant similarity to the spinach sequence upstream of the methionine shown as the start of the pea precursor in figure 1, and this methionine is probably the start of the precursor protein in pea. There is some sequence identity between pea and spinach in the remainder of the presequence. These may be signals used for the transfer of the protein from the cytoplasmic site of synthesis to the thylakoid lumen, and for possible cleavage to an intermediate, as happens in *Neurospora*[12].There is little similarity to the common amino acid framework that has been suggested for nuclear encoded chloroplast proteins[13].The presequence is not as well conserved between pea and spinach as the mature protein. The *C*-terminal region of the mature protein containing a cysteine rich region is extremely well conserved between all known Rieske Fe-S proteins.This is presumed to be the active site where the 2Fe/2S prosthetic groups are bound. In particular the motifs CTHLGCV and CPCHGS are very strongly conserved, and the four cysteines are possible Fe ligands[14]. The histidine residues have also been implicated as possible ligands by ENDOR mearurements[15].

It is clear that the *C*-terminal Fe-S centre must be located on the lumenal side of the membrane, since it is involved in electron transfer to cytochrome *f*. The *N*-terminal hydrophobic region is proposed to be a membrane attachment domain, anchoring the protein into the membrane. This could be a transmembrane domain, either a single or double span, and would give the one of the two orientations shown in figure 2. A double span may however involve more of the mature protein than indicated by underlining in figure 1, and although the following redidues have a hydrophobic nature, they contain glycine and proline

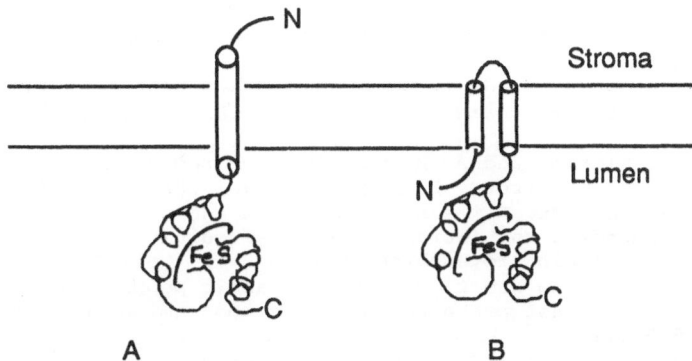

Figure 2. Possible orientations of the Rieske Fe-S protein within the thylakoid. A shows a single transmembrane span, and B a double.

residues that would destabilise an α helix. However, evidence from the laboratory of Neupert[16] indicates that *N*-and *C*-termini of the Rieske Fe-S protein are on the same side of the membrane, suggesting that the hydrophobic region may not completely span the membrane. In addition, the mitochondrial protein can be easily removed from the membrane with low concentrations of detergents. If the organisation is similar in chloroplasts, this would mean that the protein may extend only into the lumenal half of the lipid bilayer. This would mean that both the *N*-and *C*-termini would be in the lumen. These models for the orientation of the Rieske Fe-S protein will be experimentally investigated using partial proteolysis on right-side out and inside out thylakoid vesicles.

REFERENCES

1 G. Hauska, E. Hurt, N. Gabellini and W Lockau, Comparative aspects of quinol-cytochrome c/plastocyanin oxidoreductases. Biochim Biophys Acta 726:97 (1983).
2 E. Hurt and G. Hauska, Identification of the polypeptides in the cytochrome b_6-f complex from spinach chloroplasts with redox-center-carrying subunits. J Bioenerg Biomembr 14:405 (1982).
3. E. Hurt, G. Hauska and R. Malkin, Isolation of the Rieske iron-sulphur protein from the cytochrome b_6-f complex of spinach chloroplasts. FEBS Lett 134:1 (1981).
4. B. L. Trumpower, Function of the Iron-Sulphur protein of the cytochrome b-c_1 segment in electron-transfer and energy-conserving reactions of the mitochondrial respiratory chain. Biochim Biophys Acta 639:129 (1981).
5. J. Alt, P. Westhoff, B. B. Sears, N. Nelson, E. Hurt, G. Hauska and R. G. Hermann, Genes and transcripts for the polypeptides of the cytochrome b-f complex from spinach thylakoid membranes. EMBO J 2:979 (1983).

6. J. Tittgen, J. Hermans, J. Stephunn, T. Jansen, C. Jansson, B. Andersson, R. Nechushtai,N. Nelsonand R.G. Hermann,Isolation of fourteen nuclear-encoded thylakoid membrane proteins. Mol Gen Genet 204:258 (1986).

7. J. Stepphun, C. Rother, J. Hermans, T. Jansen, J. Salnikow, G. Hauska and R. G. Hermann, The complete amino-acid sequence of the Rieske Fe-S precursor protein from spinach chloroplasts deduced from cDNA analysis. Mol Gen Genet 210:171 (1987).

8. B. Pfefferkorn and H. E. Meyer, N-terminal amino acid sequence of the Rieske iron-sulfur protein from the cytochrome b_6-f complex of spinach thylakoids. FEBS Lett 206:233 (1987).

9. A. L. Phillips and J.C. Gray, Isolation and characterization of a cytochrome b-f complex from pea chloroplasts. Eur J Biochem 137:553(1983).

10. B. J. Newman and J. C. Gray, Characterisation of a full-length cDNA clone for pea ferredoxin-NADP$^+$ reductase.Plant Mol Biol 10:511 (1988).

11. F. Sanger,S. Nicklen and A. R. Coulson, DNA sequencing with chain-tertminating inhibitors. Proc Nat Acad Sci USA 74:5463(1977).

12. F. Hartl, B. Schmidt, E. Wachter, H. Weiss and W. Neupert, Transport into mitochondria and intramitochondrial sorting of the Fe/S protein of ubiquinol-cytochrome c reductase. Cell 47:939 (1986).

13. G. A. Karlin-Neumann and E. M. Tobin, Transit peptides of nuclear-encoded chloroplast proteins share a common amino acid framework. EMBO J 5:9 (1986).

14. U. Harnisch, H. Weiss and W. Sebald, The primary structure of the iron sulpur subunit of ubiquinol-cytochrome c reductase from Neurospera, determined by cDNA and gene sequencing. Eur J Biochem 149:95(1985).

15. J.F. Cline, B.M. Hoffman, W.B. Mims, E. LaHaie, D.P. Ballou and J. A. Fee, Evidence for N coordination to Fe in the [2Fe-2S] clusters of Thermus Rieske protein and pthalate dioxygenase from Pseudomonas J Biol Chem 260:3251 (1985).

16. H. Schagger, U. Borchart, W. Machleicht, T.A. Link and G. von Jagow, Isolation and amino acid sequence of the 'Rieske' iron sulfur protein of beef heart ubiquinol:cytochrome c reductase. FEBS Lett 219:161(1987).

CLONING THE GENE FOR CYTOCHROME C$_2$ FROM RHODOSPIRILLUM RUBRUM

S.J. Self, C.N. Hunter, and R.J. Leatherbarrow

Departments of Chemistry and Applied Biology

Imperial College, London SW7 2AY

Cytochrome *c2* of *Rhodospirillum rubrum* , a purple non-sulphur bacterium, is involved in the photosynthetic electron transport chain, and being the final component of the pathway, it donates electrons back to the bacterial photochemical reaction centre. The redox potentials of the various components in the electron transport system are critical to the electron flow, and are thought to be determined in part by the protein environment of the heme prosthetic groups. The crystal structure of *c2* has been resolved to 2A resolution (Salemme et al, 1973) and recently the structure of the reaction centre complex of *Rhodobacter sphaeroides* has been determined (Allen et al, 1987). This system is therefore well suited to a detailed study by protein engineering. Although the amino acid sequence of *c2* has been known for a long time (Dus et al, 1968), the DNA sequence is as yet unknown. The cloning strategy employed to isolate the cytochrome *c2* gene (designated *cycA*) from *R.rubrum* is described below.

Using the known amino acid sequence of *R.rubrum*, two families of deoxyoligonucleotides, termed P1 and P2, were designed to hybridise to the coding strand of the structural gene between amino acids Met 55 - Thr 63 and Met 91 - Glu 102 repectively. These oligonucleotides were designed as best guesses based on the codon usage of *R.rubrum*, with built in redundancies at three positions in each oligonucleotide (fig1).

477

Total restriction digests of *R.rubrum* chromosomal DNA were separated by agarose electrophoresis, and blotted onto nylon membranes. These Southern blots were then probed using the oligonucleotides end-labelled with ^{32}P. A HindIII fragment of approximately 3-4Kb was shown to contain the gene. In addition, this fragment was shown to contain a BamH1 site situated between the sites of hybridisation of the two probes, which could

Glu-Gly-Asp-Ala—Ala-Ala-Gly-Glu-Lys-Val-Ser-Lys-Lys-Cys-Leu-Ala-Cys-His-Thr-Phe-Asp-Gln-Gly-Gly-Ala-Asn-Lys-Val-Gly-Pro-Asn-Leu-Phe-Gly-Val-Phe-Glu-Asn-Thr-Ala-Ala-His-Lys-Asp-Asn-Tyr-Ala-Tyr-Ser-Glu-Ser-Tyr-Thr-<u>Glu-MET-Lys-Ala-Lys-Gly-Leu-Thr-Trp-Thr</u>-Glu-Ala-Asn-Leu-Ala-Ala-Tyr-Val-Lys-Asn-Pro-Lys-Ala-Phe-Val-Leu-Glu-Lys-Ser-Gly-Asp-Pro-Lys-Ala-Lys-Ser-Lys-<u>MET-Thr-Phe-Lys-Leu-Thr-Lys-Asp-Asp-Glu-Ile-Glu</u>-Asn-Val-Ile-Ala-Tyr-Leu-Lys-Leu-Lys

P1   3'-CTC/TTACTTC/TCGGTTCCCGGACTGG/CACCTG-5'

P2   3'-TACTGG/CAAGTTCGACTGG/CTTCCTACTACTC/TTAGCT-5'

Fig.1   Amino acid sequence of *R.rubrum* cytochrome *c*2 showing hybridisation sites of olgonucleotide probes.

be used to confirm the prescence of the gene once cloned. Size-fractionated *R.rubrum* DNA from a total *Hind*III digest was subsequently cloned into the *Hind*III site of the plasmid pACYC184 (Chang and Cohen,1978) and the mini-library generated screened using the radiolabelled probe P1, resulting in the recombinant plasmid pSS1 (fig2).

In an attempt to express the protein, plasmid pSS2 was generated by subcloning *Hind*III fragment containing the gene into the broad-host range vector pSUP202 (Simon et al, 1983). This vector contains the *mob* site necessary for transfer from the donor strain into a recipient strain by conjugative DNA transfer. Experiments are currently underway to transfer pSS2 into a strain of *R.sphaeroides* which is deficient in cytochrome *c*2 (Donohue et al, 1988). Unlike in *Rhodobacter capsulata*, cytochrome *c*2 is essential for photosynthetic growth of *R.sphaeroides* (Daldal et al, 1986:Donohue et al, 1988), enabling selection of progeny containing the plasmid under photosynthetic conditions.

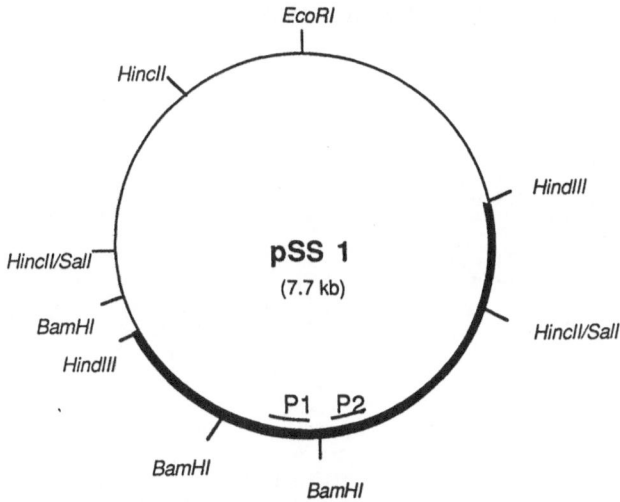

Fig.2 Restriction map of pSS1 containing the *cycA* gene from *R.rubrum*.

Sequencing the *R.rubrum cycA* gene is also currently in progress, and the initial sequence data indicates the prescence of a typical procaryotic signal sequence. This will be confirmed and homology with sequences from *R.sphaeroides* and *R.capsulatus* (Donohue et al, 1986; Daldal et al, 1986) will be determined once sequence data is complete.

479

References

Allen, J.P., Feher, G., Yeates, T.O., Kamiya, H. and Rees, D.C. (1987) *Proc. Natl. Acad. Sci.*84:6162.

Chang, A.C.Y. and Cohen, S.N. (1978) *J.Bact.* 134:1141.

Daldal, F., Cheng, S., Applebaum, J., Davidson, E. and Prince, R.C. (1986) *Proc. Natl. Acad. Sci.*83:2012.

Donohue, T.J., McEwan, A.G. and Kaplan, S. (1986) *J. Bact.* 168:962.

Donohue, T.J., McEwan, A.G., Doren, S.V., Crofts, A.R. and Kaplan, S. (1988) *Biochem* 27:1918.

Dus, K., Sletten, K. and Kamen, M.D. (1968) *J. Biol. Chem.*243:5507.

Salemme, F.R., Freer, S.T., Xuong, Ng.H., Alden, R.A. and Kraut, J. (1973) *J. Biol. Chem.* 248:3910.

Simon, R., Priefer, U. and Puhler, A. (1983) *Biotech.*784

PLANT PRODUCTION, PHOTOSYNTHESIS RATE AND RELATED CHARACTERS IN DOUBLED-

HAPLOID LINES OF <u>NICOTIANA</u> <u>TABACUM</u> SELECTED BY PHOTOSYNTHETICAL EFFICIENCY

H. Medrano, A. Pol, and E. Delgado

Lab. de Fisilogia Vegetal. Dept. de Biologia y C.S.
Universitat des les Illes Balears. 07071. Palma de Mallorca
Spain

INTRODUCTION

Because plant productivity and canopy photosynthesis are closely related, many attempts have been made to relate leaf CO_2 Exchange Rate (CER) to yield of crop plants and to select plants for high CER in order to improve yield (1, 2, 3).

Relationships between photosynthesis and yield has been object of wide discussion (4, 5) because there are both results showing good and poor correlation among plant production and CER (6, 7).

In a previous experiment, large populations of tobacco haploids were selected by survival in a low CO_2 atmosphere chamber after a 45 days selection period (8). Selected genotypes were diploidiced and self polinated. Seeds from these doubled haploid genotypes were sown to study the field performance and the photosynthetical characters.

MATERIAL AND METHODS

Material

Five doubled haploid selected genotypes (DH355, DH422, DH432, DH435 and DH451), and the source cultivar Wisconsin 38 (W-38) were studied in 1986 experiment and DH422 and DH451 in the 1987 one.

Methods.

Field assays were done in 1986 and in 1987 on the same place. In 1986 all plots were single rows of 8 plants spaced 60 cm from to another with 120 cm between rows. The experimental design consisted of three randomiced bolcks, each containing one plot of each genotype randomly distributed. The 1987 experiment was done on five randomiced plots with six plants per line in four of them for growth measurements, and with ten plants per line on the block for photosynthesis and Ribulose bisphosphate carboxilase (RuBPCase) analysis. Growth data were obtained at 26, 41, 48, and 57 days after planting.

Photosynthesis measurements were carry out on excised leaves, at the begining of flowering, in an open system of three perpex chambers in parallel (340 μE m^{-2} s^{-1}, 400 W Mazda lamps, 25°C, in water saturated air), conected with an IRGA Grubb Parsons 120.

Activities of RuBPCase were determined in crude leaf extracts. One g of leaf was weighted, clipped and inmediately frozen in liquid N_2, and stored at 80°C. Samples were homogenized in 3 ml of grinding buffer, then filtered and centrifuged (20000 g, 15 min) mantainning a temperature of 4°C. Extracts were reactivated with DTT 25 mM (2h 30 min, 30°C), and activity was measured according to the method of Lorimer et al. (9).

RESULTS AND DISCCUSION

Mean performance of selected dihaploid genotypes and Control showed in 1986 field assay that genotypes wich were selected by photosynthetic efficiency had significantly higher rates of dry matter production than the Control (tab. 1). Among ten measured traits, those more closely related to photosynthetic efficiency of the plants, leaf area, and leaf and total dry weight were significantly higher than Control in each one of the doubled haploid selected genotypes, with increases from 9.3 to 45.7 % in dry weight per plant in respect to Control.

These results confirms previously reported (8) and agree with the hypothesis of the selection criteria applied to obtain the haploids from wich the doubled haploids used in this assay were obtained.

Some selected genotypes, did not show any significant differences among them in ten measured characters (DH435 in respect DH422; DH435 in respect DH451 and DH451 in respect DH435) that agree with their common origen (W-38) and selection process of these plants.

Growth analysis of plant dry matter accumulation and leaf are per plant in 1987 experiment (tab. 2) shows that the maximun differentiation in biomass production and Net Assimilation Rate (NAR) between genotypes occurs at the 3rd growth period (fig. 1). Because in this interval, differences in capacity (Leaf Area Index, LAI) are not high enough, variations in biomass production and NAR could be understood in terms of intensity (net photosynthesis per plant) that again agree with the selection criteria applied to obtain these genotypes. DH451 shows the highest

Table 1. Mean performance and standar error of cultivar Control (W-38) and select lines, in leaf, stem, inflorescence and total dry weight (LDW, SDW, IDW and TDW), stem diameter (SD), leaf area (LA), first flower heigh (FFH), top height (TH), leaf number (LN) and days till flowering (DTF). 1986 filed assay.

Genotype	LDW*	SDW*	IDW**	TDW**	SD	LA**	FFH**	TH	LN	DTF**
			g/plant		mm	cm^2		cm		days
Control	96.6± 6.7	102.5± 8.2	20.0±3.0	219.0±15.9	37.5±0.9	1253± 48	134.4±2.9	194.8±2.3	26.1±0.9	55.5±1.3
DH355	123.5± 5.8	132.1± 8.3	16.9±2.2	272.5±14.0	38.7±0.9	1596± 48	138.5±4.2	199.0±2.3	25.4±1.0	59.3±1.0
DH422	124.2±10.1	132.4±11.6	39.6±6.7	296.2±21.5	38.2±1.0	1430± 69	117.5±5.2	197.2±3.0	26.5±1.8	50.3±1.3
DH432	111.3± 5.8	112.0± 5.4	16.1±1.7	239.4±11.4	35.9±0.8	1404± 61	137.6±3.9	195.5±2.6	23.9±0.3	58.6±1.0
DH435	139.6±13.0	150.3±13.2	29.1±5.9	319.0±29.8	40.8±1.9	1429±127	127.6±4.1	197.6±3.8	28.3±1.9	53.9±1.3
DH451	131.2±12.5	131.6±12.2	18.0±4.1	280.8±25.2	38.8±1.5	1503± 58	141.6±5.7	196.5±4.9	26.6±1.3	56.5±2.0

*, **, Significant at 0.05 and 0.01 probability level respectively according to the analysis of the variance (F test).

482

values of Crop Growth Rate (CGR), LAI and NAR. DH422 performs clearly lower, with values nearer to Control.

In respect to 1986 field assay, plant production at the last harvest shows similar values in DH422, but clearly higher in Control and DH451. In both experiments DH selected genotypes performed significantly higher than Control, with an average increase of 32 % (1986) and 25 % 1987 in respect to it. CER measurements (tab. 2) gave values closer to the Control ones obtained in previous experiments (8). This lack of correspondence with previosuly reported resouls and among plant production and CER measurements could be explained according to the thesis of Zelitch referring to the difficulty in obtaining good relationships between net CO_2 exchange rate and dry weight accumulation in the filed, from a few instantaneous measurements.

Tab. 2. Plant production parameters. Mean values of: total dry weight per plant (DW, in g) and leaf area per plant (LA in dm^2) in 1987 experiment. Mean values of dry weight and leaf area at the end of the experiment followed by the same letters are not significantly different at the 0.05 level according to Duncan's multiple range test.

	DAY 26		DAY 41		DAY 48		DAY 57	
	DW	LA	DW	LA	DW	LA	DW**	LA**
Control	17.4±5.8	28.8±8.3	138.5±24.3	106.0±4.5	197.0±17.5	222.5±10.6	258.7±9.4b	264.2±26.2b
DH422	20.1±5.5	34.2±8.4	136.3±10.1	99.1±3.2	194.9±27.8	241.9±12.1	290.9±10.2b	288.1±21.1ab
DH451	23.0±6.3	36.5±8.4	170.8±9.5	122.0±4.4	280.7±26.1	250.3±18.0	353.7±17.3a	324.5±26.0a

** Significant at 0.01 probability level according to the analysis of the variance (F test).

Tab. 3. Leaf characteristics. Total leaf chlorophyl (Chl), total leaf nitrogen (N), CO_2 exchange rate (CER), ribulose 1,5 bisphosphate carboxilase (RuBPCase) activity. Mean values in columns followed by the same letter are not significantly diffrent at the 0.05 level, according to Duncan's multiple range test. Columns without subindex correspond to mean values without statistical significant differences.

	Chl	N**	C.E.R.	RuBPCase*
	mg Chl g^{-1}DW	mg N g^{-1}DW	mg CO_2 $dm^{-2}h^{-1}$	μmolCO_2 $min^{-1}mg^{-1}$N
Control	12.1±0.3	42.6±2.3c	12.1±0.6	0.42±0.07b
DH422	13.8±0.6	48.9±1.7b	12.2±0.3	0.54±0.05a
DH451	13.7±0.6	52.4±1.6a	12.4±0.4	0.31±0.01b

*,** Significant at 0.05 and 0.01 probability level according to the analysis of the variance (F test).

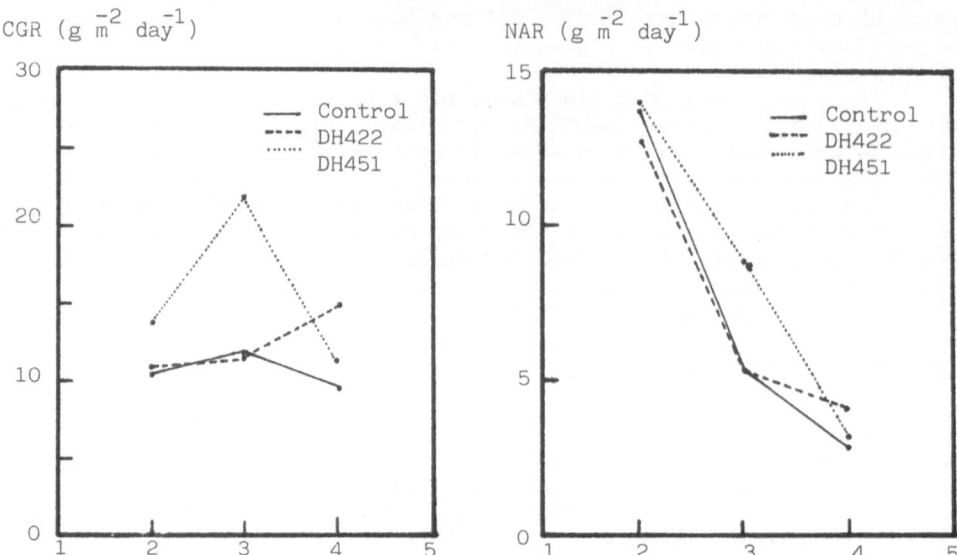

CGR (g $\overline{m}^2$ day^{-1}) NAR (g $\overline{m}^2$ day^{-1})

Figure 1. CGR and NAR values during the growth period (1987 experiment).

Leaf nitrogen content that shows highest values in selected geno-types is directly related with RuBPCase concentration in many different plants (10). This coincidence suggest that N leaf content could be rela-ted with the capacity of a plant to survive the selection coditions.

Nevertheless RuBPCase activity in DH451 is clearly lower than Con-trol, and significantly different in respect to DH422, showing a lack of correspondence with CER values, but also showing, the interest and posi-bilities of androgenic haploids as a source of genetic variability.

Reported results confirms that selected genotype yields, in two consecutive harvest (1986,1987) were significantly higher than the sour-ce cultivar, but does not clarify completely the causes that justify that increase in dry matter accumulation. Further experiments will be done to understand it and to contributed to the knowledge of photosynthesis and plant production relationships.

REFERENCES

1. C. J. Nelson, K. H. Asay and L. D. Patton, Crop Sci. 15:629/635 (1975)
2. T. M. Crosbie, R. B. Pearce and J. J. Mock, Crop Sci. 21:629/631(1981)
3. J. D. Mahon and S. L. A. Hobbs, Crop Sci. 21:616/621 (1981).
4. G. M. Dornhoff and R. M. Shibles, Crop Sci. 10:42/45 (1970).
5. A. L. Chrysty and L. A. Porter, in: "Photosynthesis Applications to Food and Agriculture", Govindjee, ed.,Academic Press, New York (1982).
6. R. M. Gifford and L. T. Evans, Ann. Rev. Plant Phys. 32:485/509 (1971).
7. I. Zelitch, BioScience 32:796/802 (1982).
8. H. Medrano and E. Primo Millo, Plant Physiol. 79:505/508 (1985).
9. G. H. Lorimer, M. R. Badger and T. J. Andrews, Anal. Biochem. 78:66/75 (1977).
10. R. C. Huffaker, in: "Encyclopedia of Plant Physiology, vol. 14A", D. Boulter, B. Parthier, ed., pag. 370/400, Springer Verlag, Berlin (1982).

CHLOROPLAST DEVELOPMENT IN T-DNA GENE 4 TRANSFORMED CALLI

R. Valcke[1], S.E. Beinsberger[2], H.A. Van Onckelen[2] and H. Clijsters[1]

[1] : Dept. S.B.M., Limburgs Universitair Centrum,Universitaire Campus, B-3610 Diepenbeek, Belgium

[2] : Dept. Biology, University of Antwerp, U.I.A., Universiteits-plein, B-2610 Wilrijk, Belgium

ABSTRACT

The first results are presented of a study on chloroplast development in transformed cell lines obtained by incorporation of chimeric T-DNA gene 4 constructions in the genome of Nicotiana tabacum cv. petit Havanna SR1 (gene 4 is coding for isopentenyltransferase). These chimeric genes were constructed in the pGV831 vector and introduced in the Ti-plasmid vector pGV 2260 of Agrobacterium tumefaciens[1]. Since pGV831 contained a selectable marker (Pnos-nptII, resistance against kanamycine) the transformed cells could be selected on a medium containing kanamycin and both auxins (IAA) and cytokinins (BAP). This procedure allowed to compare the activity of different chimeric genes in an identical background. The preliminary experiments indicate that the development of functional chloroplasts is cytokinin (ZR) dependent.

INTRODUCTION

The development of the photosynthetic apparatus in higher plants is regarded as the result of the coördination of differential gene expression of organelle constituents, encoded in nuclear and plastid genome, in dependence of the external factor light. The synthesis, transport and assembly of plastid constituents occur both in light and darkness. Some of this processes are largely stimulated by light, while others are triggered by it. The light control of gene expression of the major plastid polypeptides, the large and small subunits of Ribulose-1,5-bisphosphate Carboxylase/Oxygenase (RubisCO), the Light-Harvesting Chlorophyll a/b Protein (LHCP) and the 32 kDa herbicide binding protein, has been reviewed[2]. The involvement of endogeneous factors such as phytohormones (growth regulators) on the chloroplast development is greatly unknown[3]. Both transcription and translation have been suggested as processes stimulated by cytokinins[3]. The modulation of gene expression by phytohormones and the interaction between growth regulator and light involved in the chloroplast development has been studied by Parthier et al.[4], using developing pumpkins cotyledons and senescing leaf segments.

Our present contribution concerns the possible involvement of phytohormones on the development of the photosynthetic apparatus. The model system consists of plant cell lines obtained by incorporation of chimeric T-DNA gene 4 constructions in the genome of Nicotiana tabacum cv. petit Havanna SR1.

MATERIALS AND METHODS

Transformed cell lines were obtained by incorporation of chimeric T-DNA gene 4 constructions in the genome of Nicotiana tabacum cv. petit Havanna SR1. The gene 4 is coding for isopentenyltransferase. These chimeric genes were constructed in the pGV831 vector and introduced in the Ti-plasmid vector pGV 2260 of Agrobacterium tumefaciens[1]. Since pGV831 contained a selectable marker (Pnos-nptII, resistance against kanamycine) the transformed cells could be selected on a medium[5] containing kanamycin and both auxins (IAA) and cytokinins (BAP). The gene 4 constructions are defined as follow:
- line 831: control line, containing only the selectable marker;
- line 2449: gene 4 coupled to the octopine promotor;
- line 2488: gene 4 coupled to the promotor of the small subunit of Rubis-CO, (pSSU);
- line 2492: gene 4 coupled to the nopaline promotor.

The content of endogeneous cytokinin (ZR) and auxin (IAA) are determined as described by Weiler[6]. The chlorophyll measurements are done according to Lichtenthaler and Wellburn[7]. Low temperature in vivo fluorescence was recorded on a home made spectrofluorimeter. Photochemical activities were measured according to Allen and Holmes[8]. Samples for electron microscopy were prepared as described by J.R. Lawton[9].

RESULTS AND DISCUSSION

The expression of gene 4 was examined by analysis of the callus zeatine-riboside (ZR) content and by growth experiments on hormone-free medium. The three gene 4 containing lines were able to grow on hormone free medium. Their endogeneous ZR and IAA content increased significantly during the first 15 days compared to the control line (tabel 1, except line 2492, data not shown). The control line grows only when hormones (NAA, BAP) were supplied to the medium. The ZR and IAA content depend on the chimeric construction. The pSSU-gene 4 construction (line 2488), which is light regulated, results in a much lower ZR content (about 1/70) and a lower IAA content (1/3) compared to the octopine line (line 2449). This result clearly indicate a promotor dependent expression of gene 4. Moreover, the different activities of the chimeric genes results in a different growth pattern and macroscopical morphology (data not shown).

When grown in 16h light (30µmol $m^{-2}sec^{-1}$)/8h dark cycle on BAP/NAA containing medium, the Chl content (expressed on a fresh weight basis) increases in cell lines 831 and 2449, but decreases in lines 2488 and 2492 during the first 10 days. Afterwards, we observe a decrease in Chl content in lines 831 and 2449 during the following 10 days. The two other lines show a heterogeneous pattern. Grown on a medium supplemented with ZR, an increase in Chl content during the first 10 days is only observed in line 831. In the 2449 and 2488 lines the Chl content declines slightly during the same period, until day 20, followed by an increase noticed only in line 2488. The 2492 line shows a heterogeneous pattern. A second observation is that in the control line and in the octopine line, the presence of ZR in the culture medium induces a higher Chl content throughout the whole period. In the two other lines, this is only observable during the first 10 to 15 days. After this period the opposite occurs. The Chl a/b ratio increases only in line 831 under both culture conditions except for a small decline on day 10 (table 3). In the three other lines, irrespective of the culture conditions, the course of the Chl a/b ratio during the observed period is very irregular. The results of

Table 1. Endogeneous ZR and IAA content of three transformed cell lines as a function of time (days). ZR: expressed as pmol ZR-eq./gr.fresh weight IAA: expressed as pmol IAA/gr.fresh weight.

Callus time (days)	pGV 831		pGV 2449		pGV 2488	
	ZR	IAA	ZR	IAA	ZR	IAA
2	15	0.83	87	11.81	1043	3.61
8	34	2.19	205	19.36	1533	50.02
11	14	3.13	165	14.21	12287	42.33
21	8	1.52	6	3.56	1390	33.05
28	4	0.27	2.5	3.51	92	4.74
37	2	0.19	3.5	6.50	184	5.01

Table 2. Total Chl content, expressed as μgr./gr.fresh weight in function of time. NB: NAA/BAP medium; NB+ZR: NAA/BAP medium supplemented with ZR; (-): no data available.

Callus time (days)	pGV 831		pGV 2449		pGV 2488		pGV 2492	
	NB	NB+ZR	NB	NB+ZR	NB	NB+ZR	NB	NB+ZR
5	9+1	12+2	13+2	34+5	25+6	37+5	38+9	45+6
10	15.6+0.6	24+4	22+1	29+3	19+1	22+9	26+8	33.5+0.4
15	7.4+0.7	20+3	17+2	29+3	31+3	21+2	39+2	48+4
20	9+3	23+7	8+3	26+4	25+2	19+2	38+2	19+3
30	18+3	28+6	-	-	33+9	28+6	24+2	20+2

Table 3. Chl a/b ratio in function of time for the different cell lines. Abbreviations as in table 2.

Callus time (days)	pGV 831		pGV 2449		pGV 2488		pGV2492	
	NB	NB+ZR	NB	NB+ZR	NB	NB+ZR	NB	NB+ZR
5	1.1+0.2	1.9+0.2	2.8+0.3	2.5+0.4	2.4+0.7	3.1+0.4	2.0+0.5	3.5+0.4
10	0.82+0.07	0.89+0.02	1.7+0.4	2.5+0.7	2.0+0.2	3+1	3+2	4.7+0.8
15	1.0+0.3	2.21+0.08	2.2+0.1	2.8+0.3	4.8+0.2	3.8+0.2	4.1+0.4	4.4+0.3
20	1.0+0.2	2.9+0.4	1.9+0.2	2.9+0.3	3.7+0.1	3.23+0.08	3.74+0.07	3.4+0.3
30	3.0+0.5	3.1+0.4	-	-	3.5+0.2	3.4+0.2	2.92+0.07	3.0+0.1

the pigment analysis indicate that it is impossible to draw any conclusions about the distribution of the chlorophylls, the chlorophyll-protein complexes and the effect of supplementary added ZR.

The in vivo low temperature fluorescence spectra of the different cell lines show two emission maxima: 685 and 728 nm and a shoulder on 692 nm. Compared to the spectrum of a leaf (e.g. Phaseolus vulgaris: 685 and 735, 695 respec.) there is a shift of the long wavelenght emission band towards shorter wavelenght. The in vivo absorption spectra of the 831 and 2488 cell lines show a maximum and that of the 2492 line a minimum at 750 nm, a feature not observed in an intact leaf (data not shown). This results indicate that the organization of the photosynthetic apparatus in the transformed cell lines is probably different from this observed in normal tissue (intact leaf).

Electron microscopic observations show no qualilative differences between the different cell lines. Compared to normal leaf material, all lines contain very few plastids, small in size but rich in starch. A remarkable feature of the transformed cells is the presence of numerous peroxisomes containing large crystals of catalase.

As a general conclusion, based on this preliminary results, we can only say that there could be an effect of cytokinin on the development of the photosynthetic apparatus. The effect is most probably not a direct one. It depends clearly on the degree of the expression of gene 4 (under control of the attached promotor), in other words on the final concentration of cytokinin present in the cell.

ACKNOWLEDGMENTS

The authors wish to thank Ms B. Vanacken and E. Plaum for technical assistance. The work is supported by grant nr. 290009.87 from the F.K.F.O., a division of the National Found for Scientific Research, Belgium.

REFERENCES

1) R. Deblaere, B. Bytedier, H. De Greve, F. Deboeck, J. Schell, M. Van Montagu, and J. Leemans. (1985), Nucl.Ac.Res., 13:4777.
2) E.M. Tobin and J. Silverthorne. (1985), Ann.Rev.Plant Physiol., 36:593.
3) B. Parthier. (1979), Biochem.Physiol.Pflanzen, 174:173.
4) B. Parthier, J. Lehmann, S. Lerbs, R.A. Weidbase, and R. Woolgiehn. (1987) UCLA Symp.Mol.Biol., 44:391.
5) R.T. Fraley, S.G. Rogers, R.B. Horsch, J.S. Flick, S.P. Adams, M.C. Bittner, L.A. Brand, C.L. Fink, J.S. Fry, G.R. Galuppi, S.B. Goldberg, N.L. Hoffmann ,and S.C. Woo. (1983), Proc.Natl.Acad.Sci., 80:4803.
6) E. Weiler. (1980), Planta, 149:155.
7) H.K. Lichtenthaler and A.R. Wellburn. (1983), Biochem.Soc.Trans., 11:591.
8) J.F. Allen and N.G. Holmes. (1986), in "Photosynthesis Energy Transduction a practical approach.", H.F. Hipkins and N.R. Baker, Ed., IRL Press, Oxford,
9) J.R. Lawton. (1986), J. Microscopy, 144:201.

ABBREVIATIONS

BAP: 6-benzyl-aminopurine; IAA: indol-acetic-acid; NAA: α-naphtyl-acetic acid; ZR: zeatine riboside; Chl: chlorophyll.

PHOTOSYNTHETICALLY ACTIVE CHLOROPLASTS IN PETUNIA COROLLAS

David Weiss, Mordechay Schonfeld and Abraham H. Halevy

The Hebrew University of Jerusalem, Faculty of Agriculture
P.O.Box 12, Rehovot 76100, Israel

INTRODUCTION

Flower buds of most plants are initially green and their typical colors develop gradually during flower growth. The presence of Chl is usually not evident in mature flowers and it is assumed that chloroplasts present in the early stages are transformed into chromoplasts[1]. Photosynthetically active chloroplasts were previously reported to occur in flowers[2,3], but in all these cases the petals examined were green. In the present work we found photosynthetically active chloroplasts in pink Petunia corollas during all flower developmental stages. The photosynthetic activity of the corolla chloroplasts was compared with chloroplasts from green leaves.

Fig. 1. Different developmental stages of the Petunia flower. Sepals were removed from one side of each flower to expose the corolla

MATERIALS AND METHODS

Plants of Petunia hybrida (cv. Hit Parade Rosa) were grown in a greenhouse (18/27 night/day temperatures) at a light intensity of 500 to 600 μE m^{-2} s^{-1}. Flowers not used for experiments were removed from the plants twice a week and plants were trimmed monthly so as to keep them compact. Chl was determined spectroscopically in dimethylformamide extracts. Anthocyanin was extracted in 1% HCl in methanol and determined by the absorbance at 530 nm and 657 nm[4]. Chloroplasts from corollas and leaves were isolated essentially as previously described[5]. Electron transport in isolated chloroplasts was measured with an oxygen electrode, and Chl fluorescence was measured with a laboratory assembled apparatus[5]. The fluorescence induction time was measured essentially as described by Malkin et al[6]. Thin sections for electron microscopy were prepared and examined as previously described[7].

RESULTS AND DISCUSSION

Figure 1 shows different developmental stages of the Petunia flower, starting from a small green bud (stage 1) and ending with the fully developed pink corolla (stage 7). The Chl content of the corolla was found to increase during early developmental stages, reaching a maximum just before anthesis (stage 6). At this stage the anthocyanin content of the corolla also reached maximum (Fig. 2). Chl synthesis did not keep up with the corolla growth rate, so that its concentration started decreasing during the advanced stages of flower growth. The maximal Chl concentration in the corolla was about 40% of that of green leaves. Electron micrographs of sections made in Petunia corollas showed the presence of chloroplasts during all stages. Chloroplasts in the corolla were not observed to change into chromoplasts, and their growth and development largely paralleled the flower development. The change in corolla coloration, from light green to dark pink did not stem from Chl degradation but was due to synthesis of epidermal anthocyanins which masked the presence of green chloroplasts.

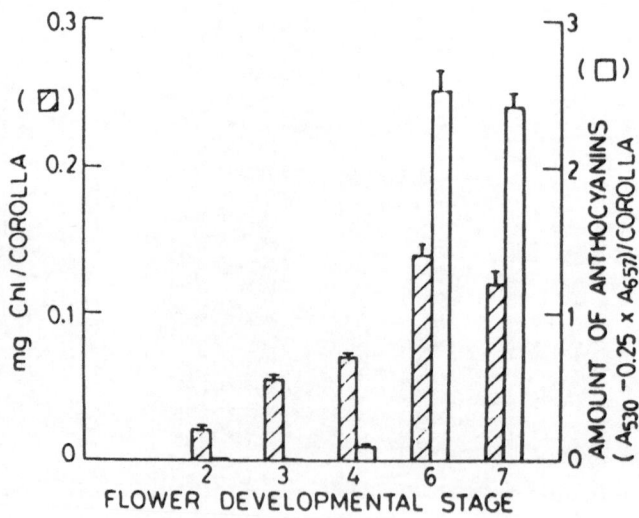

Fig. 2. Chl and anthocyanin levels in Petunia corolla during the different developmental stages

Chloroplasts isolated from pink corollas carried out light induced electron transport with methyl viologen as acceptor at 1/3 of the rate measured in chloroplasts from green leaves (Fig. 3). Both the quantum yield for electron transport and the light-saturated maximal rate were lower in corolla chloroplasts compared to chloroplasts from green leaves. Electron transport from reduced dichlorophenolindophenol to methyl viologen, which involved only PSI, amounted in corolla chloroplasts to 65% of the rate in chloroplasts from green leaves (data not shown). This higher activity might indicate a special role for PS I in corolla development and pigmentation[8].

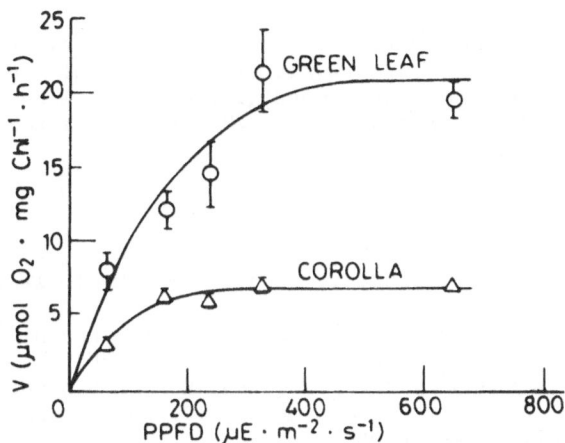

Fig. 3. Effect of Photosynthetic Photon flux density (PPFD) on the rate of whole-chain electron transport (V). Electron transport was measured in isolated chloroplasts with methyl viologen as acceptor

Reduction in quantum yield in <u>Petunia</u> corollas was also indicated by Chl fluorescence measurements in intact organs (Table 1). The relative amplitude of the variable fluorescence (F_v/F_o), which is a measure for the quantum yield[6], was lower in corollas compared to green leaves. The reduction in quantum yield was at least partially due to inhibition of electron transfer from the primary to the secondary quinone acceptor in PS II. This was evident as an increase in the "intermediate fluorescence" level, (F_i-F_o)/F_o. F_i is apparent as a shoulder in the Chl induction curve[9].

Another difference between the fluorescence patterns of pink corollas and green leaves was the rise-time from the initial (F_o) to the maximal fluorescence level (Table 1). The shorter induction time in corollas seems to indicate a larger photosynthetic unit compared to green leaves. The decrease in the light intensity required for saturation of electron transport in corolla chloroplasts (Fig. 2) might also lead to such a conclusion. A need for a large photosynthetic unit may arise at an early stage of flower development, when the corolla is still largely covered by the sepals and is similar in this respect to a shade plant[10].

Table 1. A Comparison of Chl a Fluorescence in
Petunia Corolla and Green leaves

	Green Leaves	Corollas
Fv/Fo	2.2 $\pm$ 0.1	1.6 $\pm$ 0.1
(Fi-Fo)/Fo	0.29 $\pm$ 0.01	0.76 $\pm$ 0.04
Induction time (s)	0.55 $\pm$ 0.05	0.09 $\pm$ 0.01

Means of 20 replications $\pm$ S.E.

The role of the photosynthetically active chloroplasts in the Petunia corolla is not clear yet. Growth and pigmentation of detached corollas, incubated in the light in a sugar-containing medium, were inhibited by DCMU (data to be presented elsewhere). This might indicate a role for the corolla chloroplasts in flower development.

LITERATURE

1. J. M. Whatley and F. R. Whatley, When is a chromoplast?, New Phytol. 106:667 (1987).
2. V. Sharza, Hill activity in chloroplasts from red pigmented corolla, bracts and leaves, Photosynthetica 14:79 (1980).
3. J. C. V. Vu, G. Yelenosky and M. G. Bausher, Photosynthetic activity in .the flower buds of valencia orange (Citrus sinensis [L.] osbeck), Plant Physiol. 78:420 (1985).
4. A. L. Mancinelli, C. P. H. Yang, P. Lindquist, O. R. Anderson and I. Rabino, Photocontrol of anthocyanin synthesis III. The action of streptomycin, Plant Physiol. 55:251 (1975).
5. M. Schonfeld, T. Yaacoby, O. Michael and B. Rubin, Triazine resistance without reduced vigor in Phalaris paradoxa, Plant Physiol. 83:329 (1987)
6. S. Malkin, P. A. Armond, H. A. Mooney and D. C. Fork, Photosystem II photosynthetic unit sizes from fluorescence induction in leaves. Correlation to photosynthetic capacity, Plant Physiol. 67:570 (1981).
7. E. Zamski, Structure and function of Beta vulgaris parenchima cells: Ultrstructure and sugar uptake characteristics of tissue and cell in suspension, Bot. Gaz. 147:20 (1986).
8. M. J. Schneider and W. R. Stimson, Contribution of photosynthesis and phytochrome to the formation of anthocyanin in turnip seedlings, Plant Physiol. 48:312 (1971).
9. C. J. Arntzen, K. Pfister and K. E. Steinbeck, The mechanism of chloroplast triazine resistance: alteration in the site of herbicide action, in "Herbicide Resistance in Plants," H. M LeBaron and J. Gressel, eds., John Wiley and Sons, New York, p185 (1982).
10. S. Malkin, D. C. Fork, Photosynthetic units of sun and shade plants, Plant Physiol. 67:580 (1981)

ALGAE REMOVAL IN WASTE STABILIZATION PONDS

Ayşen Türkman, and Füsun Şengül

Dokuz Eylül University
Environmental Engineering Dept.
İzmir, Turkey

INTRODUCTION

Although waste stabilization ponds are generally considered to be suitable in small communities or for the treatment of industrial wastewaters, there are cases where they are considered to be appropriate for big cities also. They are in use at least 39 countries[.] They may become feasible if land is available at a low cost. This factor, considerable sunshine and generally warm weather in combination with significant financial advantage in operation forms the basis of feasibility of waste stabilization ponds system in İzmir. İzmir is the third biggest city of Turkey with its metropolitan population of about two million. Among the treatment system alternatives, waste stabilization ponds consisting of three ponds in series has found to be most appropriate for the domestic wastewaters combined with pretreated industrial wastewaters (Figure 1). In this alternate, the wastewaters will be held in anaerobic ponds, facultative ponds, and maturation ponds for 2, 18 and 4 days respectively.

Although the presence of algae is a requirement for the growth of nearly all other forms of life, excessive growth of algae is not desired. Algal growth will be a nuisance in the summer. Algae at times may cause the BOD limit to be exceeded. Algae removal is a problem and there is no simple solution.

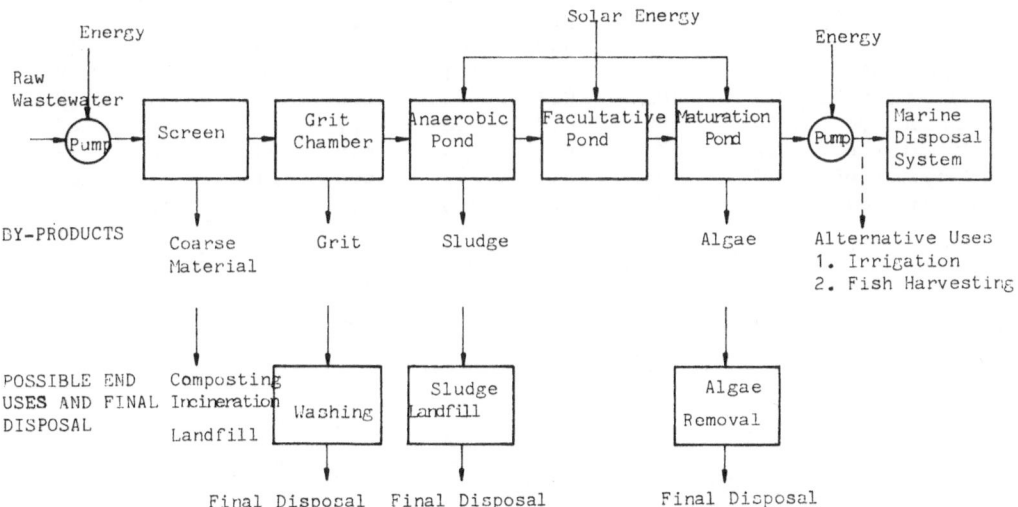

Figure 1. Flowchart of the İzmir wastewater treatment plant.

Eighteen possible effluent polishing techniques have been developed to upgrade ponds. These are 1) centrifugation, 2) microstraining, 3) in-pond removal of particulate matter, 4) coagulation, 5) bioflocculation, 6) biolo gical disks, 7) baffles and raceways, 8) in-pond chemical precipitation 9) autofloculation, 10) biological harvesting, 11) oxidation ditches, 12) land application, 13) dissolved air flotation, 14) granular media filtration, 15) intermittant sand filtration, 16) series ponds with intermediate chlori nation, 17) controlled discharge ponds, 18) application of algicides[3,4,5]

A series of reports distributed by the Canadian government has indicated success with the treatment of controlled discharge ponds by adding various coagulants from motorboat[6,7].

PILOT STABILIZATION PONDS MODEL STUDIES

Pilot treatment plant consists of three main units in the order of anaerobic, facultative and maturation ponds and arranged to work as semi-continous reactors. The volumes of anaerobic, facultative and maturation ponds are 50, 450 and 100 liters and detention times 2, 8 and 4 days respec tively. 25 liters of domestic wastewater is fed into the model and 25 l of treated water is taken every day. Pilot stabilization ponds model was under operation between December 1985 to January 1987. The samples were taken from the influent and effluent of each subsystem. Nitrogen and phosporus tests are carried out on filtered samples taken from beneath the model surface. Figure 2 indicates the model scheme. Table 1 summarizes monthly average P and N concentrations data obtained during the pilot plant studies.

Effluent samples taken from maturation pond of pilot plant are analysed for algae according to Standard Methods. Mainly euglena, chlorella and chlamydomonas are found to exist at the effluent (Figure 2).

Table 1. During the pilot plant studies obtained monthly average phosphorus and nitrogen concentrations (Figure 3).

Samp. pt. Time	I $P,g/m^3$	$N,g/m^3$	II $P,g/m^3$	$N,g/m^3$	III $P,g/m^3$	$N,g/m^3$	IV $P,g/m^3$	$N,g/m^3$
January 86	7.0	144	6.7	116	6.3	79	6.0	67
February 86	9.1	96	8.3	83	7.1	63	6.6	57
March 86	7.8	110	6.8	87	4.5	55	3.1	43
April 86	10.3	118	8.8	99	6.0	51	3.5	33
May 86	9.9	88	9.2	75	6.0	40	4.0	32
June 86	10.3	99	9.2	81	6.7	54	4.3	38
August 86	10.0	121	9.2	101	7.5	86	4.9	69
September 86	10.3	102	9.1	82	6.9	63	4.1	40
October 86	10.1	89	8.9	60	6.7	41	4.0	23
November 86	9.6	78	8.5	44	6.8	30	4.5	19
December 86	9.7	85	8.4	45	6.6	31	4.4	21
January 87	9.2	88	8.2	43	7.6	24	5.7	18

Figure 2. Algae forms determined in maturation ponds.

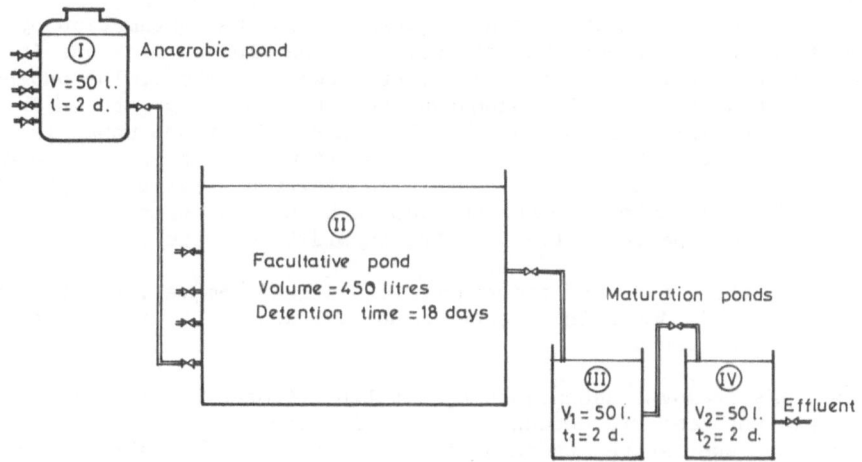

Figure 3. Pilot waste stabilization ponds model.

For the removal of algae from the maturation pond effluents, flotation coagulation and disinfection techniques are applied.

In the first series of experiments, the effluent was hold in a 40 l tank for 6 hours to allow the algae particles to setile or float. There was no noticable algae removal by either flotation or sedimentation.

In the second series of experiments, air flotation with 1.5 hours of detention time was carried out. Air was supplied by air pumps through diffusers. This method gave no noticable algae removal and considered to be unsuitable for pilot plant considered.

In the third series, jar tests have been performed in order to determine the possibility of removal by coagulation. Experimental results obtained by using lime and alum are given in Table 2.

In the fourth series, sodium hypochloride is used in order to kill algae and allowed for half an hour for settling. Disinfection with sodium hypochloride at $1-12$ g/m^3 dozages was unsuccessful for algae removal.

RESULTS

The amount of algae in waste stabilization ponds is closely related to the phsophorus concentration in wastewater. Some amount of phosphorus is

Table 2. Jar tests results for algae removal by using lime as coagulant (Nov. 12, 1986).

	Dose (g/m^3)	Algae removal (%)	pH of resulting solution
Lime	50	9.5	7.56
	100	19.0	8.10
	150	38.0	9.08
	200	38.0	9.33
	250	40.0	9.74
	200	36.0	6.60
	300	48.0	6.30
Alum	400	76.0	5.95
	500	94.0	5.40
	600	99.8	4.80
	700	99.9	4.50

removed in waste stabilization ponds system due to the consumption by algae sedimentation with sludge and adsorption by the sludge. Decrease in phosphorus content from pond tc pond may be noted in Table 1. The phosphorus content is measured as total phosphorus, but the samples are taken from the middle of the tanks since algae tend to congregate at the upper layers of the stratified pilot ponds. Surface drawoff will contain high levels of algae. Also decrease in nitrogen content is observed parallel to phosphorus decrease. This indicates the consumption of N and P to produce algae which is also observed experimentally from the turbidity of water.

Although there are many techniques for algae removal, the best technic can only be determined experimentally, depending on the species of algae existing in the ponds.

The algae removal techniques by settling, flotation and chlorination-settling was unsucessful. Because of the infrequent occurence of conditions necessary for the above techniques, they are not viable alternatives for the removal of algae from ponds at this time.

As can be seen from Table 2, use of alum was much more effective than using lime. Use of $Mg(OH)_2$ was not taken into consideration for economical reasons. Although increasing the alum dose gave very high removal efficiencies, $500 \ g/m^3$ (94 % removal) should be taken as optimum since the removal of remaining small amounts of algae becomes very expensive due to the consumption of chemicals.

In case alum is used for algae removal, additional advantage of phosporus, biochemical oxygen demand and suspended solids removal is obtained.

Although use of lime does not seem as efficient as alum, it also may be suggested since it is considerably cheaper than alum in Turkey. Use of lime will also bring the phosphorus removal, but pH correction may be necessary due to the high pH values existing in water after lime addition.

In order to decrease chemical consumption or increase the algae removal efficiency more experiments at different conditions should be performed For example use of alum in combination with polielectrolytes is expected to decrease alum dose, or use of tube setllers may increase the the removal efficency . These applications increase the plant complexity and require closer operator attention.

REFERENCES

1. E.F. Gloyna, "Waste Stabilization Ponds", World Health Organization, Geneva (1971).
2. DEÜ, İzmir Wastewater Treatment Plant Feasibility Study, Dokuz Eylül University, Faculty of Engineering and Achitecture, İzmir (1986).
3. S.S.Mcdonald and K.J.Williamson, Settling Rates of Algae from Wastewater Lagoons, Journal of Env.Eng.Div. April (1979).
4. EPA; Desing Manual Municipal Wastewater Stabilization Ponds, Center of Env.Reseach Inf, Cincinatti (1983).
5.İzmir Project, İzmir Sewerage Project, B and V Project no.14046. 301, USA (1988).
6. H. J. Graham and R. B. Hunsinger, Phosphorus Reduction from Continuous Overflow Lagoons by Addition of Coagulants to Influent Sewage, Research Report no.65, Ontario (1977).
7. Pollutech Pollution Advisory Services Ltd., Nutrient Control in Sewage Lagoons, Volume II, Project no. 72-5-12, Wastewater Technology Center, Ontario (1977).
8. Standard Methods for the Examination of Water and Wastewater, 15 th Edition, APH, AWWA, WPCF (1981).

SECTION V
STRESS PHENOMENA AND ADAPTATION

INACTIVATION OF PHOTOSYSTEM II AND TURNOVER

OF THE D1-PROTEIN BY LIGHT AND HEAT STRESSES

Gadi Schuster, Susana Shochat,
Noam Adir and Itzhak Ohad

Dept. of Biological Chemistry
The Hebrew University of Jerusalem
91904 Jerusalem, Israel

The photosynthetic electron flow activity is sensitive to physiological stress conditions such as light, heat, cold or water stress[1,2]. Interestingly, photosystem I (PS I) is not as sensitive as photosystem II (PS II), and in most cases, PS I continues to operate normally under conditions in which PS II activity is completely inhibited[1]. As plants are exposed to stress conditions in their natural habitats, the questions arise as to what are the molecular mechanisms of the stress induced inactivation of PS II and whether there are physiological processes in the chloroplasts which may protect the photosynthetic electron flow activity of PS II from the specific damage induced by environmental stresses. Among these, the inactivation of PS II by light and heat stress have been extensively studied using the green alga Chlamydomonas reinhardtii.

When photosynthetic organisms are exposed to light intensities which exceed those required to saturate photosynthesis, photosynthetic activity is inhibited (photoinhibition) due to specific damage to the reaction center of PS II (RC II)[1,2,3]. The photoinhibition phenomenon is general for all the photosynthetic oxygen evolving organisms tested so far, including cyanophytes, green algae and higher plants. The degree of photoinhibition at equal photon fluency or the light intensities required to inhibit photosynthetic electron flow to the same degree, varies for different organisms and is influenced by physiological conditions such as CO_2 concentration, water availability as well as by the structure of the photosynthetic apparatus (for example sun or shade adapted plants)[1,2,3]. When light stress is accompained by heat stress photoinhibition is induced at much lower light intensities and the process is practicaly irreversible. In this report we shall describe the phenomena occuring during the early phase of the light induced alteration of PS II under normal and heat stress conditions and discuss the mechanism involved in this process.

The initial process of photoinhibition

The initial process of PSII inactivation by light was studied using various techniques including measurements of variable fluorescence, oxygen evolution in continuous light, oxygen flash yield following exposure of cells to saturating single turnover flashes, and measurements of thermoluminescence arising from charge recombination between the S states and reduced Q_A or Q_B quinone acceptors on the donor side of PSII reaction center (RCII)[4].

The results of these measurements demonstrate that the earliest detectable alteration of RCII in cells exposed to light intensities above those required for saturation of photosynthesis as compared with non-saturating light, is a destabilization of the $S_2Q_B^-$ pair resulting in recombination at a temperature 10 - 15 ° C lower than that of the control cells (Fig. 1). This effect was accompanied by a shortening of the life time of $S_2Q_B^-$ and affected the entire population of RCII but not the $S_2Q_A^-$ pair. It preceeded the loss of O_2 evolution measured in continuous light and was corelated with a rise in the intrinsic fluorescence (Fo)[4]. Since oscillations of oxygen flash yield with flash number and reduction of Q_A were not affected at this initial stage, one can conclude that the water splitting complex, P_{680}, Pheophytin and the primary stable acceptor, Q_A, were not affected. Therefore the initial damage was found to be localized at the level of the D1 protein and is related to the Q_B binding site[4].

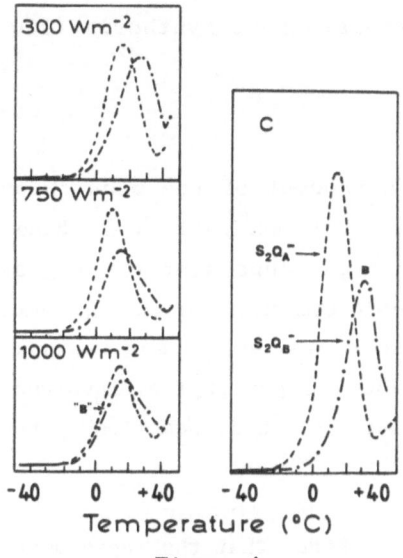

-40   0 +40   -40   0 +40

Temperature (°C)

<u>Figure 1</u>

<u>Thermoluminescence response of control and photoinhibited cells</u>

Chlamydomonas cells (45 µg Chl/ml) were incubated at light intensities as indicated for 25 min and the thermoluminescence response was measured in absence (.-.-.) or presence of DCMU (------). DCMU inhibits photosynthetic electron flow from Q_A to Q_B. The peak obtained in its presence represents the $S_2Q_A^-$ pair, whereas in its absence it represents the $S_2Q_B^-$ pair. Trace C: control cells. The values of F_v/F_o were 3.1 (control) and 2.1, 1.2 and 0.8 for the samples exposed to 300, 750 and 1000 $W \cdot m^{-2}$, respectively. The excitation flash was given at 0 - 2° C for the control and DCMU samples. The chlorophyll concentration was 45 µg per sample for the control and 35 µg for all other samples.

Turnover of the D1 protein

C. reinhardtii cells, exposed to light intensities which alter the $S_2Q_B^-$ properties but do not inactivate PS II electron flow responded by an accelerated synthesis of the D1-protein. A correlation was found between the amount of destabilization of the $S_2Q_B^-$ pair and the rate of synthesis of D1, both reaching saturation at light intensities which saturate photosynthesis (200 - 400 w.m^{-2}) (Fig. 2)[4].

The D1 protein, a product of the psbA chloroplastic gene, which appears as a diffuse poorly stainable 32kDa band in SDS-PAGE, is well known as a rapidly light dependent turning over protein[5]. This protein is an intrinsic component of RC II, forms the binding site of the secondary quinone acceptor Q_B, and together with the D2, binds the chlorophylls P-680 and pheophytin, involved in the photochemical charge separation activity of PS II reaction center.

It was suggested before that the rapid degradation and synthesis of D1 might serve as a mechanism for the protection of PS II photosynthetic activity against light induced damage (photoinhibition)[2,3,6,7]. Thus following alteration of RC II by light, the D1 protein would be removed and replaced by a newly synthesized molecule before the light induced alteration would cause further damage to the reaction center and photoinhibition of photosynthetic electron flow would occur. In such a situation, the observed photosynthetic electron flow level is a steady state that results from the balance between the light induced damage to PS II reaction centers and the rate of the "repair" process which replaces the altered D1 protein. The correlation between the destabilization of the $S_2Q_B^-$ pair and the synthesis of D1 at light intensities which do not yet cause photoinhibition, might indicate that the first event involved in the light induced alteration of PS II reaction center occurs at the Q_B site and is followed by a rapid degradation of D1 protein to prevent further damage and photoinhibition[4].

Photoinhibition and turnover of PS II proteins

In the natural habitate conditions of temperature, light intensities, CO_2 and water supplies, the mechanism that prevents photoinhibition by rapid turnover of D1 is highly efficient so that in every RC II altered by light, D1 can be degraded and replaced. However, the mechanism described above would predict that a situation

Chlamydomonas cells (35 µg Chl/ml) were incubated for 35 min at light intensities as indicated and the thermoluminescence response was measured after one excitation flash at 0 - 2° C. B, "B", the emission at 30° C and 15 - 17° C respectively, as shown in Figure 1. After the 35 min at different light intesities the cells were transferred to 40 W·m^{-2} light for 10 min and labeled for an additional 10 min with $^{35}SO_4^{2-}$. Thylakoids were isolated, seperated on SDS-PAGE and exposed to auotoradiograph. The rate of synthesis of D1 was obtained from densitometer scanning of the radioactivity of the D1 band(D1)

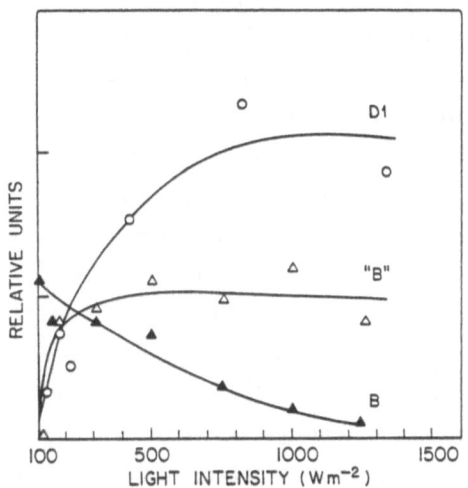

Figure 2. Changes in the thermoluminescence response and in the rate of synthesis of the D1 protein as a function of light intensity during photoinhibition

could exist, in which the rate of PS II reaction centers alteration would be faster than D1 degradation and replacement. Such a situation can be obtained in the green alga C. reinhardtii, exposed to ten fold photosynthetic saturating light (2500 w.m^{-2})[6]. In such conditions, PS II is inactivated with a half life time of 45 min, and thus cosiderably faster than the degradation of the D1 (t 1/2 90 min) (Fig. 3). Under this conditions, when 90% of PS II activity was inhibited, additional processes were observed in the chloroplasts such as reversible inhibition of the LHC II protein kinase system and changes in the size and distribution of the thylakoid intramembrane particles[9]. Under these conditions, cyclic electron flow around PS I is active as shwon by using the photoacoustic technique[10]. Using the half life time value of PS II activity under photoinhibitory conditions, one could estimate the rate of inactivation as one reaction center being inactivated for about 10^5 photon events (Fig. 3).

The question arises as whether D1 is the only PS II protein degraded during photoinhibition? Using a variety of techniques, we could demonstrate that the LHC II and CP-29 antennae polypeptides, the core complex 44 and 51 kDa chlorophyll binding proteins and the water oxidizing enhancer proteins of 17, 23 and 30kDa were not degraded during photoinhibition of C. reinhardtii cells[8]. Only the D1 and the D2 proteins were degraded (Fig. 3), although the degradation rate of the D2 was lower than that of the D1. (t 1/2 of 120 min.). The high turnover rate of the D2 occured only under extreme photoinhibitory conditions. Under saturating photosynthetic light intensity no detectable turnover of D2 could be demonstrated[8]. Unlike the D1, the D2 does not contain the amino acid sequence PEST, characteristic of some fast degraded proteins[11]. However, one should note that the amino acid sequence from 241 to 246 of the D1 amino terminus, which was considered to be the cleavage site of this protein in the light dependent turnover, is almost identical to amino acids sequence 240-245 of the D2[12].

Photoinhibition at low light intensities during heat shock

If indeed the removal of D1 from altered RC II is part of a protecting mechanism against photoinhibition the question arises as to what will happen if the chloroplast protecting mechanism, which accelerated the D1 turnover, cannot operate at an optimal rate ? Based on the above consderation one would predict that photoinhibition would be induced under lower light intensities. Indeed, such a situation was observed when C. reinhardtii cells were exposed to light at elevated temperatures (heat shock)[13]. No inhibition of photosynthetic electron flow was obtained during heat shock in the dark. Heat shock

Chlamydomonas reinhardtii cells (30 μg Chlorophyll/ml) were photoinhibited with or without addition of chloramphenicol (CAP) at 2500 W.M^{-2} light intensity.

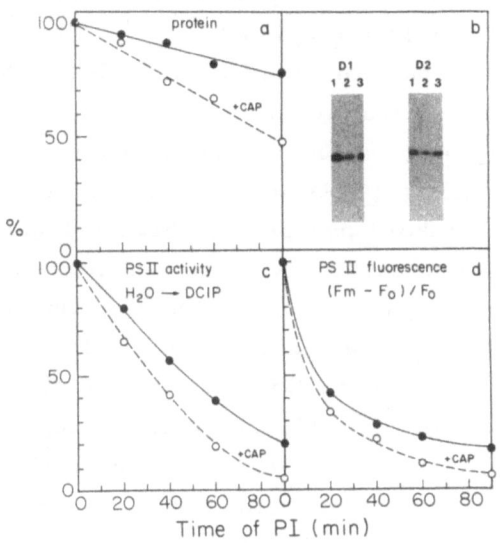

Figure 3. Loss of photosystem II activity and degradation of the D1 and D2 protein during photoinhibition of C. reinhardtii cells

a. Quantitation of the amount of the D1 protein during photoinhibition by immuno dot blotting of thylakoids as shown in Figure 2; b. Western blot, using specific antibodies to the D1 and D2 proteins, of thylakoids isolated from: control cells (1); cells photoinhibited in the presence (2) or absence (3) of chloramphenicol. Isolated thylakoids were resolved by SDS-PAGE, electrotransferred to nitrocellulose paper and decorated with D1 or D2 antibodies and ^{125}I-protein A. c. Photoinhibition of photosystem II dependent electron flow as measured in isolated thyalkoids by the light dependent, DCMU sensitive, reduction of 2-6-dichlorophenol indophenol (DCIP), with H$_2$O as electron donor. d. Photoinhibition of photosystem II measured by photosystem II fluorescence kinetics at room temperature in the presence of 5 x 10^{-6} M DCMU in whole cells.

in the light caused inactivation of photosynthetic electron flow due to damage of PS II reaction center in a way similar to the photoinhibition process at normal growth temperatures as described above[13]. The light intensities which caused photoinhibition under heat shock condition was about ten times lower than that required at growth temperature, and was similar to that required for saturating photosynthesis (approx. 250 w.m^{-2})[13]. Examination of the thylakoid polypeptide pattern by SDS-PAGE disclosed that the thylakoid proteins were aggregated (polymerized) and remained in the stacking gel (Fig. 4). The aggregation process was rather specific for the D1 at low light intensities (50-100 w.m^{-2}), and became non-specific affecting other thylakoid proteins at higher light intensities[13]. This phenomenon occured also when the D1 degradation system was inhibited during photoinhibition at low temperatures (5°C), or when isolated thylakoids, in which the degradation mechanism was not operated, were photoinhibited (Fig. 4). Thus we suggest that if D1 is not degraded in the altered RC II, it becomes aggregated or crosslinked with other PS II proteins (possibly D2), during the subsequent light exposure possibly due to interaction with free radicals. Crosslinking of proteins by free radicals is a well known process which was shown in other biological systems to induce rapid degradation processes[3,14,15,16]. Possible candidates for this effect are semiquinone anione radicals or more probably, hydroxyl radicals generated from peroxide and O_2^- in the presence of Fe^{2+} (Fenton chemistry)[3,17].

We have demonstrated before that among the heat induced proteins HSP 22 of nuclear origin is transported into the chloroplasts and becomes associated with the granna lamellae region of the thylakoids[18]. Prevention of heat shock protein synthesis by inhibitors of cytoplasmatic protein translation such as cycloheximide, accelerated the photoinhibition process and the aggregation of the D1 at elevated temperature[13]. These observations suggested that HSP 22 might be involved in protecting PS II against photoinhibition and light induced polymerization of thylakoid proteins during the heat shock process[13,18].

A. <u>C. reinhardtii</u> thylakoids were isolated and photoinhibited at 500 W·m^{-2} light intensity at 25° C in a buffer containing 50 mM Tricin pH 7.8, 10 mM NaCl, 5 mM MgCl$_2$ and 0.4 M sucrose.

1. Thylakoids incubated in the dark for one hour. 2, 3. thylakoids incubated in the light 30 and 60 min respectively. G, coomassie briliant blue R stained SDS-PAGE. Arrows indicate the accumulation of aggregated proteins in the stacking gel. W.B., Western blot using D1

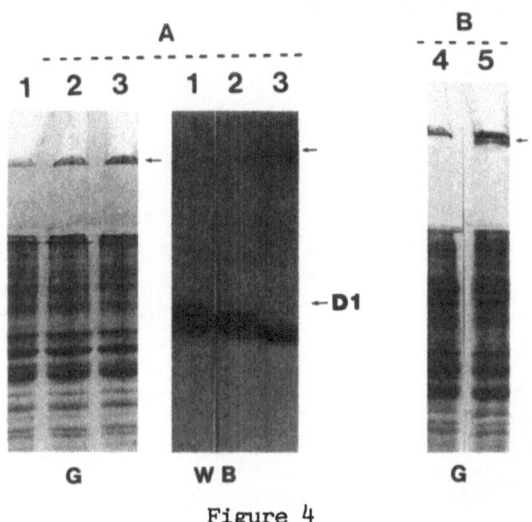

Figure 4

<u>Aggregation of thylakoid proteins during photoinhibition of isolated thylakoids and heat shock of C. reinhardtii cells</u>

antibodies. Note the disappearance of the D1 from the 32 kDa region of the SDS-PAGE and its accumulation in the aggregates.

Photosynthetic activities, as measured by photoreduction of 2,6-dichlorophenolindophenol (DCIP) were 93, 40 and 17% of sample 1, 2 and 3 respectively as compared to the time zero sample.

B. <u>C. reinhardtii</u> cells were heat shocked at 42° C in the dark (4) or at 150 W·m^{-2} light intensities (5) for 2 h. Thylakoids were isolated and analyzed by SDS-PAGE.

The sequence of processes involved in photoinhibition and turnover of the PS II reaction center are schematically presented in scheme 1. The scheme is based on the experimental results described above and others obtained in different laboratories[5,11] Many details in this picture are still open to question: What is the molecular mechanism of the initial damage? What is the triggering signal for the degradation of D1? What is the identity of the D1 proteolitic system? and what happens to the pigments that are bound to the D1 during its turnover? These questions remain open and will be the subject of future research of the PS II reaction center dynamics.

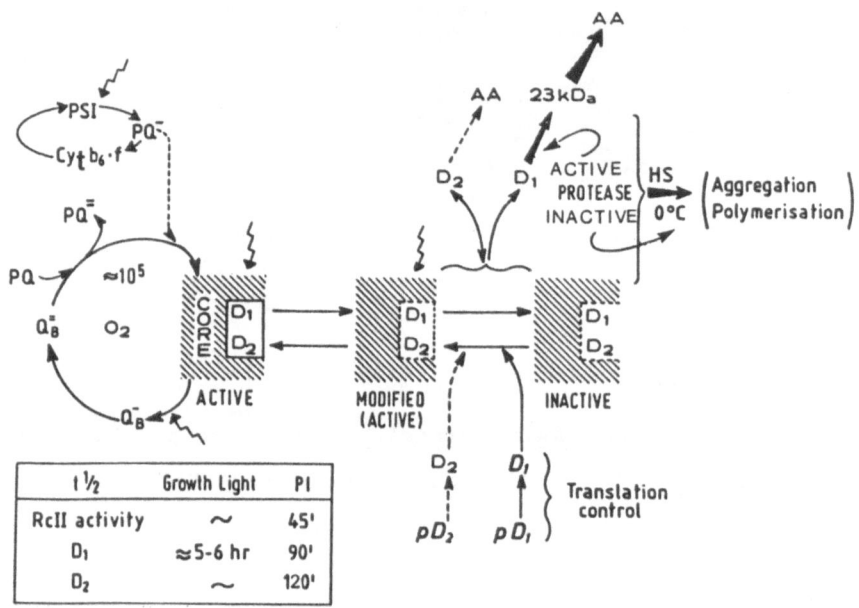

Scheme 1. Schematic representation of the processes involved in the turnover of PS II reaction center

An active PS II in electron transfer and oxygen evolution is presented on the left (ACTIVE). Following $\sim 10^5$ charge separation events RC II is modified. This modification leads to thermodynamic destabilization of the Q_B^- and probably to changes in the D1 (MODIFIED (ACTIVE)). At this stage the D1 degradation system is activated and the D1 protein turns over. Replacement of the altered D1 by a newly synthesize protein allows the PS II core to return to its initial active state. However, other light dependent modification(s) of the altered PS II core might inactivate the electron flow (INACTIVE) and accelerate the D1 protein turnover rate. Under high light intensity when the modification and inactivation processes are faster than the D1 degradation, the D2 protein is also rapidly degraded and photosynthetic activity is inhibited (photoinhibition). If the degradation system of the D1-protein is inactivated (cold or heat shock conditions) the D1 is aggregated and becomes cross-linked to other PS II proteins.

The rates of inactivation of the C. reinhardtii PS II activity and the degradation of the D1 and D2 at light intensities saturating photosynthesis (approx. 250 $w.m^{-2}$) and at photoinhibition condition (2500 $w.m^{-2}$) are indicated in the table.

References

1. Powels, S., Ann. Rev. Plant Physiol 35:15-44 (1984).

2. Kyle, D. J. and Ohad, I. in Encyclopedia of Plant Physiol. (New series), Vol. 19, Staehelin, L. A. and Arntzen, C. J., eds.), pp. 468-476. Springer, Heidelberg. (1986)

3. Kyle, D. J. in "Photoinhibition" Kyle, D. J. Osmond, C. B. and Arntzen, C J. eds. Elsevier sci. Pub. B. V. pp.197-226. (1987)

4. Ohad, I., Koike, H. K., Shochat, S. and Inoue, Y. Biochem. Biophys. Acta. 933, 288-298. (1988)

5. Kyle, D. J. Photochem. Photobiol. 41, 107-116. (1985)

6   Kyle, D. J., Ohad, I. and Arntzen, C. J. Proc. Natl.   Acad. Sci. USA 81, 4070-4074. (1985)

7. Ohad, I., Kyle, A. J. and Arntzen, C. J. J. Cell Biol. 35, 521-552. (1984)

8. Schuster, G., Timberg, R., and Ohad, I., Eur. J. Biochem. (in press) (1988).

9. Schuster, G., Dewit, M., Staehelin, L. A. and Ohad, I. J. Cell Biol. 103, 71-80. (1986)

10. Canaani, O., Schuster, G., and Ohad, I. Photosinthesis Res. (in press) (1988).

11. Greenberg, B. M., Gaba, V., Mattoo, A. K. and Edelman, M. EMBO J. 6, 2865-2869. (1987)

12. Trebst, A. and Draber, w., Photosynthesis Res. 10, 381-392. (1986).

13. Schuster, G., Even, D., Kloppstech, K. and Ohad, I. EMBO J. 7, 1-6. (1988).

14. Dean, R. T. FEBS Lett. 220: 278-282 (1987).

15. Wolff, S. P., Garner,A. and Dean, R. Trends. Biochem. Sci. 11: 27-30 (1986).

16. Davis, K. J. A., J. Biol. Chem. 262:9895-9901 (1987).

17. Asada,K. and Takahashi, M., in "Photoinhibition" Kyle, D. J. Osmond, C. B. and Arntzen, C J. eds. Elsevier sci. Pub. B. V. pp.227-287. (1987).

18. Kloppstech, K., Meyer, G., Schuster, G. and Ohad, I. EMBO J. 4, 1901-1909. (1985).

Acknowledgement

This work was supported by grants awarded to us by the Israeli Academy of Sciences and the Deutsche Forschungsgemeinschaft and by the National Council for R&D joint German-Israel Program in Biotechnology.

The data in figs. 1 and 2 are part of a research project carried out in cooperation with Dr. H. Koike and Y. Inoue from the Physical Chemical Res. Institute Solar Energy Group, RIKEN, Wako-shi, Saitama, Japan.

SUBFRACTIONAL ANALYSIS OF THYLAKOID MEMBRANE ORGANIZATION AND ADAPTATION

Bertil Andersson[1], Ulla K. Larsson[2], Cecilia Sundby[2], Pirkko Mäenpää[1]*, Sophie Bingsmark[1] and Torill Hundal[1]

[1]Department of Biochemistry, Arrhenius Laboratories, University of Stockholm and [2]Department of Biochemistry, University of Lund (Sweden)

THE SUBFRACTIONAL APPROACH

In Dutch the word for chemistry is scheikunde. This is one illustration of the central role for the separation concept in experimental chemistry. The need for efficient fractionation methods does also apply to biochemistry and photosynthetic research. Not until pure and intact chloroplasts were isolated in the fifties could the subcellular localization of CO_2 fixation be established[1]. Thylakoid preparations completely free of contaminating mitochondria were a prerequisite for proving the existence of photophosphorylation[1]. Today we see the importance of subfractionation for the isolation of thylakoid protein complexes with a defined function and composition[2]. For the recent dramatic progress in our understanding of photosystem II, subfractionation has been a real cornerstone. In this paper we describe the subfractional approach for studies on the organization and dynamics of the thylakoid membrane.

Membrane fractionation involves two principal steps, disintegration and a subsequent separation step. The disintegration can be obtained either chemically by detergents or by mechanical shearing. Differential or density gradient centrifugations are usually applied for the separation, thereby fractionating the membrane fragments according to differences in their size and density. Another separation method is partition in aqueous polymer two-phase systems[3,4]. This method separates membranes according to differences in surface properties such as charge and hydrophobicity which makes phase partition an alternative or complement to centrifugation procedures.

---

* present address: Department of Biology, University of Turku (Finland)

AQUEOUS POLYMER TWO-PHASE PARTITION

When aqueous solutions of the polymers dextran and po-lyethylene glycol are mixed above certain concentrations they give rise to two immiscible phases. The lower phase is enriched in dextran while the upper phase is enriched in polyethylene glycol. Both phases are rich in water (85-99%), a fact which makes the polymer phase systems suitable for separation of biological macromolecules and membrane partic-les. The partition of membranes in aqueous polymer two-phase systems takes place between the two phases and the interfa-ce. Several properties of the phase system, such as ionic composition, polymer concentration and temperature, influence the partition of a biological substance[4]. Finding the opti-mal phase composition for a particular separation usually involves systematic changes of the polymer concentrations or the salt composition of the phase system (as described in detail by Albertsson et al.[4]). The separation can be obtai-ned either by a few step batch procedure or by a multistage chromatographic procedure such as counter-current distribu-tion. The phase partition method has been used for the sepa-ration of macromolecules, membranes, organelles and cells from a variety of biological sources[3,4], although it has been particularly useful for the purification of plant orga-nelle membranes[5].

THYLAKOID MEMBRANE SUBFRACTIONATION: ISOLATION OF INSIDE-OUT VESICLES

By phase partition of fragmented spinach thylakoids, inside-out thylakoid vesicles could be isolated for the first time[6]. Below follows a description for one of the met-hods available for the isolation of everted thylakoids: Isolated stacked spinach thylakoids suspended in 5 mM $MgCl_2$, 5 mM NaCl, 10 mM Na-phosphate pH 7.4 and 100 mM sucrose were fragmented by two consecutive passages through a Yeda press at a nitrogen gas pressure of 10 MPa. The thylakoid frag-ments were centrifuged at 40 000 x g for 30 min to give a su-pernatant containing pure stroma thylakoid vesicles rich in photosystem I. The pellet was resuspended in a buffer compo-sed of 5 mM NaCl, 10 mM Na-phosphate pH 7.4 and 100 mM suc-rose and cycled two more times through the press. The vesic-les were fractionated in an aqueous polymer phase system composed of 5.7% (w/w) dextran T-500, 5.7% (w/w) polyethyle-ne glycol 3350, 10 mM Na-phosphate pH 7.4, 5 mM NaCl and 30 mM sucrose. The phase mixture was shaken at 2°C and allo-wed to settle. Usually the settling was facilitated by a short low speed centrifugation. About two thirds of the thy-lakoid material partitioned to the upper phase (T1) while the remaining third was recovered in the lower phase (B1). The upper phase was repartitioned one more time to give a T2 fraction and the lower phase was repartitioned twice yielding a B3 fraction. Analysis of the B3 fraction revealed that it contained predominantly vesicles that were turned inside-out. This could be shown by proton translocation mea-surements and freeze-fracture electronmicroscopy. The T2 fraction contained vesicles of normal sidedness. The use of thylakoid vesicles of opposite sidedness for the analysis of transverse thylakoid membrane asymmetry have been described elsewhere[7].

Inside-out vesicles can be isolated if the thylakoid membranes are stacked during fragmentation[7]. Under such conditions the fragments obtained from the grana thylakoids upon rupture are not single, but pairwise closely appressed to a neighbouring membrane. Resealing at the edges of two such appressed membrane pieces results in the formation of an inside-out vesicle. Thus, the everted vesicles are derived from the appressed thylakoid regions.

Comparison of the composition of inside-out vesicles with stroma thylakoid vesicles unravelled an extreme lateral heterogeneity in the organization of stacked thylakoid membranes of higher plants[7,8] (Fig. 1). Through the combination of SDS-PAGE, immunological analysis and activity measurements it was shown that inside-out vesicles were enriched in photosystem II and its main light-harvesting chlorophyll a/b proteins (LHC II) while highly depleted in photosystem I and the ATP synthase (CF_1-CF_0). The stroma thylakoids contained mainly the two latter complexes and a small portion of photosystem II. Only the cytochrome b/f complex was found in substantial amounts in both subthylakoid fractions.

Based on this subfractional analysis it became clear that the structural differentiation of higher plant thylakoids into appressed and non-appressed regions is paralleled by a functional differentiation[8,9]. The photosystem I complex and the ATP synthase are located only in non-appressed thylakoids. Only a small portion of photosystem II is normally located in these regions. The cytochrome b/f complex is distributed in both regions. In contrast, the appressed thylakoids are dominated by photosystem II and its LHC II and do not contain any photosystem I and CF_1-CF_0. More recently the introduction of immunoelectron microscopy has confirmed the extreme spatial segregation of the thylakoid membrane protein complexes[10].

More and more evidence are now accumulating to show that dynamic changes in the composition and organization of the thylakoid membrane are essential for optimizing and protecting the photosynthetic process in response to environmental changes[9,11,12] (Fig. 2). Below we describe how the subfractionation procedure based upon Yeda press fragmentation and phase partition (Fig. 3) can be used to describe

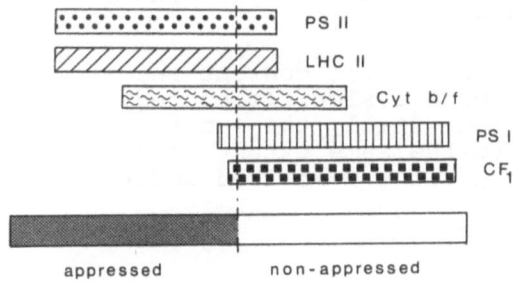

Fig. 1. Relative distribution of the main thylakoid protein complexes between appressed and non-appressed membrane regions as judged by subfractional analysis.

the lateral organization of the thylakoid membrane under conditions of (i) inoptimal excitation to the photosystems (protein phosphorylation); (ii) heat exposure and (iii) photoinhibition.

THYLAKOID DYNAMICS AND PROTEIN PHOSPHORYLATION

Thylakoid membranes have protein kinases which catalyse a rapid phosphorylation of a 25 kDa subunit of LHC II in addition to a slower phosphorylation of a 27 kDa LHC II subunit and other photosystem II subunits[13,14]. When too much light is captured into photosystem II there is an over-reduction of the plastoquinone pool leading to an activation of the kinase activity. Several observations suggest that protein phosphorylation decreases the antenna size of photosystem II while increasing that of photosystem I[11,12,13]. It has therefore been postulated that protein phosphorylation is a molecular mechanism to balance the excitation energy between the two photosystems. It may also provide a way for the plant to protect itself against photoinhibition[15].

When thylakoids that had undergone protein phosphorylation were subfractionated (Fig. 3) the following observations were made[16].

(a) The specific incorporation of [^{32}P] phosphate was considerably higher in the LHC II of stroma thylakoid vesicles than in the LHC II of inside-out vesicles.

(b) The relative amount of LHC II in stroma thylakoids nearly doubled after phosphorylation.

(c) The LHC II units imported into stroma thylakoids showed a relatively high proportion of the 25 kDa subunit as compared to bulk LHC II.

(d) Unless the phosphorylation was prolonged, photosystem II and a LHC II subpopulation containing the 27 kDa subunit remained in the appressed thylakoid region.

(e) A limited destacking (10-20%) occurred as judged by a lower yield of inside-out vesicles.

(f) The lateral rearrangement of LHC II was paralleled by a decrease in the functional antenna size of photosystem II by 15% and an increase of the functional antenna size of photosystem I by 8% (as measured by changes at 650 nm).

(g) At 0° C, no increase of the LHC II content was seen in stroma thylakoids despite a significant degree of protein phosphorylation. Still, the functional antenna size decreased to the same extent as during the room temperature conditions described under a-f.

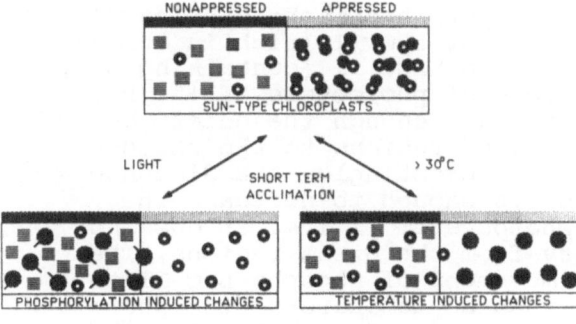

NONAPPRESSED   APPRESSED

SUN-TYPE CHLOROPLASTS

LIGHT   SHORT TERM   › 30°C
          ACCLIMATION

PHOSPHORYLATION INDUCED CHANGES   TEMPERATURE INDUCED CHANGES

○ PHOTOSYSTEM II AND INNER LHC II   ● PERIPHERAL PHOSPHO-LHC II
● PERIPHERAL LHC II   ■ PHOTOSYSTEM I

Fig. 2. Reversible changes
in the lateral organization
of the thylakoid membrane
due to protein phosphorylation and heat exposure.

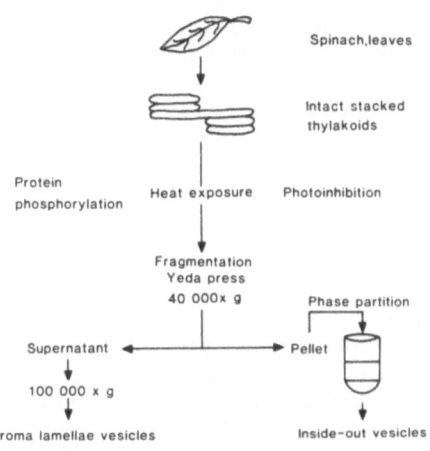

Isolation of inside –out and stroma lamellae vesicles

Spinach,leaves

Intact stacked
thylakoids

Protein   Heat exposure   Photoinhibition
phosphorylation

Fragmentation
Yeda press
40 000x g

                            Phase partition

Supernatant ◄────── ► Pellet

100 000 x g

Stroma lamellae vesicles   Inside-out vesicles

Fig. 3. Subfractional analyses of "stressed" thylakoid membranes.

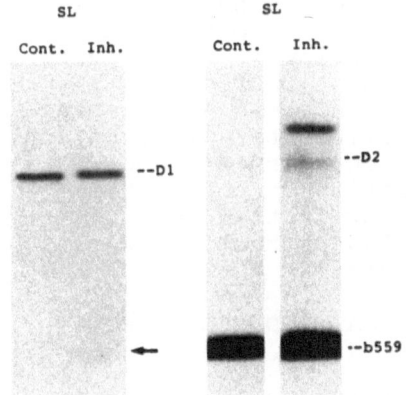

    SL   SL

Cont. Inh.   Cont. Inh.

        --D1   --D2

                          --b559

Fig. 4. Western blot analyses of stroma lamellae
vesicles isolated from
control and photoinhibited
thylakoids.

As illustrated in Fig. 2, protein phosphorylation disconnects the outer subpopulation of LHC II (rich in the 25 kDa subunit) from the inner pool of LHC II (containing only the 27 kDa subunit) and the core of photosystem II[14]. Provided the thylakoid membrane is fluid enough the outer pool of LHC II migrates from the appressed regions to the photosystem I rich stroma thylakoids. Interestingly, the disconnection of the outer LHC II pool is enough to decrease the functional antenna size of photosystem II i.e. it does not require export of LHC II away from the grana regions. The obvious significance of the movement of phospho-LHC II to the stroma thylakoids would then be to increase the photosystem I antenna size. However, as discussed above, we have only seen an 8% increase in the functional chlorophyll b antenna size of photosystem I in isolated stroma lamellae vesicles although the increase in the amount of chlorophyll b per P700 was approximately 36%[17]. One should therefore not exclude additional reasons for lateral migration of phospho-LHC II which may be related to protein turnover and more long term adaptation of light-harvesting. Still these experiments clearly illustrate the importance of a dynamic and flexibel organization of the thylakoid membrane for short term optimization of the photosynthetic process under different light conditions.

THYLAKOID DYNAMICS AND HEAT EXPOSURE

When the subfractionation approach was applied to thylakoids heated above 30°C (Fig. 3), it was observed that structural rearrangements of the photosynthetic apparatus also occurred at elevated temperatures[18] (Fig. 2) in the absence of protein phosphorylation. At such moderately elevated temperatures the outer pool of LHC II dissociates from the photosystem II core and the inner LHC II in the same manner as after protein phosphorylation[18,19]. However, with the heat treatment the pattern of lateral migration is quite the opposite. Now, the outer LHC II remains in the appressed membrane regions while the photosystem II connected to the inner LHC II becomes located in the stroma exposed thylakoids in close vicinity of photosystem I. Nevertheless, the heat induced rearrngements lead to a decreased antenna size of photosystem II. As this reorganization due to brief heat treatment is reversible at temperatures below 45°C it may be yet another regulatory mechanism to protect photosystem II from over-excitation[18]. Such an adaptation would be a useful strategy for plants to use since high temperatures are often accompanied by high light. In addition Ohad has proposed (personal communication) that the migration of photosystem II to the non-appressed regions may allow it to interact with heat-shock proteins thereby allowing protection from damages such as irreversible aggregations.

THE IMPORTANCE OF DIVALENT ANIONS FOR HEAT INDUCED REORGANIZATION OF PHOTOSYSTEM II

The heat induced reorganization of photosystem II and its light harvesting apparatus is strongly dependent on the ionic properties of the medium in which the thylakoid membranes are suspended. In a previous report[18] it was claimed that monovalent cations were required. However, as shown in

Table 1 Ionic requirement for heat induced lateral migration of photosystem II

| | Concentration (mM) | | | I.O.-vesicles |
Na$^+$	Cl$^-$	H$_2$PO$_4$$^-$	H PO$_4$$^{2-}$	chl a/b
0	10	0	0	2.19
10	10	0	0	2.17
10	10	1	4	1.98
15	15	7	3	2.23
15	15	3	7	1.96
15	15	1	9	1.72

Thylakoids were incubated in the indicated media all containing 5 mM Mg^{2+} and heated to 42°C for 5 min. Inside-out vesicles were isolated and analysed with respect to chlorophyll a/b. The ratio for inside-out vesicles isolated from non-heated thylakoids was 2.32. A lowered value represent an increased relative content of LHC II i.e. depletion of photosystem II in the appressed thylakoid regions.

Table 1 there is no correlation between the presence of Na$^+$-ions and the heat induced migration of photosystem II to the stroma thylakoids (as visualized by a lowered chlorophyll a/b value in the inside-out vesicles). Only under conditions where relatively high concentrations of the divalent anion H PO$_4$$^{2-}$ were present a significant change could be seen. Heating in alkaline buffers without divalent ions did not give a lowered chlorophyll a/b ratio indicating that it was not a general pH effect. We suggest that H PO$_4$$^{2-}$ ions destabilize an electrostatic interaction between the outer and inner subpopulations of LHC II and thereby allowing them to dissociate when the temperature becomes sufficiently high. At present, it is not clear whether other divalent anions can substitute for the H PO$_4$$^{2-}$ - ions.

THYLAKOID DYNAMICS AND PROTEIN TURNOVER

There is an obvious need for lateral migration of proteins between the appressed and non-appressed regions in connection with photosystem II protein turnover. At present we do not know much about the mechanism for such lateral trafficing. However, Mattoo and Edelman[20] have recently described transport of newly synthesized D$_1$-proteins from the stroma thylakoids into the appressed regions. The 33.5 kDa precursor of the D$_1$-protein was only found in the stroma thylakoids. While still in the stroma thylakoids it is processed to its mature 32 kDa size. Interestingly, after processing the D$_1$-protein is palmitoylated in an acylation reaction and the acylated protein moves laterally into the appressed membrane regions. The acylchain may serve to direct the protein from the stroma thylakoids to the appressed regions.

Lateral migration of proteins should also be expected during disassembly of photosystem II. Of particular interest are the events following photoinhibition which involve degradation of the reaction centre D$_1$-protein[21], release of extrinsic membrane proteins and manganese[22,23]. Photosystem II needs to be repaired. We have, using the subfractional approach (Fig. 3), studied if the repair mechanism involves lateral migration of damaged photosystem II complexes from the appressed thylakoids to the stroma exposed thylakoids where the repair may take place. Our analysis have revealed that there is an increased level of several photosystem II proteins in stroma thylakoid vesicles isolated from photoinhibited spinach thylakoids. As shown in the immunoblot of Fig. 4 there are increased levels of the D_2-protein and the 9 kDa subunit of cytochrome b-559. The band of approximately 45 kDa probably represents an oligomeric form of cytochrome b-559. This lateral migration of photosystem II subunits does not require protein phosphorylation and is therefore quite similar to that seen after brief heat exposure. As expected there is no increase of the D_1-protein in its 32 kDa form. However, we have in stroma thylakoids after photoinhibition found a 10 kDa polypeptide (Fig. 4, arrow), which cross-reacts with the D_1-protein antibody. Presumably, this is one of the primary proteolytic products of the D_1-protein[24], which previously has not been found in plant thylakoids.

We conclude that photoinhibition and D_1-protein degradation in vitro and possibly also in vivo are followed by a migration of damaged photosystem II units to stroma exposed thylakoids as part of a repair cycle. At present we can not judge whether photosystem II is disassembled and move as individial subunits or if the integral proteins of the D_1-protein depleted complex is still held together and move as one entity.

LONG-TERM ADAPTATION OF PHOTOSYSTEM II LIGHT-HARVESTING

Much of the long-term regulation of the light-harvesting of photosystem II is due to changes in the amount of LHC II. Recently, it has been possible to show that the regulation in the antenna size could be attributed to variation of the the outer 25 kDa polypeptide rich pool of LHC II[25] (Fig. 5). Thus, the same subpopulation of LHC II appears to be responsible for both the short time and long time regulation of the photosystem II antenna size. The small portion of photosystem II units residing in the stroma thylakoids (β-centres)

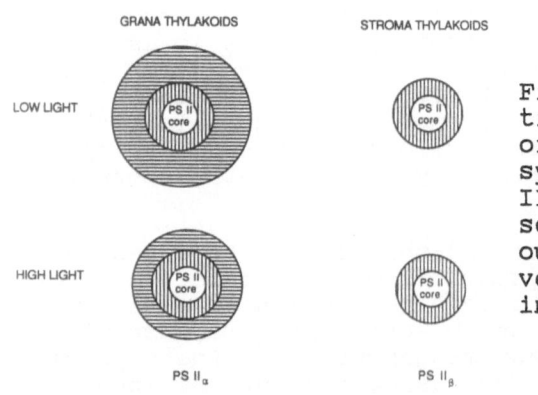

Fig. 5 Schematic illustration of long-term adaptation of the antenna size of photosystem II_α and photosystem II_β. Horizontal lines represent the outer 25 kDa rich outer pool of LHC II while vertical lines represent the inner 27 kDa rich pool.

have a smaller antenna size, which is comprised only of the inner LHC II rich in the 27 kDa polypeptide. By analyzing the antenna size of photosystem II in stroma lamellae vesicles from plants grown under different light conditions we have recently shown that photosystem II_β does not change its LHC II content. Thus, adaptive variation in the antenna size does only apply to those photosystem II units which are located in the appressed thylakoid region (α-centres). Moreover, the inner pool of LHC II does not appear to undergo adaptive changes in its size.

ACKNOWLEDGEMENTS

This project has been supported by the Swedish Natural Science Research Council and the Academy of Finland.

REFERENCES

1. D. I. Arnon, The Light Reactions of Photosynthesis, Proc. Nat. Acad. Sci. 68:2883 (1984).
2. B. Andersson and J. M. Anderson, The Chloroplast Thylakoid Membrane - Isolation, Subfractionation and Purification of Its Supramolecular Complexes, in: "Modern Methods of Plant Analysis", H. F. Linskens and J. F. Jackson, eds., Springer-Verlag, Berlin (1985).
3. P.-Å. Albertsson, "Partition of Cell Particles and Macromolecules (3rd Ed.) Wiley & Sons, New York (1985).
4. P.-Å. Albertsson, B. Andersson, C. Larsson and H.-E. Åkerlund, Phase partition - A method for purification and analyses of cells, organelles and membrane vesicles, in: "Methods in Biochemical Analysis", D. Glick, ed., Wiley & Sons, New York (1982).
5. C. Larsson, Partition in aqueous polymer two-phase system: A rapid method for separation of membrane particles according to their surface properties, in: "Isolation of membranes and organelles from plant cells", J. L. Hall and A. L. Moore, eds. Academic Press, London (1983).
6. B. Andersson, Characterization of the thylakoid membrane by subfractionation analyses, in: "Methods of Enzymology - Plant molecular biology", A. Weisbach and H. Weisbach eds., Academic Press, New York (1986).
7. B. Andersson, C. Sundby, H.-E. Åkerlund and P.-Å. Albertsson, Inside-out thylakoid vesicles - Important tools for the characterization of the photosynthetic membrane. Physiol. Plant. 65:322 (1985).
8. B. Andersson and J. M. Anderson, Lateral heterogeneity in the distribution of chlorophyll-protein complexes of the thylakoid membranes of spinach chloroplasts. Biochim. Biophys. Acta 593:427 (1980).
9. J. M. Anderson and B. Andersson, The dynamic photosynthetic membrane and regulation of solar energy conversion, Trends Biochem. Sci. 13:351 (1988).

10. O. Vallon, F. A. Wollman and J. Olive, Laterial distribution of the main protein complexes of the photosynthetic apparatus in Chlamydomonas reinhardtii and in spinach: an immunocytochemical study using intact thylakoid membranes and a PS II enriched membrane preparation, Photobiochem. Photobiophys. 12:203 (1986).

11. J. Barber, Influence of surface changes on thylakoid structure and function, Annu. Rev. Plant Physiol. 33:261 (1982).

12. L. A. Staehelin and C. J. Arntzen, Regulation of Chloroplast Membrane Function: Protein Phosphorylation Changes the Spatial Organization of Membrane Components, J. Cell. Biol. 97:1327 (1983).

13. J. Bennet, Regulation of photosynthesis by reversible phosphorylation of the light-harvesting chlorophyll a/b protein, Biochem. J. 212:1 (1983).

14. U. K. Larsson, C. Sundby and B. Andersson, Characterization of two different subpopulations of spinach LHC II: polypeptide composition, phosphorylation pattern and association with photosystem II. Biochim. Biophys. Acta, 894:58 (1987).

15. P. Horton, P. Lee, Stimulation of a cyclic electron transfer pathway around photosystem II by phosphorylation of chloroplast thylakoid proteins. FEBS Lett. 162:81 (1983).

16. U. K. Larsson, Modulation in the organization of the light-harvesting antenna of photosystem II - a molecular mechanism for light adaptation in plants. Doctoral Thesis, University of Lund (1987).

17. U. K. Larsson, E. Ögren, G. Öquist and B. Andersson Electron transport and fluorescence studies on the interaction between phospho-LHC and photosystem I in isolated stroma lamellae vesicles. Photobiochem. Photobiophys. 13:29 (1986).

18. C. Sundby and B. Andersson, Temperature-induced reversible migration along the thylakoid membrane of photosystem II regulates its association with LHC-II. FEBS Lett. 191:24 (1985).

19. K. Gounaris, A. P. R. Brain, P. J. Quinn and W. P. Williams, Structural Reorganization of chloroplast thylakoid membranes in response to heat-stress. Biochim. Biophys. Acta 766:198 (1984).

20. A. K. Mattoo, and M. Edelman, Intramembrane translocation and posttranslational palmitoylation of the chloroplast 32 kDa herbicide-binding protein. Proc. Natl. Acad. Sci. 84:1497 (1987).

21. D. J. Kyle, I. Ohad, and C. J. Arntzen, Membrane protein damage and repair: Selective loss of a quinone-protein function in chloroplast membranes, Proc. Natl. Acad. Sci. 81:4070 (1984).

22. I. Virgin, S. Styring and B. Andersson, Photosystem II disorganization and manganese release after photoinhibition of isolated spinach thylakoid membranes, FEBS Lett. 233:408 (1988).

23. I. Virgin, T. Hundal, S. Styring and B. Andersson, Disassembly of photosystem II following photoinhibition, this volume.

24. B. M. Greenberg, V. Gaba, A. K. Mattoo and M. Edelman, Identification of a primary in vivo degradation product of the rapidly-turning-over 32 kDa protein of photosystem II, EMBO J. 6:2865 (1987).
25. U. K. Larsson, J. M. Anderson and B. Andersson, Changes in the relative amount of the peripheral antenna pool of LHC II in response to light variations and thylakoid development. Biochim. Biophys. Acta 894:69 (1987).

STRESS RESPONSES IN CYANOBACTERIA

Lester Packer

Membrane Bioenergetics Group
Lawrence Berkeley Laboratory
Department of Molecular and Cellular Biology
University of California
Berkeley, California 94720

Our laboratory is trying to chart some new areas that offer attractive vistas for research in the response to stress in photosynthetic microorganisms. We use oxygenic photosynthetic cyanobacteria as a model system to examine stress responses. A useful approach has been to take a freshwater cyanobacterium suddenly exposed to seawater concentrations of sodium, and ask the question, "What are the steps involved in the adaptation of the organism to grow in high salt?" In a similar way, we have investigated other kinds of stresses, like temperature stress, toxic metal stress (change of the selenium to sulphur ratio), pH stress and photo-oxidative stress (Belkin et al., 1987). The questions that we seek answers to are: if universal mechanisms of response to stress exist, and what are the specific responses that the organisms show when they are exposed to one or another kind of stress? This is obviously a complex problem that involves changes in gene expression, bioenergetic changes and in the case of salt tolerance, of course, osmoregulatory changes (Blumwald et al., 1983a; Blumwald et al., 1983b).

Salt Stress

The idea is to try and identify the steps involved by seeing what happens to the freshwater cyanobacterium Synechococcus

<u>6311</u> when exposed to 0.5M NaCl. Most experiments have been carried out in a medium in which the cells grow. We emphasize noninvasive studies. If one examines the freshwater cyanobacterium after adding 0.5M salt medium, one finds a 50-80% inhibition of photosynthesis and an inhibition of growth rate; after two days they recover photosynthetic activity and return to normal growth rates (Fry et al.,1986). So the cells have to have some way of overcoming the energetic demand suddenly placed upon them by having to deal with pumping out sodium and also to meet the energetic demands needed to carry on biosynthesis.

We have tried to find out what the lesion is in the photosynthetic electron transport chain. We have looked at the reactions involved between water and Qa and Qb by fluorescence changes. As is known, if Qa is not working, and one adds DCMU, you get a large increase in fluorescence. We were interested to see if NaCl addition would uncouple this system. This is because photosynthesis is being driven to a large extent by the phycobiliprotein system attached to the membranes. Indeed one way of isolating phycobiliproteins is to salt wash the membranes. Thus, we reasoned maybe functional uncoupling of a phycobiliproteins would occur. However, this was found to be only a small effect and actually only a transient effect; it only occurs for a few minutes and then its gone (G. Dubertret, M. Lefort-Tran, S. Spath collaboration). Thus, we know that this is not the site where photosynthetic electron transport is inhibited. Though we still do not exactly know where it is inhibited, we think that it is at the level of the ferredoxins, because ferredoxin-linked enzyme systems are usually very salt sensitive, but we still have not proven it.

Since photosynthesis is inhibited, one would expect that cell energetics would be affected. To study this, we use the 31-PNMR. We inserted nine fiber optical light pipes into an NMR sample tube for an AM 400 spectrometer with provision by capillary tubing for gas exchange (Packer et al., 1987). In this way (figure 1), we can get very good 31-P NMR spectra of <u>Synechococcus</u> <u>6311</u> cell suspensions. One observes nicely resolved mono-ester phosphate and inorganic phosphate peaks, several nucleotide peaks and a pyrophosphate peak. After cells are exposed to 0.5M salt, a large increase in the inorganic phosphate peak occurs and one cannot any longer clearly resolve the monoester phosphate peak. The ATP peak is almost gone, but after about an hour the ATP peak returns,

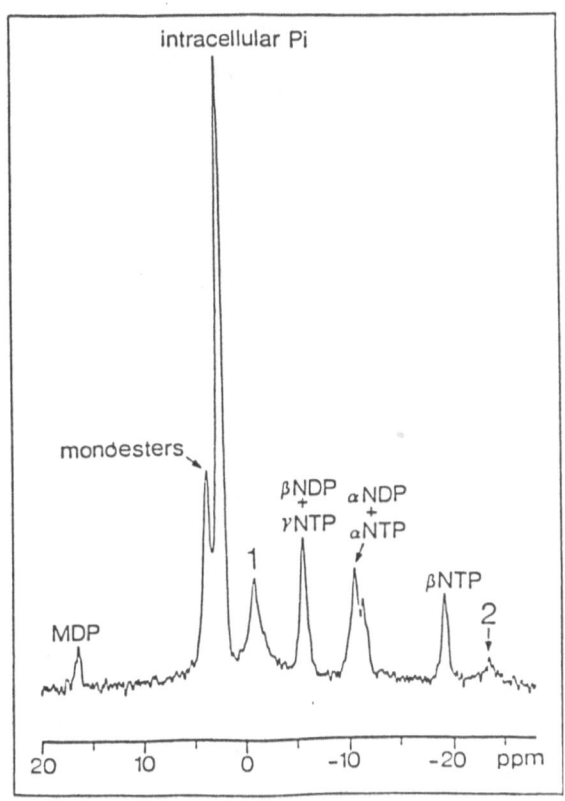

Figure 1. NMR spectra of phosphorous metabolites of intact cells of <u>Synechococcus</u> <u>6311</u> (after Packer et al., 1987)

indicating the cells begin to recover their normal energetic profile of 31P metabolites. To distinguish between the internal and external phosphate, we added an aliquot of pyrophosphate. Inorganic pyrophosphate does not enter the cells, but is hydrolyzed by an extracellular phosphatase; as the pyrophosphate peak disappears, external inorganic phosphate appears to allow one to clearly observe it from the internal peak because of the different pH outside vs. inside the cells.

NMR is also a good method to observe the entry of sodium by 23 sodium NMR. One can separate the internal signal from the external signal by adding membrane impermeable shifts reagents like dysposium salts. This splits the signal, and

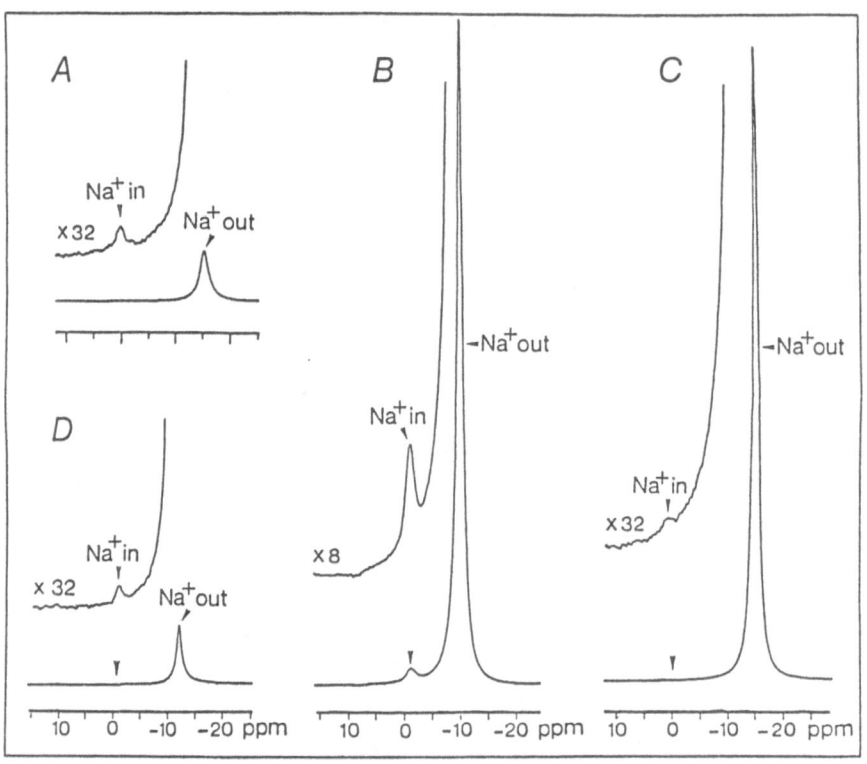

Figure 2. Time sequence of Sodium NMR spectra of saline treated intact cells of <u>Synechococus</u> <u>6311</u> (after Packer et al., 1987) A)Control Cells B) 0.3M NaCl treated cells, 10 minutes C) 0.5M NaCl grown cells, D) Salt-grown cells plus 0.5 mannitol.

one can see that initially there is very little sodium inside the cells (figure 2), but after 0.5M sodium chloride addition, a lot of sodium accumulates inside. So there is very big intrusion of sodium into the cells and subsequent spectra reveal that after some time, the cells exchange most of the sodium out. So a system for sodium exchange gets activated shortly after cells are exposed to NaCl.

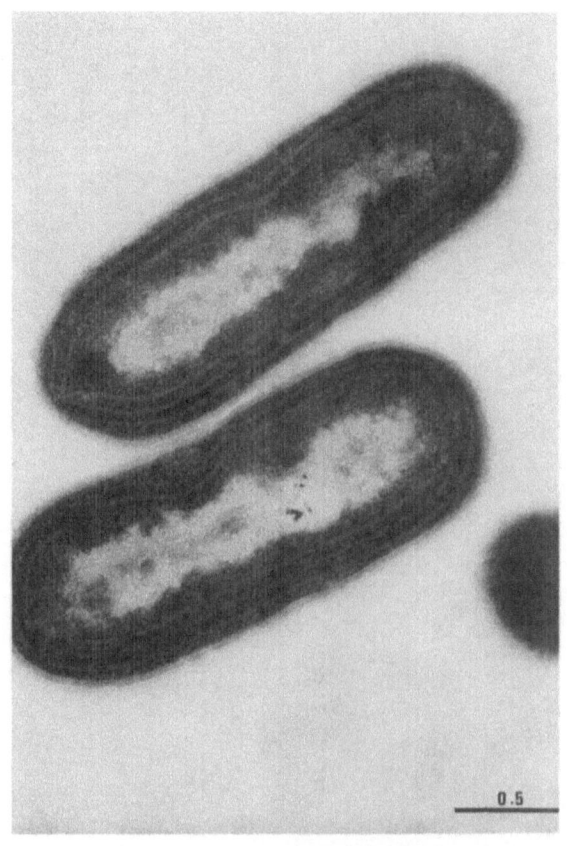

Figure 3.   Ultrastructure of <u>Synechococcus</u> <u>6311</u> at various
stages before and after exposure to 0.5M Nacl (after LeFort-
Trans et al., 1988) A) 1 minute after exposure B) Salt
adapted cells, 48 hours.

(continues)

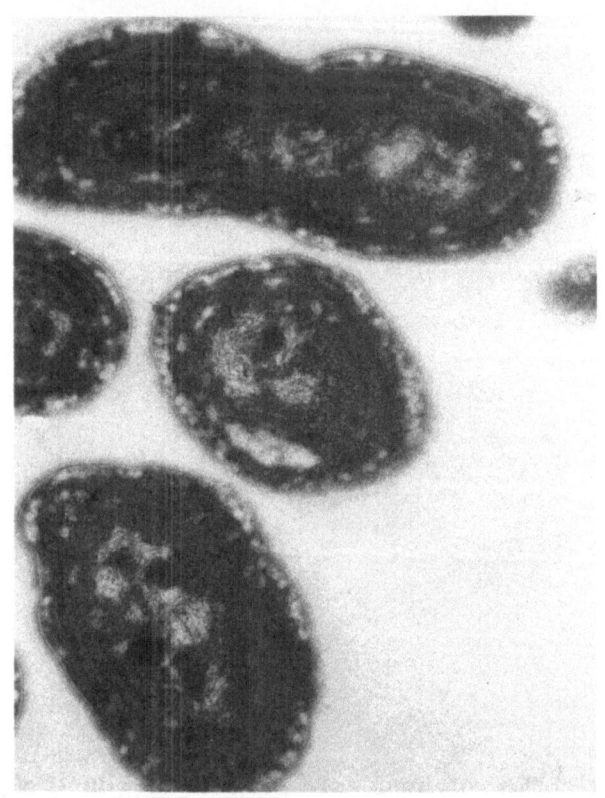

Figure 3.

Another extremely surprising observation that was made by thin section electron microscopy (Lefort-Tran et al., 1988) is that 1 minute after adding sodium to the cell suspension, all signs of the intracellular accumulation products disappear. Cyanobacteria have internal inclusion products like glycogen, carboxysomes, cyanophycin, beta-hydroxybutyrate and polyphosphate as granules. These are numerous different inclusion products that may occur under different conditions. All disappear one minute after salt exposure. Also, the DNA material which is initially coarse in appearance develops a fine more threadlike appearance one minute after salt treatment. So there appears to be some large change in the intracellular physio-chemical environment. Figure 3 A-B summarize the ultrastructural changes. Maybe it is a pH change because when the Na enters a Na+/proton exchange system is activated. We are not exactly sure what brings this about, but it is a very striking finding.

Later, glycogen begins to be resynthesized. Cells grown on 30% enriched C-13 bicarbonate show glycogen accumulation by 13C NMR. This is a good method to show accumulation and loss of glycogen under different environmental conditions (Tel-Or et al., 1986). For example, Argamellum quadruplicatum (PR6), a marine organism, that grows between 1M and 1.5M salt, glycogen accumulation and loss can be followed dynamically with time. After salt growth, glycogen accumulation is massive in Synechococcus 6311, but the chemical and NMR studies do not reveal where the glycogen is located. Using the electron microscope (as described above) one minute after exposure to salt, loss of all glycogen occurs. After 4-5 hours of salt exposure glycogen is re-formed and about 24 hours after exposure, the glycogen is primarily located in the region between the cytoplasmic membrane and the thalycoid membranes. It is almost as if it is a pearl necklace is placed around the edge of the cell (figure 3). After 48 hours, when photosynthesis is completely recovered, a more random, normal appearance of glycogen is observed. However, glycogen granules are still preferentially located between the photosynthetic membranes and the cytoplasmic membrane, but it is also found elsewhere in the cell. The distribution of the glycogen granules, when one examines lot of different cells, shows that 4-5 hours seems to be a critical point. At this time, all of the cells are filled with glycogen again. Whereas one minute after exposure, 71% of the cells do not have any glycogen as detectable by electron microcopy. This striking glycogen accumulation suggests that respiratory activity may change. Perhaps the respiratory activity is located in the cytoplasmic membrane (Lefort-Tran et al., 1988). There is, however, a dispute of C.N. Murata Laboratory with regard to location of cytochrome oxidase and other respiration components in the cytoplasmic membrane.

Indeed, an accumulation of respiratory activity is seen in the salt grown cells, usually it is up to a ten fold increase. Low temperature EPR and Cu+2 EPR signals of cytochrome oxidase are seen in Synechococcus. This signal is seen in the oxidized condition and it is not observed in the reduced sample. We have compared the oxidized Cu^{+2} signal to submitochondrial particles which showed a very similar signal to Synechococcus; not only with respect to the spectra, but also with respect to temperature dependence and power saturation (Fry et al, 1985).

The Na is pumped out and a question is "What kind of activity permits this?" We used several methods to test for the presence of the Na+/proton exchange mechanism, like that present in many bacteria (Blumwald et al., 1984). We used the fluorescence quenching of acridine dyes as one way to test for this activity. We can show that the increased respiratory activity was associated with increased Na+/proton exchange. Together with Margaret Huflejt et al. (1988), we have used the dye, bis(carboxyethyl)-carboxyfluorescein. This dye that penetrates the cell in its uncharged form; after it enters the cell, groups are cleaved by nonspecific esterases to give rise to a charged molecule that is fluorescent and pH sensitive. We have been able to use this dye loading technique to show the activity of the Na+/proton exchange system in Synechococcus 6311. So a new sensitive technique for non-invasively studying cytosolic changes in pH associated with Na+/proton exchange in intact cells is now available.

One cell structure that one would expect to be involved in any adaptation to stress should be the cytoplasmic membrane. The cytoplasmic membrane should contain pumps like Na+/proton exchange, and so we were interested in observing if the structure of the cytoplasmic membrane is altered in salt stressed cells. Together with Lefort-Trans et al. (1988), we looked at the cytoplasmic membranes in the EF and PF fracture faces. The EF fracture face has more particles. Analysis of these particles in the cytoplasmic membranes shows that there is a large shift in the particle size distribution in salt adapted cells. In control cells 54% of the particles are in the 7.5 - 9.5 nanometer diameter range. Whereas after salt adaptation, only 17% of the particles are in the small particle range. In salt adapted cells 60% of the particles are large, whereas in the control cells only 35% of the particles are in the large size class (12-14 nanometer size range). These particle changes may represent a Na+/proton exchange system, or components of the respiration system or whatever else changes in the cytoplasmic membrane in response to stress adaptation.

Toxic Metal, Temperature Stress

Changing the growth temperature or introducing toxic species into the growth medium are examples of stress responses. In toxic metal stress, we have grown Synechococcus 6311 cells in the presence of different selenium to sulfur ratios (Fry et al., unpublished results). If one increases the selenium to sulfur ratio, then one finds a growth inhibition. Also, cells growing at a selenium to sulfur ratio, of 60 to 14, show a large increase in respiration activity (figure 4A). If one adds sulfate back to change the ratio of these two substances to the control ratio, then sulfate acts like it is a competitive inhibitor of the selenium, and the selenium dependent increase in respiration is reversed. Also, if one adds sulfate back, selenium which accumulates comes out (figure 4B). Here then is another case of stress where one finds growth inhibition associated with an increased respiration, and glycogen accumulation. Temperature stress shows similar response. If cyanobacteria growing at

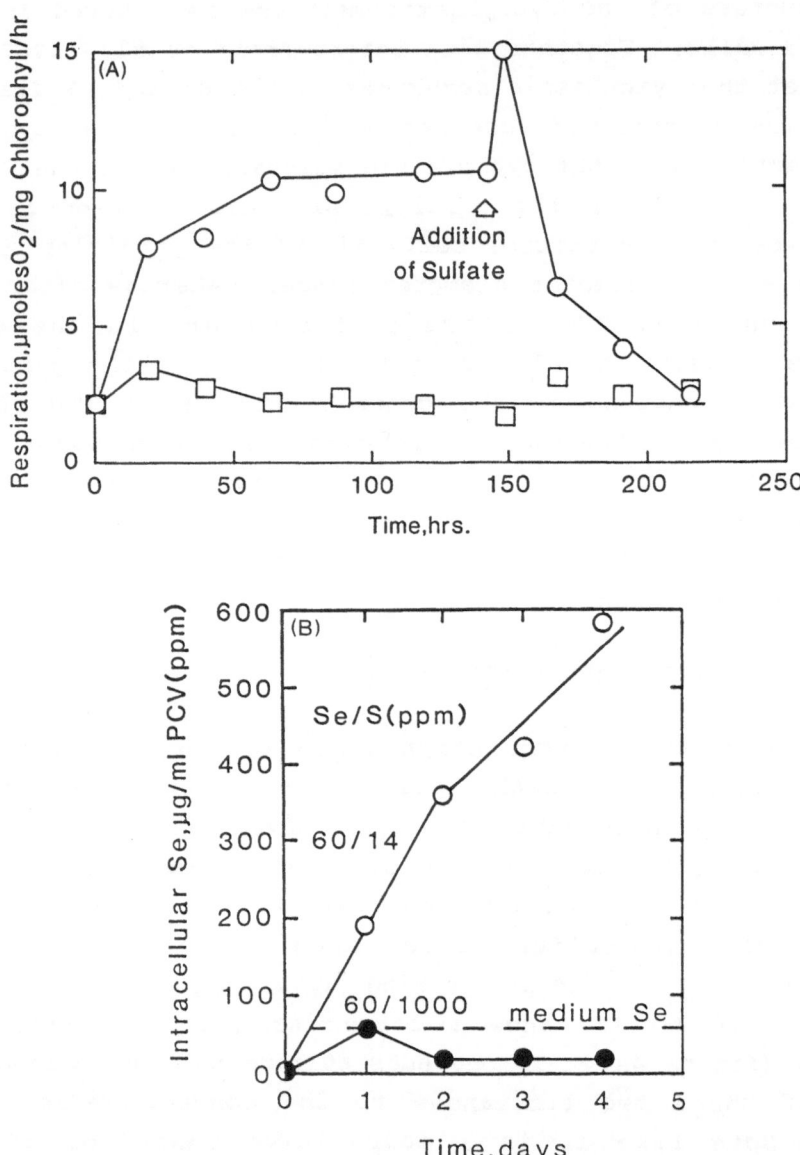

Figure 4)   Effect of changing the selenium to sulur rates on
A) respiration and B) selenium accumulation in <u>Synechococcus</u>
<u>6311</u> (after Fry et al., 1987)

40°C have their temperature decreased to 20°C the growth rate slows, respiration increases, and glycogen accumulates. Thus a universal response to stress seems to be an increase in cell respiration and increased glycogen accumulation. So the cells which are normally obligately phototrophic, come to depend upon respiratory energy and presumably the proton electrochemical potential developed by a respiratory system. We think this respiratory system is probably in the cytoplasmic membrane. Isolated cytoplasmic membranes from salt grown cells show increases in cytochrome C oxidation in the isolated cytoplasmic membranes and EPR methods also show increased Cu^{+2} signals (Fry et al., 1985). The lipid composition (Huflejt, Lang, Packer, unpublished results) and ESR lipid spin label determined order parameters (Khomatov and Packer, unpublished results) is also changed after salt adaptation. There is an increased amount and unsaturation of the fatty acids.

CONCLUDING REMARKS

There are two membrane systems in the cyanobacteria, the thalykoid and the cytoplasmic membranes. Both membranes contain proton ATPase and cytochrome oxidase. The cytoplasmic membrane systems certainly should additionally contain the ion transport systems like the Na+/proton exchange. The thalylcoid membrane contains the photosynthetic system. To what extent respiratory electron transport components in these two membranes are important in stress responses and in the energetics of these cells, still is unresolved. We believe that stress adaptation may provide a tool to be able to change the composition of these membranes perhaps selectively in the cytoplasmic membrane. This situation would allow for studies of the relative contribution to cell energetics of the photosynthetic and respiratory systems and to determine how they interact with each other through electron transport. This question is one of the unsolved problems in the oxygenic photosyntheic cyanobacteria.

ACKNOWLEDGEMENTS

Research described was supported by the NASA Closed Ecological
Life Support System (CELSS) program and the Office of
Biological Energy Research, Department of Energy.

REFERENCES

Belkin S, Mehlhorn, RJ, Packer L (1987) Determination of
Dissolved Oxygen in Photosynthetic Systems by Nitroxide Spin-
Probe Broadening. Arch Biochem Biophys 252(2):487-495

Blumwald E, Mehlhorn, RJ, Packer L (1983) Studies of
Osmoregulation in Salt Adaptation of Cyanobacteria Using ESR
Spin Probe Techniques. Proc Natl Acad Sci 80:2599-2602

Blumwald E, Mehlhorn RJ, Packer L (1983) Ionic Osmoregulation
During Salt Adaptation of the Cyanobacterium Synechococcus
6311. Plant Physiol 73:377-380

Blumwald E, Wolosin JM, Packer L (1984) Na+/H+ Exchange in the
Cyanobacterium Synechococcus 6311. Biochem Biophys Res Commun
122:452-459

Fry IV, Hrabeta J, Giauque RD, Packer L, Uptake and Cellular
Distribution of Selenium by a Primary Producer, the
Cyanobacterium Synechococcus 6311. In Preparation

Fry IV, Huflejt M, Erber WWA, Peschek GA, Packer L (1986) The
Role of Respiration During Adaptation of the Freshwater
Cyanobacterium Synechococcus 6311 to Salinity. Arch Biochem
Biophys 244:686-691

Fry IV, Peschek GA, Huflejt M, Packer L (1985) EPR Signals of
Redox Active Copper in EDTA Washed Membranes of the
Cyanobacterium Synechococcus 6311. Biochem Biophys Res Commun
129:106-116

Huflejt M, Negulescu P, Machen T, Packer L (1988) Na^+ and
Light-Dependent Regulation of Cytoplasmic pH in Cyanobacterium
Synechococcus 6311. Biophys J 53(2):616a

Lefort-Tran M, Pouphile M, Spath S, Packer L, Cytoplasmic
Membrane Changes During Adaptation of the Fresh Water
Cyanobacterium Synechococcus 6311 to Salinity. Plant
Physiology, in press.

Packer L, Spath S, Martin J, Roby C, Bligny R (1987) 23Na and
31P NMR Studies of the Effects of Salt Stress on the Fresh
Water Cyanobacterium Synechococcus 6311. Arch Biochem Biophys
256(1):354-361

Tel-Or E, Spath S, Packer L, Mehlhorn R.J. (1986) Carbon-13
NMR Studies of Salt Shock-Induced Carbohydrate Turnover in the
Marine Cyanobacterium Agmenellum quadruplicatum. Plant
Physiol. 82:646-652

DISASSEMBLY OF PHOTOSYSTEM II FOLLOWING PHOTOINHIBITION

Ivar Virgin, Torill Hundal, Stenbjörn Styring and Bertil Andersson

Department of Biochemistry, Arrhenius Laboratories, University of Stockholm, S-106 91 Stockholm

INTRODUCTION

Photosystem II(PS II) of higher plants and cyanobacteria is a multi-protein complex, where the D1 and D2 proteins are thought to comprise the reaction center. According to this model the D1 and D2 proteins are arranged as a heterodimer analogous to the L and M reaction center subunits of photosynthetic bacteria, carrying all the redox components required for the primary photochemistry of PS II(1,2). In addition, circumstantial evidence suggests that the manganese ions required for water oxidation are associated with the D1 and D2 proteins(3,4). When the photosynthetic apparatous is exposed to excess light the PS II activity is inhibited, and an increased turnover of the D1 protein is induced. In this study the inhibition of PS II activity and the degradation of the D1 protein has been correlated to other changes in the organization of the PS II complex.

MATERIALS AND METHODS

Thylakoid membranes were isolated from spinach leaves essentially as described in(5), and suspended to a concentration of 150 μg chlorophyll/ml in 10 mM sodium phosphate, pH 7.4/5 mM NaCl/5 mM MgCl$_2$/100 mM sucrose. The thylakoid samples were exposed to white light(7000 μE x m^{-2}x s^{-1}) under aerobic conditions at 20°C for the indicated periods of time. The content of individual proteins were determined by quantitative immunoblotting(6). Oxygen evolution were determined polarographically(6). Manganese and cytochrome b$_{559}$ were characterized by EPR(6).

RESULTS

The steady state PS II activity decreased after photoinhibitory treatment(fig 1). The inhibition after 90 minutes of illumination was 95 % and the inhibition half time was approximately 20 minutes. This inhibition pattern was compared to changes in several other PS II parameters.

In order to investigate the effect on the oxidizing side of PS II, the formation of the S$_2$-state multiline EPR signal was studied. This signal originates from the manganese cluster and can be used as a spectroscopic probe for its functional intactness. Photoinhibition results in impairment of the ability to generate the multiline signal. This inhibition correlates well with the inhibition of steady state oxygen evolution(fig 1).

The EPR spectra demonstrates that photoinhibition induces a conversion of the high potential form of cytochrome b_{559} to its low potential form(fig 2). This conversion closely correlates with the inhibition of the activity(fig 1). Furthermore, no electron donation from cytochrome b_{559} to the reaction center could be seen, although the electron donation from the manganese cluster was inhibited. Immunoblotting of the 9 kDa subunit of cytochrome b_{559} revealed unchanged levels following photoinhibition.

The disappearance of the D1 protein, as measured by quantitative immunoblotting, was considerably slower than the inhibition of the PS II activity(fig 1). After 90 minutes, when the inhibition of oxygen evolution was near completion, there was still approximately 50 % of the D1 protein remaining. The level of the D2 protein also decreased, although at a considerably slower rate.

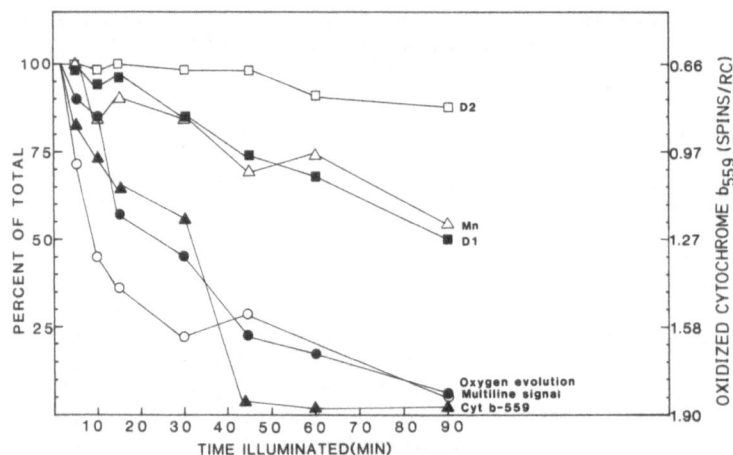

Figure 1. Effects on PS II after illumination of isolated spinach thylakoids.

Since it has been suggested that the D1 and the D2 proteins are involved in manganese binding(3,4), an analysis of total manganese in the thylakoids was of interest. Strikingly, the release of manganese from the membranes closely correlated to the disappearance of the D1 protein(fig 1). The total loss of manganese represented 2 manganese ions/reaction center of PS II(assuming 500 chlorophyll molecules/PS II) corresponding to 4 manganese ions released per D1 protein degraded.

Another consequense of photoinhibition is that the extrinsic 33, 23 and 16 kDa proteins are released from the inner thylakoid surface. In inside-out thylakoid vesicles prepared from illuminated thylakoids there was a loss of the extrinsic proteins in relative amount which correlated to the degree of D1 protein degradation(table 1). The same release could be seen by direct photoinhibitory studies on inside-out thylakoid vesicles.

Preliminary subfraction studies on photoinhibited thylakoids suggests that there are increased levels of PS II proteins in the non-appressed thylakoid regions. Thus, we have been able to detect about three times higher relative amounts of cytochrome b_{559} compared to normal in stroma lamellae vesicles isolated from photoinhibited thylakoids. This shows that lateral migration of proteins is essential for the degradation/-reassembly of PS II following photoinhibition and D1-protein turnover.

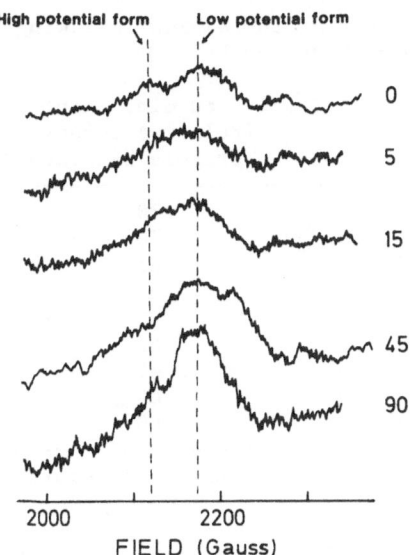

Figure 2. EPR spectra showing the high and low potential form of cytochrome b559 following photoinhibition.

DISCUSSION

Our results demonstrate that photoinhibition primarily destroys the photochemical reactions, and that protein degradation is a subsequent event in agreement with previous studies(7).

Of special interest was the finding that the impairment of the manganese multiline signal correlated with the decrease of the steady state activity. This suggests that the inhibition resulted from an inability to perform stable charge separation rather than by slowing down several steps in the electron transfer chain. It should be noted that since the illumination during measurements of the multiline signal was done at 200 K there was no involvment of the Q_B-site in the inhibited reaction.

Our results also show a conversion of cytochrome b559 high potential form to its low potential form and that this conversion correlates with the inhibitory process. This demonstrates that photoinhibition results in changes in the protein conformation of PS II prior to the breakdown of the D1 protein.

The fact that there is a strong correlation between the release of manganese from the thylakoid membrane and the degradation of the D1 protein is intriguing. One likely interpretation is that the D1 protein binds manganese. Due to the degradation of the D1 protein, manganese is lost from the active site on the lumenal side of the thylakoid membrane and leaks subsequently through the membrane. In addition, the stoichiometry of the manganese release shows that all four manganese ions are associated with the D1 protein. Alternatively, the release of manganese could be due to a general disassembly of the PS II complex following D1 degradation. Indeed, the release of the three extrinsic proteins from the inner thylakoid surface points to additional changes in the protein organization of PS II. However, it is unlikely that the loss of the extrinsic proteins causes the release of the manganese. When the 33 kDa protein is released from thylakoids e.g by $CaCl_2$ treatment, there is only a slow release of manganese which do not involve all four manganese ions(8,9).

CONCLUSIONS

These data describe several sequential events following photoinhibition of PS II.

Table 1. Polypeptide content of thylakoids following 60 minutes of illumination as revealed by immunoblotting.

		60 min. light (relative amount of protein retained in %)
Experiment A	D1	70
	33 kDa	70
	23 kDa	65
	16 kDa	60
Experiment B	D1	55
	33 kDa	65
	23 kDa	60
	16 kDa	60

Experiment A: Inside-out thylakoid vesicles prepared from illuminated intact thylakoids.

Experiment B: Inside-out thylakoid vesicles directly illuminated.

i)   Photochemical damage leading to loss of stable charge separation.

ii) Conformational changes of PS II reflected in an altered potential of cytochrome b_{559}.

iii) D1 protein degradation and manganese release.

iv) Release of extrinsic proteins from the inner thylakoid surface.

v)   D2 protein degradation.

vi) Increased levels of PS II proteins in stroma exposed thylakoids.

REFERENCES

1. H. Michel and J. Deisenhofer, in: "Photosynthesis III," L.A Staehelin and C.J. Arntzen, eds. pp. 371-381, Springer-Verlag, Heidelberg(1986).
2. A. Trebst, Z. Naturforsch. 41C, 240-243(1986)
3. B. Andersson, R.T. Sayre and L. Bogorad, Chemica Scripta. 27 B, 195-200 (1987).
4. G.C. Dismukes, Chemica Scripta, in press.
5. B. Andersson, H.-E. Åkerlund and P.Å Albertsson, FEBS Lett. 77. 141-145 (1977).
6. I. Virgin, S. Styring and B. Andersson, FEBS Lett. 233, 408-412(1988).
7. I. Ohad, D.J. Kyle and J. Hirschberg, EMBO J. 4, 1655-1659(1985).
8. M. Miyao and N. Murata, FEBS Lett. 170, 350-354(1984).
9. T.-A. Ono and Y. Inoue, FEBS Lett. 168, 281-286(1984).

DAMAGING PROCESSES IN THE REACTION CENTER AND ON ITS ACCEPTOR SIDE DURING PHOTOINACTIVATION OF PSII PARTICLES

Ladislav Nedbal, and Jiři Masojídek

Institute of Microbiology, Czechoslovak Academy of Sciences, 379 81 Třebon
Czechoslovakia

The photoinactivation of photosystem II is a very complex process which gradually affects all of its parts. A damage has been reported to occur in the reaction center itself /P680-Pheo/[1,2,3,4] as well as on its donor /Z,D,Mn/S-states/[5,6] and acceptor /Q_A,Q_B/[7,8,9,10] sides. It is not clear, yet, where the photoinactivation starts, in what order the secondary processes occur and what the underlying mechanism of the damage is. These questions are addressed also in the present paper.

Oxygen evolving photosystem II particles were isolated from peas as in Ref.11. The particles /15 µg Chl/ml/ were exposed to 250 W/m^2 white light at 20 °C in a cuvette 20 mm thick under aerobic /further O-conditions/ and under anaerobic strongly reducing /R-conditions, 0.5 mg dithionite/ml/ conditions. The exposure medium and the photoinactivation treatment have been described in Ref.8. The samples taken during the light treatment were spun down, resuspended in a fresh medium containing oxygen and only then the fluorescence and absorption measurements were done. The variable and constant components of the chlorophyll fluorescence were measured in a phosphoroscope. The difference absorption measurements were done using the dual-wavelength spectrophotometer Shimadzu UV-3000 with a cross illumination unit. The pheophytin photoreduction was measured in samples with added dithionite /E_h=-250 mV/ as a light-dark difference of the absorption at 685 nm /ΔA_{685}/. To ensure that the signal is not due to fluorescence or to chlorophyll bleaching, we measured ΔA_{650} and ΔA_{676} in parallel. The ratio between these signals remained the same during the photoinactivation. Chlorophyll-protein analysis of the particles was carried out under non-denaturing conditions[12] using the buffer system developed by Laemmli[13].

Three types of photoinactivation processes with different kinetics could be distinguished: /1/ the fast process /$t_{1/2}$=1min/, /2/ the slow process /$t_{1/2}$=15min/ and /3/ the very slow process /$t_{1/2}$=110min/. /1/ occurs only under the R-conditions, while /2/ and /3/ take place under both O- and R-conditions.

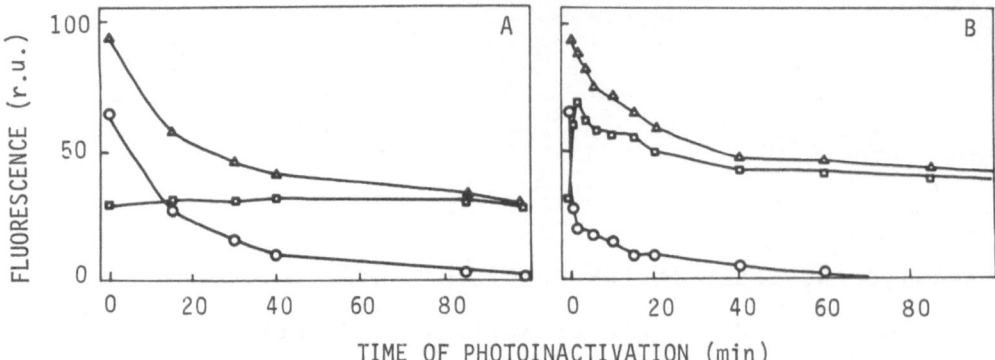

Fig.1. Fluorescence behaviour of the particles during
       photoinactivation under aerobic /A/ and anaerobic
       strongly reducing /B/ conditions.
       /□/ - constant fluorescence Fo; /O/ - variable
       fluorescence Fv; /△/ - maximal fluorescence
       Fmax=Fo+Fv.

The fast process /Fig. 1B/ results in a parallel decline
of variable fluorescence Fv /curve O/ and of Hill reaction
rate /see Refs.7,8/ which is accompanied by an antiparallel
increase of constant fluorescence Fo /curve □/. We assume
that trapping of Q_A in a negatively charged stable state is
responsible for the effects observed.
     The slow process is characterized by a decline of
maximal fluorescence Fmax /curves △ in Figs.1A,B/. The rate
of the decline is similar under both O- and R-conditions.
Under O-conditions /Fig.1A/, this decline is due to the well
known disappearance of Fv /curve O/. Under R-conditions
/Fig.1B/, the slow decline of Fmax /curve △/ represents the
disappearance of the increment of Fo /curve □/ generated by

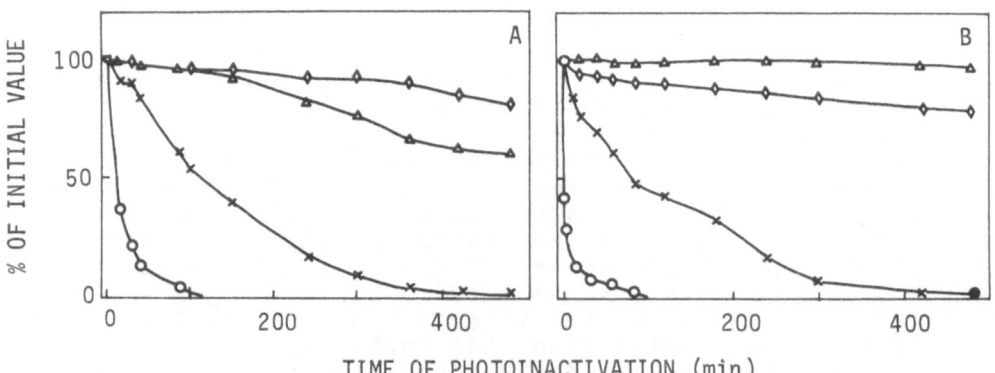

Fig.2. Relative changes of variable fluorescence /O/, of
       photoaccumulation of reduced pheophytin - ΔA685 /x/
       and of chlorophyll content /△/ of the particles during
       photoinactivation under aerobic /A/ and anaerobic
       strongly reducing /B/ conditions. The variable
       fluorescence of dark control is also shown /◊/.

TIME OF PHOTOINACTIVATION (hours)

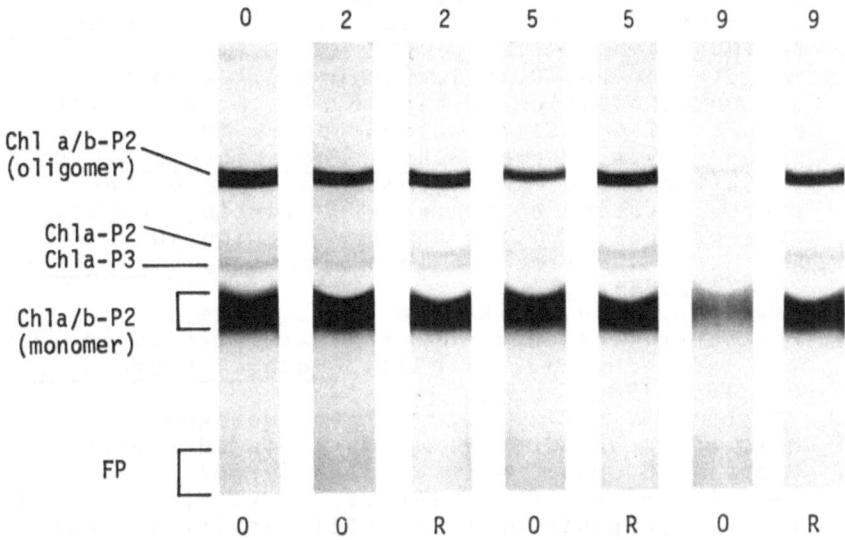

Fig.3. Changes of chlorophyll protein composition of the particles during photoinactivation under aerobic /O/ and anaerobic strongly reducing conditions /R/. Nomenclature of the complexes as in Ref.12: Chla-P2 and Chla-P3 correspond to CP43 and CP47, respectively.

the fast process. We assume that the slow process consists in neutralization of the negative charge in the domain of Q_A in a manner that renders Q_A non-functional. The alternative interpretation suggesting the damage of PSII donor side seems improbable as a cause for the fluorescence decline because dithionite which is an effective PSII electron donor does not restore the initial Fmax when added during the fluorescence measurements.

The very slow process proceeds also with the same rate under both O- and R-conditions. It is seen as a decline of the ΔA_{685} signal /Figs.2A,B, curves x/ in presence of dithionite which results from the photoaccumulation of the reduced pheophytin. We assume that it is due to an impairment of the primary charge separation.

The photoinactivation is not necessarily related to chlorophyll bleaching /Figs.2A,B, curves △/ or to degradation of chlorophyll proteins /Fig.3/ since these were observed only under O-conditions. Also, the changes in polypeptide pattern occured only under O-conditions /not shown/. It can be concluded that the protein degradation <u>in vitro</u> is only a consequence of the photoinactivation.

References

1. B. Arntz and A. Trebst, On the role of the Q_B protein of PSII in photoinhibition, <u>FEBS Lett.</u> 194:43 /1986/
2. S. Demeter, P. J. Neale and A. Melis, Photoinhibition: Impairment of the primary charge separation between P680 and pheophytin in PSII of chloroplasts, <u>FEBS Lett.</u> 214:370 /1987/

3. K. Satoh and D. C. Fork, Photoinhibition of reaction centers of photosystems I and II in intact Bryopsis chloroplasts under anaerobic conditions, Plant Physiol. 70:1004 /1982/

4. R. E. Cleland and C. Critchley, Studies on the mechanism of photoinhibition in higher plants. Inactivation by high light of PSII reaction center function in isolated chloroplasts, Photobiochem. Photobiophys. 10:83 /1985/

5. F.E. Callahan and G.M. Cheniae, Studies on the photoactivation of the water-oxidazing enzyme. I. Processes limiting photoactivation in hydroxylamin-extracted leaf segments, Plant Physiol. 79:777 /1985/

6. S. M. Theg, L. J. Filar and R. A. Dilley, Photoinactivation of chloroplasts already inhibited on the oxidizing side of PSII, Biochim. Biophys. Acta 849:104 /1986/

7. L. Nedbal, E. Šetlíková, J. Masojídek and I. Šetlik, The nature of photoinhibition in isolated thylakoids, Biochim. Biophys. Acta 848:108 /1986/

8. A. Allakhverdiev, E. Šetlíková, V. V. Klimov and I. Šetlik, In photoinhibited PSII particles pheophytin photoreduction remains unimpaired, FEBS Lett. 226:186 /1987/

9. D. J. Kyle, I. Ohad and C. J. Arntzen, Membrane protein damage and repair: selective loss of a quinone protein function in chloroplast membranes, Proc. Natl. Acad. Sci. USA 81:4070 /1984/

10. A. K. Mattoo, H. Hoffman-Falk, J. B. Marder and M. Edelman, Regulation of protein metabolism: coupling of photosynthetic electron transport to in vivo degradation of the rapidly metabolized 32-kD protein, Proc. Natl. Acad. Sci. USA 81:1380 /1984/

11. N. I. Shutilova, V. V. Klimov, V. A. Shuvalov and V. M. Kutyurin, Isolation and investigation of photochemical and spectral properties of PSII subchlorplast fragments which are highly purified from PSI admixtures, Biofizika /russ./ 20:844 /1975/

12. O. Machold, Relationship between the 43 kDa chlorophyll protein of PSII and the rapidly metabolized 32 kDa Q_B protein, FEBS Lett. 204:363 /1986/

13. U. K. Laemmli, Cleavage of structural proteins during the assembly of the head of bacteriophage T4, Nature 227:680 /1970/

PHOTOINHIBITION OF REACTION CENTRE ACTIVITY IN PHOTOSYSTEM TWO PREPARATIONS WITH REDUCED RATES OF WATER OXIDATION

Weiqiu Wang, David J. Chapman and James Barber

AFRC Photosynthesis Research Group, Department of Pure & Applied Biology, Imperial College of Science, Technology & Medicine London SW7 2BB, UK

INTRODUCTION

Significant reduction in photosynthetic rate and subsequent loss of crop yield can occur as a result of increases in light intensity in field environments, particularly when there is an additional stress factor such as that of low temperature (Powles, 1984). Although attempts to understand the molecular basis of this photoinhibition have been largely unsuccessful, it has been clearly shown that the primary sites of damage are the photosystems which drive the electron transfer reactions of photosynthesis (Kyle, 1987). Inactivation of the photosystems arises largely because the many pigment molecules in the leaf absorb and then transfer light energy to the reaction centre of the photosytems and this focus of energy gives the potential for initiation of damaging reactions. Damage is likely to occur if dissipation of energy is not achieved by the maintenance of sufficiently high rates of electron flow and it has been suggested that factors which enhance photoinhibition, such as low CO_2 levels, could act through reducing rates of electron transfer (Powles, 1984). In this context there are many points at which electron flow could be inhibited although the main locus for photodamage seems to be photosystem two (PS2) (Kyle, 1987). Here we report our work on the significance of impaired water oxidation and electron donation to the PS2 reaction centre.

MATERIALS AND METHODS

Previously described methods (Chapman et al., 1988) were used to prepare PS2 enriched membrane preparations from chloroplasts of Pisum sativum plants and measure rates of oxygen evolution using a Clark type oxygen electrode. In the latter case 0.5 mM phenylbenzoquinone (PBQ) has been used as an artificial electron acceptor. Electron transfer was also assayed by reduction of 0.01 mM 2,6-dichlorophenol-indophenol (DCPIP) with or without 0.5 mM diphenylcarbazide (DPC) as an artificial electron donor. PS2 preparations were photoinhibited by treatment at 1200 μE m^{-2}s^{-1}, 10°C and 30 μg chlorophyll.ml^{-1} in a buffer of 20 mM MES pH 6.5, 5 mM MgCl$_2$, 15 mM NaCl and 10% glycerol. Alkaline pH treatment was by resuspension of PS2 membrane pellets (centrifuged at 40,000 xg, 4°C, 20 min) in 20 mM Bis Tris Propane (BTP) buffer, 5 mM MgCl$_2$, 15 mM NaCl and 10% glycerol; high salt washes were by resuspension in either 1.2 M NaCl or 1 M CaCl$_2$ in a buffer as above except using MES pH 6.5 in place of BTP; and the citrate buffer treatment was by resuspension in 10 mM citric acid-sodium citrate buffer, 400 mM sucrose and 20 mM NaCl, pH 3.0.

RESULTS

PS2 enriched membrane preparations were washed at a range of different pH values and then resuspended at pH 6.0 for assay of electron transfer activity. As reported previously (Chapman et al., 1988) inhibition of water to DCPIP, or water to PBQ, activity by alkaline pH was observed. However, as shown in Fig.1 this was not paralleled by an inactivation of the PS2 reaction centre as monitored by reduction of DCPIP in the presence of the artificial electron donor DPC.

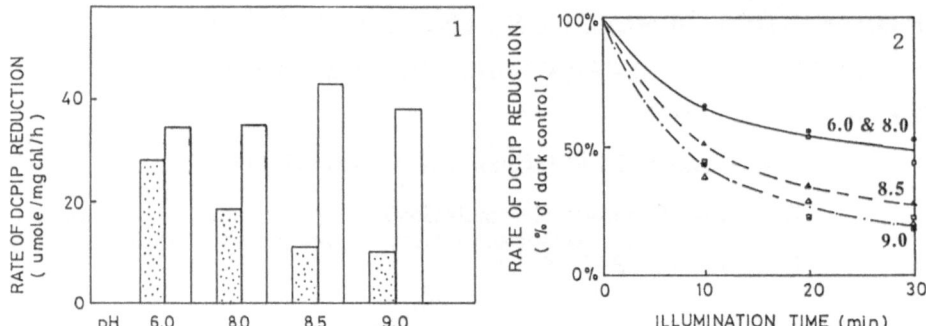

Fig.1   Effect of pH treatment on photosynthetic electron transport activity of PS2 enriched membrane preparations. After incubation at the pH indicated for 30 min in the dark at 10°C, samples were centrifuged, resuspended at pH 6.5 and asayed in the absence (shaded) or presence (unshaded) of DPC as an electron donor.

Fig.2   Time course of the photoinactivation of PS2 photosynthetic electron transport activity in samples pretreated at different pH values. Samples were incubated at different pH values, centrifuged and resuspended as in Fig.1 and then exposed to light of 1200 $\mu E.m^{-2}.s^{-1}$ at 10°C before assay of DCPIP photoreduction with DPC as the electron donor. Rates are given as percentages of rates in the dark controls which were incubated for equivalent times at 10°C.

In Fig.2 results are given for the effect of light on the preparations after different pH wash treatments. In all cases exposure to light inhibited rates of electron transfer from DPC to DCPIP. However, the effect was most pronounced after washes at pH 8.5 and 9.0. Thus, the greatest photoinhibition of reaction centre activity was in the samples treated so as to inhibit water oxidation activity. Measurements of rates of reduction of DCPIP in the absence or presence of DPC demonstrated that 1.2 M NaCl, 1 M $CaCl_2$ or citrate buffer at pH 3.0 were all effective means of inhibiting water oxidation while maintaining full reaction centre activity (Fig.3). After all three of these treatments exposure to light resulted in photoinhibition of the DPC to DCPIP rate (Fig.4). In the case of NaCl and $CaCl_2$ wash treatments the inhibition of water oxidation was reversed by adding $CaCl_2$ prior to the assay (Fig.3) and in samples treated in this way there was a reduced effect of exposure to light (Fig.4). The artificial electron donor DPC was used to further assess the role of electron donation to the reaction centre in protecting against photoinhibition. Addition of DPC prior to illumination significantly reduced the initial rate of light induced loss of reaction centre activity (Fig.5).

DISCUSSION

The four different methods used for inhibition of water oxidation in this paper are known to have different effects on the molecular structure of the PS2 complex. Extrinsic 17 and 23 kDa proteins are removed by alkaline pH up to 8.5 (Chapman et al., 1988) and 1.2 M NaCl (Miyao and Murata, 1983), the 17, 23 and 33 kDa proteins are removed by pH 9.0 (Kuwabara and Murata, 1982; Chapman et al., 1988) and 1 M $CaCl_2$ (Ono and Inoue, 1983). Citrate buffer at pH 3.0 inhibits water oxidation without the loss of extrinsic proteins although Ca^{2+} depletion does occur (Ono et al., 1988). We have shown that inactivation of water oxidation by all four methods

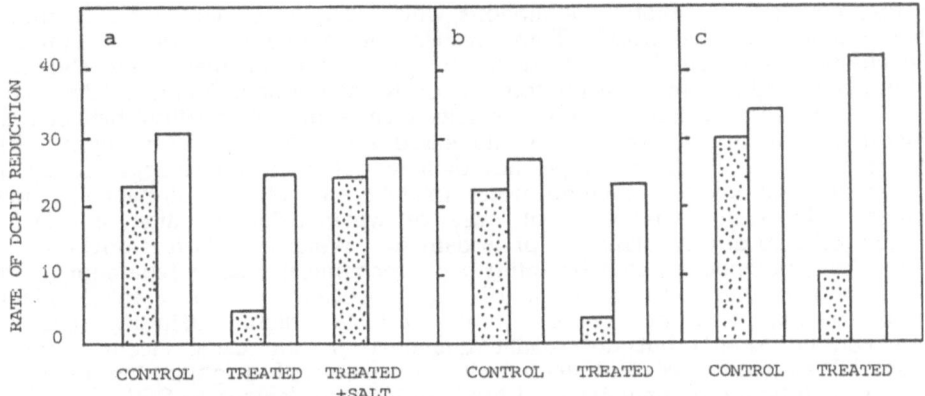

Fig.3 Photosynthetic electron transport activity of PS2 enriched membranes treated by incubation in a) 1.2 M NaCl, b) 1 M CaCl$_2$ and c) citrate buffer pH 3.0. After treatment the samples were centrifuged and resuspended at pH 6.5 for assay of DCPIP reduction in the absence (shaded) or presence (unshaded) of DPC as an electron donor. Prior to assay a salt addition (10 mM CaCl$_2$ + 30 mM NaCl) was made as indicated (+ salt). Rates are given as μ moles DCPIP.mg chl^{-1}.h^{-1}.

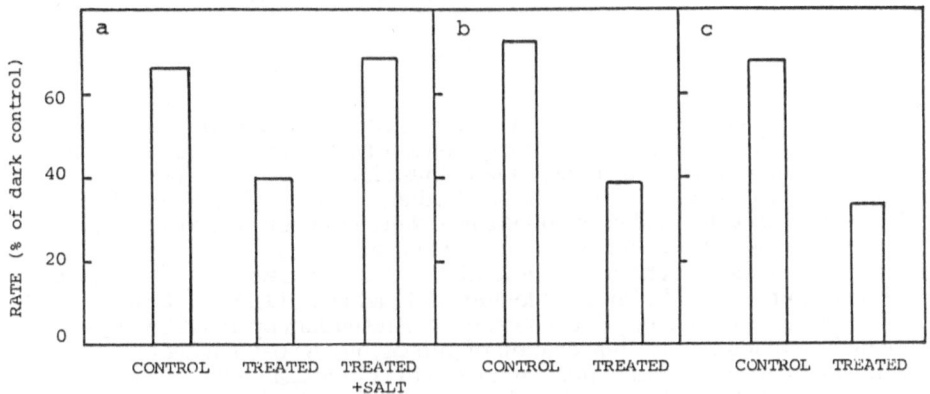

Fig.4 Photoinactivation of PS2 photosynthetic electron transport activity in samples pretreated by incubation in a) 1.2 M NaCl, b) 1 M CaCl$_2$ and c) citrate buffer pH 3.0. Treated samples were centrifuged and resuspended at pH 6.5 as in Fig.3 and then exposed to light of 1200 μE.m^{-2}.s^{-1} at 10°C for 20 min before assay of DCPIP photoreduction with DPC as the electron CaCl$_2$ + 30 mM NaCl) was made as indicated (+ salt). Rates are given as percentages of rates in dark controls also incubated at 10°C for 20 min.

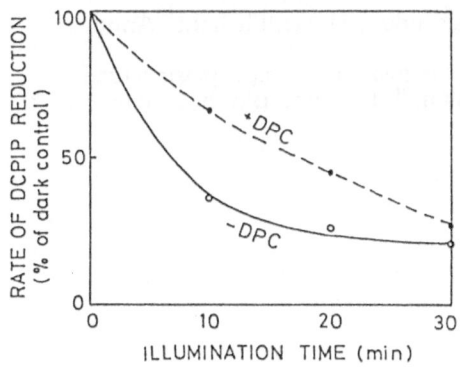

Fig.5 Effect of addition of DPC on the photoinactivation of PS2 enriched membranes pretreated by washing at pH 9.5. After incubation at pH 9.5 samples were centrifuged, resuspended at pH 6.5 and exposed to light, as in Fig.2, in the absence (continuous line) or presence (dashed line) of 1 mM DPC. DCPIP photoreduction was assayed after addition of DPC to all samples and rates are given as percentages of dark controls incubated at 10°C for the same times as light treatments.

gave PS2 enriched membrane preparations which were more vulnerable to photo-inhibition than active controls. These results corroborate the work of others in which chloride depletion, Tris washing or hydroxylamine treatment were shown to result in similar enhancement of photoinhibition (Callahan and Cheniae, 1986; Theg et al., 1986). Taken together these results give a firm correlation between the inhibition of water oxidation and the increased sensitivity of PS2 to light. The existence of a causal relationship is not definite but is strongly suggested by the fact that all methods of inhibiting water oxidation resulted in enhanced photoinhibition. In addition recovery of rates of water oxidation after inactivation also gave recovery of a resistance to photodamage. Water oxidation is driven by the high oxidising potential created by primary photochemical charge separation in the PS2 reaction centre. As a result of the inactivation of the water splitting reaction, without loss of reaction centre activity, highly oxidising chemical species can be formed but not readily quenched by the usual electron transfer process. Thus destructive and uncontrolled oxidation of neighbouring molecules could arise and be the primary cause of the light induced damage to PS2. Indeed the involvement in photoinhibition of a more highly oxidising component in PS2 ($\sim +1.2$ V) than in PS1 ($\sim +0.5$ V) could be one of the main reasons why PS2 has a greater susceptibility to photodamage.

ACKNOWLEDGEMENTS

We are grateful to the Agricultural and Food Research Council and one of us (W.W.) is the recipient of a Chinese Academy of Sciences Award.

REFERENCES

Callahan, F.E., Becker, D.W. and Cheniae, G.M. Studies on the photoactivation of the water-oxidising enzyme II. Characterisation of weak light photoinhibition of PSII and its light-induced recovery. Plant Physiology 82;261-269 (1986)

Chapman, D.J., De Felice, J., Davis, K. and Barber, J. Effect of alkaline pH on photosynthetic water oxidation and the association of extrinsic proteins with photosystem two. Biochem. J. 258 in press (1989)

Kyle, D.J. The biochemical basis for photoinhibition of photosystem II. In: Topics in Photosynthesis Vol.9. Photoinhibition ed. Kyle, D.J., Osmond, B. and Arntzen, C.J., Pub. Elsevier Science Publishers, B.V., Amsterdam pp 197-226 (1987)

Kuwabara, T. and Murata, N. Inactivation of photosynthetic oxygen evolution and concomitant release of three polypeptides in the photosystem II particles of spinach chloroplasts. Plant & Cell Physiol. 23;533-539 (1982)

Miyao, M. and Murata, N. Partial disintegration and reconstitution of the photosynthetic oxygen evolution system. Binding of 24 kD and 18 kD polypeptides. Biochim. Biophys. Acta 725;87-93 (1983)

Ono, T. and Inoue, Y. Mn-preserving extraction of 33-, 24- and 16-kDa proteins from O_2-evolving PSII particles by divalent salt-washing. FEBS Lett. 164;255-260 (1983)

Ono, T. and Inoue, Y. Discrete extraction of the Ca atom functional for O_2 evolution in higher plant (Photosystem II) by a simple low pH treatment. FEBS Lett. 227;147-152 (1988)

Powles, S.B. Photoinhibition of photosynthesis induced by visible light. Annu. Rev. Plant Physiol. 35;15-44

Theg, S.M., Filar, L.J. and Dilley, R.A. Photoinactivation of chloroplasts already inhibited on the oxidising side of photosystem II. Biochim. Biophys. Acta 849;104-111

PHOTOINHIBITION OF ELECTRON TRANSPORT ACTIVITY OF PHOTOSYSTEM II IN ISOLATED THYLAKOIDS STUDIED BY THERMOLUMINESCENCE AND DELAYED LUMINESCENCE

Imre Vass, Narendranath Mohanty and Sándor Demeter

Institute of Plant Physiology, Biological Research Center
of the Hungarian Academy of Sciences
P.O.Box 521, H-6701 Szeged, Hungary

INTRODUCTION

Photosynthetic organisms exposed to higher light intensity than that required to saturate photosynthesis gradually lose their photosynthetic capacity. The phenomenon is called photoinhibition and related to a damage in photosystem II (PS II) [1-5]. However, the opinions differ concerning the exact site of photodamage. Works using algae cells suggest that the Q_B binding protein is injured [2,9]. On the other hand the majority of experiments on isolated chloroplasts advocate a primary site of photoinhibition in the P_{680}-Pheo-Q_A section of the electron transport chain [3-5]. Recently thermoluminescence (TL) and delayed luminescence (DL) proved to be useful methods in the investigation of PS II photochemistry. Well characterized TL and DL signals are arising from the $S_2Q_A^-$ and $S_2Q_B^-$ charge recombinations [6-8]. Thus it is expected that the effect of photoinhibition on the Q_A and Q_B acceptors can be easily followed by these techniques. Recent TL investigation of photoinhibition in Chlamydomonas reinhardii cells led to the conclusion that in the first stage of photoinhibition the Q_B binding site is modified while the Q_A acceptor is only slightly influenced [9]. Considering that in isolated thylakoids the process of photoinhibition may differ from that occurring in intact cells we carried out TL and DL measurements of photoinhibited spinach thylakoids.

MATERIALS AND METHODS

Thylakoids were isolated from market spinach as usual and suspended in a medium composed of 50 mM HEPES (pH 7.5)/5mM $MgCl_2$/10 mM NaCl/0.4 M sorbitol to give 1-2 mg Chl/ml and stored on ice until use. For photoinhibitory treatment samples were diluted with the suspension medium to 100 μg Chl/ml and illuminated with white light of 500 W/m^2 intensity in a flat Petri dish at 4 $^{\circ}$C for various periods of time. Thermoluminescence was excited with a single flash given at 5 $^{\circ}$C. Glow curves were recorded from -40 $^{\circ}$C to 80 $^{\circ}$C at a heating rate of 20 $^{\circ}$C/min [7]. The single-flash induced delayed light emission was detected at 20-25 $^{\circ}$C at 30 μg Chl/ml [8]. The rate of photosynthetic oxygen evolution was measured at saturating light intensities using a Clark-type electrode at 25 $^{\circ}$C in the presence of 1.5 mM dimethylbenzoquinone as electron acceptor. Fluorescence induction was measured at 10 μg Chl/ml.

RESULTS AND DISCUSSION

Untreated thylakoids excited by a saturating flash at 5°C exhibited a TL band at about 28°C (Fig. 1. left side). This band originates from the $S_2Q_B^-$ charge recombination [6,7]. In the presence of DCMU, the B band was replaced by a band at about 2°C (Fig. 1. right side). This band arises from $S_2Q_A^-$ recombination and is called the Q or D band [6,7]. When thylakoids were exposed to higher light intensity (500 W/m^2) than that required to saturate photosynthesis the amplitude of the B and Q bands gradually decreased (Fig. 1.). However, their peak positions and consequently their half life-times remained constant (Fig. 1.). In the decay of delayed luminescence two components could be distinguished in the seconds to minutes time scale (Fig. 2.). The slower component of 28–30 s half life-time (Q_B component) corresponds to the B TL band and originates from the $S_2Q_B^-$ recombination [8]. The faster component, decaying with a half life-time of about 3s (Q_A component), is attributed, like the Q TL band, to the $S_2Q_A^-$ charge recombination [8]. In the presence of DCMU the Q_B component disappeared and the Q_A component was intensified (not shown). Corresponding to the constant peak positions of the Q and B TL bands, the half life-times of the Q_A and Q_B DL components did not change with the duration of the photoinhibitory treatment, although their intensities decreased. The decay course of the emission intensities of the B and Q bands was identical during photoinhibition (Fig. 3.). Similarly, the amplitude of the Q_A and Q_B DL components exhibited a parallel decrease as the photoinhibition proceeded (not shown). The light saturated

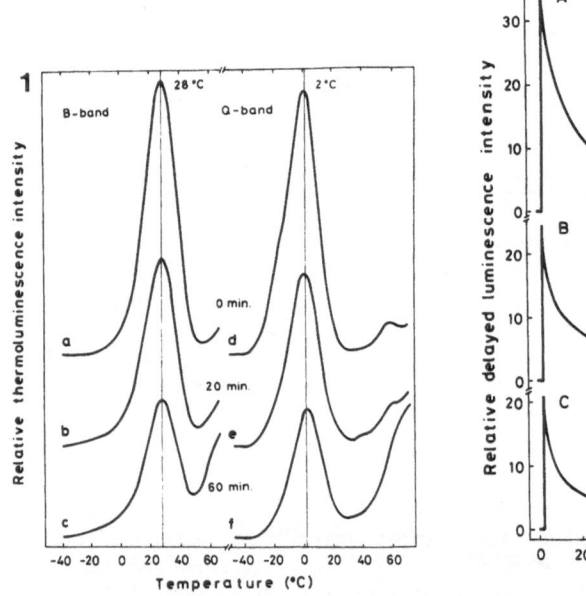
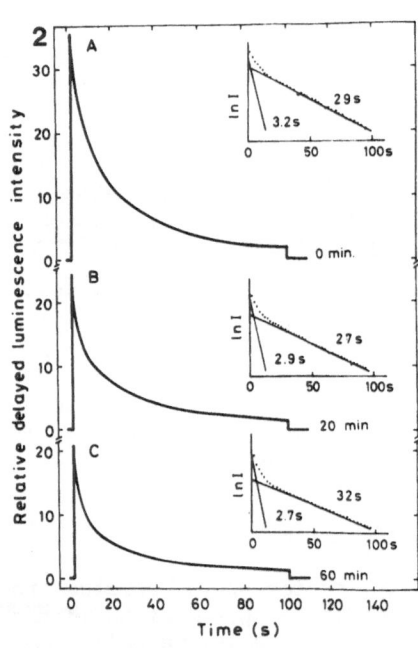

Figure 1. Changes in thermoluminescence characteristics of isolated thylakoids during photoinhibition. Samples were illuminated with 500 W/m^2 white light for 0, 20 and 60 minutes. The B and Q thermoluminescence bands were excited by a single flash in the absence (curves a, b, c) and presence (curves d, e, f) of 1 μM DCMU, respectively.

Figure 2. Changes in delayed luminescence characteristics of isolated thylakoids exposed to photoinhibitory conditions for various periods of time: A, 0 min; B, 20 min; C, 60 min. The insets show the resolved exponential components with half life-times.

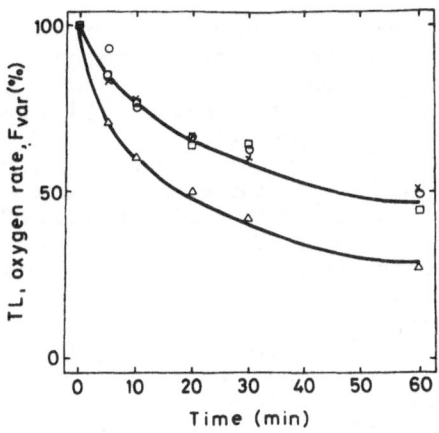

Figure 3. Time course of photoinactivation of PS II activity. c, intensity of the B TL band ; □ , intensity of the Q TL band measured in the presence of 1 μM DCMU; x, rate of oxygen evolution measured in the presence of 1.5 mM DMQ as electron acceptor; $\triangle$, relative variable fluorescence $((F_{max}-F_o)/F_o$). The initial rate of oxygen evolution was 110 μM O_2/mg Chl/h.

electron transport rate of PS II measured from H_2O to dimethylbenzoquinone diminished by the same extent as the amplitude of the B and Q bands (Fig. 3.). However, the loss of the variable fluorescence which is a generally used test of photoinhibitory damage to photosynthetic organelles, ran ahead of the inhibition of the electron transport rate (Fig. 3.).

According to our present knowledge, the PS II reaction center complex consists of a heterodimer composed of D_1 and D_2 subunit polypeptides and cytochrome b-559. D_1 provides binding site for Q_B while Q_A is very probably located on D_2 [10]. Investigations performed with algae cells led to the conclusion that the primary effect of photoinhibitory attack is a damage of the Q_B binding site [2,9]. At variance with this conclusion experiments carried out with isolated chloroplasts or PS II membranes suggested that the primary photodamage impairs electron transfer at the level of Pheo or Q_A [3-5].

Taking the advantage of TL in studying the effect of photoinhibition on the Q_A and Q_B acceptors, recently, Ohad et al. [9] carried out a detailed photoinhibitory study of <u>Chlamydomonas reinhardii</u> cells. They found that in the first stage of photoinhibition the Q_B binding site was modified. Exposure of the cells to a light intensity of 500 W/m^2 resulted in a shift of the B TL band from 30°C to 15-17°C. The appearance of this modified B band has been attributed to a change in the conformation of the D_1 protein which caused a destabilization of the $S_2Q_B^-$ state. The amplitude of the TL band ascribed to the Q_A acceptor was decreased only by 30-40% under similar photoinhibitory treatment.

Our results are contradicting a primary photoinhibitory attack on the Q_B binding site. The parallel time courses of the decrease in the amplitudes of the Q and B TL bands as well as that of the Q_A and Q_B DL components indicates that the reduction of the Q_A and Q_B acceptors is simultaneously impaired by photoinhibitory treatment. The constant TL peak positions as well as DL half life-times also contradict any preferential photo-induced change in the redox state of the Q_B acceptor.

On the basis of our TL and DL measurements we conclude that in isolated thylakoids the Q_B binding site is not the primary target of photoinhibition. We propose two alternatives to reconcile the data presented here and reported previously in the literature.

I. Photoinhibition impairs electron transfer at a site in the $Z-P_{680}-Pheo-Q_A$ section of the electron transport chain. Our methods do not make possible a more precise localization of the site of action. However, based on data in recent literature [5] a damage at Q_A or between Pheo and Q_A would be the most probable.

II. It was suggested that the primary cause of photoinhibition is the light-induced accumulation of the reactive quinone species Q_B^{2-} which damages the Q_B binding site [2]. The accumulation of the reduced form of the primary quinone acceptor (Q_A^-) might be responsible for a similar damage to the Q_A binding site. In isolated thylakoids which does not have CO_2 fixation capability high light intensity reduces fully all of the acceptor pools. Due to the equal amount of permanently reduced Q_B and Q_A both the Q_B and Q_A binding sites are damaged to the same extent. In intact algae cells electrons are continuously drained off the plastoquinone pool towards PS I resulting in a partially reduced electron transport chain i.e. there is a smaller population of reduced Q_A than that of Q_B. This can explain the slower photodamage of the Q_A binding site compared to that of Q_B as observed in algae cells [9].

We prefer the second alternative because it can explain the primary photodamage to the Q_B binding site in algae cells as well as the parallel impairment of Q_A and Q_B reduction in isolated thylakoids.

ACKNOWLEDGEMENTS

This work was supported by the Research Funds of the Hungarian Academy of Sciences AKA (219/86) and OKKFT (TT 310/86).

REFERENCES

1   S.B. Powles, Photoinhibition of photosynthesis induced by visible light, Annu. Rev. Plant. Physiol. 35:15 (1984)

2   D.J. Kyle, I. Ohad and C.J. Arntzen, Membrane protein damage and repair:Selective loss of a quinone-protein function in chloroplast membranes, Proc. Natl. Acad. Sci. USA 81:4070 (1984)

3   B. Arntz and A. Trebst, On the role of the Q_B protein in PS II in photoinhibition, FEBS Lett. 194:43 (1986)

4   R.E. Cleland, A. Melis and P.J. Neale, Mechanism of photoinhibition: photochemical reaction center inactivation in system II of chloroplasts, Photosynthesis Res. 9:79 (1986)

5   S.I. Allakhverdiev, E. Setlikova, V.V. Klimov and I. Setlik, In photoinhibited photosystem II particles pheophytin photoreduction remains unimpaired, FEBS Lett. 226:186 (1987)

6   A.W. Rutherford, A.R. Crofts and Y. Inoue, Thermoluminescence as a probe of photosystem II photochemistry. The origin of the flash-induced glow peaks, Biochim. Biophys. Acta 682, 457-465 (1982)

7   S. Demeter and I. Vass, Charge accumulation and recombination in photosystem II studied by thermoluminescence. I. Participation of the primary acceptor Q and secondary acceptor B in the generation of thermoluminescence of chloroplasts, Biochim. Biophys. Acta 764:24 (1984)

8   É. Hideg and S. Demeter, Binary oscillation of delayed luminescence: Evidence of the participation of Q_B^- in the charge recombination, Z. Naturforsch. 40c:827 (1985)

9   I. Ohad, H. Koike, S. Shochat and Y. Inoue, Changes in the properties of reaction center II during the initial stages of photoinhibition as revealed by thermoluminescence measurements, Biochim. Biophys. Acta 933:288 (1988)

10 O. Nanba and K. Satoh, Isolation of a photosystem II reaction center consisting of D-1 and D-2 polypeptides and cytochrome b-559, Proc. Natl. Acad. Sci. USA 84:109 (1987)

CHARACTERIZATION OF THE HEAT INDUCED EFFECTS ON THE PHYSICAL PROPERTIES AND ORGANIZATION STATE OF PSII SUBCHLOROPLAST MEMBRANES

Alexander G. Ivanov and Mira C. Busheva

Bulgarian Academy of Sciences
Central Laboratory of Biophysics
1113 Sofia, Bulgaria

INTRODUCTION

The inhibitory effect of heating on the PSII-associated photochemical activity and heat-induced changes in the structural organization and electrical properties of thylakoid membranes are well established[1,2,3]. Although the heat inactivation of oxygen evolution in isolated PSII particles and the concomitant alterations in the electron donor site of PSII (partial protein and Mn^{2+} release from the particles) have been also reported[4,5], little is known about heat-induced effects on the physical properties and organization state of PSII membranes. In the present study we have reported a decrease of the net negative surface charge density in heat treated PSII membranes. In addition, 77K fluorescence measurements have indicated that structural reorganization of chlorophyll-protein complexes (LHCa/b and PSII core) also occur under the same conditions.

MATERIAL AND METHODS

PSII subchloraplast particles (BBY particles) were prepared from pea chloroplasts using Triton X-100 - treatment following the procedure described by Berthold et al.[6]. Before use the PSII particles were washed twice in a medium containing 25 mM MES buffer (pH 6.5), 0,3 M sucrose, 10 mM NaCl and collected by centrifugation at 40000 x g for 10 min. The resulted pellet was resuspended in 0.01 M Tricine buffer (pH 8.0), 0.33 M sucrose, 0.01 M KCland 5 mM $MgCl_2$ where indicated. Heat treatment was carried out at $50°C$ for 5 min in the dark. The surface charge density (Θ) of control and heat-treated samples was measured by the method of Chow and Barber[7] using the fluorescence of a 9-aminoacridine (9AA) probe. 77K fluorescence emission and excitation spectra were recorded with a Jobin Yvon JY3 spectrofluorimeter. Chlorophyll (10 µg/ml in the probe) fluorescence was excited at 436 nm (slits width=4 nm).

RESULTS AND DISCUSSION

The data summarized in Tabl.1. indicated that net negative electrical charge calculated for isolated PSII (BBY) particles is higher as compared to the average surface charge density of the thylakoid membranes reported earlier[3]. This result is in full agreement with the Θ value estimated by Mansfield et al.[8] for the "inside out" PSII vesicles and confirms the charge asymmetry existing between inner and outer surfaces of the thylakoid membranes. It has been recently supposed[9] that the charge asymmetry might be partially due to the presence of lumenal exposed and negatively charged[10] polypeptides (24 and 33 kDa) of the oxygen evolving system.

As can be seen in Tabl.1. heat treatment led to an increase of the surface charge density in chloroplast membranes, whilst in PSII particles remarkable decrease of the $\ominus$ value was observed. It is evident that the opposite effects of heat treatment reflected different heat-induced events

Table 1. Effects of heat treatment on the net negative surface charge densities of envelope-free chloroplasts and PSII subchloroplast membranes. Means and standard errors were calculated from 5 independent experiments.

Fraction	Surface charge density $(C.m^{-2}) \pm$ S.E.
Chloroplasts	$- 0.028 \pm 0.001_3$
Heat-treated chloroplasts	$- 0.037 \pm 0.002_3$
PSII membranes	$- 0.034 \pm 0.002$
Heat-treated PSII membranes	$- 0.020 \pm 0.003$

probably due to the different structural organization of the two type of membranes. In a previous paper Ivanov et al.[3] suggested that the increase of the net negative charge in heat-treated chloroplast membranes might be attributed to the well characterized[1,2] heat-induced physical separation of the LHCa/b protein complex from the core complex of PSII. Obviously, the decrease of $\ominus$ in heat-treated PSII membranes could not be explaned by this mechanism, although the same structural changes can not be excluded in the PSII preparations.

It must be noted that similar decrease of the net negative charge has been reported[3] for 18, 24 and 33 kDa polypeptides[4] depleted PSII particles. Moreover, Yamamoto and Nishimura[4] and Nash et al.[5] demonstrated that heat-treatment of PSII particles also resulted in partial release of 18, 24 and 33 kDa proteins, the 33 kDa polypeptide being mostly affected. Bearing in mind that: 1) 24 and 33 kDa polypeptides are negatively charged[10], and 2) that the same proteins are directly bound to the membrane[11], it seems very likely that the decrease of surface charge of PSII membranes could be attributed to the heat-induced polypeptides release (mainly of the 33 kDa protein). Furthermore, our data confirm the suggestion of Isogai et al.[12] that the binding site of the 33 kDa polypeptide on the inner surface of PSII is positively charged.

The fluorescence emission spectra of higher plant chloroplasts at 77K exhibited three main peaks located around 685, 695 and 735 nm, which are generally thought to be associated with well characterized chlorophyll-protein complexes of LHCa/b, PSII and PSI respectively. The isolated PSII subchloroplast membrane fractions exhibited only characteristic peak at 685 nm accompanied by a secondary peak or shoulder of variable intensity at 695 nm. The 77K fluorescence measurements have indicated that Mg^{2+} causes remarkable increase of the F695/F685 ratio in PSII preparations (Fig.1. and Tabl.2.). Yamamoto and Ke[13] proposed earlier that this could be due to the cation-induced enhancement of energy transfer from LHCa/b complex to PSII reaction center complex. However. there is some strong evidence[14] that part of F695 originate from LHCa/b complex, especially from its aggregated form.

It was found that heating at $50^{\circ}C$ of isolated PSII membranes lead to considerable increase of the intensity of secondary fluorescence peak, which is accompanied by 2.5-5.0 nm shift toward longer wavelengths, thus implying remarkable heat-induced alterations in the interactions between LHCa/b complex and PSII core[2] (Fig.1. and Tabl.2.). This effect is more pronounced in the presence of Mg^{2+} and does not depend on the presence of DCMU (data not shown). No evidence was found for any heat-induced enhancement of the energy transfer from LHCa/b complex to PSII core complex. Moreover, fluorescence

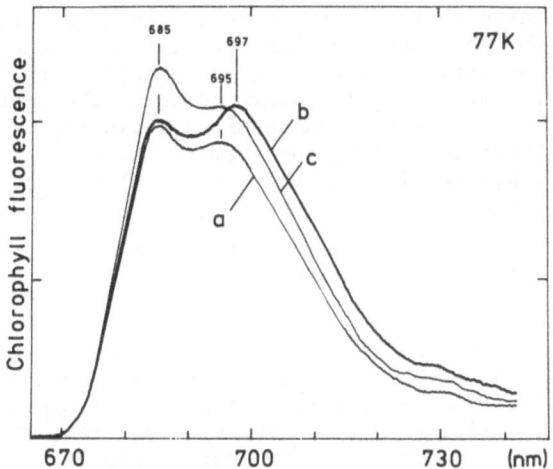

Figure 1. 77K fluorescence spectra of PSII particles:
a)- in the presence of 5 mM Mg^{2+}, b)- in the
absence of $MgCl_2$, c)- heat-treated (50°C, 5
min) in the presence of Mg^{2+}.

excitation peak at 475 nm was decreased in heat-treated particles (E436/
E475=0.81 for F685) as compared to the control ones (E436/E475=0.72), indicating that LHCa/b transfer less effectively excitation energy to Chl a
of the PSII core complex as a result of heating. Thus, the increase of 695-
700 amplitude could be related to an aggregated form of LHCa/b complex[14].
The spectral shift observed in heat-treated PSII particles also confirm this
suggestion. The formation of longer wavelength fluorescence band at 699 nm
has been observed by Larkum and Anderson[14] when the LHCa/b was reconstituted
alone into liposomes at low lipid/protein ratious, i.e. aggregated form (s)

Table 2. Low temperature (77K) fluorescence parameters of PSII
subchloroplast membranes. Means and standard errors
were calculated from 5 independent experiments.

Fraction	λ^1_{max}	λ^2_{max}	F_λ^2/F_λ^1
PSII membranes (+Mg)	685	695	0.930 ± 0.025
PSII membranes (−Mg)	685	695	0.870 ± 0.008
Heat-treated PSII (+Mg)	685	697.5	1.130 ± 0.028
Heat treated PSII (−Mg)	683	700	1.050 ± 0.011

of LHCa/b complex exist. Furthermore, our assumption is in agreement and
can be fully understood in terms of heat-induced rearrangements of chlorophyll-protein complexes within the thylakoid membranes. Gounaris et al.[1]
and more recently Sundby et al.[2] have demonstrated that part of LHCa/b
dissociate from the PSII core and form aggregates, which remain free and
separated from both photosystems.

Hence, our results demonstrated that heat-induced effects on the isolated PSII particles at the level of PSII-LHCa/b organization are similar,
if not identical to those reported for thylakoid membranes. At the same
time, the decrease of the surface charge density observed under the same
conditions raise the question to what extent the heat-induced polypeptides
release is responsible for the 77K fluorescence changes. Our preliminary

experiments indicated that both the surface charge and fluorescence changes are partially reversible when the heat-treated PSII membranes are reconstituted with crude extract containing the extrinsic polypeptides of the oxygen evolving complex. A full understanding of the problem under discussion requires freeze-fracture and detailed rebinding studies. Such studies will be important for localozing the primary effect of heat stress and the role of extrinsic polypeptides in the structural organization of PSII-LHCa/b supramolecular complex.

Acknowledgements - This work was supported by the Ministry of Culture, Science and Education under Research contract 519.

REFERENCES

1. K. Gounaris, A.P.R. Brain, P.J. Quinn and W.P. Williams, Structural and functional changes associated with heat-induced phase-separations of non-bilayer lipids in chloroplast membranes, FEBS Lett.153:47 (1983).
2. C. Sundby, A, Melis, P. Maenpaa and B. Andersson, Temperature-dependent changes in the antenna size of Photosystem II. Reversible conversion of Photosystem II to Photosystem II , Biochim. Biophys. Acta 851: 475 (1986).
3. A. G. Ivanov, M.Velitchkova and D.Kafalieva, Multiple effects of trypsin- and heat-treatments on the ultrastructure and surface charge density of pea chloroplast membranes. Influence on P700$^+$ parameters, Progr. Photosynth. Res.2:741 (1987).
4. Y. Yamamoto and M.Nishimura, Organization of the O_2-evolution enzyme complex in a highly active O_2-evolving photosystem-II preparation, in: "The Oxygen Evolving System af Photosynthesis", Y.Inoue, A.R. Crofts, Govindjee, N.Murata, G.Renger and K.Satoh, eds., Academic Press Japan, Tokyo (1983).
5. D. Nash, M.Miyao and N.Murata, Heat inactivation of oxygen evolution in Photosystem II particles and its acceleretion by chloride depletion and exogenous manganese, Biochim. Biophys. Acta 807:127 (1985).
6. D. A.Berthold, G.T.Babcock and C.F. Yokum, A highly resolved oxygen-evolving photosystem II preparation from spinach thylakoid membranes. ESR and electron-transport properties, FEBS Lett. 134:231 (1981).
7. W. C.Chow and J. Barber, 9-aminoacridine fluorescence changes as a measure of surface charge density of the thylakoid membrane, Biochim. Biophys. Acta 589:346 (1980).
8. R. W.Mansfield, H.Y.Nakatani, J.Barber, S.Mauro and R.Lannoye, Charge density on the inner surface of pea thylakoid membranes, FEBS Lett. 137:133 (1982).
9. M. C.Busheva and A.G.Ivanov, Effects of 33, 24 and 18 kDa polypeptides depletion on the surface charge density and structure of PSII particles, Compt. rend. Acad. bulg. Sci. 41:121 (1988).
10. T. Kuwabara and N.Murata, Chemical and physicochemical characterization of the proteins involved in the oxygen evolution system, Adv. Photosynth. Res. 1:371 (1984).
11. B. Andersson, C.Larsson, C.Jansson, U.Ljungberg and H.-E.Akerlund, Immunological studies on the organization of proteins in photosynthetic oxygen evolution, Biochim. Biophys. Acta 766:21 (1984).
12. Y. Isogai, M.Nishimura and S.Itoh, Charge distribution on the membrane surface of the donor site of photosystem II. Effects of the free radical relaxing agent dysprosium on the power saturation of EPR signal II in the PSII-particles, Plant Cell Physiol. 28:1493 (1987).
13. Y. Yamamoto and B.Ke, Regulation of excitation energy distribution in photosystem II fragments by magnesium ions, Biochim. Biophys. Acta 592:296 (1980).
14. A. W. D.Larkum and J.M.Anderson, The reconstitution of a photosystem II protein complex, P-700-chlorophyll a-protein complex and light- harvesting chlorophyll a/b protein, Biochim. Biophys. Acta 679:410 (1982).

IN BROADLEAVED EVERGREENS VARIATIONS IN Fv/Fm INDUCED BY PHOTOINHIBITION ARE COUPLED TO REDUCTIONS IN PSII RC UNIT SIZE

Guido Bongi and Francesco Loreto

C.N.R. Olivicoltura

Via Madonna Alta, 06100 Perugia, Italy

Different sources of stress, when coupled to high light fluence, produce photoinhibitive damage on broadleaved evergreens at irradiations which are 4-5 times higher than reported on annual leaves[1]. Quantum yield of PSII photochemistry Qy(p) as indicated by the ratio Fv/Fm at 692nm (Kitajima and Butler[2]) was found to be quite a convenient probe of photoinhibitive damage; Bjorkman[3] on Hedera canariensis, and same other authors on different species found highly signifiant relationship between quantum yield for oxygen evolution, Qy(ox) and Qy (p). The same procedure when applied to our plants resulted however in a trend lowered by a constant factor in respect to those experiences (fig.1).

Evergreen leaves exposed to high light are thick and heteromorphous (bifacial). Hence, whole leaf quantum yield of oxygen evolution includes, at the light flux herewith used for Qy(ox), substantial amounts of respiratory metabolism by inner layers and, moreover, light response of intermediate layers is steeper producing a further increase in the apparent Qy(ox). Qy(p), on the other hand, is taken just from sub-epidermal layers; in our limited experience this discrepancy corresponds empirically to a large increase in photoinhibition tolerance.

These heteromorphous leaves also show room temperature emission spectra substantially different between the upper and the lower part for the adaxial face showing remarquable reduction of F684 (fig.2). This can be explained considering that altough both F743 and F684 are in the saturated portion of the F/|Chl| relationship, on the upper layer F684 is furthermore quenched by self-absorption produced by an higher chlorophyll density.

In photoinhibited samples Demming and Bjorkman [4] found that the kinetic constant for thermal decay and process other than photochemistry, Kdt, was greatly increasing in respect to Kf, the first order kinetic constant for fluorescence, perhaps in relation to an increase of activity of violaxantin-zeaxantin cycle. Conversely they found a decrease in Kp/Kf, being Kp the second order kinetic constant for PSII photochemistry (v(PSII) = Kp |Chl*| Q/Qox).

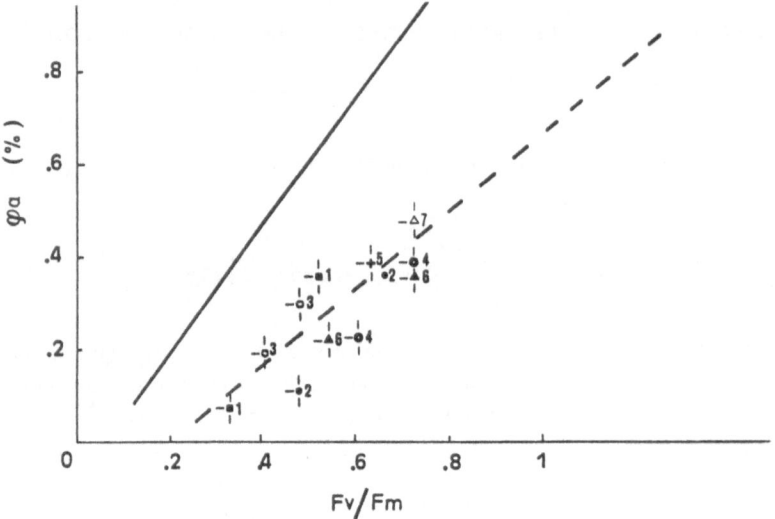

Fig.1 Relationship between the photon yield of O2 evolution (φa) and the quantum yield of PSII photochemistry (Fv/Fm). Solid line depicts data given in [1]. 1= olive chilled under high light; 2= olive under salt stress; 3= olive chilled; 4= olive under nitrogen starvation; 5= olive control; 6= Prunus laur. chilled; 7= carob.

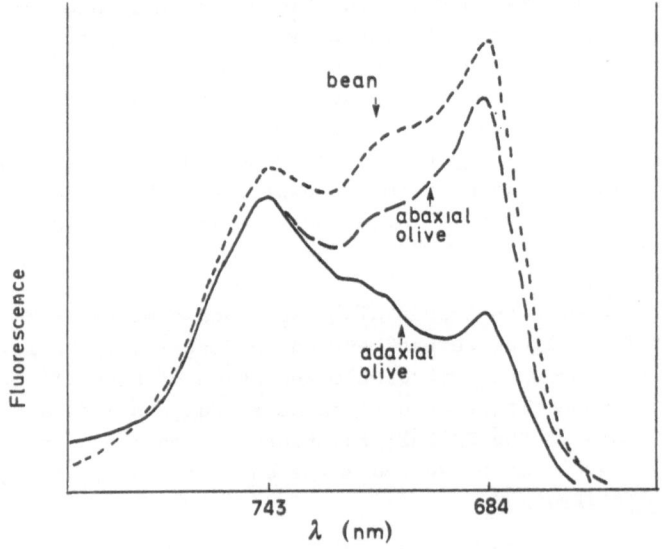

Fig.2 Fluorescence emission spectra at room temperature in the abaxial (lower) and adaxial (upper) face of olive leaves compared to emission observed in bean.

In detail, in this experiment photoinhibitive conditions where applied to whole plants under salt stress, nitrogen starvation or chilling at an irradiance of 1400 uE m-2 s-1 for several hours. Shaded plants were relatively undamaged on Qy(ox); to avoid partial recovery after treatments, the irradiation was prolonged until constant Qy(ox) was obtained. No photoxydation of chlorophyll was however found.
As species we used the mediterranean evergreens Olea europea, Prunus lauroceraso and Ceratonia siliqua. Fluorescence meaurements were carried out with a modulated fluorimeter (PAM Walz, Effeltrich).

Basically we considered a Qy(Fo) = 0.02 ([2]) at room temperature, quenched by leaf absorbance at 692 nm in variable proportions according to the optical properties of intact leaf, by an amount which is experimentally between 75.6 and 83.2% (in an Ulbricht sphere on leaf fragments infiltrated overnight by water DCMU saturated solution). Kp/Kf was then obtained by the relation (Fv/Fm)/Qy(Fo). Normalizing Fm yield to Qy(Fo), one can eventually obtain Kdt/Kf by the relation 1/Qy(Fm)-1 (derived after Kitajima and Butler[2]).

However Butler model about kinetic ratios in primary photochemistry, relies on the assumption that the amount of acceptors Q is constant among different treatments, as verified by the above induction area in electron-transport-inhibited samples, or by the amount of atrazine-binding sites.

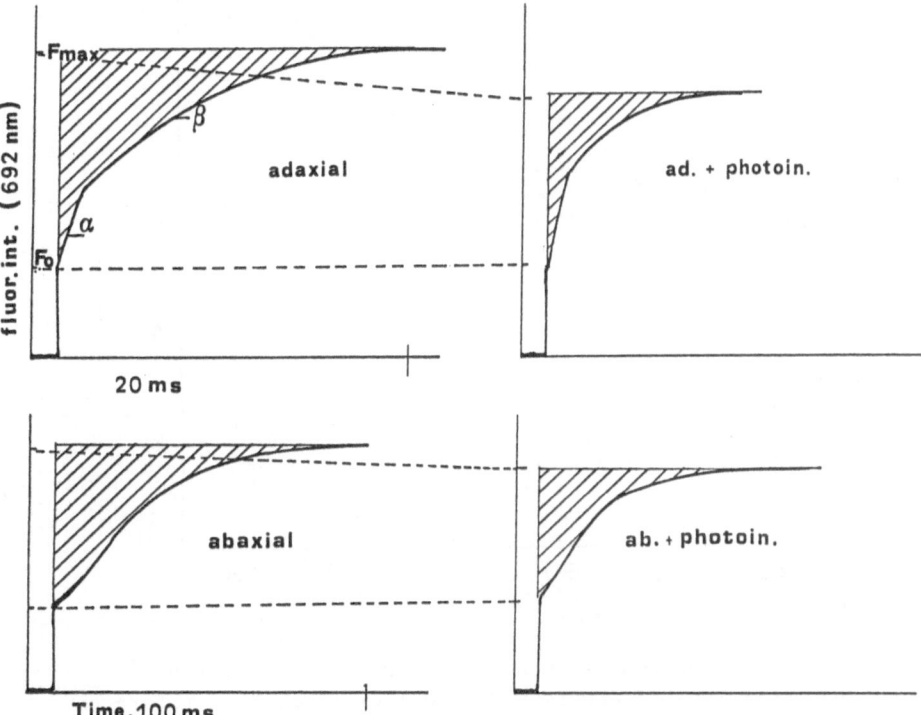

Fig.3 Fluorescence induction kinetics in the adaxial (upper) and abaxial (lower) side of evergreen leaves either be fore (left) or after photoinhibition (right). DCMU inhibited samples.

To estimate Kp relative variations one should therefore first verify a constancy of |Q| among treatments; infact it is clear that any change in |Q| among treatments not taken into account will produce an apparent variation in Kp/Kf just related to different amounts of PSII reaction centres |Chl*|.

That was infact the case in our samples which showed a clear decrease in |Q| as stress went by. Moreover a linear relationship was found between |Q| and Kp/Kf contrasting the fundamental statement of indipendency between a kinetic constant and its substrate concentration (fig.4).

As results of photoinhibitive treatments we thus obtained a reduction in the "active" area of the leaves rather than a modified light conversion which is quite common at higher organization levels as in patches of closed stomata or even in paraheliotropic leaf movements. Kdt/Kf variations are probably linked to a lesser sensitivity of dispersive phenomena to stress treatment which resulted in a varied proportion to Kp.

Whitin certain stress limits the fast kinetic of recovery (unpubl.res.) indicates probably the assembly rate of reaction centres as the sensitive site of photoinhibition in these tolerant plants.

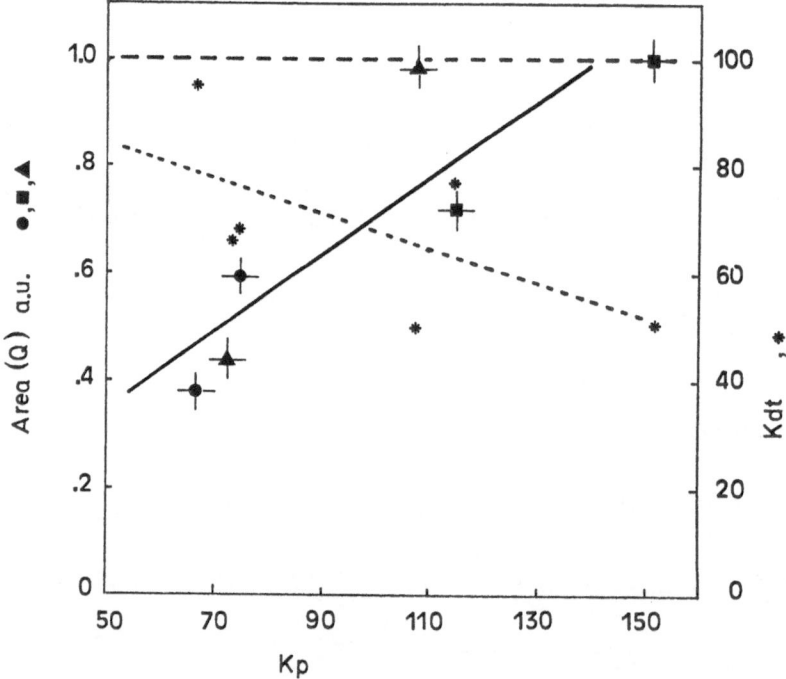

Fig. 4   Reaction centres concentration |Q| and Kdt for different values of apparent Kp; different symbols refers to treatments in fig. 1 (Squares= low nitrogen, circles= chilling, triangles= salt photoinhibited samples). The horizontal, dashed line on the top represents the expected response of Kp, whereas the solid line represents the result of this experiment.

REFERENCES

1) S.P.Long, A.Nugawela, G.Bongi, and P.K.Farage. Chilling dependent photoinhibition of photosynthetic CO2 uptake. in Progress in Photosynt.Res. vol IV, J.Biggins ed.,M.Nijhoff Publ. Dordrecht (1987).

2) M.Kitajima and W.L.Butler. Quenching of chlorophyll fluorescence and primary photochemistry in chloroplasts by dibromothymoquinone. Bioch.Biophys.Acta 376: 105-115 (1975).

3) O.Bjorkman. High irradiance stress in higher plants and interaction with other stress factors. in Progress in Photosynt.Res. vol IV, J.Biggins ed. M.Nijhoff Publ. Dordrecht (1987).

4) B.Demming and O.Bjorkman. Comparison of the effect of excessive light on chlorophyll fluorescence (77K) and photon yield of O2 evolution in leaves of higher plants. Planta 171: 171-184 (1987).

THE PRIMARY LIMITATION TO PHOTOSYNTHESIS FOLLOWING LOW TEMPERATURE STRESS
UNDER HIGH LIGHT

Gretchen F. Sassenrath[1], Donald R. Ort[1,3] and Archie R. Portis[2,3]

Depts Plant Biology[1] and Agronomy[2], and Agricultural Res Serv[3]

U.S. Dept of Agriculture, Univ Illinois, Urbana, IL 61801

INTRODUCTION

Certain plants which are accustomed to growing under temperate
conditions show a distinct inhibition of function when exposed for brief
periods to temperatures that are well below that optimal for growth yet
above freezing. This chill-induced impairment of metabolism persists for an
extended period of time after plants are rewarmed to growth temperatures.
In many instances, photosynthesis is the foremost physiological process
that displays a loss of activity following a low temperature exposure.
Damage to photosynthesis is exacerbated when low temperatures are sustained
in the presence of high light intensities.

We are interested in understanding the nature of inhibition to photosynthesis following chilling stress, particularly the complicitous damage
resulting from low temperature exposure at high light intensities. Net
photosynthesis is inhibited more than 60% in intact tomato leaves upon
subsequent rewarming following a 6 hour exposure at 4^0C under high intensity
illumination (1000 $\mu E/m^2 \cdot s$). The majority of the reduction is non-stomatal
in nature and results from direct inhibition of chloroplastic function.
However, whole chain electron transport rates are reduced less than 10% in
chloroplasts isolated from prestressed tomato leaves[1]. Additionally, in
situ measurements of the decay kinetics of the electrochromic shift indicate
that phosphorylation capacity imposes no significant limitation to net
photosynthesis after light chilling[2]. These observations implicate a
process other than those involved in light utilization and energy conservation in the loss of chloroplast activity following chilling in the light.
Light chilling may induce a lesion in an enzymatic reaction involved in
carbon reduction, leading to an inhibition of CO_2 assimilation. It is the
aim of this study to ascertain the inhibitory affect of light-chill stress
on the in vivo enzymatic performance of the photosynthetic carbon reduction
cycle.

RESULTS AND DISCUSSION

Measurements of the photosynthetic carbon reduction cycle intermediates
from intact tomato leaves under steady state photosynthetic conditions
reveals a significant buildup of fructose-1,6-bisphosphate following
chilling in the light (Fig 1). Modest increases in levels of dihydroxy-

acetonephosphate and sedoheptulose-1,7-bisphosphate also result from a light chill exposure. Similar increases in the levels of these substrates are not observed in control plants illuminated at subsaturating light that have nearly identical rates of net photosynthesis as light-chilled plants, eliminating the possible that the changes in substrate levels arise merely due to a diminished uptake of CO_2. Rather, the increased substrate levels indicate a lower in vivo turnover rate, possibly due to a limitation of phosphatase activity. A slower rate of catalysis of FBP would be expected to increase the levels of DHAP due to the near equilibrium of the aldolase reaction in situ[5]. The reduction in levels of phosphoglycerate and ribulose-1,5-bisphosphate observed in tomato leaves after a light chill exposure would be consistent with a limitation of phosphatase activity.

In order to determine if enzyme activity is indeed rate limiting following light chilling, in vivo activation states of ribulose-1,5-bisphosphate carboxylase/oxygenase (Rubisco) and fructose-1,6-bisphosphatase (FBPase) were measured in whole leaf extracts. A rapid freeze quench apparatus was used to quickly isolate the leaf discs and trap the enzymes in the state in which they occurred in vivo. FBPase activates with a t1/2 of about 4 min upon illumination of intact tomato leaves, declining slightly

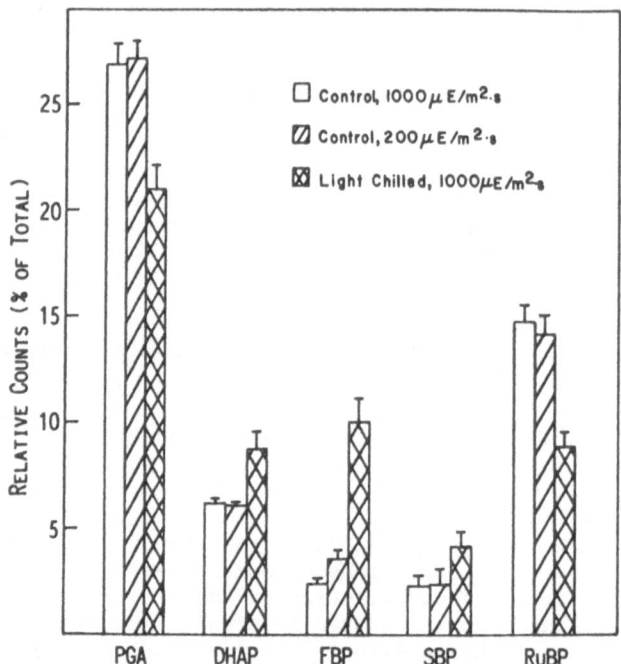

Figure 1. Relative amounts of intermediates of the photosynthetic carbon reduction cycle. Incorporation of $^{14}CO_2$ into intact tomato leaves and separation of labelled compounds by HPLC was conducted as previously described[3,4]. Relative incorporation of label into photosynthetic carbon reduction cycle intermediates following a 2 min label with $^{14}CO_2$ at steady state rates of photosynthesis at 1500 µl/l CO_2 is compared for control leaves illuminated at high (1000 µE/m²·s) and low (200 µE/m²·s) light and prestressed light-chilled leaves illuminated at high light. Prestressed leaves were chilled in the light for 6 hours at 4°C under 1000 µE/m²·s incident light. Reported values are means of 4 separate leaf samples ± s.e. Photosynthetic rates at 1500 µl/l CO_2 for the three treatments are: high light control = 25.4 µmol CO_2/m²·s; low light control = 10.9 µmol CO_2/m²·s; light chill = 9.4 µmol CO_2/m²·s.

from the maximal activity to reach a steady state rate after 30 min illumination (Fig 2). The measured enzymatic capacities are well in excess of those required to support control photosynthetic rates. Following rewarming of light-chilled tomato leaves, FBPase fails to reactivate even after 1 hour illumination. The low level of FBPase activity is sufficient to support the measured rate of net photosynthesis in light-chilled plants, yet is far below the level of activity necessary to support control rates of photosynthesis. The substantial reduction in FBPase activity suggests an inability to reactivate FBPase is the primary rate limitation to net CO_2 uptake following a light chill stress. Although FBPase activity is substantially inhibited in leaf extracts from rewarmed plants, full catalytic activity can be regained upon reductive activation with dithiothreitol (Table 1). Light-chill stress apparently does not directly affect enzymatic integrity, but more likely disrupts the endogenous activation mechanism of FBPase, the thioredoxin:ferredoxin system.

Rubisco activity is slightly inhibited following a light chill stress, although the reduced rates are sufficient to support control rates of photosynthesis. Rubisco activity is thus not limiting to net carbon uptake following chilling in the light.

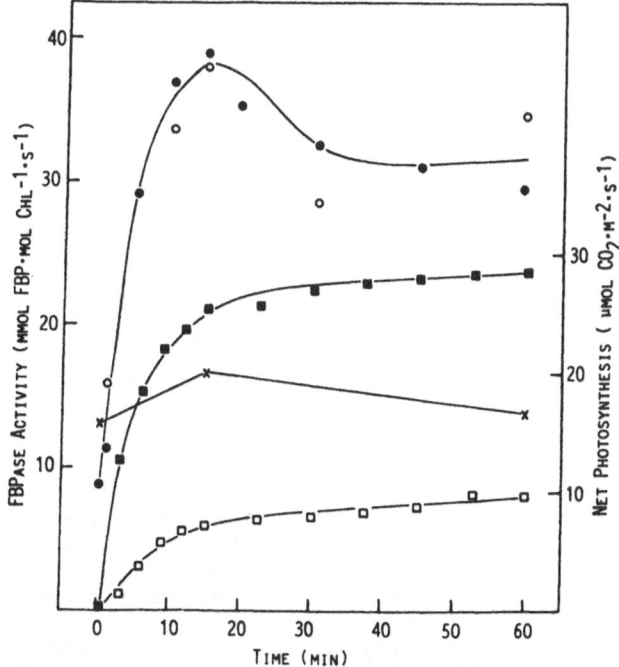

Figure 2. Activation of Photosynthesis and FBPase in Control and Chilled Tomato Leaves. Net photosynthesis with increasing time of illumination was measured in an open gas exchange system on intact leaves of previously dark adapted control (■) and light-chilled (□) plants. FBPase activity was measured in whole leaf extracts of leaves isolated at different times of light exposure from control (●), dark-chilled (o), and light-chilled (x) leaves, by the procedures described in Table 1. Dark chilled plants were incubated at 1 C for 16 hours; light chilling was performed as described in Fig 1. Rates of photosynthesis are plotted directly across from the minimum enzymatic activity necessary to support that level of photosynthesis.

Although Rubisco activity is reduced following a light chill stress, the limitation to CO_2 assimilation is more likely a result of lower phosphatase activity. Interestingly, the FBPase activity of light chilled plants prior to rewarming is comparable to that observed in control plants. This supports the notion that the enzyme itself is catalytically competent, but cannot be maintained in an active configuration once the light-chilled plant is allowed to rewarm. The loss of FBPase activity at warm temperatures could arise from an alteration of the redox state of the chloroplast following light chilling. Since FBPase is dependent on reductive activation via the thioredoxin:ferredoxin system, the presence of chemical oxidants, such as H_2O_2 and O_2^- , that may be produced during the low temperature exposure at high light intensities[8], would be particularly detrimental to activation of the enzyme. Alternatively, a particular step of the thioredoxin:ferredoxin system could be damaged during light chill stress, resulting in a loss of FBPase activity upon rewarming.

Although several steps in CO_2 assimilation are dependent on reductive activation by thioredoxin, only activity of the phosphatases appears to be limiting following light chill stress. Chilling in the light could cause loss of enzymatic capacity at any step involved in photosynthetic carbon reduction. However, diminished capacity at a step which is normally rate-limiting to photosynthesis would be particularly deleterious due to the regulatory nature of the reaction. Since the phosphatases are likely important control points of photosynthesis[5,9], inhibition of activity at this point would result in a substantial reduction in CO_2 assimilation.

Table 1. In vivo and Total Enzymatic Activities of Rubisco and FBPase Following Low Temperature Stress at High Light Intensity. In vivo and total enzymatic activities of Rubisco and FBPase were measured in light- and dark-adapted control leaves, and leaves pretreated for 6 hours at 4^0C under 1000 $\mu E/m^2 \cdot s$ light. Light-chilled leaves were sampled at 4^0C immediately after the light chill, or after 1 hour reillumination at 30^0C following a 1 hour dark rewarm. Plants were illuminated at 1000 $\mu E/m^2 \cdot s$ and 1500 $\mu l/l$ CO_2 until steady state rates of photosynthesis were reached. Enzymatic activities in whole leaf extracts were then determined using radiolabelled substrates[6,7]. Total Rubisco activity was determined in whole leaf extracts preincubated in 6-phosphogluconate. Total FBPase activity was determined by in vitro activation using dithiotheitol. The minimum required enzymatic rate is calculated from the measured photosynthetic rates of each particular treatment, and gives the minimum turnover required to support that rate of photosynthesis. Reported values are means of 3 determinations $\pm$ s.e.

Sample	Rubisco Activity mmol CO_2/mol Chl·s			FBPase Activity mmol FBP/mol Chl·s		
	in vivo	Total	Minimum	in vivo	Total	Minimum
Control:						
Light	128 ± 10.8	195 ± 11.5	75.0	32.9 ± 2.93	27.9	25.0
Dark	90 ± 5.2	182 ± 8.8	–	9.5 ± 0.92	29.5	–
Light Chill:						
Light	160 ± 14.2	192 ± 17.0	15.0	29.3 ± 4.1	29.4	5.0
1 hr dark + 1 hr light	100 ± 11.8	155 ± 9.0	25.0	13.8 ± 1.8	28.3	8.3

CONCLUSION

A primary limitation to net photosynthesis following light chill stress in intact tomato leaves appears to result from an inability to maintain the stromal FBPase in a fully active configuration. Although SBPase may be similarly inhibited, no other enzymatic step involved in photosynthetic carbon reduction is limiting to net carbon uptake following light chill stress. The loss of phosphatase activity is not due to damage directly to the catalytic competence of the enzyme, but rather is dependent on an inability to activate the enzyme. The inactivation of the enzyme may result from damage to a component of the thioredoxin:ferredoxin system, or from production of oxidants during chilling in the light.

REFERENCES

1. S. C. Kee, B. Martin, and D. R. Ort, The effects of chilling in the dark and in the light on photosynthesis of tomato: electron transfer reactions, Photosyn Res 8:41 (1986)
2. R. R. Wise and D. R. Ort, Effects of light chilling of photophosphorylation in cucumber, in "Molecular Biology and Bioenergetics," Govindjee, ed., In press (1988)
3. C. A. Atkins and D. T. Canvin, Photosynthesis and CO_2 evolution by leaf discs: Gas exchange, extraction, and ion-exchange fractionation of ^{14}C-labelled photosynthetic products, Can J Bot. 49:1225 (1971)
4. S. Boag and A. R. Portis, Jr., Metabolite levels in the stroma of spinach chloroplasts exposed to osmotic stress: Effects of the pH of the medium and exogenous dihydroxyacetone phosphate, Planta 165:416 (1985)
5. J. A. Bassham and G. H. Krause, Free energy changes and metabolic regulation in steady state photosynthetic carbon reduction, Biochim Biophys Acta 198:207 (1969)
6. G. F. Sassenrath, D. R. Ort, and A. R. Portis, Jr., Effect of chilling on the activity of enzymes of the photosynthetic carbon reduction cycle, in: "Prog in Photosyn Res., Vol IV," J. Biggens, ed., Martinus Nijhoff Publ, Dordrecht (1987)
7. A. R. Portis, Jr., M. E. Salvucci, W. L. Ogren, Activation of ribulose-bisphosphate carboxylase/oxygenase at physiological CO_2 and ribulosebisphosphate concentrations by rubisco activase, Plant Physiol. 82:967 (1986)
8. B. Halliwell, "Chloroplast Metabolism" Clarendon Press, Oxford (1981)
9. H. W. Heldt, C. J. Chon, R. McC. Lilley, and A. Portis, The role of fructose- and sedoheptulosebisphosphatase in the control of CO_2 fixation, in "Proc Fourth Internatl Congr on Photosyn" (1977)

COMPARISON OF CHILL-INDUCED AND ROOM TEMPERATURE PHOTOINHIBITION

Esa Tyystjärvi, Jari Ovaska, Eva-Mari Aro and Pirjo Karunen

Department of Biology
University of Turku
Finland

INTRODUCTION

Bright light induces photoinhibition of photosynthesis. When plants are subjected to chilling temperature, symptoms of photoinhibition are often observed to appear even in moderate light (for a review, see Öquist & al.[1]). It is often thought, that the role of low temperature is only to accelerate photoinhibition. In this study, we distinguish between high light induced photoinhibition and the inhibition caused by the combination of low temperature and light.

Table 1. Different combinations of duration, photon flux density and temperature of treatments. RT = room temperature

Photon flux density $\mu mol\ m^{-2}s^{-1}$	Time, min.			
	30	60	120	240
750	RT 1°C	RT 1°C	RT 1°C	
1500	RT	RT 1°C	RT 1°C	
2500			RT	RT
Darkness			1°C	

MATERIALS AND METHODS

Pumpkin (Cucurbita pepo L.) plants were treated under various photon flux densities either at room temperature (RT) or 1°C (Table 1.). Full leaf turgor was maintained with saturated humidity. The temperature was measured with a thermoelement inserted into the leaf. Fluorescence induction (77 K) at 687 nm and the apparent quantum yield of photosynthetic oxygen evolution were measured from leaf disks.

RESULTS

All light treatments caused a decrease in the ratio of variable to maximum fluorescence (F_V/F_{max}) at 77 K (Fig. 1.). The decrease in F_V/F_{max} was mainly dependent on the duration and photon flux density, and only slightly on leaf temperature. At the beginning of the treatments, the decrease in F_V/F_{max} was due to a decrease in F_{max}, later mainly to an increase in F_0. The treatments also led to a decrease in the apparent quantum yield of photosynthetic oxygen evolution. At room temperature, the high light induced changes in the quantum yield of oxygen evolution correlated well with the changes in F_V/F_{max} (Fig. 2., circles). However, light treatment at 1°C had a much greater effect on the quantum yield than on F_V/F_{max} (Fig. 2., squares). Chilling in darkness (2h, 1°C) did not influence fluorescence parameters, and had only a small effect on the apparent quantum yield.

Measurements of light reaction activities from isolated pumpkin chloroplast thylakoids showed that the loss of PS II activity depended mainly on the light level and duration of the treatment, not on the temperature. On the contrary, the loss of whole chain electron transport capacity occurred only when the leaves were subjected to 1°C in light. PS I was not affected by any of the treatments.

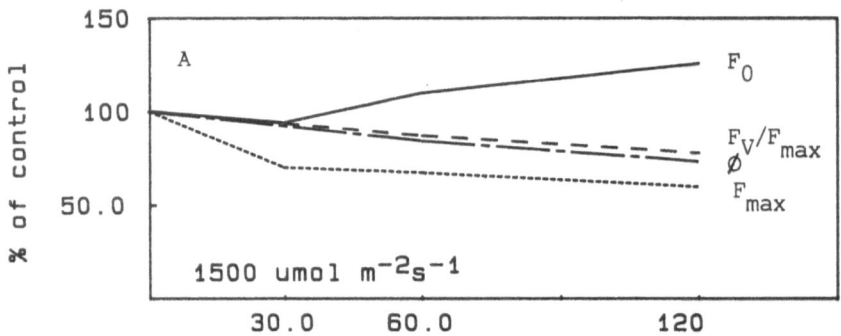

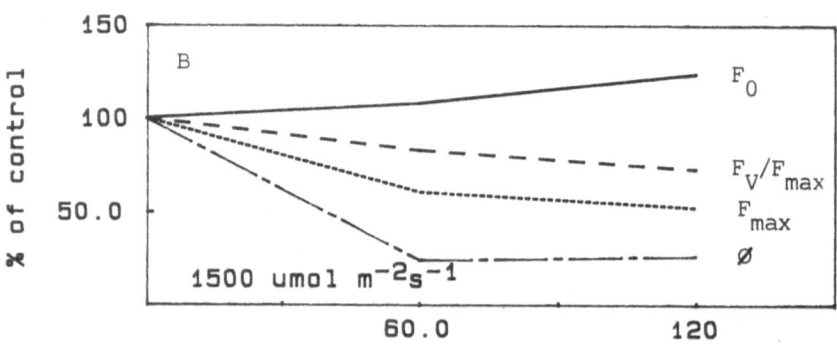

Fig. 1. Effect of high light (PPFD 1500 umol m^{-2}s^{-1}) on 77 K fluorescence induction and apparent quantum yield of photosynthetic oxygen evolution (∅) of pumpkin leaves (% of control). Means of three independent experiments. (A) 30, 60 and 120 min. light treatment at room temperature. (B) 60 and 120 min. light treatments at 1°C.

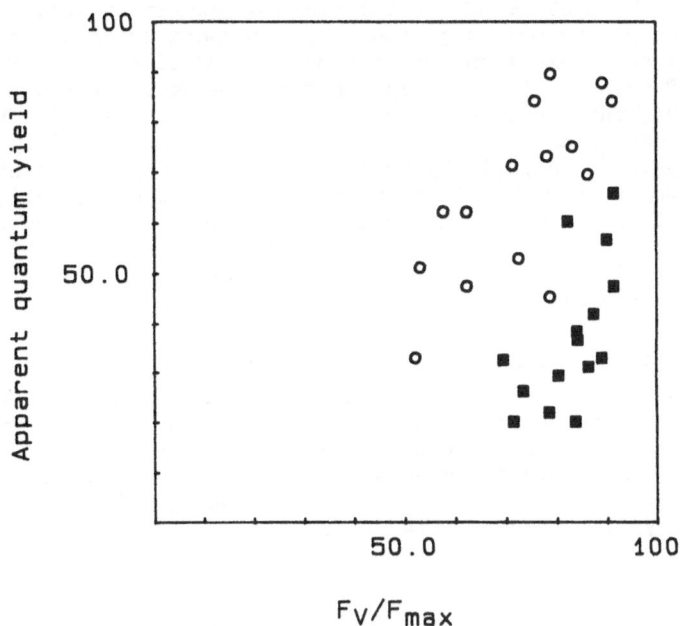

Fig. 2. The relationship between the ratio of variable to maximum fluorescence at 77 K (F_V/F_{max})
and apparent quantum yield of photosynthetic
oxygen evolution of pumpkin leaves measured
after light treatments at RT (o) and 1°C (■).
All values are expressed as % of control.
Each point corresponds to a single experiment. The various combinations of light level and duration of the treatment are listed
in Table 1.

DISCUSSION

There are two main hypotheses explaining the accelerating effect of
low temperature on photoinhibition of photosynthesis. Low temperature is
explained to reduce the amount of light energy that can be safely handled
by photosynthesis[2]. Secondly, the recovery from photoinhibition is
temperature-dependent, its rate being low at chilling temperatures[3]. In
the present study, we have shown that the acceleration of photoinhibition
of PS II in chilling temperatures may be a consequence of a more primary
inhibitory event, requiring both light and low temperature. Our results
from pumpkin leaves support the following conclusions:

1. Light alone – irrespective of temperature – is responsible for photoinhibition of PS II.
2. High or moderate light combined with low temperature causes an
   inhibition of electron transfer between the two photosystems.
3. The inhibition of electron transfer between PS II and PS I is mainly
   responsible for the decline in quantum yield of photosynthetic oxygen
   evolution, when pumpkin plants are subjected to chilling temperature
   in the light.

The inhibition of electron transfer between the two photosystems may contribute to the decline of quantum yield of oxygen evolution often observed, when plants are subjected to low temperature and moderate light. Further, a block in electron transfer between PS II and PS I reduces the proportion of absorbed light energy that can be used for the photochemistry in PS II. Thus, the acceleration of PS II photoinhibition in chilling temperatures may be a consequence of this phenomenon. Possibly, the PS II in pumpkin leaves is especially well protected against photoinhibition, and that is why we saw that PS II photoinhibition was only slightly accelerated at low temperature. This is supported by our previous results[4] with pumpkins: a four days' treatment at 5°C and a PPFD of 300 umols m^{-2} s^{-1} caused only a slight inhibition of PS II, while the whole-chain electron transport was severely inhibited.

One consequence of our results is that 77 K fluorescence can not be used as the only method to monitor the inhibitory effects of light and chilling temperature on photosynthesis. The results obtained at room temperature are in agreement with those of Björkman & Demmig[5], showing a linear relationship between changes in photon yield of oxygen evolution and F_V/F_{max} ratio of 77 K fluorescence in leaves treated with different photoinhibitory light levels (Fig. 2.).

REFERENCES

1. G. Öquist, D. H. Greer, and E. Ögren, Light stress at low temperature, in: "Topics in Photosynthesis", Vol. 9., D. J. Kyle, B. Osmond, D. J. Arnzen, eds., Elsevier Publishers, B.V., Amsterdam (1987).
2. E. Ögren, G. Öquist, and J-E. Hällgren, Photoinhibition of photosynthesis in Lemna gibba as induced by the interaction between light and temperature. I. Photosynthesis in vivo, Physiol. Plant. 62: 181-186 (1984).
3. D. H. Greer, J. A. Berry, and O. Björkman, Photoinhibition of photosynthesis in intact bean leaves: role of light and temperature, and requirement for chloroplast-protein synthesis during recovery, Planta 168: 253-260.
4. P. Mäenpää, E-M. Aro, S. Somersalo and E. Tyystjärvi, The effect of chilling in light on the thylakoid membrane. Rearrangement of the thylakoid at chilling temperature in the light, Plant Physiol., in press.
5. O. Björkman and B. Demmig, Photon yield of O_2 evolution and chlorophyll fluorescence characteristics at 77 K among vascular plants of diverse origins, Planta 170: 489-504 (1987).

THE EFFECT OF SALINITY, PHOTOINHIBITION AND INTERACTION OF BOTH ON

PHOTOSYNTHESIS IN BARLEY

Prabhat K. Sharma and David O. Hall

Department of Biology, King's College London, Campden Hill Road
London W8 7AH

ABSTRACT

The effect of salinity, photoinhibition and interaction of the two on various photosynthetic parameters was investigated in leaves and isolated chloroplasts of barley. Plants were grown in a controlled environment and irrigated with different concentrations (0-200 mM) of NaCl + CaCl2 salts. Photoinhibition treatment was given at 1600 uE/M2/S for four hour at 5 C and 20 C. The light saturated rate of CO2 uptake and maximum quantum yield decreased with increasing salt concentrations. Photoinhibition also decreased the CO2 assimilation rate and apparent quantum yield. Stomatal conductance was more sensitive to salt stress than photoinhibition. Salt stress and photoinhibition both modified the amplitude of leaf variable fluorescence (Fm) at 20 C in the presence of DCMU. Electron transport activity measured in isolated chloroplasts from plants grown at low salinity levels showed a slight increase in photosystem II (PS II) and photosystem I (PS I) activity in comparison with control; higher salt concentrations did not affect the electron transport activity. In control plants photoinhibition significantly decreased the PS II activity but no changes were observed in PS I activity. Interaction of photoinhibition and salt stress increases the inhibition of PS II activity in comparison with control photoinhibited plants.

INTRODUCTION

Salinity of the soil and irrigation water is a serious problem in agriculture especially in arid and semi-arid regions usually leading to severe crop losses. Non-halophyte species vary greatly in their response to short and long term salinity. The reduction of growth may result from salt effects on dry matter allocation, ion relations, water status, physiological processes, biochemical reactions or combinations of such factors (1). The earliest response observed in non-halophytes to high salinity is that their leaves grow more slowly; root growth is usually less affected than shoot growth thus increasing the root:shoot ratio. At low salinity, however, root growth may not decrease at all while shoot growth declines ,eg. barley (2) or root growth may even increase ,eg. sorghum (3). These effects are clearly observed even in the short term, before ion concentrations in the shoot have built up to high levels. The reduction in growth of the leaf is often accompanied by a decreased rate of photosynthesis but the extent of decrease depends upon the concentration and type of salt, the manner in which the stress is imposed and the relative sensitivity of the species (4). Salinity in the growth medium results in a disarrangement in leaf pigment composition which induces a decrease in chlorophyll content (5).

Plants growing under salt stress often experience photoinhibition including other environmental stresses, such as low and high temperatures and drought, which usually amplify the salinity effect. A series of studies have attempted to define the conditions under which chilling damage occurs to the photosynthetic capacity (6, 7). These studies confirm the importance of interactions between light and temperature to produce an inhibitory effect on photosynthesis. It has been suggested that excess radiant energy damages mainly photosystem II (PS II) or PS II related reactions (8).

There have been relatively few photosynthetic studies of the interaction of different stresses. Our work analyses the interaction of salt, photoinhibition and temperature stresses on photosynthesis.

MATERIAL AND METHODS

Seeds of barley (H. vulgare L.; Golden Promise) were obtained from Miln Masters, King's Lynn, UK, and stored in the dark at room temperature.

GERMINATION AND GROWTH CONDITIONS: Prior to sowing seeds were soaked in tap water for an hour. 8 g seeds were planted in 10 cm diameter plastic pots containing vermiculite (Micafil) soaked with different salt concentrations: 50, 100, 150 and 200 mM NaCl, 50 mM NaCl + 50 mM CaCl2 and 100 mM NaCl + 100 mM CaCl2; tap water was used as the control. Pots were watered alternate days with the solutions. Plants were grown for 8-12 days in a controlled environment cabinet with 16 hours photoperiod (day/night temperature 22 C/ 16 C ± 1 C), illuminated with a combination of fluorescent tubes and incandescent bulbs to provide a photosynthetic photon flux density (PPFD) of 100 uE/M2/S at plant level.

PHOTOINHIBITION CONDITIONS: An apparatus for exposing plants to 1600 uE/M2/S PPFD at 5 C and 20 C temperature was constructed in a 4 C cold room. Three linear 2 kW tungsten halogen lamps were mounted in a curved aluminium reflector. The light passed through a glass tank with 18 cm flowing water to absorb ultra-violet and infra-red radiation. The photoinhibition treatment was given for 4 hour.

ELECTRON TRANSPORT ASSAYS: Chloroplasts (50 percent intact type "A") were mechanically isolated from 8 day old barley leaves following the method of Richards and Hall (9). Electron transport activity was assayed using various electron donors and acceptors in a Rank (Clark type) oxygen electrode at 20 C and 2000 uE/M2/S PPFD illuminated using a Quartz iodine 300 W projector. The light was filtered through an orange Cinemoid filter (No. 105, Lee Ltd, Andover, England) with a cutoff wavelength of 500 nm and 4 cm of water in a flat glass bottle.

FLUORESCENCE MEASUREMENT: Leaf fluorescence was monitored with a leaf disc electrode (Hansatech, King's Lynn, UK; 10). All measurements were taken at 20 C. 12 day old leaves were infiltrated with a medium containing 0.35 M sorbitol, 20 mM tricine (PH 7.5), 10 mM NaCl (with or without DCMU) in the dark under vacuum for 45 minutes. Measurements of fluorescence emission at 685 nm, indicative of photosystem II, were made using a fibre optic light source, the spectrum of which showed a single sharp peak at 675 nm providing 200 uE/M2/S PPFD at the leaf surface.

GAS EXCHANGE MEASUREMENTS: Net photosynthesis, stomatal conductance, transpiration rate, vapour pressure deficit and internal CO2 were measured in an open type gas exchange system, with a pair of detached

leaves with their cut ends in water, using an infra-red gas analyser
(ADC, Hoddesdon, UK; type LCA; 6).

RESULTS AND DISCUSSION

As evident from fig 1, a decrease of 54% in CO_2 assimilation rate (Fc)
was observed in 200 mM NaCl treated plants. The photoinhibitory effect
of light is enhanced at chilling temperatures. Approximately a 66%
decrease was observed in plants grown at 200 mM NaCl and
photoinhibited at 5 C; the assimilation rate was less affected at 20
C. A combination of Na + Ca salts proved more deleterious.

Stomatal conductance (gs) was relatively more sensitive than the
assimilation rate or transpiration rate (fig 2). Approximately 76%
reduction in gs was observed at 200 mM NaCl. Photoinhibition at 5 C
and 20 C with 100 mM NaCl results in approximately the same amount of
inhibition in stomatal conductance. No further decrease in gs was
observed with photoinhibition at higher concentrations of NaCl. Vapour
pressure deficit increased with higher concentrations of salt and was
further increased with combinations of salt and photoinhibition (data
not shown).
Fig 3 shows the light response curve of CO_2 fixation. Maximum apparent
quantum yield was decreased by 77% in plants treated with 50 mM NaCl
and photoinhibited at 5 C with 1600 uE/M2/S, followed by 200 mM NaCl
(68% decrease) and than by 50 mM NaCl plus photoinhibition at 20 C
(64% decrease).
The decrease in the reduction in assimilation rate may be explained by
lower stomatal conductance. Lower activity of the photosynthetic
enzyme RuBP carboxylase and disturbed inorganic metabolism may also
contribute to the lower activity of Fc in stressed plants (11). The
suboptimal temperature also restricts the photosynthetic rate. The
reduction in apparent quantum efficiency at higher salinities implies
damage to the effectiveness of photon capture by photosynthetic
pigments and/or damage to photosynthetic electron transport (6, 7).

Fig 4 depicts the time course of chlorophyll fluorescence at 685 nm at
20 C. These induction curves exhibit characterstic phases of
fluorescence levels Fo, Fp, Fs and Ft (12). An induction phase (Fi)
was observed in salt stressed plants. Phase Fp was delayed along with
phase Fs in stressed plants. A decrease of 59% in Fm in plants treated
with 50 mM NaCl + 50 mM $CaCl_2$ and photoinhibited at 20 C was observed.
No change in Fo level was observed which implies that interaction
between salt and photoinhibition has not caused any structural
alterations. A decrease in Fm suggested that the extent of reduction
of QA is decreased; this may result from photoinhibitory damage to the
oxidizing side of the PS II centre and this resulting in a lower
reducing capacity. A hump was also observed between phase M and T in
salt stressed plants which probably indicate some changes in dark
reaction cycle.
The photosynthetic electron transport activity measured in
chloroplasts isolated from leaves of salt treated barley plants showed
an increase (10-40%) in PS II at low salt concentrations as assayed
from water to phenylenediamine (PD). Ps I activity also showed a
similar increase at low salt concentrations as assayed using reduced
2,6-DichlorophenolIndophenol (DCPIP),
N,N,N´,N´,-Tetramethyl-p-phenylenediamine (TMPD), or
2,3,5,6-Tetramethyl-1,4-phenylenediamine (DAD) as electron donors to
Methyl Viologen (MV). No inhibition in either PS II or PS I activities
was observed at higher concentration of salt in comparison with
control (Table 1). Barley plants irrigated with salt solutions or
water (control) and photoinhibited at 5 C and 1600 uE/M2/S PPFD showed

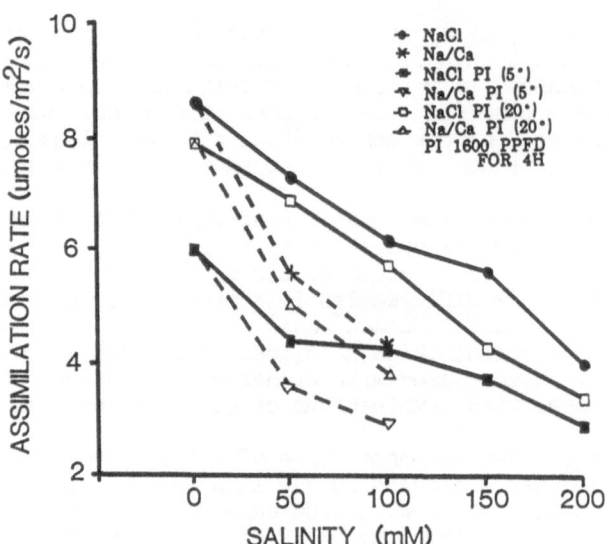

fig 1- Effect of interaction between
salt, photoinhibition and temperature
on CO2 assimilation rate.

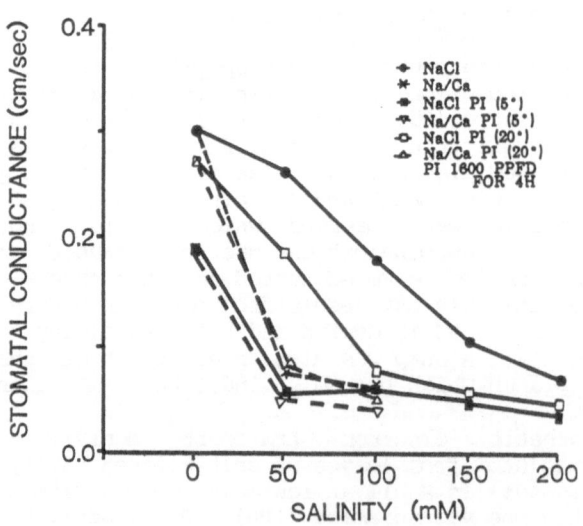

fig 2- Effect of interaction between
salt, photoinhibition and temperature
on stomatal conductance.

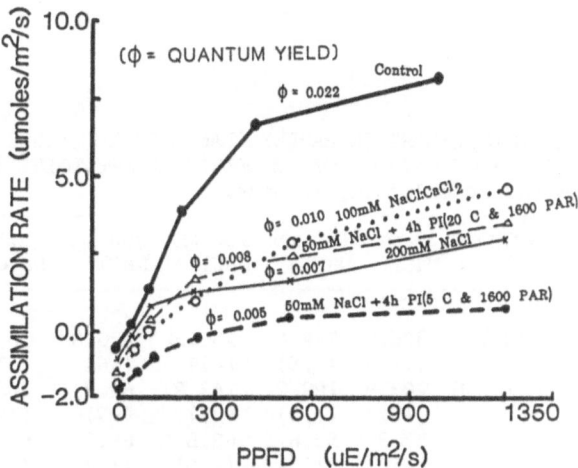

fig 3- Effect of interaction between
salt, photoinhibiton and temperature
on maximum apparent quantum efficiency.

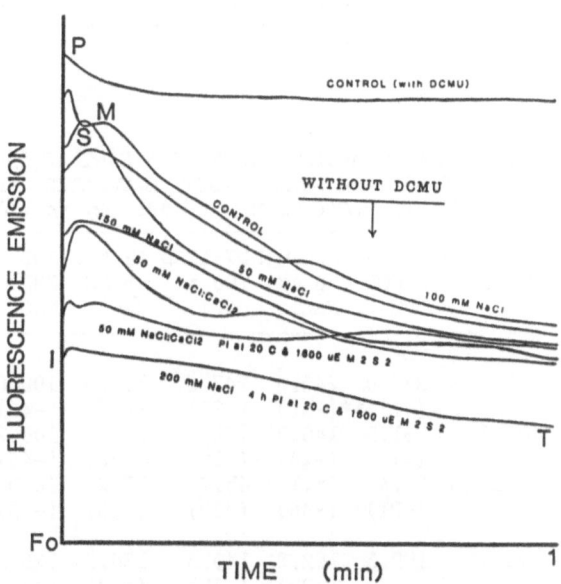

fig 4-Effect of interaction between
salt, photoinhibition and temperature
on fluorescence induction curve at 20 C

a 46% decrease in combined PS I plus PS II activity and a 44% decrease in PS II activity in comparison with photoinhibited plants. The increase in electron transport activity at lower salt concentrations

TABLE 1   EFFECT OF SALT STRESS IN BARLEY PLANTS ON THE LIGHT SATURATED ELECTRON TRANSPORT ACTIVITY (UNCOUPLED) OF ISOLATED CHLOROPLASTS (NO PHOTOINHIBITORY TREATMENT) umoles O2/ mg chlorophyll/ hour.

ASSAYS	CONTROL H2O	50 mM NaCl	100 mM NaCl	150 mM NaCl	200 mM NaCl	50:50mM Na:Ca*	100:100mM Na:Ca
PS I + PS II							
H2O--> MV	301.6	326.9 (+8)**	379.0 (+25)	296.3 (-2)	320.8 (+6)	303.3 (+1)	322.0 (+6)
H2O--> FeCN	154.5	203.9 (+32)	183.2 (+19)	143.9 (-7)	64.7 (+7)	164.3 (+6)	149.2 (-3)
DPC--> MV	51.8	57.3 (+10)	59.6 (+15)	62.5 (+20)	66.2 (+28)	61.9 (+20)	68.9 (+35)
PS II							
H2O--> PD	153.4	210.7 (+37)	221.0 (+44)	150.4 (-2)	181.5 (+18)	187.2 (+22)	170.8 (+11)
PS I							
DCPIP/ASC--> MV	66.7	96.2 (+44)	77.8 (+16)	66.5 (0)	71.1 (+6)	64.9 (-3)	71.1 (+7)
TMPD/ASC--> MV	311.1	395.0 (+27)	391.7 (+26)	310.9 (0)	438.9 (+41)	315.5 (+1)	307.2 (-1)

TABLE 2   THE EFFECT OF 4 HOUR PHOTOINHIBITORY TREATMENT AT 5 C AND 1600 uE/M2/S IN SALT STRESSED BARLEY PLANTS ON THE LIGHT SATURATED ELECTRON TRANSPORT ACTIVITY (UNCOUPLED) OF IOLATED HLOROPLASTS. umoles O2/ mg chlorophyll/ hour.

| ASSAYS | CONTROL H2O | <----- 4h PHOTOINHIBITED AT 5 C AND 1600 uE/m2/s -----> | | | | | | |
		CONTROL H2O	50 mM NaCl	100 mM NaCl	150mM NaCl	200 mM NaCl	50:50 Na:Ca*	100:100 Na:Ca
PS I + PS II								
H2O--> MV	355.5	272.0 (-23)**	240.0 (-32)	247.6 (-30)	191.5 (-46)	190.6 (-46)	222.6 (-37)	193.6 (-45)
H2O--> FeCN	185.1	161.5 (-13)	140.9 (-24)	151.3 (-18)	114.9 (-38)	100.2 (-46)	125.7 (-32)	113.4 (-39)
DPC--> MV	65.2	51.4 (-21)	35.1 (-46)	45.7 (-30)	37.2 (-43)	40.0 (-39)	37.5 (-42)	38.9 (-40)
PS II								
H2O--> PD	215.1	159.5 (-26)	152.3 (-29)	140.3 (-35)	136.5 (-37)	125.5 (-42)	127.6 (-41)	120.0 (-44)
PS I								
DCPIP/ASC--> MV	86.4	78.1 (-9)	69.2 (-20)	74.0 (-14)	76.5 (-11)	80.2 (+7)	77.5 (-10)	74.6 (-14)
DAD/ASC--> MV	865.7	900.4 (+4)	858.1 (-1)	842.1 (-3)	785.5 (-9)	978.6 (+13)	776.1 (-10)	745.6 (-14)
lTMPD/ASC--> MV	400.0	399.9 (0)	352.8 (-12)	416.8 (+4)	343.2 (-14)	400.8 (0)	336.4 (-16)	346.5 (-13)

* = NaCl + CaCl2 IN mM (1:1 RATIO).   **= PERCENT.

is unexplained but may be due to ionic changes within the chloroplasts or as a result of cation like Ca++ binding to the 23 kD protein and faciliting electron transport at higher rate.

REFERENCES

1. H. Greenway and R. Munns. Mechanisms of salt tolerence in non-halophytes. Ann. Rev. Plant Physiol. 31: 149-190 (1980).
2. R. delane, H. Greenway, R. Munns and J. Gibbs. Ion concentration and carbohydrate status of elongating leaf tissue of H. vulgare at high external NaCl. I. relationship between solute concentration and growth. J. Exp. Bot. 33: 557-573 (1982).
3. R. Weimberg, H.R. Lermer and M.A. Poljakaff. Changes in growth and water soluble concentration in Sorghum bicolor stressed with sodium and potassium. Plant Physiol. 62: 472-480 (1984).
4. D.J. Longstreth and P.S. Nobel. Salinity effects on leaf anatomy. Plant Physiol. 63: 700-703 (1979).
5. M. Stiborova, S. Ksinska and A. Brezinova. Effect of NaCl on the growth and biochemical characteristics of photosynthesis of barley and maize. Photosynthetica 21: 320-328 (1987).
6. S.P. Long, T.M. East and N.R. Baker. Chilling damage to photosyntesis in young Zea mays. J. Exp. Bot. 34: 177-188 (1983).
7. O. Bjorkman. High irradince stress in higher plants and interaction with other stress factor. in: progress in photosynthesis research. ed. J. Biggins, Martinus Nijhoff Publ., Dordrecht. 11-18 (1987).
8. S.B. Powles. Photoinhibition of photosynthesis induced by visible light. Ann. Rev. Plant Physiol. 35: 15-44 (1984).
9. G.E. Richards and D.O. Hall. Photoinhibition at chilling temperature in intact leaves and isolated chloroplasts. In: Progress in photosynthesis research. ed J. Biggins, Martinus Nijhoff Publ., Dordrecht. 39-43 (1987).
10. T. Delieu and D.A. Walker. Simultaneous measurement of oxygen evolution and chlorophyll fluorescence from leaf pieces. Plant physiol. 72: 534-541 (1982).
11. N.A. Rajmane and B.A. Karadge. Photosynthesis and photorespiration in winged been (Psophocarpus tetragonolobus L.) grown under saline conditions. Photosynthetica 20: 139-145 (1986).
12. G. Papageorgiou. Chlorophyll fluorescence: An intrinsic probe of photosynthesis. In: Bioenergetics of photosynthesis. ed. Govindjee. Academic Press, New York. pp 319-371 (1975).

PHOTOINHIBITION IN OUTDOOR CULTURES OF THE CYANOBACTERIUM SPIRULINA PLATENSIS

Avigad Vonshak and Rachel Guy

The Microalgal Biotechnology Laboratory
The Jacob Blaustein Institute for Desert Research
Ben-Gurion University, Sede-Boker Campus 84993, Israel

ABSTRACT

The photosynthetic activity, degree of photoinhibition and the output rate of a dense culture of Spirulina platensis strain Art-B, grown outdoors in an open raceway pond was studied. In the morning, when the photon flux density (PFD) was 1,100 uE m^{-2} s^{-1}, no photoinhibition was detected and the photosynthetic activity was 109 umols h^{-1} mg cha. At noon time, when PFD was 2,100 uE m^{-2} s^{-1}, a photoinhibition level of 25 was estimated and the photosynthetic activity of the culture was 84 umols h^{-1} mg cha. Shading the outdoor cultures by shade nets resulted in a decrease in photoinhibition of the shaded cultures, followed by an increase in the productivity from 11.5 gr m^{-2} d^{-1} at the non-shaded culture to 14.7 gr m^{-2} d^{-1} at the 30% shaded one.

Cultures grown under different shade conditions or in the laboratory under continuous illumination varied in their chlorophyll to cartenoids ratio. Exposing each of these cultures to high PFD in the lab indicated a change in the sensitivity to the high light stress, where the unshaded cultures were the most resistant to high PFD and the laboratory cultures being the most sensitive ones.

INTRODUCTION

The main problem limiting development of industrial production of algal biomass, is the relatively low output rates obtained in large commercial production sites [1,2]. In previous papers [3-5], we have suggested that photoinhibition would be one of the main limiting factors for obtaining high production rates when environmental conditions i.e., light and temperature should favor high photosynthetic rates.

The response of higher plants as well as microalgae to high photon flux densities (PFD), has been intensively studied under laboratory conditions [6,7] It is clear that the damage caused to the photosynthetic apparatus by high PFD is repairable and recovery back to the original photosynthetic activity occurs once light intensity is reduced [8,9]. We have recently studied the response of Spirulina to high PFD under controlled laboratory conditions [10], demonstrating that Spirulina cultures underwent a process of photoinhibition and, recovery when incubated in low light condition.

Although much information is available on the effect of high PFD on photosynthesis in laboratory grown cultures, very limited information is available on photoinhibition in densely grown microalgal cultures outdoors. Those cultures are exposed to a diurnal change in PFD, from under saturation levels for photosynthesis in the morning up to over saturation and inhibitorty levels at noon time. In this work, an attempt was made to quantify the level of photoinhibition imposed on outdoor algal cultures and to estimate to what extent this inhibition is responsible to reduced output rates (productivity) measured in those cultures.

MATERIALS AND METHODS

Strains and culture conditions; Spirulina platensis strain SP-RB [10] were used for this study. Outdoor cultures were grown in a modified Zarouk's medium [11] in 2.5 m^2 ponds. Depth of culture was 10 cm, stirring was provided by a paddle-wheel at 15 RPM, and biomass concentration was kept constant at 0.5 - 0.6 $O.D_{560nm}$ by daily removal of biomass using a 350 mesh screen. All nutritional requirements were supplied in excess, as previously described [12], in order to eliminate nutrient deficiency.

Shade was provided by standard shade nets used in greenhouses ("Ben- Tzur & Drouianoff", Herzelia, Israel). The cut-off of the nets was 15%, 25-30% and 50% of solar radiation. Light intensity was measured by using a Li-Cor-185 photometer and quantum sensor. O_2 evolution activity and photoinhibition were measured as previously described[10].

Photosynthetic activity in the outdoor ponds was measured by following the increase in O_2 concentration in the pond above saturation level after reducing the O_2 concentration in the culture by bubbling N_2 through the cultures for 5-10 minutes [13].

Recovery experiments were carried out by taking samples from the ponds and diluting them with fresh medium to a final concentration of 2.3 ug/ml chlorophyll[10]. Thereafter, they were incubated in dim light for 1 and 2 hrs at 30°C. At each time interval O_2 evolution was measured[10]. Biomass was measured as dry weights[14] and chlorophyll was determined according to Bennet and Bogorad[15].

RESULTS AND DISCUSSION

Spirulina SP-RB was grown outdoors at mid summer when radiation at noontime was beyond 2000 uE m^{-2} s^{-1}. Cultures were exposed to different light intensities by covering the ponds with shade nets that cut off a fraction of the solar radiation without any spectral change.
When the rates of O_2 evolution in the cultures were compared, a decrease in O_2 evolution activity was observed in the unshaded ponds at noon time, when light intensity was at the highest. A different response was observed in all the shaded ponds i.e. an increase in the O_2 evolution activity was measured at noontime, indicating that shading was most likely providing protection from the inhibitory effect of the high light intensity to which the unshaded ponds were exposed.(Tab-1).

TABLE 1
Photosynthetic activity of Spirulina in outdoors ponds under different shade conditions

Shade	Time sampled	Photosynthetic Activity u mol O_2 h^{-1} mg ch^{-1}
None	Morning	109
	Noon	84
25-30%	Morning	80
	Noon	120
50%	Morning	78
	Noon	116

Morning: temp. 19°-21°C. light, 1100 uE m^{-2} s^{-1}
Noon: temp. 30°-33°C. light, 2100 uE m^{-2} s^{-1}

The difference in water temperature in the various ponds was at the most 2-3°C, thus the different O_2 evolution activity rates in the ponds could not be attributed to fluctuations in temperature. As optimum growth temperature of Spirulina is 33-35°C, the decrease in O_2 evolution activity observed in the uncovered ponds at noon could not be the result of the increase in pond temperature.

We have already demonstrated, in a previous work, that when Spirulina was photoinhibited under laboratory conditions and transferred to low light intensities, a recovery process occurred so that the original O_2 evolution activity was reached after 2-3 hours of incubation[10]. This characteristic was used as a tool to estimate the degree of photoinhibition of the Spirulina cultures grown outdoors. Samples were taken from the ponds early in the morning when radiation did not exceed 900 uE m^{-2} s^{-1} and incubated in the lab at dim light. O_2 evolution activity of the samples was measured before the incubation and after one and two hours of incubation. The same procedure was followed at noontime (2200 uE m^{-2} sec^{-1}).

By comparing the photosynthetic activity values before and after the recovery treatments, in the morning and noon samples, the degree of photoinhibition could be determine. No significant increase in photosynthetic activity of the morning sampled cultures could be observed after the recovery treatment. In contrast the cultures sampled from the unshaded ponds at noon had an higher photosynthetic activity after the recovery treatment. In the samples from the 25-30% shaded ponds the recovery treatment had no effect (table 2). This shows that the Spirulina cultures were photoinhibited during part of the day and, by providing shading the cultures could be protected from the photoinhibitory effect of light.

TABLE 2
Photoinhibition in outdoors Spirulina cultures

Shade	Time sampled	Percent of Photoinhibition SP-RB
None	Morning	3.2
	Noon	25
25-30%	Morning	0
	Noon	0

* Percent of photoinhibition calculated from the increase in the photosynthetic activity after a recovery treatment

The output rates of the Spirulina cultures grown under the different shade conditions used are summarized in table 3. A significant increase in the productivity was observed in SP-RB cultures grown under 25-30% shading. Even a 50% cut-off in the light intensity still resulted in a somewhat higher production rate as compared with the the unshaded culture.

TABLE 3
Output rates of _Spirulina_ culture grown outdoors under different shade conditions.

Shade	Productivity ($gr \, m^{-2} \, day^{-1}$)
None	11.5
25-30%	14.7
50%	12.7

Results are an average of at least 30 days of operation.

The fact that the unshaded ponds were exposed to photoinhibition during part of the day is clearly reflected in their productivity. The 25-30% and 50% shading provided an sufficient protection from photoinhibition during noontime so that high productivity could be maintained at the high light intensities without photoinhibition. In the 50% shaded ponds, light intensity became too low and limited productivity.

The data presented here indicate that the algal cultures in the open ponds were subjected to photoinhibition for about 6 hours each day in the summer season. The three sets of experiments (recovery treatment, O_2 evolution measurements in the ponds and productivity measurements) complement each other and clearly demonstrate that SP-RB was photoinhibited in the open pond by 25-30% and that the shade net provided the conditions for maximal O_2 evolution activity and maximal production. One way to improve the productivity of outdoor grown cultures is to provide protection from photoinhibition. While this can be accomplished to some extent by shading, a better approach will be to screen for strains of higher resistance to photoinhibition, which utilize high PFD advantageously.

REFERENCES

1. Richmond, A., Microalgal culture. CRC Critical Reviews in Biotechnology, 1986, 4, 369-438.

2. Richmond, A., <u>Spirulina</u>, In: <u>Microalgal Biotechnology</u>, eds. M. Borowitzka and L. Borowitzka, Cambridge Univ. Press, 1988, 84-121.

3. Vonshak, A. and A. Richmond., Problems in developing the biotechnology of algal biomass production. <u>Plant & Soil</u>, 1985, <u>89</u>, 129-135.

4. Vonshak, A., Strain selection of <u>Spirulina</u> suitable for mass production. <u>Hydrobiologia</u>, 1987, <u>151</u> 75-77.

5. Vonshak, A., Biological limitations in developing the biotechnology for algal mass cultivation. <u>Sciences de l'Eau</u>, 1987, <u>6</u>, 99-103.

6. Powles, S.B., Photoinhibition of photosynthesis induced by visible light. <u>Ann. Rev. Plant Physiol.</u>, 1985, <u>35</u>, 15-44.

7. Kyle, D.J. and I. Ohad, The mechanism of photoinhibition in higher plants and green algae. In: <u>Encyclopedia of Plant Physiology New Series</u>, eds. L.A. Staehelin and C.J. Arntzen. Photosynthesis III, <u>19</u>, Springer-Verlag Berlin, 1986, pp. 468-475..

8. Ohad, I., D.J. Kyle and C.J. Arntzen. Membrane protein damage and repair removal and replacement of inactivated 32 kD polypeptides in chloroplast membranes, <u>J. Cell Biol.</u> ,1984, <u>99</u>, 481-485.

9. Samuelsson, G., A.L. Lonneborg, E. Rosenqvist, P. Gustafsson and G. Oquist, Photoinhibition and reactivation of photosynthesis in the cyanobacteria Anacystis nidulans. <u>Plant Physiol</u>, 1985, <u>79</u>, 992-995.

10. Vonshak, A., R. Guy, R. Poplawsky and I. Ohad., Photoinhibition and its recovery in two strains of the cyanobacteria <u>Spirulina</u> platensis. <u>Plant & Cell Physiol.</u>, <u>29</u> 721-726.

11. Vonshak, A., A. Abellovich, S. Boussiba, and A. Richmond. Production of <u>Spirulina</u> biomass: effects of environmental factors and population density. <u>Biomass,</u> 1982 <u>2</u>, 175-185.

12. Vonshak, A., S. Boussiba, A. Abellovich, and A. Richmond, On the production of <u>Spirulina</u> biomass: The maintenance of pure culture under outdoor conditions. <u>Biotech. and Bioeng.</u>, 1983, <u>25</u>, 341-351.

13. Ben-Yaakov, S., H. Guterman, A. Vonshak, and A. Richmond., An automatic method for on-line estimation of the photosynthetic rate in open algal ponds. <u>Biotech. and Bioeng.</u>, 1985, <u>27</u>, 1136-1145.

14. Vonshak, A., Laboratory Techniques for the Culturing of Microalgae. In: _Handbook for Microalgal Mass Culture_. ed. Richmond, A. CRC Press, Boca Raton, Florida, 1986, pp. 117-145.

15 Bennet, A. and L. Bogorad. Complementary chromatic adaptation in a filamentous blue-green alga. _J. Cell Biol_, 1973, 58, 419-435.

MECHANISM OF HEAT INDUCED STIMULATION OF PSI ACTIVITY IN PEA CHLOROPLASTS

Maya Velitchkova, Alexander G. Ivanov and Rosa Petkova[*]

Bulgarian Academy of Sciences, Central Laboratory of
Biophysics, Institute of Plant Physiology, 1113 Sofia
Bulgaria

INTRODUCTION

It was well documented[1,2] that short time heat stress (5 min at 50°) results in an increase of PSI-mediated electron transport rate. An explanation of this phenomenon has been given by Thomas et al[?]. , who proposed an exposure of new electron donor sites within cyt b6/f complex for the artificial electron donor. Considerable increase (50%) of light-induced ESRI signal (P700[+]) in heat stressed chloroplasts was also reported[3]. Recently, it was demonstrated[4], that even at very low light intensity the amplitude of P700[+] remains higher and that under limiting light conditions P700 photo-oxidation occurs faster in heat-treated chloroplasts. It was supposed[3,4], that the mechanism of heat-induced stimulation of P700 photoconversion includes a redistribution of the excitation light energy in favor to PSI via an increase of PSI absorption cross section and enhancement of the spillover. This assumption was made on the basis of the well established[5,6] heat-induced structural reorganization of the thylakoid membranes - randomization of the pigment-protein complexes of PSII and PSI, dissociation of LHCa/b protein complex from core complex of PSII and its possible lateral migration along the thylakoid membrane to the PSI-enriched stroma lamellae. In the present study the effects of heat-stress on the energy distribution and PSI electron transport rate in total thylakoids and stromal thylakoid fractions were compared.

MATERIALS AND METHODS

Fresh pea chloroplasts and stromal subchloroplast fractions were obtained as described by Ford et al.[7]. The high salt envelope-free chloroplasts were used for sonic fractionation of stromal thylakoids using a sonicator type UZDN-2T (USSR) at appropriate power setting. Chlorophyll a/b ratio of stromal membranes varied between 4.7-6.0 at different experiments. The Chl concentration and Chl a/b ratio were estimated using the method of Wellburn and Lichtenthaler[8]. Chloroplasts and stromal thylakoids were incubated at designated temperature for 5 min than cooled to 20° for further measurements. The PSI-mediated electron transport were estimated from rates of O_2 uptake using Rank Brothers oxygen electrode assembly. Standard assay medium contained: phosphate buffer (pH 7.5), 0.33 M sucrose, 130 mM KCl, 0.4 mM DCPIP, 0.1 mM methylviologen, 0.04 mM DCMU, 4 mM ascorbate, 0.4 mM sodium azide and 25 µg Chl/ml. Measurements were carried out under saturating light conditions. 77K fluorescence emission spectra were recorded by an Jobin Yvon JY3 spectrofluorimeter equipped with red sensitive photomultiplier (Hamamatsu R928) and a liquid nitrogen device. Slits width=4 nm. Mg^{2+}-dependent chlorophyll fluorescence rise was monitored by home made apparatus under rapid stirring. Chlorophyll fluorescence was excited at 436 nm and measured at 685 nm. Chlorophyll fluorescence under F_o conditions was monitored using extremly weak excitation light as described by Schreiber and Armond[2].

RESULTS AND DISCUSSION

The thermal stress of isolated chloroplasts was known[1,2] to stimulate the PSI activity when $DCPIP.H_2$ was used as an artificial electron donor. Our results presented on Tabl. 1. also indicate consoderable increase of PSI electron transport rate (42%) in envelope-free chloroplasts as a result of heat-treatment. Thomas et al.[2] proposed that this stimulation is due to the heat-induced exposure of new electron donor sites within cyt b6/f complex in the electron transport chain. Cyt b6/f is generaly believed to be distributed throughout the complete membrane at an approximately uniform density[10]. In this way it is reasonable to expect that heat-treatment could

Table 1. The effects of heat stress (5 min at 50°) on PSI activity, F_o level and 77K chlorophyll fluorescence ratio-F735/F685 (F_K) inenvelope-free chloroplasts and stromal thylakoids. The data are presented as a ratio of corresponding values observed for heated (h) and control unheated (c) membranes.

Sample	PSI^h/PSI^c	F_o^h/F_o^c	F_K^h/F_K^c
Envelope-free chloroplasts	1.42 ± 0.06	1.45 ± 0.11	1.46 ± 0.09
Stromal thylakoids	0.64 ± 0.18	1.09 ± 0.01	1.15 ± 0.04

increase PSI activity in stromal thylakoids also.

Surprisingly, our results demonstrate (see Tabl.1.) that no stimulation is observed in stromal thylakoids, where PSI complexes are preferentially situated. Moreover, thermal action inhibits the PSI activity by 35% in the latter sample. It must be noted that stromal thylakoids posses the same initial PSI electron transport rates as total thylakoids (data not shown), which is consistent with the data reported by Ford et al.[7]. This indicates that the opposingeffect of heating on the PSI photochemical activity in stromal fraction could not be due to any damage at the cyt b6/f level as a result of fractionating procedure.

More recently it has been supposed[4], that heat-induced redistribution of excitation energy infavor to PSI (enhancement of spillover) can not be as a possible mechanism for explanation of the increase of PSI activity. Since the Mg^{2+}-associated increase ofchlorophyll fluorescence and the ratio of F735/F685 in the 77K chlorophyll fluorescence emissionis currently believed to reflect changes in spillover of excitation energy from PSII to PSI such experiments were performed.

The results presented on Fig. 1. show that the effect of heat treatment above $40^\circ C$ is to reduce the size of Mg^{2+}-induced fluorescence rise. This could be connected with an increased ability for interaction between chlorophyll-protein complexes of PSII and PSI at elevated temperatures. A second effectof heat treatment at the same temperature region is to decrease the rate ($t_{\frac{1}{2}}$ is increased) of fluorescence rise. These changes reflect presumably the effect of heating on the lateral mobility of the supramolecular complexes of both photosystems.

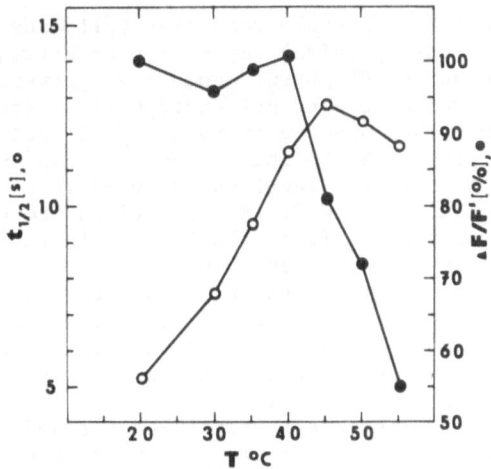

Fig. 1. Temperature dependence curves of Mg^{2+}-indiced chlorophyll fluorescence rise. Reaction medium contained:
0.01 M tricine (pH 8,0), 0.33 M sucrose, 10 mM KC1,0.0
5 mM EDTA and 10 µg Chl/ml. $\Delta F/F'$ refers to the size of
the fluorescence increase, where F' is the initial
fluorescence level observed before addition of Mg^{2+}
and ΔF is the maximum changes observed following addition of 5 mM Mg^{2+}. 100% is the value for $\Delta F/F'$ at $20^{\circ}C$.
The $t_{\frac{1}{2}}$ corresponds to the half time for fluorescence
increase after Mg^{2+} addition.

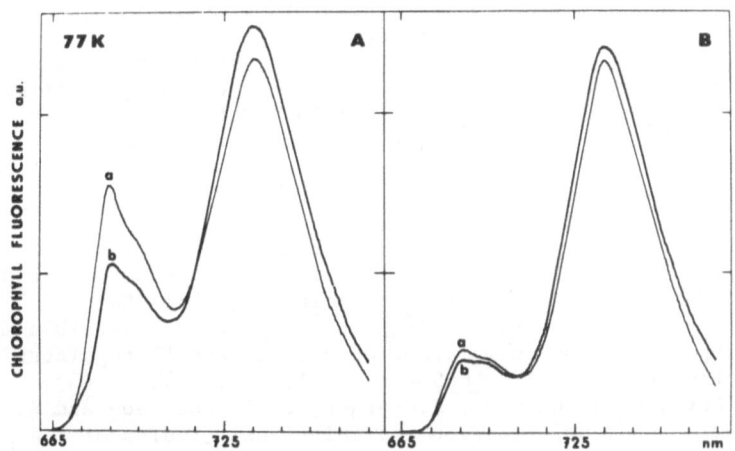

Fig. 2. Low temperature (77K) chlorophyll fluorescence emission
spectra of envelope-free chloroplasts (A) and stromal
thylakoids (B). b/ - heat-treated membranes, a/ - control membranes.

Furthermore, heat stress at $50^{\circ}C$ leads to the considerable increase of
the ratio F735/F685 when measured for envelope-free chloroplasts (Fig. 2.
and Tabl. 1.). For stromal thylakoids this increase is only 15%.
Thus, it is evident that remarkable enhancement of spillover take place
in heat treated chloroplasts. This is full agreement and could be fully understood within the framework of earlier reported[5,6] data for general reor-

ganization of the thylakoid protein complexes following heat treatment.

It must be noted that physical separation of stromal mlmbranes, i.e. their partial depletion of LHCa/b-PSII complexes prevents the stimulation effect of thermal treatment on the PSI photochemical activity. Hence, it seems very likely that the observed increase of PSI activity in whole chloroplasts is related presumably to the changes in LHCa/b-PSII complex rather than to the changes at cyt b6/f level as proposed in[2].

Fluorescence measurements under F_o conditions are known to reflect the alterations at the pigment level. As can be seen in Tabl. 1. F_o level is increased by about 45% in heat treated chloroplasts, whilst no remarkable changes is observed in stromal fraction. This could be attributed to the earlier established[5,6] heat-induced physical dissociation of LHCa/b complex from the core complex of PSII. As proposed recently[3,4] a part of this dissociated LHCa/b might serve as additional antenna for PSI resulting in an increase of its absorption cross section.

Thus, the well established structural changes and rearrangement of the membrane complexes could be responsible for the heat-induced stimulation of PSI activity by means of at least three possible mechanisms: an enhancement of spillover, enlargement of PSI absorption cross section as well as an exposure of new electron donor sites.

ACKNOWLEDGEMENTS - This work was supported by the National Science Commettee, Research project 519.

REFERENCES

1. M. A. Stidham, E. G. Uribe and G. J. Williams III, Temperature dependence of photosynthesis in Agropyron smithii Ridb. II Contribution from electron transport and photophosphorylation, Plant Physiol. 69:929 (1982).
2. P. G. Thomas, P. J. Quinn and W. P. Williams, The origin of photosystem-I-mediated electron transport stimulation in heat-stressed chloroplasts, Planta 167:133 (1986).
3. A. G. Ivanov, M. Y. Velitchkova and D. N. Kafalieva, Heat-induced changesin photosystem I reaction centre in pea chloroplast membranes, Compt. rend. Acad. bulg. Sci,39:123 (1986).
4. M. Velitchkova and A. Ivanov, Light intensity dependence of P700 photooxidation in heat stressed pea chloroplasts, in: "Electromagnetic Fields and Biomembranes", M. Markov and M. Blank, eds., Plenum Press, New York and London.(1988).
5. K. Gounaris, A. P. R. Brain, P. J. Quinn and W. P. Williams, Structural reorganization of chloroplast thylakoid membranes in response to heat-stress, Biochim. Biophys. Acta 766:198 (1984).
6. C. Sundby and B. Andersson, Temperature-induced reversible migration along the thylakoid membrane of photosystem II regulates its association with LHC-II, FEBS Lett. 191:24 (1985).
7. R. C. Ford, D. J. Chapman, J. Barber, J. Z. Pedersen and R. D. Cox, Fluorescence polarization and spin-label studies of the fluidity of the stromal and granal chloroplast membranes, Biochim. Biophys. Acta 681:145 (1982).
8. A. R. Wellburn and H. Lichtenthaler, Formulae and program to determine total carotenoids and chlorophyll a and b. Leaf extracts in different solvents, Adv. Photosynth. Res.2:9 (1984).
9. U. Schreiber and P. A. Armond, Heat-induced changes of chlorophyll fluorescence in isolated chloroplasts and related heat-damage at the pigment level, Biochim. Biophys. Acta 502:138 (1978).
10. D. R. Allred and L. A. Staehelin, Implications for cytochrome b6/f location for thylakoid electron transport, J. Bioenerg. Biomembr.18: 419 (1986).

REGULATION OF PHOTOSYNTHETIC CARBON METABOLISM IN WATER-STRESSED

SUNFLOWER

D.W. Lawlor, Carmen Gimenez*, D.A. Ward and A.T. Young

AFRC Institute of Arable Crops Research, Rothamsted
Experimental Station Harpenden, Herts., AL5 2JQ,
UK and *Department of Agronomy, ETSIA, Cordoba 14071, Spain

INTRODUCTION

Water stress limits plant production in many parts of the world. When leaf mesophyll cells lose water and turgor their capacity for photosynthesis declines[1,2] and the rate of photosynthesis per unit leaf area decreases. This is not caused by stomatal closure only, but by inhibition of metabolism [1,2,3]. The mechanisms by which cell water deficit affects metabolism and slows CO_2 assimilation are not well understood[4,5]. Light harvesting and electron transport in thylakoid membranes are less sensitive to water stress than is photophosphorylation [4,5,6], which takes place in the chloroplast stroma and may be affected by increased concentrations of stromal components. Without an understanding of the effects of water stress on the mechanisms of photosynthesis, selection of plants with characteristics better adapted to drought will be hindered, as will direct manipulation of the photosynthetic system.

Here we present some data on photosynthesis in two varieties of sunflower, selected for differences in productivity, including aspects of leaf composition, ribulose bisphosphate (RuBP) carboxylase activity and RuBP amounts. The effects of water stress on photosynthesis are discussed and analysed in relation to previously observed changes of adenylates, energy charge and reduced and oxidised nucleotides in wheat caused by water stress[7].

MATERIALS AND METHODS

Sunflower (Helianthus annuus L.) varieties Sungro and SH were grown in 10 l containers of potting compost. Plants were stressed by replacing only a fraction of the water transpired so that intermediate and severe stresses (leaf water potentials, ψ, of ca. -1.5 and -2.5 MPa respectively, measured with a pressure chamber) were reached after ca. 3 and 7 days. Photosynthesis (P) was measured as a function of CO_2 pressure in the intercellular spaces of the leaf, in a multi-channel open-gas exchange system, at a leaf temperature of 25°C, with a vapour pressure deficit of ca. 1 kPa and 21 kPa oxygen at 1500 μmol quanta photosynthetically active radiation m^{-2} s^{-1}. Leaves in the chambers were rapidly frozen (within 0.1 s) with a freeze clamp and used for enzyme and metabolite analysis. The initial activity of RuBP carboxylase was measured immediately upon grinding the tissue in buffer and compared with the total activity obtained after 10 min incubation with CO_2 and Mg^{++} to estimate the activation state <u>in vivo</u>. Total soluble protein was measured with Coomassie brilliant blue; RuBP carboxylase protein was separated by DISC PAGE and the amount of protein determined colorimetrically by comparison with standard RuBP carboxylase protein. Metabolites were extracted from leaf tissue with perchloric acid and measured by standard methods[4]. Cell numbers were measured on macerated tissue with a Coulter Counter and cell volume estimated from tissue water content/cell number.

RESULTS AND DISCUSSION

Leaves of well-watered Sungro contained (Table 1) the same amounts of total soluble protein (TSP) and chlorophyll (Chl) per unit area, and had the same Chl a/b ratio as SH. Sungro had fewer, but larger, cells per area with more TSP per cell but slightly less TSP per volume of cell when fully hydrated. The difference in cell size may be important as small cells with large surface area to volume ratio would be advantageous for photosynthesis, which was indeed greater in unstressed SH leaves. Water stress decreased mesophyll cell volume substantially. Therefore, the number of cells and TSP per unit area of leaf increasd and the TSP per cell was unaffected by stress, but the TSP per unit of cell volume increased with stress.

The ratio of RuBP carboxylase protein to TSP was 0.44 in Sungro and 0.45 in SH and was unaffected by stress. RuBP carboxylase activity per unit leaf area was ca. 0.2 μmol min^{-1} cm^{-2} in unstressed leaves and was only slightly greater in Sungro than SH and decreased with stress. The RuBP carboxylase activity per unit of protein also decreased with stress in both varieties. Initial enzyme activity was ca. 90% of the total activity in well-watered plants but was greater in stressed leaves, probably because compounds produced in stressed tissue inhibited the enzyme during the longer incubation. The specificity factor of RuBP carboxylase was the same in Sungro, SH and wheat (Dr A.J. Keys pers. comm.).

Photosynthetic rate (P) of young, mature, unstressed leaves of SH was greater than that of Sungro when measured at CO_2 saturation (Pmax; Fig. 1, 2a). The difference in P was not a consequence of stomatal conductance (gs) which was greater in SH than Sungro (Fig. 2b) because Pmax is independent[3] of gs. The CO_2 limited rate of photosynthesis (carboxylation efficiency) was greater (Fig. 2b) in Sungro than SH. Water deficit decreased Pmax (Fig. 1, 2a) similarly in both varieties and also carboxylation efficiency (Fig. 2b). The CO_2 compensation point also increased with stress (Fig. 1) as observed in earlier experiments with sunflower[5].

Table 1. Leaf characteristics of sunflower varieties Sungro and SH with different water potentials (ψ). Abbreviations in the table and units are: Cell No. x 10^{10} m^{-2}; Chlorophyll (Chl) and total soluble protein (TSP) in g m^{-2}; Protein/cell volume (PCV) in pg pl^{-1}.

| ψ(MPa) | Sungro | | | | | | SH | | | | | |
	Cell No.	Cell vol.	Chl	Chl a/b	TSP	PCV	Cell No.	Cell vol.	Chl	Chl a/b	TSP	PCV
-0.6	3.8	5.6	0.5	2.8	6.4	30	4.7	3.9	0.5	2.8	6.0	33
-1.5	5.6	3.3	0.6	2.6	7.9	43	5.3	3.6	0.5	2.7	7.7	40
-2.6	5.7	2.5	0.5	2.4	8.2	58	5.5	2.5	0.4	2.7	7.4	54

Inspection of the data for RuBP carboxylase showed that changes in the amount and activity _in vitro_ were not the cause of the decrease in P with stress. Therefore, factors other than the enzyme must have been limiting assimilation and so we measured the amounts of RuBP, the substrate for CO_2 assimilation in leaves (Fig. 2a). The amount of RuBP ws greater in SH than Sungro in control and stressed plants; in both varieties it decreased with stress in parallel to the decrease in P. If the reduction in RuBP was due to inhibition of the photosynthetic carbon reduction cycle between 3-phosphoglyceric acid (3PGA) and RuBP then 3PGA should accumulate. However, this was not observed (data not give); 3PGA decreased similarly to RuBP.

A major limitation to RuBP synthesis in water-stressed mesophyll cells may be in photophosphorylation[1,4,5], as other cycle enzymes (as well as RuBP carboxylase) appear not to be inhibited by stress[4]. Evidence from water stressed wheat[7] shows that the amounts of adenylates change with decreasing water potential (Table 2); ATP decreases, ADP rises and consequently Energy Charge falls as expected if photophosphorylation were inhibited. An effect of slow synthesis of ATP would be inhibition of RuBP production as shown by the decreased amount of RuBP, and this would inhibit photosynthesis. The observed decrease in carboxylation efficiency is noteworthy as it is not always observed in stressed cells[4]. Also Pmax and carboxylation efficiency decrease in proportion. These results suggest that RuBP is not saturating the enzyme at low intercellular CO_2 in stressed leaves. Therefore, the efficiency of photophosphorylation and RuBP regeneration is decreased by stress. Coupling factor, the enzyme complex responsible for ATP synthesis, is inhibited in stressed mesophyll cells[1,8], possibly by large ion (principally Mg^{++}) concentrations[8,9]. However, the mechanism whereby ions decrease coupling factor activity is not fully understood[8].

In addition to changing the pools of adenylates, water stress alters the nucleotide concentrations in photosynthetic tissue (Table 2)[7]. The

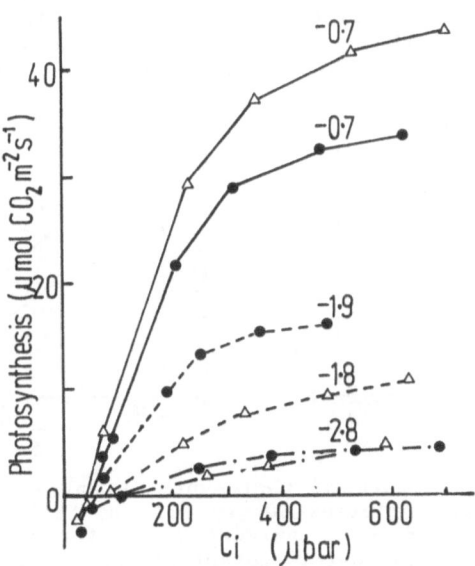

Fig. 1. Photosynthesis by leaves of sunflower varieties Sungro (●) and SH (▲) of different water potential, as a function of intercellular CO_2 pressure (Ci) (ψ, in MPa on curves).

Table 2. Effect of water stress (ψ in MPa) over 12 h on the amounts of adenylates and energy charge (EC) and amounts of nucleotides and the ratio of reduced to oxidised nucleotides (R/O) in wheat leaves[7].

ψ	ATP	ADP	EC	NADPH	NADP$^+$	NADPH	NAD$^+$	R/O
	(μmol m^{-2})		–		(μmol m^{-2})			–
−0.4	170	60	0.80	3	9	2	14	0.22
−0.7	90	100	0.67	5	14	3	16	0.27
−1.3	60	100	0.61	3	15	7	7	0.45
−1.8	50	160	0.57	4	4	8	6	1.20

amount of NADPH is maintained and the amount of NADH substantially increased, so that the ratio of total reduced to oxidised nucleotides increased with stress in wheat leaves. The decrease in P with stress is therefore more closely related to changes in ATP than in NADPH. The greater amounts of NADH may be related to the increased dark respiration and catabolism observed in stressed tissues [2,5]. The increased amounts of reduced nucleotides and the decreased ATP and Energy Charge will greatly affect many metabolic processes in the cell, for example the accumulation of amino acids (particularly proline) observed in stressed leaves (data not given).

SUMMARY

Sunflower variety SH had larger photosynthetic rates than Sungro in CO_2 saturated conditions; the main difference in leaf composition and structure related to increased photosynthesis was the smaller mesophyll cells of SH. Water stress had little effect on the amount, activity or activation state of RuBP carboxylase, which is therefore not considered to be the limiting factor for photosynthesis in water-stressed tissues. The decrease in

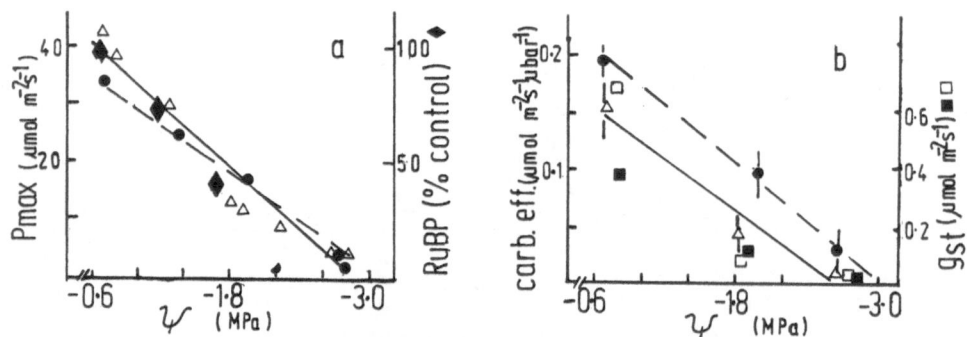

Fig. 2. a) The maximum rate of photosynthesis at saturating CO_2 (P max) for sunflower varieties Sungro (●) and SH (▵) and the amount of RuBP (◆) in SH as a percentage of the unstressed control in relation to leaf water potential (ψ). b) The carboxylation efficiency (± s.e.) of Sungro (●) and SH (▵) leaves, measured at limiting CO_2 together with the stomatal conductance (gs) of Sungro (■) and SH (□), in relation to leaf water potential (ψ).

amount of RuBP, in parallel to the decrease in photosynthesis, suggests that regeneration of RuBP regulates CO_2 assimilation. Relating these changes to the effect of water stress on adenylates and Energy Charge in stressed wheat suggests that inhibition of RuBP synthesis in dehydrated sunflower mesophyll cells is caused by inhibited photophosphorylation. An increase in the ratio of reduced nucleotides to ATP may be related to changes in photosynthetic metabolism and respiration.

REFERENCES

1. J. S. Boyer, Nonstomatal inhibition of photosynthesis in sunflower at low leaf water potentials and high light intensities, Plant Physiol. 48: 532 (1971).
2. D. W. Lawlor, Water stress induced changes in photosynthesis, photorespiration, respiration and CO_2 compensation concentration of wheat, Photosynthetica 10: 378 (1976).
3. G. Cornic, J-L Prioul and G. Louason, Stomatal and non-stomatal contribution in the decline in leaf net CO_2 uptake during rapid water stress, Physiol. Plant. 58: 295 (1983).
4. T. D. Sharkey and M. R. Badger, Effects of water stress on photosynthetic electron transport, photophosphorylation, and metabolite levels of Xanthium strumarium mesophyll cells, Planta 156: 199 (1982).
5. D. W. Lawlor, Effects of water and heat stress on carbon metabolism of plants with C_3 and C_4 photosynthesis, in: "Stress Physiology in Crop Plants", H. Mussell and R.C. Staples eds; Wiley Interscience, New York, (1979).
6. R. W. Keck and J. S. Boyer, Chloroplast response to low leaf water potential, III Differing inhibition of electron transport and photophosphorylation, Plant Physiol. 53: 474 (1974).
7. D. W. Lawlor and R. Khanna-Chopra, Regulation of photosynthesis during water stress, in: "Advances in Photosynthesis Research", C. Sybesma, ed., Martinus Nijhoff/Dr W. Junk Publishers, The Hague (1984).
8. H. M. Younis, J. S. Boyer and Govindjee, Confirmation and activity of chloroplast coupling factor exposed to low chemical potential of water in cells, Biochim. Biophys. Acta 54: 328 (1979).
9. J. S. Boyer and H. M. Younis, Molecular aspects of photosynthesis at low leaf water potentials, in: "Effects of Stress on Photosynthesis", R. Marcelle et al., ed., Martinus Nijhoff/Dr W. Junk Publishers, The Hague (1983).

GALACTOLIPASE ACTIVITY AND FREE FATTY ACIDS IN CHLOROPLASTS
AS INDICATORS OF CHILLING SENSITIVITY OF CLOSELY RELATED
PLANT SPECIES *

Joanna Gemel and Zbigniew Kaniuga

Institute of Biochemistry, University of Warsaw
Al. Żwirki i Wigury 93, 02-089 Warszawa, Poland

INTRODUCTION

Cold and dark treatment of leaves of chilling-sensitive(CS) plants results in both accumulation of free fatty acids(FFA) (cf.1) and degradation of galactolipids [2] in chloroplasts. Moreover both galactolipase activity and FFA levels appear to be higher in CS than chilling-resistant plants [3,4]. In this communication data concerning these two parameters in chloroplasts of closely related plants species but of different chilling sensitivity are presented.

PLANT MATERIALS AND METHODS

Less and more chilling-sensitive plants in this study are designated here as CT and CS, respectively. Differences with respect to their chilling sensitivity known either from horticultural practice, breeding selection or literature were additionally checked by the extent of Photosystem II inactivation following chilling stress of leaves [5]. Decrease of Hill reaction activity or oxygen evolution was measured(cf.6).

In all experiments young, fully expanded leaves were used. Leaves of cucumber and melon were stored (cf.5) at $4.5^{\circ}C$ and those of maize, tomato and potato at $0^{\circ}C$ in dark for few days as indicated.

Chloroplast galactolipase was isolated and its activity measured as described by Anderson et al.[7] while for determination of FFA content gas-liquid chromatography was applied as previously[4].

RESULTS

The data of table 1 indicate that in each group of plants studied FFA levels in chloroplasts of more CS plants are higher than in less CS species. Similarly, following chilling stress the extent of FFA accumulation is also much higher in more CS than less CS plants. This relationship is characteristic for breeding lines and cultivars of cucumber, melon, maize, wild potato as well as for domestic and wild tomatoes originally growing at different altitudes

---

* This is the 22nd paper of series: Photosynthetic apparatus of chilling-sensitive plants

Table 1. Free Fatty Acid Levels in Chloroplasts of Fresh and Cold Treated Leaves of Closely Related Plant Species with Different Chilling Sensitivity

Type	Plant species	Time of cold stress[a]	Free fatty acid[b] Control	Chilled	Ratio[c]
	Cucumber				
CS	Skierniewicki	2	0.94	2.14	2.4
	Warszawski	2	1.17	5.10	4.3
CT	Line 303	3	0.44	0.70	1.6
	Borszczagowski	3	0.44	1.10	2.5
	Melon				
CS	Hearts of Gold	2	0.66	0.94	1.4
	Line 191	3	0.85	1.19	1.4
	Super Market	-	1.10	----	---
CT	Line 14	2	0.17	0.54	3.2
	Line 190	3	0.12	0.66	5.5
	Line 9/6	-	0.36	----	---
	Maize				
CS	Line F7 RpIII	3	0.43	3.72	8.6
CT	Line S72	3	0.17	0.43	2.5
	Potato				
CS	*S. tuberosum*	6	0.34	1.22	3.6
	S. toralapanum	6	0.39	1.90	4.8
CT	*S. chaucha*	6	0.39	0.39	1.0
	S. ajanhuirii	6	0.36	0.57	1.6
	Tomato				
CS	*L. esculentum*	4	0.52	0.87	1.6
	L. hirsutum 700 m	8	0.53	0.77	1.5
CT	*L. hirsutum* 3100 m	8	0.41	0.57	1.4
	L. peruvianum	6	0.27	0.33	1.2

[a] Days
[b] μmol * mg Chl^{-1}
[c] Ratio of FFA content in chilled/control samples

in Peru. Higher FFA ratio in less CS lines of melon than in more CS cultivars is not contradictory with general relationship, since FFA level in control and their increase in chilled samples of melon is in agreement with data obtained for other plants.

Galactolipase activity in chloroplasts is higher in plants of more CS than of less CS ones (Table 2). As reported previously [4] galactolipase activity could not be estimated directly in chloroplast fraction of *Cucurbitaceae* members. However, its activity can be detected and evaluated indirectly by determination of galactolipids degradation and FFA accumulation following cold stress of leaves.

CONCLUSIONS

The evaluation of chilling tolerance due to measurements of Hill reaction activity decrease in chilled leaves [5,7] can be now complemented

Table 2. Galactolipase Activity in Chloroplasts of Closely
Related Plant Species with Different Chilling
Sensitivity

Type	Plant species	Galactolipase activity[a]
	Maize	
CS	Line F7 RpIII	1.59
CT	Line S72	0.49
	Potato	
CS	*S. tuberosum*	0.64
	S. toralapanum	0.51
CT	*S. chaucha*	0.09
	S. ajanhuirii	0.00
	Tomato	
CS	*L. esculentum*	1.41
	L. hirsutum 700m	0.90
CT	*L. hirsutum* 3100m	0.64
	L. peruvianum	0.36

[a] μmol FFA released $*$ min^{-1} $*$ (mg $protein^{-1}$)

by assay of chloroplast galactolipase activity and FFA levels in fresh leaves
and extent of FFA accumulation in chloroplasts of chilled leaves.

The present experiments with the use of several closely related plant
species including breeding forms selected for greater chilling tolerance support our postulate[3] that determination of galactolipase activity and FFA
levels in chloroplasts may be useful parameters for evaluation of chilling
sensitivity of plants. Therefore it is suggested that both parameters may
have application in physiological studies and breeding programmes for selection of chilling tolerant species.

ACKNOWLEDGEMENTS

We are grateful to Dr. A.Korzeniewska (Chair of Genetics and Plant
Breeding, Agricultural University, Warsaw) for providing cucumber and melon
leaves, to Dr. A.Michalska (Breeding Station of Horticultural Crops ,Warsaw)
for leaves of domestic and wild tomatoes, to Prof. J.Bojanowski (Institute of
Plant Breeding and Acclimatization, Radzików) for maize seeds and to Prof.
D.Kleinhempel (Institute of Potato Research, Gross Lusewitz,GDR) for series
of wild potatoes. This work was supported by grant from CPBP 05.02, project
1.09.

REFERENCES

1. Z. Kaniuga and W.P.Michalski, Photosynthetic apparatus of chilling-sensitive plants.II.Change in free fatty acid composition and photoperoxida-
tion in chloroplasts following cold storage and illumination of leaves
in relation to Hill reaction activity, Planta 140:129(1978).
2. W. P.Michalski and Z.Kaniuga, Photosynthesic apparatus of chilling-sensitive plants.VII.Comparison of the effect of galactolipase treatment of
chloroplasts and cold-dark storage of leaves on photosynthetic electron
flow, Biochim.Biophys.Acta 589:84(1980).

3. Z. Kaniuga and J.Gemel, Galactolipase activity and free fatty acid level in chloroplasts. Novel approach to characteristics of chilling sensitivity of plants, FEBS Lett. 171:55(1984).

4. J. Gemel and Z.Kaniuga, Comparison of galactolipase activity and free fatty acid levels in chloroplasts of chill-sensitive and chill-resistant plants, Eur.J.Biochem. 166:229(1987).

5. Z. Kaniuga, B.Sochanowicz, J.Ząbek and K.Krzystyniak, Photosynthetic apparatus of chilling-sensitivity plants.I.Reactivation of Hill reaction inhibited on the cold and dark storage of detached leaves and intact plants, Planta 140:121(1978).

6. Z. Kaniuga, J.Gemel and B.Zabłocka, Fatty-acid induced release of manganese from chloroplasts, Biochim.Biophys.Acta 850:473(1986).

7. M. M.Anderson, .R.E.McCarty and E.A.Zimmer, The role of galactolipase in spinach chloroplasts lamellar membranes.I.Partial purification of bean leaf galactolipid lipase and its action on subchloroplast particles, Plant Physiol. 53:699(1974).

8. R. M.Smillie and R.Nott, Assay of chilling injury in wild and domestic tomatoes based on photosystem activity of chilled leaves, Plant Physiol. 63:796(1979).

ASSEMBLY OF CHL-PROTEIN COMPLEXES IN MEMBRANES OF IRON-STRESSED SYNECHOCOCCUS SP. PCC7942 PROCEEDS IN THE ABSENCE OF CHLOROPHYLL SYNTHESIS

T.A. Troyan, G.S. Bullerjahn, and L.A. Sherman

Division of Biological Sciences, University of Missouri
Columbia, MO   65211 USA

Introduction

Iron-limited growth of plants and cyanobacteria results in a dramatic decrease in chlorophyll and thylakoid content. Iron deficient cells of the cyanobacterium Synechococcus sp. PCC7942 demonstrate significant changes in the organization of their photosynthetic membranes. One of the most dramatic changes is evidenced by the depletion of PSI and PSII chlorophyll-protein complexes and the accumulation of an accessory chlorophyll-protein complex, CPVI-4, which is present in very low amounts in unstressed cells. CPVI-4 harbors the majority of the chlorophyll in the iron-stressed cells. Its function may be that of a weak antenna or perhaps that of a chlorophyll reservoir which facilitates recovery when iron becomes available (Riethman and Sherman, 1988). We examined the recovery of cells from iron limitation by monitoring low temperature chlorophyll fluorescence of whole cells and chlorophyll-protein changes in thylakoid membranes under conditions in which chlorophyll synthesis or protein synthesis is inhibited.

Materials and Methods

Synechococcus sp. PCC 7942 was grown in iron-limiting media under the conditions previously described (Sherman and Sherman, 1983). Recovery was initiated with the addition of 6 mg of ferric ammonium citrate per liter of medium. Chlorophyll synthesis was inhibited with 50 µM gabaculine (Fluka Chemical Corp.) and protein synthesis was inhibited with 0.25 mg per ml chloramphenicol (Sigma Chemical Co.).

Cells were harvested at several time intervals after the addition of iron and inhibitors. Recovery was monitored spectrally and electrophoretically. Whole cell 77K chlorophyll fluorescence emission spectroscopy was performed using an SLM spectrofluorimeter as previously described by Pakrasi et al. (1985a). For electrophoretic analysis, membranes were prepared from frozen cells harvested at various time points during recovery. These cells were thawed, broken using a chilled French pressure cell, and the membranes isolated by differential centrifugation as described previously (Pakrasi et al., 1985b). The membrane pellet was re-suspended in 50 mM MES (pH 6.5), 1 mM phenylmethylsulfonylfluoride, benzamidine and E-aminocaproic acid and the chlorophyll concentration determined by the method of Arnon (1949). Membrane suspensions of 0.5 - 0.8 mg chlorophyll per ml were solubilized with 1% dodecyl-ß-D-maltoside (w/v) purchased from Calbiochem-Behring.   The solubilized suspension was centrifuged at 14,000 rpm for 10 minutes in an Eppendorf microfuge at 4°C to pellet the unsolubilized material. The supernatant was stored frozen at -70°C. Electrophoretic analysis of chlorophyll-protein complexes was performed as previously described (Pakrasi et al., 1985b).

Results and Discussion

Unstressed cells have fluorescence emission peaks at 685 nm, 696 nm and 716 nm, which represent PSII (685 and 696 nm) and PSI (716 nm) associated chlorophyll complexes. Low iron cells have one emission peak at 685 nm, attributed to the accumulation of the chlorophyll-protein complex CPVI-4 (Riethman and Sherman, 1988). During iron recovery, the intensity of the 685 nm fluorescence peak decreases while peaks at 696 nm and 716 nm appear (Figure 1, panel A). This recovery pattern is still apparent under conditions where no new chlorophyll is being synthesized (Figure 1, panel B).

The effect of gabaculine on chlorophyll concentration of iron-reconstituted cultures is shown in Table I. Note that the chlorophyll concentration of gabaculine-treated cultures decreases with time, probably due to photodestruction of the stable chlorophyll pool. During the time course of these experiments, the chlorophyll concentration of the control culture doubles.

Chlorophyll-protein gels of cells during recovery corroborate the fluorescence results by demonstrating an increase in the number of chlorophyll protein complexes associated with PSI and PSII (data not shown) with a corresponding decrease in CPVI-4 in the presence or absence of de novo chlorophyll synthesis (Figure 2). Chloramphenicol treated cells, however, exhibit no recovery as monitored by low temperature fluorescence, which indicates the necessity of new protein synthesis during recovery. These results strongly suggest that recovery of photosynthetic membranes from iron-stress is dependent upon new protein synthesis, but that newly formed chlorophyll-protein complexes are made utilizing chlorophyll originally associated with the CPVI-4 complex.

We have previously suggested (Riethman et al., 1988) that CPVI-4 acts as a reservoir of chlorophyll molecules dedicated to de novo PSI and PSII reaction center assembly. Since CPVI-4 is present in low amounts in normally-grown cells, we extended this model to suggest that CPVI-4 is a complex required for chlorophyll transfer to PSI and PSII centers under all growth conditions. This work demonstrates that chlorophyll molecules can be mobilized from CPVI-4 to newly synthesized reaction centers, and that the coordinate assembly of pigment and protein into PSI and PSII centers can occur by the turnover of an existing chlorophyll pool. Such a mechanism may also occur in higher plants, as Akoyunoglou and co-workers have presented evidence that LHCII can donate chlorophyll to reaction centers in developing chloroplasts (Akoyunoglou and Akoyunoglou, 1985). This chlorophyll-transfer phenomenon may represent a general event occurring during thylakoid biogenesis.

TABLE 1 - Effects of Gabaculine on Chlorophyll Concentration

Time (h) after Fe addition	Chlorophyll concentration (µg/ml)	
	+ gabaculine	- gabaculine
0	1.06	1.06
3	1.06	1.26
6	0.96	1.31
12	0.95	2.03

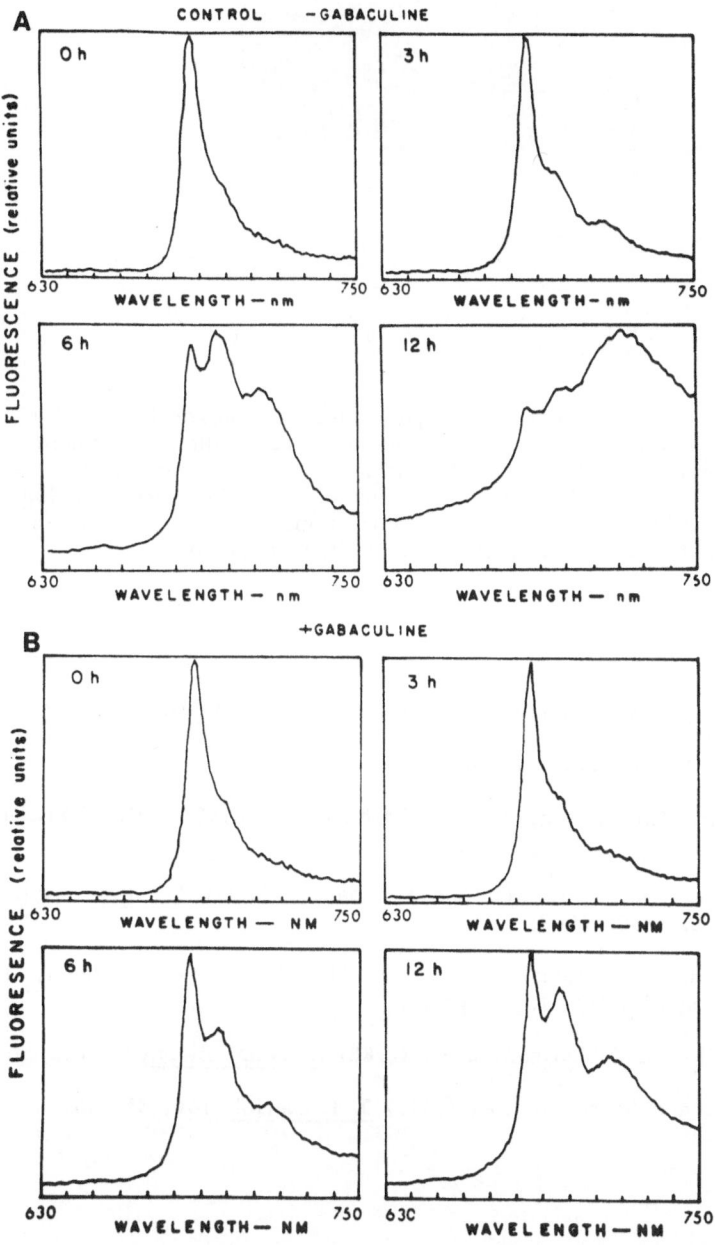

Figure 1. 77 K fluorescence emission spectra of cells 0, 3, 6, and 12 hours after iron reconstitution. Panel A follows recovery with no inhibition and Panel B follows recovery without de novo chlorophyll synthesis (gabaculine treatment).

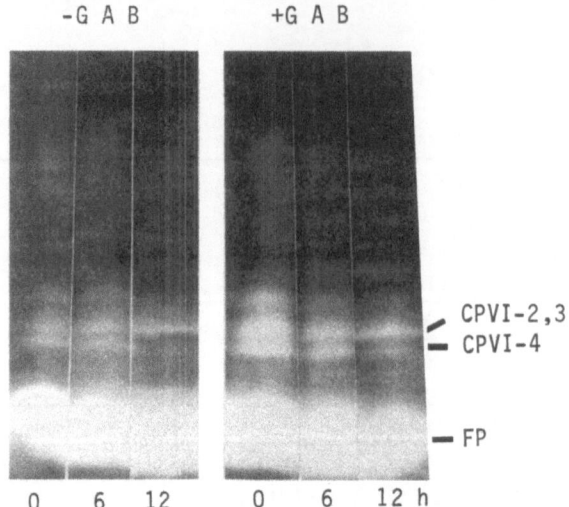

Figure 2. UV fluorescent chlorophyll protein complexes 0, 6 and 12 hours after iron reconstitution with or without gabaculine treatment. The gel is pictured with UV transillumination. CPVI-2 and CPVI-3 represent the major chlorophyll-binding proteins of the PSII core homologous to CP47 and CP43 of chloroplasts. Note their accumulation in the gabaculine-treated and control cultures. FP, free pigment

References

Akoyunoglou, A. and Akoyunoglou, G. (1985) Plant Physiol. 79, 425-431.

Arnon, D.I. (1949) Plant Physiol. 24, 1-15.

Pakrasi, H.B., Goldenberg, A. and Sherman, L.A. (1985a) Plant Physiol. 79, 290-295.

Pakrasi, H.B., Riethman, H.C. and Sherman, L.A. (1985b) Proc. Natl. Acad. Sci. USA 82, 6903-6907.

Riethman, H.C., Bullerjahn, G.S., Reddy, K.J. and Sherman, L.A. (1988) Photosynth. Res. 16, in press.

Riethman, H.C. and Sherman, L.A., (1988) Biochim. Biophys. Acta, in press.

Sherman, D. and Sherman, L.A. (1983) J. Bacteriol. 156, 393-401.